McGraw-Hill
networks™
A Social Studies Learning System

GO online
connected.mcgraw-hill.com

**MEETS YOU ANYWHERE —
TAKES YOU EVERYWHERE**

McGraw-Hill
netw⊙rks™

**MEETS YOU ANYWHERE —
TAKES YOU EVERYWHERE**

It's **ALL** online

1. Go to *connected.mcgraw-hill.com.*

2. Get your User Name and Password from your teacher and enter them.

3. Click on your **Networks** book.

4. Select your chapter and lesson.

HOW do you learn?

Read • Reflect • Watch • Listen • Connect • Discover • Interact

start **netw rk**ing

WHAT do you learn?

History • Geography • Economics • Government • Culture

start **network**ing

WHAT do you use?

Graphic Organizers • Primary Sources • Videos • Games • Photos

start**networ**king

McGraw-Hill
netw⬤rks™
A Social Studies Learning System

GEOGRAPHY

THE HUMAN AND PHYSICAL WORLD

Richard G. Boehm, Ph. D.

Mc Graw Hill Education

Bothell, WA • Chicago, IL • Columbus, OH • New York, NY

AUTHORS

Senior Author

Richard G. Boehm, Ph.D., was one of the original authors for *Geography for Life: National Geography Standards,* which outlined what students should know and be able to do in geography. He was also one of the authors of the *Guidelines for Geographic Education,* in which the Five Themes of Geography were first articulated. Dr. Boehm has received many honors, including "Distinguished Geography Educator" by the National Geographic Society (1990), the "George J. Miller Award" from the National Council for Geographic Education (NCGE) for distinguished service to geographic education (1991), "Gilbert Grosvenor Honors" in geographic education from the Association of American Geographers (2002), and the NCGE's "Distinguished Mentor Award" (2010). He served as president of NCGE, has twice won the *Journal of Geography* award for best article, and also received the NCGE's "Distinguished Teaching Achievement." Presently, Dr. Boehm holds the Jesse H. Jones Distinguished Chair in Geographic Education at Texas State University in San Marcos, Texas, where he serves as director of The Gilbert M. Grosvenor Center for Geographic Education. His most current project includes the production of the video-based professional development series, *Geography: Teaching With the Stars.* Available programs may be viewed at www.geoteach.org.

Contributing Author

Jay McTighe has published articles in a number of leading educational journals and has coauthored 10 books, including the best-selling *Understanding By Design* series with Grant Wiggins. McTighe also has an extensive background in professional development and is a featured speaker at national, state, and district conferences and workshops. He received his undergraduate degree from The College of William and Mary, earned a Masters degree from The University of Maryland and completed post-graduate studies at The Johns Hopkins University.

Cover Photo Credits: Main Image: Bruno Morandi/Robert Harding World Imagery/Getty Images; **Background:** Bjorn Holland/The Image Bank/Getty Images. **Thumbnails:** (l to r, top to bottom) Peter Zelei/Getty Images; John Coletti/Getty Images; Matt Champlin/ Getty Images; ©Jacom Stephens/Getty Images; NASA/NOAA/SPL/Getty Images; BananaStock/PictureQuest; ©Ocean/Corbis; Ingram Publishing; ©Pete Atkinson/Getty Images; Getty Images/Digital Vision; ©Image Source, all rights reserved; Jack Goldfarb/Getty Images; Getty Images; ©Image Source/Getty Images; Beanstock Images/Radius Images/Getty Images; Stephen Waldron.

McGraw-Hill networks ™ meets you anywhere—takes you everywhere. It's All Online. 1. Go to connected.mcgraw-hill.com. 2. Get your User Name and Password from your teacher and enter them. 3. Click on your *networks* book. 4. Select your chapter and lesson.

www.mheonline.com/networks

Copyright © 2015 McGraw-Hill Education

All rights reserved. No part of this publication may be reproduced or distributed in any form or by any means, or stored in a database or retrieval system, without the prior written consent of McGraw-Hill Education, including, but not limited to, network storage or transmission, or broadcast for distance learning.

Send all inquiries to:
McGraw-Hill Education
8787 Orion Place
Columbus, OH 43240

ISBN: 978-0-07-664288-5
MHID: 0-07-664288-7

Printed in the United States of America.

5 6 7 8 9 DOW 17 16 15 14

REVIEWERS AND CONSULTANTS

ACADEMIC CONSULTANTS

Frederick L. Bein, Ph.D.
Professor of Geography
Indiana University-Purdue University
Indianapolis Adjunct Professor of Geography
Moi University
Indianapolis, Indiana

David Berger, Ph.D.
Ruth and I. Lewis Gordon Professor of Jewish
   History
Dean, Bernard Revel Graduate School
Yeshiva University
New York, New York

Randy Bertolas, Ph.D.
Professor of Geography
Wayne State College
Wayne, Nebraska

Lara Bryant, Ph.D.
Assistant Professor of Geography
Keene State College
Keene, New Hampshire

Tom Daccord
Educational Technology Specialist
Co-Director EdTech Teacher
Boston, Massachusetts

Charles Gritzner, Ph.D.
Professor of Geography, Retired
South Dakota State University
Brookings, South Dakota

P.P. Karan, Ph.D.
Professor of Geography
University of Kentucky
Lexington, Kentucky

Jeff Lash, Ph.D.
Associate Professor of Geography
University of Houston-Clear Lake
Houston, Texas

Elizabeth Leppman, Ph.D.
Writer
Map and Book Editor
Lexington, Kentucky

Joseph Manzo, Ph.D.
Professor of Geography
Concord University
Athens, West Virginia

Kent McGregor, Ph.D.
Associate Professor of Geography
University of North Texas
Denton, Texas

Olga Medvedkov, Ph.D.
Professor of Geography
Wittenberg University
Springfield, Ohio

Jerry Mitchell, Ph.D.
Research Associate Professor of Geography
University of South Carolina
Columbia, South Carolina

Shannon O'Lear, Ph.D.
Associate Professor of Geography
University of Kansas
Lawrence, Kansas

Justin Reich
Educational Technology Specialist
Co-Director EdTech Teacher
Boston, Massachusetts

David Rutherford, Ph.D.
Associate Professor of Geography & Public Policy
University of Mississippi
Oxford, Mississippi

Richard Sambrook, Ph.D.
Professor of Geography
Department Head
Eastern Michigan University
Ypsilanti, Michigan

Ginger Schmid, Ph.D.
Assistant Professor of Geography
Minnesota State University
Mankato, Minnesota

Joseph Stoltman, Ph.D.
Professor of Geography
Western Michigan University
Kalamazoo, Michigan

William Strong, Ph.D.
Professor Emeritus of Geography
University of North Alabama
Florence, Alabama

TEACHER REVIEWERS

Kimberly Coffelt
Social Studies Teacher
Southeast High School
Wichita, Kansas

Sara Heck
Social Studies Teacher
Teays Valley Local School District
Ashville, Ohio

Kenny Lee
Social Studies Curriculum
   Coordinator
Westerville City Schools
Westerville, Ohio

Barry Leonard
Social Studies Teacher
Graves County High School
Mayfield, Kentucky

Lorena McMenomy
Social Studies Educator
Hillcrest High School
Dallas, Texas

Jason Orendi
Social Studies Teacher
Boonsboro High School
Martinsburg, West Virginia

Jamie Reese
Social Studies Teacher
Buckhorn High School
New Market, Alabama

Adam Schwartz
Social Studies Teacher
Montgomery County Public
   Schools
Rockville, Maryland

Judi Shortt
Social Studies Teacher
Edmond Public Schools
Edmond, Oklahoma

Natalie Wojinski
Geography Teacher
Hercules High School
San Francisco, California

CONTENTS

UNIT ONE

Orlando Sentinel/McClatchy-Tribune/Getty Images

CHAPTER 1

How Geographers Look at the World 11

ESSENTIAL QUESTION

How does geography help us interpret the past, understand the present, and plan for the future?

Harry Kikstra/Flickr/Getty Images

CHAPTER 2

The Physical World 37

ESSENTIAL QUESTION

How do physical processes shape Earth's surface?

CONTENTS

CONTENTS

Viviane Ponti/Lonely Planet Images/Getty Images

Tomasz Tomaszewski/National Geographic Stock

CONTENTS

UNIT FIVE

x

CONTENTS

CONTENTS

CONTENTS

UNIT NINE

FEATURES

FEATURES

Global Connections

Connecting Geography to ....

PRIMARY SOURCES

Analyzing PRIMARY SOURCES

Analyzing PRIMARY SOURCES

MAPS

Interactive Charts, Graphs, and Tables

netw rks

Explore these assets as well as all assets from the Student Edition online!

Interactive Images and Slide Shows

ONLINE RESOURCES

networks Explore these assets as well as all assets from the Student Edition online!

xxix

⌄ Time Lines

How do I study Geography?

Geographers have tried to understand the best way to teach and learn about geography. In order to do this, geographers created the Five Themes of Geography. The themes acted as a guide for teaching the basic ideas about geography to students like yourself.

People who teach and study geography, though, thought that the Five Themes were too broad. In 1994, geographers created 18 national geography standards, *Geography for Life*. These standards were more detailed about what should be taught and learned. The Six Essential Elements act as a bridge connecting the Five Themes with the standards.

These pages show you how the Five Themes are related to the Six Essential Elements and the 18 standards.

5 Themes of Geography

1 Location

Location describes where something is. Absolute location describes a place's exact position on the Earth's surface. Relative location expresses where a place is in relation to another place.

2 Place

Place describes the physical and human characteristics that make a location unique.

3 Regions

Regions are areas that share common characteristics.

4 Movement

Movement explains how and why people and things move and are connected.

5 Human-Environment Interaction

Human-Environment Interaction describes the relationship between people and their environment.

(t to b)Thorsten Henn/Cultura/Getty Images, David S. Boyer/Getty Images, John Lamb/Stone/Getty Images, Panoramic Images/Glow Images

6 Essential Elements

18 Geography Standards

I. The World in Spatial Terms

Geographers look to see where a place is located. Location acts as a starting point to answer "Where Is It?" The location of a place helps you orient yourself as to where you are.

1 How to use maps and other tools

2 How to use mental maps to organize information

3 How to analyze the spatial organization of people, places, and environments

II. Places and Regions

Place describes physical characteristics such as landforms, climate, and plant or animal life. It might also describe human characteristics, including language and way of life. Places can also be organized into regions. Regions are places united by one or more characteristics.

4 The physical and human characteristics of places

5 How people create regions to interpret Earth's complexity

6 How culture and experience influence people's perceptions of places and regions

III. Physical Systems

Geographers study how physical systems, such as hurricanes, volcanoes, and glaciers, shape the surface of the Earth. They also look at how plants and animals depend upon one another and their surroundings for their survival.

7 The physical processes that shape Earth's surface

8 The distribution of ecosystems on Earth's surface

9 The characteristics, distribution, and migration of human populations

10 The complexity of Earth's cultural mosaics

IV. Human Systems

People shape the world in which they live. They settle in certain places but not in others. An ongoing theme in geography is the movement of people, ideas, and goods.

11 The patterns and networks of economic interdependence

12 The patterns of human settlement

13 The forces of cooperation and conflict

V. Environment and Society

How does the relationship between people and their natural surroundings influence the way people live? Geographers study how people use the environment and how their actions affect the environment.

14 How human actions modify the physical environment

15 How physical systems affect human systems

16 The meaning, use, and distribution of resources

VI. The Uses of Geography

Knowledge of geography helps us understand the relationships among people, places, and environments over time. Applying geographic skills helps you understand the past and prepare for the future.

17 How to apply geography to interpret the past

18 How to apply geography to interpret the present and plan for the future

SCAVENGER HUNT

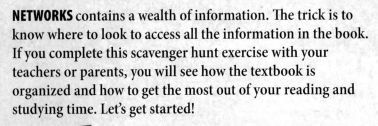

NETWORKS contains a wealth of information. The trick is to know where to look to access all the information in the book. If you complete this scavenger hunt exercise with your teachers or parents, you will see how the textbook is organized and how to get the most out of your reading and studying time. Let's get started!

1 How many chapters are in Unit 1?

2 What does Unit 3 cover?

3 Where can you find the Essential Question for each chapter?

4 Where can you find primary sources in your textbook?

5 How can you identify content vocabulary and academic vocabulary in the narrative?

6 Where do you find graphic organizers in your textbook?

7 You want to quickly find a map in the book about climate in Europe. Where do you look?

8 Where would you find the latitude and longitude for Dublin, Ireland?

9 If you needed to know the Spanish term for *primate city*, where would you look?

10 Where can you find a list of all the features in the book?

(t) Image Source / Punchstock,
(c) ScotStock/Alamy,
(b) Ingram Publishing/Alamy

REFERENCE ATLAS

ATLAS KEY

Ice Cap

Tundra

Evergreen Forest

Mixed Forest

Mountains

Grassland

Oceans

Seas

Desert

SYMBOL KEY

.... Claimed boundary	✪ National capital	Dry salt lake
—— International boundary (political map)	○ State/Provincial capital	Lake
—— International boundary (physical map)	• Towns	Rivers
	▼ Depression	Canal
	▲ Elevation	

GEOGRAPHIC DICTIONARY

Ocean

Archipelago

Gulf

Reservoir

Volcano

Isthmus

Plateau

Highlands

Canyon

Cliff

Cape

Bay

Harbor

Reef

Island

Channel

Peninsula

---

archipelago a group of islands

basin area of land drained by a given river and its branches; area of land surrounded by lands of higher elevations

bay part of a large body of water that extends into a shoreline, generally smaller than a gulf

canyon deep and narrow valley with steep walls

cape point of land that extends into a river, lake, or ocean

channel wide strait or waterway between two landmasses that lie close to each other; deep part of a river or other waterway

cliff steep, high wall of rock, earth, or ice

continent one of the seven large landmasses on the Earth

delta flat, low-lying land built up from soil carried downstream by a river and deposited at its mouth

divide stretch of high land that separates river systems

downstream direction in which a river or stream flows from its source to its mouth

escarpment steep cliff or slope between a higher and lower land surface

glacier large, thick body of slowly moving ice

gulf part of a large body of water that extends into a shoreline, generally larger and more deeply indented than a bay

harbor a sheltered place along a shoreline where ships can anchor safely

highland elevated land area such as a hill, mountain, or plateau

hill elevated land with sloping sides and rounded summit; generally smaller than a mountain

island land area, smaller than a continent, completely surrounded by water

isthmus narrow stretch of land connecting two larger land areas

lake a sizable inland body of water

lowland land, usually level, at a low elevation

mesa broad, flat-topped landform with steep sides; smaller than a plateau

mountain land with steep sides that rises sharply (1,000 feet or more) from surrounding land; generally larger and more rugged than a hill

Labels on image: Desert, Oasis, Mountain Peak, Sound, Basin, Mountain Range, Source of River, Glacier, Tributary, Hills, Strait, Valley, Upstream, Lake, Downstream, River, Mouth of River, Escarpment, Lowland, Plain, Delta, Seacoast

mountain peak pointed top of a mountain

mountain range a series of connected mountains

mouth (of a river) place where a stream or river flows into a larger body of water

oasis small area in a desert where water and vegetation are found

ocean one of the four major bodies of salt water that surround the continents

ocean current stream of either cold or warm water that moves in a definite direction through an ocean

peninsula body of land jutting into a lake or ocean, surrounded on three sides by water

physical feature characteristic of a place occurring naturally, such as a landform, body of water, climate pattern, or resource

plain area of level land, usually at low elevation and often covered with grasses

plateau area of flat or rolling land at a high elevation, about 300 to 3,000 feet (90 to 900 m) high

reef a chain of rocks, coral or sand at or near the surface of the water

river large natural stream of water that runs through the land

sea large body of water completely or partly surrounded by land

seacoast land lying next to a sea or an ocean

sound broad inland body of water, often between a coastline and one or more islands off the coast

source (of a river) place where a river or stream begins, often in highlands

strait narrow stretch of water joining two larger bodies of water

tributary small river or stream that flows into a large river or stream; a branch of the river

upstream direction opposite the flow of a river; toward the source of a river or stream

valley area of low land usually between hills or mountains

volcano mountain or hill created as liquid rock and ash erupt from inside the Earth

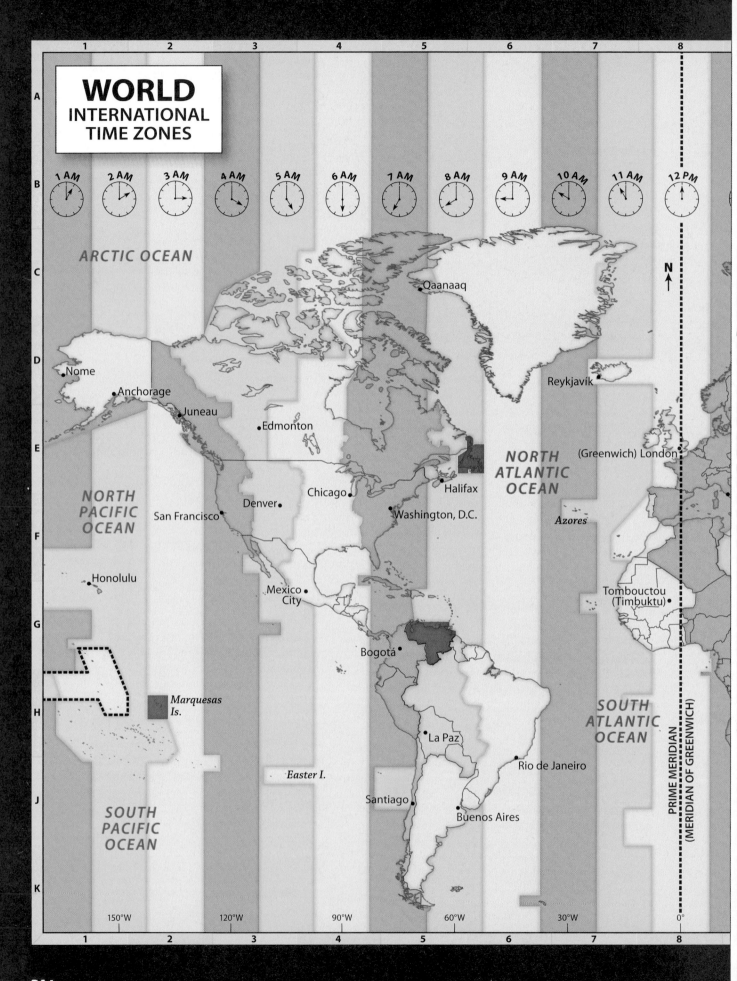

WORLD
INTERNATIONAL TIME ZONES

1 AM 2 AM 3 AM 4 AM 5 AM 6 AM 7 AM 8 AM 9 AM 10 AM 11 AM 12 PM

ARCTIC OCEAN

N

Qaanaaq

Nome

Anchorage

Juneau

Reykjavík

Edmonton

NORTH
ATLANTIC
OCEAN

(Greenwich) London

NORTH
PACIFIC
OCEAN

Chicago

Denver

Halifax

San Francisco

Washington, D.C.

Azores

Honolulu

Mexico
City

Tombouctou
(Timbuktu)

Bogotá

Marquesas
Is.

SOUTH
ATLANTIC
OCEAN

La Paz

Easter I.

Rio de Janeiro

Santiago

Buenos Aires

SOUTH
PACIFIC
OCEAN

PRIME MERIDIAN
(MERIDIAN OF GREENWICH)

150°W   120°W   90°W   60°W   30°W   0°

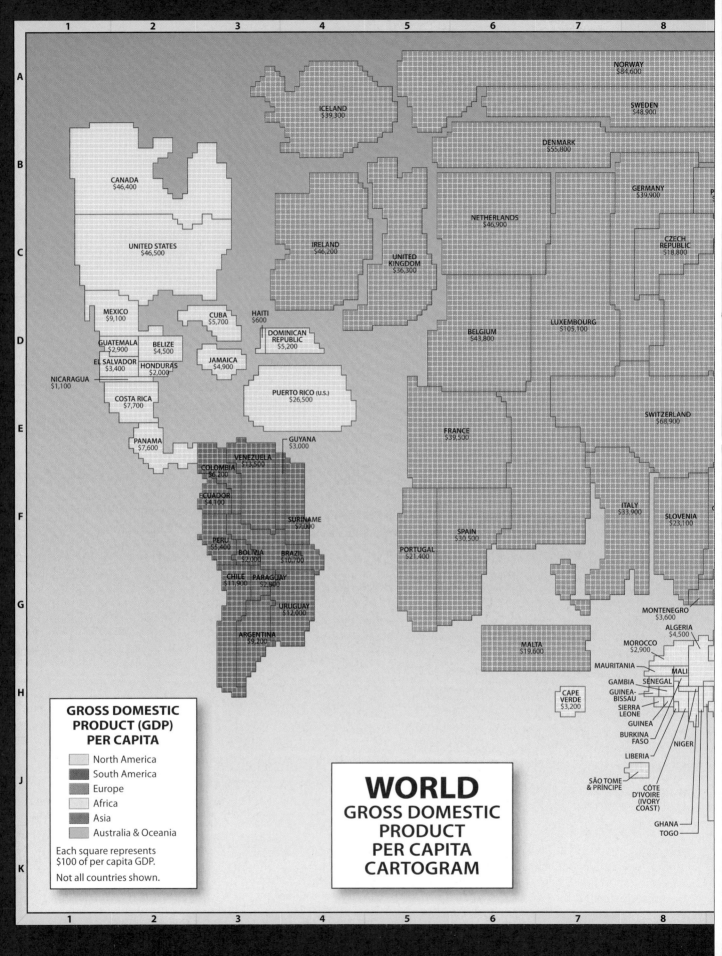

WORLD
GROSS DOMESTIC PRODUCT PER CAPITA CARTOGRAM

GROSS DOMESTIC PRODUCT (GDP) PER CAPITA

- North America
- South America
- Europe
- Africa
- Asia
- Australia & Oceania

Each square represents $100 of per capita GDP.

Not all countries shown.

A

NORWAY $84,600

ICELAND $39,300

SWEDEN $48,900

DENMARK $55,800

B

CANADA $46,400

GERMANY $39,900

NETHERLANDS $46,900

CZECH REPUBLIC $18,800

C

UNITED STATES $46,500

IRELAND $46,200

UNITED KINGDOM $36,300

BELGIUM $43,800

LUXEMBOURG $105,100

D

MEXICO $9,100

CUBA $5,700

HAITI $600

DOMINICAN REPUBLIC $5,200

GUATEMALA $2,900

BELIZE $4,500

EL SALVADOR $3,400

HONDURAS $2,000

JAMAICA $4,900

SWITZERLAND $68,900

NICARAGUA $1,100

COSTA RICA $7,700

PUERTO RICO (U.S.) $26,500

E

PANAMA $7,600

GUYANA $3,000

VENEZUELA $13,500

COLOMBIA $6,200

FRANCE $39,500

F

ECUADOR $4,100

SURINAME $7,000

PERU $5,400

BOLIVIA $2,000

BRAZIL $10,700

SPAIN $30,500

PORTUGAL $21,400

ITALY $33,900

SLOVENIA $23,100

CHILE $11,900

PARAGUAY $2,800

G

URUGUAY $12,000

ARGENTINA $9,200

MONTENEGRO $3,600

ALGERIA $4,500

MOROCCO $2,900

MALTA $19,600

MAURITANIA

H

CAPE VERDE $3,200

GAMBIA

SENEGAL

MALI

GUINEA-BISSAU

SIERRA LEONE

GUINEA

BURKINA FASO

NIGER

LIBERIA

J

SÃO TOME & PRÍNCIPE

CÔTE D'IVOIRE (IVORY COAST)

GHANA

TOGO

K

RA6

A

FINLAND
$44,500

ESTONIA
$14,100

B

LATVIA
$10,700

JAPAN
$43,100

POLAND
$12,300

LITHUANIA
$11,000

BELARUS
$5,700

UKRAINE
$3,000

C

KYRGYZSTAN
TAJIKISTAN
AFGHANISTAN

NORTH KOREA
$500

SLOVAKIA
$16,000

RUSSIA
$10,400

AUSTRIA
$45,200

GEORGIA

KAZAKHSTAN
$9,200

MONGOLIA

MYANMAR
(BURMA)

SOUTH KOREA
$21,100

MOLDOVA

CHINA
$4,400

LAOS

D

ROMANIA
$7,500

AZERBAIJAN

UZBEKISTAN

BHUTAN
$2,000

VIETNAM

HUNGARY
$12,900

ARMENIA

TURKMENISTAN
$4,600

NEPAL

THAILAND
$4,600

CAMBODIA

IRAN
$5,200

PAKISTAN
$1,000

BANGLADESH
$700

TURKEY
$10,100

INDIA
$1,400

SRI LANKA

MALAYSIA
$8,400

BULGARIA
$6,400

SERBIA
$5,100

E

PHILIPPINES
$2,100

KOSOVO

KUWAIT
$45,400

MACEDONIA

BOSNIA &
HERZEGOVINA
$4,500

IRAQ
$900

KIRIBATI
$1,500

CROATIA
$13,800

GREECE
$26,500

SYRIA
$2,900

QATAR
$72,400

F

LEBANON
$9,300

BRUNEI
$32,600

JORDAN

ALBANIA

ISRAEL
$29,300

UNITED ARAB
EMIRATES
$39,600

SINGAPORE
$43,800

G

SOLOMON
ISLANDS

INDONESIA
$2,900

PAPUA
NEW GUINEA

SAUDI
ARABIA
$15,800

TUNISIA
$4,200

EGYPT
$2,700

OMAN
$20,800

EAST TIMOR
(TIMOR-LESTE)
$700

FIJI ISLANDS
$3,500

LIBYA
$11,300

YEMEN
$1,400

CHAD

SUDAN
$1,800

H

CAMEROON
$1,200

ERITREA

ETHIOPIA

DJIBOUTI
$1,300

CENTRAL AFRICAN
REPUBLIC

SOMALIA

KENYA

EQUATORIAL
GUINEA
$16,900

UGANDA

RWANDA

DEM. REP. OF
THE CONGO

BURUNDI

NEW
ZEALAND
$32,400

TONGA

GABON
$12,500

TANZANIA

CONGO

ZAMBIA

J

ANGOLA
$4,300

MALAWI
$400

NIGERIA

ZIMBABWE

BENIN

MOZAMBIQUE

MAURITIUS
$7,500

AUSTRALIA
$57,100

NAMIBIA
$5,100

BOTSWANA
$7,400

SWAZILAND
$3,300

MADAGASCAR
$400

LESOTHO
$1,000

K

SOUTH AFRICA
$7,300

SOURCE: The United Nations Statistics Division online, 2010

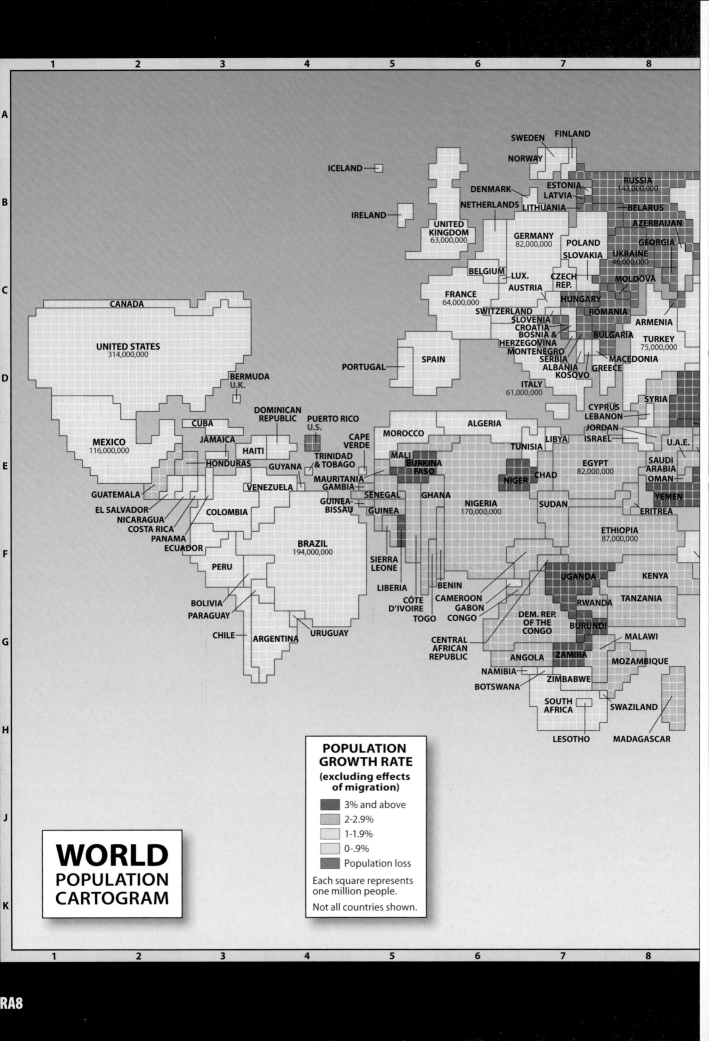

WORLD
POPULATION
CARTOGRAM

**POPULATION
GROWTH RATE**
**(excluding effects
of migration)**

- 3% and above
- 2-2.9%
- 1-1.9%
- 0-.9%
- Population loss

Each square represents
one million people.

Not all countries shown.

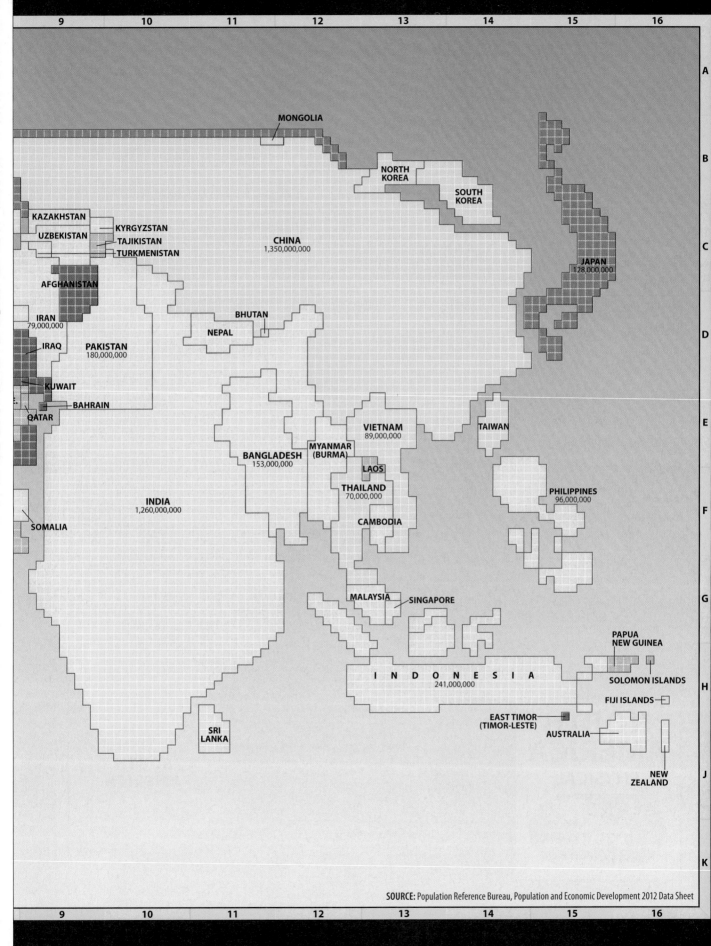

9　　10　　11　　12　　13　　14　　15　　16

A

MONGOLIA

B

NORTH
KOREA

SOUTH
KOREA

KAZAKHSTAN

KYRGYZSTAN

UZBEKISTAN

TAJIKISTAN

TURKMENISTAN

CHINA
1,350,000,000

C

JAPAN
128,000,000

AFGHANISTAN

BHUTAN

IRAN
79,000,000

NEPAL

D

IRAQ

PAKISTAN
180,000,000

KUWAIT

BAHRAIN

VIETNAM
89,000,000

TAIWAN

E

QATAR

BANGLADESH
153,000,000

MYANMAR
(BURMA)

LAOS

PHILIPPINES
96,000,000

F

SOMALIA

INDIA
1,260,000,000

THAILAND
70,000,000

CAMBODIA

MALAYSIA

SINGAPORE

G

PAPUA
NEW GUINEA

I N D O N E S I A
241,000,000

SOLOMON ISLANDS

H

FIJI ISLANDS

SRI
LANKA

EAST TIMOR
(TIMOR-LESTE)

AUSTRALIA

NEW
ZEALAND

J

K

SOURCE: Population Reference Bureau, Population and Economic Development 2012 Data Sheet

9　　10　　11　　12　　13　　14　　15　　16

NORTH AMERICA
PHYSICAL

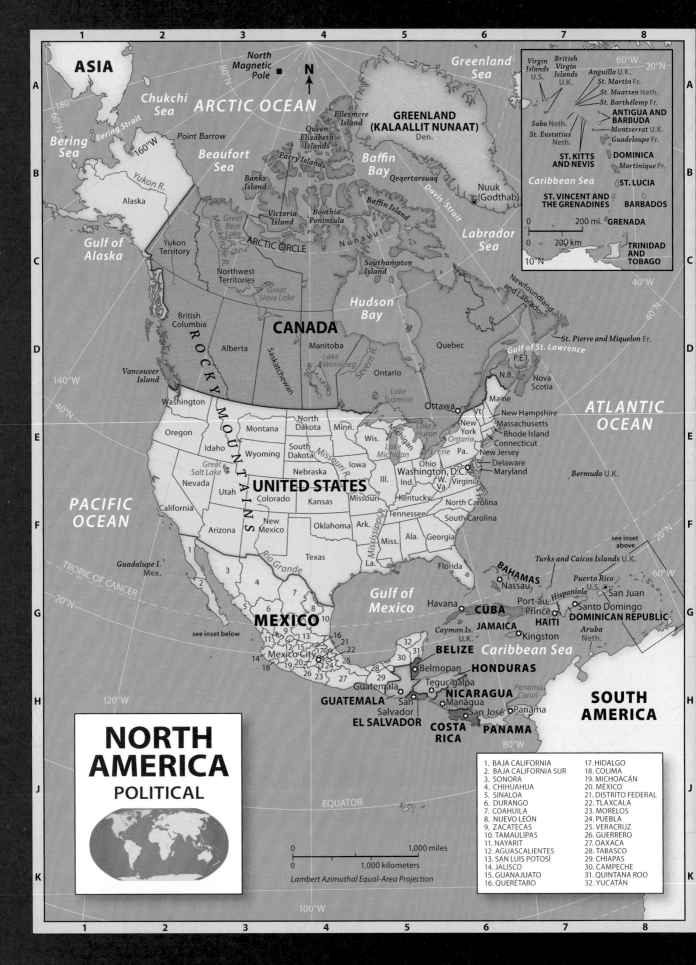

NORTH AMERICA POLITICAL

1 2 3 4 5 6 7 8
Grid reference rows: A B C D E F G H J K

ASIA

ARCTIC OCEAN

North Magnetic Pole

N

Chukchi Sea

Bering Sea

Bering Strait

Point Barrow

Beaufort Sea

Parry Islands

Banks Island

Yukon R.

Alaska

160°W

180°

60°N

Gulf of Alaska

Yukon Territory

Mackenzie R.

Great Bear Lake

Victoria Island

Boothia Peninsula

Northwest Territories

ARCTIC CIRCLE

Ellesmere Island

Queen Elizabeth Islands

Baffin Bay

Baffin Island

Davis Strait

Qeqertarsuaq

Nuuk (Godthab)

GREENLAND (KALAALLIT NUNAAT) Den.

Greenland Sea

80°W

20°N

Great Slave Lake

British Columbia

Vancouver Island

Washington

140°W

40°N

CANADA

Alberta

Saskatchewan

Manitoba

Severn R.

Lake Winnipeg

Ontario

Quebec

Southampton Island

Nunavut

Hudson Bay

Labrador Sea

Newfoundland and Labrador

Gulf of St. Lawrence

St. Pierre and Miquelon Fr.

P.E.I.

N.B.

Nova Scotia

Maine

40°N

ATLANTIC OCEAN

Lake Superior

Lake Huron

Lake Michigan

Lake Ontario

L. Erie

Montana

North Dakota

Minn.

Wis.

Michigan

Oregon

Idaho

Wyoming

South Dakota

Iowa

Ill.

Ind.

Ohio

Pa.

New York

New Hampshire

Vt.

Massachusetts

Rhode Island

Connecticut

New Jersey

Delaware

Maryland

W. Va.

Virginia

Washington, D.C.

Bermuda U.K.

PACIFIC OCEAN

Missouri R.

Great Salt Lake

Nevada

Utah

Colorado

Nebraska

Kansas

Missouri

Kentucky

Tennessee

North Carolina

South Carolina

California

UNITED STATES

ROCKY MOUNTAINS

Arizona

New Mexico

Oklahoma

Ark.

Miss.

Ala.

Georgia

Texas

Rio Grande

Mississippi R.

La.

Florida

Gulf of Mexico

Guadalupe I. Mex.

TROPIC OF CANCER

120°W

20°N

1

2

3

4

5

6

7

8

9

10

11

12

13

15

16

17

21

22

14

18

19

20

23

24

25

26

27

28

29

30

31

32

MEXICO

Mexico City

see inset below

Havana

CUBA

BAHAMAS

Nassau

Cayman Is. U.K.

JAMAICA

Kingston

HAITI

Port-au-Prince

Hispaniola

Santo Domingo

DOMINICAN REPUBLIC

San Juan

Puerto Rico U.S.

Turks and Caicos Islands U.K.

see inset above

60°W

20°N

BELIZE

Belmopan

Caribbean Sea

Aruba Neth.

GUATEMALA

Guatemala

San Salvador

EL SALVADOR

HONDURAS

Tegucigalpa

NICARAGUA

Managua

COSTA RICA

San José

PANAMA

Panama

Panama Canal

80°W

SOUTH AMERICA

EQUATOR

100°W

0   1,000 miles
0   1,000 kilometers
Lambert Azimuthal Equal-Area Projection

Virgin Islands U.S.

British Virgin Islands U.K.

Anguilla U.K.

St. Martin Fr.

St. Maarten Neth.

St. Barthélemy Fr.

ANTIGUA AND BARBUDA

Saba Neth.

St. Eustatius Neth.

Montserrat U.K.

Guadeloupe Fr.

ST. KITTS AND NEVIS

DOMINICA

Martinique Fr.

Caribbean Sea

ST. LUCIA

ST. VINCENT AND THE GRENADINES

BARBADOS

GRENADA

TRINIDAD AND TOBAGO

60°W

20°N

10°N

40°W

0   200 mi.
0   200 km

Legend:
1. BAJA CALIFORNIA
2. BAJA CALIFORNIA SUR
3. SONORA
4. CHIHUAHUA
5. SINALOA
6. DURANGO
7. COAHUILA
8. NUEVO LEÓN
9. ZACATECAS
10. TAMAULIPAS
11. NAYARIT
12. AGUASCALIENTES
13. SAN LUIS POTOSÍ
14. JALISCO
15. GUANAJUATO
16. QUERÉTARO
17. HIDALGO
18. COLIMA
19. MICHOACÁN
20. MÉXICO
21. DISTRITO FEDERAL
22. TLAXCALA
23. MORELOS
24. PUEBLA
25. VERACRUZ
26. GUERRERO
27. OAXACA
28. TABASCO
29. CHIAPAS
30. CAMPECHE
31. QUINTANA ROO
32. YUCATÁN

Caribbean Sea

1 2 3 4 5 6 7 8

N

0   1,000 miles
0   1,000 kilometers
Lambert Azimuthal Equal-Area Projection

40°W

VENEZUELA
Caracas

GUYANA
SURINAME

80°W

Malpelo I.
Bogotá

COLOMBIA

LLANOS
Orinoco R.

Angel Falls
Total drop
3,212 ft. 979 m

Georgetown
Paramaribo
Cayenne
FRENCH GUIANA

GUIANA HIGHLANDS

Boundary claimed
by Suriname

*Marajó
Island*

Quito
ECUADOR

Rio Negro
EQUATOR

0°

A   M   A   Z   O   N

Marañón R.
Amazon R.
Amazon R.

B   A   S   I   N

Ucayali R.

PERU

S e l v a s

Purus R.
Madeira R.
Topajos R.
Xingu R.
Araguaia R.
Tocantins R.
São Francisco R.

BRAZIL

Lima

*Machu
Picchu*

*Lakes
Titicaca*

La Paz
BOLIVIA

*MATO GROSSO
PLATEAU*

B R A Z I L I A N

Brasília

Sucre

Altiplano

*Salar
de Uyuni*

CHACO

H I G H L A N D S

20°S

TROPIC OF CAPRICORN

GRAN

PARAGUAY

Paraguay R.
Paraná R.

*Iguazú
Falls*

San Ambrosio I.

San Félix I.

Asunción

P A M P A S

Paraná R.

Uruguay R.

ATLANTIC
OCEAN

CHILE
*Aconcagua
22,834 ft.
6,960 m*

Juan Fernández Is.
Santiago

Buenos Aires

URUGUAY
Montevideo

ARGENTINA

Rio de la Plata

Colorado R.

Negro R.

Chiloé Island

*Valdés Peninsula
-131 ft.
-40 m*

*Taitao
Peninsula*

*Gulf of
San Jorge*

**PACIFIC
OCEAN**

Wellington I.

P A T A G O N I A

*Falkland Islands
(Islas Malvinas)*

Stanley

Tierra del Fuego

*Strait of
Magellan*

Cape Horn

South Georgia Island

SOUTH AMERICA
PHYSICAL

80°W   60°W   40°W   20°W

100°W

20°S

40°S

Caribbean Sea

1 2 3 4 5 6 7 8

0 1,000 miles
0 1,000 kilometers
Lambert Azimuthal Equal-Area Projection

N

Santa Marta
Barranquilla
Maracaibo
Caracas
Cartagena
Valencia
VENEZUELA
Ciudad Guayana
GUYANA
SURIName
Bucaramanga
San Cristóbal
Paramaribo
Medellín
Orinoco R.
Georgetown
Cayenne
Bogotá
FRENCH GUIANA
Maracaibo
Fr.
Malpelo I.
Cali
Col.
COLOMBIA
Boa Vista
Boundary claimed by Suriname
Marajó Island
Esmeraldas
Rio Negro
Quito
EQUATOR
ECUADOR
A M A Z O N
Amazon R.
Guayaquil
Manaus
Belém
São Luís
Iquitos
Santarém
B A S I N
Fortaleza
Marañón R.
Amazon R.
Teresina
Purus R.
Madeira R.
Tapajós R.
Natal
PERU
Xingu R.
Campina Grande
Rio Branco
Pôrto Velho
Tocantins R.
Araguaia R.
Recife
Callao
São Francisco R.
Lima
Machu Picchu
BRAZIL
Salvador
Ayacucho
Cuzco
Trinidad
Lake Titicaca
Brasília
La Paz
Goiânia
Arequipa
BOLIVIA
Santa Cruz
Uberlândia
Arica
Oruro
Sucre
Campo Grande
Uberaba
Belo Horizonte
Iquique
Tarija
Paraguay R.
Paraná R.
20°S
20°S
Antofagasta
PARAGUAY
Londrina
Campinas
Nova Iguaçu
TROPIC OF CAPRICORN
Salta
São Paulo
Rio de Janeiro
CHILE
Asunción
Santos
San Félix I.
San Ambrosio I.
Curitiba
Chile
San Miguel de Tucumán
Uruguaiana
ATLANTIC OCEAN
La Serena
Paraná R.
Santa Maria
Pôrto Alegre
Coquimbo
Córdoba
Uruguay R.
Juan Fernández Is.
Rosario
Chile
Valparaíso
Mendoza
URUGUAY
Santiago
Buenos Aires
Montevideo
La Plata
Rio de la Plata
Concepción
ARGENTINA
Mar del Plata
Colorado R.
Bahía Blanca
Negro R.
Puerto Montt
40°S
40°S
SOUTH AMERICA
POLITICAL
Comodoro Rivadavia
PACIFIC OCEAN
Falkland Islands (Islas Malvinas)
Stanley
Administered by United Kingdom
Claimed by Arg.
Río Gallegos
Punta Arenas
South Georgia Island
U.K.
Ushuaia
Strait of Magellan
Cape Horn
100°W
80°W
60°W
40°W
20°W

EUROPE
PHYSICAL

N

| | 1 | 2 | 3 | 4 | 5 | 6 | 7 | 8 |

A

Reykjavík
ICELAND

Norwegian Sea

B

S
C
A
N
D
I
N

Faeroe Islands

C

Shetland Islands

NORWAY

Oslo
SWEDEN

Gulf of

D

Outer Hebrides

Orkney Islands

British Isles

Highlands

Stockholm

Gotland

Edinburgh
Belfast
IRELAND
UNITED KINGDOM

Irish Sea
Dublin

North Sea

Skagerrak

Jutland
Zealand

Kattegat

Baltic

E

Great Britain

Celtic Sea

Cardiff

Thames R.

London

NETHERLANDS
Amsterdam

Berlin

DENMARK
Copenhagen

Elbe R.

N
O
R
T
H

ATLANTIC OCEAN

Land's End

English Channel

Brussels
BELGIUM

Rhine R.

GERMANY

POLAND

F

Brittany

Seine R.

Paris
LUXEMBOURG
Luxembourg

Prague
CZECH REPUBLIC

Oder R.

Loire R.

FRANCE

Danube R.

Bratislava
SLOVAKIA

G

Bay of Biscay

Cantabrian Mountains

LIECHTENSTEIN
Bern
SWITZERLAND
Vaduz

Vienna
AUSTRIA

Budapest
HUNGARY

Mont Blanc
15,771 ft.
4,807 m

Massif Central

A
L
P
S

SLOVENIA
Ljubljana

Po R.

Zagreb
CROATIA

Douro R.

Ebro R.

Pyrenees

Andorra
la Vella

Rhône R.

H

MONACO

Riviera

SAN MARINO

BOSNIA & HERZEGOVINA
Sarajevo

ANDORRA

A
p
e
n
n
i
n
e
s

Adriatic Sea

I B E R I A N

Madrid

Lisbon
PORTUGAL

Tagus R.

Corsica

ITALY

MONTENEGRO
Podgorica

SPAIN
P E N I N S U L A

Rome

Tiranë
ALBANIA

J

Cape
St. Vincent

Baetic Mountains

Balearic Islands

Sardinia

VATICAN CITY
(within Rome)

Strait of Gibraltar
GIBRALTAR

M e d i t e r r a n e a n

Tyrrhenian Sea

Ionian Sea

Sicily
Etna
10,902 ft.
3,323 m

K

AFRICA

MALTA
Valletta

| | 1 | 2 | 3 | 4 | 5 | 6 | 7 | 8 |

North Cape

Barents Sea

A

80°E

60°N

ASIA

B

U R A L   M O U N T A I N S

Europe/Asia boundary

Pechora R.

Kola Peninsula

White Sea

L A P L A N D

S C A N D I N A V I A

Northern Dvina R.

FINLAND

Lake Region

Lake Onega

C

50°N

70°E

Helsinki ✪

Lake Ladoga

Gulf of Finland

RUSSIA

D

Tallinn ✪

ESTONIA

Sea

Bothnia

LATVIA

Riga ✪

Moscow ✪

Ural R.

E

LITHUANIA

Vilnius ✪

RUSSIA

Minsk ✪

BELARUS

C E N T R A L

Volga R.

KAZAKHSTAN

Warsaw ✪

Vistula R.

R U S S I A N

Don R.

E A S T E R N   E U R O P E A N   P L A I N

F

40°N

60°E

(Kyiv) Kiev ✪

U P L A N D

Dnieper R.

Caspian   Depression

Dniester R.

Carpathian Mts.

UKRAINE

Tisza R.

MOLDOVA

Chișinău ✪

Sea of Azov

Mt. Elbrus
18,510 ft.
5,642 m ▲

Caspian Sea

G

60°E

ROMANIA

Belgrade ✪ Bucharest ✪

Danube R.

Crimea

C a u c a s u s   Mountains

GEORGIA

AZERBAIJAN

Baku ✪

SERBIA

B A L K A N

KOSOVO

Priština ✪ Sofia ✪

BULGARIA

Balkan Mts.

Black Sea

H

Skopje ✪

P E N I N S U L A

MACEDONIA

Bosporus

T U R K E Y

GREECE

Dardanelles

Sea of Marmara

0 |———| 400 miles
0 |———| 400 kilometers
Lambert Azimuthal Equal-Area Projection

J

30°N

Aegean Sea

Athens ✪

Peloponnese

ASIA

Rhodes

Nicosia ✪

Crete

CYPRUS

K

50°E

Sea

30°E

40°E

EUROPE
POLITICAL

N

	1	2	3	4	5	6	7	8

30°W 20°W 10°W 70°N 0° 10°E 20°E

Tromsø

A

Reykjavík • Akureyri
ICELAND

ARCTIC CIRCLE

PRIME MERIDIAN

60°N

Norwegian Sea

Åre

B

Faeroe Islands
Denmark • Tórshavn

Trondheim
Ålesund

SWEDEN

Sundsvall

NORWAY

Gulf of

C

Shetland Islands
Lerwick

Bergen

Oslo

Uppsala
Stockholm

50°N

Isle of Lewis

Orkney Islands

Stavanger

Göteborg *Gotland*

Inverness
SCOTLAND
Aberdeen

North Sea

Århus

Baltic

30°W

Glasgow • Edinburgh
NORTHERN IRELAND Belfast

DENMARK
Copenhagen • Malmö

D

IRELAND
Dublin
Cork

UNITED KINGDOM

Irish Sea
Liverpool
Manchester

Kiel
Hamburg

Gdańsk
Bydgoszcz

E

WALES
Cardiff

ENGLAND
Birmingham

London
Southampton

The Hague • Amsterdam
NETHERLANDS

Elbe R.

Berlin

POLAND

Łódź

Celtic Sea

Land's End

Thames R.

Brussels
BELGIUM

Rhine R.

GERMANY
Bonn
Frankfurt

Wrocław

Kraków

ATLANTIC OCEAN

English Channel

Le Havre

Prague
CZECH REPUBLIC

F

Brest

Rennes

Seine R.

Paris

LUXEMBOURG
Luxembourg

Bratislava

SLOVAKIA

Nantes

Loire R.

Strasbourg

Danube R.

Munich

LIECHTENSTEIN Vienna

Budapest

40°N

La Rochelle

FRANCE
Limoges

Zürich
Bern

Vaduz

AUSTRIA

HUNGARY

G

La Coruña

Bay of Biscay

Bordeaux

Geneva
SWITZERLAND

SLOVENIA
Ljubljana
Zagreb

20°W

Vigo

Donostia-
San Sebastián

Lyon

A L P S

Milan

Venice

CROATIA

Porto

Bilbao

Po R.

Turin

BOSNIA & HERZEGOVINA

Coimbra

Valladolid

Pyrenees

Andorra
la Vella

Nice

Genoa

Rhône R.

SAN MARINO

Sarajevo

H

PORTUGAL

Madrid
Zaragoza

ANDORRA

Marseille

MONACO

Adriatic Sea

MONTENEGRO

Lisbon

Tagus R.

Ebro R.

Barcelona

Corsica
France

ITALY

Podgorica

Tiranë

Cape St. Vincent

SPAIN
Córdoba

Valencia

Murcia

Palma

Sardinia
Italy

VATICAN CITY
(within Rome)

Rome

Naples

ALBANIA

J

Cádiz
Seville
Málaga

Cartagena

Balearic Islands
Spain

Cagliari

Tyrrhenian Sea

Ionian Sea

Strait of Gibraltar
GIBRALTAR
U.K.

M e d i t e r r a n e a n

Palermo

Sicily
Italy

Messina
Catania

30°N

MALTA Valletta

K

AFRICA

10°W 30°N 0° 10°E 20°E

	1	2	3	4	5	6	7	8

A commonly accepted division between Asia and Europe—here marked by a gray line—is formed by the Ural Mountains, Ural River, Caspian Sea, Caucasus Mountains, and the Black Sea with its outlets, the Bosporus and the Dardanelles.

Europe/Asia boundary

ASIA

Barents Sea

Tobseda

Pechora

Kola Peninsula

Murmansk
Ivalo Kirovsk
Kiruna
Umba
White Sea

L A P L A N D

Kemi
Luleå
Oulu
Umeå
FINLAND

Kem'
Arkhangel'sk
Severodvinsk

Syktyvkar

Northern Dvina R.

U R A L   M O U N T A I N S

Vaasa
Kuopio
Lake Onega

Perm'

Pori
Tampere
Lake Ladoga

Kirov

Ufa

Turku
Helsinki
St. Petersburg

RUSSIA

Kazan'

Bothnia

Tallinn
ESTONIA

Novgorod
Yaroslavl'

Nizhniy Novgorod

Orenburg

Sea

LATVIA
Riga
Daugavpils

Tver'
Moscow

Samara

Ural R.

LITHUANIA
Vilnius
Vitsyebsk
Smolensk
Ryazan'
Penza
Saratov

Oral

Kaunas
Kaliningrad
BELARUS
Minsk
Homyel'
Bryansk
Kursk

Don R.

Volga R.

KAZAKHSTAN

Warsaw
Vistula R.

Chernihiv
Sumy
Kyiv (Kiev)
Kharkiv
Poltava

Volgograd

L'viv
Vinnytsya
UKRAINE
Dnipropetrovs'k
Donets'k

Astrakhan

Dnieper R.

Dniester R.

Carpathian Mts.

MOLDOVA
Chişinău

Rostov

Stavropol'

Caspian Sea

Odessa
Crimea
Simferopol'
Sea of Azov
Kerch
Grozny

Belgrade
Bucharest
ROMANIA
Sevastopol'
Yalta
Caucasus Mountains
AZERBAIJAN

SERBIA
Constanţa
GEORGIA
Baku

KOSOVO
Priština
Balkan Mts.
Varna
BULGARIA
Sofia
Black Sea

Skopje
MACEDONIA
Bosporus
İstanbul

Thessaloniki
T U R K E Y

GREECE
Dardanelles
Sea of Marmara

Aegean Sea

ASIA

Athens
Peloponnese

0   400 miles
0   400 kilometers
Lambert Azimuthal Equal-Area Projection

Rhodes
Nicosia

Iraklíon
Crete
Greece
CYPRUS

Sea

30°E   30°E   40°E   50°E   60°E   70°E   80°E

60°N   50°N   40°N   30°N

70°E   70°E   60°E

AFRICA
PHYSICAL

AFRICA
POLITICAL

ASIA
PHYSICAL

9   10   11   12   13   14   15   16

NORTH AMERICA

OCEAN

Chukchi
Sea
Wrangel I.
New
Siberian
Islands
Severnaya
Zemlya
Taymyr
Peninsula
Laptev
Sea
East
Siberian
Sea
Chukchi
Peninsula
Gulf of
Anadyr
Bering
Strait
Bering
Sea
Kolyma Mts.
Kolyma R.
Aleutian Islands
Commander Is.
Kamchatka
Peninsula

CENTRAL
SIBERIAN
PLATEAU
Cherski Range
Verkhoyanski Mts.
Angara R.
Lena R.
Aldan R.
Lena R.
Amur R.
Sea of
Okhotsk
Sakhalin
Kuril Islands

B E R I A
S I A
Lake
Baikal
Yablonovyy Range
Greater Khingan Range
Manchurian Plain
Sikhote-Alin' Range
Hokkaidō

Shan
Ulaanbaatar
MONGOLIA
GOBI
Beijing
Yablonovyy Range
Amur R.
**NORTH
KOREA**
P'yŏngyang
Seoul
**SOUTH
KOREA**
Sea of
Japan
(East
Sea)
Honshū
JAPAN
Tokyo
Nampo Shoto
TROPIC OF CANCER

Qilian Shan
Qaidam
Basin
Huang He
(Yellow R.)
North
China
Plain
Yellow
Sea
Shikoku
Kyūshū

C H I N A
Sichuan
Basin
Chang Jiang (Yangtze)
Gongga Shan
24,790 ft.
7,556 m
East
China
Sea
Ryukyu Islands
Okinawa
PACIFIC OCEAN

Mekong R.
Salween R.
Xi R.
Taipei
TAIWAN
Mariana Islands
Brahmaputra
Irrawaddy R.
Hanoi
Philippine
Sea
Caroline Islands

**MYANMAR
(BURMA)**
LAOS
Vientiane
Hainan
Luzon
Nay Pyi
Taw
THAILAND
Bangkok
South China
Sea
Manila
Mindoro
Samar
EQUATOR
Andaman
Islands
CAMBODIA
Phnom Penh
VIETNAM
PHILIPPINES

Andaman
Sea
Nicobar
Islands
Gulf of Thailand
Palawan
Mindanao
New Guinea

Bandar Seri Begawan
Sulu Sea
Celebes
Sea
Ceram
Aru
Is.
Kuala Lumpur
MALAYSIA
BRUNEI
Molucca
Buru
Tanimbar
Is.
Dolak
SINGAPORE
Borneo
Sulawesi
(Celebes)
Arafura Sea
Mentawai
Islands
I N D O N E S I A
Sumatra
Greater Sunda Islands
Dili
AUSTRALIA
Jakarta
Java
Java Sea
**EAST TIMOR
(TIMOR-LESTE)**
Timor
Timor Sea

Malay Peninsula

9   10   11   12   13   14   15   16

0   1,000 miles
0   1,000 kilometers
Two-Point Equidistant Projection

ASIA
POLITICAL

ASIA

N

TROPIC OF CANCER

20°N

M I C R O N E S I A

NORTHERN
MARIANA
ISLANDS
U.S.

Saipan

GUAM U.S.

Agana

Wake Island
U.S.

MARSHALL ISLANDS

Bikini
Atoll

Ralik Chain

Ratak Chain

Yap
Islands

Chuuk
Islands

Caroline Islands

Palikir

Pohnpei
(Ponape)

Majuro

Melekeok

PALAU

FEDERATED STATES OF MICRONESIA

Baker
Island
U.S.

0°

EQUATOR

Howland
Island
U.S.

New
Guinea

M E L A N E S I A

Tarawa
(Bairiki)

Gilbert Islands

Yaren
NAURU

Mt. Wilhelm
14,762 ft.
4,500 m

PAPUA
NEW GUINEA

New
Britain

Solomon Is.

SOLOMON
ISLANDS

TUVALU

Funafuti

Torres Strait

Port Moresby

Honiara

Santa
Cruz Is.

Wallis Is.
Fr.

Darwin

Gulf of
Carpentaria

CORAL SEA
ISLANDS
TERRITORY
Austral.

VANUATU

Port-Vila

Suva

FIJI
ISLANDS

INDIAN
OCEAN

Kimberley
Plateau

NORTHERN
TERRITORY

GREAT DIVIDING RANGE

NEW
CALEDONIA
Fr.

Nouméa

Coral Sea

Nuku'alofa

20°S

GREAT SANDY
DESERT

AUSTRALIA

Macdonnell
Ranges

QUEENSLAND

TONGA

WESTERN
AUSTRALIA

GREAT VICTORIA
DESERT

SOUTH
AUSTRALIA

Lake Eyre (dry)
-52 ft.
-16 m

NEW
SOUTH
WALES

Brisbane

Norfolk Island
Austral.

Lord Howe Island
Austral.

Kermadec
Islands
N.Z.

Perth

Great Australian
Bight

Dorling R.

Adelaide

Murray R.

AUSTRALIAN
CAPITAL
TERRITORY

Canberra

Sydney

Auckland

North
Island

40°S

VICTORIA

Melbourne

Mt. Kosciusko
7,310 ft.
2,228 m

Tasman
Sea

NEW
ZEALAND

Wellington

TASMANIA

Hobart

Mt. Cook
12,349 ft.
3,764 m

Christchurch

Chatham
Islands
N.Z.

South
Island

Stewart Island

INTERNATIONAL DATE LINE

120°E   140°E   160°E   180°

NORTH PACIFIC OCEAN

NORTH AMERICA

TROPIC OF CANCER

20°N

HAWAII
U.S.

Johnston Atoll
U.S.

Kingman
Reef
U.S.

Palmyra Atoll U.S.

P O L Y

Kiritimati
(Christmas I.)

EQUATOR   0°

Jarvis I.
U.S.

KIRIBATI

Line Islands

Phoenix Is.

TOKELAU
N.Z.

SAMOA
Apia

AMERICAN
SAMOA
U.S.
Pago Pago

Samoa Is.

COOK ISLANDS
N.Z.

Niue
N.Z.

Tahiti
Society Is.

Papeete

FRENCH POLYNESIA
Fr.

Austral Is.

Marquesas Is.

Tuamotu Archipelago

Henderson Island
U.K.

Pitcairn
Island
U.K.

20°S

TROPIC OF CAPRICORN

Easter I.
Chile

P O L Y N E S I A

INTERNATIONAL DATE LINE

SOUTH PACIFIC OCEAN

160°W   140°W   120°W

OCEANIA
PHYSICAL /
POLITICAL

0   1,000 miles
0   1,000 kilometers
Mercator Projection

ARCTIC OCEAN
PHYSICAL

500 miles
500 kilometers
Lambert Azimuthal Equal-Area Projection

RUSSIA

Gulf of Ob

Ob' R.
Yenisey R.
White Sea
FINLAND
GERMANY
NETH.
LUX.
DENMARK
FRANCE
SWEDEN
BELGIUM
NORWAY
North Sea
UNITED KINGDOM
IRELAND

Taymyr Peninsula
Kara Sea
Novaya Zemlya
Barents Sea
Norwegian Sea
Svalbard
90°E
60°E
30°E

Lena R.
Laptev Sea
120°E
Severnaya Zemlya
Franz Josef Land
Greenland Sea
ICELAND
Denmark Strait
30°W
70°N
60°N

New Siberian Islands
Oodaaq Island
NORTH POLE +
GREENLAND
60°W
80°N
NORTH ATLANTIC OCEAN

Sea of Okhotsk
East Siberian Sea
150°E
ARCTIC OCEAN
Lincoln Sea
Queen Elizabeth Islands
Ellesmere Island
Hayes Peninsula
Cape Farewell
ARCTIC CIRCLE

Kamchatka Peninsula
180°
Wrangel Island
150°W
Baffin Bay
Davis Strait

Chukchi Sea
120°W
Devon I.
Melville I.
Somerset I.
Baffin Island

Chukchi Peninsula
Point Barrow
North Slope
Banks Island
Prince of Wales I.
Boothia Peninsula
Melville Peninsula
Foxe Basin
Hudson Strait

St. Lawrence Island
Bering Strait
Seward Peninsula
Brooks Range
Beaufort Sea
Victoria Island
Southampton Island

NORTH PACIFIC OCEAN
Aleutian Islands
Bering Sea
Nunivak Island
Bristol Bay
Yukon R.
ALASKA
Mackenzie R.
Great Bear Lake
CANADA
Hudson Bay

ANTARCTICA
PHYSICAL

South Orkney Is.
SOUTHERN OCEAN
60°W
30°W
Riiser-Larsen Ice Shelf
Fimbul Ice Shelf
0°
30°E
SOUTHERN OCEAN
60°E

South Shetland Islands
ANTARCTIC PENINSULA
GRAHAM LAND
Larsen Ice Shelf
Weddell Sea
COATS LAND
QUEEN MAUD LAND
ENDERBY LAND

Mt. Jackson
13,745 ft.
4,189 m
Filchner Ice Shelf
Valkyrie Dome
Amery Ice Shelf

Alexander I.
PALMER LAND
Ronne Ice Shelf
Berkner Island
TRANSANTARCTIC MOUNTAINS
AMERICAN HIGHLAND

Bellingshausen Sea
ELLSWORTH LAND
Vinson Massif
16,066 ft.
4,897 m
Ellsworth Mts.
ANTARCTICA
EAST ANTARCTICA

SOUTH PACIFIC OCEAN
90°W
ANTARCTIC CIRCLE
WEST ANTARCTICA
POLAR PLATEAU
+ SOUTH POLE
90°E
West Ice Shelf
Shackleton Ice Shelf

Bentley Subglacial Trench
-8,383 ft.
-2,555 m
Dome Circe
WILKES LAND

MARIE BYRD LAND
Amundsen Sea
Ross Ice Shelf
80°S
Roosevelt I.
Ross Sea
Ross I.
Mt. Erebus
12,448 ft.
3,794 m
VICTORIA LAND
Talos Dome

SOUTHERN OCEAN
INDIAN OCEAN
60°S
70°S
120°E

500 miles
500 kilometers
Lambert Azimuthal Equal-Area Projection

The World

UNIT **1**

World
Physical

ARCTIC OCEAN

Beaufort Sea

Chukchi Sea

Mt. McKinley (Denali) 20,320 ft. (6,194 m)

Alaska Range

60°N

Bering Sea

Gulf of Alaska

Aleutian Islands

Coast Mts.

Cascade Range

ROCKY MTS.

NORTH AMERICA

Death Valley ▼ -282 ft. (-86 m)

30°N

TROPIC OF CANCER

Baja California

Hawaiian Islands

NORTH PACIFIC OCEAN

Mackenzie R.

Canadian Shield

Lake Winnipeg

Great Lakes

Mississippi R.

Rio Grande

Gulf of Mexico

Appalachian Mts.

Central America

West Indies

Caribbean Sea

EQUATOR

Amazon Basin

Amazon R.

SOUTH AMERICA

São Francisco R.

A N D E S

Samoa Islands

SOUTH PACIFIC OCEAN

TROPIC OF CAPRICORN

30°S

Aconcagua 22,834 ft. (6,960 m)

Paraná R.

60°S

ANTARCTIC CIRCLE

Ross Sea

150°W TRANSANTARCTIC MOUNTAINS 60°W

90°W

30°W

0°

ARCTIC OCEAN

Baffin Bay

Greenland

Greenland Sea

ARCTIC CIRCLE

Iceland

Hudson Bay

Labrador Sea

NORTH ATLANTIC OCEAN

Atlas

S A

SOUTH ATLANTIC OCEAN

PRIME MERIDIAN

Falkland Islands

Strait of Magellan

Scotia Sea

Elevations

10,000 ft.	(3,000 m)
5,000 ft.	(1,500 m)
2,000 ft.	(600 m)
1,000 ft.	(300 m)
0 ft.	(0 m)
Below sea level	

— National boundary

▲ Mountain peak

▼ Lowest point

150°W   120°W   90°W   60°W   30°W

ARCTIC OCEAN

Laptev Sea

Barents Sea

Kara Sea

East Siberian Sea

Norwegian Sea

Central Siberian Plateau

West Siberian Plain

Northern European Plain

Ural Mts.

Volga R.

Ural R.

Ob' R.

Lena R.

North Sea

Baltic Sea

EUROPE

ALPS

ASIA

Altay Shan

Lake Baikal

Sea of Okhotsk

Kamchatka Peninsula

60°N

GOBI

Black Sea

Caucasus Mts.

Caspian Sea

Zagros Mts.

Taklimakan Desert

HIMALAYA

Huang He (Yellow R.)

Sea of Japan (East Sea)

Japan

NORTH PACIFIC OCEAN

Mts.

Mediterranean Sea

Dead Sea −1,312 ft. (−400 m)

Arabian Peninsula

Chang Jiang (Yangtze R.)

Yellow Sea

East China Sea

30°N

Mt. Everest 29,028 ft. (8,848 m)

TROPIC OF CANCER

H A R A

Libyan Desert

Red Sea

Nile R.

Gulf of Aden

Arabian Sea

Deccan Plateau

Ganges R.

Bay of Bengal

Indochina Peninsula

South China Sea

Philippine Sea

AFRICA

Philippine Islands

Gulf of Guinea

Congo Basin

Lake Victoria

Malay Peninsula

Kilimanjaro 19,341 ft. (5,895 m)

EQUATOR

0°

Lake Tanganyika

Indonesia

New Guinea

INDIAN OCEAN

Arafura Sea

Fiji Islands

Kalahari Desert

Coral Sea

Orange R.

TROPIC OF CAPRICORN

Great Sandy Desert

AUSTRALIA

Great Dividing Range

Lake Eyre −52 ft. (−16 m)

Great Victoria Desert

30°S

2,000 miles at Equator

0

0   2,000 kilometers at Equator

Murray R.

Darling R.

New Zealand

Robinson projection

Tasman Sea

Tasmania

N
W   E
S

The Atlantic, Indian, and Pacific Oceans merge around Antarctica. Some define this as an ocean, calling it the Antarctic Ocean, Austral Ocean, or Southern Ocean. While most accept four oceans (including the Arctic Ocean), there is little international agreement on the name and extent of a fifth ocean.

60°S

Ross Sea

SOUTHERN OCEAN

ANTARCTICA

30°E   60°E   90°E   120°E   150°E

MAP STUDY

1. **Physical Systems** Which part of North America has high elevations? Are they higher than the Andes of South America?

2. **Environment and Society** What types of physical features make Asia a crossroads for trade?

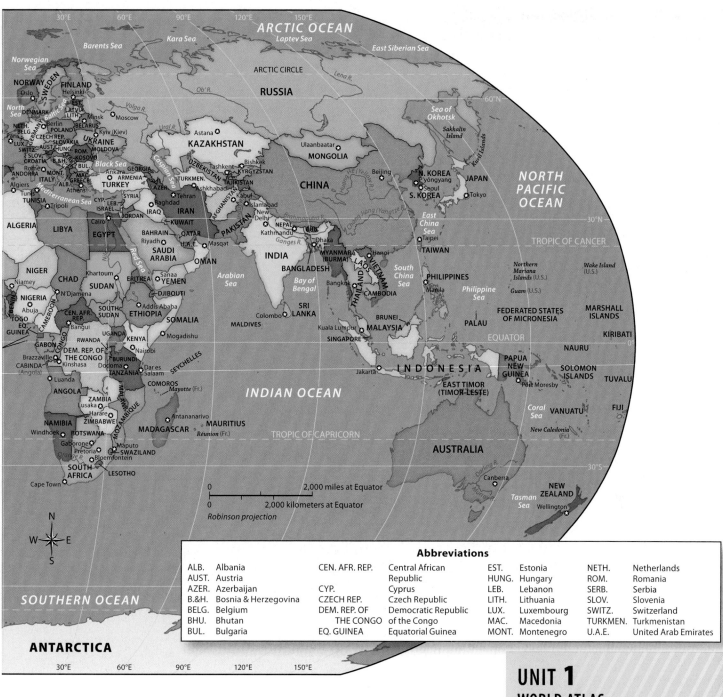

ARCTIC OCEAN

Barents Sea   Kara Sea   Laptev Sea   East Siberian Sea

Norwegian Sea

ARCTIC CIRCLE

NORWAY SWEDEN FINLAND
Oslo Helsinki

North Sea DENMARK
EST.
LATVIA
LITH. Minsk
NETH. BELG. GERMANY Berlin POLAND Kyiv (Kiev) UKRAINE
Paris LUX. CZECH REP. SLOVAKIA MOLDOVA
FRANCE AUST. HUNG.
SWITZ. SLOV. ROM.
Rome CROATIA B.&H. KOSOVO
ANDORRA ITALY MONT. BUL.
Algiers ALB. GREECE
Tunis MAC.
TUNISIA Tripoli

Moscow

RUSSIA

Volga R.   Ob' R.   Lena R.

60°N

Sea of Okhotsk

Sakhalin Island

Kuril Islands

Astana KAZAKHSTAN Ulaanbaatar MONGOLIA

Ural R.

Bishkek UZBEKISTAN KYRGYZSTAN
Tashkent
TURKMEN. TAJIKISTAN
Ashkhabad

Beijing

N. KOREA P'yŏngyang Seoul S. KOREA

JAPAN

Tokyo

NORTH PACIFIC OCEAN

Mediterranean Sea

GEORGIA
Ankara ARMENIA
TURKEY AZER.
CYP. LEB. SYRIA Tehran
ISRAEL IRAQ Baghdad IRAN
JORDAN
Cairo

AFGHANISTAN Kabul
Islamabad
PAKISTAN New Delhi

Black Sea Caspian Sea

Huang He (Yellow R.)

CHINA

Chang Jiang (Yangtze R.)

East China Sea

30°N

TAIWAN Taipei

TROPIC OF CANCER

ALGERIA LIBYA EGYPT

NIGER CHAD SUDAN
Niamey N'Djamena Khartoum
NIGERIA Abuja
BENIN TOGO CAMEROON CEN. AFR. REP. SOUTH SUDAN
EQ. GUINEA Bangui
GABON CONGO DEM. REP. OF THE CONGO RWANDA UGANDA KENYA
Brazzaville Kinshasa BURUNDI Nairobi
CABINDA (Angola) Dodoma TANZANIA Dar es Salaam
Luanda

Red Sea

BAHRAIN QATAR
Riyadh U.A.E. Masqat
SAUDI ARABIA OMAN

KUWAIT

Sanaa YEMEN
ERITREA DJIBOUTI
Addis Ababa ETHIOPIA SOMALIA Mogadishu

Arabian Sea

INDIA

NEPAL Kathmandu BHU.
Ganges R. BANGLADESH Dhaka
Brahmaputra R.

Bay of Bengal

MYANMAR (BURMA) LAOS VIETNAM Hanoi
Bangkok THAILAND CAMBODIA

South China Sea

PHILIPPINES Manila

Northern Mariana Islands (U.S.)

Wake Island (U.S.)

Philippine Sea Guam (U.S.)

FEDERATED STATES OF MICRONESIA

MARSHALL ISLANDS

SRI LANKA Colombo
MALDIVES

Kuala Lumpur MALAYSIA
SINGAPORE

BRUNEI
PALAU

EQUATOR

KIRIBATI

NAURU

SEYCHELLES

ANGOLA ZAMBIA MALAWI
Lusaka Harare
NAMIBIA ZIMBABWE MOZAMBIQUE
Windhoek BOTSWANA
Gaborone SWAZILAND
Pretoria Maputo
SOUTH AFRICA Bloemfontein LESOTHO
Cape Town

COMOROS Mayotte (Fr.)

Antananarivo MADAGASCAR MAURITIUS
Réunion (Fr.)

INDIAN OCEAN

Jakarta INDONESIA

EAST TIMOR (TIMOR-LESTE)

PAPUA NEW GUINEA Port Moresby

SOLOMON ISLANDS

TUVALU

Coral Sea VANUATU

New Caledonia (Fr.)

FIJI

TROPIC OF CAPRICORN

AUSTRALIA

Darling R.

Murray R.

Canberra

30°S

Tasman Sea Wellington

NEW ZEALAND

0   2,000 miles at Equator
0   2,000 kilometers at Equator
Robinson projection

N W E S

SOUTHERN OCEAN

ANTARCTICA

30°E   60°E   90°E   120°E   150°E

Abbreviations

ALB.	Albania	CEN. AFR. REP.	Central African Republic	EST.	Estonia
AUST.	Austria			HUNG.	Hungary
AZER.	Azerbaijan	CYP.	Cyprus	LEB.	Lebanon
B.&H.	Bosnia & Herzegovina	CZECH REP.	Czech Republic	LITH.	Lithuania
BELG.	Belgium	DEM. REP. OF THE CONGO	Democratic Republic of the Congo	LUX.	Luxembourg
BHU.	Bhutan			MAC.	Macedonia
BUL.	Bulgaria	EQ. GUINEA	Equatorial Guinea	MONT.	Montenegro

NETH.	Netherlands
ROM.	Romania
SERB.	Serbia
SLOV.	Slovenia
SWITZ.	Switzerland
TURKMEN.	Turkmenistan
U.A.E.	United Arab Emirates

UNIT 1
WORLD ATLAS

MAP STUDY

1. *Environment and Society* How do Earth's physical features appear to have shaped the borders of countries in Southeast Asia?

2. *Human Systems* List three countries whose capital cities are located on the coast.

The World **5**

World

Population Density

ARCTIC OCEAN

Beaufort Sea

Chukchi Sea

Greenland Sea

Bering Sea

60°N

Yukon R.

Mackenzie R.

⊙ Edmonton

NORTH AMERICA

Vancouver

Seattle

Labrador Sea

Glasgow ⊙

London

Minneapolis

Montreal

Toronto

Madrid

San Francisco

Denver

Chicago

Detroit

New York

Lisbon

Phoenix

St. Louis

Philadelphia

Washington, D.C.

NORTH ATLANTIC OCEAN

Casablanca

Los Angeles

Dallas

Atlanta

San Diego

Houston

30°N

Rio Grande

NORTH PACIFIC OCEAN

TROPIC OF CANCER

Monterrey

Tampa

Guadalajara

Havana

Miami

Mexico City

San Juan

Santo Domingo

Mississippi R.

Dakar

Caribbean Sea

Maracaibo

San José ⊙

Caracas

Medellín

Abidjan

Accra

Bogotá

Belém

Fortaleza

EQUATOR

0°

Guayaquil

Amazon R.

SOUTH AMERICA

Recife

Lima

Cities
(Statistics reflect metropolitan areas.)

■ Over 5,000,000

□ 2,000,000–5,000,000

⊙ 1,000,000–2,000,000

La Paz

Brasília

Salvador

São Francisco R.

Belo Horizonte

TROPIC OF CAPRICORN

São Paulo

Rio de Janeiro

SOUTH ATLANTIC OCEAN

Asunción ⊙

Curitiba

30°S

Paraná R.

Pôrto Alegre

Santiago

SOUTH PACIFIC OCEAN

Buenos Aires

PRIME MERIDIAN

60°S

ANTARCTIC CIRCLE

150°W

120°W

90°W

60°W

30°W

0°

150°W

120°W

90°W

60°W

30°W

0°

POPULATION

Per sq. mi.	Per sq. km
1,250 and over	500 and over
250–1,249	100–499
63–249	25–99
25–62	10–24
2.5–24	1–9
Less than 2.5	Less than 1
Uninhabited	Uninhabited

2,000 miles at Equator

2,000 kilometers at Equator

Robinson projection

MAP STUDY

1. *Human Systems* What cities in the United States have populations over five million people?

2. *Environment and Society* What generalizations can you make about the parts of the world that are least populated?

Resources

Coal		Nickel	
Petroleum		Copper	
Natural gas		Lead	
Iron ore		Manganese	
Tin		Gold	
Zinc		Silver	
Bauxite		Platinum	
Cobalt		Gems	
Uranium			

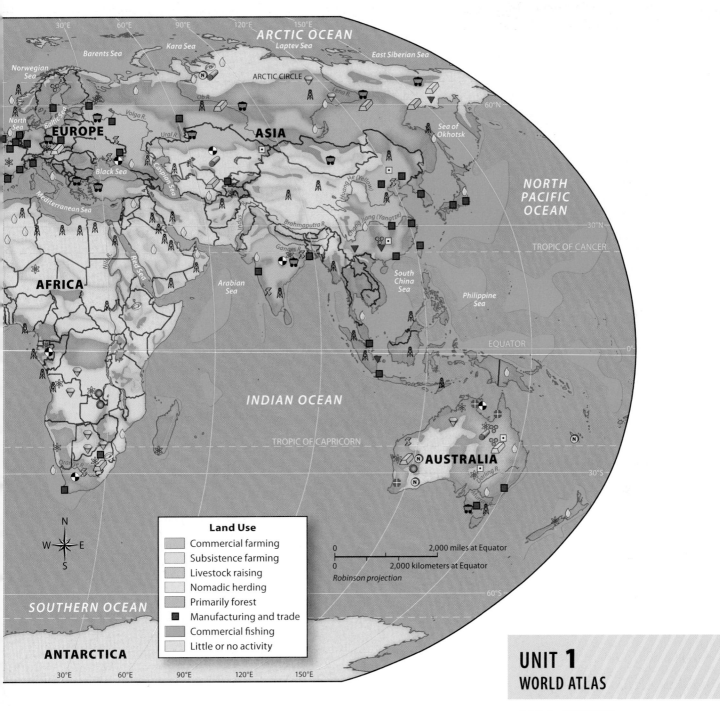

Land Use

- Commercial farming
- Subsistence farming
- Livestock raising
- Nomadic herding
- Primarily forest
- Manufacturing and trade
- Commercial fishing
- Little or no activity

0   2,000 miles at Equator

0   2,000 kilometers at Equator

Robinson projection

ARCTIC OCEAN

Barents Sea   *Kara Sea*   *Laptev Sea*   *East Siberian Sea*

Norwegian Sea

ARCTIC CIRCLE

Lena R.   *Sea of Okhotsk*

North Sea   *Baltic Sea*   **EUROPE**   *Volga R.*   *Ural R.*   **ASIA**

Black Sea   *Caspian Sea*   *Ob R.*

Mediterranean Sea

NORTH PACIFIC OCEAN

Huang He (Yellow)

TROPIC OF CANCER

Brahmaputra R.   *Zhang Jiang (Yangtze)*

AFRICA   *Nile R.*   *Red Sea*   *Arabian Sea*   *Indus R.*   *Ganges R.*

South China Sea

Philippine Sea

EQUATOR

INDIAN OCEAN

TROPIC OF CAPRICORN

AUSTRALIA

Darling R.

SOUTHERN OCEAN

ANTARCTICA

30°E   60°E   90°E   120°E   150°E

MAP STUDY

1. *Environment and Society* What are the primary land use activities in high latitude climate regions?

2. *Human Systems* In which areas of the world is commercial fishing a dominant economic activity?

World
Climate

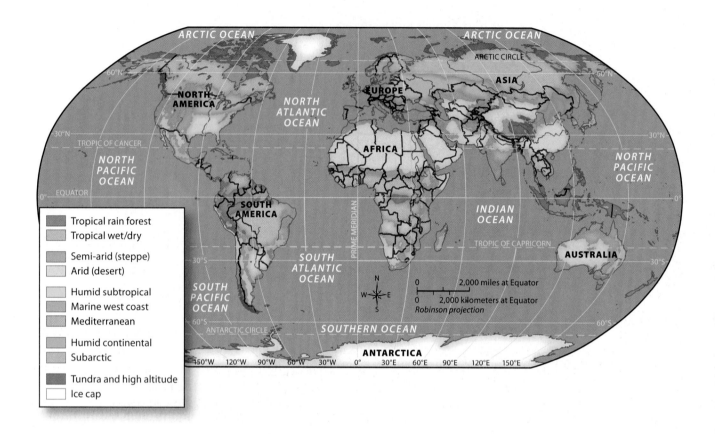

ARCTIC OCEAN · ARCTIC OCEAN · ARCTIC CIRCLE · ASIA · NORTH AMERICA · NORTH ATLANTIC OCEAN · EUROPE · 60°N · 30°N · TROPIC OF CANCER · NORTH PACIFIC OCEAN · AFRICA · NORTH PACIFIC OCEAN · EQUATOR · 0° · SOUTH AMERICA · PRIME MERIDIAN · INDIAN OCEAN · 30°S · SOUTH ATLANTIC OCEAN · TROPIC OF CAPRICORN · AUSTRALIA · SOUTH PACIFIC OCEAN · 60°S · ANTARCTIC CIRCLE · SOUTHERN OCEAN · ANTARCTICA · 150°W 120°W 90°W 60°W 30°W 0° 30°E 60°E 90°E 120°E 150°E

Map Legend:
- Tropical rain forest
- Tropical wet/dry
- Semi-arid (steppe)
- Arid (desert)
- Humid subtropical
- Marine west coast
- Mediterranean
- Humid continental
- Subarctic
- Tundra and high altitude
- Ice cap

0   2,000 miles at Equator
0   2,000 kilometers at Equator
Robinson projection

MAP STUDY

1. *Physical Systems* Which two climate regions occur on and around the Equator?

2. *Physical Systems* What types of climate occur in the high latitudes?

How Geographers Look at the World

networks

There's More Online about how geographers look at the world.

CHAPTER 1

ESSENTIAL QUESTION • *How does geography help us interpret the past, understand the present, and plan for the future?*

Geography Matters...

Geographers are not the only ones who find value in geography skills. In fact, you use geography skills every day. The online maps you use to find the nearest grocery store, the GPS technology in your smartphone that tells your friends where you are, the weather reports that help you decide what to wear each day, the transportation systems that get you from place to place, and even your understanding of the people around you—their beliefs, values, and ways of life— are all examples of how geography is integrated into your daily life. Understanding the elements of geography in greater detail can help you become a better decision maker, planner, and citizen of the United States and the world.

◄ Philippe Cousteau, Jr., grandson of the famous marine explorer Jacques Cousteau, paddles in the waters of Blue Spring State Park in Florida while working on a documentary about Blue Spring and its manatees.

Orlando Sentinel/McClatchy-Tribune/Getty Images

distribution *of* political power

Political geography deals with the ways in which political processes and spatial environments interact and affect one another. Political boundaries such as the borders of countries, states, cities, and electoral districts are all part of political geography. Even election outcomes can be affected by the interaction of politics and geography.

Redistricting in Ohio

Ohio Districts, 2010

SOURCE: Ohio House of Representatives

Some counties can be divided between districts as a result of gerrymandering. For example, eastern Hamilton County is part of the Ohio 2nd district, while the rest of the county is part of the 1st district.

Gerrymandering can be used in ways that some perceive as positive. For example, districts are created in which a majority of the voters are a racial or ethnic minority, as in Ohio's 11th district.

Proposed Districts in Ohio

SOURCE: The Columbus Dispatch, September 14, 2011

During the process of reapportionment based on the 2010 Census, Ohio Republicans proposed a new congressional map that left them a good chance to hold many seats for the next decade.

The approved map of districts reflects some, but not all, of the Republicans' proposals.

The proposed 9th district, which would have snaked along Lake Erie from Toledo to Cleveland, packs concentrations of Democratic voters into one district.

The proposed 16th district moved a Democratic representative into a district that favored a Republican representative.

What is redistricting?

In the 1960s, after decades of inequalities, the U.S. Supreme Court ruled that each person's vote should be worth as much as any other person's vote in both federal and state elections. The government divided the states into districts, with the number of districts in each state determined by the state's population. In this way, states with larger populations get more votes than states with smaller populations, establishing what is called the "one person, one vote" requirement. In order to draw the boundary lines of congressional districts, the U.S. government uses the census, or official count of all the people living in the country. When new census data is collected, sometimes district boundaries must be redrawn to reflect changes in population size and distribution. This process is known as *redistricting*, and it ensures that each person's vote receives equal weight in government.

1. Human Systems What issues might have prompted the Supreme Court's ruling in the 1960s to establish the "one person, one vote" requirement?

How are electoral districts drawn?

Apportionment is the process used to decide how many representatives each state will have to represent it in the U.S. Congress. Census data is used to determine how many representatives each state will have. After the number of representatives for each state is determined, the state congressional districts are drawn so that there is one district for each representative. The government uses U.S. Census data along with geographic information systems (GIS) to create maps that draw district lines in such a way that each district in a state has a roughly equal population. For example, a district with people living fairly spread apart will cover a larger geographic area than a district that is densely populated, such as in urban areas. However, they both can be defined as districts with equal representation. In addition to population size, redistricting takes into account factors such as maintaining city boundaries, preserving boundaries from previous districts, and even avoiding contests between existing representatives.

2. Human Systems What term is used to describe the process the government uses in deciding how many representatives each state will have? What tools are used by the government to help make these decisions?

How does redistricting affect the distribution of political power?

Unfortunately, the process of redistricting is not perfect. This creates an opportunity for misuse by those who seek to manipulate the outcome of elections to favor particular candidates. *Gerrymandering* is a term used to describe the act of drawing political district boundary lines in such a way that one party or candidate has an advantage over another. *Packing* and *cracking* are methods of gerrymandering used to minimize the power of a particular group of voters. Packing concentrates members of a group in a single district, which helps the opposing party win other districts that are nearby. Cracking splits the voters among multiple districts, which dilutes their impact and prevents them from being a majority when they vote in the elections. Packing and cracking are commonly used together in such a way that they influence voting.

3. Places and Regions According to the maps, which districts would be significantly changed by the Ohio Republicans' plan? Which would remain the same?

THERE'S MORE ONLINE

READ about the effect of the census on redistricting • WATCH a video about the redistricting process

Winkel Tripel Projection

Mercator Projection

Goode's Interrupted Equal-Area Projection

Robinson Projection

GEOGRAPHY CONNECTION

Each type of map projection has its advantages as well as some degree of inaccuracy. In addition to the major categories of map projections—planar, cylindrical, and conic—many other map projections can be used depending on the information the mapmaker wishes to show.

1. **THE WORLD IN SPATIAL TERMS** Which projection appears to have the least amount of distortion of distances and size of landmasses?

2. **THE USES OF GEOGRAPHY** Which projection is preferred for sea navigation? Why?

Goode's Interrupted Equal-Area projection resembles a globe that has been cut apart and laid flat. The process of creating this *interrupted projection* can be compared to slicing an orange peel in order to lay it flat on a page. Although this projection shows the true size and shape of Earth's landmasses, distances between land features are generally distorted.

✔ **READING PROGRESS CHECK**

Explaining Why is a trip from Tokyo to Los Angeles a longer distance than it appears to be on a map?

Determining Location

GUIDING QUESTION *How is location determined?*

Geography addresses the question of *where*. To answer this question, a geographer identifies a location. Both globes and maps use a grid system in order to form a pattern of lines that cross one another. These patterns are used to help find the location of places on the Earth's surface.

Lines of latitude, or parallels, circle the Earth parallel to the Equator. Although they run in an east-to-west direction, they measure distance to the north and south of the Equator. The measurements are in degrees. Parallels north of the Equator are called north latitude. Parallels south of the Equator are called south latitude. The Equator is defined as 0° latitude, the North Pole as 90° N, and the South Pole as 90° S.

Longitude lines, also called meridians, are lines that connect the North and South Poles. They run in a north-to-south direction, but they measure distance east and west of the Prime Meridian, which is identified as 0° longitude. Meridians run perpendicular to the lines of latitude, and they also use the measurement of degrees. Meridian lines located east of the Prime Meridian are identified as east longitude, and lines located west are known as west longitude. The longitude line located 180° from the Prime Meridian, on the opposite side of the Earth, is called the International Date Line.

The Equator divides the Earth in half, creating Northern and Southern Hemispheres. The Northern Hemisphere includes any location north of the Equator up to 90° N, while the Southern Hemisphere includes any location south of the Equator up to 90° S. Just as the Equator splits the Earth into Northern and Southern Hemispheres, the Prime Meridian and International Date Line split the globe into east and west halves. Locations east of the Prime Meridian are identified as part of the Eastern Hemisphere, and locations west of the Prime Meridian as part of the Western Hemisphere. All points on Earth are located in two of the four hemispheres: north or south and east or west.

An **absolute location** is an exact global address derived from the latitude and longitude lines that intersect at that place. For example, Tokyo, Japan, is located at approximately 36° N latitude and 140° E longitude. For a more precise reading of a location, a degree is divided into 60 minutes ('). Each minute is then divided into 60 seconds (") just as hours and minutes on a clock are divided to provide a more exact time. For example, the absolute location of the famous Tokyo Tower is 35°39′30.96″ N latitude and 139°44′43.59″ E longitude.

While absolute location identifies exact points using latitude and longitude, **relative location** uses a reference point to identify one place in relation to another. To find relative location, find a reference point—a location you already know—on a map. Then look in the appropriate direction for the new location. For example, locate the city of Paris on the map of France and use this as your reference point. The relative location of the city of Lyon can be described as southeast of Paris.

absolute location the exact position of a place on the Earth's surface

relative location location in relation to other places

✔ **READING PROGRESS CHECK**

Listing List the four hemispheres of the Earth.

GEOGRAPHY CONNECTION

Lines of latitude and longitude create a grid system on Earth's surface.

1. *THE WORLD IN SPATIAL TERMS* Which lines of latitude and longitude divide the Earth into hemispheres?

2. *THE USES OF GEOGRAPHY* How are latitude and longitude used to determine absolute location?

Latitude and Longitude

Latitude

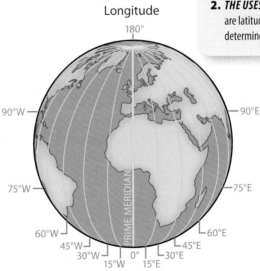
Longitude

Mathematics plays an integral role in the study of geography, especially the process of projecting Earth's spherical features onto a flat map. Early cartographers hand-calculated their map projections using geometry, trigonometry, and even calculus. Today, computer programs take much of the mathematical legwork out of the process of mapmaking. However, geographers still use mathematics for many tasks, such as calculating the area of a landmass, determining the volume of a body of water, and computing distances between locations.

MAKING CONNECTIONS How does mathematics allow cartographers to create flat depictions of a spherical Earth?

elevation the height of a land surface above the level of the sea

Using Maps

GUIDING QUESTION *How do maps work?*

In addition to lines of latitude and longitude, maps include other important tools to help you understand the information they provide. Learning to use map tools will help you interpret the language of maps more easily.

Parts of a Map

The purpose of a map is identified by the map's title. For example, a map titled "Housing Developments in Washington, D.C." would show different details than a map titled "Topography of the Washington, D.C., area." The time period of a map is another important clue to understanding what the map shows. For example, a map titled "Europe Before World War I" would show country borders and national capitals of Europe that are quite different from Europe's current political borders. The map title is the first thing you should look at when reading a map because it provides context for the map's content.

An effective map will provide a legend, or key, to explain the meaning of various symbols used on the map. Geographic features represented on the map are identified by symbols, also called icons. Icons vary by map, depending on the details that are the focus of the map. Roads, highways, railroads, landmarks, parks, and buildings are all human-made features shown by icons. Dots are often used to represent cities. Sometimes the relative sizes of cities are shown using dots of different sizes. Capital cities can be identified by a star within a circle.

The compass rose indicates direction or orientation of a map. North, east, south, and west are the four cardinal directions. The intermediate directions—northeast, northwest, southeast, and southwest—may also be shown. The compass rose looks like intersecting arrows or points of a star.

Line symbols on a map emphasize various features of human activity, such as boundary lines, roads, streets, or routes of trade and transportation. On political maps, boundary lines highlight the borders between different countries and states. Line symbols can also represent physical features such as rivers, earthquake faults, and ocean shorelines.

Colors can be used to distinguish elements on a map. For example, a political map might make each country a different color. On a physical map, colors may indicate the various ranges of **elevation,** or the height above sea level. Colors are used for a variety of other purposes, including identifying water features such as oceans, lakes, or rivers; land features such as deserts, valleys, plains, or mountain ranges; and human-made features such as roads, parks, or streets. The meaning of a color can be identified by the map title or is provided in the legend.

All maps are drawn to a certain scale. Scale represents the consistent, proportional relationship between the measurements shown on a map and the actual measurements of the Earth's surface. Maps use scale to shrink what would be large distances and features of a region to a manageable size. When a map is scaled to fit on paper, every feature of the map is scaled by the same amount so that each feature will have the same proportion to every other feature on the map. However, not all parts of a map will be perfectly to scale because flat maps are subject to some distortion. A map's scale is identified by a scale bar, which compares distances shown by a map to actual distances on the Earth. For example, a scale bar might indicate that one inch (2.5 cm) on the map represents 100 miles (160.9 km).

The amount of scale portrayed on a map depends on its depiction as a *small-scale* or *large-scale* map. A small-scale map shows a larger area with fewer details. For instance, a small-scale map can focus on a specific country and its

Map labels (Paris map): RNES · BLVD DE BATIGNOLLES · BLVD DE CLICHY · Gare St. Lazare · ace Charles e Gaulle oile) · Arc de Triomphe · BLVD HAUSSMANN · CHAMPS ÉLYSÉES · Opéra · HAILLOT · Grand Palais · Place de la Concorde · Tuileries Gardens · RUE DE RIVOLI · Seine R. · QUAI DES TUILERIES · QUAI D'ORSAY · National Assembly · INVALIDES · Eiffel Tower · Esplanade des Invalides · ST. GERMAIN · BLVD SAINT GERMAIN · Champ de Mars · Hôtel des Invalides · Rodin Museum · BLVD RASPAIL · Luxembourg Palace · 1 mile · 1 kilometer

Map labels (France map): National boundary · Regional boundary · National capital · Major city · U.K. · 50°N · English Channel · Lille · BELG. · GERMANY · LUX. · Paris · SWITZERLAND · Nantes · Orléans · FRANCE · ATLANTIC OCEAN · Lyon · 45°N · Bay of Biscay · Bordeaux · ITALY · Toulouse · Nice · Marseille · SPAIN · 0 200 miles · 0 200 kilometers · ANDORRA · Mediterranean Sea · Corsica · Lambert Azimuthal Equal-Area projection · 0°E

neighboring countries to show boundary lines, major cities and capitals, important land or water features, or regional topography. For example, the scale bar on a map of France and its bordering countries could show a relationship of one inch (2.5 cm) as equal to 200 miles (321.9 km) in actual distance. On the other hand, a large-scale map can show a small area with a great amount of detail. It narrows in on an identified region to show more specific details. The map measurements of a large-scale map use much smaller distances than on a map of France. For example, a large-scale map of the city of Paris shows the layout of streets, major roads, bridges, parks, and important landmarks such as museums, hotels, and churches. The scale bar of a map of Paris could show a relationship of one inch (2.5 cm) as equal to one mile (1.6 km) in actual distance. This measurement is much more specific than for a small-scale map of France.

Types of Maps

A cartographer can choose from several types of maps in order to convey geographic information. Physical maps, political maps, and thematic maps each serve a unique purpose and are suited to showing different types of information. A physical map shows location and topography, or shape, of the Earth's land features. A study of a country's land and other physical features can help to explain the historical development of the country. For example, mountains may be barriers to transportation, and rivers and streams can provide access to a country's interior. Physical maps show water features such as rivers, streams, and lakes. They also show landforms such as mountains, plains, plateaus, and valleys.

Physical maps highlight general **relief** through shading and texture. Relief shows the differences in elevation between the various landforms of an area. An elevation key can use colors to indicate specific, measured differences in elevation above sea level.

GEOGRAPHY CONNECTION

Small-scale maps, like this political map of France, show a large area but little detail. Large-scale maps, like the map of the city of Paris, can show a small area with a great amount of detail.

1. **PLACES AND REGIONS** Using the scale bar on the map of France, what is the distance from Paris to Nice in miles?

2. **HUMAN SYSTEMS** What types of human-made features does the map of Paris include that the map of France does not?

relief the variation in elevation across an area of Earth's land

A political map provides the boundaries and locations of political units such as countries, states, counties, cities, and towns. It can show the networks and links that exist within and between political units. While some boundaries are distinguished by natural features such as bodies of water or landforms, the majority of features on a political map are human-made. Such human-made features can include boundary lines, cities and capitals, railroads, roads, highways, streets, buildings, and other landmarks.

thematic map a map that emphasizes a single idea or a particular kind of information about an area

A **thematic map** emphasizes a particular theme or subject. Thematic maps can show various natural and human-made features. While a general map can cover various topics, thematic maps focus on a single topic such as climate, natural vegetation, population density, or economic activities.

A type of thematic map called a flow-line map displays the movement of people, animals, goods, and ideas. It also illustrates physical processes such as hurricanes and glaciers. Arrows illustrate the flow and direction of movement for the map subject. Arrows can be distinguished by different colors to represent the varied directions of movement, and the thickness of the arrows can show the amount that is moving.

Mental Maps

While the diverse parts of a map are crucial elements for identifying a map's purpose, a map reader must also learn how to mentally organize the spatial

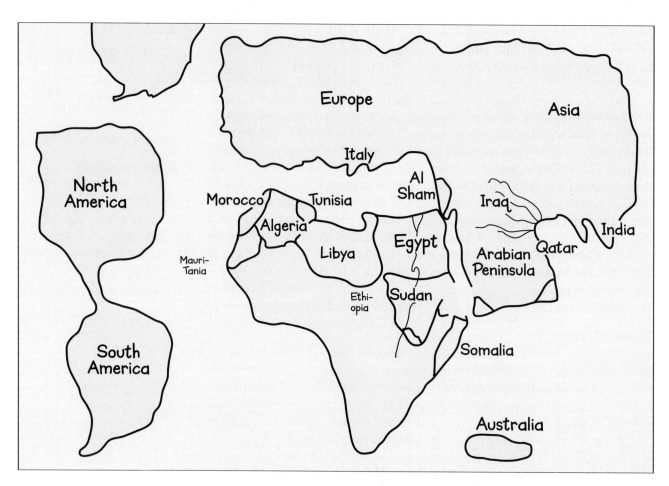

Mental maps reflect a person's personal point of view or perception about a place.

▲ CRITICAL THINKING

1. **Drawing Inferences** Based on this mental map, do you think the person who created it knows a lot or a little about the world? Explain.

2. **Hypothesizing** Do you think your mental map of the world would be more or less detailed than this one? Why?

information emphasized by a map. Mental maps provide an effective method for answering and remembering important geographical questions about locations, characteristics, patterns, and relationships of places and regions. Mental maps describe an individual's **internal** map—his or her perception of features of the Earth's surface. Mental maps can be applied to local or global scales, from the orientation of an individual within a classroom or home, to the visualization one could see from an airplane on a transcontinental flight.

internal existing or lying within

A person's mental map helps geographers understand how individuals view their relationship to the space around them. Mental maps mix precise, objective data with an individual's subjective perception or existing knowledge of places. For example, a student may be asked to memorize locations and names of the original 13 colonies to contextualize his or her experience and understanding of those places today. A biologist's mental map of major climate regions could include his or her knowledge of the distribution of various types of animals and plants within each region. Using mental maps of the world's regions over time allows a geographer to analyze patterns of human settlement in the world and how settlements relate to land features and availability of resources. This information can be used to determine why particular routes of settlement, trade, and transportation developed.

Mental maps can change according to an individual's experience and perception of people, places, regions, and environments. Because mental mapping is based on the individual, one individual's memory of a place differs from another individual's memory. A new experience in an unfamiliar location would alter the accuracy and number of details recalled by a person's mental map of that place. For example, traveling through a country would help a person more easily recall city names and locations of that country than by memorizing such places by only looking at a map. The pictures of two students asked to sketch from memory their view of the Grand Canyon would vary depending on what features the individuals noticed at the time, as well as how much their memory of the place has changed since they saw the canyon.

Recalling a place related to emotion, such as a city that a student likes or dislikes, would also alter the details depicted in a mental map. An individual's personal connection to a place or event through personal experiences or exposure to written and visual accounts can influence what is part of a mental map. As a result, mental mapping is beneficial for geographers to understand the significance of information related to a specific place and how it is perceived by individuals.

✔ **READING PROGRESS CHECK**

Explaining What do relief maps show?

Geospatial Technologies

GUIDING QUESTION *How are geospatial technologies used to learn about the world?*

Geospatial technologies assist geographers and other professionals in answering geographic questions. Global positioning systems (GPS), geographic information systems (GIS), and remote sensing from satellites and aircraft incorporate multiple information sources to construct detailed geographic representations of specific aspects of the Earth's surface.

Global Positioning System

A **global positioning system (GPS)** is used to determine the exact, or absolute, location of something on Earth. Made up of a network of satellites and receiving instruments, GPS provides an accurate location with respect to latitude,

global positioning system (GPS) a navigational system that can determine absolute location by using satellites and receivers on Earth

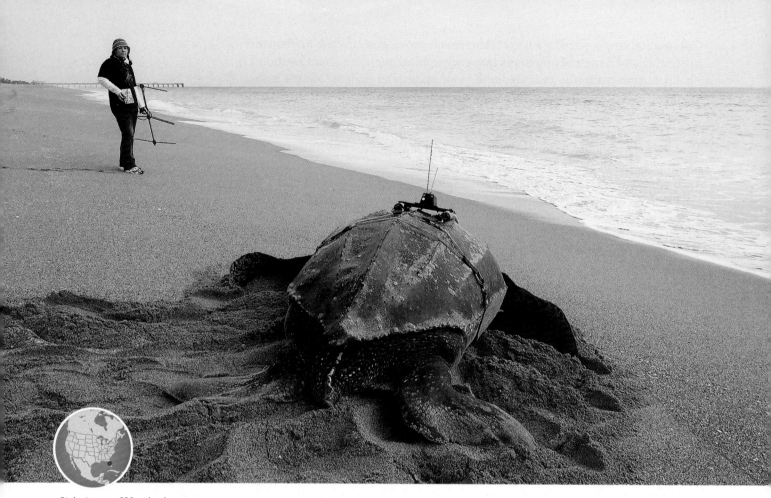

Biologists use GPS technology to track the movement and behavior of animals in the wild.

▲ CRITICAL THINKING

1. *Drawing Conclusions* How might wildlife biologists use GPS technology to protect endangered species such as the African elephant?

2. *Classifying* What other fields use GPS technology to gather information?

longitude, and even altitude. GPS technology in the United States relies on a system of 24 satellites that make 6 full orbits around the Earth every 12 hours. The European Union, as well as some individual countries such as Russia and China, have their own satellites that support GPS systems.

The satellites in all these systems send out radio signals that are picked up by GPS receivers on Earth. In a process called triangulation, a GPS receiver measures the precise time taken for radio signals from four or more satellites to travel to the receiver. The receiver then multiplies the time by the speed of a radio wave to calculate the distance between it and the satellite. When signals from the four or more satellites are processed in the same manner, the receiver's built-in computer determines the point at which at least four satellite signals intersect on the Earth. This intersection then identifies the receiver's latitude, longitude, and altitude. The more satellites that are used, the more accurate the location that is pinpointed.

GPS technology serves a commercial purpose for military machinery, space shuttles, aircraft, ships, submarines, trucks, trains, and ambulance fleets. Yet it can also aid in multiple forms of personal navigation. The GPS receiver in a car, for example, tracks the car's changing location on an electronic map to provide constantly updated directions based on where the car is located and where it is headed. Many current GPS receivers are battery-powered and are no larger than the palm of your hand, while GPS computer chips are smaller than your fingernail.

Many fields of science employ GPS technology. For example, seismologists, the scientists who study earthquakes, can use GPS to determine the size of earthquakes. Scientists first plant GPS receivers in the ground in regions vulnerable to tectonic, or earthquake-prone, activity. Once an earthquake hits,

(t)Chris Johnson/Alamy

seismologists can quickly measure the strength of an earthquake by calculating how far the planted GPS receivers move. This measurement allows scientists to predict how likely the earthquake is to produce large ocean waves called tsunamis. Because tsunamis can cause devastating destruction to coastal communities, early warning would diminish loss of human life by advising people to flee as soon as possible.

Another function of GPS technology in the field of science is to track the migration of animals to determine any changing patterns within the animals' ecosystems. Biologists tag animals with GPS receivers so they can track their movement due to seasonal changes, changes of food or shelter, or threats to their habitat by human activity or by other animals.

Geographic Information Systems

Advances in technology have changed the way maps are made. An important tool in mapmaking today involves computers with software programs called **geographic information systems (GIS)**. But more than simply making maps, GIS can be used to perform advanced geographical analysis.

Many types of data can be entered into a GIS. These data come from a wide variety of sources such as maps, satellite images, printed text, and statistical databases. The primary and most important function of a GIS, however, is to link the location of a place with the characteristics, or attributes, found at that location. That function helps us not only identify and list the characteristics of places, but also analyze how places compare to one another and interact with one another. These patterns of interactions are known as spatial organization, and the study of them is called spatial analysis.

The locational data of places is stored in a GIS as latitude and longitude coordinates. These coordinates can be obtained from existing maps, GPS receivers, and satellite images. The attribute data come from a wide range of

geographic information systems (GIS) computer programs that process and organize details about places on Earth and integrate those details with satellite images and other pieces of information

GIS Layers

Topographic base

Zoning

Floodplains

Wetlands

Land Cover

Soils

Survey control

Composite overlay

⌄ DIAGRAM SKILLS

Geographic information systems allow different kinds of information to be saved on separate layers.

◂ CRITICAL THINKING

1. **Formulating Questions** What questions could a geographer answer using the layers of information in this sample GIS file?

2. **Evaluating** How does GIS allow cartographers to create maps and make changes quickly and easily?

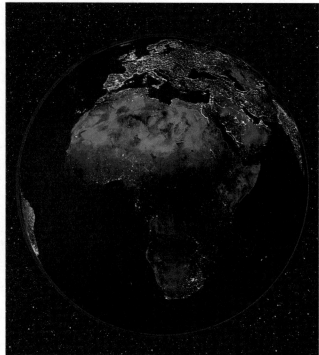

The stark differences between North America and Africa at night are clearly seen with satellite images.

▲ **CRITICAL THINKING**

1. *Drawing Conclusions* What do these images tell you about North America? About Africa?

2. *Analyzing* How might the information from these images be used by geographers?

transmit to send from one place to another

sources. The GIS stores all these data in a digital database. Cartographers then select an appropriate map projection and program the GIS to produce thematic maps of the data. Each of the various types of attribute data in the database can be displayed on the map as a theme. The different themes can be saved as separate electronic layers that can be turned on or off. The GIS can show just one layer of information or multiple layers at the same time.

In this way, maps can be made—and changed—quickly and easily to display various complex types of information with a single map source. GIS is used by a diverse range of professional fields with work that relies on maps, including environmental and urban planners, marketing researchers, retail developers, environmentalists, and other professionals.

Satellites

A satellite is a natural or human-made object that orbits a planet or other large astronomical body. The first human-made satellite was launched in 1957 by the Soviet Union, and today hundreds of human-made satellites orbit the Earth. These satellites are useful for many purposes, such as navigation, collecting atmospheric data to make weather predictions, and communications for cellular phones and the Internet. Some satellites collect visual information of the Earth's surface as they orbit the Earth. Others carry instruments to observe the presence of and interaction between the land, ocean, air, and living things. Because scientists can gather very specific information regarding atmospheric phenomena, they can compare satellite images with ground research and knowledge of Earth's natural history to analyze how the environment has changed over time.

Different types of satellites are used to collect different types of information about the Earth. Once the data are collected, computers on a satellite store and then **transmit** those data by radio signals to receiving stations on Earth. Scientists on Earth receive the transmitted data and use their specialized knowledge to interpret the meaning of these data. The received data sometimes serve as inputs to computer models using mathematical formulas. Just as cartographers produce maps from data processed by GIS, so do scientists who use the converted data from satellite imagery to study Earth's natural and human-made processes.

NASA/NOAA/SPL/Getty Images

Remote sensing is any technique used to measure, observe, or monitor a subject or process without physically touching the object under observation. For example, scientists use remote sensing when they analyze images from satellites, telescopes, and cameras in airplanes and spacecraft. Often, remote sensing collects images of things that could not be seen with the unaided human eye. It is also a useful process for obtaining information from locations that would otherwise be dangerous or difficult to reach, such as estimating precipitation rates in a desert region. The immediate and frequent flow of images from remote sensing allows cartographers to create detailed and relevant maps to estimate constant and changing environmental conditions, such as sediment buildup, air pollution, ocean surface roughness, surface temperatures, biomass volumes, mineral resources, and changes created by storms and floods.

remote sensing the science of obtaining information about an object or an area from a distance, typically from instruments in aircraft or satellites

Quality and Limitations of Geospatial Technologies

Geospatial technologies are excellent sources of information because they provide actual images and data related to a location and can provide a great amount of detail. While scientists can use observational and historical data to gather information about a place, geospatial technology acts as a primary source for compiling raw data. Because they are a relatively new innovation in comparison to traditional forms of mapmaking, the current uses of geospatial technologies can be limited. These informational technologies are constantly changing, and they will improve with the advancement of computer, aerospace, and Internet-based technology. Accelerated development in geospatial technology offers a number of possibilities for its use with government, private industry, scientists, and the general public. Because the economic, cultural, and political activities of the world's regions have become increasingly interconnected, information related to the world's physical and human systems needs to be readily available, consistent, and up-to-date. The combination of mental mapping with GPS, GIS, and aerial imagery can create a very detailed picture of places and regions.

Geospatial technologies allow access to a wealth of information about what sorts of features and objects are in the world and where those features and objects are located. This "geospatial information" can be very helpful for identifying and navigating, but by itself, does not help much in answering the "why" or the "why care" questions that lie at the heart of understanding and making decisions about the world in which we live. It is important to go beyond geospatial information to geographical understanding of peoples, places, and environments—and the connections among them—that are interesting as well as useful.

☑ **READING PROGRESS CHECK**

Identifying What type of data do geospatial technologies provide?

LESSON 1 REVIEW

Reviewing Vocabulary
1. *Identifying* What is the difference between absolute location and relative location?

Using Your Notes
2. *Describing* Using your graphic organizer, list and describe the four common map projections.

Answering the Guiding Questions
3. *Comparing* How are globes and map projections related?

4. *Explaining* How is location determined?

5. *Summarizing* How do maps work?

6. *Analyzing* How are geospatial technologies used to learn about the world?

Writing Activity
7. *Informative/Explanatory* If you were planning to open a sporting goods store, in what ways could GIS technologies help you choose a good location? Discuss the types of layers that might be helpful to your decision.

netw⊙rks

There's More Online!

- ☑ **CHART** Skills for Thinking Like a Geographer
- ☑ **IMAGE** Hiker Reading a Map
- ☑ **IMAGE** Rice Paddy in Southern China
- ☑ **IMAGE** Scientist Studying a Glacier
- ☑ **MAP** Perceptual Regions of the United States
- ☑ **INTERACTIVE SELF-CHECK QUIZ**
- ☑ **VIDEO** The Geographer's Craft

Reading **HELP**DESK

Academic Vocabulary

- **primary**
- **obtain**
- **fluctuation**

Content Vocabulary

- **spatial perspective**
- **site**
- **situation**
- **formal region**
- **functional region**
- **perceptual region**

TAKING NOTES: *Key Ideas and Details*

IDENTIFYING As you read the lesson, use a graphic organizer like the one below to identify the basic elements of geography.

Elements of Geography

LESSON 2
The Geographer's Craft

ESSENTIAL QUESTION · *How does geography help us interpret the past, understand the present, and plan for the future?*

IT MATTERS BECAUSE

The root of the word geography *is an ancient Greek word meaning "earth description." Geographers study the location and relationships of Earth's physical and living features. They look for links between places and people and identify patterns in order to learn why those patterns exist or occur.*

A Geographic Perspective

GUIDING QUESTION *What is the spatial perspective?*

Geographers focus on understanding the world and answering questions about it. An important part of the geographical perspective is the spatial perspective. A **spatial perspective** focuses on how individual places, people, or objects are related to one another across the surface of the Earth. Using a spatial perspective, geographers examine why things are located where they are. Geographers also use the spatial perspective to examine what makes regions distinct based on changes and movements of people over time and changes in the physical environment. This spatial perspective can also be viewed from a local perspective, such as where to select a new home, or the global perspective, such as the economic activities of countries in the form of trade.

Other aspects of the geographic perspective include the ecological perspective and the perspective of experience. The ecological perspective focuses on understanding Earth as a complex set of interacting living and nonliving components. It involves thinking about the connections and interactions that operate among ecosystems and human societies. The perspective of experience considers how people make meaning from the world in which we live. As people live in locations on Earth, they have experiences and build memories that give places on Earth unique characteristics. Awareness and sensitivity to these uniquenesses are an important part of the perspective of experience.

Not all people who study or incorporate geography into their profession are labeled geographers. Geography skills can be applied to a variety of fields, including government, business, and education.

(l to r)Thorsten Henn/Cultura/Getty Images, ©BambooSIL/SuperStock, Ashley Cooper/Alamy, Paul Chesley/Stone/Getty Images

One broad cluster of career opportunities in geography is teaching and education. Teaching opportunities exist at all levels, from elementary to high school to university levels of education. Teachers with a background or training in geography topics are in demand in the United States for elementary and high schools. Students with formal geographic training from a university can find work in many diverse businesses, industries, and professional fields.

Because geography itself has many specialized fields, there are many ways that people use geography in their work. Those with a knowledge of physical geography can work as meteorologists who study the atmosphere and weather patterns; as emergency management officials dealing with natural hazards, such as earthquakes or hurricanes; as ecologists who study the interrelationships of organisms and their environment; as soil scientists; and as environmental managers. Work in the environmental field includes assessing the environmental impact of proposed development projects regarding air and water quality and wildlife.

spatial perspective a way of looking at the human and physical patterns on Earth and their relationships to one another

Skills for Thinking Like a Geographer

Skill	Examples	Tools and Technologies
Asking Geographic Questions helps you pose questions about your surroundings	• Why has traffic increased along this road? • What should be considered when building a new community sports facility?	• Maps • Globes • Internet • Remote sensing • News media
Acquiring Geographic Information helps you answer geographic questions	• Compare aerial photographs of a region over time. • Design a survey to determine who might use a community facility.	• Direct observation • Interviews • Reference books • Satellite images • Historical records
Organizing Geographic Information helps you analyze and interpret information you have collected	• Compile a map showing the spread of housing development over time. • Summarize information obtained from interviews.	• Field maps • Databases • Statistical tables • Graphs • Diagrams • Summaries
Analyzing Geographic Information helps you look for patterns, relationships, and connections	• Draw conclusions about the effects of road construction on traffic patterns. • Compare information from different maps that show available land and zoning districts.	• Maps • Charts • Graphs • GIS • Spreadsheets
Answering Geographic Questions helps you apply information to real-life situations and problem solving	• Present a report showing the results of a case study. • Suggest locations for a new facility based on geographic data gathered.	• Sketch maps • Reports • Research papers • Oral or multimedia presentations

One of the most important geographic tools is the ability to think geographically. The five skills identified above are key to geographic understanding.

▲ CRITICAL THINKING

1. *Analyzing* What types of patterns might you recognize by comparing aerial photographs of a region over a specific time period?

2. *Identifying the Central Issue* Provide a real-life example of how you might apply three of the skills described in the chart. Explain the issue, including which skills you chose, and how they would be used.

Those with knowledge of human geography find work in many areas such as health care, transportation, population studies, economic development, public policy, and international economics. Human geographers with a background in urban planning are hired by local and state government agencies to focus on projects such as housing and community development and parks and recreation planning. An economic geographer examines human economic activities. He or she may work at such tasks as market analysis and site selection for stores, factories, and restaurants. A regional geographer studies the features of a particular region and may assist government and businesses in making decisions about land use. Geographers can also find employment as writers and editors for publishers of textbooks, maps, atlases, news and travel magazines, and Web sites.

One of the most important tools of a geographic perspective is the ability to think geographically about the world. Geographers use five basic skills that are key to geographic understanding: asking, acquiring, organizing, analyzing, and answering geographic questions. Asking geographic questions provides information that can be used to better understand one's surroundings. As knowledge is acquired, it can be applied to recognize patterns and relationships that will help in real-life situations.

☑ **READING PROGRESS CHECK**

Listing List at least three fields outside of geography that use geography skills.

The Elements of Geography

GUIDING QUESTION *What are the elements of geography?*

Geography uses the geographic perspective to study the peoples, places, environments, histories, and cultures of the world's regions. Geographers study the interactions between peoples, places, and environments to explain why and how patterns of interaction occur. There are six overall elements geographers consider in their work: the world in spatial terms, places and regions, physical systems, human systems, environment and society, and the uses of geography.

Maps can be used to determine site and situation.

▼ **CRITICAL THINKING**

1. *Speculating* What could this hiker's map tell her about the site and situation of the landscape she is viewing?

2. *Describing* Explain how a map, such as the one shown, could be used to identify spatial relationships.

28

Thorsten Henn/Cultura/Getty Images

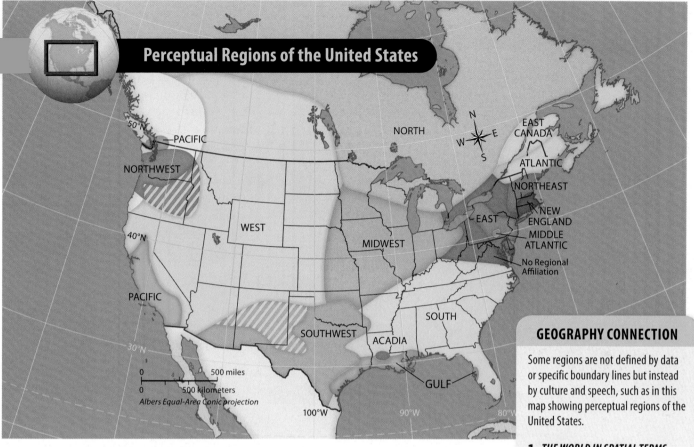

Perceptual Regions of the United States

NORTH

EAST CANADA

PACIFIC

NORTHWEST

50°N

ATLANTIC

NORTHEAST

WEST

EAST

NEW ENGLAND

40°N

MIDDLE ATLANTIC

MIDWEST

No Regional Affiliation

PACIFIC

SOUTH

30°N

SOUTHWEST

ACADIA

0   500 miles

0   500 kilometers

GULF

Albers Equal-Area Conic projection

100°W   90°W   80°W

The World in Spatial Terms

Spatial relationships link people and places based on their locations and relationships to each other. Location is a reference point for geographers in the same way that dates serve as reference points for historians.

One way of locating a place is by describing its absolute location—the exact spot at which the place is found on the Earth. To determine absolute location, geographers use the system of latitude and longitude. On a day-to-day basis, humans tend to identify a place based on relative location—a place's location in relation to another place. For example, New Orleans is located near the mouth of the Mississippi River. Knowing the relative location of a place helps you to think spatially. By creating a mental map based on relative location, you can orient yourself in space and develop an awareness of the world around you.

The broad or specific definition of a geographic location based on relative or absolute location also takes into account a place's site and situation. **Site** is the specific location of a place, including its physical setting. For example, the site of San Francisco is its location at the end of a peninsula in northern California. **Situation** refers to a more general location, defined by a place's geographic position in relation to other places and its connections to other regions. San Francisco's situation is a port city on the Pacific coast, close to California's agricultural lands.

Places and Regions

A place has physical and human significance. It has distinguishing characteristics defined by its features and surroundings. Geographers study and assess the similarities and differences between places to express what features are unique to each place. To interpret the Earth's complexity, geographers group places with similar characteristics into regions. A region can be defined by physical traits

GEOGRAPHY CONNECTION

Some regions are not defined by data or specific boundary lines but instead by culture and speech, such as in this map showing perceptual regions of the United States.

1. **THE WORLD IN SPATIAL TERMS** According to the map, in which region do you live?

2. **PLACES AND REGIONS** What aspects of culture do you think define a perceptual region in the United States?

site the specific location of a place, including its physical setting

situation the geographic position of a place in relation to other places or features of a larger region

Rice paddy fields in southern China reflect both human systems and physical systems at work.

▲ CRITICAL THINKING

1. *Categorizing* What type of region would you categorize the rice paddy fields as being? Explain your choice.

2. *Formulating Questions* Create one question you would ask the people living in the area that could help you determine whether the rice paddy is a formal or perceptual region.

formal region a region defined by a common characteristic, such as production of a product

primary of first rank, importance, or value

functional region a central place and the surrounding territory linked to it

such as climate, landforms, soils, vegetation, animal life, and natural resources. A region can also have human significance, as defined by characteristics such as language, religion, political or economic systems, and population distribution. Geographers identify three types of regions: formal, functional, and perceptual.

A **formal region** features a unifying characteristic, such as a product produced in that region. For example, the Corn Belt is a band of farmland stretching from Ohio to Nebraska in the United States. It is a formal region because corn is its **primary** crop. A **functional region** incorporates a central node and a surrounding area that is connected to the node by some defined function. For example, a cell tower provides the central node for a surrounding area in which cell phone users can obtain phone reception. A **perceptual region** uses a looser standard for characterization, defined more by commonly accepted tradition or value than by objective data. For example, the term "heartland" refers to a central area in the United States in which traditional values of family and hospitality are believed to predominate. A perceptual region could also be labeled a vernacular region. This refers to patterns native to a particular region in spite of boundary lines. The Creole dialect that is spoken in southern Louisiana is an example of a vernacular region. It is defined more by the culture and speech of the region than by a designation of state and city boundaries.

Physical Systems and Human Systems

Because geography can cover a broad range of themes, geographers divide their focus into major branches: physical geography and human geography. Physical geography—climate, land, water, plants, and animal life—looks at these processes

©BambooSIL/SuperStock

and their significance to humans. Human geography, or cultural geography, analyzes human activities and their relationship to the cultural and physical environments. Political, economic, social, and cultural factors can include themes such as urban development, economic production and consumption, and population change. Because physical and human geography are still very broad in their focus, they can be further divided into subject areas. For example, climatology is the study of climate and long-term atmospheric conditions and their impact on ecology and society. Historical geography is the study of places and human activities over time based on the geographic factors that have shaped them.

Geographers study how physical features and processes of land, water, and climate interact with plants and animals to create, support, or change ecosystems. An ecosystem is a community of plants and animals that depend upon one another and their surroundings for survival. Geographers also study the processes by which people operate across Earth's surface—how they settle the Earth, form societies, and create permanent features. A recurring theme in geography is the ongoing movement of people, goods, and ideas. Human migration and settlement, as well as the exchange of ideas and practices among cultures, can over time transform societies, traditions, and the landscape in which humans live. In studying human systems, geographers look at how people compete or cooperate to change or control aspects of the Earth to meet their needs.

Environment and Society

The relationship between people and their physical environment is a theme embodied by human-environment interaction. Ways in which people use their surroundings, ways in which they change it voluntarily and involuntarily, and the consequences that result from such human-environment interaction are very important themes for geographers. Pollution, construction, human population growth, conservation of parks, and reintroduction of species into the wild are just a few of the ways humans change the physical environment. Yet the physical environment can also have an effect on humans. For example, physical barriers such as rivers, mountains, and deserts limit human movement and growth. Natural phenomena such as hurricanes, earthquakes, and heavy storms or droughts force humans to adapt their activities and lifestyles to the changing environment. By understanding how the Earth's physical features and processes shape and are shaped by human activity, geographers help societies make informed decisions about their relationship with their surrounding physical environment.

The Uses of Geography

Geography provides insight into how physical features and living things developed in the past. It also takes into account current trends regarding the physical and human environment in order to plan for future needs. Planning and policy making must account for interactions between humans and the natural environment. Data regarding physical features and processes can highlight suitable sites for resource extraction or for human habitation. Urban planners analyze trends in human growth within a specified region to determine where and what systems, such as schools, roads, public services, and businesses, are necessary for supporting a growing population. Geographers analyze past data in order to determine effective future actions to sustain and support both the natural environment and human development. Although people trained in geography are in great demand in the workforce, many of them do not have geographer as a job title. Geography skills are useful in so many different situations that geographers have more than a hundred different job titles.

Analyzing
PRIMARY SOURCES

The Island

"The island seems to have a tenacious hold on the human imagination. Unlike the tropical forest or the continental seashore it cannot claim ecological abundance, nor—as an environment—has it mattered greatly in man's evolutionary past. Its importance lies in the imaginative realm. Many of the world's cosmogonies, we have seen, begin with the watery chaos: land, when it appears, is necessarily an island…. In numerous legends the island appears as the abode of the dead or of immortals."

—Yi-fu Tuan, *Topophilia: a Study of Environmental Perception, Attitudes, and Values,* 1974

DBQ *PLACES AND REGIONS*
What emotions do you feel when you think of an island? Explain whether it is based on your imagination or on real experiences.

perceptual region a region defined by popular feelings and images rather than by objective data

Scientists study Russell Glacier in Greenland and the causes of its rapid melting.

▲ **CRITICAL THINKING**

1. *Formulating Questions* List three questions the scientist might be asking about the melting glacier.

2. *Evaluating* What words would you use to describe the site where the scientist is located?

Geographers work in a variety of jobs in government, business, and education. They often combine the study of geography with other areas of study. For example, an ecologist must know the geographic characteristics of a place or region in which he or she studies living organisms. Similarly, a travel agent must have knowledge of the physical and human geography of a place in order to plan trips for clients.

✔ **READING PROGRESS CHECK**

Identifying What are the three types of regions?

Research Methods

GUIDING QUESTION *What methods do geographers use to conduct their work?*

To do their work, geographers use several research methods. Direct observation and measurement, mapping, interviewing, production and use of statistics, and the use of technology are all specialized research methods used by geographers.

Direct observation and measurement involves analysis of patterns of human activity that take place on the Earth's surface. Geographers using this method will visit a place to gather information about it from what they observe of the place and its geographic features. Geographers also employ remote sensing from satellite images and aerial photographs to locate specific information without having to visit the site in person. For example, aerial photographs or satellite images can be used to locate mineral deposits, to determine the size of freshwater sources, or to see the extent of urban sprawl.

Mapping is essential to geographers. Many findings from geography research can be shown visually and spatially on maps better than they can be explained through statistical methods or written documents. Complex information can be collected and shown in more easily understood terms through using maps that highlight features, patterns, and relationships of people, places, and things. Maps

Ashley Cooper/Alamy

also are useful for making comparisons. For example, a geographer might compare population density maps or transportation networks maps of two counties in order to determine where to build new schools.

Interviewing requires a geographer to ask questions rather than just collect data, images, and on-site observations. Specifically, for human geographic studies, geographers may want to find out how people think or feel about certain places. They may also want to examine the ways in which people's beliefs and attitudes have affected the physical environment. To **obtain** such information, geographers interview their subjects. They can do this by selecting a particular group of people for study. Rather than contacting every individual in the group, however, geographers use a carefully selected sample of people whose answers represent the larger group.

Geographers also analyze *statistics*. Numerical data, such as temperature and snowfall, can provide insight into a region's climate trends. Geographers use computers to organize and present this information in an understandable manner, as well as to look in detail at the data for patterns and trends. For example, studies that identify age, ethnicity, and gender of specified regions can emphasize possible trends within a human population. After identifying such patterns and trends, geographers use statistical tests to see whether their ideas are valid.

obtain to gain or acquire, usually by planning or effort

✔ **READING PROGRESS CHECK**

Defining What do geographers do?

Geography and Other Subjects

GUIDING QUESTION *How is geography related to other subjects?*

Geography has important relationships to other subjects. Geographers use geographic tools and methods to understand historical patterns, economies, politics and political patterns, and the impact of societies and cultures on the landscape.

To visualize what places could have looked like in the past, geographers take into account historical perspectives of the place. For example, to gather information about how a city has changed over time, geographers can collect information from historical sources regarding census data, economic output, birth and death rates, natural disasters, disease, and major **fluctuations** in population size. Such data can address questions concerning how human activities have changed the natural vegetation, or how waterways are different today than in the past. Such historical perspectives provide insight as to which institutions or development should be constructed to avoid repeating past complications between human growth and the physical environment.

fluctuation a shift from a previous condition

Additionally, analysis of historical and current political patterns emphasizes changing boundary lines and government systems. Geographers are also interested in how the natural environment has influenced political decisions and how governments change natural environments. For example, in the 1960s the Egyptian government built the massive Aswān High Dam on the Nile River to help irrigate the land. The dam altered the Nile River valley and significantly impacted the region's people.

Human geographers, also called cultural geographers, use the ideas of sociology and anthropology to study human tendencies and past cultures and their influence on current traditions and social norms. Because people come from diverse cultural backgrounds, their interpretations of information and experiences differ depending on their frame of reference. For example, residents of a particular neighborhood may define boundaries based on location of activities, such as stores they frequent and people with whom they have contact.

The natural shelter and deep waters of Victoria Harbor make Hong Kong one of the world's largest shipping ports.

▲ **CRITICAL THINKING**

1. *Analyzing* What factors might a geographer study to learn about the economy of a particular country?

2. *Making Connections* How is interdependence important to economic activity?

However, the local governments create neighborhood boundaries to facilitate services and maintenance. Furthermore, other neighborhood boundaries may be created by police departments and school districts that have different needs to meet and different reasons for creating spatial divisions. Human geography can be used to study the relationships between people, places, and environments by mapping information about them into a spatial context using GIS and other geospatial technologies. Human geographers study the way people are rooted in particular places and how they have constructed various types of regions. Some geographers specialize in studying the feelings one has about a place, which is very closely connected with theories in the fields of psychology and behavioral science.

Geographers study economies to understand how the locations of resources affect the ways people make, transport, and use goods, and how and where services are provided. Geographers are interested in how locations are chosen for various economic activities. Where and what human groups choose to produce and consume depend on a variety of factors: location of natural resources for mining and extracting, fertile soil for farming, suitable climates for living and producing, and proximity to good transport routes and other cultures to establish trade relations.

Economic activity relies on not only a society's production and use of goods, but also on the transport of goods between cultures in the form of trade. Such interdependence between global economies is part of what defines relationships and communication between various cultures. The growth of technological and communication systems in today's world also affects these relationships. The ability to call a client halfway across the world, reduce production time by mechanical production instead of human labor, use the Internet to communicate ideas instantly, and send goods overnight via air delivery are examples of how innovations in technology have increased the speed and efficiency of the movement of information and goods.

☑ **READING PROGRESS CHECK**

Naming What is another name used for human geographers?

LESSON 2 REVIEW

Reviewing Vocabulary

1. *Understanding Relationships* How are site and situation different?

Using Your Notes

2. *Listing* Using your graphic organizer, write a definition of *geography* in your own words.

Answering the Guiding Questions

3. *Identifying* What is the spatial perspective?

4. *Categorizing* What are the elements of geography?

5. *Organizing* What methods do geographers use to conduct their work?

6. *Evaluating* How is geography related to other subjects?

Writing Activity

7. *Informative/Explanatory* Write a paragraph explaining how geography helps us interpret the past, understand the present, and plan for the future.

Paul Chesley/Stone/Getty Images

Directions: On a separate sheet of paper, answer the questions below. Make sure you read carefully and answer all parts of the questions.

Lesson Review

Lesson 1

1 **Describing** Describe the problems that arise when the curves of a globe become straight lines on a map.

2 **Comparing and Contrasting** Explain the similarities and differences between the Winkel Tripel projection and the Mercator projection.

3 **Describing** What is the importance of scale in reading maps?

4 **Listing** List three examples of things a map can show.

Lesson 2

5 **Explaining** Why is the U.S. Corn Belt considered a formal region?

6 **Discussing** What are two research methods used by geographers?

7 **Differentiating** What is the difference between physical geography and human geography?

8 **Summarizing** Why is human-environment interaction an important theme for geographers?

Applying Map Skills

Refer to the Unit 1 Atlas to answer the following questions.

9 **Places and Regions** Which continent has the most countries with the highest population densities?

10 **Environment and Society** Look at the map of world economic activities. Name three forms of land use for the continent of Africa.

11 **Physical Systems** What physical features in Africa might explain areas with little or no economic activity?

12 **The World in Spatial Terms** Look at the political map. What is the absolute location of Houston, Texas? What is the relative location of Houston?

DBQ Analyzing Primary Sources

Use the cartoon to answer the following questions.

PRIMARY SOURCE

13 **Analyzing Visuals** Why do we see the globe as being upside down? How is the penguin's point of view different? Why?

14 **Finding the Main Idea** What does this cartoon say about the broad definition of geography?

Exploring the Essential Question

15 **Understanding Historical Interpretation** Imagine you are studying an archaeological site to determine why a civilization came to end. What kind of geographic information might you want to gather in order to determine what happened to the civilization? Consider the skills for thinking geographically. Write a paragraph containing your response.

Need Extra Help?

If You've Missed Question	**1**	**2**	**3**	**4**	**5**	**6**	**7**	**8**	**9**	**10**	**11**	**12**	**13**	**14**	**15**
Go to page	14	15	18	18	30	32	30	30	2	2	2	2	35	35	32

Leo Cullum/The New Yorker Collection/www.cartoonbank.com

Directions: On a separate sheet of paper, answer the questions below. Make sure you read carefully and answer all parts of the questions.

Critical Thinking

16 *Assessing* What are the different types of geospatial technologies? How has this advanced technology improved the way maps are created?

17 *Making Connections* What type of information would help you determine why traffic has increased on a certain road?

18 *Evaluating* How could geography be interpreted differently based on changing human perspectives?

19 *Drawing Conclusions* How can we use geography to make decisions for the future?

20 *Compare and Contrast* What differentiates mental mapping from other forms of visual mapping?

College and Career Readiness

21 *Examining Information* Describe three ways geographic knowledge or tools are used in professions other than geography.

Research and Presentation

22 *Research Skills* Using the Internet, conduct research to learn about how geography is used in the film industry. Write one paragraph describing a specific example of this use.

Writing About Geography

23 *Informative/Explanatory* Imagine you are a geographer working on a plan for a new community center. Using the different types of research methods as your guide, what factors would you use to develop the plan? Explain your choices in a paragraph.

21st Century Skills

Use the excerpt below to answer the questions that follow.

PRIMARY SOURCE

"*Geography—real-world geography—is the art and science of location, or place. It is about spatial patterns and spatial processes. It is about which way the wind blows from Chernobyl, the Pacific 'ring of fire,' AIDS, terrorists, and refugees. It is about acid rain, El Niño, ocean dumping, cultural censorship, droughts and famines. . . .*

Real-world geography also explores things in locations: why something is where it is and what processes change its distribution. Geography is the why of where of an ever-changing universe. Its surpassing objective is to discover the processes that move over space and connect places and continually transform the location and character of everything."

—George J. Demko, *Why in the World: Adventures in Geography,* 1992

24 *Geography Skills* What are some of the world issues that George Demko lists as concerns of geography?

25 *Primary and Secondary Sources* According to the excerpt, what is the main objective of geography?

26 *Identifying Cause & Effect* Identify how the physical environment affects human activity, and likewise how human activity affects the physical environment.

27 *Economics* How might the economic activities of a region affect its physical and human geography?

Need Extra Help?

If You've Missed Question	**16**	**17**	**18**	**19**	**20**	**21**	**22**	**23**	**24**	**25**	**26**	**27**
Go to page	21	22	28	31	20	33	36	32	36	36	36	36

The Physical World

networks

There's More Online about the physical world.

ESSENTIAL QUESTION · *How do physical processes shape Earth's surface?*

Geography Matters...

Because much of our world had not been explored before 1800, geography prior to that time focused on discovery, basic data collection, and observation. Geographers were busy determining the elevation of land surfaces, describing landforms, identifying weather and climate patterns, and classifying soils and ecosystems. The geographers of today use sophisticated tools and research methods to monitor the processes that shape Earth's surface, the natural resources essential to our survival, and the ways in which humans affect and are affected by the physical environment. Understanding the planet we call home helps us make better decisions about how to utilize, preserve, and protect the Earth and adapt to its changing environments.

◄ A climber rests alone on Mount Everest.

Harry Kikstra/Flickr/Getty Images

economics *and* resources: water scarcity

Water is a basic need for all humans. Providing freshwater to support human activity is a significant challenge in drought-prone areas around the world. Agriculture, industry, and thriving population centers all depend on a reliable supply of safe, fresh water to support human endeavors.

THERE'S MORE ONLINE

SEE a diagram of the desalination process • *READ* an infographic about freshwater resources (bkgd)Ami Vitale/Alamy, (tcl)Nir Alon/Alamy, (tcl)Richard Allenby-Pratt/arabianEye/Corbis, (tcr)Jose Luis Roca/AFP/Getty Images

How does water scarcity impact economic activity?

North Africa and Southwest Asia are two areas of the world where there is a scarcity—lack of an adequate supply—of freshwater for agriculture, industry, drinking, bathing, and other human uses of water due to limited resources. A scarcity of freshwater limits the ability of a city, region, or country to engage in many economic pursuits that allow people to live comfortably and thrive. Without water, agricultural fields cannot be irrigated, industrial production cannot occur, and hospitals cannot provide the necessary quality of care.

1. **Environment and Society** Why is water scarcity a problem? Explain what the term *water scarcity* means and discuss how it affects human activities.

How is technology being used to solve the problem of water scarcity?

The town of Ashqelon is located on Israel's Mediterranean coast with an abundant supply of salt water lapping at its shore. Desalination (salt removal) technology is being used to turn the sea water into freshwater for Israeli citizens and industries that have faced recent droughts and population growth. Ashqelon, 40 miles (64.4 km) south of Tel Aviv, has the largest desalination plant of its kind in the region and one of the largest in the world. Seawater is pumped through a series of filters and reverse-osmosis membranes which allow the smaller water molecules to pass through while excluding the larger molecules of salt and other impurities. The plant provides approximately five percent of Israel's freshwater and is predicted to soon make Israel an exporter of water.

2. **Human Systems** How has desalination technology affected Israel's economy? What countries may import freshwater from Israel?

What are the environmental effects of desalination?

The freshwater provided by the Ashqelon desalination plant supports Israel's thriving agricultural and high-tech industries, both heavily dependent on a steady supply of freshwater. However, this water comes with several costs to the environment. Fossil fuels are used to power the pumps that circulate the water through the filter plant, contributing to pollution. The brine, or leftover material that the filters remove, is pumped back into the sea, resulting in a salty desert on the sea floor around the outflow pipes. The plant also occupies a coastal area that might otherwise harbor wildlife and offer recreation options.

3. **Environment and Society** Explain the costs and benefits of providing freshwater using reverse-osmosis desalination.

CHAPTER 2 Assessment

Directions: On a separate sheet of paper, answer the questions below. Make sure you read carefully and answer all parts of the questions.

Critical Thinking

13 **Making Generalizations** Explain how weathering and erosion help create soil.

14 **Identifying Cause and Effect** How would life on Earth be different if there were no precipitation?

15 **Comparing and Contrasting** What is the difference between groundwater and aquifers, and why is the distinction important?

16 **Defending** How can wind erosion actually be helpful in some areas?

Applying Map Skills

Refer to the Unit 1 Atlas to answer the following questions.

17 **Physical Systems** Which coast of South America appears to have the most signs of plate tectonic movement due to its landforms?

18 **The World in Spatial Terms** Which continents are located along the Ring of Fire?

19 **Places and Regions** Use what you know of the world's oceans to draw a mental map showing the continents and oceans. Be sure to label the five oceans.

Exploring the Essential Question

20 **Making Connections** Write a one-page essay explaining how knowledge of plate tectonics can help governments prepare for natural disasters. Provide at least two specific examples.

Writing About Geography

21 **Informative/Explanatory** Describe at least three landforms found in the region of the world where you live. Include the name of each landform and the processes that created it.

DBQ Analyzing Primary Sources

Use the document to answer the following questions.

The center of the Earth is filled with intense heat and pressure. These natural forces drive numerous changes such as volcanoes and earthquakes that renew and enrich Earth's surface. The physical processes can also disrupt, and often destroy, human life. As a result, scientists are working to learn how to predict them.

PRIMARY SOURCE

❝*Scientists are doing everything they can to solve the mysteries of earthquakes. They break rocks in laboratories, studying how stone behaves under stress. They hike through ghost forests where dead trees tell of long-ago tsunamis. They make maps of precarious, balanced rocks to see where the ground has shaken in the past, and how hard. They dig trenches across faults, searching for the active trace. They have wired up fault zones with so many sensors it's as though the Earth is a patient in intensive care.*❞

—Joel Achenbach, "The Next Big One," *National Geographic,* April 2006

22 **Analyzing** How does Achenbach describe the work done to solve earthquake mysteries?

23 **Interpreting** What did Achenbach mean by "wired up fault zones...as though the Earth is a patient in intensive care"?

Research and Presentation

24 **Research Skills** Using the Internet, research the U.S. Geological Survey and the specific role it plays in contributing to our knowledge of earthquakes. Create a multimedia presentation to share your findings.

Need Extra Help?

If You've Missed Question	**13**	**14**	**15**	**16**	**17**	**18**	**19**	**20**	**21**	**22**	**23**	**24**
Go to page	50	51	54	49	4	4	4	45	43	56	56	48

Climates of the Earth

ESSENTIAL QUESTION · *Why is climate important to life on Earth?*

Geography Matters...

The relationships between the Earth and the sun influence climate and weather patterns all over the Earth. The various climates on Earth support different communities of plants and animals that are called biomes. Human activity is also affected by Earth's climates, such as knowing that only certain vegetables can grow in our backyard gardens. Fortunately, our planet is just the right distance from our sun to make life on Earth possible.

◄ A young Inuit boy on Canada's Baffin Island wears warm clothing to protect against the cold Arctic conditions.

Sue Flood/The Image Bank/Getty Images

climate change:
the impacts *on* humans

Climate scientists have amassed data on weather patterns and climate since the 1980s. Most agree that the average surface air temperature across the globe has increased about 1.4°F (0.8°C) since then. Predictions show that warming of the planet, or global warming, will accelerate through the end of the twenty-first century.

THERE'S MORE ONLINE

WATCH a video of a melting glacier • *SEE* a map of rising global temperatures

/AFP/Getty Images

(c)Arctic-Images/Iconica/Getty Images, (c)John Short"/Design Pics/Corbis, (cr)Dondi Tawatao/Getty Images News/Getty Images

What is climate change and global warming?

Climate change is a significant change in temperature, precipitation, and wind patterns that lasts for at least several decades. These changes may be difficult to notice, but there are several signs of climate change that are observable, such as heat waves, flooding, droughts, melting glaciers, and extreme weather events. Global warming refers to warming of Earth's average temperature that has occurred since the mid-1800s. It represents only one aspect of climate change. Global warming does not simply mean that the climate everywhere on Earth will become warmer. An important part of what it means is that some weather patterns will become more extreme.

1. Physical Systems How is global warming different from climate change?

What are the causes of climate change?

Although average global surface temperature does fluctuate due to natural causes, most climate scientists agree that human activity is a major contributing factor to recent warming of the planet. As global population and industrialization increase, a variety of human activities increase the presence of so-called "greenhouse gases" in Earth's atmosphere. In contrast, the greenhouse effect is a naturally occurring process in which these gases act as a blanket, trapping the sun's heat within the atmosphere. The human-produced greenhouse gases amplify the effect and increase the warming of Earth. These gases—primarily carbon dioxide and methane—are released into the atmosphere through the burning of fossil fuels and wood, along with some industrial processes. Agricultural practices and the decay of organic waste in landfills also contribute to greenhouse gases.

2. Physical Systems Describe two factors contributing to climate change.

What are the potential impacts of climate change on people?

As Earth's surface temperature warms and its climate becomes less stable, people will be affected in several ways. Melting glaciers will cause sea levels to rise, likely rendering some coastal regions and islands uninhabitable. Increased droughts, heat waves, and wildfires will affect some regions, while other areas will be impacted by severe flooding. Storms and other extreme weather events may increase in their frequency and intensity as well. As a result, people may face increased health and safety risks along with threats to their homes, cities, and critical infrastructure systems.

3. Environment and Society How will climate change affect humans? Consider water supplies, agriculture, power and transportation systems, the environment, and human health.

networks

There's More Online!

☑ **DIAGRAM** The Earth's Seasons
☑ **INFOGRAPHIC**
The Greenhouse Effect
☑ **INTERACTIVE**
SELF-CHECK QUIZ
☑ **VIDEO** Earth-Sun Relationships

Sun

LESSON 1
Earth-Sun Relationships

Reading **HELP**DESK

Academic Vocabulary

- **predictable**
- **eventually**

Content Vocabulary

- **weather**
- **climate**
- **axis**
- **revolution**
- **equinox**
- **solstice**
- **midnight sun**
- **greenhouse effect**

TAKING NOTES: *Key Ideas and Details*

DESCRIBING As you read the lesson, use a chart like the one below to list the characteristics of Earth-sun relationships and describe their effects on climate.

Earth-Sun Relationships	Effects on Climate

ESSENTIAL QUESTION · *Why is climate important to life on Earth?*

IT MATTERS BECAUSE
Daily life on Earth is influenced by the dynamic relationship between the Earth and the sun. The amount of direct sunlight reaching Earth's surface plays an important role in affecting the temperature of different places. The Earth's rotation determines when we receive sunlight, giving us day or night. The Earth's tilt and its revolution around the sun result in the four seasons we experience.

Climate and Weather

GUIDING QUESTION *How do the relationships between the Earth and the sun affect climate?*

There is an important difference between climate and weather. **Weather** is the condition of the atmosphere in one place over a short period of time, such as hours or days. For example, when you look outside the window in the morning to decide what to wear that day, you are checking the weather. **Climate**, on the other hand, refers to the average weather conditions as measured over many years. Climate is the reason why you decide to buy certain types of clothing to wear based on where you live.

Earth's Tilt and Rotation
The relationship between the Earth and the sun directly affects climate. An important aspect of the Earth-sun relationship is that the Earth's **axis** is tilted. The axis runs from the North Pole to the South Pole through the center of the planet. Currently, the Earth is tilted at about 23½°.

Because of the Earth's tilted axis, not all places on Earth receive the same amount of direct sunlight at the same time. For this reason, Earth's tilt affects the temperature of a particular place. Temperature is the measure of how hot or cold a place is. Temperature is measured in degrees on a set scale. The most common scales for measuring temperature are Fahrenheit (°F) and Celsius (°C).

Why is it usually warmer during the day than it is during the night? This depends on which side of the planet is facing the sun. The Earth rotates on its axis, making one complete rotation every 24 hours, or one day. The Earth's rotation from west to east ensures that every part of the world receives sunlight in a **predictable** pattern during those 24 hours. The side of the planet not facing the sun is colder, and the side of the planet facing the sun is warmer.

Earth's Revolution

While the Earth rotates on its axis, it also revolves around the closest star to us, the sun. It takes the Earth one year, approximately 365 days, to complete one **revolution** around the sun. The Earth's revolution, combined with its tilted axis, affects the amount of sunlight that reaches different locations on the Earth at different times of the year. People who live in the Northern Hemisphere experience summer when the Northern Hemisphere is tilted toward the sun and is receiving the most direct sunlight. The seasons are reversed north and south of the Equator. When it is summer in the Northern Hemisphere, it is winter in the Southern Hemisphere because the Southern Hemisphere is tilted away from the sun and receives less direct sunlight. Likewise, when it is fall in the Northern Hemisphere, it is spring in the Southern Hemisphere.

Twice a year (around March 21 and September 23), the direct sunlight falls on the Equator. This day is called an **equinox**, meaning "equal night," because daytime and nighttime hours are equal. On the equinox, equal amounts of light reach the Northern and Southern Hemispheres. The two equinoxes mark the shift in seasons between winter and spring and between summer and fall.

In addition to the Equator, there are two other lines of latitude that run parallel to the Equator and mark important changes in the Earth's seasons. As the Earth proceeds in its revolution around the sun, the direct rays of the sun

weather condition of the atmosphere in one place during a short period of time

climate weather patterns typical for an area over a long period of time

axis an imaginary line that runs through the center of the Earth between the North and South Poles

predictable expected or able to be foreseen

revolution in astronomy, the Earth's yearly trip around the sun, taking 365¼ days

equinox one of two days (about March 21 and September 23) on which the sun is directly above the Equator, making day and night equal in length

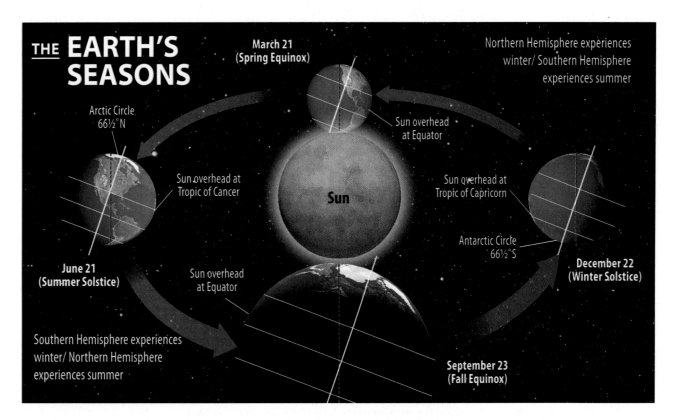

THE EARTH'S SEASONS

March 21 (Spring Equinox)

Northern Hemisphere experiences winter/ Southern Hemisphere experiences summer

Arctic Circle 66½°N

Sun overhead at Equator

Sun overhead at Tropic of Cancer

Sun

Sun overhead at Tropic of Capricorn

Antarctic Circle 66½°S

June 21 (Summer Solstice)

Sun overhead at Equator

December 22 (Winter Solstice)

Southern Hemisphere experiences winter/ Northern Hemisphere experiences summer

September 23 (Fall Equinox)

The angle of the sun's rays as they strike the Earth affects the Northern and Southern Hemispheres differently.

▲ **CRITICAL THINKING**

1. *Contrasting* Explain the difference between solstices and equinoxes.

2. *Drawing Conclusions* In what months do the sun's rays directly strike the Equator? The Tropics of Cancer and Capricorn?

Climates of the Earth **61**

solstice one of two days (about June 21 and December 22) on which the sun's rays strike directly on the Tropic of Cancer or Tropic of Capricorn, marking the beginning of summer or winter

eventually strike the Tropic of Cancer, the latitude line at 23½° N that passes through Mexico, North Africa, and India. The sun usually hits the Tropic of Cancer around June 21, bringing the longest day of sunlight to the Northern Hemisphere. This date is known as the summer **solstice** and marks the beginning of the summer season in the Northern Hemisphere. By about September 23, the Earth has revolved so that the direct rays of the sun hit the Equator again. This equinox marks the end of summer and the beginning of the fall season in the Northern Hemisphere.

As the Earth continues in its revolution, the direct rays of the sun eventually strike the Tropic of Capricorn—the latitude line at 23½° S running through South America, the southern tip of Africa, and Australia—around December 22. This marks the winter solstice, bringing the shortest day of sunlight to the Northern Hemisphere and signaling the beginning of the winter season.

The most dramatic variation in the amount of sunlight occurs near the Poles. For six months of the year, one Pole gets continuous sunlight while the other Pole receives none. From about March 20 to about September 23, the polar area north of the Arctic Circle (66½° N) experiences continuous daylight or twilight. The polar area south of the Antarctic Circle (66½° S) experiences continuous daylight or twilight for the other six months of the year. Continuous daylight, a phenomenon also known as **midnight sun**, is caused by the tilt of the Earth's axis as it revolves around the sun. The Poles are very sparsely populated, so many people remain unaffected by midnight sun. However, parts of northern North America and northern Europe—such as Alaska, Sweden, Denmark, Norway, Finland, and others—have become popular tourist destinations particularly because of midnight sun.

midnight sun continuous daylight, a time when the sun is visible at midnight during the summer in either the Arctic or Antarctic Circle

☑ **READING PROGRESS CHECK**

Contrasting What factor distinguishes weather from climate?

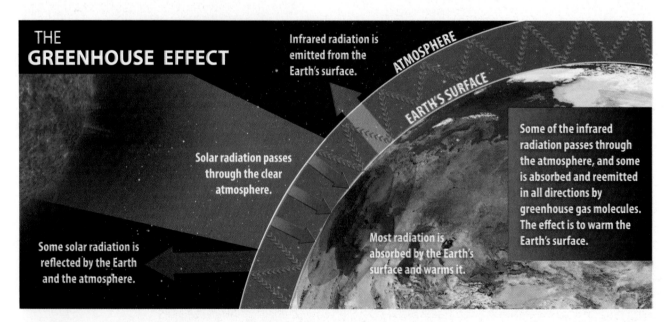

THE
GREENHOUSE EFFECT

ATMOSPHERE

EARTH'S SURFACE

Infrared radiation is emitted from the Earth's surface.

Solar radiation passes through the clear atmosphere.

Some solar radiation is reflected by the Earth and the atmosphere.

Most radiation is absorbed by the Earth's surface and warms it.

Some of the infrared radiation passes through the atmosphere, and some is absorbed and reemitted in all directions by greenhouse gas molecules. The effect is to warm the Earth's surface.

Solar radiation, when combined with greenhouse gases like carbon dioxide and methane, results in the warming of the Earth's surface. Without this greenhouse effect, Earth would be too cold for most living things.

▲ **CRITICAL THINKING**

1. Interpreting What processes of the greenhouse effect contribute to the warming of the Earth's surface?

2. Sequencing What happens to the solar radiation that is not absorbed by the Earth?

The Greenhouse Effect

GUIDING QUESTION *What is the greenhouse effect?*

Even on the warmest days, only some of the sun's rays pass through the Earth's atmosphere. The atmosphere reflects some of the radiation back into space. Enough radiation reaches Earth's surface, however, to warm the air, land, and water. Once Earth's surface absorbs the radiation from the sun and is warmed, it radiates this heat energy back again into the atmosphere.

Normally, the atmosphere provides just the right amount of insulation to promote life on the planet. The 50 percent of the sun's radiation that reaches the Earth is converted into infrared radiation, or heat. As shown in the infographic, clouds and greenhouse gases—atmospheric gases such as water vapor, methane, and carbon dioxide (CO_2)—absorb the reradiated heat energy and trap it so that most of it cannot escape back into space. The atmosphere is therefore like a greenhouse. It traps enough radiation to warm the land, water, and air and help plants grow while reflecting some radiation to ensure that the Earth does not overheat.

This **greenhouse effect** is the warming of the Earth that occurs when the sun's radiation passes through the atmosphere, is absorbed by the Earth, and is radiated as heat energy back into the atmosphere where it cannot escape into space. Without the greenhouse effect, Earth's average temperature would be below 0°F (-17°C) and life as we know it could not exist.

To understand the planet's natural greenhouse effect, consider that according to the laws of physics the radiation Earth receives from the sun must be equally balanced by the heat Earth radiates back out to space. If Earth gave back less energy than it received, the planet would **eventually** become too warm to support life. Likewise, if Earth gave back more energy than it received, the planet would be too cold for life.

greenhouse effect the capacity of certain gases in the atmosphere to trap heat, thereby warming the Earth

eventually taking place later; in the end

PRIMARY SOURCE

❝ The greenhouse effect is one of the fundamental facts of atmospheric science. It is real; that fact is beyond dispute. Without it, the entire surface of the ocean would be frozen solid. Life—at least the kind that depends on liquid water and warmth—could not survive. We owe our existence to the greenhouse effect. So why are we worried about it? . . . The answer is that some of the heat radiation leaks out through the atmosphere, because there is not enough water vapor, carbon dioxide, and other gases to absorb all of the IR [infrared heat radiation]. Think of the atmosphere as a leaky blanket. ❞

—Richard A. Muller, from the Columbia Forum, "Physics for Future Presidents," January/February 2009

☑ **READING PROGRESS CHECK**

Assessing How does the greenhouse effect influence Earth's surface temperature?

LESSON 1 REVIEW

Reviewing Vocabulary
1. *Making Connections* What is the relationship between weather, climate, axis, temperature, revolution, equinox, and solstice?

Using Your Notes
2. *Discussing* Using your graphic organizer, discuss how temperature is affected by the tilt of the Earth.

Answering the Guiding Questions
3. *Identifying Cause and Effect* How do the relationships between the Earth and the sun affect climate?

4. *Explaining* What is the greenhouse effect?

Writing Activity
5. *Informative/Explanatory* Write a paragraph explaining the differences in weather you would expect in Alaska and Florida.

networks

There's More Online!

- ☑ **DIAGRAM** The Rain Shadow Effect
- ☑ **MAP** World Zones of Latitude and Wind Patterns
- ☑ **MAP** World Ocean Currents
- ☑ **INTERACTIVE SELF-CHECK QUIZ**
- ☑ **VIDEO** Factors Affecting Climate

Reading **HELP**DESK

Academic Vocabulary

- **crucial**
- **derive**

Content Vocabulary

- **current**
- **prevailing wind**
- **Coriolis effect**
- **doldrums**
- **El Niño**
- **windward**
- **leeward**
- **rain shadow**

TAKING NOTES: *Key Ideas and Details*

DESCRIBING As you read the lesson, use a web diagram like the one below to list the factors that cause both wind and ocean currents.

Wind and Ocean Currents

LESSON 2
Factors Affecting Climate

ESSENTIAL QUESTION • *Why is climate important to life on Earth?*

IT MATTERS BECAUSE

The climate of a particular place may have extreme weather and temperature ranges that are caused by several geographic features. Both the latitude and the elevation of a place, along with wind and ocean currents, influence its climate. Two sides of a mountain range may also have two different climates. While one side receives more precipitation as air rises, the opposite side has drier, warmer air as air descends.

Latitude, Climate, and Elevation

GUIDING QUESTION *How are climate patterns related to each zone of latitude?*

The Earth's annual revolution around the sun creates predictable climate patterns. These patterns correspond with bands, or zones, of latitude.

The low latitude zone is between 30° S and 30° N. This zone includes the Tropic of Capricorn, the Equator, and the Tropic of Cancer. Portions of the low latitude zone receive direct rays from the sun year-round and therefore have warm to hot climates.

The high latitude zone includes the Earth's polar areas, which stretch from 60° N to 90° N and from 60° S to 90° S. When either the Northern or the Southern Hemisphere is tilted toward the sun, its polar area receives nearly continuous, but indirect, sunlight.

The midlatitude zone, between 30° N and 60° N in the Northern Hemisphere and between 30° S and 60° S in the Southern Hemisphere, is home to climates that have the most variable weather on Earth. The midlatitudes generally have a temperate climate, or one that varies from fairly hot to fairly cold, with dramatic seasonal weather changes. Warm/hot and cool/cold air masses move across the midlatitudes. These movements and the interactions between the different air masses affect weather in this latitude zone throughout the year.

At all latitudes, elevation influences climate because of the relationship between the elevation of a place and its temperature. The Earth's atmosphere thins as altitude increases. Less dense air retains less heat. As elevation increases, temperatures decrease by about 3.5°F (1.9°C) for each 1,000 feet (305 m). This effect occurs at all latitudes.

Cool moist air drops moisture

For example, in Ecuador, the city of Quito (KEE•toh) is nearly on the Equator. However, Quito is located in the Andes at an elevation of more than 9,000 feet (2,743 m), so average temperatures there are more than 20 degrees cooler than in the coastal lowlands. Sunlight is bright in places with high elevation because the thinner atmosphere filters fewer rays of the sun. But even in bright sunlight, the world's highest mountains, such as the Andes, are cold and snowy.

☑ **READING PROGRESS CHECK**

Explaining What happens to temperature as elevation increases?

Winds and Ocean Currents

GUIDING QUESTION *How do winds and ocean currents affect climate?*

How do wind and water work together to affect weather? Air moving across the surface of the Earth is called wind. Winds occur because sunlight heats the Earth's atmosphere and surface unevenly. Warm temperatures cause air to rise and create areas of low pressure. Cool temperatures cause air to sink, which creates areas of high pressure. Air moves along the pressure gradient from areas of high pressure to low pressure, so the cool air then flows in to replace the warm rising air. These movements cause winds to distribute the sun's energy around the planet. For this reason, wind patterns are **crucial** to a region's climate.

Ocean **currents** also help distribute energy around the planet. As they circulate, cold water from the polar areas moves slowly toward the Equator. These are cold ocean currents because they consist of cooler water flowing into warmer water. The opposite is also true: warm water moves away from the Equator and these are warm currents because they consist of warmer water flowing into cooler water.

crucial vitally important

current cold or warm stream of seawater that flows in the oceans, generally in a circular pattern

GEOGRAPHY CONNECTION

Global wind patterns are affected by latitude.

1. *PHYSICAL SYSTEMS* In what direction are winds deflected in the Northern Hemisphere?

2. *PLACES AND REGIONS* In what latitude zone are the warm winds deflected toward the west?

World Zones of Latitude and Wind Patterns

The rain shadow effect is influenced by landforms and climate.

▶ **CRITICAL THINKING**

1. *Interpreting* Why does air lose moisture as it rises over mountains?

2. *Speculating* Winds are blowing from the west toward the east of a mountain range running from north to south. Which side of the mountain range is greener and has more trees, the western side or the eastern side?

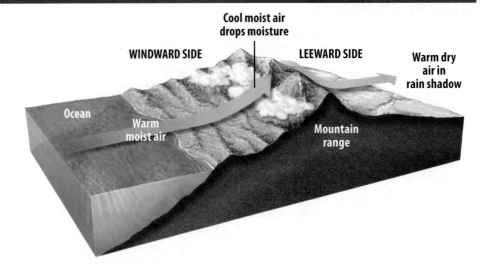

Landforms and Climate

GUIDING QUESTION *How can landforms and bodies of water affect climate?*

While one can generalize about the climates of places located in the same latitude zones, they vary based upon the presence or absence of certain physical features. Large bodies of water, for example, are slower to heat and cool, so they tend to keep temperatures in surrounding lands moderate. Coastal lands receive the benefit of this influence and experience less changeable weather. Conversely, the interiors of the continents tend to experience extremes in seasonal temperatures.

Yet another physical feature that affects the climates in the latitude zones are the mountain ranges. As the diagram shows, mountain ranges push wind upward and, as a result, the rising air cools and releases moisture in the form of precipitation. Most of the precipitation falls on the **windward** side of the mountain, or the side of the mountain range facing the wind. After the precipitation is released, winds become warmer and drier as they descend on the opposite, or **leeward**, side of the mountains. The hot, dry air produces little precipitation in an effect known as a **rain shadow**. The rain shadow effect often causes dry areas—and even deserts—to develop on the leeward sides of mountain ranges.

☑ **READING PROGRESS CHECK**

Discussing What happens to winds after they release precipitation?

windward being in or facing the direction from which the wind is blowing

leeward being in or facing the direction toward which the wind is blowing

rain shadow result of a process by which dry areas develop on the leeward sides of mountain ranges

---

LESSON 2 REVIEW

Reviewing Vocabulary

1. *Describing* How are prevailing winds influenced by the Coriolis effect?

Using Your Notes

2. *Making Connections* Which of the factors from your graphic organizer do you think have the strongest effect on the climate where you live?

Answering the Guiding Questions

3. *Explaining* How are climate patterns related to each zone of latitude?

4. *Identifying Cause and Effect* How do wind currents and ocean currents affect climate?

5. *Evaluating* How can landforms and bodies of water affect climate?

Writing Activity

6. *Informative/Explanatory* Suppose you are on a ship sailing in the low latitudes. Write a paragraph explaining what might happen as you drift near the Equator.

netw⊙rks

There's More Online!

☑ **MAP** World Biomes

☑ **PRIMARY SOURCE**
Controversy Over Climate Change

☑ **INTERACTIVE**
SELF-CHECK QUIZ

☑ **VIDEO** World Climate Patterns

Reading **HELP**DESK

Academic Vocabulary

- **overlap**
- **distinct**

Content Vocabulary

- **biome**
- **natural vegetation**
- **average daily temperature**
- **oasis**
- **prairie**
- **coniferous**
- **deciduous**
- **mixed forest**
- **permafrost**

TAKING NOTES: *Key Ideas and Details*

IDENTIFYING Use a web diagram like the one below to take notes about the Earth's four climate zones.

LESSON 3
World Climate Patterns

ESSENTIAL QUESTION · *Why is climate important to life on Earth?*

IT MATTERS BECAUSE
Climate patterns vary from region to region. However, factors such as wind and air pressure can create zones where a climate becomes quite dramatic. For example, most of Australia has a dry climate, but when trade winds meet during the summer months, Australia's northern coast sees intense thunderstorms.

Climate Regions and Biomes

GUIDING QUESTION *How are world climates and biomes organized?*

Climates in the world are organized into four climate zones: tropical, dry, midlatitude, and high-latitude climates. These climates support different kinds of biomes. A **biome** is a major type of ecological community defined primarily by distinctive **natural vegetation** and animal groups. The characteristics of biomes may **overlap** with one another.

Tropical Climates

Tropical climates are found in or near low latitudes in areas otherwise referred to as the Tropics. The two most widespread kinds of tropical climate regions are wet climates and dry climates.

Tropical rain forest climates have an **average daily temperature** of 80°F (27°C), and since the warm air is humid, or saturated with moisture, it rains almost daily. Annual rainfall averages from 50 to 260 inches (125 to 660 cm). This continual rain tends to strip the soil of nutrients. The biome in these climates are the tropical rain forests, characterized by thick vegetation that grows in layers. Tall trees form a canopy over shorter trees and bushes, and shade-loving plants grow on the completely shaded forest floor. The world's largest tropical rain forest is in the Amazon River basin. Similar climate and vegetation exist in other parts of South America, the Caribbean, Asia, and Africa. Due to the vast amount of plant food, wildlife is also abundant, and scientists estimate that more than half of all of the plant and animal species exist in the tropical rain forests. As with all biomes, plants and animals vary within a given rain forest, with differentiation often occurring. For example, trees that grow in the mountains of the Amazon rain forest do not grow in the lowlands of the same rain forest.

biome major type of ecological community defined primarily by distinctive plant and animal groups

natural vegetation plant life that grows in a certain area if people have not changed the natural environment

overlap to partly cover

average daily temperature the average of the daily high temperature and the overnight low; often used for comparison across climate regions

distinct recognizably different

oasis small area in a desert where water and vegetation are found

Tropical wet/dry climates have pronounced dry and wet seasons, with high year-round temperatures. These regions, also called savannas, have fewer plants and animals than the tropical rain forest climates. One distinguishing characteristic of a tropical wet/dry climate is that sunlight is not blocked by trees and is able to reach much of the ground surface. This makes for more highly specialized plant and animal species. Tropical savannas are found in Africa, Central and South America, Asia, and Australia. Each maintains **distinct** types of plant and animal life. A key factor determining the types of plants and animals is the length and severity of the dry season.

Dry Climates

The two main types of dry climates are semi-arid (or steppe) and arid (or desert), both of which occur in low latitudes and midlatitudes. Geographers distinguish these dry climates based on the amount of rainfall and the vegetation in each.

Steppes are usually located away from oceans or large bodies of water and therefore are less humid. However, they do receive an average of 10 to 30 inches (25 to 76 cm) of rainfall per year. Steppes experience warm summers and harshly cold winters. Some steppes have heavy snowfall, while others are susceptible to droughts and violent winds. Steppes are found on almost every continent and are home to a diverse variety of grasses.

Deserts are extremely dry areas that receive about 10 inches (25 cm) of rainfall or less per year and support a very small amount of plant and animal life. Only plants that can live without much water and tolerate unreliable precipitation and extreme temperatures live in the desert. Deserts are usually hot and dry, although some deserts experience snowfall in the winter. In some desert areas, underground springs support an **oasis**, an area of lush vegetation. Temperatures tend to vary widely from day to night, as well as from season to season.

World Biomes

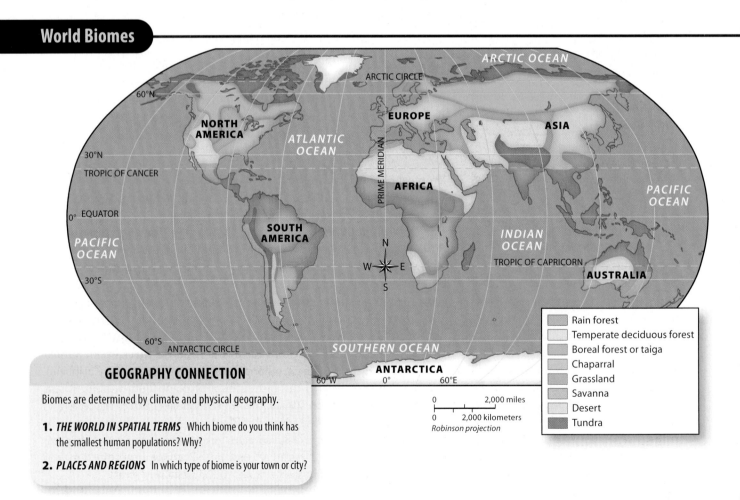

GEOGRAPHY CONNECTION

Biomes are determined by climate and physical geography.

1. **THE WORLD IN SPATIAL TERMS** Which biome do you think has the smallest human populations? Why?

2. **PLACES AND REGIONS** In which type of biome is your town or city?

Legend:
- Rain forest
- Temperate deciduous forest
- Boreal forest or taiga
- Chaparral
- Grassland
- Savanna
- Desert
- Tundra

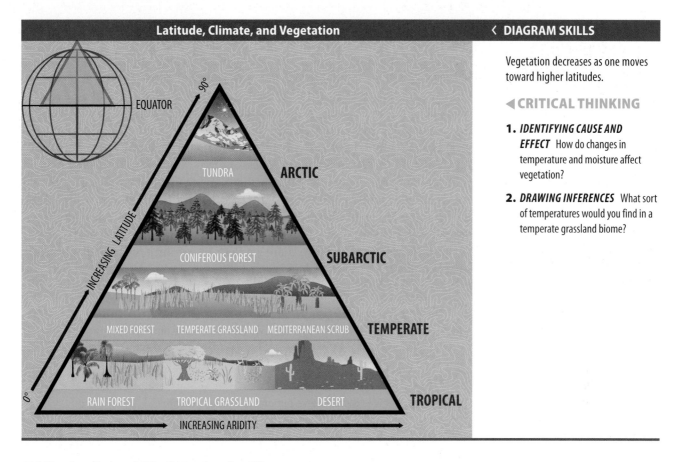

Latitude, Climate, and Vegetation

EQUATOR

90°

INCREASING LATITUDE

0°

TUNDRA — **ARCTIC**

CONIFEROUS FOREST — **SUBARCTIC**

MIXED FOREST · TEMPERATE GRASSLAND · MEDITERRANEAN SCRUB — **TEMPERATE**

RAIN FOREST · TROPICAL GRASSLAND · DESERT — **TROPICAL**

INCREASING ARIDITY

‹ **DIAGRAM SKILLS**

Vegetation decreases as one moves toward higher latitudes.

◄ **CRITICAL THINKING**

1. *IDENTIFYING CAUSE AND EFFECT* How do changes in temperature and moisture affect vegetation?

2. *DRAWING INFERENCES* What sort of temperatures would you find in a temperate grassland biome?

Midlatitude and High-Latitude Climates

The midlatitude climates include four temperate climate regions: humid subtropical, marine west coast, Mediterranean, and humid continental. Midlatitude climates experience variable weather patterns due to two conflicting air masses. Tropical air masses move from the Equator to the Poles, and polar air masses move in the opposite direction from the Poles to the Equator.

Humid subtropical climates, which include the southeastern United States as well as parts of Brazil, China, Japan, Australia, and India, are characterized by short, mild winters and nearly year-round rain. The wind patterns and high pressure from nearby oceans keep humidity levels high. Vegetation consists of **prairies** and evergreen and deciduous forests. **Coniferous** trees, most of which are evergreens, have cones. **Deciduous** trees, most of which have broad leaves, change color and drop their leaves in autumn.

Marine west coast climates, including the southern coast of Chile, parts of Australia, the British Isles, and the Pacific coast of North America, are mainly between the latitudes of about 30° N and 60° N and about 30° S and 60° S. Ocean winds bring cool summers and cool, damp winters. Abundant rainfall supports both coniferous and deciduous trees, often resulting in **mixed forests**.

Lands surrounding the Mediterranean Sea, in addition to the southwestern coast of Australia and central California, have mild, rainy winters and hot, dry summers. The natural vegetation includes thickets of woody bushes and short trees known as Mediterranean shrubs. Geographers classify as *Mediterranean* any coastal midlatitude area with similar climate and vegetation.

In some midlatitude regions of the Northern Hemisphere, landforms influence climate more than winds, precipitation, or ocean temperatures do. *Humid continental* climate regions do not experience the moderating effect of ocean winds because of their northerly continental, or inland, locations. The farther north one travels, the longer and more severe are the snowy winters, and the shorter and

prairie an inland grassland area

coniferous referring to vegetation having cones and needle-shaped leaves, including many evergreens, that keep their foliage throughout the winter

deciduous falling off or shed seasonally or periodically; trees such as oak and maple, which lose their leaves in autumn

mixed forest forest with both coniferous and deciduous trees

permafrost permanently frozen layer of soil beneath the surface of the ground

"Scientists analyzed summertime storm activity in the eastern U.S. from 1995 to 2009…. They discovered that tornadoes and hailstorms occurred at a rate of about 20 percent above average during the middle of the week. In contrast, the phenomena occurred at a rate of roughly 20 percent below average on the weekend….The team then investigated Environmental Protection Agency air-quality monitoring data and noted that human-made, summertime air pollution over the eastern U.S. peaks midweek. The cycle is linked to more human-made pollution created during the five-day workweek, such as commuters driving to and from work."

—Charles Q. Choi,
National Geographic News,
December 29, 2011

DBQ *ANALYZING* Do you think there is a cause-and-effect relationship between increased air pollution and increased tornado activity? Explain.

cooler are the summers. Vegetation is similar to that found in marine west coast areas, with evergreens outnumbering deciduous trees in the northernmost areas.

In high-latitude climates, freezing temperatures are common all year because of the lack of direct sunlight. As a result, the amount and variety of vegetation is limited here. Just south of the Arctic Circle are the *subarctic* climate regions. Winters here are bitterly cold, and summers are short and cool. Subarctic regions have the world's widest temperature ranges. In parts of the subarctic, only a thin layer of surface soil thaws each summer. Below is permanently frozen subsoil, or **permafrost**. Brief summer growing seasons may support needled evergreens.

Closer to the Poles are *tundra* climate regions. Winter darkness and bitter cold last for months, while summer has only limited warming. The layer of thawed soil is even thinner than in the subarctic. Trees cannot establish roots, so vegetation is limited to low bushes, very short grasses, mosses, and lichens (LY• kuhns).

Snow and ice, often more than 2 miles (3 km) thick, constantly cover the surfaces of *ice cap* regions. Lichens are the only form of vegetation that can survive in these areas where monthly temperatures average below freezing.

☑ **READING PROGRESS CHECK**

Making Connections Why do high-latitude climates have limited vegetation?

Climate Change

GUIDING QUESTION *What causes climates to change over time?*

Climate change refers to major changes in the factors used to measure climate over an extended period of time. For example, scientists have concluded that the average global temperature has increased by 1.4°F (0.8°C) over the last century. Some indicators of climate change include rising global temperatures, severe weather changes such as intense heat waves and changes in precipitation, an increase in severe weather events, and rising sea levels. Scientists search for answers by studying the interrelationships among ocean temperatures, greenhouse gases, wind patterns, and cloud cover. While scientists continue to disagree about the causes of climate change, we do know that the Earth undergoes natural and predictable cycles of cooling and warming caused by factors such as solar flares and volcanic activity.

However, increased global temperatures can also be attributed to greenhouse gas emissions. Burning fossil fuels releases gases that mix with water in the air, forming acids that fall in rain and snow. Acid rain can destroy forests. Fewer forests may result in climate change. The exhaust released from burning fossil fuels in automobile engines and factories is heated in the atmosphere by the sun's ultraviolet rays, forming smog, a visible chemical haze in the atmosphere.

☑ **READING PROGRESS CHECK**

Explaining How does the burning of fossil fuels create smog?

LESSON 3 REVIEW

Reviewing Vocabulary

1. *Identifying* Use the following three terms in one sentence that describes humid subtropical climate regions: prairie, coniferous, deciduous.

Using Your Notes

2. *Describing* Use your graphic organizer to describe the Earth's four climate zones.

Answering the Guiding Questions

3. *Categorizing* How are world climates organized?

4. *Drawing Conclusions* What causes climates to change over time?

Writing Activity

5. *Informative/Explanatory* Write a paragraph detailing how the four major climate regions are related to the three zones of latitude.

Directions: On a separate sheet of paper, answer the questions below. Make sure you read carefully and answer all parts of the questions.

Lesson Review

Lesson 1

① *Listing* What are the two types of solstices?

② *Explaining* Why are the seasons different in the Northern and Southern Hemispheres?

③ *Describing* Describe what is meant by "midnight sun."

Lesson 2

④ *Defining* How are winds named?

⑤ *Describing* Describe the Coriolis effect. How does it differ based on latitude?

⑥ *Discussing* What effects do mountain ranges have on climates in a region?

Lesson 3

⑦ *Identifying* What are the four temperate climate regions into which midlatitude climates are classified?

⑧ *Explaining* How do geographers distinguish between the types of dry climate?

⑨ *Listing* What are three indicators of climate change?

Critical Thinking

⑩ *Making Generalizations* Has human activity had a positive or negative effect on the global climate? Explain.

⑪ *Comparing and Contrasting* How are tropical rain forest and tropical wet/dry climates alike? How are they different?

⑫ *Drawing Conclusions* How might human activities negatively affect plants and animals?

21st Century Skills

⑬ *Creating and Using Graphs, Charts, Diagrams, and Tables* All of the cities in the table are located in the Tropics except for Telluride. Why does Quito have a lower average temperature than the other tropical cities?

⑭ *Compare and Contrast* Use the information in the table to compare and contrast how elevation and latitude affect temperatures on Earth. Provide two examples.

The Influence of Elevation on Temperature

	Elevation	Latitude & Longitude	Average Temperature
Quito, Ecuador	9,233 ft. (2,811 m)	0°09′ S 78°29′ W	58°F (14°C)
Nairobi, Kenya	5,327 ft. (1,623 m)	1°19′ S 36°55′ E	67°F (19°C)
Bjumbura, Burundi	2,568 ft. (782 m)	3°19′ S 29°19′ E	77°F (25°C)
Manaus, Brazil	276 ft. (84 m)	3°09′ S 59°59′ W	81°F (27°C)
Telluride, Colorado	8,760 ft. (2,670 m)	37°57′ N 107°49′ W	39°F (3°C)

Source: www.weatherbase.com

College and Career Readiness

⑮ *Clear Communication* Use the Internet to research careers in climatology and meteorology. Then write a job posting for one of the careers you learned about. Be sure that the job posting describes the required skills and experience necessary for the position.

Need Extra Help?

If You've Missed Question	①	②	③	④	⑤	⑥	⑦	⑧	⑨	⑩	⑪	⑫	⑬	⑭	⑮
Go to page	62	61	62	65	66	68	71	70	72	72	69	72	73	73	66

Directions: On a separate sheet of paper, answer the questions below. Make sure you read carefully and answer all parts of the questions.

Applying Map Skills

Refer to the Unit 1 Atlas to answer the following questions.

16 *Environment and Society* How is human activity affected by specific climate regions? Using the maps in the unit atlas, choose a climate region and provide an example of how it influences human activity.

17 *Human Systems* Using the maps in the unit atlas, what connections can you make between densely populated areas and climate regions?

Exploring the Essential Question

18 *Making Connections* Choose one of the biomes discussed in this chapter: tropical rain forest, tropical wet/dry, semi-arid (steppe), arid (desert), humid subtropical, marine west coast, Mediterranean, humid continental, subarctic, tundra, and ice cap. Use what you have learned about the biome to create a poster illustrating the plants and animals in the biome. Conduct additional research to include statistics on average rainfall and amount of direct sunlight. Posters should also illustrate the interactions of human systems within this biome. Posters should be highly visual and can include photos of typical vegetation, graphs/charts, and maps.

Research and Presentation

19 *Research Skills* Use Internet and library resources to gather information about the leatherback sea turtle. Create a multimedia presentation outlining the leatherback sea turtle's migration and mating patterns in map and graph/chart forms. Compare and contrast historical and current data regarding how many leatherback sea turtles existed in the past and exist today. Be sure to (1) explain why the turtle follows these migratory and mating patterns and (2) how climate change and human activity have impacted these patterns.

DBQ Analyzing Primary Sources

Use the document to answer the following questions.

PRIMARY SOURCE

"*It has been demonstrated that climate changes at millennial, centennial, and even decadal scales, as many studies in the twentieth century have revealed. Such studies were largely done by examining environmental data such as tree rings, pollen assemblages, lake sediment, and ice cores. These data are objective and usually continuous, but they are often difficult to interpret. For example, a narrow tree ring could mean either a dry spell or cold spring, or both. Consequently, conclusions obtained this way are often associated with considerable uncertainties or ambiguities.*"

—Pao K. Wang, "Chinese historical documents and climate change", www.accessscience.com

20 *Analyzing* According to the excerpt, how did scientists learn that climates have changed in the past?

21 *Interpreting* What example is provided to explain why climate change data is often difficult to interpret?

Writing About Geography

22 *Informative/Explanatory* Use standard grammar, spelling, sentence structure, and punctuation to write a one-page essay explaining how the Earth's rotation on its axis and its revolution around the sun creates certain types of climate conditions in low, mid-, and high latitude regions. Be sure to cite specific examples in your essay.

Need Extra Help?

If You've Missed Question	**16**	**17**	**18**	**19**	**20**	**21**	**22**
Go to page	4	4	69	72	74	74	60

The Human World

ESSENTIAL QUESTION · *How do the characteristics and distribution of human populations affect human and physical systems?*

Geography Matters...

Have you ever traveled far from home and had new experiences of sight, sound, and taste? Examining these varieties in the human experience is the work of the human geographer. Human geographers study relationships between humans and their natural environment and analyze geographic patterns. They look at present-day events with an eye for trends, knowing that human actions and decisions can have major impacts on people near and far.

◄ A woman works on a tea plantation in Kerala, India.

©Tuul/Robert Harding World Imagery/Corbis

the human development index

The relative social and economic status of countries is often described in the media and among academics as either "less developed," "newly industrialized," or "more developed." What precisely is meant by the term development? And how is development measured?

Human Development

ARCTIC OCEAN

W — E
S

Norway
Human Development Index Rank: 1
Life expectancy at birth: 81.1 years
Education index*: 0.985
Population: 4,924,800

United States
Human Development Index Rank: 4
Life expectancy at birth: 78.5 years
Education index*: 0.939
Population: 313,085,400

China
Human Development Index Rank: 101
Life expectancy at birth: 73.5 years
Education index*: 0.623
Population: 1,347,565,300

60°N

PACIFIC OCEAN

ATLANTIC OCEAN

30°N

PACIFIC OCEAN

Haiti
Human Development Index Rank: 158
Life expectancy at birth: 62.1 years
Education index*: 0.406
Population: 10,123,800

0°

INDIAN OCEAN

30°S

Human Development Index, 2011
- Very High
- High
- Medium
- Low
- No data

Sierra Leone
Human Development Index Rank: 180
Life expectancy at birth: 47.8 years
Education index*: 0.304
Population: 5,997,500

0   4,000 miles
0   4,000 kilometers
Mercator projection

60°S

*Education index is derived from mean years and expected years of schooling.
The maximum observed value of 0.978 (New Zealand) represents the highest level of education.

60°W   30°W   0°   30°E   60°E   90°E   120°E   150°E

SOURCE: United Nations Development Progamme, Human Development Report, 2011.

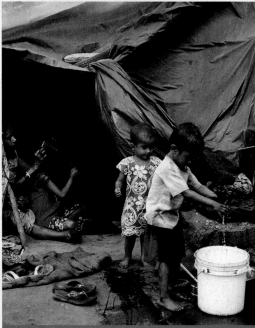

What is development?

Although different societies and cultures have their own perspectives and values that shape the types of human development that occur, most agree that development includes technological progress that leads to a better standard of living for people. The modern notion of development as a means of improving people's lives gained favor during the Industrial Revolution. Advances in science and technology led to increased agricultural and industrial production, which in turn provided a greater food supply and more manufactured products that people needed. Such new inventions and products led to easing some people's workloads, providing comfort and convenience, and allowing many to begin acquiring wealth. Broadly speaking, development is improvement in people's living situations and economic prospects.

1. **Human Systems** Why is human development an important topic? Define the term *human development* and describe the role of technology as a component of it.

Why do we measure development?

Governments rely on measures of development when making important policy decisions and addressing the specific needs of both their own citizens and those of other countries. For example, local governments use measures of development as a way to bring the needs of people to the attention of the central government in the hopes of having funds and resources allocated appropriately. Geographers and other social scientists are also interested in measures of development because they study the spatial patterns of human need and well-being, looking for cause-and-effect relationships. Identifying these types of relationships helps governments, aid agencies, and others make changes to improve the economic, social, or environmental conditions in which people live. Although there are many ways to measure development, many governments today use the Human Development Index (HDI), a statistic used to rank countries by their level of human development. The index, devised by Pakistani economist Magbub-ul-Haq in 1990, was designed to include measurements of improvement in people's well-being, rather than focusing solely on national economic indicators.

2. **Places and Regions** How is measuring development useful? Explain how measures of development are used.

How does the Human Development Index (HDI) measure development?

The HDI ranks countries' level of human development by measuring three dimensions of development: health, education, and living standards. The measure of health is determined by life expectancy at birth. Access to education is measured by examining mean, or average, years of schooling and expected years of schooling. Living standards are measured by gross national income per capita. The Human Development Index uses the data from these three dimensions of development to calculate one composite statistic for each country. The HDI is published annually by the United Nations Development Programme, and it is used to measure how economic policy decisions affect people's quality of life.

3. **The Uses of Geography** Why do you think the HDI incorporates more than one dimension of development? What do these dimensions indicate about the level of development in a country? Provide examples to support your answer.

(tl)View Stock/Getty Images, (tc)©Hill Street Studios/Matthew Palmer/Blend Images, (tr)Sajjad Hussain/AFP/Getty Images

THERE'S MORE ONLINE

VIEW a graph of changes in HDI ranking over time • **COMPARE** population indicators of several countries

networks

There's More Online!

☑ **CHART** Cultural Universals

☑ **MAP** World Language Families

☑ **MAP** World Culture Hearths

☑ **INTERACTIVE SELF-CHECK QUIZ**

☑ **VIDEO** Global Cultures

LESSON 1
Global Cultures

Reading **HELP**DESK

Academic Vocabulary

- **similar**
- **major**

Content Vocabulary

- **culture**
- **language family**
- **ethnic group**
- **culture region**
- **cultural diffusion**
- **culture hearth**

TAKING NOTES: *Key Ideas and Details*

IDENTIFYING Use a graphic organizer like the one below to take notes as you read about global cultures in this lesson.

Elements of Culture	Cultural Change

ESSENTIAL QUESTION · *How do the characteristics and distribution of human populations affect human and physical systems?*

IT MATTERS BECAUSE

The world's people organize communities, develop ways of life, and adjust to the differences and similarities they experience. As the world becomes increasingly interconnected, cultures spread and are shared. To help understand this cultural diversity, geographers divide the Earth into culture regions, which are defined by the presence of common cultural elements such as language and religion.

Elements of Culture

GUIDING QUESTION *What are the elements of culture?*

Geographers study **culture**, the way of life of a group of people who share **similar** ways of thinking, believing, and living, expressed in common elements or features. For example, a particular culture can be understood by looking at language, religion, daily life, history, art, government, technology, and economy.

Language is a key element in a culture's development. Through language, people communicate information and experiences and pass on cultural values and traditions. Even within a culture, however, there are language differences. Some people may speak a dialect, or a local form of a language, that differs from the main language. These differences may include variations in the pronunciation and meaning of words.

Linguists, scientists who study languages, organize the world's languages into **language families**—large groups of languages having similar roots. Seemingly diverse languages may belong to the same language family. For example, English, Spanish, and Russian are all members of the Indo-European language family.

Religion is another important element of a culture. Religious beliefs vary significantly around the world. For many people, religion provides an important sense of identity. It also influences many aspects of daily life, from the practice of moral values to the celebration of holidays and festivals. Throughout history, religious symbols and stories have shaped cultural expressions such as literature, painting and sculpture, architecture, and music.

SuperStock/age fotostock

A social system develops to help the members of a culture work together to meet basic needs. In all cultures, the family forms an important group. Most cultures are also made up of social classes, groups of people ranked according to ancestry, wealth, education, or other criteria. Moreover, cultures may include people who belong to different ethnic groups. An **ethnic group** is made up of people who share a common language, history, or place of origin.

Geographers also analyze governments to help understand a culture. Governments of the world share certain features, such as maintaining order within the country, providing protection from outside dangers, and supplying other services to the people. Governments can be categorized by levels of power—national, regional, and local—and by type of authority—a single ruler, a small group of leaders, or a body of citizens and their representatives.

Economic activities also influence and shape a culture. People must make a living, whether in farming, industry, or by providing services. Geographers study how a culture utilizes its natural resources to meet such needs as food and shelter. They also analyze the ways in which people produce, obtain, use, and sell goods and services.

To organize their understanding of cultural development, geographers divide the Earth into culture regions. Each **culture region** includes areas that have certain traits in common. They may share similar economic systems, forms of government, or social groups. Their histories, religions, and art forms may share similar influences.

☑ **READING PROGRESS CHECK**

Explaining Why are social groups important to the development of a culture?

culture way of life of a group of people who share similar culture traits, including beliefs, customs, technology, and material items

similar comparable

language family group of related languages that have all developed from one earlier language

ethnic group group of people who share common ancestry, language, religion, customs, or place of origin

GEOGRAPHY CONNECTION

Most languages spoken throughout the world belong to one of 12 language families. Languages within a family have a common origin.

1. ***PLACES AND REGIONS*** Where are the Dravidian languages spoken?

2. ***HUMAN SYSTEMS*** Do people within the same language family necessarily speak the same language? Explain your response.

World Language Families

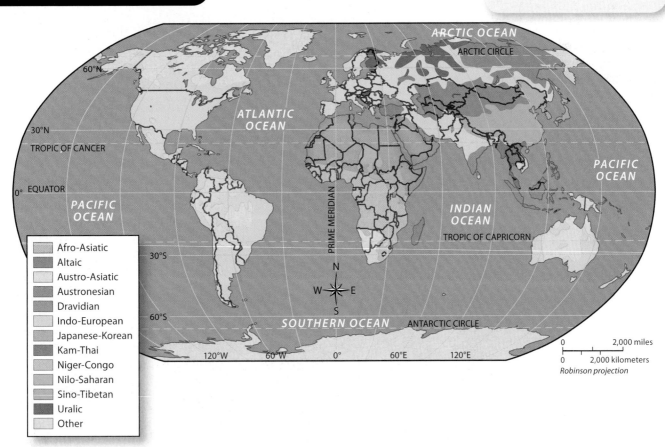

- Afro-Asiatic
- Altaic
- Austro-Asiatic
- Austronesian
- Dravidian
- Indo-European
- Japanese-Korean
- Kam-Thai
- Niger-Congo
- Nilo-Saharan
- Sino-Tibetan
- Uralic
- Other

culture region division of the Earth in which people share a similar way of life, including language, religion, economic systems, and values

cultural diffusion the spread of culture traits, material and non-material, from one culture to another

major greater in importance or interest

culture hearth a center where cultures developed and from which ideas and traditions spread outward

Cultural Change

GUIDING QUESTION *What are two major ways in which cultures change over time?*

Cultures are dynamic and continually changing. Internal factors—new ideas, lifestyles, and inventions—create change within cultures. Change can also come through spatial interaction such as trade, migration, and war. The spread of new knowledge from one culture to another is called **cultural diffusion**. Cultural diffusion has been a **major** factor in cultural development since the dawn of human history, and the pace of cultural change has accelerated in contemporary times. The earliest humans were small groups of hunters and gatherers, who moved from place to place in search of animals to hunt, plants to gather, water, and useful materials. As they migrated, they helped spread culture traits from one group and place to another.

Cultural Change in History

The world's first civilizations arose in **culture hearths**—early centers of civilization whose ideas and practices spread to surrounding areas. The map shows that some of the most influential culture hearths developed in areas that make up the modern countries of Egypt, Iraq, Pakistan, China, and Mexico.

These five culture hearths had certain geographic features in common. They all emerged from farming settlements in areas with a mild climate and fertile land. In addition, all five culture hearths were located near a major river or source of water. Making use of favorable environments, the people dug canals and ditches to irrigate the land. All of these factors contributed to what is known as the agricultural revolution, a major shift from food gathering to food production that enabled people to grow surplus crops.

Surplus food set the stage for the rise of cities and civilizations and the development of long-distance trade. The increased wealth from trade led to the rise of cities and complex social systems. These new cities needed a well-organized

GEOGRAPHY CONNECTION

Culture hearths are centers where the world's first civilizations arose.

1. *PLACES AND REGIONS* Where in Asia were the first major settlements located?

2. *HUMAN SYSTEMS* What water feature do most of the culture hearths have in common?

World Culture Hearths

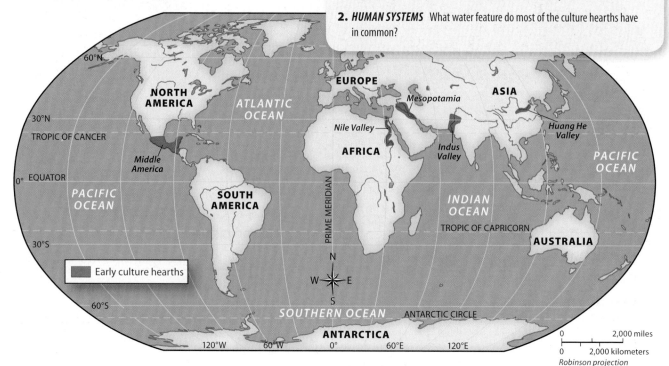

government to coordinate harvests, plan building projects, and manage an army for defense. Officials and merchants created writing systems to record and transmit government and trade information.

Cultural diffusion has increased rapidly during the last 250 years. In the 1700s and 1800s, some countries began to industrialize, using power-driven machines and factories to mass-produce goods. This period is known as the Industrial Revolution. With new production methods, these countries produced goods quickly and cheaply, and their economies changed dramatically. These developments also led to social changes. As people left farms for jobs in factories and mills, cities grew larger.

Cultural Change in the Contemporary World

At the end of the twentieth century, the world experienced a new turning point— the information revolution. Computers now make it possible to store huge amounts of information and instantly send it all over the world, thus allowing more rapid spread of ideas and traditions among the cultures of the world. The Internet has been responsible for communication and socialization around the world via social networking sites and other sites that allow users to share many types of information and stay connected with others. Consequently, the world feels much smaller than it might have previously.

Cultural contact among different peoples promotes cultural change as ideas and practices spread. Computer technology certainly accelerates the spread of cultural change, but other connections among people do as well. Trade and travel are important avenues by which cultural change occurs. Migration has also fostered cultural diffusion. People migrate for many reasons. Positive factors—better social and economic conditions and religious or political freedoms—may draw people from one place to another. Most people move from one place to another in search of better economic opportunity. Negative factors— wars, persecution, and famines—also motivate people to migrate. In some instances, as in the case of enslaved Africans brought to the Americas, mass migrations have been forced. Regardless of the reasons, migrants carry their cultures with them, and their ideas and practices often blend with those of the people already living in the migrants' adopted countries.

☑ **READING PROGRESS CHECK**

Identifying Where were the five earliest culture hearths located?

Contact between different cultures usually leads to change in both systems.

▲ **CRITICAL THINKING**

1. Describing In what way is this picture an example of cultural contact?

2. Speculating What types of ideas and practices might be exchanged due to this example of cultural contact?

SuperStock/age fotostock

LESSON 1 REVIEW

Reviewing Vocabulary

1. Explaining Explain the relationships between culture, ethnic group, culture region, cultural diffusion, and culture hearth.

Using Your Notes

2. Describing Use your graphic organizer to describe the external factors that change cultures.

Answering the Guiding Questions

3. Identifying What are the elements of culture?

4. Drawing Conclusions What are two major ways in which cultures change over time?

Writing Activity

5. Explanatory Research to find three different definitions of culture. Write a paragraph comparing the definitions.

networks

There's More Online!

☑ **DIAGRAM** Demographic Transition Model

☑ **GRAPH** Projected Populations

☑ **INTERACTIVE SELF-CHECK QUIZ**

☑ **VIDEO** Population Geography

Reading **HELP**DESK

Academic Vocabulary

- **community**
- **trend**

Content Vocabulary

- **birthrate**
- **death rate**
- **natural increase**
- **migration**
- **demographic transition**
- **doubling time**
- **population pyramid**
- **population distribution**
- **population density**

TAKING NOTES: *Key Ideas and Details*

DESCRIBING As you read, use a graphic organizer like the one below to take notes on the demographic transition model and challenges of growth.

Population Geography
├─ The Demographic Transition Model
└─ Challenges of Growth

LESSON 2
Population Geography

ESSENTIAL QUESTION · *How do the characteristics and distribution of human populations affect human and physical systems?*

IT MATTERS BECAUSE

Earth's human population increased dramatically during the nineteenth century and much of the twentieth century. Although that growth has begun to slow in recent decades, there are far more people on Earth today than ever before. The result is that issues of population growth or decline are crucial to all countries. Geographers play an important role in examining ways to plan for the future and solve problems of tomorrow.

Population Growth

GUIDING QUESTION *What factors influence population growth?*

More than 7 billion people now live on Earth, and most of the population inhabits about 30 percent of the planet's land area. Global population continues to grow and is expected to level off at 9 billion by the year 2050. Such rapid growth was not always the case. From the year 1000 until 1800, the world's population increased slowly. Then the number of people on Earth more than doubled between 1800 and 1950. It doubled again between 1950 and about 2000.

The Demographic Transition Model

Scientists in the field of *demography*, the study of populations, use statistics to learn about population growth. The **birthrate** is the number of births per year for every 1,000 people. The **death rate** is the number of deaths per year for every 1,000 people. **Natural increase**, or the growth rate of a population, is the difference between an area's birthrate and its death rate. **Migration**, or the movement of people from place to place, must also be considered when examining population changes.

The **demographic transition** model uses birthrates and death rates to show how populations in countries or regions can change over time. The model was first used to show the relationship of declining birthrates and death rates to industrialization in Western Europe. Death rates can fall quickly as a result of more abundant and reliable food supplies, improved health care, access to medicine and technology, and better living conditions. Birthrates decline more slowly because the declines result from changes in cultural traditions that can often take longer.

(tl)Mitchell Funk/Photographer's Choice/Getty Images, (tc tr)Mitchell Funk/Photographer's Choice/Getty Images

Today, most of the world's industrialized and technologically developed countries have experienced the transition from high birthrates and death rates to low birthrates and death rates. These countries have reached what is known as *zero population growth,* in which the birthrate and death rate are equal. When this balance occurs, a country's population does not grow as a result of natural increase, although it can still grow as a result of migration.

Although birthrates have fallen significantly in many countries in Asia, Africa, and Latin America over the past 40 years, they are still higher than in the industrialized world. Families in these regions traditionally are large because of cultural beliefs about marriage, family, and the value of children. For example, a husband and wife in a rural agricultural area may choose to have several children who will help farm the land. The high number of births often continues after death rates decrease as a result of improved living conditions. This causes the population to greatly increase. As a result, the **doubling time**, or the number of years it takes a population to double in size, has been reduced to below 50 years in some parts of Asia, Africa, and Latin America. In contrast, the average doubling time of a more developed country can be more than 300 years.

Challenges of Growth

Rapid population growth presents many challenges that affect individual countries and the global **community**. As the number of people increases, so does the difficulty of producing enough food to feed them. In Africa, for example, food shortages in

birthrate number of births per year for every 1,000 people

death rate number of deaths per year for every 1,000 people

natural increase the growth rate of a population; the difference between birthrate and death rate

migration the movement of people from place to place

demographic transition the model that uses birthrates and death rates to show how populations in countries or regions change over time

doubling time the number of years it takes for a population to double in size

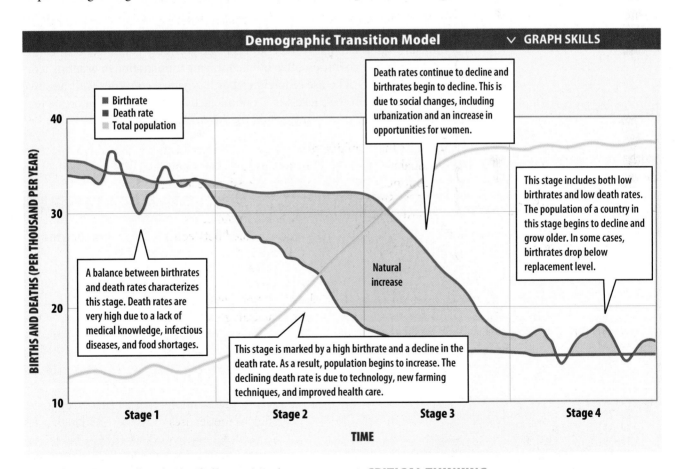

Demographic Transition Model ⌄ **GRAPH SKILLS**

Death rates continue to decline and birthrates begin to decline. This is due to social changes, including urbanization and an increase in opportunities for women.

This stage includes both low birthrates and low death rates. The population of a country in this stage begins to decline and grow older. In some cases, birthrates drop below replacement level.

A balance between birthrates and death rates characterizes this stage. Death rates are very high due to a lack of medical knowledge, infectious diseases, and food shortages.

Natural increase

This stage is marked by a high birthrate and a decline in the death rate. As a result, population begins to increase. The declining death rate is due to technology, new farming techniques, and improved health care.

BIRTHS AND DEATHS (PER THOUSAND PER YEAR)

■ Birthrate
■ Death rate
■ Total population

Stage 1   Stage 2   Stage 3   Stage 4

TIME

Changes in population trends can be identified by examining the relationships between birthrates and death rates.

▲ **CRITICAL THINKING**

1. *Analyzing* What characteristics of Stage 2 create an increase in population?

2. *Identifying* What effect does an increase in opportunities for women have on populations? Of what stage is it a characteristic?

community people with common interests living in a particular area

population pyramid a diagram that shows the distribution of a population by age and gender

trend a general movement

production and availability continue to be an issue. In better circumstances, Africa's agricultural sector would respond to rising prices by increasing food supply. In Africa, however, lack of government investment and other support of agriculture—along with warfare, poor access to rural areas, and weather and pests that can ruin crops—have combined to bring hunger to the region.

In addition, populations that grow rapidly use resources more quickly. Some countries face shortages of water, housing, and clothing, for instance. Rapid population growth strains these limited resources. Another concern is that the world's population is unevenly distributed by age, with the majority of some countries' populations being infants and young children who cannot contribute to food production. This population structure can be seen with a **population pyramid**.

While some experts are pessimistic about the long-term effects of rapid population growth, others are optimistic that, as the number of humans increases, the levels of technology and creativity will also rise. For example, scientists continue to study and develop ways to boost agricultural productivity. Fertilizers can improve crop yields. Irrigation systems can help increase the amount of land available for farming. New varieties of crops have been created to withstand severe conditions and yield more food.

In the late 1900s some countries in Europe began to experience a **trend** called *negative population growth,* in which the annual death rate exceeds the annual birthrate. Hungary and Germany, for example, both show change rates of about -0.2. This situation has economic consequences different from, but just as serious as, those caused by high growth rates. In countries with negative population growth, it is difficult to find enough workers to keep the economy going. Labor must be recruited from other countries, often by encouraging immigration or granting temporary work permits. The use of foreign labor has helped countries with negative change rates maintain their levels of economic activity. But it also can create tensions between the "host" population and the communities of newcomers.

☑ READING PROGRESS CHECK

Identifying Where was the demographic transition model first used?

Stages Of Growth: Population Pyramids

CHAD, 2012 (Rapid Growth)

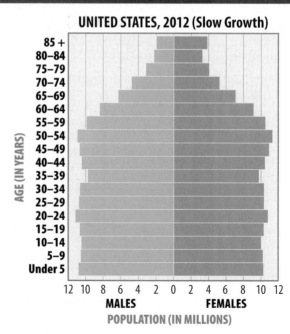

UNITED STATES, 2012 (Slow Growth)

84

Population Distribution

GUIDING QUESTION *What influences population distribution?*

Not only do population growth rates vary among Earth's regions, but the pattern of human settlement, or **population distribution**, is uneven as well. Population distribution is related to the Earth's physical geography. Only about 30 percent of Earth's surface is made up of land, and much of that land is inhospitable. High mountain peaks, barren deserts, and frozen tundra make human activity difficult in many places. Almost everyone on Earth lives on a relatively small portion of the planet's land—a little less than one-third. Most people live where fertile soil, available water, and a climate without harsh extremes make human life possible.

Of all the continents, Europe and Asia are the most densely populated. Asia alone contains more than 60 percent of the world's people. Many people throughout the world live in metropolitan areas—cities and their surrounding urbanized areas—where populations are highly concentrated. Today, most people in Europe, North America, South America, and Australia live in or around urban areas.

Geographers determine how crowded a country or region is by measuring **population density**—the number of people living on a square mile or square kilometer of land. To determine population density in a country, geographers divide the total population of the country by its total land area.

Population density varies widely from country to country. Canada, with a low population density of about 8 people per square mile (3 people per sq. km), offers wide-open spaces and the choice of living in thriving cities or quiet rural areas. In contrast, Bangladesh has one of the highest population densities in the world—about 2,721 people per square mile (1,046 people per sq. km).

Countries with populations of about the same size do not necessarily have similar population densities. For example, both Tunisia and the Czech Republic have about 10.5 million people. Tunisia, with a larger land area, has only 169 people per square mile (65 people per sq. km). However, the Czech Republic has 349 people per square mile (134 people per sq. km).

population distribution
the variations in population that occur across a country, a continent, or the world

population density the average number of people living on a square mile or square kilometer of land

✓ GRAPH SKILLS

ITALY, 2012 (Decline)

Source: U.S. Census Bureau, Statistical Abstract of the United States: 2012

Population pyramids include information about the age and gender distribution of a country's population. The population structure of a country provides insight into population growth trends and the country's stage in the demographic transition.

◀ **CRITICAL THINKING**

1. **Comparing and Contrasting** How does the shape of the population pyramid for a country undergoing rapid growth differ from that of a country experiencing population decline?

2. **Analyzing** What does the shape of the U.S. population pyramid tell you about the country's population structure?

"The United States is strongly committed to protecting and assisting refugees and we offer resettlement to more refugees each year than all other countries in the world combined. Since 1975, more than three million refugees have made new homes in the United States, and nearly half of them have become U.S. citizens.

Refugees are contributing in ways large and small to business, academia, the arts, science and technology. Today we celebrate the success of refugees who have built new lives here and in other resettlement countries, but we also recognize the millions of refugees who remain displaced in camps, cities, and rural settlements around the world."

—Hillary Rodham Clinton, press statement, June 20, 2012

DBQ *SPECULATING* Why might refugees choose to migrate to the United States? What difficulties might refugees face?

Because the measure of population density includes all the land area of a country, it does not account for uneven population distribution within a country. In Egypt, for example, overall population density is 216 people per square mile (83 people per sq. km). In reality, over 90 percent of Egypt's people live along the Nile River. The rest of Egypt is desert. Thus, some geographers describe a country's population density in terms of land that can be used to support the population rather than total land area. When Egypt's population density is measured this way, it is about 6,993 people per square mile (2,688 people per sq. km)!

The Earth's population is moving in great numbers. People are moving from city to city or from rural villages to cities. The resulting growth of city populations brought about by such migration and the changes that come with this increase are called *urbanization.* The primary cause of urbanization is the desire of rural people to find jobs and a better life in more prosperous urban areas. Rural populations have certainly grown in some countries, but the amount of farmland has not increased to meet the growing number of people. As a result, many rural migrants find urban jobs in manufacturing and service industries.

About half of the world's people live in cities. Between 1960 and 2012, the population of metropolitan Mexico City rose from about 5 million to more than 19 million. Other cities in Latin America, as well as in Asia and Africa, have seen similar growth. Some cities contain a large part of their country's entire population. For example, about one-third of Argentina's population lives in Buenos Aires.

PRIMARY SOURCE

❝Never in human history have so many people changed their locations and lifestyles so quickly. Each month, there are 5 million new city dwellers created through migration or birth in Africa, Asia, and the Middle East....❞

—Doug Saunders, "The great shift from farm to city," *Los Angeles Times,* June 19, 2011

Population movement also occurs between countries. Some people emigrate from the country of their birth and move to another. They are known as emigrants in their homeland and immigrants in their new country. In the past 40 years, millions of people have left Africa and Asia to find jobs in the wealthier countries of Europe. These types of economic factors that attract people to a place are called *pull factors.* Religious and political freedoms are also examples of pull factors that attract people to another country. Other factors of migration, called *push factors*, trigger out-migration; these also include lack of economic opportunities and religious or political persecution. Some people are forced to flee their country because of wars, food shortages, or other problems. They are refugees, or people who flee to escape persecution or disaster.

☑ **READING PROGRESS CHECK**

Explaining What factors influence the migration of people from one country to another?

LESSON 2 REVIEW

Reviewing Vocabulary

1. *Explaining* Define demographic transition model and doubling time. How are they related to each other?

2. *Calculating* Define population density and how it is calculated.

Using Your Notes

3. *Describing* Use your graphic organizer to explain why population growth varies from place to place.

Answering the Guiding Questions

4. *Discussing* What factors influence population growth?

5. *Stating* What influences population distribution?

Writing Activity

6. *Narrative* What human-made structures might be present in countries that have large numbers of people concentrated in relatively small areas? Write a paragraph with supporting details to explain your answer.

networks

There's More Online!

☑ **CHART** WTO Members

☑ **IMAGE** Rhine River Valley

☑ **IMAGE** Four Corners

☑ **IMAGE** Pakistan-India Border

☑ **IMAGE** Terrorist Attack
on Mumbai

☑ **INTERACTIVE SELF-CHECK QUIZ**

☑ **VIDEO** Political Geography

LESSON 3
Political Geography

ESSENTIAL QUESTION · *How do the characteristics and distribution of human populations affect human and physical systems?*

Reading HELPDESK

Academic Vocabulary

- **unique**
- **authority**

Content Vocabulary

- **unitary system**
- **federal system**
- **autocracy**
- **monarchy**
- **oligarchy**
- **theocracy**
- **democracy**
- **natural boundary**
- **cultural boundary**
- **geometric boundary**

TAKING NOTES: *Key Ideas and Details*

DESCRIBING As you read, use a graphic organizer like the one below to take notes on the features of government.

Features of Government	
Levels of Government	Types of Government
Geography and Government	Conflict and Cooperation

(l to r) J.D. Pooley/Getty Images News/Getty Images; Guenter Fischer/Alamy; Jupiterimages/Comstock Images/Getty Images; Marketa Jirouskova/Oxford Scientific/Getty Images; Bhaskar Paul/The India Today Group/Getty Images

IT MATTERS BECAUSE
Governments and economies of countries around the world are becoming increasingly interconnected. Some countries or groups of countries, such as the European Union, have strong economies that allow them to help improve standards of living in other countries.

Features of Government

GUIDING QUESTION *What influences the level and type of a country's government?*

Today the world includes nearly 200 independent countries that vary in size, military might, natural resources, and world influence. Each country is defined by characteristics such as territory, population, and sovereignty, or freedom from outside control. These elements are brought together under a government. A government must make and enforce policies and laws that are binding upon all people living within its territory.

Levels of Government

The government of each country has **unique** characteristics that relate to that country's historical development. To carry out their functions, governments are organized in a variety of ways. Most large countries have several different levels of government. These usually include a national or central government, as well as the governments of smaller internal divisions such as provinces, states, counties, cities, towns, and villages.

A **unitary system** of government gives all key powers to the national or central government. This structure does not mean that only one level of government exists. Rather, it means that the central government creates state, provincial, or other local governments and gives them limited sovereignty. The United Kingdom and France both developed unitary governments as they emerged from smaller territories during the late Middle Ages and early modern times.

A **federal system** of government divides the powers of government between the national government and state or provincial governments. Each level of government has sovereignty in some areas. The United States developed a federal system after the thirteen colonies became independent from Great Britain.

President Barack Obama is the leader in the United States federal system.

▲ CRITICAL THINKING

1. *Identifying* Over what type of government does the president of the United States preside?

2. *Contrasting* How is this form of government different from an oligarchy?

unique being the only one; without a like or an equal

unitary system form of government in which all key powers are given to the national or central government

federal system form of government in which powers are divided between the national government and state or provincial governments

autocracy system of government in which one person rules with unlimited power and authority

authority power to influence or command thought, opinion, or behavior

monarchy a form of autocracy with a hereditary king or queen exercising supreme power

oligarchy system of government in which a small group holds power

Another similar type of government structure is a confederation, or a loose union of independent territories. The United States at first formed a confederation, but this type of political arrangement failed to provide an effective national government for the new nation. As a result, the U.S. Constitution established a strong national government while preserving some state government powers. Today, other countries with federal or confederate systems include Canada, Switzerland, Mexico, Brazil, Australia, and India.

Types of Governments

Governments can be classified by asking the question: Who governs the state? Under this classification system, all governments belong to one of the three major groups: (1) autocracy, or rule by one person; (2) oligarchy, or rule by a few people; or (3) democracy, or rule by many people.

Any system of government in which the power and authority to rule belongs to a single individual is an **autocracy** (aw•TAH•kruh•see). Autocracies are the oldest and one of the most common forms of government. Most autocrats achieve and maintain their position of **authority** through inheritance or by the ruthless use of military or police power.

Several forms of autocracy exist. One is an absolute or totalitarian dictatorship in which the decisions of a single leader determine government policies. The government under such a system can come to power through a revolution or an election. The totalitarian dictator seeks to control all aspects of social and economic life. Examples of totalitarian dictatorships include Adolf Hitler in Nazi Germany, Saddam Hussein in Iraq, Raul Castro of Cuba, and Kim Jong Un of North Korea.

Monarchy (MAH•nuhr•kee) is another form of autocratic government. In a monarchy, a king or queen exercises the supreme powers of government. Monarchs usually inherit their positions. Absolute monarchs have complete and unlimited power to rule. The king of Saudi Arabia, for example, is an absolute monarch. Absolute monarchs are rare today, but from the 1400s to the 1700s, kings or queens with absolute power ruled most of Western Europe.

Today, some countries, such as the United Kingdom, Sweden, Japan, Jordan, and Thailand, have constitutional monarchies. Their monarchs share governmental powers with elected legislatures or serve as ceremonial leaders.

An **oligarchy** (AH•luh•GAHR•kee) is any system of government in which a small group holds power. The group derives its power from wealth, military power, social position, or a combination of these elements. Today the governments of communist countries, such as China, are mostly oligarchies. Leaders in the Communist Party and the military control the government.

Sometimes religion is the source of power in an oligarchy. A **theocracy**, for example, is a government of officials believed to be divinely inspired. In a theocracy, a divine power is thought to be the head of the government. Government officials receive their inspiration, guidance, and authority to rule from this divine power. For example, Islamic sharia law is imposed in parts of North Africa and Southwest Asia today.

Both dictatorships and oligarchies sometimes claim they rule for the people. Such governments may try to give the appearance of control by the people. For example, they might hold elections but offer only one candidate. Such governments may also have some type of legislature or national assembly elected by or representing the people. These legislatures, however, only approve policies and decisions already made by the leaders. As in a dictatorship, oligarchies usually suppress all political opposition.

J.D. Pooley/Getty Images News/Getty Images

A **democracy** is any system of government in which leaders rule with the consent of the citizens. The term *democracy* comes from the Greek *demos* (meaning "the people") and *kratia* (meaning "rule"). The ancient Greeks used the word *democracy* to mean government by the many in contrast to government by the few. The key idea of democracy is that people hold sovereign power.

Direct democracy, in which citizens themselves decide on issues, exists in some places at local levels of government. No country today has a national government based on direct democracy. Instead, democratic countries have representative democracies, in which the people elect representatives with the responsibility and power to make laws and conduct government. An assembly of the people's representatives may be called a council, a legislature, a congress, or a parliament.

Many democratic countries, such as the United States and France, are republics. In a republic, voters elect all major officials, who are responsible to the people. The head of state—or head of government—is usually a president elected for a specific term. Not every democracy is a republic. The United Kingdom, for example, is a democracy with a monarch as head of state. This monarch's role is ceremonial, however, and elected officials hold the actual power to rule.

theocracy system of government in which those who rule are regarded as divinely inspired

democracy system of government in which leaders rule with consent of the citizens

✔ READING PROGRESS CHECK

Contrasting How does an autocracy differ from an oligarchy?

Geography and Government

GUIDING QUESTION *How does geography influence a country's government?*

Governments can be greatly influenced by geography. Geographic areas can actually determine how political and administrative units are drawn up and how they will be governed. Democratic countries have entities based on location, which are divided into local bodies that might have different laws.

A government must consider the cultural and religious beliefs of its citizens in order to govern effectively. In autocracies, governments frequently suppress

Types of Boundaries

Natural Boundary: The Rhine River Valley near Ruggell forms the border between Switzerland (left bank) and Liechtenstein (right bank).

Geometric Boundary: The Four Corners site in the United States occurs where the states of Utah, New Mexico, Arizona, and Colorado all meet.

Cultural Boundary The Wagah border post that separates northern India from eastern Pakistan is an example of a cultural boundary.

▲ CRITICAL THINKING

1. Identifying Which of the boundary types might change slowly over time without the consent of the governments involved? Explain your choice.

2. Analyzing Describe the Four Corners site and explain why it is considered a geometric boundary.

(bl)Guenter Fischer/Alamy, (bc)Jupiterimages/Comstock Images/Getty Images, (br)Marketa Jirouskova/Oxford Scientific/Getty Images

Firefighters extinguish the flames coming out from the Taj Palace Hotel in Mumbai, India, during a terrorist attack in the city in November 2008.

▲ **CRITICAL THINKING**

1. *Defining* What elements of the definition of terrorism are shown in the photo?

2. *Speculating* What are some of the long-term effects that people might have who survived but witnessed the bombing?

natural boundary

a fixed limit or extent defined along physical geographic features such as mountains and rivers

cultural boundary

a geographical boundary between two different cultures

geometric boundary

a boundary that follows a geometric pattern

cultural and religious groups in order to maintain order and power. In democracies, governments usually take account of cultural and religious beliefs in order to protect their people's freedoms and ensure their well-being.

Geography influences governments as they develop policy to provide people with goods and services. Governments must also know where their citizens are moving, why they are moving there, and how that affects their relationship with the environment. Infrastructures, such as roads, bridges, and power plants, must be built based on the geographic distribution of people using both current demographic data and future projections.

Several geographic factors influence the development of political boundaries. A **natural boundary** follows physical geographic features such as mountains and rivers. For example, the Mississippi River forms the borders between several U.S. states. Natural boundaries are often more defensible and easy to identify.

Other boundaries develop to separate areas with cultural differences, such as places with different religions or languages. These **cultural boundaries** geographically divide two identifiable cultures. For example, when Britain partitioned India and created Pakistan, it created a religious cultural boundary. Muslims were reorganized into Pakistan and Hindus into India.

At other times, cultural and natural landforms are not considered when boundaries are drawn. In these cases, treaties might create **geometric boundaries** to separate countries or nations. Geometric boundaries—which often follow straight lines and do not account for natural and cultural features—exist between Libya, Egypt, and Algeria.

Political boundaries, referred to as borders, are not always permanent. Many areas of the world have seen changing borders as the result of wars and territorial disputes. Border disputes arise from unsettled territorial claims or as a result of one state desiring the resources of a neighboring state. In February 1848, Mexico and the United States signed a treaty which ended the war between them and gave large portions of the Southwest, including present-day California, to the United States. Several days earlier, gold had been discovered near the present-day capital of Sacramento. This started the gold rush and sped up California's statehood.

☑ **READING PROGRESS CHECK**

Identifying What is the cause of many boundary disputes?

Conflict and Cooperation

GUIDING QUESTION *How do cooperation and conflict shape the division of Earth's surface?*

Cooperation and conflict have contributed to and resulted from the political geographic divisions of the world. Global cooperation is frustrated by many factors, including border disputes, tensions over larger territories, multiple ethnic groups within one state, competition for fewer resources, and control of strategic sites.

Nationalism often contributes to political conflicts. Nationalism is a belief that the individual's loyalty and devotion to the nation or state surpasses other individual or group interests. After passing through the new countries of Latin America, nationalism spread in the early nineteenth century to central Europe and from there, toward the middle of the century, to eastern and southeastern Europe. This period is considered the age of nationalism in Europe. Asia and Africa saw a rise in nationalism at the beginning of the twentieth century as powerful movements took place. Nationalism can breed conflict if it reaches fanatical levels. This can and often does lead to war.

Terrorism is also a type of political conflict. Terrorism inspires fear and is any violent and destructive act committed to intimidate a people or a government.

Bhaskar Paul/The India Today Group/Getty Images

Terrorist attacks are usually carried out in such a way as to maximize the severity and length of the psychological impact. Not usually government supported, each act of terrorism is devised to have an impact on many large audiences. Terrorists also attack national symbols to show power and to attempt to shake the foundation of the country or society they are opposed to. For example, there was a series of terrorist attacks on September 11, 2001, at the World Trade Center in New York City, the Pentagon near Washington, D.C., and in the sky over western Pennsylvania. In 2012 there was an attack on the U.S. embassy in Libya. Terrorist acts frequently have a political purpose. They desire change so badly that failure to achieve change is seen as a worse outcome than the deaths of civilians.

Terrorism can be influenced by geographic factors, as in the Israeli-Palestinian conflict in which many innocent lives were lost. In 1947 Palestine was divided to establish the nation of Israel, resulting in two separate homelands for the Palestinian and Jewish peoples. This division of the land has polarized Israelis and Palestinians for over 60 years, resulting in ongoing violent conflicts in the region.

Alliances and cooperation can also be explored from a geographic perspective. Treaties and international organizations are examples of how countries work together to resolve conflicts and establish ways to share resources. For example, much of the acid rain in Canada comes from pollution in the United States. As a result, in the early 1990s the two countries signed a cooperative agreement that would reduce acid rain.

The United Nations (UN) is an international organization whose stated aims are facilitating cooperation in international law, international security, economic development, social progress, human rights, and aspirations to achieve world peace. The UN was founded in 1945 after World War II to stop wars between countries and to provide a platform for international dialogue. The North Atlantic Treaty Organization (NATO) is an alliance of 16 sovereign Euro-Atlantic countries dedicated to maintaining democratic freedom by means of collective defense. The World Trade Organization (WTO) is an international body that oversees trade agreements and settles trade disputes among countries. A country's membership in the WTO is an important step in its development, and less developed countries strive to become members. In 2000, hoping that trade might open China to democratic change, the United States granted full trading privileges to China and supported its entrance into the WTO. The following year China was admitted to the WTO.

✔ **READING PROGRESS CHECK**

Discussing What is the function of the United Nations?

Connecting Geography to GOVERNMENT

Geopolitics

The *how* and *why* of the creation of political divisions—such as zones, countries, states within countries, and territories—are examples of the connection between geography and politics. *Geopolitics* examines how political units are influenced by geographic factors, such as a country's size, location, and resources. Geopolitical issues influence government and foreign policies, and guide political and economic decisions. It is crucial in a world with terrorism, globalization, and technological advances that governments understand how people and places are interconnected.

EXPLORING ISSUES Discuss a specific example of how geography and politics are related.

LESSON 3 REVIEW

Reviewing Vocabulary

1. *Contrasting* Describe the difference between a unitary system and a federal system.

Using Your Notes

2. *Displaying* Use your graphic organizer to summarize features of government. Include both the levels of government and types of governments.

Answering the Guiding Questions

3. *Identifying* What influences the level and type of a country's government?

4. *Discussing* How does geography influence a country's government?

5. *Expressing* How do cooperation and conflict shape the division of Earth's surface?

Writing Activity

6. *Informative/Explanatory* Write a one-page essay explaining the human and physical geographic characteristics that can influence a country's foreign policy.

HOW HAS GLOBALIZATION CHANGED MODERN CULTURE?

Globalization—the widening exchange of culture traits such as trade, technology, and ideas—is a process that has been taking place for as long as trade has existed. As the world becomes increasingly interconnected economically and socially, largely thanks to advances in communication and transportation, the process of globalization has reached every corner of the Earth.

There are many benefits to globalization. For example, technology developed in one country and shared with others helps to increase economic efficiency and standards of living. International trade has also allowed countries to specialize in the goods they produce well and trade for the goods they do not, enabling economies to grow more rapidly and giving people access to products and resources they may not otherwise have. Globalization has helped us become more aware of the cultures and lifestyles of others around the world, giving us new and unique perspectives on the diversity of life.

While globalization has many benefits, it also has drawbacks. As ideas, products, and even lifestyles are shared between cultures, traditional cultural heritage can become diluted by outside influences. Language, artistic traditions, clothing styles, and even behaviors can all be altered through interactions with other cultures. Language, for example, is one way in which globalization can lead to the permanent loss of some of the world's rare languages, such as Xyzyl, a language spoken in northwest Mongolia. Experts predict that there may be more than 500 languages that are spoken by fewer than 10 people. Many groups, including the United Nations Educational, Scientific and Cultural Organization (UNESCO), urge the protection of traditional culture in the face of increasing globalization.

Protect Cultural Heritage

PRIMARY SOURCE

❝ Today, with the rapid advance of globalization, the loss of intangible cultural heritage . . . can now be observed throughout the world. The threat of extinction to intangible cultural heritage is particularly noticeable in developing countries in Asia, Africa and the Middle East, today. Therefore, while modernization and industrialization remain urgent issues, it is at the same time essential to preserve and transmit these traditional cultures.

Taking into consideration the fact that every culture has been more or less influenced by others, and has forged a cultural identity within history, it goes without saying that the openness of one culture to others is very significant. However, the rapid flow of people, products and information—or rapid cultural interpenetration caused by globalization—menaces minority cultures, especially their intangible cultural heritage, which should be handed down from generation to generation. It is therefore most necessary that measures be taken to prevent this loss. ❞

—Shogo Arai, Parliamentary Secretary for Foreign Affairs, "Japan and the Preservation of Intangible Cultural Heritage," 2004

Increase Cultural Awareness

PRIMARY SOURCE

❝ Paradoxically, it is precisely in the context of increasing globalization that more and more peoples and communities of the world have begun to recognize the importance of their cultural heritage—whether tangible or intangible—as a contribution to the world's cultural diversity. Communities in every land have come to realize that their cultural heritage, which is by nature fragile, plays a crucial role in their identity and that their engagement in safeguarding activities contributes to a sense of continuity. As a result, while globalization has undeniably contributed to the dissemination of cultures, its effects on cultural diversity can, if we are not careful, be negative. . . .

But more is needed in order to respond to peoples' growing awareness of the importance of their culture . . . ❞

—Koïchiro Matsuura, Director-General of UNESCO, "Globalization, Intangible Cultural Heritage and the Role of UNESCO," 2004

What do you think?

1. ***Drawing Conclusions*** Arai states that the loss of cultural heritage is most noticeable in developing countries. Why might the culture of more developed countries have such a strong impact on less developed countries?

2. ***Identifying Central Issues*** According to Matsuura, how has globalization increased cultural awareness?

3. ***Hypothesizing*** What steps do you think can be taken to ensure that cultural heritage is not diminished as a result of globalization? How might international organizations like UNESCO play a role?

(l)Tori Koene/Alamy, (tl)Foto 24/Alamy, (tr)Borderlands/Alamy

networks

There's More Online!

- ☑ **IMAGE** New York Stock Exchange
- ☑ **IMAGE** Alaska Oil Marine Terminal
- ☑ **MAP** World GDP
- ☑ **TABLE** Economic Activities and Economic Development
- ☑ **INTERACTIVE SELF-CHECK QUIZ**
- ☑ **VIDEO** Economic Geography

Reading **HELP**DESK

Academic Vocabulary

- **regulate**
- **incentive**

Content Vocabulary

- **traditional economy**
- **market economy**
- **free enterprise**
- **capitalism**
- **mixed economy**
- **command economy**
- **more developed country**
- **newly industrialized country**
- **less developed country**

TAKING NOTES: *Key Ideas and Details*

ORGANIZING As you read about economic geography, complete a web diagram like the one below to list the major concepts of economic systems, development, and world trade.

LESSON 4
Economic Geography

ESSENTIAL QUESTION · *How do the characteristics and distribution of human populations affect human and physical systems?*

IT MATTERS BECAUSE
The growth of the global economy continues to make the world's peoples increasingly interdependent, or reliant on each other. Natural resources are extracted and traded around the world. Other trade items could be goods, services, and even labor. Countries with varying levels of economic development have become increasingly interdependent through this world trade.

Economic Systems

GUIDING QUESTION *What are the three main types of economic systems?*

All economic systems must make three basic economic decisions: (1) what and how many goods and services should be produced, (2) how they should be produced, and (3) who gets the goods and services that are produced. These decisions are made differently in the three major economic systems—traditional, market, and command.

In a **traditional economy**, habit and custom determine the rules for all economic activity. Individuals are not free to make decisions based on what they would like to have. Instead, their behavior is defined by the customs of their elders and ancestors. For example, it was a tradition in the Inuit society of northern Canada that a hunter would share the food from the hunt with the other families in the village. Today, traditional economies exist in very limited parts of the world. One of the few advantages in a traditional economy is that the roles of individuals are clearly defined. There are also many disadvantages to this type of society. These societies are often very slow to change. When new technologies are introduced, these ideas and techniques are discouraged.

In a **market economy**, individuals and private groups make decisions about what things to produce. People, as shoppers, choose what products they will or will not buy, and businesses produce more of what they believe consumers want. A market economy is based on the concept of **free enterprise**, the idea that private individuals or groups have the right to own property or businesses and make a profit with only limited government interference. In a free enterprise system,

(l to r)Stan Honda/AFP/Getty Images, Bernd Rohrschnei/age fotostock, Roberto Escobar/AFP/Getty Images, Alaska Stock Images/National Geographic Stock (tl)Stan Honda/AFP/Getty Images, (tr)Bernd Rohrschnei/age fotostock

All economic systems must make decisions about the types and quantity of goods and services, including how they are going to be produced and sold.

◀ **CRITICAL THINKING**

1. *Identifying* Which type of system does the image on the left represent?

2. *Describing* What features in the image on the right represent characteristics of a traditional economy?

people are free to choose what jobs they will have and for whom they will work. People have the ability to make as much money as they can and do what is in their best interest. Another positive aspect of market economies is that the government tries to stay out of the way of businesses and there is a great variety of goods and services for consumers. An economic system organized in this way is referred to as **capitalism**.

One major problem with this type of economy is that it does not always provide the basic needs to everyone in the society. The weak, sick, disabled, and old sometimes have trouble providing for themselves and often slip into poverty.

No country in the world, however, has a pure market economy system. Today, the U.S. economy and others like it are described as mixed economies. A **mixed economy** is one in which the government supports and **regulates** free enterprise through decisions that affect the marketplace. In this arrangement, the government's main economic task is to preserve the free market by keeping competition free and fair and by supporting public interests. Governments in modern mixed economies also influence their economies by spending tax revenues to support social services.

In a **command economy**, the government owns or directs the means of production—land, labor, capital (machinery, factories), and business managers—and controls the distribution of goods. Believing that such economic decision making benefits all of society and not just a few people, countries with command economies try to distribute goods and services equally among all citizens. Public taxes, for example, are used to support social services, such as housing and health care, for all citizens. However, citizens have no voice in how this tax money is spent.

Socialism and communism are examples of command economies because they involve heavy government control. However, in practice these two economic systems are mixed economies. In a socialist economy, the government owns some, but not all, of the basic productive resources. The government also provides for some of the basic needs of the people, such as education and health care. Communism is an extreme form of socialism in which all property is collectively, not privately, owned. Under communism, the government decides how much to produce, what to produce, and how to distribute the goods and services produced. One political party—the Communist Party—makes decisions and may even use various forms of coercion to ensure that the decisions are carried out at lower political and economic levels.

traditional economy
a system in which tradition and custom control all economic activity; exists in only a few parts of the world today

market economy
an economic system based on free enterprise, in which businesses are privately owned and production and prices are determined by supply and demand

free enterprise a system in which private individuals or groups have the right to own property or businesses and make a profit with limited government interference

capitalism a system in which factors of production are privately owned

mixed economy a system of resource management in which the government supports and regulates enterprise through decisions that affect the marketplace

regulate to govern or direct according to rule

command economy a system of resource management in which decisions about production and distribution of goods and services are made by a central authority

incentive something that motivates one to act

Recent history has demonstrated, however, that communist economies lack the free decision making and **incentives** that foster business innovation and generation of products that people need and want. Customers can be limited in their choices, and economies can stagnate. As a result of these problems, command economies often decline. The Soviet Union, as described below by a Russian observer, provided an example of this situation:

PRIMARY SOURCE

❝In 1961, the [Communist] party predicted . . . that the Soviet Union would have the world's highest living standard by 1980. . . . But when that year came and went, the Soviet Union still limped along, burdened by . . . a stagnant economy.❞

—Dusko Doder, "The Bolshevik Revolution," *National Geographic*, October 1992

By 2000, Russia and the other countries that were once part of the Soviet Union were developing market economies. Communist China and Vietnam have also allowed some free enterprise to promote economic growth, although their governments tightly control political affairs.

Socialism allows a wider range of free enterprise. It has three main goals: (1) an equitable distribution of wealth and economic opportunity; (2) society's control, through its government, of decisions about public goods; and (3) public ownership of services and factories that are essential. Some socialist countries, like those in Western Europe, are democracies. Under democratic socialism people have basic human rights and elect their political leaders.

☑ **READING PROGRESS CHECK**

Assessing On what idea is a market economy based?

Economic Development

GUIDING QUESTION *What influences economic development?*

Most natural resources are not evenly distributed throughout the Earth. This uneven distribution affects the global economy. As a result, countries specialize in the economic activities best suited to their resources.

Geographers and economists classify all of the world's economic activities into four types. Primary economic activities—such as farming, grazing, fishing, forestry, and mining—involve taking or using natural resources directly from the Earth. Such activities take place near the natural resources that are being gathered or used. For example, coal mining occurs at the site of a coal deposit.

Secondary economic activities use raw materials to make a tangible product that is new and more valuable than the original raw material. Such activities include manufacturing automobiles, assembling electronic goods, producing electric power, or making pottery. These activities occur close to the resource or to the market for the finished good.

Tertiary economic activities do not involve directly acquiring and remaking natural resources. Instead, these activities provide services to people and businesses. Doctors, teachers, lawyers, bakers, truck drivers, and store clerks all provide professional, wholesale, or retail services.

Quaternary economic activities are concerned with the processing, management, and distribution of information. They are vitally important to modern economies that have been transformed in recent years by the information revolution. Just as with tertiary economic activities, people performing these activities include "white collar" professionals working in education, government, business, information processing, and research.

Economic Activities and Economic Development			CHART SKILLS
Country	**Level of Economic Development**	**Major Economic Activities**	**GDP per Capita (purchasing power parity)**
United States	More Developed	service industries, commercial agriculture, industrial supplies	$49,000
Sweden	More Developed	service industries, iron and steel, precision machinery	$40,900
South Korea	More Developed	electronics, telecommunications, automobiles	$32,100
Mexico	Newly Industrialized	service industries, food and beverages, consumer goods manufacturing	$14,800
South Africa	Newly Industrialized	service industries, mining, automobile assembly	$11,100
China	Newly Industrialized	mining and ore processing, textiles, petroleum	$8,500
Palau	Less Developed	tourism, subsistence agriculture, fishing	$8,100
Pakistan	Less Developed	agriculture, textiles, crude oil production	$2,800
Zimbabwe	Less Developed	mining, steel production, wood products	$500

Source: *CIA World Factbook*, 2012

The major economic activities of a country have a direct relationship to the country's level of development. The level of development also has a relationship to standard of living, as shown in this table with GDP per capita (gross domestic product per person).

▲ **CRITICAL THINKING**

1. Classifying Which countries in the table have a major economic activity that is considered a primary economic activity?

2. Evaluating What information in the table illustrates the standard of living in each country? What is the relationship between economic development and standard of living?

Economic activities, including *industrialization*, or the spread of industry, help influence a country's level of economic development. Those countries having more technology and manufacturing, such as the United States and Canada, are called **more developed countries**. Most people work in service or information industries and enjoy a high standard of living. Because of modern techniques, only a small percentage of workers in more developed countries is needed to grow enough food to feed entire populations. For similar reasons, relatively small percentages of the people are employed in manufacturing industries in more developed countries.

Newly industrialized countries have moved from primarily agricultural activities to primarily manufacturing and industrial activities. This transition to manufacturing and industry often brings improvements in socioeconomic development. Examples of newly industrialized countries are Mexico, Malaysia, and Turkey.

Those countries that, according to the United Nations, exhibit the lowest indicators of socioeconomic development are **less developed countries**. In many less developed countries, which are primarily in Africa, Asia, and Latin America, agriculture remains dominant. Even though some commercial farming occurs, most farmers in these countries engage in subsistence farming, growing only enough food for family needs. Some countries' involvement in light industry grows out of a history of cottage industries, businesses that employ workers in their homes. As a result, most people in less developed countries remain poor, as economic development typically reduces poverty.

✓ **READING PROGRESS CHECK**

Listing List the four types of economic activities and explain how these economic activities relate to a country's level of development.

more developed country
a country that has a highly developed economy and advanced technological infrastructure relative to other less developed nations

newly industrialized country a country that has begun transitioning from primarily agricultural to primarily manufacturing and industrial activity

less developed country
a country that, according to the United Nations, exhibits the lowest indicators of socioeconomic development

Economies and World Trade

GUIDING QUESTION *What stimulates and supports world trade?*

World trade is the exchange of capital, labor, goods, and services across international borders or territories, involving the import and export of goods. In most countries, such trade represents a significant share of gross domestic product (GDP). Trade among countries has been present throughout history, but its economic, social, and political importance has increased in recent centuries.

The unequal distribution of natural resources is one factor that promotes a complex network of trade among countries. Countries export their specialized products, trading them to other countries that cannot produce those goods. When countries cannot produce as much as they need of a certain good, they import it, or buy it from another country. That country, in turn, may buy the first country's products, making the two countries trading partners.

Other factors affecting world trade are differences in labor costs and education. Multinationals often base their business decisions on these factors. They locate their headquarters in a more developed country and locate their manufacturing or assembly operations in less developed or newly industrialized countries with low labor costs. In recent decades, many less developed countries have allowed multinationals to build factories or form partnerships with local companies.

ANALYZING PRIMARY SOURCES

Coffee Country

Coffee has a long tradition as an important export crop for both Guatemala and Mexico, employing thousands of workers and bringing income to the region. However, a coffee crisis is causing families that have grown coffee for generations to leave their fields and head for the city or for the border, leaving entire coffee estates abandoned.

❝The problem is what is known as the international coffee crisis. Simply put, there's too much cheap coffee flooding the market these days. It comes from countries such as Brazil, and more recently Vietnam, which have been using massive agribusiness techniques.

However, some small family farmers have found a way to prosper by following the environmental guidelines for what is known as fair trade. They sell their coffee directly to buyers, thereby cutting out the middleman. The difficulty is that the success depends upon the taste of the coffee.

And so the tasting process becomes a critical make-or-break step for many farmers, and failure of the taste test can mean taking a loss on an entire season.

'In the current crisis,' [reporter Sam] Quinones observes, 'peasant coffee growers have to learn the Starbucks lesson and focus on quality. Consumers, meanwhile, have to be willing to pay extra for the best coffee, searching out regional coffees the way they do with wine.'

Even those consumers who like good coffee don't know where it comes from. And many haven't even heard of the fair trade concept. Until all that changes, the international coffee crisis may not be going away anytime soon.❞

—PBS Frontline/World, "Guatemala/Mexico- Coffee Country," May 2003

A woman working on a large coffee estate separates the green coffee seeds from the ripe ones.

DBQ ▲ **CRITICAL THINKING**

1. ***Identifying Cause and Effect*** In what ways could coffee purchased in countries far from Guatemala or Mexico affect people living there?

2. ***Problem Solving*** How does fair trade help the coffee farmers in this region?

Roberto Escobar/AFP/Getty Images

Russell Illig/Getty Images

Oil is a major commodity of trade in the world markets.

◀ **CRITICAL THINKING**

1. *Contrasting* How is an import different from an export?

2. *Evaluating* Explain why a country would need to import oil.

International trade is, in principle, not different from domestic trade. The main difference is that international trade is typically more costly due to additional costs such as tariffs, time costs due to border delays, and costs associated with country differences such as language, the legal system, or other cultural barriers.

Another difference between domestic and international trade is that factors of production such as capital and labor are typically more mobile within a country than across countries. Thus, international trade is mostly restricted to trade in goods and services, and only to a lesser extent to trade in capital, labor, or other factors of production. Trade in goods and services can serve as a substitute for trade in factors of production.

Sometimes, a country can import goods that make extensive use of that factor of production and thus embody it. An example is the import of labor-intensive goods by the United States from China. Instead of importing Chinese labor, the United States imports goods that were produced with Chinese labor.

Emerging markets are nations with social or business activity in the process of rapid growth and industrialization. The economies of China and India are considered to be the largest. The seven largest emerging and developing economies by either nominal GDP or GDP (PPP) are China, Brazil, Russia, India, Mexico, Indonesia, and Turkey. The ASEAN–China Free Trade Area, launched on January 1, 2010, is the largest regional emerging market in the world.

☑ **READING PROGRESS CHECK**

Contrasting Explain the differences between international trade and domestic trade.

LESSON 4 REVIEW

Reviewing Vocabulary

1. *Identifying* What are the advantages and disadvantages to a command economy?

Using Your Notes

2. *Describing* Use your graphic organizer to describe the three basic economic decisions that are made by economic systems.

Answering the Guiding Questions

3. *Listing* What are the three main types of economic systems?

4. *Describing* What influences economic development?

5. *Discussing* What stimulates and supports world trade?

Writing Activity

6. *Informative/Explanatory* Write an essay explaining the advantages and disadvantages a less developed country might experience by joining a free trade agreement.

PATTERNS OF RESOURCE
DISTRIBUTION

Earth's natural resources are unevenly distributed across the globe. As a result, a country might have abundant access to one resource but may have a limited supply of another resource. Saudi Arabia, for example, is a country rich in petroleum, but it has very limited access to freshwater. This uneven distribution of resources is a major reason countries trade with one another, exchanging the resources they have for the ones they need. Access to natural resources can also be a cause of conflict in some parts of the world.

CANADA
- 20,680,000,000 BBL
- 3,040,220 sq km
- 3,069 cu km
- 6,988 kg per hectare

UNITED STATES
- 175,200,000,000 BBL
- 3,101,340 sq km
- 3,300 cu km
- 3,490 kg per hectare

BRAZIL
- 12,860,000,000 BBL
- 5,195,220 sq km
- 8,233 cu km
- 4,055 kg per hectare

LEGEND

 Proven Oil Reserves

 Renewable Water Resources

 Forest Area (timber)

 Cereal Yield (grains)

GLOBAL SCALE
Relative amount compared to total world resources.

VERY LOW LOW MED HIGH VERY HIGH

GERMANY
276,000,000 BBL
110,760 sq km
188 cu km
6,716 kg per hectare

SAUDI ARABIA
262,600,000,000 BBL
9,770 sq km
2.4 cu km
5,621 kg per hectare

CHINA
14,800,000,000 BBL
2,068,610 sq km
2,829.6 cu km
5,521 kg per hectare

INDIA
5,682,000,000 BBL
684,340 sq km
1,907.8 cu km
2,537 kg per hectare

SOUTH AFRICA
15,000,000 BBL
56,660 sq km
50 cu km
4,162 kg per hectare

Making Connections

1. *Analyzing* Which of the four resources on this map do you think Canada supplements with imports? From which country might Canada import this resource?

2. *Drawing Conclusions* Using one of the countries featured on this map as an example, explain how a country's physical geography, including climate, location, and landforms, influences its access to natural resources.

networks

There's More Online!

☑ **IMAGE** Urban Landfill

☑ **DIAGRAM** Central Place Theory

☑ **INFOGRAPHIC** Urban Land Use Models

☑ **GRAPH** Ten Largest Cities

☑ **INTERACTIVE SELF-CHECK QUIZ**

☑ **VIDEO** Urban Geography

City

Town

Market Tov

○ Village

LESSON 5
Urban Geography

ESSENTIAL QUESTION · *How do the characteristics and distribution of human populations affect human and physical systems?*

Reading HELPDESK

Academic Vocabulary

- **function**
- **structure**

Content Vocabulary

- **urban sprawl**
- **connectivity**
- **metropolitan area**
- **central place theory**
- **world cities**

TAKING NOTES: *Key Ideas and Details*

DESCRIBING Complete a graphic organizer similar to the one below describing the nature of cities, patterns of urbanization, and challenges of urban growth.

Urban Geography		
The Nature of Cities	Patterns of Urbanization	Challenges of Urban Growth

IT MATTERS BECAUSE

Urban geography is a branch of human geography concerned with cities and the people who live in them. An urban geographer analyzes patterns of settlement and growth in urban areas and evaluates the impact of cities on people and the environment.

The Nature of Cities

GUIDING QUESTION *How does a city's function influence its structure?*

The Industrial Revolution ushered in a new age of urbanization in the world's history. As the focus shifted from agricultural production to industrial production, people began moving to cities in large numbers. As more people moved to cities, the physical size of cities also began to grow. This spreading of urban areas onto undeveloped land near cities is called **urban sprawl**. Currently, the world's urban population is growing at a much faster rate than that of the rural population. Over half of the world's people now live in cities, and this proportion is highest in the developed regions of the world. Eighty-two percent of Americans now live in urban areas, and more than two-thirds of the people of Europe, Russia, Japan, and Australia do as well. This growth is due to **connectivity**, the directness of routes and communication linking pairs of places.

The Function of Cities

Only recently have people gathered in the densely populated and highly structured settlements we call cities. The first cities were established about 5,000 years ago, but it has only been in the last 200 years—with the expansion of industrialization, economic growth, and global population at exponential rates—that cities have grown significantly in size and number. At the start of the twentieth century, only about one person in ten lived in a city. Today, the proportion of urban and rural dwellers is approximately equal. By 2025, it is expected that nearly two-thirds of the world's population will live in urban areas.

All cities serve a variety of **functions**. For example, manufacturing, retail, and service centers are often located in urban areas. These functions are the economic base of a city, generating employment and wealth. The

Kevin Leigh/Photolibrary/Getty Images

larger a city is, the more numerous and highly specialized its functions are likely to be. Smaller cities and towns have fewer functions, which tend to be of a more general nature. In the field of health care, for example, clinics are found in a wide range of places, but specialized teaching hospitals tend to be located only in larger cities.

Cities also tend to be centers of culture and creativity. Artists, musicians, architects, philosophers, scientists, and writers gravitate toward cities where there are patrons, communities of other artists, universities, clients, and a skilled workforce. Today's urban centers of culture have changed over time, mostly based on their economic or political strength with the outside world.

There are several reasons cities can support a variety of functions. The large population of a city means there are plenty of workers available to support a variety of industries. The large population also means there is a large market of consumers to sustain the demand for specialized functions. From a functional perspective, infrastructure facilitates the production of goods and services, and also the distribution of finished products to markets. In addition, it provides basic social services such as schools and hospitals. Roads provide adequate transportation, and safe buildings provide secure housing.

Urban areas have both advantages and challenges. The diversity of peoples and activities encourages innovation and creativity, but overcrowding, crime, poverty, social conflict, and pollution can become challenges. An urban area differs from country to country, but each is considered a **metropolitan area**—a region that includes a central city and its surrounding suburbs.

urban sprawl spreading of urban developments on land near a city

connectivity the directness of routes linking pairs of places

function a special purpose

metropolitan area region that includes a central city and its surrounding suburbs

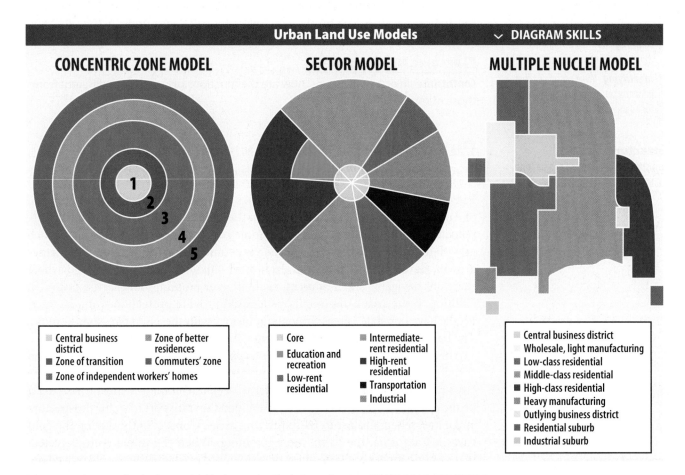

Urban Land Use Models ⌄ **DIAGRAM SKILLS**

CONCENTRIC ZONE MODEL

1
2
3
4
5

- Central business district
- Zone of transition
- Zone of independent workers' homes
- Zone of better residences
- Commuters' zone

SECTOR MODEL

- Core
- Education and recreation
- Low-rent residential
- Intermediate-rent residential
- High-rent residential
- Transportation
- Industrial

MULTIPLE NUCLEI MODEL

- Central business district
- Wholesale, light manufacturing
- Low-class residential
- Middle-class residential
- High-class residential
- Heavy manufacturing
- Outlying business district
- Residential suburb
- Industrial suburb

Urban geographers use urban land use models like these to describe the internal structure of cities. They examine location patterns of people and businesses within the urban setting.

▲ **CRITICAL THINKING**

1. Identifying What district is typically largest in the multiple nuclei model?

2. Describing What zone is located between the commuters' zone and the zone of independent workers' homes in the concentric zone model?

Landfills are examples of the type of infrastructures a city must have in order to support large populations.

▲ CRITICAL THINKING

1. **Defining** How does a landfill fit the definition of an infrastructure?

2. **Identifying** What are a few other infrastructures a city must maintain for its citizens?

structure something constructed or arranged in a definite pattern of organization

central place theory geographical theory that seeks to explain the number, size, and location of human settlements in an urban system

The Structure of Cities

Urban **structure** is the arrangement of land use in urban areas. Sociologists, economists, and geographers have developed several models explaining where different types of people and businesses tend to exist within the urban setting. Urban structure can also refer to the urban spatial structure, which concerns the arrangement of public and private space in cities and the degree of connectivity and accessibility.

The *concentric zone model* was the first to explain distribution of social groups within urban areas. Based on a single city, Chicago, it was created by sociologist Ernest Burgess in 1924. According to this model, a city grows outward from a central point in a series of rings. A second theory of urban structure was proposed in 1939 by economist Homer Hoyt. The *sector model* proposed that a city develops in sectors instead of rings. Certain areas of a city are more attractive for various activities, whether by chance or geographic and environmental reasons. As the city grows and these activities flourish and expand outward, they do so in a wedge shape and become a sector of the city.

Geographers C. D. Harris and E. L. Ullman developed the *multiple nuclei model* in 1945. According to this model, a city contains more than one center around which activities revolve. Some activities are attracted to particular nodes while others try to avoid them. For example, a university node may attract well-educated residents, pizzerias, and bookstores, whereas an airport may attract hotels and warehouses. Other businesses may also form clusters for automobile repair, tire stores, or arts districts. Incompatible activities will avoid clustering in the same area, thus explaining why heavy industry and high-income housing rarely exist together in the same neighborhood.

☑ **READING PROGRESS CHECK**

Contrasting In economic terms, how are the functions of larger cities different from those of smaller cities?

Patterns of Urbanization

GUIDING QUESTION *What influences the location and growth of cities?*

Factors that led to the early growth of cities are still influential today. The growth of American cities, for example, began in the late 1700s. This growth was directly related to certain influential factors. Some of the same factors that led to such growth hundreds of years ago are the very same factors influencing the further growth, development, and urbanization of cities today. People will go where there are navigable water sources, such as river crossings and fertile deltas. In addition, if the area is mountainous it could provide protection from enemies. People have populated areas throughout history with these same reasons in mind. The basics of survival are food and water sources and security from enemies.

In addition to factors that promote population, there are factors that can shrink population in a region. For example, if an industry is no longer needed, the city's population will move on in order to seek a livelihood. Ghost towns are a prime example of this. The railroads provided the transportation for prospectors and entrepreneurs to get to these thriving areas. Consequently, during the gold rush in California, the towns were booming. When the mines were depleted, there was no longer any reason for these towns to exist and the people moved on.

The **central place theory** is a spatial theory in urban geography that attempts to explain the reasons behind the distribution patterns, size, and number of cities and towns around the world. It attempts to illustrate how settlements locate in relation to one another, the amount of market area a central

Kevin Leigh/Photolibrary/Getty Images

place can control, and why some central places function as hamlets, villages, towns, or cities. It also attempts to provide a framework by which those areas can be studied both for historical reasons and for the locational patterns of areas today.

A **world city** is a city generally considered to play an important role in the global economic system. World cities possess such features as having international diverse cultures, an active influence on and interaction in world affairs, a large population, a major international airport, and an advanced transportation system. The world city concept comes from geography and urban studies. It can be seen as a type of "point of entry" for studying the changes that come from globalization.

One important world city that has resulted from the geography of its location is İstanbul. It straddles a strait, thus placing it on vital land and sea trade routes. In addition, it can easily defend itself against enemy factions because of its water location. Farming is good because of its fertile soil, and consequently İstanbul has been under attack and conquered many times. Its desirable location is the very thing that has made it vulnerable.

New trends in cities have emerged, including the development of suburban business districts and major diversified centers, to name just two. Typically, suburbia refers to an outlying community around a city. These communities are business districts in their own right. The name that is now most commonly used to describe these terms is "edge cities."

These new suburban cities have sprung up all over and are home to glistening office towers and huge retail complexes. They are always located close to major highways. "Boomers" are the most common type of edge cities, having developed around a shopping mall or highway interchange such as Pasadena, California. On the suburban fringe of Phoenix, Arizona, Sun City is a "greenfield"—a new, master-planned city built on undeveloped land. In contrast, "uptown" edge cities are historic activity centers built over an older city or town, such as the Rosslyn-Ballston Corridor in Virginia.

world cities cities generally considered to play an important role in the global economic system

☑ **READING PROGRESS CHECK**

Summarizing What does the central place theory attempt to explain?

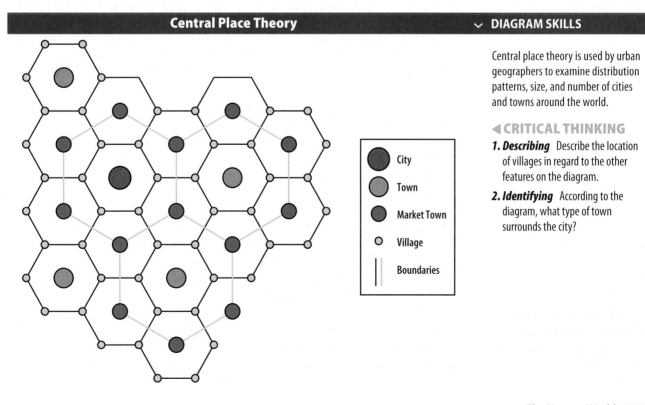

Central Place Theory

Legend:
- City
- Town
- Market Town
- Village
- Boundaries

∨ DIAGRAM SKILLS

Central place theory is used by urban geographers to examine distribution patterns, size, and number of cities and towns around the world.

◄ CRITICAL THINKING

1. ***Describing*** Describe the location of villages in regard to the other features on the diagram.

2. ***Identifying*** According to the diagram, what type of town surrounds the city?

Challenges of Urban Growth

GUIDING QUESTION *What problems do urban areas face?*

The urbanization process refers to much more than simple population growth. It involves changes in the economic, social, and political structures of a region. Rapid urban growth is responsible for many environmental and social changes, and its effects are strongly related to issues of pollution and economics. These changes are not always positive. The rapid growth of cities strains their capacity to provide services such as energy, education, health care, transportation, sanitation, and physical security.

The more developed countries experienced urbanization during the nineteenth and twentieth centuries along with the Industrial Revolution. During this time, urbanization resulted from and contributed to industrialization. New job opportunities in the cities motivated people to migrate from rural areas to cities. At the same time, migrants provided cheap, plentiful labor for the emerging factories. Today, the circumstances are rather different in less developed countries. People are forced out of rural areas because of insufficient land on which to grow subsistence crops. Meanwhile, there are not enough jobs to accommodate the many migrants looking for employment, creating a large surplus labor force. This influx keeps wages low and can lead to poverty in many urban areas. Foreign investment companies from more developed countries see such situations as attractive. By employing these workers they can produce goods for far less.

Modern cities all over the world face many of the same problems: poor housing, homelessness, pollution, and social problems such as addiction, crime, and gang violence. People often live in old houses or other structures without electricity or sanitation. Some live on the streets with little access to adequate food or shelter. In addition, cars and industries pollute city air and water. Unemployment continues to grow. Larger multiethnic cities continue to face conflicts between different cultural groups.

One of the major effects of rapid urban growth is urban sprawl—scattered development that increases traffic, saps local resources, and destroys open space. Urban sprawl is responsible for changes in the physical environment and can also diminish the local character of the community. Small local businesses find it difficult to compete with larger stores and restaurants. Some cities are trying to be proactive and establish measures aimed at fighting urban sprawl by limiting construction and using innovative land-use planning techniques or community cooperation. One new form of land use is called "smart growth" or "New Urbanism," in which cities plan the communities' growth in a strategic way for livable and walkable neighborhoods.

✔ **READING PROGRESS CHECK**

Defining In what ways does rapid growth strain cities?

LESSON 5 REVIEW

Reviewing Vocabulary

1. *Defining* What is urban sprawl, and what is it responsible for?

Using Your Notes

2. *Listing* Use your graphic organizer to list the three models of urban structures.

Answering the Guiding Questions

3. *Explaining* How does a city's function influence its structure?

4. *Describing* What influences the location and growth of cities?

5. *Discussing* What problems do urban areas face?

Writing Activity

6. *Argument* Consider the negative aspects of urban sprawl. What are some possible solutions the government could provide to bolster infrastructure and services? Develop an argument to support your ideas.

Assessment

Directions: On a separate sheet of paper, answer the questions below. Make sure you read carefully and answer all parts of the questions.

Lesson Review

Lesson 1

1 *Describing* Describe how culture affects the daily lives of people.

2 *Explaining* Explain the elements of culture geographers use to organize the world into culture regions.

3 *Drawing Conclusions* How did the agricultural revolution influence cultural diffusion?

Lesson 2

4 *Evaluating* What factors contribute to the uneven distribution of the world's population?

5 *Analyzing* How is the demographic transition model used to explain a country's population growth?

6 *Summarizing* How do the effects of zero population growth and negative population growth differ? How are they similar?

Lesson 3

7 *Comparing and Contrasting* What different roles might local citizens have in government decision making under a unitary system, a federal system, and a confederation?

8 *Explaining* Explain the different ways in which an autocracy, an oligarchy, and a democracy exercise authority.

9 *Defining* List and describe the three types of political boundaries.

Lesson 4

10 *Listing* What are the characteristics of a mixed economy?

11 *Contrasting* Contrast a more developed country with a less developed country. What are the primary differences?

12 *Analyzing* What are the advantages and disadvantages of a capitalist economy?

Lesson 5

13 *Explaining* Explain two functions of urban areas.

14 *Inferring* What are some issues associated with urban sprawl? Discuss some solutions to address these issues.

15 *Identifying Central Issues* What factors have influenced the site and growth of cities? Give examples.

21st Century Skills

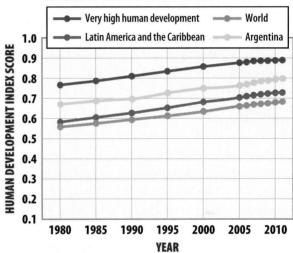

RECENT TRENDS IN DEVELOPMENT

Source: United Nations Development Programme

16 *Using Graphs, Charts, Diagrams, and Tables* According to the graph, how does the Human Development Index score of Argentina compare to that of the world?

College and Career Readiness

17 *Economics* Imagine that you are interviewing people for a position studying economic geography. Write a series of interview questions (with sample answers) that assesses the interviewees' understanding of economic activities and economic development.

Need Extra Help?

If You've Missed Question	1	2	3	4	5	6	7	8	9	10	11	12	13	14	15	16	17
Go to page	78	78	80	85	82	83	87	88	90	95	97	95	102	106	104	107	94

Directions: On a separate sheet of paper, answer the questions below. Make sure you read carefully and answer all parts of the questions.

Writing About Geography

18 *Informative/Explanatory* Use standard grammar, spelling, sentence structure, and punctuation to write a one-page essay discussing problems in today's urban areas. Include examples suggesting solutions and their feasibility.

DBQ Analyzing Primary Sources

Use the document to answer the following questions.

Governments and economies of countries around the world are becoming increasingly interconnected. Some countries, such as members of the European Union, have had strong economies that allowed them to help improve standards of living in other countries.

PRIMARY SOURCE

" *I drove east on the highway that connects the capital of Talinn with Narva, on the Russian border. Nearly everywhere I looked I saw the handiwork of the European Union, starting with the road itself. The EU has already invested millions of euros to improve the highway, which serves as the main link to St. Petersburg, Russia. This highway passes the town of Sillamäe, once a 'closed' city run by the Soviet military, which enriched uranium for weapons programs in a huge factory overlooking the sea. The EU is here, too, kicking in more than a million dollars to help prevent the radioactive waste from leaching into the Baltic Sea.* "

—Don Belt, "Europe's Big Gamble," *National Geographic,* May 2004

19 *Speculating* Why would the EU want to invest in the improvement of other countries?

20 *Interpreting* What did Belt mean by "I saw the handiwork...starting with the road itself"?

Applying Map Skills

Use the Unit 1 Atlas to answer the following questions.

21 *Places and Regions* The Ural Mountains are a natural border that divides Asia from what other continent?

22 *Human Systems* How would you describe the population density pattern of Australia?

23 *The Environment and Society* Which of the countries in South America have petroleum resources?

Exploring the Essential Question

24 *Making Connections* Recall what you have learned about the demographic transition model. In which types of economic activities would a country in Stage 2 most likely be involved?

25 *Researching* Research and compare two countries in the ways they depend on the environment for the products they export.

Critical Thinking

26 *Making Generalizations* Explain the factors that influence a country's ability to control territory.

27 *Identifying Cause and Effect* What cultural changes have resulted from the information revolution?

28 *Comparing and Contrasting* What is the difference between a culture region and a culture hearth?

Research and Presentation

29 *Research Skills* Use Internet and library resources to gather information about a city plagued by urban sprawl. Your research should focus on the past, present, and future of this city. Create a multimedia presentation outlining that city's experiences along the road to urbanization. Be sure to (1) describe the human systems that affected the city's growth and (2) explain how other cities in the region might learn from its experience.

Need Extra Help?

If You've Missed Question	**18**	**19**	**20**	**21**	**22**	**23**	**24**	**25**	**26**	**27**	**28**	**29**
Go to page	106	108	108	2	6	8	82	97	89	81	79	102

Europe's Big Gamble by Don Belt. National Geographic Magazine, May 2004. All rights reserved.

The United States and Canada

©Blaine Harrington III/Corbis

1 Culture Immigration to the United States and Canada from other areas of the world has had a dramatic effect on the cultures of the two countries. Parades celebrating these cultural roots are common throughout the region.

Robedzy Cronik tta/Lonely Planet Images/Getty Images

EXPLORE the REGION

The region of the **UNITED STATES** and **CANADA** stretches from the Pacific Ocean in the west to the Atlantic Ocean in the east. The two countries share many physical features—mountains frame their eastern and western edges, cradling a central area of vast plains. The region is a land of immigrants. Many made this land their home by choice. Others were forced to come as exiles or enslaved workers. Along with native peoples, these groups have shaped the cultures of the region.

THERE'S MORE ONLINE

(cr)Jon Spaull/Dorling Kindersley/Getty Images, (bl)S. Greg Panosian/Vetta/Getty Images, (br) Brand X Pictures/PunchStock

③ Lakes and Rivers Long rivers, such as the Mississippi River, have played an important role in trade and industry in both the United States and Canada.

② Mountains The Rocky Mountains are the longest mountain range in North America, stretching from British Columbia in Canada to New Mexico in the United States.

④ Economy Today the service industry employs most of the workers in the United States and Canada. Many of these jobs are located in urban centers such as Toronto, Canada's largest city.

The United States and Canada **111**

The United States and Canada
Physical

RUSSIA

ARCTIC OCEAN

Greenland Sea

GREENLAND

Chukchi Sea

Point Barrow

Bering Sea

Bering Strait

Ellesmere Island

ARCTIC CIRCLE

Beaufort Sea

Banks Island

Queen Elizabeth Islands

Devon I.

Baffin Bay

Brooks Range

Yukon R.

Aleutian Is.

Mt. McKinley (Denali) 20,320 ft. (6,194 m)

Alaska Range

Baffin Island

Kodiak I.

Yukon Plateau

Victoria Island

Davis Strait

Labrador Sea

Gulf of Alaska

Mt. Logan 19,551 ft. (5,959 m)

Mackenzie Mts.

Great Bear Lake

Southampton I.

Hudson Strait

Ungava Peninsula

Labrador

Coast Mts.

Mackenzie R.

Great Slave Lake

C A N A D I A N   S H I E L D

Queen Charlotte Islands

Peace R.

Lake Athabasca

Athabasca R.

Churchill R.

Nelson R.

Hudson Bay

Newfoundland

Gulf of St. Lawrence

Cape Breton I.

Fraser Plateau

R O C K Y   M O U N T A I N S

Saskatchewan R.

Lake Winnipeg

Laurentian Mts.

Sable I.

Vancouver I.

Columbia R.

Fraser R.

G R E A T   P L A I N S

Lake Superior

Lake Huron

St. Lawrence R.

Lake Ontario

Cape Cod

PACIFIC OCEAN

Columbia Plateau

Snake R.

Black Hills

Missouri R.

Mississippi R.

Lake Michigan

Lake Erie

Appalachian Mts.

Mt. Shasta 14,162 ft. (4,317 m)

Cascade Range

Sierra Nevada

Great Basin

Great Salt Lake

Platte R.

Central Lowlands

Ohio R.

Piedmont

Chesapeake Bay

Cape Hatteras

Mt. Whitney 14,495 ft. (4,418 m)

Coast Ranges

Colorado Plateau

Colorado R.

Ozark Plateau

Arkansas R.

ATLANTIC OCEAN

Death Valley -282 ft. (-86 m)

Red R.

Rio Grande

C O A S T A L   P L A I N

The Everglades

MEXICO

Gulf of Mexico

TROPIC OF CANCER

Elevations

10,000 ft. (3,000 m)	
5,000 ft. (1,500 m)	
2,000 ft. (600 m)	
1,000 ft. (300 m)	
0 ft. (0 m)	
Below sea level	

—— National boundary

—— State or provincial boundary

▲ Mountain peak

▼ Lowest point

Hawaiian Islands

Kauai
Niihau   Oahu   Molokai
Lanai   Maui
Kahoolawe
Hawaii

PACIFIC OCEAN

0   100 miles
0   100 kilometers
Albers Equal-Area Conic projection

0   600 miles
0   600 kilometers
Lambert Azimuthal Equal-Area projection

The United States and Canada
Political

160°E

RUSSIA

Bering Strait

80°N

Bering Sea

180°

60°N

ARCTIC OCEAN

Beaufort Sea

Yukon R.

Alaska

Yukon

Gulf of Alaska

160°W

40°N

PACIFIC OCEAN

Mackenzie R.

Northwest Territories

British Columbia

Alberta

Sask.

CANADA

Manitoba

GREENLAND (DENMARK)

20°W

Baffin Bay

ARCTIC CIRCLE

Nunavut

Labrador Sea

40°W

Newfoundland and Labrador

Hudson Bay

Quebec

Ontario

Gulf of St. Lawrence

P.E.I.

N.B.

Nova Scotia

60°W

St. Lawrence R.

Ottawa ✪

Maine

Wash.

Oregon

Montana

Idaho

Wyoming

Nevada

Utah

California

Arizona

New Mexico

N. Dak.

S. Dak.

Nebraska

Colorado

Kansas

Okla.

Texas

Missouri R.

Minn.

Iowa

Mo.

Ark.

Miss.

La.

Wis.

Mississippi R.

Ill.

Michigan

Ind.

Ky.

Tenn.

Ala.

Ga.

Fla.

Ohio

W. Va.

Va.

N.C.

S.C.

Vt.

N.H.

Mass.

R.I.

Conn.

N.J.

Del.

Md.

N.Y.

Pa.

Washington, D.C. ◉

UNITED STATES

20°N

TROPIC OF CANCER

140°W

ATLANTIC OCEAN

Rio Grande

MEXICO

Gulf of Mexico

◉ National capital

Hawaii (inset)

160°W   155°W

Kauai

Niihau   *Oahu Molokai*

Hawaii   *Lanai* *Maui*
          Kahoolawe

20°N

Hawaii

PACIFIC OCEAN

0   100 miles

0   100 kilometers

Albers Equal-Area Conic projection

0   600 miles

0   600 kilometers

Lambert Azimuthal Equal-Area projection

120°W

100°W

UNIT 2
REGIONAL ATLAS

MAP STUDY

1. **Physical Systems** Compare the overall elevation of the western and eastern parts of the region.

2. **Human Systems** Which U.S. state is on the farthest northern latitude?

The United States and Canada **113**

The United States and Canada
Climate and Vegetation

RUSSIA

ARCTIC OCEAN

Bering Sea

Bering Strait

ARCTIC CIRCLE

Anchorage

Hudson Bay

Gulf of St. Lawrence

PACIFIC OCEAN

0   600 miles
0   600 kilometers
Lambert Azimuthal Equal-Area projection

Vancouver

Seattle

Winnipeg

Ottawa

Toronto

New York City

Chicago

Washington, D.C.

Denver

ATLANTIC OCEAN

PACIFIC OCEAN

Honolulu

Los Angeles

Atlanta

Albers Equal-Area Conic projection

MEXICO

Gulf of Mexico

New Orleans

TROPIC OF CANCER

100°W

ARCTIC CIRCLE

TROPIC OF CANCER

EQUATOR

Vegetation

- Tropical rain forest
- Tropical grassland (savanna)
- Desert scrub and desert waste
- Temperate grassland
- Mediterranean scrub
- Deciduous forest
- Coniferous forest
- Mixed forest (deciduous and coniferous)
- Tundra
- Ice cap

Climate

- Tropical rain forest
- Tropical wet/dry
- Semi-arid (steppe)
- Arid (desert)
- Humid subtropical
- Marine west coast
- Mediterranean
- Humid continental
- Subarctic
- Tundra and high altitude
- Ice cap

The United States and Canada
Economic Activity

RUSSIA

ARCTIC OCEAN

160°E

Bering Sea

Bering Strait

ARCTIC CIRCLE

Hudson Bay

Gulf of St. Lawrence

PACIFIC OCEAN

ATLANTIC OCEAN

MEXICO

Gulf of Mexico

Land Use
- Commercial farming
- Livestock raising
- Primarily forest
- Manufacturing and trade
- Commercial fishing
- Little or no activity

Resources
- Coal
- Petroleum
- Natural gas
- Iron ore
- Tin
- Zinc
- Cobalt
- Uranium
- Nickel
- Copper
- Lead
- Gold
- Silver
- Platinum
- Gems

0   600 miles
0   600 kilometers
Lambert Azimuthal Equal-Area projection

160°W   155°W
PACIFIC OCEAN
Hawaii
20°N
0   100 miles
0   100 kilometers
Albers Equal-Area Conic projection

UNIT 2
REGIONAL ATLAS

MAP STUDY

1. **Physical Systems** What climate regions are found in the United States?

2. **Environment and Society** What economic activities are found in the area around the Great Lakes?

RUSSIA

ARCTIC OCEAN

GREENLAND (DENMARK)

ARCTIC CIRCLE

Bering Sea

Bering Strait

PACIFIC OCEAN

Hudson Bay

Gulf of St. Lawrence

Edmonton

Vancouver

Calgary

Seattle

Portland

Montreal

Ottawa

Boston

Toronto

Buffalo

Minneapolis-St. Paul

Detroit

New York City

Milwaukee

Pittsburgh

Philadelphia

Chicago

Baltimore

Washington, D.C.

Sacramento

Indianapolis

San Francisco

St. Louis

Cincinnati

San Jose

ATLANTIC OCEAN

Las Vegas

Denver

Kansas City

Los Angeles

San Bernardino

Memphis

Atlanta

San Diego

Phoenix

Dallas

Ft. Worth

San Antonio

Austin

Orlando

Houston

Tampa

Miami

MEXICO

Gulf of Mexico

CUBA

POPULATION

Per sq. mi.	Per sq. km
1,250 and over	500 and over
250–1,249	100–499
63–249	25–99
25–62	10–24
2.5–24	1–9
Less than 2.5	Less than 1
Uninhabited	Uninhabited

Cities
(Statistics reflect metropolitan areas.)

■ Over 5,000,000
□ 2,000,000–5,000,000
⊙ 1,000,000–2,000,000

PACIFIC OCEAN

Hawaii

0   100 miles
0   100 kilometers
Albers Equal-Area Conic projection

0   600 miles
0   600 kilometers
Lambert Azimuthal Equal-Area projection

EQUATOR

UNIT 2
REGIONAL ATLAS

MAP STUDY

1. **Environment and Society** In what area is Canada's greatest population density located?

2. **Places and Regions** What generalizations can you make about the region's major population centers?

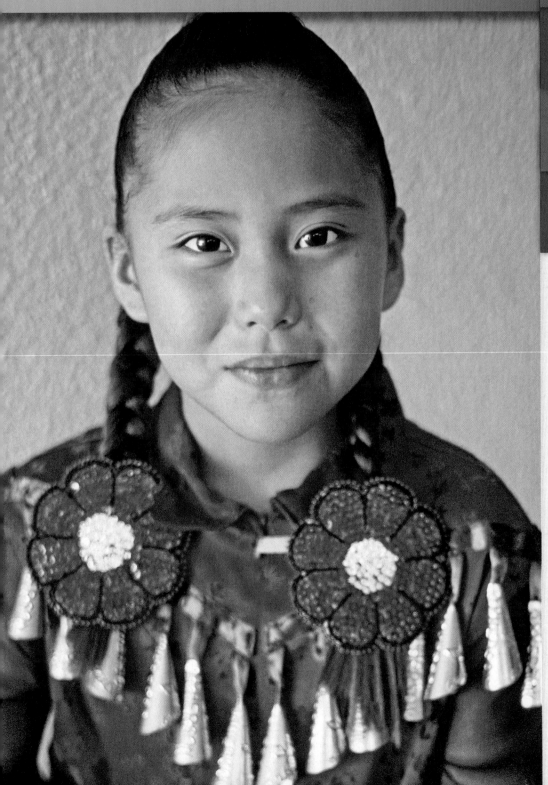

The United States

ESSENTIAL QUESTION · *How do physical systems and human systems shape a place?*

networks

There's More Online about the United States.

CHAPTER 5

Geography Matters...

What do the words *the United States* bring to mind? The images are endless for a land that offers so many opportunities. Some people hope to travel across its four million miles of highways, while others seek the "American Dream" of economic opportunity or religious and cultural freedom. It is not easy to characterize the American experience because the United States is a land of many cultures and peoples. The country is also known for its varied landscape of big cities, rugged mountains, wide plains, and forests that cover vast stretches of land.

◀ Native Americans such as this Navajo girl keep their traditions alive through dance, dress, and spiritual practices.

National Geographic Image Collection/Alamy

patterns
of immigration

One of the defining attributes of the United States is that it is largely a country of immigrants and their descendants. About 13 percent of people in the United States are foreign born, while Native Americans, Alaska Natives, and Native Hawaiians make up about 2 percent of the population. The remaining population is descended from immigrants.

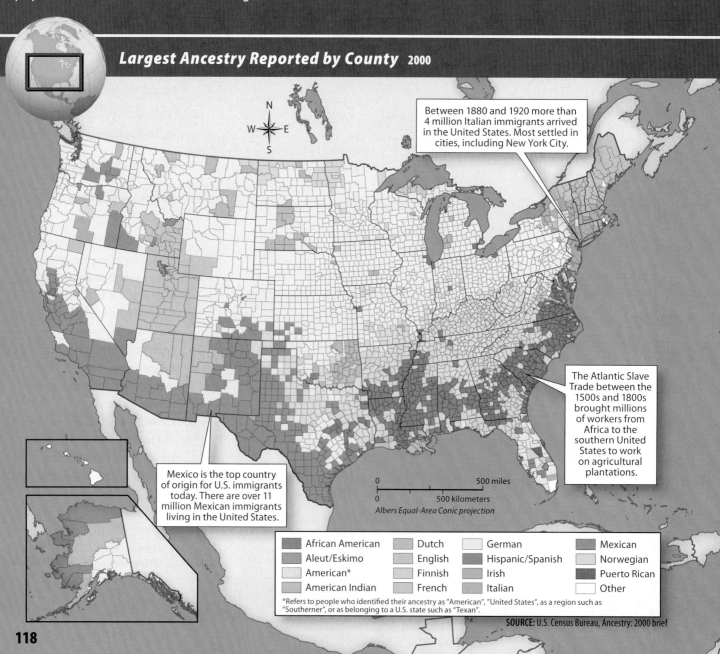

Largest Ancestry Reported by County 2000

Between 1880 and 1920 more than 4 million Italian immigrants arrived in the United States. Most settled in cities, including New York City.

The Atlantic Slave Trade between the 1500s and 1800s brought millions of workers from Africa to the southern United States to work on agricultural plantations.

Mexico is the top country of origin for U.S. immigrants today. There are over 11 million Mexican immigrants living in the United States.

0   500 miles
0   500 kilometers
Albers Equal-Area Conic projection

African American	Dutch	German	Mexican
Aleut/Eskimo	English	Hispanic/Spanish	Norwegian
American*	Finnish	Irish	Puerto Rican
American Indian	French	Italian	Other

*Refers to people who identified their ancestry as "American", "United States", as a region such as "Southerner", or as belonging to a U.S. state such as "Texan".

SOURCE: U.S. Census Bureau, Ancestry: 2000 brief

Where have immigrants to the United States come from?

The history of immigration to the United States can be divided into four distinct periods. The earliest period, the colonial period, was characterized by an influx of English and other European peoples, many of whom came as indentured servants to work for little or no pay. During the mid-nineteenth century, the majority of immigrants originated from the countries of northern and western Europe. The third major period of immigration was primarily from southern and eastern Europe. The start of the fourth wave of immigrants is marked by the abolishment in 1965 of the quotas that limited immigration from the Eastern Hemisphere based on country of origin. This opened the doors to a wave of immigrants from Asia and Africa. At the same time, worsening economic conditions in many Latin American countries led to a rise in the number of Latino immigrants to the United States. Today, Mexicans are the largest single immigrant group in the United States.

1. Places and Regions Describe the four periods into which immigration to the United States can be divided.

Why did they immigrate to the United States?

Two sets of factors influence immigration patterns: push factors and pull factors. Push factors motivate people to leave their countries of origin. They can be quite varied, including religious and ethnic persecution, lack of economic prospects or personal security, civil strife, and war. Environmental conditions such as drought, famine, and flooding and events such as earthquakes and volcanic eruptions can also push people to migrate to a new country. In the nineteenth century, for example, many Irish immigrants came to the United States to escape the Irish potato famine of 1845 to 1852. Similarly, many Chinese began migrating to the United States during this same time because of the population explosion and food shortages in China. Pull factors are those which draw people to a new country, such as employment opportunities, political and religious freedom, and better living conditions, including access to education and health care. Pull factors drew people to the United States during all four major periods of immigration, as people came seeking work, freedom, and a better way of life.

2. Human Systems What are the reasons for immigration to the United States? Define push factors and pull factors and list at least three examples of each.

How has immigration shaped human geography in the United States?

Successive waves of immigrants have absorbed, challenged, and ultimately reshaped American culture. As people move, their cultural traits and ideas move with them, creating and modifying the cultural landscape in their new home. These imprints on the United States include spoken languages, religious beliefs and institutions, family and ethnic traditions, architectural styles, art, music, and food. These cultural modifications can be seen in the ethnic neighborhoods of both small and large cities where grocery stores, churches and other places of worship, schools, restaurants, and other businesses reflect the immigrant culture of each community. Examples of ethnic neighborhoods include Little Tokyo in Los Angeles, California; Little Havana in Miami, Florida; Polish Hill in Pittsburgh, Pennsylvania; and Arabian Village in Detroit, Michigan.

3. Human Systems Write a paragraph describing three examples of how immigrants have left their imprint on the cultural landscape of your community or a place you have visited.

(tl)©Bob Sacha/Corbis, (tc)©Howard Davies/Corbis, (tr)Jim West/Alamy

THERE'S MORE ONLINE

VIEW a graph of the ethnic makeup of the United States • **WATCH** a video about U.S. ethnic diversity

networks

There's More Online!

- ☑ **DIAGRAM** Western Topography
- ☑ **INFOGRAPHIC** Hurricanes in the United States
- ☑ **MAP** The Continental Divide and the Fall Line
- ☑ **INTERACTIVE SELF-CHECK QUIZ**
- ☑ **VIDEO** Physical Geography of the United States

Reading HELPDESK

Academic Vocabulary

- **shift**
- **alter**

Content Vocabulary

- **tributary**
- **headwaters**
- **divide**
- **fall line**
- **hurricane**
- **fossil fuel**

TAKING NOTES: *Key Ideas and Details*

IDENTIFYING Use a graphic organizer like the one below to take notes on the physical geography of the United States as you read.

Physical Geography of the United States		
Landforms	Water Systems	Climate, Biomes, and Resources

LESSON 1
Physical Geography of the United States

ESSENTIAL QUESTION · *How do physical systems and human systems shape a place?*

IT MATTERS BECAUSE

With over 3.5 million square miles (9 million sq. km) of land, the United States is the third-largest country in the world. Its natural environment is diverse and makes the country one of the world's most productive regions.

Landforms

GUIDING QUESTION *How has tectonic activity helped create so many of the landforms in the United States?*

Many of the landforms of the United States can be traced back to glacial activity and the tectonic plate movement of the Earth's crust. The Pacific and Rocky Mountain ranges in the west and the Appalachian Mountains in the east are the result of powerful tectonic plate activity. The tectonic forces **shifted** giant rock slabs upward.

Considered young in geologic terms, the Pacific Ranges of the United States consist of the Sierra Nevada, the Cascade Range, the Coast Range, and the Alaska Range. Mount McKinley in the Alaska Range, at 20,320 feet (6,194 m), is the highest point in the United States. The Rocky Mountains begin in New Mexico and stretch northward over 3,000 miles (4,828 km). Between the Pacific Ranges and the Rocky Mountains is an area of plateaus and dry basins that was formed by volcanic lava seeping upward through cracks in the Earth's crust. These lava flows **altered**, or changed, the land forming the Columbia Plateau. Farther south are the flat-topped mesas of the Colorado Plateau and the spectacular gorge of the Grand Canyon, which plunges more than a mile into the Earth at its deepest points.

Extending eastward from the Rockies, the landscape flattens considerably to form the Great Plains, which stretch from 300 miles (483 km) to over 700 miles (1,126 km) wide. The land of the plains is higher in the west and slopes downward until it reaches the Central Lowlands. The plains continue eastward to the base of the Appalachian Mountains, the oldest mountain range on the North American continent. This range extends 1,500 miles (2,414 km), from Canada into the state of Alabama. As tectonic plates within the Earth's crust collided and pushed upward, they formed the Appalachians. The resulting peaks were shaped further by ice and water. Between the Appalachian Mountains and the Atlantic Coastal Plain lies the Piedmont, a low-rolling, fertile plateau cut by many rivers.

The Hawaiian Islands, located about 2,400 miles (3,862 km) off the western coast of the mainland United States, were formed when magma erupted from a spot on the seafloor, called a hot spot. This hot spot created a string of 8 major and 124 smaller islands that make up the Hawaiian Island chain.

shift to change the place, position, or direction of

alter to change partly

☑ **READING PROGRESS CHECK**

Explaining How were the Pacific Ranges formed?

Water Systems

GUIDING QUESTION *How have rivers and lakes been important to the economic development of the United States?*

Many lakes, rivers, and **tributaries** play a crucial role in many aspects of life. The Mississippi River is one of the longest rivers in North America. It flows 2,350 miles (3,782 km) from its **headwaters**, or source, in Minnesota and reaches a width of 1.5 miles (2.4 km) at its mouth, where it empties into the Gulf of Mexico. The Colorado River and the Rio Grande both have their headwaters in the Rocky Mountains where many tributaries merge to form these two major waterways.

A physical feature called a **divide** determines the direction of river flow. The Continental Divide is a high ridge in the Rocky Mountains. Waterways to the west of the divide flow into the Pacific Ocean. Waterways to the east of the divide flow toward the Arctic Ocean, Hudson Bay, Atlantic Ocean, and Mississippi River system.

In the eastern United States, the **fall line** marks the place where the higher land of the Piedmont drops to the lower Atlantic Coastal Plain. Along the fall line, eastern rivers break into rapids and waterfalls, preventing ships from the Atlantic Ocean from traveling farther inland. Many cities, such as Philadelphia, Baltimore, and Washington, D.C., were established along the fall line.

tributary a smaller river or stream that feeds into a larger river

headwaters the source of a stream or river

divide a high point or ridge that determines the direction rivers flow

fall line a boundary in the eastern United States where the higher land of the Piedmont drops to the lower Atlantic Coastal Plain

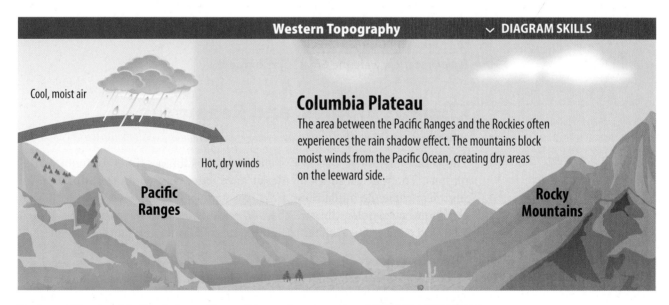

Western Topography ⌄ **DIAGRAM SKILLS**

Cool, moist air

Hot, dry winds

Pacific Ranges

Columbia Plateau
The area between the Pacific Ranges and the Rockies often experiences the rain shadow effect. The mountains block moist winds from the Pacific Ocean, creating dry areas on the leeward side.

Rocky Mountains

The shape and location of the Cascade Range affect conditions on the Columbia Plateau.

▲ **CRITICAL THINKING**

1. Drawing Conclusions What type of vegetation would you expect to find in the flat plateau areas between mountain ranges?

2. Analyzing How does the rain shadow effect impact the Columbia Plateau?

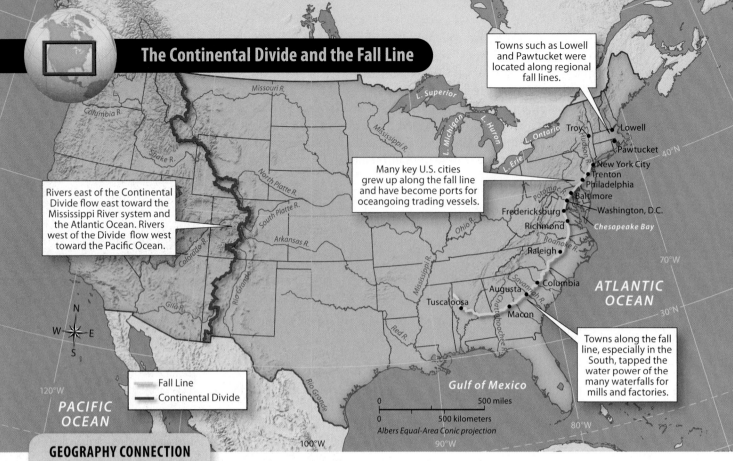

Towns such as Lowell and Pawtucket were located along regional fall lines.

Rivers east of the Continental Divide flow east toward the Mississippi River system and the Atlantic Ocean. Rivers west of the Divide flow west toward the Pacific Ocean.

Many key U.S. cities grew up along the fall line and have become ports for oceangoing trading vessels.

Towns along the fall line, especially in the South, tapped the water power of the many waterfalls for mills and factories.

Fall Line
Continental Divide

0   500 miles
0   500 kilometers
Albers Equal-Area Conic projection

GEOGRAPHY CONNECTION

The Continental Divide and the fall line are important to the flow of rivers and river traffic in the United States.

1. **PHYSICAL SYSTEMS** How does the Continental Divide affect the flow of rivers in the western United States?

2. **PLACES AND REGIONS** Describe how the fall line influenced economic development in the eastern United States.

Formed when glacier basins filled with water, Lake Superior, Lake Huron, Lake Erie, Lake Ontario, and Lake Michigan make up the Great Lakes. The glaciers uncovered major deposits of natural resources, including iron ore and coal, that later spurred explosive economic growth. The Great Lakes serve many economic and recreational purposes, but none more valuable than the Great Lakes–St. Lawrence Seaway System, a series of canals, rivers, and waterways linking the Great Lakes with the Atlantic Ocean. The seaway helped make cities along the Great Lakes, such as Chicago, powerful trade and industrial centers.

☑ **READING PROGRESS CHECK**

Explaining How were the Great Lakes formed?

Climate, Biomes, and Resources

GUIDING QUESTION *What factors cause variations in climate and vegetation in the United States?*

The United States has a variety of climates. The climates differ for a number of reasons. The high latitudes of Alaska have long, cold winters and brief, mild summers while the midlatitude areas have temperate climates. Places with high elevation have cooler climates than do those with low elevation. The United States even has tropical climates in the states of Hawaii and Florida.

Climate Regions and Biomes

The Southeast has a humid subtropical climate that is rainy with long, muggy summers and mild winters. Since it borders large bodies of water—the Atlantic Ocean and Gulf of Mexico—there is no dry season. Maple, oak, and pine trees are plentiful, and many types of mammals, reptiles, and amphibians are common.

Wetlands and swamps such as Florida's Everglades shelter a great variety of vegetation and wildlife. In late summer and early autumn, **hurricanes**—ocean

hurricane a large, powerful windstorm that forms over warm ocean waters storms hundreds of miles wide with sustained winds of about 74 miles per hour (119 km per hour) or more—can pound the region's coastlines.

The climate of the Great Plains reflects its location in the center of the continent. Because this area is far from the moderating influences of the ocean waters, it experiences very cold winters and hot summers. This is known as a continental climate. Moreover, parts of this area have a humid continental climate because they receive significant precipitation. This interior climate also extends into the hills and plateaus between the Mississippi River and the Appalachian Mountains. In the Great Plains and the eastern United States, violent spring and summer thunderstorms called supercells often spawn tornadoes with winds that can reach 200 miles (339 km) per hour.

Some areas west of the Great Plains have a semiarid climate with a mixture of vegetation, depending on latitude and elevation. These are transitional climates that occur between the humid continental climates and the arid climates of the Colorado Plateau. Animals in semiarid regions include deer, bison, coyotes, and wolves.

To the west of the semiarid regions, dry air moves down the leeward side of the mountains, creating an arid climate. This is called the rain shadow effect. Plants in arid climates, such as scrub bushes and cacti, have developed long root systems and other adaptations that allow them to survive with little water.

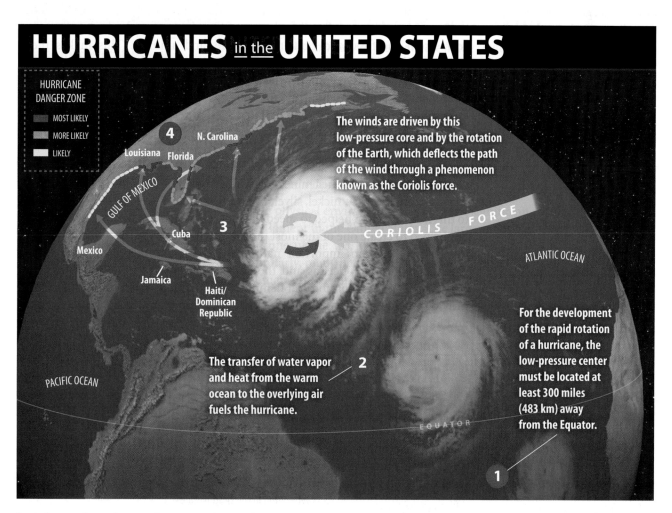

HURRICANES in the UNITED STATES

HURRICANE DANGER ZONE

- MOST LIKELY
- MORE LIKELY
- LIKELY

4 N. Carolina

Louisiana Florida

GULF OF MEXICO

Cuba **3**

Mexico

Jamaica

Haiti/ Dominican Republic

The winds are driven by this low-pressure core and by the rotation of the Earth, which deflects the path of the wind through a phenomenon known as the Coriolis force.

CORIOLIS FORCE

ATLANTIC OCEAN

For the development of the rapid rotation of a hurricane, the low-pressure center must be located at least 300 miles (483 km) away from the Equator.

The transfer of water vapor and heat from the warm ocean to the overlying air fuels the hurricane.

2

PACIFIC OCEAN

EQUATOR

1

Tropical storms that are known as hurricanes in the United States are called typhoons in the western North Pacific Ocean and cyclones in the South Pacific and Indian Oceans.

▲ **CRITICAL THINKING**

1. **Interpreting** Describe the air pressure inside the core of a hurricane and how proximity to the Equator affects the development of this type of storm.

2. **Analyzing Visuals** Which parts of the United States are most vulnerable to damage from hurricanes?

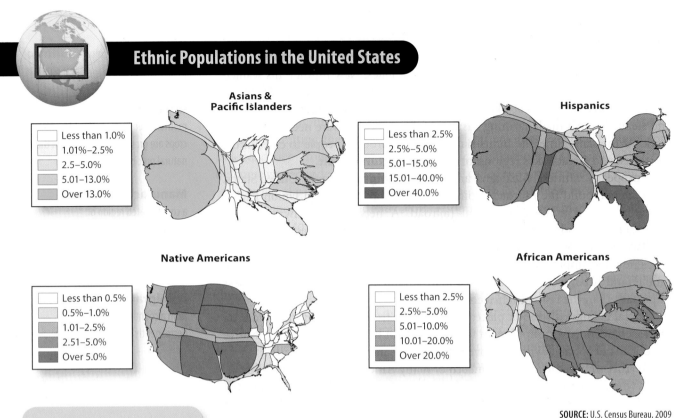

Asians & Pacific Islanders

Less than 1.0%
1.01%–2.5%
2.5–5.0%
5.01–13.0%
Over 13.0%

Hispanics

Less than 2.5%
2.5%–5.0%
5.01–15.0%
15.01–40.0%
Over 40.0%

Native Americans

Less than 0.5%
0.5%–1.0%
1.01–2.5%
2.51–5.0%
Over 5.0%

African Americans

Less than 2.5%
2.5%–5.0%
5.01–10.0%
10.01–20.0%
Over 20.0%

SOURCE: U.S. Census Bureau, 2009

GEOGRAPHY CONNECTION

The United States Census captures the country's ethnic diversity.

1. **PLACES AND REGIONS** Which region of the United States has the greatest percentage of Native Americans, the Central Plains or the East?

2. **PLACES AND REGIONS** How would you characterize the distribution of the Asian American population in the United States?

Sunbelt a mild climate region in the southern and southwestern portions of the United States

Population Patterns

GUIDING QUESTION *What factors influence population patterns in the United States?*

More than 315 million people live in the United States today. While about 2.5 million are Native Americans, a majority are immigrants or the descendants of immigrants from Europe, Asia, Africa, and Latin America. Some arrived only recently, while others belong to families whose ancestors came to the region centuries ago.

The average population density of the United States is about 87 people per square mile (33 people per sq. km). Outside of large urban areas, however, the population is widely distributed. The Northeast and Great Lakes regions are densely populated because they are the historic centers of industry and commerce. The Pacific coast attracts people looking for a mild climate and economic opportunities, resulting in a population cluster there. The least densely populated areas of the country include the subarctic region of Alaska, the dry Great Basin, and parts of the arid and semiarid Great Plains.

The population structure of the United States is changing. The U.S. Census Bureau projects that the population aged 65 and older will likely grow from some 40.2 million in 2010 to about 88.5 million by the year 2050. This increase presents challenges to the federal government as costs for Social Security and Medicare rise. The health care sector, the business sector, and families will also be affected.

Since the 1960s, the Manufacturing Belt has suffered a decline in population and economic strength as manufacturers relocated. Many businesses moved to **Sunbelt** states in the South and Southwest. Over the years, as mechanized agriculture has required fewer workers, the United States also has experienced urbanization, the movement of people from rural areas to cities. Today, most people in the United States live in the country's 366 metropolitan areas. A metropolitan area is a city with a population of at least 50,000 people including outlying communities, called suburbs.

Many U.S. population clusters lie in coastal areas with strong economies linked to world trade. Pacific coast cities provide important links to the rest of the world, especially to the growing Asian economies. The **megalopolis** that stretches from Santa Barbara, California, to Mexico is also an important corridor for world trade. Along the Atlantic coast, a chain of closely linked metropolitan areas from Boston, Massachusetts, to Washington, D.C., form the Boswash megalopolis. The Great Lakes region has three megalopolises: one centered in Chicago, Illinois, another in Buffalo, New York, and a third in Detroit, Michigan.

megalopolis a large population concentration made up of several large and many smaller cities, such as the area between Boston and Washington, D.C.

☑ **READING PROGRESS CHECK**

Speculating Provide a specific example of how the aging of the population might affect Americans.

Society and Culture Today

GUIDING QUESTION *How has immigration influenced the culture of the United States?*

Today, the United States has one of the most diverse populations in the world. Some immigrants have come to the United States to seek political and religious freedom or to find economic opportunities. Others are fleeing wars or natural disasters. Rich natural resources, industry, and economic wealth make the United States an attractive destination.

Throughout history, immigrants have often faced discrimination, but they have invariably enriched their new country through their hard work and talents and by bringing greater cultural diversity to the country. In 2012 the Census Bureau reported that 13 percent of the total U.S. population was foreign born and that more than half of the foreign-born population came from Latin America.

Immigrants have contributed to the country's diverse religious beliefs. Since the country's founding, religious freedom has been a core value in the United States. Today, most Americans who are members of an organized religion are Christian, with the majority being Protestant. Judaism, Islam, and Buddhism are among the other religions practiced in the United States. About 16 percent of the U.S. population today is not affiliated with any organized religion.

Family and Status of Women

Although population patterns in the United States continue to change, the family remains a vital institution. About half of adults are married, but some are single people living alone, or single mothers or fathers living with their children. More and more women work outside the home. Women have also continued to make gains in college completion rates, exceeding the graduation rate of men for the people between the ages of 25 and 34.

The Arts

The history of music in the United States can be traced back to Native American traditions. Europeans later brought their own folk and religious music. At the start of the 1900s, a distinct form of music known as **jazz** developed in African American communities throughout the United States. Jazz blended African rhythms with European harmonies. By the second half of the century, country music and rock-and-roll had become popular, not only in North America but around the world. Blues, punk rock, and hip-hop all have their origins in the United States, though hip-hop also traces its roots to the dance hall musicians of Jamaica.

jazz musical form that developed in the United States in the early 1900s, blending African rhythms and European harmonies

Many styles of art can also be found in the United States. In the early 1900s, a group of American artists known as the Ashcan School painted the grim realities of urban America. In the mid-1900s, many artists adopted European abstract styles, which express artists' emotions and attitudes without depicting recognizable images.

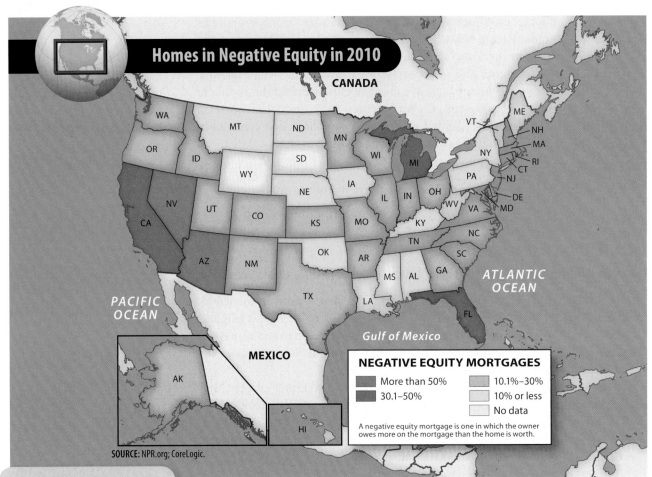

Homes in Negative Equity in 2010

CANADA

WA
MT
ND
MN
OR
ID
SD
WY
WI
MI
VT
ME
NH
NY
MA
NE
IA
IL
IN
OH
PA
CT
RI
NJ
NV
UT
CO
KS
MO
WV
VA
MD
DE
CA
AZ
NM
OK
AR
KY
TN
NC
SC
MS
AL
GA
TX
LA
FL

PACIFIC
OCEAN

AK

HI

MEXICO

Gulf of Mexico

ATLANTIC
OCEAN

NEGATIVE EQUITY MORTGAGES

- More than 50%
- 30.1–50%
- 10.1%–30%
- 10% or less
- No data

A negative equity mortgage is one in which the owner owes more on the mortgage than the home is worth.

SOURCE: NPR.org; CoreLogic.

GEOGRAPHY CONNECTION

During the foreclosure crisis, millions of Americans were unable to pay their home mortgages.

1. *PLACES AND REGIONS* Which states were least affected by negative equity mortgages?

2. *HUMAN SYSTEMS* Which states are shown with a more than 50% negative equity statistic?

The American-born graffiti art movement, begun by disenfranchised urban youth, has grown into a worldwide phenomenon. Graffiti is now created by artists who are commissioned by governments and private citizens to create enormous and intricate works of art.

✓ **READING PROGRESS CHECK**

Identifying What is the most common religion practiced in America today?

Economic Activities

GUIDING QUESTION *How is the U.S. economy an important part of the global economy?*

The United States has always been based on a free market economy and experiences ups and downs. Between the two world wars in the twentieth century, a long and devastating depression affected tens of millions of Americans. As the 1950s began, the manufacturing sector became a driving force of the economy. Manufacturers helped the economy by turning from wartime production to the manufacturing of cars, televisions, and appliances.

Today, the U.S. economy is a free market economy that allows people to profit from owning their own businesses. This freedom—coupled with laws that protect private property rights, employment opportunities, and the health and safety of workers—has created a great economic power. The country's wealth, measured in terms of gross national product (GNP), is due to universal education, technology and innovation, abundant natural resources, high agricultural output, and highly developed industries. The country has important reserves of natural gas and petroleum and is also a world leader in coal exports.

Resources, Power, and Industry

Agriculture in the United States has undergone many changes since the 1950s. The average size of farms has grown and continues to grow. Manufacturing has also evolved. The number of manufacturing jobs has declined while time efficiency and productivity have increased for most U.S. factories.

While agriculture and manufacturing are still important, the **postindustrial** economy is dominated by high-tech, biotechnology, and service industries. The service sector has grown more than any other part of the economy in recent decades. In the high-tech industry, California's Silicon Valley and cities such as Seattle, Washington, and Austin and Dallas, both in Texas, are leaders in software development. The North Carolina cities of Raleigh, Durham, and Chapel Hill form the Research Triangle region, known for attracting biotechnology companies.

In recent years, businesses in the United States have turned to offshoring, the practice of setting up plants abroad to produce parts or products for domestic use and international sale. While offshoring decreases the costs of goods, some people argue that it takes jobs away from American workers.

Good transportation and reliable communications are crucial to the economy of the United States. The automobile is still the most commonly used personal transportation method in the country. Its use has resulted in large investments in the building and maintenance of highways, roads, and bridges. The country also relies on air travel as a major method of transportation. A large percentage of the freight in the United States is transported by truck. The country's long-distance communications are carried via wireless, microwave, and satellite relays. Cellular and digital services have made mobile communication the norm, with fewer and fewer households using traditional landline telephones.

postindustrial economy that emphasizes services and technology rather than industry and manufacturing

The Economic Downturn

In 2008 the United States entered a serious economic downturn caused by an excessive number of ill-advised home mortgage loans. This resulted in a record number of homes going into **foreclosure**. The downturn was called the subprime mortgage crisis. At the same time, the stock market became unstable and unemployment rose. The bad loans led to the failure of some large banks and required large government investments to save many other banks. Because the U.S. economy and financial system are so important to the global economy, many other countries also spiraled into an economic downturn. The U.S. and global economies are improving but still growing less than is needed for robust health.

foreclosure legal proceeding in which a borrower's rights to a property are relinquished due to his or her inability to make payments on the loan

☑ **READING PROGRESS CHECK**

Describing What are the characteristics of the U.S. free market economy?

LESSON 2 REVIEW

Reviewing Vocabulary

1. *Defining* Define Sunbelt and megalopolis, and describe the locations of each.

Using Your Notes

2. *Identifying* Use the notes from your graphic organizer to write a paragraph describing the economic activities of the United States.

Answering the Guiding Questions

3. *Describing* How did physical geography and a spirit of independence influence the development of the United States?

4. *Explaining* What factors influence population patterns in the United States?

5. *Discussing* How has immigration influenced the culture of the United States?

6. *Expressing* How is the U.S. economy an important part of the global economy?

Writing Activity

7. *Informative/Explanatory* Write an essay describing the postindustrial economy of the United States.

Directions: On a separate sheet of paper, answer the questions below. Make sure you read carefully and answer all parts of the questions.

Applying Map Skills

Use the Unit 2 Atlas to answer the following questions.

16 *Physical Systems* Describe the general differences in elevation you encounter as you travel from California east to Oklahoma.

17 *Places and Regions* Using your mental map of the United States, list the bodies of water that surround Florida.

18 *Human Systems* What vegetation type is found in the most central part of the United States?

Exploring the Essential Question

19 *Making Connections* Imagine you are writing a science fiction novel about a person who has the power to change history by going back in time to sources that contributed substantially to two types of major pollution in the United States. Write a summary of the plot in your novel to explain the changes you would make to humans' actions—and the impact of these changes.

Research and Presentation

20 *Research Skills* Use the Internet and other resources to gather information about the effects of NAFTA on the physical environment of member countries. What cooperation issues other than trade does NAFTA cover? Create a multimedia presentation to share your findings.

Writing About Geography

21 *Informative* Use standard grammar, spelling, sentence structure, and punctuation to write a one-page essay that explains the effects of a specific type of pollution on your town or state. Discuss steps your town or state is taking to alleviate the problems.

DBQ Analyzing Primary Sources

Use the document to answer the following questions.

In the United States today, years of industrial emissions, automobile exhaust gases, and the extraction of natural resources have taken their toll on the environment. Curbing dependency on nonrenewable energy sources is an important issue. More and more people are exploring alternative energy sources that are both clean and renewable.

PRIMARY SOURCE

"*Freedom! I stand in a cluttered room surrounded by the debris of electrical enthusiasm: wire peelings, snippets of copper, yellow connectors, insulated pliers. For me these are the tools of freedom. I have just installed a dozen solar panels on my roof, and they work. A meter shows that 1,285 watts of power are blasting straight from the sun into my system, charging my batteries, cooling my refrigerator, humming through my computer, liberating my life.*

The euphoria of energy freedom is addictive. . . . Maybe that's because for me, as for most Americans, one energy crisis or another has shadowed most of the past three decades."

—Michael Parfit, "Powering the Future," *National Geographic,* August 2005

22 *Analyzing* What is the author installing and what will they be used for?

23 *Interpreting* What did Parfit mean by " the euphoria of energy freedom"?

Need Extra Help?

If You've Missed Question	16	17	18	19	20	21	22	23
Go to page	112	112	114	136	138	135	140	140

Canada

networks

There's More Online about Canada.

ESSENTIAL QUESTION • *How do physical systems and human systems shape a place?*

Geography Matters...

Although Canada borders the United States, many Americans do not know much about our neighbor. The world's second-largest country in land area, Canada stretches across the northern half of North America. The vast land is rich in resources, especially energy resources. It has many climate regions, each with different landscapes and peoples. Most Canadians, however, live in the southern part of the country, near the border with the United States.

◄ A Canadian woman plays ice hockey on a pond. Hockey is a popular sport in Canada.

Alexander Nicholson/Taxi/Getty Images

energy resources
and indigenous rights

*As an energy-rich country, **Canada** has massive coal reserves and enough petroleum and natural gas to supply its population. Extracting these resources, however, affects local populations. In Canada's northern frontier, this often pits the rights of the indigenous peoples who live there against those involved in the extraction.*

What are the economic benefits of extracting energy resources?

Canada has the third-largest proven oil reserves in the world after Saudi Arabia and Venezuela. Canada's economy depends on the United States as an export market. Pipelines connect Canadian oil production regions with refining and export centers in the United States. Canada also has a surplus of natural gas and electricity available to export.

Energy reserves fuel economic growth. Besides profiting from exports, the Canadian economy benefits from the jobs created by the extraction of fossil fuels. Canadian households and businesses also enjoy access to an abundant supply of affordable energy. All of these factors boost the country's economic growth.

1. **Environment and Society** Write a paragraph making the case for the extraction of an energy resource.

What are the human and environmental impacts of extracting energy resources?

The three main sources of oil production in Canada are the oil sands of Alberta, the conventional resources in western Canada, and the offshore oil fields in the Atlantic. The oil sands currently produce the majority of Canadian oil. These "tar sands" require traditional pit mining for surface deposits, but extracting deeper reserves involves the injection of steam underground. Oil sands extraction is very energy intensive and poses environmental risks, such as oil spills from pipelines.

The cost of extraction to indigenous communities has generally exceeded any benefit they receive. Land usually cannot be used for agriculture after extraction. The wildlife upon which many indigenous peoples rely for food may also be harmed by extraction. Indigenous peoples have even been forced to leave their land so companies could use the land for resource extraction.

2. **Human Systems** Describe the causes of conflict between resource extraction companies and indigenous peoples.

How have groups worked together to address this issue?

The Canadian government has addressed land rights of some indigenous communities for years. Currently, the government has ownership of their reserves. Indian Oil and Gas Canada is a government organization that manages and regulates oil and gas resources on indigenous peoples' lands.

The government has also cooperated with indigenous peoples to address conflicts and concerns over land rights. In 2012 it held hearings over a challenge by the Athabasca Chipewyan to a tar sands project proposed by Shell Oil Canada. The hearings provided an opportunity for the Athabasca Chipewyan to voice their concerns about the environmental damage the project might cause to their lands.

3. **Human Systems** How can groups work together to ease the negative impacts of resource extraction?

(tl)Norm Betts/Bloomberg/Getty Images; (tc)Jeff Malet Photography/Newscom; (tr)Rick Madonik/Newscom

A CLOSER LOOK:
ALBERTA

The Canadian province of Alberta is the source of most of Canada's mineral wealth. In addition to oil sands, it leads the nation in production of coal, natural gas, and crude oil. Alongside its resources, Alberta also has one of the largest indigenous populations in Canada. Nearly 200,000 indigenous people call Alberta home.

- ■ **CRUDE OIL & TAR SANDS**
- ■ **COAL**
- □ **NATURAL GAS**
- ▲ **INDIGENOUS SETTLEMENTS**

THE INDIGENOUS POPULATION

BLOOD TRIBE	SADDLE LAKE FIRST NATION	SAMSON CREE NATION	SIKSIKA NATION	LITTLE RED RIVER CREE NATION
7,555	5,883	5,550	3,634	3,341

Energy and Indigenous Rights in Alberta

There are over 5,000 square miles of indigenous reserves and settlements in Alberta. This, along with Alberta's vast mineral reserves, has led to some disputes over land rights and resource ownership. To attempt to resolve these disputes fairly, Alberta developed Canada's first mineral rights consultation policy in 2005.

VERY LOW   LOW   MED   HIGH   VERY HIGH

CANADA'S ENERGY PRODUCTION

TAR SANDS
1.5 MILLION BARRELS PER DAY
#1 RANK IN THE WORLD

OIL
1.2 MILLION BARRELS PER DAY
#20 RANK IN THE WORLD

COAL
68.2 MILLION METRIC TONNES PER YEAR
#14 RANK IN THE WORLD

NATURAL GAS
14.7 BILLION CUBIC FEET PER DAY
#3 RANK IN THE WORLD

netw**o**rks

There's More Online!

- ☑ **IMAGE** Grazing Bull Moose
- ☑ **MAP** Glaciers of the Last Ice Age
- ☑ **MAP** St. Lawrence Seaway System
- ☑ **INTERACTIVE SELF-CHECK QUIZ**
- ☑ **VIDEO** Physical Geography of Canada

LESSON 1
Physical Geography of Canada

ESSENTIAL QUESTION · *How do physical systems and human systems shape a place?*

Reading **HELP**DESK

Academic Vocabulary

- **series**
- **controversy**

Content Vocabulary

- **timberline**
- **chinook**
- **tar sands**
- **fishery**
- **overfishing**
- **aquaculture**

TAKING NOTES: *Key Ideas and Details*

CATEGORIZING As you read the lesson, use a graphic organizer like the one below to take notes on the physical geography of Canada.

Physical Geography of Canada		
Landforms	Water Systems	Climate, Biomes, and Resources

IT MATTERS BECAUSE

Canada is a country of great physical variety and natural wealth. This wealth includes breathtaking landforms shaped by water, wind, ice, and tectonic activity over millions of years. It also includes abundant water and energy resources as well as a variety of wildlife.

Landforms

GUIDING QUESTION *How do landforms link the geography of Canada and the United States?*

Canada covers approximately the northern third of North America, stretching from the Pacific to the Atlantic. It is made up of 3 territories and 10 provinces. A province is a political unit similar to a state.

Mountains on Canada's eastern and western edges cradle a central region of plains. When people first arrived on the plains, they found a sea of grass and dark, fertile soil that later became some of the world's most productive farmland. To the east of the plains stand the rocks of the Laurentian Highlands, the Canadian Shield, and the ancient, rounded Appalachian Mountains. To the west are the younger Rocky Mountains. A variety of climates are found in Canada, from the frozen tundra and cold, subarctic climates in the north to the steppe and humid continental climates along the border with the United States.

The western third of Canada is mountainous. Collisions between tectonic plates millions of years ago thrust up a **series** of sharp-peaked mountains called the Pacific Ranges. These ranges include the Cascade Range and the Coast Range. Mount Fairweather, on the border of Alaska and western British Columbia, at 15,300 feet (4,663 m) is one of the tallest coastal mountains in the world.

Like the Pacific Ranges, the Rocky Mountains farther east grew as geologic forces heaved slabs of rock upward. The Rocky Mountains link the United States and Canada, stretching more than 3,000 miles (4,828 km) from New Mexico to Alaska. The Canadian Rockies extend northward from the Rockies in the United States.

East of the Rockies, the land falls in elevation and flattens into the Great Plains and Interior Lowlands, which extend across the center of

(l)Chris Harris/age fotostock, (c)MPI/Archive Photos/Getty Images, (r)Streeter Lecka/Getty Images Sport/Getty Images the region and south into the United States. These plains and lowlands are generally flat, but diverse in terms of landscape types. They can include hills, escarpments, low mountains, forests, and river valleys. Water from these lowlands drains into the Atlantic Ocean through the St. Lawrence River. The region was shaped by glacial activity during the last ice age. It therefore has deep arable soil that makes it important for agriculture.

At the eastern edge of the Interior Lowlands lies the Canadian Shield. The Canadian Shield is a giant core of rock anchoring the continent and centered on Hudson Bay and James Bay. This stony land makes up the eastern half of Canada and parts of the northeastern United States. Erosion and glacial ice from the last ice age have smoothed the surface and created lakes, rivers, and streams.

The heavily eroded Appalachian Mountains are North America's oldest mountains. They extend from Canada's Maritime Provinces south through the eastern United States to the state of Alabama. The Appalachians were formed by tectonic plate movements. Over time they were shaped by ice and running water. Coastal lowlands lie to the east of the Appalachians.

☑ **READING PROGRESS CHECK**

Making Connections Why are landforms an important part of Canada's geography?

Water Systems

GUIDING QUESTION *How are water systems important to the Canadian economy?*

Freshwater lakes and rivers have helped make Canada prosperous. Abundant water satisfies the needs of cities and rural areas. It also provides power for homes and industries and moves resources across the country. The Mackenzie River, which flows from the Great Slave Lake to the Arctic Ocean, drains much of Canada's northern interior. The Fraser River flows southwest and drains into the Pacific Ocean just south of Vancouver.

series a number of things of the same type following one after the other in space or time

Glaciers of the Last Ice Age

GEOGRAPHY CONNECTION

Glaciers have dramatically changed the landscape of Canada. Due to Canada's location, it experienced widespread glaciation during the last ice age.

1. *THE WORLD IN SPATIAL TERMS* Which glacier was the largest during the last ice age?

2. *PHYSICAL SYSTEMS* What are the lasting results of the glaciers, as seen in the landscape today?

St. Lawrence Seaway System

0 — 300 miles
0 — 300 kilometers
Lambert Azimuthal Equal-Area projection

The St. Lawrence Seaway connects the St. Lawrence River and the Great Lakes with a series of 6 canals and 19 locks.

The Moses-Saunders Power Dam spans the St. Lawrence River. Half of the dam is owned by the U.S. and the other half by Canada.

More than 4,000 ships travel the St. Lawrence Seaway into and out of the Great Lakes each year carrying nearly 45 million tons of cargo.

- Port
- Canadian lock
- U.S. lock

1. St. Lambert & Côte Ste. Catherine
2. Lower Beauharnois & Upper Beauharnois
3. Snell
4. Eisenhower
5. Iroquois
6. Welland Canal (8 locks)
7. Soo Locks

GEOGRAPHY CONNECTION

The St. Lawrence Seaway System is an important link in Canada's transportation network.

1. **PLACES AND REGIONS** List the bodies of water that are connected by the St. Lawrence Seaway System.

2. **HUMAN SYSTEMS** How does the system contribute to the economy of the region?

The St. Lawrence River flows about 760 miles (1,223 km) from Lake Ontario to the Gulf of St. Lawrence in the Atlantic. It forms part of Canada's border with the United States. The cities of Quebec, Montreal, and Ottawa grew up along the St. Lawrence and its tributaries. Niagara Falls, on the Niagara River, forms another part of the Canada-U.S. border. The falls are a key source of hydroelectric power.

In northern Canada, naturally occurring dams created by glacial ice formed Great Bear Lake and Great Slave Lake. Elsewhere, moving glaciers tore at the earth, leaving behind basins that became the Great Lakes. Deposits of coal, iron ore, and other minerals nearby favored industrial development and urban growth here.

☑ **READING PROGRESS CHECK**

Identifying Central Issues How does water help the Canadian economy?

Climate, Biomes, and Resources

GUIDING QUESTION *What factors cause variations in climate and vegetation in Canada?*

Due to its great expanse of latitude, Canada experiences a large variation in climate and vegetation types. It varies from the bitter cold of the high-latitude areas to the radically changing seasons of the interior regions. Canada is characterized by biomes common to the midlatitude and high latitude.

Canada is rarely affected by natural weather hazards. The waters off the Pacific and Atlantic coasts are too cold to support hurricanes. Also, the extreme temperatures needed for thunderstorms, tornadoes, and hail are less common.

Russ Heinl/age fotostock

Climate Regions and Biomes

Ocean currents play a key role in Canada's climates. The Gulf Stream, for example, is a warm, northward-flowing ocean current off the southeastern coast of Canada. It moderates coastal temperatures and carries nutrients as far north as Newfoundland. Merging with the colder air and water of the southward-flowing Labrador Current off eastern Canada, the potential for fog-bound coastal conditions is always present. Off the Pacific coast, the cold Alaska Current flows southward and parallel to British Columbia. It churns up nutrients from the ocean floor, inviting whales to follow a moving feast during their migrations.

Climate and vegetation are variable in the southern third of Canada—from about 40° N to 50° N latitude. The area's humid continental climate ranges from hot and humid to cool and wet. The farther north one travels, the more severe and snowy the winters become, with shorter and cooler summers. Coniferous evergreen trees tend to outnumber deciduous trees in the northernmost areas.

Canada's Pacific coast has a marine west coast climate. The Pacific Ranges force moist ocean air upward, where it cools and releases moisture, causing heavy rainfall. Winters are overcast and rainy. Summers are cloudless and cool, appealing to many varieties of warblers and other small birds that go north to nest. Ferns, mosses, and coniferous forests grow here. The soils in Canada's forests are quite acidic. This is due to the leaching (removal) of minerals out of the topsoil during rainfall. With fewer minerals, the soils are unfit for agriculture. The exceptions are the soils in the mixed and deciduous forests.

Areas between the Pacific Ranges and Rocky Mountains have a drier climate due to the rain shadow effect. This area has a semi-arid (steppe) climate, with grasslands and coniferous forests. Spruce and fir trees cover the middle elevations of the ranges. Beyond the **timberline**, the elevation above which trees cannot grow, lichens (LY• kuhns) and mosses are found. These higher elevations of the Canadian Rockies and Pacific Ranges are high-latitude climate regions. Rocky Mountain sheep, mountain goats, elk, mule deer, and black bears are common in the southern latitudes of these mountains. The semi-arid (steppe) climate predominates east of the Rocky Mountains, again because of the rain shadow effect. In addition, in spring, a warm, dry wind called the **chinook** (shuh•NUK) blows down the eastern slopes of the Rockies and melts the snow.

In the north, there are subarctic, tundra, and ice cap climates. In these high-latitude climates, freezing temperatures are common all year because of a lack of direct sunlight. The amount and variety of vegetation is limited.

Boreal forest, also known as taiga, occupies the bulk of Canada's northern areas. It is the world's second-largest area of uninterrupted forest. The boreal forest has long winters and moderate to high precipitation each year. Covered mostly with coniferous evergreen trees, the region is an important source of pulpwood and lumber. The forest is also home to moose, black bears, beavers, Canada lynx, wolves, snowshoe hares, Canada jays, blue jays, crows, and ravens.

Just south of the Arctic Circle lie the subarctic climate regions. These regions have Earth's widest temperature ranges, varying by as much as 120°F (49°C) from their bitterly cold winters to cool, short summers. In parts, only a thin layer of surface soil thaws each summer. Below is permanently frozen subsoil, or permafrost.

timberline elevation above which it is too cold for trees to grow

chinook a seasonal warm wind that blows down the Rockies in late winter and early spring

tar sands sand or sandstone naturally impregnated with petroleum

fishery an area in which fish or sea animals are caught

A bull moose grazes in Bowron Lake Park in British Columbia.

▼ **CRITICAL THINKING**

1. **Analyzing Visuals** What features of the moose make it suited for its biome?

2. **Speculating** How might large populations of moose be a threat to the forests in some parts of Canada?

Chris Harris/age footstock

Biology

Using aquaculture, fish can be raised in concrete or earthen ponds, coastal pools, or floating cages in the open ocean, where they are protected from predators and environmental hazards. Aquaculture is an important industry in Canada. People raise Atlantic salmon, cod, tilapia, rainbow trout, and more. The industry provides income and jobs and can help meet the increasing global demand for seafood and restore threatened wild populations of species.

HYPOTHESIZING How might aquaculture provide economic benefits for countries lacking land-based resources for international trade?

overfishing harvesting fish to the point that species are depleted and the value of the fishery reduced

aquaculture the cultivation of seafood

controversy the presence of opposing views

A few needled evergreens survive here, along with herds of woodland caribou that are endangered in some areas.

Closer to the North Pole is the tundra climate region. Winter darkness and bitter cold last for months, and the brief summer period brings only limited warming effects. The layer of thawed soil is even thinner here than in the subarctic, with vegetation limited to low bushes, short grasses, mosses, and lichens. Winter temperatures can fall to –70°F (–57°C).

Natural Resources

Abundant natural resources such as energy, minerals, timber, and fish have made Canada wealthy. Their extraction, however, has led to depletion and environmental problems. Energy resources include coal, petroleum, and natural gas. Canada's petroleum and natural gas reserves lie largely in or near Alberta. The Athabasca **Tar Sands** contain large deposits of extremely heavy crude oil, much of which is in semisolid form. Large inputs of energy and water are required to transform these substances into synthetic crude oil for use by humans.

Mineral resources are also plentiful in Canada. The Rocky Mountains yield gold, silver, and copper. Parts of the Canadian Shield are rich in iron ore and nickel. Canada's mineral inventory includes 33 percent of the world's production of potash (a mineral salt used in fertilizers), 4 percent of its copper and gold, and 5 percent of its silver. Because mining involves heavy equipment, uses large quantities of water, and moves a great deal of rock and other natural materials, it can damage land, water, and air systems.

Timber is a vital resource in Canada. Trees once covered much of Canada. Today, forests cover only about 34 percent of the country. Trees are a renewable resource, but only if people take steps to protect forests and their ecosystems. Cutting down trees can have a negative effect on the other plants and animals. Many species have been lost. Positive efforts to preserve forests include replanting trees, cooperating to protect the many species of native forest animals facing extinction, and preserving old-growth forests.

The coastal Atlantic and Pacific waters have long been important **fisheries**, or places for catching fish and other sea animals. The Grand Banks, once one of the world's richest fishing grounds, covers about 139,000 square miles (360,000 sq. km) of Canada's southeast coast. In recent years, **overfishing** has caused fish stocks to drop dangerously. Canada is now working to protect key species. **Aquaculture**, or fish farming, is a growing economic activity. This has caused some **controversy**, as crowded conditions can lead to disease among farmed species. The industry is working to minimize risks to consumers.

✔ **READING PROGRESS CHECK**

Specifying What are Canada's natural energy resources?

LESSON 1 REVIEW

Reviewing Vocabulary
1. *Summarizing* Explain the significance of timberline, chinook, fishery, overfishing, aquaculture, and tar sands.

Using Your Notes
2. *Listing* Using your notes from the graphic organizer, describe the major Canadian landforms.

Answering the Guiding Questions
3. *Applying* How do landforms link the geography of Canada and the United States?

4. *Analyzing* How are water systems important to the Canadian economy?

5. *Identifying* What factors cause variations in climate and vegetation in Canada?

Writing Activity
6. *Informative/Explanatory* Write a short paragraph explaining how tectonic forces have influenced landscapes in Canada.

networks

There's More Online!

☑ **GRAPH** Annual Growth Rate of Canadian Industries

☑ **IMAGE** Inuit People

☑ **TIME LINE** Canada: Expansion and Diversity

☑ **INTERACTIVE SELF-CHECK QUIZ**

☑ **VIDEO** Human Geography of Canada

LESSON 2
Human Geography of Canada

ESSENTIAL QUESTION · *How do physical systems and human systems shape a place?*

Reading HELPDESK

Academic Vocabulary

- **trace**
- **advocate**

Content Vocabulary

- **Inuit**
- **First Nations**
- **dominion**
- **Quebecois**
- **separatism**
- **Loyalist**
- **emigrate**

TAKING NOTES: *Key Ideas and Details*

IDENTIFYING As you read the lesson, use a graphic organizer like the one below to take notes on the human geography of Canada.

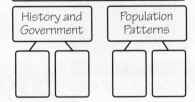

Human Geography of Canada

History and Government

Population Patterns

(tl)©Photo Collection Alexander Alland, Sr./Corbis, (tcl)©Jeff Greenberg/Alamy, (tcr)Eric Baccega/age fotostock, (tr)©Bettmann/Corbis, (tc)Boomer Jerritt/All Canada Photos/Getty Images,

IT MATTERS BECAUSE

Diversity has been important in shaping Canada. The indigenous peoples who have lived on the land for centuries, the French and British colonists who settled the area later, and the waves of immigrants from Europe, Asia, Africa, and Latin America have all influenced the human geography of this country.

History and Government

GUIDING QUESTION *How did British and French cultures influence Canada's history?*

Many Canadians can **trace** their roots back to countries from around the world. Some are descendants of the country's native peoples. Their earliest ancestors arrived in North America thousands of years ago.

Unity, Expansion, and Diversity

About 200,000 native peoples were living in what is now Canada when Europeans arrived off the coast of Newfoundland in 1497. In the next 200 years the native populations declined, as Europeans claimed their lands and diseases from Europe spread. After 1950 high birthrates and access to improved medical care contributed to population growth.

The main indigenous groups that exist in Canada today are the **Inuit**, the Métis, and **First Nations** peoples. The Inuit are the indigenous people of the Canadian Arctic. Métis are persons who have both indigenous and French Canadian ancestry. *First Nations* is a term that refers to indigenous peoples of Canada who are neither Inuit nor Métis. The current population of First Nations peoples is about 700,000 of the 1.25 million indigenous people in Canada today.

French explorers helped establish claims to land in the region in the early 1600s. This land was named New France, and part of it became the province of Quebec. Territorial rivalry between Great Britain and France began in 1670, when the British chartered the Hudson's Bay Company to seek a northwest passage to the Pacific Ocean. A clash of French and British interests along the Atlantic coast led to wars.

Britain and France were first drawn to the North American continent to gain riches from precious metals and beaver pelts that came mainly from the Atlantic coast. Exploration inland soon followed, however.

trace to follow or study in detail or step by step

Inuit a member of the Arctic native peoples of North America; once known as Eskimo

First Nation one of the indigenous peoples of Canada who are neither Inuit nor Métis

dominion a partially self-governing country with close ties to another country

Quebecois Quebec's French-speaking inhabitants

Both European powers sought to claim more resources and more colonies. The British eventually drove the French from the Hudson Bay area, capturing Quebec in 1759 and winning control of New France in 1763. The Quebec Act, passed by the British in 1774, gave French settlers the right to keep their language, religion, and laws. The act also extended Canadian territory south to the Ohio River. This angered American colonists and brought them closer to war with the British.

During the early 1800s, English- and French-speaking communities feuded over colonial government policies. Fears of a takeover by the United States, however, forced both to work together. In 1867 the colonies of Quebec, Ontario, Nova Scotia, and New Brunswick united as provinces of the **Dominion** of Canada, a new country within the British Empire. Manitoba, British Columbia, Alberta, Saskatchewan, Prince Edward Island, and Newfoundland became provinces over the next 100 years.

Canada was created as a dominion, a partially self-governing country with close ties to Great Britain. It gained full independence in 1931, but the British government kept the right to approve changes to Canada's constitution. This legislative link to Great Britain finally ended in 1982 with passage of the Constitution Act. Today, Canada is a constitutional monarchy.

The executive part of Canada's federal government includes the governor-general, the prime minister, and the cabinet. The British monarch serves as the head of state and appoints a governor-general to act in his or her place. The national legislature, or Parliament, includes the Senate and the House of Commons. Canada's prime minister is the actual head of government. Nine judges sit on the Supreme Court of Canada, the country's highest court.

In the 1800s, Canada acquired lands stretching from the Atlantic to the Pacific and from the Arctic to the U.S. border. The British government encouraged immigration to Canada. Between 1815 and 1855, a million people arrived from Great Britain. This made the French-speaking citizens a minority and fueled French nationalism among **Quebecois** (kay•beh•KWAH), Quebec's French-speaking inhabitants. This nationalism has been present throughout Canada's history.

Widespread immigration from other parts of the world began in Canada in the late 1800s. Some came for the Klondike Gold Rush, but many more were attracted by the fertile soil of the prairies of Alberta, Saskatchewan, and Manitoba.

©Photo Collection Alexander Alland, Sr./Corbis

TIME LINE ⌄

CANADA
Expansion and Diversity ➔

Canada's history features steady expansion and multiculturalism.

▶ CRITICAL THINKING

1. Identifying What are two events that show how or why Canada's settlement expanded?

2. Explaining How has Canada attempted to address past mistreatment of native peoples?

1776 ➔ .... **➔1900** ..

1800s — Thousands of immigrants from England, Scotland, and Ireland arrive each year.

1776 — Loyalist refugees from the American War of Independence settle in Canada.

1885 — Canadian Pacific Railroad is completed.

1898 — Gold rush along upper Yukon River; Yukon Territory is given separate status.

Immigrants from Germany, Scandinavia, Ukraine, Japan, and China arrived to settle the land. In the 1800s Canada also sheltered African Americans who had escaped slavery in the United States. Canada never practiced slavery. These refugees, many of whom escaped via the Underground Railroad, were safely beyond the reach of American laws once they arrived in Canada.

Westward expansion in Canada came at a price, however, as immigrants pushed First Nations peoples off their lands. The injustice was formally recognized in 1998, when the Canadian government apologized to native peoples for their mistreatment. The government established a "healing fund" to make reparations.

In the 1900s Canada became an industrialized, urban country. Mineral resources were utilized, and hydroelectric projects and transportation systems were developed. World War II stimulated the economy, making it a crucial military and industrial power. After the war, Canada sought to improve federal assistance to its citizens through pensions, unemployment insurance, and medical care.

Modern Challenges

The United States has more trade with Canada than any other country. Being neighbors and sharing a border has permitted international trade that is beneficial to the economies of both countries. In 1994 the North American Free Trade Agreement (NAFTA) eliminated tariffs and other trade barriers between Canada, Mexico, and the United States. Although the open border and a history of cooperation have benefited both countries, some Canadians continue to dislike the effect free trade has had on their culture. Canadians struggle to maintain a separate identity while being bombarded by U.S. popular culture.

Conflicts continue as French-speaking Canadians seek greater protection for their language and culture. Many desire Quebec's independence and strongly support **separatism**—the breaking away of one part of a country to create a separate, independent country. In 2012 members from the Parti Québécois, a political party in Quebec that **advocates**—or publicly supports—separatism, were elected to political office in Quebec.

separatism the breaking away of one part of a country to create a separate, independent country

advocate to publicly recommend or support

✓ **READING PROGRESS CHECK**

Identifying What factors led to the decline of the native populations of Canada?

(c)Jeff Greenberg/Alamy, (c)Boomer Jerritt/All Canada Photos/Getty Images, (b)©Bettmann/Corbis

Great Britain transfers final legal powers to Canada. A new constitution is adopted.

1982

1998 A referendum rejects Quebec's independence from Canada by 1 percent.

Truth and Reconciliation Commission begins hearings about past policies harmful to indigenous peoples and their cultural identities.

2010

➔**2012**

1992 North American Free Trade Agreement (NAFTA) enacted

1999 Nunavut is formed, becoming the first Canadian territory with a majority indigenous population.

2012 Prime Minister Harper calls for fresh start in government relations with First Nations peoples.

Population Patterns

GUIDING QUESTION *What explains Canada's diverse mix of people and settlement patterns?*

Canada is a highly developed country with bustling cities. It also has sparsely populated areas of beautiful, pristine wilderness. These rugged natural areas sometimes have difficulty supporting communities. But they are as much a part of the country's cultural identity as are the busy metropolitan areas.

Nunavut is the cultural and economic homeland for the Inuit people. Today, about 25,000 Inuit from the Arctic live mostly in settlements of 25 to 500 people, with just a few larger towns. In recent decades, mining, oil exploration, and pipeline construction, along with a decline in the demand for fur, have increased their reliance on government services.

Immigrants to Canada came in search of political and religious freedom, economic and educational opportunities, and refuge from wars. For example, **Loyalists**, or American colonists who remained loyal to the British government, fled to Canada after the American Revolution. They settled in the Maritime Provinces of Nova Scotia, New Brunswick, and Prince Edward Island. Some immigrant groups settled in areas that let them keep their familiar ways of life.

The ethnic origins of Canadians vary from province to province. Today, more than one-fourth of Canadians identify themselves as being of mixed ethnic origins. In addition, more than 1 million or 4 percent of all Canadians identify themselves as of North American Indian, Inuit, or Métis ancestry.

Rugged terrain and a cold arctic climate with a short growing season make much of Canada difficult for human settlement. About 90 percent of the population lives within 100 miles (160 km) of the U.S.-Canada border. Average population density is about 9 people per square mile (3 people per sq. km). More densely populated areas are clustered near the coasts, the Great Lakes, and in places that support agriculture, fishing, and trade. Over the past 100 years, most internal migration has been westward to the Prairie Provinces of Manitoba, Saskatchewan, and Alberta. This move was due in part to the discovery of oil and natural gas since the 1960s.

According to 2012 estimates, 70 percent of the country's population is between the ages of 15 and 64. Since Canadians enjoy a long life expectancy and low infant mortality rates, Canada has an aging population. This presents economic challenges, as a larger share of the population reaches retirement and requires more government-funded health care.

Approximately 81 percent of Canada's 34 million inhabitants live in urban areas. These urban areas developed in places with intensive commercial agriculture in the Great Lakes and St. Lawrence lowlands areas. Smaller settlements were merged into larger metropolitan areas. Today, these cities are important centers for commerce, education, and transportation. As Canada's largest city, Toronto is an industrial and financial center. Montreal is an industrial and shipping center. On the Pacific coast, Vancouver handles nearly all of the trade between Canada and Asia. Edmonton grew with the development of the petroleum industry.

☑ READING PROGRESS CHECK

Paraphrasing How has Canada's physical geography influenced its settlement patterns?

Loyalist an American colonist who remained loyal to the British government

Nunavut, home to many Inuit people, was established as a separate territory in 1999.

▼ CRITICAL THINKING

1. ***Describing*** What kinds of settlements do many Inuit people live in?

2. ***Drawing Conclusions*** Why do you think it was important for the Inuit people to have Nunavut declared a separate territory?

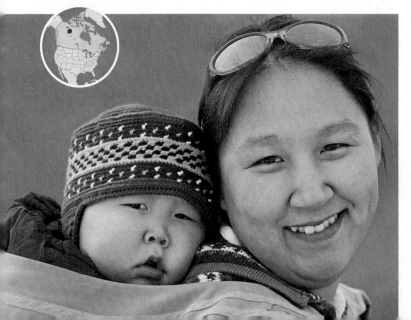

Eric Bacega/age fotostock

Society and Culture Today

GUIDING QUESTION *Why is Canada often called a multicultural society?*

Canada has a truly multicultural society, largely due to immigration. There are many reasons people decide to **emigrate**, or leave their countries to settle in another. Canada's high standard of living and government policies protecting multiculturalism are appealing to people of diverse ethnic origins. Historically these immigrants have arrived primarily from Europe. Recent migration patterns, however, feature large increases in immigration from Asia and Central and South America, and an overall decrease in immigration from Europe.

Language and religion reflect Canada's diverse immigrant population. There are two official languages—English and French. The British introduced English to most of Canada. In the province of Quebec, however, French was established as the dominant language. Languages spoken also include German, Italian, and Chinese. Native languages include Cree and Inuktitut, the language of the Inuit. Christians make up the largest religious group in Canada. Other religions practiced in Canada by small minorities include Islam, Buddhism, Hinduism, Judaism, and Sikhism.

Education and health care are supported by the government. The literacy rate in Canada is 99 percent, and there is an extensive network of public and private schools. Attending school is required for children ages 6 to 16. Each province is responsible for organizing and administering public education. The federal government sets standards, with each province responsible for financing and managing its own system. In some cases, rising costs resutling in part from an aging population have created a need to limit benefits or raise taxes.

Family and Status of Women

Today, the average Canadian family contains three members. Two parents live in 67 percent of Canadian families, but the percentage is decreasing just as in the United States. The overall composition of the Canadian family is changing and families have fewer children. Since women are increasingly joining the Canadian workforce, there are a growing number of dual-income households.

The government of Canada has passed several laws protecting the rights of women. The Status of Women in Canada (SWC) was created in 1971 to increase women's participation and equality in all aspects of life. In 1982 the Canadian Charter of Rights and Freedom ensured gender equality in employment, public life, and education. Canadian women have the same literacy rate as men and outnumber men in getting secondary and university-level educations. They still, however, face barriers to equality in the labor market.

The Arts

The arts in Canada have been influenced primarily by native and European cultures. For example, First Nations artists combine old and modern techniques with traditional color schemes and motifs that often relate to the natural world. In the twentieth century, museums and scholars began to focus on art created by native peoples. Immigration has also added distinctive features to Canadian art.

French explorers, missionaries, and settlers wrote the earliest Canadian literature. Their writings had strong historical and religious themes. Popular modern Canadian fiction writers include Margaret Atwood, author of *The Handmaid's Tale,* and Yann Martel, author of *Life of Pi.*

Toronto is highly regarded in the realms of theater and music. It is the third-largest production center in the English-speaking world, after London and New York City. The world-renowned Toronto Symphony Orchestra and the top-ranked National Ballet of Canada call Toronto home.

Analyzing PRIMARY SOURCES

Canada Slips in Gender Equality Rankings

"Canada is losing ground in a key global measure of gender equality, sliding out of the world's top 20 chiefly due to a lack of female representation in politics.

'In the future, talent will be more important than capital or anything else,' said Klaus Schwab, the forum's [World Economic Forum] founder and executive chairman. 'To develop the gender dimension is not just a question of equality; it is the entry card to succeed and prosper in an ever more competitive world.' "

—Tavia Grant, *The Globe and Mail,* October 24, 2012

DBQ *IDENTIFYING CENTRAL ISSUES* What is the chief cause of Canada's slide from the world's top 20 in gender equality?

emigrate to leave one's own country to settle permanently in another

Canadian industries generally faced some difficult times as a result of the 2008 global recession, but have since shown signs of bouncing back.

▶ **CRITICAL THINKING**

1. Contrasting Which industry's performance is most unlike the others?

2. Summarizing What has been the general trend in Canadian industry in recent years?

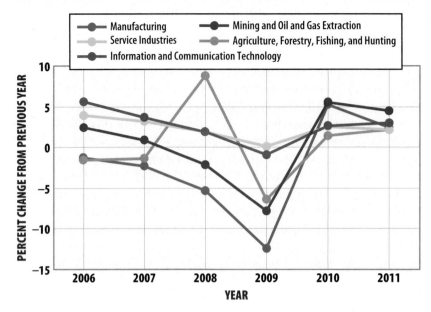

Source: Statistics Canada, 2012

Canada is also at the forefront of contemporary art. The "Vancouver School" artists focus on photoconceptualism, and the Nova Scotia College of Art and Design is known for innovation in photography and graphic design.

☑ **READING PROGRESS CHECK**

Explaining How has the government worked to improve the status of women?

Economic Activities

GUIDING QUESTION *What factors define Canada as a postindustrial market economy?*

Like the United States, Canada maintains a market economy. In 2009, 58 percent of Canada's exports were made up of agricultural, energy, forestry, and mining products. Thus, natural resources are essential to the Canadian economy. Canada also has the second-largest oil reserves in the world and makes a substantial amount of money from exporting oil and natural gas.

Resources, Power, and Industry

Canada is a leader in the production of uranium, iron ore, coal, petroleum, copper, and silver. Canada is the world's third-largest producer of hydroelectricity, relies heavily on coal-fueled power, and can meet its own petroleum needs while still having a surplus of oil and natural gas. These surpluses are exported worldwide. The tar sands in Alberta are a major source of crude oil.

Natural resources and agriculture make up over 60 percent of Canada's exports. Despite this, those economic activities provide Canada with only about 4 percent of its GDP. Services such as transportation and communication, retail, health care, and others provide about 70 percent of GDP. The Canadian economy produces goods and services at home as well as being linked to the global economy through imports, exports, and trade agreements such as NAFTA.

Canada continues to be one of the world's important suppliers of agricultural products, specifically wheat, corn, and other grain crops. Canada's Prairie Provinces are major grain producers. They are also home to many cattle ranches.

The fishing industry used to heavily support the economy of the Atlantic coast. In the past few decades, however, overfishing has reduced this industry's impact. Forestry is important for certain provinces, such as British Columbia.

The last few decades have seen a rapid growth of high-tech and electronics industries. Canada has 28 million Internet users—82 percent of its population. Another high-tech industry in Canada is its space and aircraft industry.

The development of reliable transportation systems has been essential to the economic growth of Canada due to its large land area. The majority of Canada's transportation systems are located in the southern part of the country. The Trans-Canada Highway runs 4,860 miles (7,821 km) from Victoria, British Columbia, to St. John's, Newfoundland. Communication networks have also promoted development of the economy. Cellular and digital services using satellites have made telephone communication more accessible in distant places. Business transactions and personal communications can be completed instantaneously using e-mail and the Internet.

The Economy Today

Since Canada's economy is linked to the global market, it suffered some losses in the 2008 global recession. However, it was not as seriously affected as the United States. This was largely because Canadian banks were more conservative in extending credit. The housing market stayed healthy as a result. Also, Canada's federal deficit is one of the smallest in the Western world. Unemployment rates have been generally lower than those in the United States and much of Europe.

Trade is still central to the Canadian economy. In addition to forestry products, leading exports are automobiles, auto parts, crude petroleum, natural gas, electricity, aluminum, machinery, and equipment. Canada tends to import manufactured goods, such as chemical products, textiles, and foods. Exports to the United States amount to over half of all Canadian trade. Canada exports more to the United States than it imports from it, resulting in a trade surplus.

While natural resources and trade are important to Canada's economy, the service sector employs more people than all other sectors combined, with 70 percent of the over 18 million Canadians in the labor force. For this reason, Canada is considered a postindustrial society. The service industry includes retail jobs in the increasingly common large chain stores such as Walmart and Best Buy. Tourism is also a fast-growing service industry because Canada offers natural and cultural diversity in addition to national parks and historic sites.

☑ **READING PROGRESS CHECK**

Analyzing Why was Canada less affected than the United States by the 2008 global recession?

LESSON 2 REVIEW

Reviewing Vocabulary
1. *Making Connections* Use the following vocabulary terms in a paragraph about the early history of Canada: First Nation, Inuit, Loyalist, and dominion.

Using Your Notes
2. *Describing* Use your graphic organizer to describe the ethnic makeup of Canada today.

Answering the Guiding Questions
3. *Sequencing Information* How did British and French cultures influence Canada's history?

4. *Outlining* What explains Canada's diverse mix of people and settlement patterns?

5. *Inferring* Why is Canada often called a multicultural society?

6. *Determining Importance* What factors define Canada as a postindustrial market economy?

Writing Activity
7. *Narrative* Suppose you are an immigrant writing a letter to relatives about your new home in Canada. Explain your reasons for settling where you live.

Two Decades of NAFTA

In 1994 Canada, the United States, and Mexico signed the North American Free Trade Agreement (NAFTA) to reduce or eliminate tariffs on many goods traded among the three countries. Since its implementation, NAFTA has had mixed results. While it has helped to increase trade throughout North America, the gains for the three countries have not been equal. Supporters of NAFTA point to lower product prices and increased industrial integration among the countries. Opponents of the agreement argue it has led to job displacement and has done little to improve the quality of life in Mexico.

CANADA–U.S.
$315.3 Billion

U.S.–CANADA
$280.9 Billion

MEXICO–CANADA
$24.6 Billion

U.S.–MEXICO
$198.4 Billion

MEXICO–U.S.
$262.9 Billion

CANADA–MEXICO
$5.5 Billion

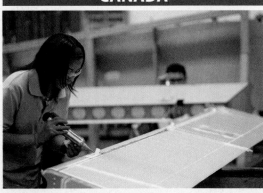

CANADA

> Few Canadians speak up for enhancing ties with Mexico. But before leaving Ottawa this month, Emilio Goicoechea, Mexico's ambassador, wrote a rebuttal urging Canada to stay the trilateral course. Trade between the two has grown fivefold since 1994.... Some Canadian companies have invested in Mexico: Bombardier has factories making aircraft parts and trains, while Scotiabank is Mexico's seventh-biggest bank.
>
> —The Economist
> "No mariachis, please," February 12, 2009

UNITED STATES

> [A] worker ... whose 16.6% tariff was eliminated by Nafta, saw wage growth rate about 11 percentage points less than a worker in a business that didn't depend on high tariffs. So ... Nafta did hurt workers at the lower end of education spectrum. Blue-collar workers in vulnerable industries suffered large absolute declines in real wages as a result of the agreement....
>
> —Bob Davis
> "The Battle Over NAFTA Continues,"
> Wall Street Journal, November 22, 2010

MEXICO

> ...NAFTA has made it easier to buy American products in Mexico. I can shop at H-E-B and other stores in Mexico and buy Procter & Gamble and other American-made products that we could not buy in Mexico before NAFTA. When I lived in Mexico and wanted American products before NAFTA, the taxes imposed on American products made them unaffordable.
>
> —Eduardo Bravo
> "NAFTA Has Fueled Job Growth,"
> SanAntonio Express-News, November 19, 2012

Making Connections

1. **Places and Regions** Which NAFTA member country exports the largest volume of goods to the other two member countries?

2. **Human Systems** How might a person's role in the economy, such as an industrial worker, a business owner, or a policy maker, impact his or her views on NAFTA?

3. **The Uses of Geography** Using what you know about NAFTA, write a one-page letter to the editor of a local newspaper explaining why you think the overall effect of NAFTA has been positive or negative.

*Interact with **Global Connections** Online*

(t)Chico Sanchez/Alamy, (c)H. Mark Weidman Photography/Alamy, (b)Russell Gordon/Alamy

netw⊙rks

There's More Online!

☑ **IMAGE** Human Impact on Wildlife

☑ **INFOGRAPHIC** Canada's Boreal Forest

☑ **INTERACTIVE SELF-CHECK QUIZ**

☑ **VIDEO** People and Their Environment: Canada

LESSON 3
People and Their Environment: Canada

Reading HELPDESK

Academic Vocabulary

- **extract**
- **diminish**

Content Vocabulary

- **old-growth forest**

TAKING NOTES: *Key Ideas and Details*

IDENTIFYING As you read, use a web diagram like the one below to outline issues that arise as a result of interactions between the Canadian people and their environment.

ESSENTIAL QUESTION · *How do physical systems and human systems shape a place?*

IT MATTERS BECAUSE
While Canada once had an abundance of wildlife and natural resources, human activity has caused many of these resources to decline and some to even disappear. Human activities also cause pollution and contribute to climate change.

Managing Resources

GUIDING QUESTION *How do economic activities in Canada put the country's natural resources at risk?*

Canada engages in numerous economic activities that involve **extracting**, or removing, natural resources. Logging, which involves cutting down trees for human use, has resulted in more trees being cut down in a given amount of time than can grow. For example, a logger can cut down 100 trees in one day. However, it will take decades for the same number of trees to grow back. Logging, when left unchecked, can lead to the complete destruction of forests and the animals that live in those forests.

Canada's boreal forest is one of the largest forest and wetland ecosystems remaining on Earth. It is home to some of the biggest populations of wolves, grizzly bears, and caribou. It also has lakes that support many types of fish and trees that shelter billions of birds.

The boreal forest has been shrinking because of logging, mining, and oil and gas extraction. As of 2011, only 10 percent of it was federally protected. Many nature advocates and conservation groups are calling on the Canadian government to protect more of the boreal forest from these devastating activities, which can deplete the animal and plant populations. Parts of the boreal at high risk are old-growth forests.

Old-growth forests are complex forests that have developed over a long period of time and are relatively untouched by human activity. They are increasingly rare and also need to be protected.

The same is true for wetlands. Home to large numbers of plant and animal species, wetlands are considered one of the most productive ecosystems in the world. Wetlands cover about 14 percent of Canada,

(tc)Dennis Macdonald/Photolibrary/Getty Images, (tcr)©Jenny E. Ross/Corbis, (tr)Education Images/UIG/Getty Images mainly in Ontario, Manitoba, and the Northwest Territories. Just like old-growth forests, Canadian wetlands are at risk. Wetlands have been drained so the land can be used for industrial, commercial, and agricultural purposes. If wetlands are not protected, even more will be destroyed.

Overfishing is caused by catching more fish than an ecosystem can replenish naturally. As with logging, if the fish cannot multiply at the same rate as they are caught, then the fish supply will seriously **diminish**, or become less. Overfishing can lead to the extinction of fish species if it is not stopped or regulated. This could have happened with Pacific salmon, which migrate in a route that takes them through waters in both Canada and the United States. Overfishing by both countries seriously depleted the salmon stock, and fishing industries in both countries suffered. In 1999 Canada and the United States signed an agreement to promote salmon conservation and harvest-sharing principles. While conservation efforts and fishery regulation have been implemented, some species of Pacific salmon are still endangered.

extract to remove

old-growth forest complex forest that has developed over a long period of time and is relatively untouched by human activity

diminish to make less or cause to appear less

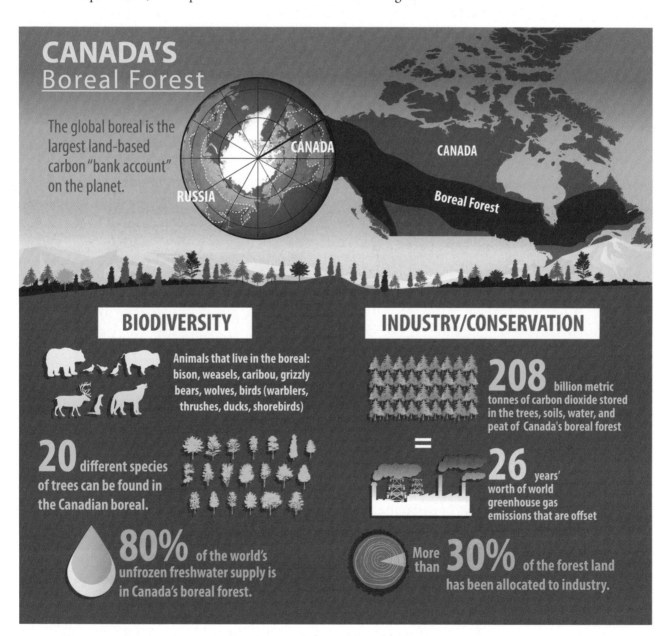

CANADA'S Boreal Forest

The global boreal is the largest land-based carbon "bank account" on the planet.

RUSSIA
CANADA
CANADA
Boreal Forest

BIODIVERSITY

Animals that live in the boreal: bison, weasels, caribou, grizzly bears, wolves, birds (warblers, thrushes, ducks, shorebirds)

20 different species of trees can be found in the Canadian boreal.

80% of the world's unfrozen freshwater supply is in Canada's boreal forest.

INDUSTRY/CONSERVATION

208 billion metric tonnes of carbon dioxide stored in the trees, soils, water, and peat of Canada's boreal forest

=

26 years' worth of world greenhouse gas emissions that are offset

More than **30%** of the forest land has been allocated to industry.

Canada's boreal forest in the far north latitudes is situated just south of the treeless tundra of the polar region.

▲ **CRITICAL THINKING**

1. Analyzing Explain how Canada's boreal forest affects all life on Earth.

2. Assessing What are some threats facing Canada's boreal forest today?

Canada is the world's third-largest producer of hydroelectricity—electricity generated from flowing or falling water. While the cost of hydroelectricity is low, making it a great source of renewable energy, it often requires damming. Many environmental groups and indigenous communities protest dam building because damming interrupts the flow of rivers and impacts local ecosystems. In some cases, these protests have succeeded in persuading companies to change their plans and work toward projects that do less harm to local ecosystems.

The continued extraction of natural resources—particularly coal, oil, and natural gas from the tar sands in Alberta—fuels a growing concern among many people over climate change. It requires a large amount of energy to convert the heavy crude oil found in the tar sands into the more useful synthetic crude oil. Furthermore, the process of converting synthetic crude oil into usable substances, such as gasoline and kerosene, is also energy intensive and releases large amounts of carbon dioxide into the air. Additionally, exporting these substances via airplanes and ships releases still more carbon dioxide. Carbon dioxide is believed to contribute directly to climate change and the rising of the Earth's temperatures. Like people around the world, many Canadians are looking for ways to develop renewable sources of energy that do not contribute to climate change.

☑ **READING PROGRESS CHECK**

Assessing What natural resources are most at risk in Canada?

ANALYZING PRIMARY SOURCES

Acid Rain: Sugar Maples at Risk

As acid rain and other effects of air and water pollution escalate, the risk to plant life is increasing. In addition to damaging existing plants, acid rain can slow the growth of new plant life or, as scientists are learning by studying sugar maple trees in Canada, stop new growth altogether. Over time, entire forests can be obliterated as a result of acid rain destruction.

❝Sugar maple abundance already has dropped in parts of the northeastern U.S. and southeastern Canada over the past 40 years, primarily because of high acid levels in soils.

The upper Great Lakes region has mostly escaped the damage because its soils are rich in calcium, which provides a buffer against acid. But in an article published this month in the *Journal of Applied Ecology,* scientists say they've discovered another way that acid rain harms sugar maple seedlings in upper Great Lakes forests.

Donald Zak of the University of Michigan says acid rain prevents dead maple leaves from decaying on the forest floor, creating a barrier that hampers growth of new trees.❞

—"Acid Rain Could Kill Maples Near Great Lakes," Associated Press, December 16, 2011

Industrial smokestacks emit pollutants, including sulfur and nitrogen oxides, that rise into the upper atmosphere, leading to acid rain.

DBQ ▲ **CRITICAL THINKING**

1. ***Identifying Central Issues*** According to Donald Zak, how does acid rain interfere with seedling development?

2. ***Speculating*** Maple syrup from sugar maple trees is an important export product in parts of Canada. How might acid rain affect the economy in areas where maple syrup is harvested?

PHOTO: Dennis Macdonald/Photolibrary/Getty images; TEXT: Acid rain could kill maples near Great Lakes by The Associated Press. © The Associated Press. All rights reserved.

©Jenny E. Ross/Corbis

Canadian Wildlife Service scientists evaluate the condition of a tranquilized mother polar bear and her young triplet cubs near western Hudson Bay. This is part of an effort to help protect animals living in Canada's boreal forest.

▲ **CRITICAL THINKING**

1. *Speculating* What data do you think the scientists are checking to assess the health of the polar bear and her cubs?

2. *Hypothesizing* How might the bears' health be an indicator of the health of their habitat?

Human Impact

GUIDING QUESTION *How do human activities impact the environment in Canada?*

Human activities, such as overfishing, logging, and mining, can seriously impact the environment in Canada. Agricultural, industrial, mining, and forestry activities have also contributed to pollution of the ocean waters. Acid rain is another serious threat that can even contaminate lakes, streams, and rivers far from the source of the pollution. These acids can come from natural sources, such as volcanoes, but human activities such as burning fossil fuels also release such acids into the air. Acid rain causes lakes, streams, and rivers to be contaminated. Since Canada is connected to the United States through a number of wind and water systems, pollution in the United States is negatively affecting the environment in Canada. For example, emissions in the United States can result in acid rain in Canada, threatening timber and water resources.

Acid rain contributes to water pollution, but it is not the only threat to water supplies. Many companies and factories put their waste, such as heavy metals and pesticides, into water sources. Water pollution affects plants, animals, and small organisms living in the water. Contaminated water can kill trees, plants, and animals. It can also be harmful to humans. The United States and Canadian governments worked together to produce the Great Lakes Water Quality Agreement and the Clean Water Act in the 1970s to address the serious issue of water pollution that threatens the Great Lakes. The result has been greatly improved water quality and the resurgence of some fish populations.

☑ **READING PROGRESS CHECK**

Describing How have human activities affected water systems in Canada?

Addressing the Issues

GUIDING QUESTION *How is the Canadian government working to address environmental issues?*

In Canada, climate change is an issue that the provinces, not the federal government, address. In 2009 Canada was ranked seventh in total greenhouse gas emissions, behind China, the United States, India, Russia, Japan, and Germany.

The main sources of greenhouse gas emissions are transportation, electricity generation, and producing and refining fossil fuels and petroleum. In 2011 Canada pulled out of the Kyoto Protocol. The Kyoto Protocol is an international agreement signed by 37 industrialized countries that agreed to reduce their greenhouse gas emissions. The Canadian government failed to meet its goals and chose to remove itself from the agreement.

Individual provinces have taken actions to address the climate change issue. In 2009 Ontario passed sweeping legislation to support renewable energy and promote conservation. The new law—the Green Energy and Green Economy Act (GEGEA)—promotes the development of renewable "green" energy projects within the province. The act also advocates for efficient energy and energy conservation efforts in homes, schools, and offices. GEGEA offers all of Ontario's residents—including home owners and large companies—financial incentives to develop small- and large-scale renewable energy sources. As an added benefit, it was projected to produce some 50,000 jobs, both directly and indirectly, in its first three years. Ontario is also the leader in Canadian wind and solar energy projects. So far, Nova Scotia is the only province besides Ontario to set formal goals for wind power projects. Such projects can attract billions of dollars in international investments and also help create jobs.

Hydroelectricity is the most efficient source of renewable energy in Canada and accounts for more than half of all the electricity produced. It is far from a perfect solution, however. The places suitable for the installation of hydroelectric facilities are limited. In addition, the damming of rivers can seriously disrupt the surrounding ecosystems.

The Canadian government has begun to explore wind and solar power options for energy production in the southern provinces. Organizations outside the government are also working hard to address environmental issues in Canada—and to put pressure on the Canadian government to address them more effectively.

There are positive efforts to report. Canada is taking steps to protect its renewable resources. Protection of Pacific salmon is one example, although it remains to be seen whether those efforts will enjoy ultimate success. Still, the effort illustrates how the government and others can take aggressive action to protect and preserve precious natural resources.

☑ **READING PROGRESS CHECK**

Describing What kinds of renewable energy sources is the Canadian government exploring?

LESSON 3 REVIEW

Reviewing Vocabulary
1. *Defining* Provide a definition and example for the following term: old-growth forest.

Using Your Notes
2. *Listing* Using your web diagram, identify a way in which Canadians are impacting their environment and one way in which they are trying to manage that impact.

Answering the Guiding Questions
3. *Identifying* How do economic activities in Canada put natural resources at risk?

4. *Evaluating* How do human activities impact the environment in Canada?

5. *Identifying Central Issues* How is the Canadian government working to address environmental issues?

Writing Activity
6. *Informative/Explanatory* With a partner, select one natural resource from the following list: water, oil, natural gas, and coal. Write a paragraph describing why this resource is valuable. How has human activity affected this resource, and what are the ways the Canadian government can do more to address the conservation of this resource?

Directions: On a separate sheet of paper, answer the questions below. Make sure you read carefully and answer all parts of the questions.

Lesson Review

Lesson 1

① *Describing* Describe how tectonic forces have played a role in shaping Canada's physical geography.

② *Discussing* Provide two examples of Canadian lakes that were formed by glacier activity and describe the process.

③ *Explaining* How does Canada's climate change as one travels farther north?

Lesson 2

④ *Assessing* Describe the ways in which physical geography and natural resources have influenced Canadian culture and economy.

⑤ *Finding the Main Idea* How have the cultures of various indigenous groups influenced Canada's cultural diversity?

⑥ *Describing* Describe some ways that the Canadian family has changed in recent years.

Lesson 3

⑦ *Summarizing* What efforts have been made to promote renewable "green" energy by the Canadian government?

⑧ *Making Inferences* Why is it in the best interest of industries to use natural resources responsibly?

⑨ *Evaluating* Why do many environmental groups and indigenous communities feel that hydroelectricity does more harm than good?

Exploring the Essential Question

⑩ *Diagramming* Create a cause-and-effect chart that shows how physical features and climate have influenced the human settlement of Canada. How have these relationships shaped Canada today?

21st Century Skills

Review the circle graph. Then answer the questions that follow.

CHINESE IMPORTS FROM CANADA BY SECTOR, 2011 (as a share of total imports from Canada)*

Source: Global Trade Atlas (Chinese statistics)
*May not add up to 100% due to rounding.

⑪ *Using Graphs, Charts, Diagrams, and Tables* China imports the most products from Canada in which two categories?

⑫ *Geography Skills* What does the high percentage of exports of wood imply about Canada's physical geography?

Applying Map Skills

Use the Unit 2 Atlas maps to answer the following questions.

⑬ *Physical Systems* List all the climate regions found in Canada and describe their location in relation to latitude.

⑭ *The World in Spatial Terms* Which parts of Canada are the most densely populated and which are the least densely populated? Why do you think this is the case?

⑮ *Human Systems* Use your mental map of Canada to describe the relative location of Ottawa.

Need Extra Help?

If You've Missed Question	❶	❷	❸	❹	❺	❻	❼	❽	❾	❿	⓫	⓬	⓭	⓮	⓯
Go to page	144	146	146	149	153	153	162	158	162	147	163	163	116	112	147

Directions: On a separate sheet of paper, answer the questions below. Make sure you read carefully and answer all parts of the questions.

Critical Thinking

16 *Making Connections* What is the connection between the semi-arid (steppe) climate region and the Pacific Ranges?

17 *Drawing Inferences* What can you infer from the fact that the Canadian government has recently begun working with indigenous populations to ensure that their culture and ways of life are preserved?

18 *Exploring Issues* What are different methods for preventing overfishing and the destruction of old-growth forests? Be sure to include examples of what a government can do and what a citizen can do.

19 *Evaluating Counter Arguments* Given the threat of environmental damage from reliance on fossil fuels, why do you think Canada continues to rely on fossil fuels for energy and as a commodity for trade?

College and Career Readiness

20 *Decision Making* You are a recent college graduate seeking a teaching job in Canada. Based on your knowledge of the landforms, climates, and cultural features of Canada, in which area of the country would you choose to live and why? Explain your answer using details from the chapter.

Writing About Geography

21 *Argument* Environmentalists working on resource management in Canada face many challenges, such as overfishing, loss of old-growth forests, acid rain, and climate change. Which challenge do you feel is the most significant? Write a one-page essay explaining your position. Be sure to describe the problem and to explain why you feel it is the most significant.

DBQ Analyzing Primary Sources

Use the document to answer the following questions.

PRIMARY SOURCE

"A conservation group is warning against allowing too much fishing of sockeye salmon on the Fraser River despite expectations that this year's run will be one of the biggest in 100 years.

Society director Craig Orr said in a release headlined 'Don't succumb to sockeye fever' that most of this year's sockeye are from one place—the Adams River—while stocks from many other sources are severely depleted.

'We should all rejoice in this year's bounty, but remember that returns of Fraser River sockeye in this decade have been extremely low for reasons not yet understood,' said Orr.

Environmentalist Vicky Husband, an advisor to Watershed Watch, said the future of the whole fishery ecosystem has to be taken into account.

'We've endured a century of over-fishing and collapse of smaller sockeye populations,' said Husband."

—"Salmon Overfishing Warning Issued," CBC News, September 1, 2010

22 *Identifying Cause and Effect* Why does Husband feel that overfishing threatens the entire fishery ecosystem?

23 *Making Inferences* Why does Orr use the phrase "sockeye fever" to warn against overfishing of salmon on the Fraser River?

Research and Presentation

24 *Research Skills* Use the Internet to research the major immigrant groups who came to Canada during different periods of its history. As you read, take notes. Use your notes to share what you learned with a classmate.

Salmon over-fishing warning issued by The Canadian Press; © The Canadian Press, 2010. All rights reserved.

Need Extra Help?

If You've Missed Question	**16**	**17**	**18**	**19**	**20**	**21**	**22**	**23**	**24**
Go to page	147	151	158	142	146	149	164	164	152

Latin America

UNIT **3**

©Imageplus/Corbis

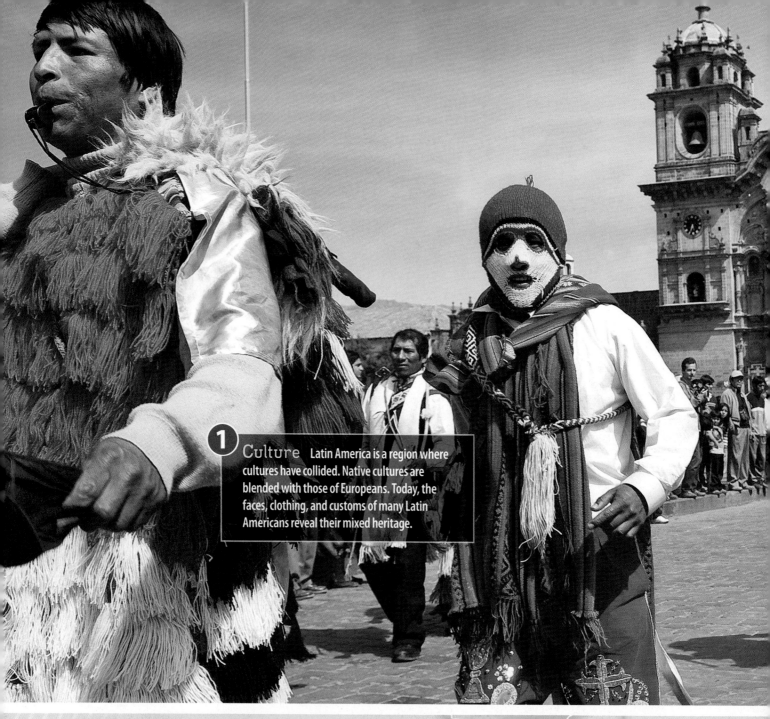

① Culture Latin America is a region where cultures have collided. Native cultures are blended with those of Europeans. Today, the faces, clothing, and customs of many Latin Americans reveal their mixed heritage.

EXPLORE the REGION

Spanning more than 85 degrees of latitude, **LATIN AMERICA** encompasses Mexico, Central America, the Caribbean, and South America. It is a region of startling physical contrasts, from the high peaks of the Andes to the lush rain forests of the Amazon. Latin America's human geography reflects a shared colonial legacy, but does present some contrasts between urban and rural, rich and poor, more developed and less developed.

THERE'S MORE ONLINE

HUGHES HervA©/hemis.fr/Getty Images

(cr)Paul Harris/AWL Images/Getty Images,
(bl)Luiz Felipe Castro/Flickr/Getty Images,
(br)Ken Hornbrook/Flickr/Getty Images

3 Rain Forests Like a snake slithering through the grass, the Nanay River meanders through the Peruvian rain forest.

4 Mountains The jagged peaks of Chile's Torres del Paine are part of the Andes, the world's longest mountain chain.

2 Cities Many people migrate to urban centers such as Mexico City where work offers the possibility of economic advancement. This rapid growth forces cities to look for ways to provide their growing populations with necessary resources.

Latin America
Physical

UNITED STATES

Bermuda Islands

ATLANTIC OCEAN

40°N

Gulf of California

SIERRA MADRE OCCIDENTAL

Rio Bravo

SIERRA MADRE ORIENTAL

MEXICAN PLATEAU

Gulf of Mexico

Baja California

Yucatán Peninsula

SIERRA MADRE DEL SUR

TROPIC OF CANCER

20°N

Bahamas

Cuba

Greater Antilles

Hispaniola

WEST INDIES

Jamaica

Puerto Rico

Guadeloupe

Martinique

Caribbean Sea

Lesser Antilles

Mosquito Coast

Lake Nicaragua

Lake Maracaibo

Trinidad

Isthmus of Panama

Orinoco R.

Angel Falls

ANDES

Llanos

Guiana Highlands

N W E S

1,000 miles

1,000 kilometers

Lambert Azimuthal Equal-Area projection

EQUATOR

Galápagos Islands

Rio Negro

AMAZON BASIN

SELVAS

Amazon R.

Madeira R.

Tapajós R.

Xingu R.

Araguaia R.

Tocantins R.

Marajó Island

Cape São Roque

Caatinga

São Francisco R.

Elevations

10,000 ft. (3,000 m)
5,000 ft. (1,500 m)
2,000 ft. (600 m)
1,000 ft. (300 m)
0 ft. (0 m)
Below sea level

—— National boundary
▲ Mountain peak
Lowest point

La Montaña

Lake Titicaca

Altiplano

MATO GROSSO PLATEAU

BRAZILIAN HIGHLANDS

Campos

PACIFIC OCEAN

Atacama Desert

ANDES

Gran Chaco

Paraguay R.

Paraná R.

Cape São Tomé
Cape Frio

20°S

TROPIC OF CAPRICORN

Aconcagua 22,834 ft. (6,960 m)

PAMPAS

Uruguay R.

Colorado R.

Río de la Plata

UNIT 3
REGIONAL ATLAS

MAP STUDY

1. Environment and Society What physical features could present barriers to the development of Latin America?

2. Human Systems What European countries still control territory in Latin America?

Juan Fernández Islands

Chiloé Island

PATAGONIA

Valdés Peninsula -131 ft. (-40 m)

Falkland Islands (Islas Malvinas)

ATLANTIC OCEAN

Strait of Magellan

Tierra del Fuego

South Georgia Island

Cape Horn

140°W   120°W   100°W   80°W   60°W   40°W   20°W

60°S

Latin America
Political

UNITED STATES

Ciudad Juárez
Tijuana
Chihuahua
Monterrey
Gulf of California
MEXICO
Guadalajara
Mexico City
Puebla
Veracruz
Orizaba

Gulf of Mexico

BAHAMAS
Nassau
Havana
CUBA
*Cayman Is.
(U.K.)*
Port-au-Prince
DOMINICAN REPUBLIC
Santo Domingo
BELIZE
Belmopan
JAMAICA
HAITI
Kingston
Guatemala
HONDURAS
GUATEMALA
Tegucigalpa
San Salvador
NICARAGUA
Managua
EL SALVADOR
San José
Panama
COSTA RICA
PANAMA

Bermuda (U.K.)

TROPIC OF CANCER

Caribbean Sea

Aruba (Neth.)   *Neth. Antilles (Neth.)*
Caracas
Medellín
VENEZUELA
Bogotá
Cali
COLOMBIA
Port-of-Spain
TRINIDAD & TOBAGO
GUYANA
Georgetown
Paramaribo
Cayenne
SURINAME
FRENCH GUIANA (Fr.)

Orinoco R.

Quito
ECUADOR
Galápagos Islands (Ecuador)

EQUATOR

Río Negro   *Amazon R.*
Manaus
Marajó Island
Belém

PERU
Lima
Madeira R.
Tocantins R.
São Francisco R.
Fortaleza
BRAZIL
Recife

BOLIVIA
Lake Titicaca
La Paz
Santa Cruz
Brasília
Salvador
Arequipa
Sucre
Belo Horizonte

PACIFIC OCEAN

TROPIC OF CAPRICORN

Paraguay R.
Paraná R.
PARAGUAY
Asunción
São Paulo
Rio de Janeiro
Curitiba

ARGENTINA
Uruguay R.
Pôrto Alegre
Valparaíso
Rosario
URUGUAY
Santiago
Buenos Aires
Montevideo
Juan Fernández Islands (Chile)
Río de la Plata
CHILE
Bahía Blanca

ATLANTIC OCEAN

Strait of Magellan

Falkland Islands (Islas Malvinas)
Administered by United Kingdom (Claimed by Arg.)

South Georgia Island (U.K.)

| 0 | 1,000 miles |
| 0 | 1,000 kilometers |

Lambert Azimuthal Equal-Area projection

◈ National capital
○ Department capital
• Major city

Inset (Caribbean)

Virgin Islands (U.S.)   *British Virgin Islands (U.K.)*
Puerto Rico (U.S.)
Anguilla (U.K.)
St. Martin (Fr.)
St. Maarten (Neth.)
St. Barthélemy (Fr.)
ANTIGUA AND BARBUDA
Saba (Neth.)
St. Eustatius (Neth.)
Montserrat (U.K.)
Guadeloupe (Fr.)
ST. KITTS AND NEVIS
DOMINICA
Martinique (Fr.)
Caribbean Sea
ST. LUCIA
ST. VINCENT AND THE GRENADINES
BARBADOS
| 0 | 200 mi. |
| 0 | 200 km |
GRENADA
TRINIDAD & TOBAGO

Latin America
Climate and Vegetation

Climate

- Tropical rain forest
- Tropical wet/dry
- Semi-arid (steppe)
- Arid (desert)
- Humid subtropical
- Marine west coast
- Mediterranean
- Tundra and high altitude

Vegetation

- Tropical rain forest
- Tropical grassland (savanna)
- Desert scrub and desert waste
- Temperate grassland
- Mediterranean scrub
- Deciduous forest
- Coniferous forest
- Mixed forest (deciduous and coniferous)
- Highland (vegetation varies with elevation)

UNITED STATES

Gulf of Mexico

Tijuana

Mexico City

San Salvador

Panama

Nassau

Caribbean Sea

Caracas

Bogotá

Paramaribo

Manaus

Belém

Lima

La Paz

Brasília

Asunción

Rio de Janeiro

Santiago

Buenos Aires

PACIFIC OCEAN

ATLANTIC OCEAN

ATLANTIC OCEAN

TROPIC OF CANCER

EQUATOR

40°N

20°N

0°

20°S

N
W E
S

0   1,000 miles
0   1,000 kilometers
Lambert Azimuthal Equal-Area projection

TROPIC OF CANCER

EQUATOR

TROPIC OF CAPRICORN

20°N

0°

20°S

40°S

60°S

140°W   120°W   100°W   80°W   60°W   40°W   20°W

Latin America
Economic Activity

Land Use

- Commercial farming
- Subsistence farming
- Livestock raising
- Primarily forest
- ■ Manufacturing and trade
- Commercial fishing
- Little or no activity

UNITED STATES

Gulf of Mexico

TROPIC OF CANCER

Caribbean Sea

ATLANTIC OCEAN

0   1,000 miles
0   1,000 kilometers
Lambert Azimuthal Equal-Area projection

N
W ✦ E
S

Resources

Coal		Cobalt	
Petroleum		Nickel	
Natural gas		Copper	
Iron ore		Lead	
Tin		Manganese	
Zinc		Gold	
Bauxite		Silver	
Uranium		Gems	

PACIFIC OCEAN

TROPIC OF CAPRICORN

UNIT 3
REGIONAL ATLAS

MAP STUDY

1. **Physical Systems** What are the predominant types of natural vegetation in the Tropics?

2. **Places and Regions** What generalizations can you make about the locations of the region's manufacturing areas?

Latin America
Population Density

ATLANTIC OCEAN

40°N

Mexicali
Tijuana
Hermosillo
Torreón
Culiacán
Monterrey
Matamoros
Gulf of Mexico
Mazatlán
Guadalajara
León
Tampico
Mérida
Havana
TROPIC OF CANCER
Mexico City
Puebla
Veracruz
Acapulco
Port-au-Prince
San Juan
Kingston
Santo Domingo
20°N
Guatemala
Tegucigalpa
San Salvador
Managua
Caribbean Sea
Fort-de-France
San José
Barranquilla
Maracaibo
Caracas
Cartagena
Port-of-Spain
Panama
Valencia
Mérida
Medellín
Ciudad Guayana
Georgetown
Bogotá
Paramaribo
Cali
Quito
Manaus
Belém
São Luís
Guayaquil
Iquitos
Fortaleza
Piura
Teresina
Natal
EQUATOR
0°
PACIFIC OCEAN
Trujillo
Pucallpa
Rio Branco
Pôrto Velho
Recife
Lima
Cuzco
Maceió
Ica
Salvador
Arequipa
La Paz
Santa Cruz
Cuiabá
Goiânia
Brasília

0   1,000 miles
0   1,000 kilometers
Lambert Azimuthal Equal-Area projection

Uberlândia
Belo Horizonte
Ribeirão Prêto
Vila Velha
TROPIC OF CAPRICORN
Antofagasta
Campinas
São Paulo
Rio de Janeiro
Salta
Asunción
Curitiba
San Miguel de Tucuman
Florianópolis
20°S
Córdoba
Pôrto Alegre
Valparaíso
Mendoza
Rosario
Santiago
Buenos Aires
Montevideo
Concepción
Bahía Blanca
Mar del Plata

Punta Arenas

UNIT 3
REGIONAL ATLAS

MAP STUDY

1. *Human Systems* Which parts of Latin America are the most densely populated? What might account for this?

2. *Places and Regions* What generalizations can you make about the locations of South America's cities?

POPULATION

Per sq. mi.	Per sq. km
1,250 and over	500 and over
250–1,249	100–499
63–249	25–99
25–62	10–24
2.5–24	1–9
Less than 2.5	Less than 1
Uninhabited	Uninhabited

Cities
(Statistics reflect metropolitan areas.)

- ■ Over 5,000,000
- □ 2,000,000–5,000,000
- ◉ 1,000,000–2,000,000
- ● 500,000–1,000,000
- ○ Under 500,000

172

140°W   120°W   100°W   80°W   60°W   40°W

Mexico

ESSENTIAL QUESTION · *How do physical systems and human systems shape a place?*

Geography Matters...

Places reflect their relationship between humans and their environment. Mexico today is a result of history, geography, and increased globalization. Cultures have collided in Mexico for centuries. Indigenous civilizations flourished, followed by Europeans who brought new laws, languages, and religions. Today, the faces and customs of many Mexican people reflect their mixed heritage. Economic forces have pulled Mexico into the global economy with a promise of prosperity dependent on natural resources and industrial growth.

◄ Traditional Mexican dress is colorful and reflects the style of the region.

©Alison Wright/Corbis

173

challenges
of urbanization

Rapid urban growth brings challenges to city governments around the world as they struggle to provide housing, services, infrastructure, and jobs, as well as curb pollution. Governments have limited funds to spend on basic upkeep and services. As a result, cities like **Mexico City** *experience challenges such as environmental problems and poverty.*

Growth of Mexico City
- 1910
- 1960
- 2000

Why has Mexico City grown so fast?

Today's Mexico City is the result of years of rural-to-urban migration by people looking for better economic opportunities. The first influx of these economic migrants coincided with rapid industrialization in the late nineteenth century. The pull forces of industrial jobs and the push forces of rural land policies drew people to Mexico City. The rural poor moved to the city as land was purchased around them by wealthy landowners. Similar factors continue to bring economic migrants to Mexico City seeking a better life for themselves and their families.

1. Human Systems What are the pull factors influencing migration to Mexico City? How are these different from the push factors that bring people to the city?

What is the social impact of rapid growth?

Economic migrants move to the city expecting to find jobs. Unfortunately, unemployment is common. Some migrants find temporary jobs or work in the informal sector—"underground economies" that are not taxed or regulated by the government. People often do not have access to health care and education. Lack of infrastructure—housing, electrical grids, sewer facilities, and roads—to support the growing population leads to the development of shantytowns. The influx of people to Mexico City puts enormous pressures on the natural environment. Underground water aquifers are being depleted, causing the city to sink. Inadequate sewer facilities lead to polluted land and water. Full of rubbish, landfills have been closed. Unregulated by the government, shantytowns are built in environmentally sensitive areas such as hill slopes.

2. Environment and Society What challenges has rural-to-urban migration created for the government of Mexico City?

What can be done?

Government agencies and other groups continue to establish initiatives and special projects to address these challenges. The government of Mexico and public-private partnerships are investing in sustainable and environmentally friendly housing development. Plan Verde (Green Plan) includes a range of programs to promote environmental sustainability by easing traffic congestion, reducing greenhouse gas emissions, and encouraging public transportation, cycling, and walking options. The Mexico City Climate Action Program provides funding for sustainable housing as well as renewable energy programs.

3. Human Systems Write a paragraph explaining how environmentally friendly policies could improve life in Mexico City.

(cl)Jacques Descloitres, MODIS Land Rapid Response Team, NASA/GSFC. URL: http://visibleearth.nasa.gov/view.php?id=64140, (c)Florian Kopp/age fotostock, (cr)Jose Luis Magana/AP Images

THERE'S MORE ONLINE

VIEW a map of the world's megacities • *WATCH* a video of Mexico's urban sprawl

networks

There's More Online!

☑ **IMAGE** Land Stripped by Copper Mine in Chihuahua, Mexico

☑ **DIAGRAM** Mexico's Location on the Tectonic Plates

☑ **INFOGRAPHIC** Vertical Climate Zones of Mexico

☑ **INTERACTIVE SELF-CHECK QUIZ**

☑ **VIDEO** Physical Geography of Mexico

Reading **HELP**DESK

Academic Vocabulary

- **feature**
- **inevitable**

Content Vocabulary

- **land bridge**
- **seismic**
- **vertical climate zone**

TAKING NOTES: *Key Ideas and Details*

SUMMARIZING As you read the lesson, use a graphic organizer like the one below to take notes on the physical geography of Mexico.

Physical Geography of Mexico		
Landforms	Water Systems	Climates, Biomes, and Resources

LESSON 1
Physical Geography of Mexico

ESSENTIAL QUESTION · *How do physical systems and human systems shape a place?*

IT MATTERS BECAUSE

Mexico has rich farmland, abundant access to the ocean, and freshwater resources. With diverse climates and biomes it supports a variety of food crops. The country also has substantial mineral resources and has developed strong manufacturing and service sectors.

Landforms

GUIDING QUESTION *Why is the Mexican Plateau considered the heartland of Mexico?*

Mexico is the southernmost country in North America. It shares its entire northern border with the United States. Mexico, along with Central America, joins the continents of North America and South America. This physical feature is called a **land bridge** because it connects two geographic landforms.

The western side of Mexico is part of the Ring of Fire. The Ring of Fire is the area where the Pacific tectonic plate collides with other tectonic plates, creating areas of **seismic** activity with earthquakes. This activity helps shape the landforms of Mexico. Seismic activity opens parts of the Earth's crust and triggers the formation of volcanoes.

The mountains that stretch along the northwestern edge of Mexico are called the Sierra Madre Occidental. These mountains have been deeply cut by westward-flowing rivers and streams, which have formed deep gorges. The largest is *Barranca del Cobre* (Copper Canyon), a beautiful natural wonder in the state of Chihuahua, Mexico.

The Sierra Madre Oriental is considered the southern extension of the Rocky Mountains of Canada and the United States. The average mountain height is between 8,000 and 9,000 feet (2,438 and 2,743 m). A few exceptions include mountains that reach heights of 12,000 feet (3,657 m).

Between these two mountain ranges is the inland Mexican Plateau. Moderate, consistent temperatures make this area an attractive place to live. It is the largest and most densely populated region of

(tl)Royalty-Free/Corbis, (tc)Beth Wald/Aurora/Getty Images, (tr)Lawrence Goldman Photography/Flickr/Getty Images

Mexico. The Mexican Plateau is broken into two parts, the huge *Mesa del Norte* (Northern Plateau) and the smaller but heavily populated *Mesa Central* (Central Plateau). The dry Northern Plateau is home to several large cities. The Central Plateau is considered the breadbasket—or major grain-producing region—of Mexico. It is less arid than the Northern Plateau and **features** several smaller valleys. Most of the food grown in Mexico comes from this area.

The Gulf Coastal Plain is a wide stretch of land east of the Sierra Madre Oriental. These mountains extend from the Texas-Mexico border along the Gulf of Mexico to the Yucatán Peninsula. In the south, a series of mountain ranges and plateaus called the Southern Highlands reach from just south of Mexico City to the southwest edge of Mexico's border with Guatemala.

The variety of landforms in Mexico—from large plateaus and valleys to long mountain ranges and highlands—has made it possible to support large communities of people. There are ample grazing and farmlands on the North and Central Plateaus, in the Southern Highlands, and along the coastlines. The population density in Mexico is greater near parts that have the most agriculture, especially on the Central Plateau and in the Southern Highlands.

✔ **READING PROGRESS CHECK**

Interpreting How has the geography of Mexico affected the way people use the land?

land bridge a strip of land that connects two larger landmasses, enabling migration of plants and animals to new areas

seismic relating to or caused by an earthquake

feature to have as a characteristic or as a prominent attribute

Plate Tectonics in Mexico   ⌄ **DIAGRAM SKILLS**

NORTH AMERICAN PLATE

AFRICAN PLATE

CARIBBEAN PLATE

The Caribbean Plate borders the North American Plate, the South American Plate, the Nazca Plate, and the Cocos Plate. These borders are regions of intense seismic activity, including earthquakes and volcanoes.

PACIFIC PLATE

COCOS PLATE

SOUTH AMERICAN PLATE

NAZCA PLATE

Mexico's location near the edges of tectonic plates makes it a hot spot for volcanoes and earthquakes.

▲ **CRITICAL THINKING**

1. Classifying On which tectonic plates is Mexico located?

2. Drawing Conclusions Why do you think this area of tectonic activity is called the Ring of Fire?

Royalty-Free/Corbis

Water Systems

GUIDING QUESTION *Why does Mexico have few major rivers and natural lakes?*

Northern Mexico is generally characterized by a dry climate. This makes permanent waterways rare. The high mountain ranges and plateaus create temperate **vertical climate zones** that do not collect the volume of water that is more common in tropical regions. The few rivers and natural lakes that exist are found in the central part of the country and are generally small. One important exception is the Rio Grande. Known as the Río Bravo del Norte in Mexico, it forms part of the border between Mexico and the United States.

The Lerma River is one of Mexico's most important rivers. It begins in the Toluca Basin, on the Central Plateau west of Mexico City. The Lerma River feeds into Lake Chapala, the largest natural lake in Mexico.

The Gulf of Mexico is the large body of water that forms Mexico's east coast. It supports diverse sea life including an ancient sea creature known as the manatee. The Gulf of Mexico is famous for shrimp and supplies the fishing industry in both the United States and Mexico. The waters in the Gulf of Mexico are relatively sheltered from ocean currents, so the beaches are calm and the waters are warm.

On the western side of Mexico, the Gulf of California divides the Baja Peninsula from the northern coast of Mexico. This body of water supports a remarkable diversity of aquatic animals. These include several types of whales, the giant Pacific manta ray, endangered leatherback sea turtles, and great white sharks.

✓ **READING PROGRESS CHECK**

Describing What is the importance of the Río Bravo del Norte to Mexico?

Climate, Biomes, and Resources

GUIDING QUESTION *How does climate affect human activities in Mexico?*

The climate of a particular region **inevitably** affects the way of life that people have in each place. For example, people who graze cattle on the Northern Plateau anxiously await rain each year. The farmers in the valleys of the Central Plateau

---

vertical climate zone
a climate zone that occurs as elevation increases, with its own natural vegetation and crops

inevitable incapable of being avoided or evaded

Differences in elevation create distinct climate zones in Mexico and other high-altitude areas in Latin America.

CRITICAL THINKING ▶

1. Analyzing Visuals Which climate zones are found above 6,000 feet (1,829 m)?

2. Synthesizing How might increasing elevation affect the type of resources found in each vertical climate zone?

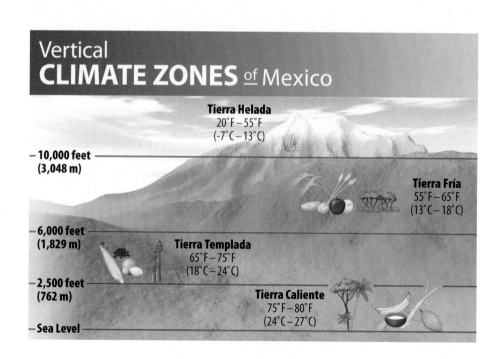

Vertical CLIMATE ZONES of Mexico

Tierra Helada
20°F – 55°F
(-7°C – 13°C)

Tierra Fría
55°F – 65°F
(13°C – 18°C)

— 10,000 feet
(3,048 m)

Tierra Templada
65°F – 75°F
(18°C – 24°C)

— 6,000 feet
(1,829 m)

Tierra Caliente
75°F – 80°F
(24°C – 27°C)

— 2,500 feet
(762 m)

— Sea Level

depend on their climate to help them get enough water for their crops. The three factors that influence Mexico's climates are the regional high-pressure systems, the northeast trade winds, and the vertical climate zones.

Regional high-pressure systems keep the north and central parts of Mexico dry and cause occasional droughts. The northeast trade winds are responsible for the pattern of tropical storms. Vertical climate zones create the temperate or mild climates found throughout the southern part of Mexico. The elevation of the highlands keeps them at a comfortable temperature that is also helpful for growing certain crops.

Northern Mexico is defined as a chaparral biome. It has mild, rainy winters and hot, dry summers. The plant life consists of cacti, shrubs, and shrub oak. High winds and low-growing plants make the soil in this region good for grasslands. The soil is also thin and rocky, however, so it is not good for crops.

Southern Mexico has a variety of biomes. In the lower altitudes along the coasts, the climate is hot. On the east coast, daily rainfall and high humidity occur. This biome supports very diverse plant and animal life and is essentially a rain forest.

Minerals are an important part of Mexico's economy, especially silver. Mexico is the world's leading producer of silver. In the area called the "Silver Belt" on the Mexican Plateau, both industrial and precious minerals are mined. Zinc, bauxite (the ore of aluminum), lead, gold, mercury, cadmium, and such trace minerals as antimony, manganese, and copper are also important. Timber, fish, and agricultural products are also a significant part of Mexico's economy.

Mexico is a leading petroleum-producing country. Petroleum exports account for a large share of foreign-exchange earnings. Mexico ranks thirteenth in the world for crude oil exports. About three-fourths of Mexico's electricity is generated by thermal power plants that are fired mainly by oil and natural gas. Another one-tenth of Mexico's electric power is created by nuclear power and renewable resources of wind, solar energy, and biomass (plant materials and animal waste used as a source of fuel).

Beth Wald/Aurora/Getty Images

☑ **READING PROGRESS CHECK**

Assessing How do vertical climate zones affect the economic activity of the Southern Highlands?

Copper mining takes place in the Mexican copper belt in the west part of the country.

▲ **CRITICAL THINKING**

1. *Analyzing Visuals* Describe the negative impacts of strip mining as depicted in this photo.

2. *Comparing* Where is Mexico's copper belt located in comparison to the "Silver Belt"?

LESSON 1 REVIEW

Reviewing Vocabulary

1. *Discussing* Write a paragraph that discusses the geography of vertical climate zones.

Using Your Notes

2. *Describing* Use your graphic organizer from the lesson to describe three of Mexico's water systems.

Answering the Guiding Questions

3. *Drawing Conclusions* Why is the Mexican Plateau considered the heartland of Mexico?

4. *Interpreting* Why does Mexico have few major rivers and natural lakes?

5. *Making Connections* How does climate affect human activities in Mexico?

Writing Activity

6. *Informative/Explanatory* Write a paragraph describing Mexico's location along the Ring of Fire and how that creates natural hazards.

Reading HELPDESK

Academic Vocabulary

- culture
- diverse

Content Vocabulary

- **mestizo**
- **conquistador**
- **cash crop**
- **syncretism**
- **megacity**
- **primate city**
- **extended family**
- **gross domestic product**
- **maquiladora**
- **free trade zone**

TAKING NOTES: *Key Ideas and Details*

PARAPHRASING Use a graphic organizer like the one below to describe the human geography of Mexico.

Human *Geography* of Mexico	
History and Government	Population Patterns

LESSON 2
Human Geography of Mexico

ESSENTIAL QUESTION · *How do physical systems and human systems shape a place?*

IT MATTERS BECAUSE
Mexico's human geography reflects influences from the Maya and Aztec civilizations, the introduction of Spanish culture during the colonial era, and cultural and social elements shared from recent interaction with the United States and other countries.

History and Government

GUIDING QUESTION *What influenced Mexico's political and social structures?*

Variations in the physical geography of Mexico led to the development of diverse **cultures,** languages, and civilizations among the indigenous peoples of Mexico. These differing peoples developed cultures to suit the environments in which they lived. These cultures can be seen in the regional distinctions of Mexico today.

The northern half of Mexico, on the inland plateau and in the mountains, originally had a small population of mostly independent groups of nomadic people. Agriculture was used to supplement hunting, herding, and gathering of food. Some of these seminomadic groups still live in their traditional homelands, separated from most outside influences. The Tarahumara people in the Sierra Madre Occidental are one example of an indigenous group who still live in northern Mexico.

The southern half of Mexico was geographically more **diverse.** It could support large-scale agriculture and produce the variety and abundance of foods necessary to maintain empires and cities. Centered in the Yucatán Peninsula, the Maya civilization was one of the earliest and largest civilizations in Mexico. The Maya built huge stone cities, which were abandoned a few hundred years before the arrival of the first Spanish explorers. The Maya ruled a vast territory and engaged in long-distance trade with other cultures, including Teotihuacán and the Zapotec. Their descendants still live in and around the areas of their former empire. Many of these Mayan people maintain their culture, speak their ancestral languages, and practice the same cultural traditions.

The Aztec Empire arose in central Mexico. The Aztec ruled from their capital, Tenochtitlán (tay•NAWCH•teet•LAHN), the site of present-day Mexico City. They had conquered other peoples in the area when the

(tl)Courtesy Everett Collection/Alamy, (tc)©Jack Kurtz/ZUMA/Corbis, (tr)David Maung/Bloomberg/Getty Images

Spanish arrived in 1519. Although **mestizos**, people of mixed Spanish and indigenous heritage, now densely populate the region, there are groups that trace their ancestry to the Aztec.

After the conquest of the Aztec by Spanish **conquistador** Hernán Cortés and his men, the Spanish took the wealth of Mexico's gold and silver resources. They also found value in the variety of food available to the local people, quickly taking corn, tomatoes, chocolate, and other native crops on the return trips to Spain. Large tracts of land in Mexico were given to the Spanish settlers of Mexico. These landowners began growing **cash crops** such as cacao (chocolate) and maize (corn) in large quantities, which they exported to Spain. This further enriched the Spanish. Mexico remained a part of the Spanish Empire for nearly three centuries. It was governed by Spain under a highly structured political system ruled by officials called viceroys who were appointed by the Spanish monarch.

In the late 1700s, throughout Mexico and the rest of Latin America, people started to protest European rule. In 1821 Mexico became the first Spanish territory to win its independence. Mexico was free from Spain, but the political system was ruled by a small group of wealthy landowners, army officers, and Catholic clergy who remained in power. Power struggles, public dissatisfaction, and civic revolts made the new republic fragile and chaotic. During this time a new type of leader emerged, the caudillo (kow•DEE•yoh), or military dictator. For brief periods in the 1800s, the government moved toward democratic principles. However, the caudillos found ways to return to power.

The long and bloody Mexican Revolution overthrew the caudillos and established a new constitution in 1917. This brought reforms and established the current Mexican government as a federal republic. Power was divided into three branches of government—legislative, executive, and judicial—and a president could only be elected for one six-year term. However, the rule of law did not last for long. In 1929 one political party, the *Partido Revolucionario Institucional* (PRI), was elected and established a corrupt monopoly on the political system of Mexico. The PRI went on to control the political establishment for nearly 70 years. Not until 2000 was the opposition party, *Partido Acción Nacional* (PAN), able to win the presidency. In 2012, however, the PRI was reelected.

Over the past few decades, drug cartels have come to control different regions of Mexico. New cartels have been forming or breaking away from older and larger cartels. These new cartels compete with old cartels for power and control of drug-producing territories. The result is internal warfare in Mexico. The cartels have increasingly incited street gun battles, massacres in the mountains, and other acts of violence and terror.

Struggles for additional reforms in the government continue. Indigenous communities, small farmers, and groups of underpaid laborers are continuing to pressure the government for greater inclusion in the political system. Corruption remains a common reality in the government. A small group of very wealthy landowners still controls most of Mexico's wealth.

☑ READING PROGRESS CHECK

Exploring the Issues Why do drug cartels have such a powerful influence in Mexico?

Population Patterns

GUIDING QUESTION *What factors have shaped Mexico's population patterns?*

For unknown reasons, the Maya had abandoned their cities in the Yucatán and southern Mexico by the time the Spanish conquistadors arrived. They were mainly living as subsistence farmers in small communities, where many still live today.

culture the customary beliefs, social forms, and material traits of a racial, religious, or social group

diverse differing from one another

mestizo refers to people of mixed indigenous and European descent

conquistador Spanish for "conqueror"; Spanish soldier who participated in conquest of indigenous peoples of Latin America

cash crop farm product grown to be sold or traded rather than used by the farm family

The Aztec were the growing military and governing power in central Mexico, conquering their neighbors and establishing a powerful central urban region in the Valley of Mexico.

The indigenous peoples of Mexico were a mixture of different groups and empires, having multiple languages and belief systems. Intermarriage between these indigenous peoples and the Spanish resulted in the mestizo population. The Spanish arrival brought other changes to Mexico. It resulted in war and disease outbreaks. The Spanish also established the Catholic Church in Mexico. A unique melding of traditional indigenous beliefs and the newly introduced Catholic culture occurred in a process called **syncretism**.

The Spanish established a solid footing in Mexico by using Catholic missions as religious outposts to help spread Christian ideas. They centralized power around presidios, or military outposts, to protect Spanish towns and missions. Communities expanded near the missions, and Spanish estate farming and ranching were introduced. Some Spanish settlements grew into large cities, using a variety of local resources. For example, Oaxaca de Juárez grew rich on the cochineal (KAH•chuh•NEEL) insect used to make a permanent red dye.

Spanish landowners took advantage of Mexico's large and impoverished indigenous communities. Workers on large landowners' estates, known as haciendas, harvested cash crops but were paid very little for their labor. It was, and in many places still is, very difficult for poor workers to earn enough to become independent of their landlord. After independence from Spain, members of poor rural communities sought independence from the landowners.

Rural-to-urban migration has greatly altered the population's distribution. Today, more than 78 percent of Mexican citizens live in urban areas. Mexico City is a **megacity** that has absorbed surrounding towns and smaller cities into the urban sprawl of the dominant city. It has received millions of migrants as well. The poor seek work in urban areas, migrating to regional capitals or to Mexico City. Mexico City is important because it is the **primate city**. It is more than just

syncretism a blending of beliefs and practices from different religions into one faith

megacity a great city that is made up of several large and small cities

primate city a city that dominates a country's economy, culture, and government and in which population is concentrated; usually the capital

(t)Courtesy Everett Collection/Alamy, (b)©Danny Lehman/Corbis

---

MEXICAN
Independence
and Change ➜

The story of Mexico's history is a fascinating one. Starting with ancient civilizations and tragic conquest, it is followed by independence and revolution, resulting in the Mexico of today.

▶ **CRITICAL THINKING**

1. *Analyzing* Why do you think Benito Juárez is considered a hero to Mexicans?

2. *Sequencing* What political party did Vicente Fox oppose and when did it regain power?

1800 ➜

1810
Father Miguel Hidalgo sparks the decade-long struggle for independence from Spain.

1821
Treaty recognizes Mexican independence from Spain.

1824
Mexico becomes a republic and adopts a federal constitution.

1833

Caudillo Antonio López de Santa Anna becomes president and halts reforms.

1846–1848
Mexico engages in war with the United States over Texas territory. Treaty of Guadalupe Hidalgo ends war in 1848.

the political capital and largest city. It is also the cultural and economic center of Mexico. Mexico City has the best schools, hospitals, and housing in the country. It also has serious overcrowding, the most impoverished neighborhoods, and high levels of air pollution and water pollution.

✓ **READING PROGRESS CHECK**

Analyzing How did Spanish landowners take advantage of the indigenous people?

Society and Culture Today

GUIDING QUESTION *How does Mexican society and culture reflect the country's colonial past?*

Colonial rule lasted for nearly 300 years in Mexico and gave the country's different cultures an important point of unifying identity. The majority of Mexicans identify themselves as Catholic. The rituals practiced in churches, however, are a blend of the diverse internal influences that drive Mexican culture.

Family and the Status of Women

Mexican culture highly values family. Today, many upper-class families are moving into more nuclear living arrangements, with only two generations in one house. The majority of Mexican homes, however, still hold three or four generations. The **extended family**, including cousins and grandparents, acts as a social support network with older generations raising younger ones. This is especially common in rural areas and in poorer communities.

Spanish culture established the tradition of machismo, or male dominance, in Mexico. As in other Latin American countries, women in Mexico have traditionally been limited to culture-specific roles in society. Their first obligation is to the family. However, women have made huge strides toward equality in the last few decades. More women are entering the workforce and are involved in traditionally male professions, as well as in the political arena.

extended family household made up of several generations of family members

(bl)©Jack Kurtz/ZUMA/Corbis, (br)Odd Andersen/AFP/Getty Images

La Reforma ("The Reform")
President Benito Juárez brings great changes to Mexico.

1858–1872

➜1900

Corrupt dictatorial rule of General Porfirio Díaz leads to Mexican Revolution.

1910–1920

1917

New constitution brings additional reforms and establishes Mexico as a federal republic.

One political party, the *Partido Revolucionario Institucional* (PRI), begins its domination of Mexican politics.

1929

PRI's control ends when Vicente Fox Quesada of the opposition party *Partido Acción Nacional* (PAN) is elected president.

2000

➜2012

2012

PRI regains control with Enrique Peña Nieto winning the presidency.

PERCENTAGE OF FEMALE EMPLOYEES PER SECTOR, MEXICO, 2010

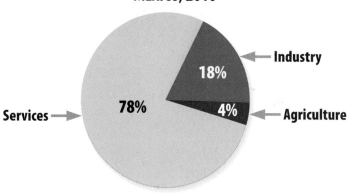

Industry 18%

Services → 78%

4% ← Agriculture

Source: http://data.worldbank.org/, World Bank, *World Development Indicators*

Women in Mexico have been entering the labor force in increasing numbers in recent years.

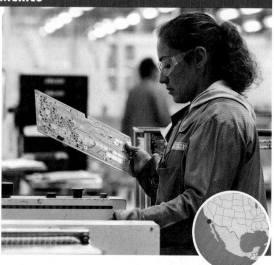

▲ **CRITICAL THINKING**

1. Analyzing   What economic sector employs the most women in Mexico?

2. Analyzing Visuals   Which sector is the photo depicting?

The Arts

Mexico's arts have been greatly influenced by its indigenous cultures and Spanish heritage. Since the 1930s, Mexican culture has been seeking a unified image through the promotion of traditional folk culture. Art forms that were once region-specific or limited to a particular cultural group are now universally recognized as Mexican. Early indigenous architecture includes Maya pyramids and Aztec temples and palaces. Many of these buildings were decorated with murals (wall paintings) and mosaics (wall pictures made with stones or tiles). Murals were also popular in the European tradition, and the Spanish encouraged this art form.

Murals continue to be an important art form in Mexico. The most famous Mexican muralist was Diego Rivera. He became known for his elaborate and often culturally and politically themed wall paintings and frescoes (paintings done in wet plaster). His art is displayed in museums and government and civic buildings in Mexico City and in the United States. In the central arch detail of his fresco *From Conquest to 1930,* Rivera used images of peasants and workers to show foreign influence and Mexico's struggle for independence.

☑ **READING PROGRESS CHECK**

Depicting   How does the art of Diego Rivera reflect Mexico's cultural heritage?

Economic Activities

GUIDING QUESTION   *How has Mexico's place in the global economy changed over time?*

In the 1500s Mexico became the center of the Spanish Empire in the Americas. It was a rich colony because of its many natural resources. Because it was the first Spanish colony to win independence, Mexico became a cultural leader and example for other colonies ruled by Spain and Portugal. Mexico is still trying to create an economy that benefits all Mexicans, not just the wealthy upper class. In fact, Mexico is still struggling to be a completely free and equal republic. Colonialism still influences Mexico's economy.

David Maung/Bloomberg/Getty Images

Central America
and the Caribbean

ESSENTIAL QUESTION · *How do physical systems and human systems shape a place?*

netw⊕rks

There's More Online about Central America and the Caribbean.

CHAPTER 8

Geography Matters...

Central America and the Caribbean is one of the most biologically diverse regions in the world. Many types of landforms and climates support abundant plant and animal species. Cultures here have been influenced by ancient civilizations, colonial pasts, political upheavals, and by a modern mix of peoples migrating in and out of the region. No single influence, culture, or history can characterize this region of diversity and biological and cultural wealth.

◀ Caribbean music is a diverse blend of African, European, and indigenous influences.

Christian Aslund/Lonely Planet Images/Getty Images

spatial diffusion:
the columbian exchange

Christopher Columbus's voyage to the Americas began a dramatic process of interchange of peoples, animals, plants, cultures, ideas, and diseases between the Eastern and Western Hemispheres. This exchange had both positive and negative consequences. The Columbian Exchange, also known as the Grand Exchange, profoundly altered the course of human development across the globe.

The Columbian Exchange

AMERICAS TO EUROPE, ASIA, AND AFRICA

EUROPE, AFRICA, AND ASIA TO AMERICAS

NORTH AMERICA

EUROPE

AFRICA

SOUTH AMERICA

Turkeys
Pumpkins
Squash
Peppers
Sweet Potatoes
Tobacco
Pineapples
Peanuts
Cacao
Potatoes
Beans
Tomatoes
Corn
Vanilla

Citrus Fruits
Grapes
Bananas
Onions
Olives
Turnips
Coffee Beans
Sugarcane
Honeybees
Pears & Peaches
Grains
• Wheat
• Rice
• Barley
• Oats
Livestock
• Cattle
• Sheep
• Pigs
• Horses

Diseases
• Smallpox
• Influenza
• Typhus
• Measles
• Malaria
• Diphtheria
• Whooping Cough

EQUATOR 0°

30°N

0   1,000 miles

0   1,000 kilometers

Mercator projection

N W E S

90°W   60°W   30°W   0°

194

How did the Columbian Exchange affect agriculture?

New foods were exchanged between Europe and the Americas as a major part of the Columbian Exchange. Europeans introduced new food plants and domesticated animals to the Americas in an attempt to "Europeanize" the region. The plants involved in the exchange altered the economies of peoples in the Americas, Europe, and other continents, leaving a lasting imprint on all. Europeans introduced livestock such as cattle, pigs, and sheep as well as grains such as wheat. Sugarcane, coffee, and citrus fruits from Asia were also brought by the Europeans. Yams, bananas, and rice were brought from Africa to the Americas. Europeans also took food crops native to the Americas back across the Atlantic Ocean. Europeans discovered a variety of valuable Native American crops, including beans, squash, chili peppers, sunflowers, tomatoes, sweet potatoes, avocados, and cacao; but potatoes and corn were most important.

1. **Human Systems** Describe how the exchange of food left a lasting imprint in the Americas and Europe.

What was an unintended consequence of the Columbian Exchange?

The unintentional transmission of infectious diseases had serious and long-lasting effects on the peoples and cultures on both sides of the Atlantic Ocean. Human populations that develop in isolation from one another are often highly vulnerable to new infectious diseases introduced by outsiders. Lack of antibodies to the new diseases and lack of access to medicine can result in widespread illness and death. Smallpox, malaria, typhus, cholera, and a host of other diseases were introduced to the Americas by the Europeans. The native populations had no experience with these diseases and therefore had no resistance to them. As a result, native populations were decimated. Although it is difficult to determine the full extent of the population loss, estimates suggest that 80–95 percent of the indigenous peoples may have died in the first 100 to 150 years following the arrival of the Europeans in 1492. Other unintentional consequences included transmission of diseases from the Americas to Europe, although much fewer in number.

2. **Environment and Society** Write a summary of how the culture and history of Central America and the Caribbean might have been different if diseases had not been transmitted during the Columbian Exchange.

How were the lifestyles of indigenous peoples altered?

Contact with Europeans not only wiped out most of the native population of the Americas through disease, but also permanently altered the culture of those who survived. For example, because the Americas lacked large animal species, indigenous peoples had no beasts of burden. Asians and Europeans had long since domesticated several large animals for use in agriculture, trade, and warfare. Horses, long extinct from the Americas, were reintroduced by the Spanish, whose horse-mounted soldiers quickly overran and dominated indigenous forces. The horse transformed the Americas through warfare, hunting, and culture. Of even greater significance was that people themselves crossed the Atlantic. The presence of new peoples and their impact on the indigenous peoples of the Americas, due to the Columbian Exchange, initiated one of the largest cultural transformations in human history.

3. **Human Systems** What advantages did horses give Spanish invaders in the Americas?

(tl)Luis Marden and Helen S. Williams/National Geographic Stock, (tc)Eye of Science/Science Source, (tr)Javier Larrea/AGE Fotostock

THERE'S MORE ONLINE

VIEW a slide show of goods that were exchanged • **READ** a quote about the effects of the Columbian Exchange

netw◉rks

There's More Online!

☑ **MAP** Physical Geography: Central America and the Caribbean

☑ **INFOGRAPHIC** Biodiversity in Central America

☑ **INTERACTIVE SELF-CHECK QUIZ**

☑ **VIDEO** Physical Geography of Central America and the Caribbean

Reading **HELP**DESK

Academic Vocabulary

- **energy**
- **exhibit**

Content Vocabulary

- **isthmus**
- **archipelago**
- **biodiversity**

TAKING NOTES: *Key Ideas and Details*

SUMMARIZING Use a graphic organizer like the one below to take notes on the natural resources of Central America and the Caribbean.

Natural Resources: Central America and the Caribbean

LESSON 1
Physical Geography of Central America and the Caribbean

ESSENTIAL QUESTION · *How do physical systems and human systems shape a place?*

IT MATTERS BECAUSE

Central America acts as a land bridge connecting North and South America. The region boasts dense rain forests, coastal plains, and high mountains, and is one of the world's great biodiversity hotspots. The more than 7,000 islands of the Caribbean exhibit their own unique landforms and living things. Shaped in part by the Ring of Fire, this region is a volatile zone of earthquakes and volcanic eruptions.

Landforms

GUIDING QUESTION *Why are the majority of Central America's people concentrated in the Central Highlands?*

Much of Central America is hilly or mountainous, although swamps and lowlands extend along both coasts. These landforms create three distinct belts: the Pacific Lowlands, the Caribbean Lowlands, and the Central Highlands. The narrow plains of the Pacific Lowlands extend from Guatemala to Panama. The Caribbean Lowlands are also narrow, except in Nicaragua and Honduras. Central America's most distinctive landforms—mountains—form the Central Highlands. The region climbs steadily higher west of the Caribbean Lowlands. It rises up to the western plateau highlands where mountains and some 40 volcanic cones attain elevations of more than 12,000 feet (3,700 m). These are the volcanic highlands, or the Volcanic Axis. These mountains are an extension of the Sierra Madre of Mexico. Volcanic eruptions and earthquakes are not uncommon. The weathered lava produces fertile soil, making these highlands rich agricultural zones and areas of dense population.

Its location on the Ring of Fire brings potential hazards to living there. Yet humans have thrived in the mountains, valleys, and plateaus of the Central Highlands for thousands of years. The region's cooler climates, adequate rainfall, and rich natural resources—water, volcanic soil, timber, and minerals—attracted the area's earliest peoples and is where the majority of the people live today.

(tl) J Marshall/Alamy, (tr) ©Corbis

Another distinctive feature is the **Isthmus** of Panama. It extends west to east, connecting North and South America and separating the Caribbean Sea from the Gulf of Panama. The mountains, swampy coastal lands, and dense rain forests make contact between people difficult in this area.

In the Caribbean, many of the more than 7,000 islands are the tops of mountains that are part of the mainland's Central Highlands. The islands of the Greater and Lesser Antilles, however, are part of an **archipelago.** This archipelago is composed of the crests and peaks of a mountain range formed from collisions between the Caribbean plate and other tectonic plates. Tectonic activity continues to change the landscape. In 2010, for example, an earthquake struck Haiti's capital, Port-au-Prince, collapsing buildings and killing large numbers of people.

isthmus a narrow strip of land connecting two larger land areas

archipelago a group or chain of islands

☑ **READING PROGRESS CHECK**

Identifying What factors attracted people to settle in the highlands?

Water Systems

GUIDING QUESTION *How are Central American rivers and lakes important to the human systems of the area?*

Inland lakes and waterways play a vital role in Central America, aiding growth and development. The water systems also provide transportation, drinking water, drainage, irrigation, and a source of hydroelectric power. Lake Nicaragua is Central America's largest freshwater lake. It is the only one in the world to contain oceanic animal life such as sharks, swordfish, and tarpon. Nearby Lake Managua has commercially viable fish and alligators. It is drained by a river that flows into Lake Nicaragua and fed by streams from the highlands. Nicaragua's capital city, Managua, is located on the southern shore of Lake Managua.

GEOGRAPHY CONNECTION

A wide array of landforms and varied elevations give Central America and the Caribbean a diverse physical landscape.

1. *PHYSICAL SYSTEMS* What is the name of the most dominant mountain range in Central America that runs from north to south?

2. *PLACES AND REGIONS* Name the bodies of water that border the islands of the Caribbean.

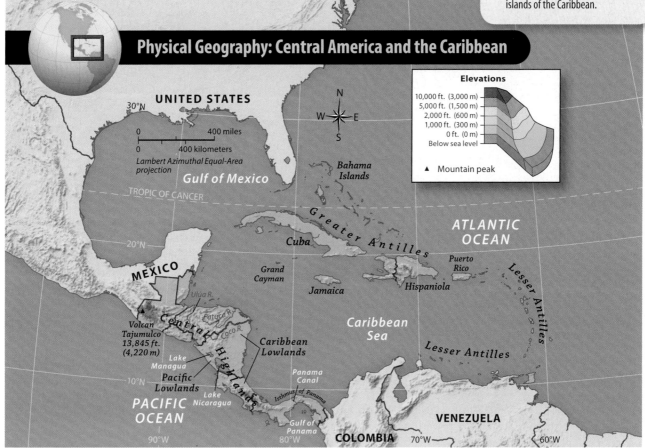

Physical Geography: Central America and the Caribbean

The Panama Canal bisects the Isthmus of Panama. It is one of the world's most important human-made waterways. The canal allows ships to travel between the Atlantic and Pacific Oceans without making the long trip around South America's Cape Horn. International traffic through the Canal Zone is dominated by ships traveling between East Asia and the eastern United States. Approximately two-thirds of trading ships are sailing to or from U.S. ports.

Many of Central America's rivers provide commercial water routes because they are short and steep. The San Juan River, an outlet for Lake Nicaragua, drains into the Caribbean Sea. El Salvador's only navigable river, the Lempa River, generates hydroelectricity from the **energy** of moving water.

energy usable power

exhibit to demonstrate or show openly

The warm, clear waters of the Antillean regions of the Caribbean Sea contain coral reefs. These reefs **exhibit** a wide variety of reef fish. Other common types of characteristic marine life include manatees, manta rays, spiny lobsters, dolphins, and numerous species of sea turtles. Commercial fishing of sardines and tuna and the utilization of other marine resources have increased international trade.

☑ **READING PROGRESS CHECK**

Describing Describe the importance of the Panama Canal to international trade.

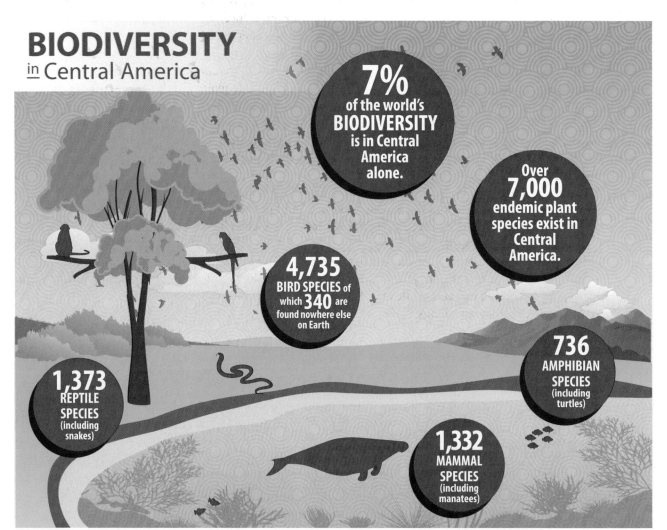

BIODIVERSITY
in Central America

7% of the world's **BIODIVERSITY** is in Central America alone.

Over **7,000** endemic plant species exist in Central America.

4,735 BIRD SPECIES of which **340** are found nowhere else on Earth

736 AMPHIBIAN SPECIES (including turtles)

1,373 REPTILE SPECIES (including snakes)

1,332 MAMMAL SPECIES (including manatees)

The natural richness of Central America is evident in its diverse animal and plant life.

▲ **CRITICAL THINKING**

1. ***Analyzing*** In what way might Central America's position as a land bridge linking two continents affect its biodiversity?

2. ***Drawing Conclusions*** What might be one effect of the loss of biodiversity in Central America?

Climate, Biomes, and Resources

GUIDING QUESTION *How does its location in the Tropics affect the climate of Central America and the Caribbean?*

Climates in the region are dependent on many factors, such as proximity to the sea, elevation, latitude, and local topography. For much of the region, its location in the Tropics and the prevailing winds that carry warm, moist air from the Atlantic Ocean result in high temperatures and abundant rainfall year-round. A tropical rain forest climate and rain forests dominate much of Central America. The tropical forest biome has a continuous canopy of trees and diversity of species. These species include tall trees, ferns, and mosses, as well as birds, bats, small mammals, and insects. Costa Rica and Panama are global **biodiversity** hotspots, rich in natural resources that may one day provide medicines and other products.

A tropical wet/dry climate is typical of Caribbean islands. Moist winds sweep in from the east. This leaves leeward island areas—those not exposed to wind—dry. Thus some islands may have high temperatures and rainfall but also an extended dry season. Grasslands flourish, although soils may not be very fertile or suitable for large-scale agriculture. On mountainous islands, windward slopes facing moist winds help cool the warm, humid air and increase precipitation. Intense hurricanes strike each year from June to November in the northern Caribbean and the Gulf of Mexico.

Elevation affects the climate and ecosystems of some parts of Central America more than distance from the Equator. Such areas have vertical climate zones in which animals, plants, and climates change as altitude increases. The *tierra caliente*, or "hot land," lies at elevations below 2,500 feet (762 m). Bananas and sugarcane are grown here. The *tierra templada*, or "temperate land," lies between 2,500 to 6,000 feet (762 to 1,829 m) and is the most densely populated. Ecosystems with broadleaf evergreens at lower altitudes give way to evergreens at upper elevations. Land at 6,000 to 10,000 feet (1,829 to 3,048 m) is known as the *tierra fria*, or "cold land." Winter frosts are common here, and potatoes and barley are grown. Above the tree line is the *tierra helada*, or "frozen land." This is an area of permanent ice and snow.

Natural resources extracted or used throughout Central America include nickel, iron ore, fish, timber, and petroleum. Guatemala and certain islands refine petroleum, while Belize extracts crude oil. Much commercial fishing also takes place around the Caribbean islands.

☑ **READING PROGRESS CHECK**

Categorizing How is the climate of Central America different from that of the Caribbean islands?

Connecting Geography to **SCIENCE**

Climatology

Predicting when hurricanes will form and where they will land is often difficult. In recent years, climatologists have used El Niño to help predict hurricanes. El Niño is a climatic change that affects the equatorial Pacific. It includes uncommonly warm waters off the western coast of Peru and Ecuador. Often the increased wind levels during El Niño prevent hurricanes from forming, so there tend to be fewer Atlantic hurricanes during El Niño. Knowing when El Niño winds will develop can help predict when hurricanes are a threat and when they are not.

IDENTIFYING How do climatologists use El Niño to help them predict the behavior of hurricanes?

biodiversity biological diversity in an environment as indicated by numbers of different species of plants and animals

---

LESSON 1 REVIEW

Reviewing Vocabulary
1. *Making Connections* Write a sentence to describe the difference between an isthmus and an archipelago.

Using Your Notes
2. *Listing* Use your graphic organizer to list the natural resources of Central America and the Caribbean and their importance to the region.

Answering the Guiding Questions
3. *Describing* Why are the majority of Central America's people concentrated in the Central Highlands?

4. *Examining* How are Central American rivers and lakes important to the human systems of the area?

5. *Interpreting* How does its location in the Tropics affect the climate in Central America and the Caribbean?

Writing Activity
6. *Informative/Explanatory* Write a paragraph describing how altitude affects the climates and biodiversity of Central America and the Caribbean.

netw⊙rks

There's More Online!

☑ **IMAGE** Kuna Molas

☑ **IMAGE** The Panama Canal

☑ **INTERACTIVE SELF-CHECK QUIZ**

☑ **TIME LINE** Central America and the Caribbean: Paths to Independence

☑ **VIDEO** Human Geography of Central America and the Caribbean

LESSON 2
Human Geography of Central America and the Caribbean

ESSENTIAL QUESTION · *How do physical systems and human systems shape a place?*

Reading **HELP**DESK

Academic Vocabulary

- **attribute**
- **obvious**

Content Vocabulary

- **population pressure**
- **dialect**
- **patois**
- **matriarchal**
- *latifundia*
- *minifundia*
- **cottage industry**
- **ecotourism**

TAKING NOTES: *Key Ideas and Details*

IDENTIFYING Use a graphic organizer like the one below to take notes as you read the lesson.

Human Geography of Central America and the Caribbean

History and Government

Population Patterns

Society and Culture Today

IT MATTERS BECAUSE

A study of the human geography of Central America and the Caribbean highlights how history, geography, and the blending of native and outside cultures have shaped the region. The region's mixing of languages, religions, economic practices, and social customs combine traditional and modern. Such diversity provides both benefits and challenges as the region works toward increased economic and social development.

History and Government

GUIDING QUESTION *How did colonialism influence the history and government of Central America and the Caribbean?*

The voyages of Christopher Columbus to the Americas from 1492 to 1504 triggered a period of conquest and colonization in Central America and the Caribbean. Spain founded the region's first permanent European settlement on the island of Hispaniola in 1493. Rodrigo de Bastidas made Spain's first claim to Central America in March 1501. Other Spanish conquistadors followed. They attempted to subdue the local people and establish permanent colonies. Vasco Nuñez de Balboa explored extensively, crossing the isthmus, claiming land for Spain, and finding enough gold and pearls to establish the first profitable colony in the Americas, called Castilla del Oro.

The king of Spain replaced Balboa with Pedro Arias Dávila, known as Pedrarias. Pedrarias expanded the colony, but was notorious for enslaving and murdering indigenous people. In 1519, as governor of Panama, he established Panama City on the Pacific coast and later moved the capital there. In 1524 Pedrarias sent Francisco Hernández de Córdoba to Nicaragua to conquer the region. Córdoba established Granada on Lake Nicaragua. He also founded León not far from Lake Managua and attempted to rule Nicaragua himself. Pedrarias had Córdoba executed and named himself governor of Nicaragua. While Pedrarias and Córdoba fought to conquer southern Central America, Cristóbal de Olid set sail

(tl)Kip Ross/National Geographic/Getty Images, (tcl)©Bettmann/Corbis, (tcl)©Corbis, (tcr)Sven Creutzmann/Mambo Photography/Getty Images News/Getty Images, (tr)©Bob Krist/Corbis to Honduras. Pedro de Alvarado, who had served with Cortés in the conquest of Mexico, set out overland to conquer Guatemala and El Salvador. When gold was discovered in Honduras, various Spanish forces fought each other for control of the region. In Costa Rica, the native population strenuously resisted Spanish efforts at conquest. Spain did not establish a permanent colony there until 1561.

The physical geography influenced colonization of the region. Remote areas of Central America remained outside of Spanish control. This allowed Great Britain to colonize Belize and the Mosquito Coast of Nicaragua. British Honduras (Belize) was the only non-Spanish colony in Central America. In time, however, France, the Netherlands, and Portugal also established colonies in the Caribbean and other parts of the Americas.

By the mid-1600s, forced labor, starvation, and European diseases had nearly killed the entire indigenous population. For the Europeans, the drastically reduced numbers of indigenous people meant a labor shortage. They began bringing Africans, who had been forcibly captured, enslaved, and transported by ship, to the Caribbean to meet the demand for workers.

In the late 1700s, Africans and indigenous people began to take organized action to free themselves from slavery and European control. François Toussaint-Louverture, a soldier born to enslaved parents, led a revolt of enslaved Africans in Haiti. By 1804, Haiti had won its independence from France. Haitian independence inspired downtrodden people throughout Latin America, but frightened many of the elite who had much to lose. Most Caribbean colonies did not gain independence until the 1900s. Cuba gained self-rule in 1898 as a result of the Spanish-American War, but remained under the protection of the United States until 1902. Some islands remain under foreign control or continue to have foreign ties today.

During the 1800s, several Central American colonies struggled for independence from Spain. In 1823 the federation of the United Provinces of Central America was formed. Eventually it divided into five separate countries: Costa Rica, El Salvador, Guatemala, Honduras, and Nicaragua. In 1903 Panama declared its independence from Colombia and signed a treaty with the United States creating the Panama Canal Zone. The Panama Canal opened in 1914 and was controlled by the United States. Today, under Panamanian control, the canal is being upgraded to reduce traffic congestion and to allow larger ships to pass.

In the 1900s, many Central American and Caribbean countries faced political, social, and economic upheaval. In Panama, the Canal Zone and related industries brought new wealth to the upper classes. Most Panamanians benefited little from the canal, however. In Cuba, a revolution in 1959 produced a communist state. Fidel Castro ruled until 2008 when he handed power to his brother Raúl, who initiated some economic reforms. Armed conflict, civil wars, unstable economies, and poor social conditions were all problems in the region. By the end of the twentieth century, however, Central American governments sought ways to increase international trade and diversify their economies. In the twenty-first century, many Central Americans have exercised their right to vote, demanding positive change. For example, Guatemala ended its history of rule by military regime and continues its political and economic recovery, including regular elections since 1996.

☑ **READING PROGRESS CHECK**

Explaining What role did Haiti play in the movement for independence from European control in the region?

The Panama Canal provides a passage through the isthmus by means of a system of canal locks.

▼ **CRITICAL THINKING**

1. *Speculating* What major challenges would builders of the Panama Canal have faced?

2. *Identifying* Which bodies of water does the Panama Canal link?

Kip Ross/National Geographic/Getty Images

Settlement of Central America and the Caribbean did not begin with the arrival of Christopher Columbus. Before 1492, indigenous groups such as the Arawaks and Caribs called the Caribbean home. In Central America, the Maya civilization flourished between A.D. 300 and 900. It began to weaken around A.D. 800, and the native population of the Caribbean began a swift decline with the arrival of Europeans. However, rather than disappearing, indigenous cultures blended with European cultures and influence the region today.

SPECULATING What cultural influences may be active in Central America and the Caribbean today?

Population Patterns

GUIDING QUESTION *How have migration and rapid growth affected the population?*

The populations of Central America and the Caribbean reflect, to varying degrees, a blending of ethnic groups. In Central America, about 60 percent are mestizos and 20 percent indigenous, with other groups of mixed ancestry making up the rest. For example, in Guatemala about half the population is Maya; in the Bahamas, the majority is of African descent.

Until about 1900, more people immigrated to Central America and the Caribbean than left. Since then outmigration has increased as people sought economic opportunities or escaped civil wars and instability. Many people also migrate within the region for similar reasons, mostly moving to cities.

Population density varies throughout Central America and the Caribbean. About 79 percent of Central Americans live in highland cities and towns, however. Most countries have a single primate city, usually the capital. El Salvador is Central America's most densely populated country. The population density of the Caribbean is among the highest in Latin America and greatest near cities and ports. Caribbean countries combine small land areas with large populations that tend to grow at a rapid rate, causing **population pressure**.

The population growth rate of Central America is one of the highest in the world. In 2007 Central America had more than 40 million people, quadruple its population in the 1950s. Fertility rates have decreased overall but are high relative to more developed countries, particularly given high infant mortality rates and low life expectancy. Projected rapid population growth may strain the environment and the ability of the region to accommodate the number of people.

✔ **READING PROGRESS CHECK**

Determining Importance Why have countries in Central America and the Caribbean experienced a dramatic increase in migration to cities?

©Bettmann/Corbis

CENTRAL AMERICA
and the CARIBBEAN
Paths to Independence ➜

Central America and the Caribbean have been influenced by struggles for independence and modern challenges.

▶ **CRITICAL THINKING**

1. ***Analyzing*** What territories did Spain lose after the Treaty of Paris of 1898?

2. ***Predicting*** What conditions might have led to the Haitian revolt against France?

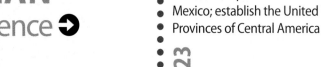

➜**1800**

1823

• Central American provinces declare independence from Mexico; establish the United Provinces of Central America

1804

Following revolt led by François Toussaint-Louverture, Haiti declares independence.

1838–1841

United Provinces of Central America separates into five independent countries: Costa Rica, El Salvador, Honduras, Guatemala, and Nicaragua.

Society and Culture Today

GUIDING QUESTION *How have the traditions and beliefs of indigenous peoples, Africans, and Europeans shaped society and culture in Central America and the Caribbean?*

The merging of indigenous, European, African, and Asian cultures has created a distinct Latin American identity. This is expressed through unique blendings of language, traditions, and art. In Central America, the primary language of most countries is Spanish. In the Caribbean, primary languages include Dutch, English, French, and Spanish. Each country has its own **dialects**, or local languages, and millions of people speak indigenous languages and are bilingual. Some speak a form of **patois** (PA•TWAH), dialects that blend languages. Haitian Creole, for example, blends French vocabulary with African and Spanish words.

Many types of religions are practiced in the region. In Central America, 80 percent of people are Roman Catholic while in the Caribbean, similarly high populations of Roman Catholics occur mainly where French or Spanish is spoken. Protestant religions are common where English is spoken. Other faiths include Hinduism, Islam, and indigenous and African religions, often mixed with Christianity and other faiths. Santería in Cuba and voodoo in Haiti are examples.

The quality of education and health care varies greatly from country to country, as well as within rural and urban areas. Generally, children are required to complete elementary school, but many do not, lacking money for clothing and supplies or the means to travel long distances to school. Countries with less-developed economies and a lower standard of living tend to have more disease and malnutrition and a low life expectancy.

Family and Status of Women

In the Caribbean, family structure is often **matriarchal**, or headed by a woman. This type of family is characteristic of West Africa, from where many local people trace their roots. This is not the case in Central America, where families are mostly male dominated. Current trends show alterations in the

population pressure the sum of factors within a population that reduce the ability of an environment to support the population, therefore resulting in migration or population decline

dialect local form of a language used in a particular place or by a certain group

patois a dialect used in everyday speech that blends elements of several languages

matriarchal family ruled by a woman such as a mother, grandmother, or aunt

(c)Sven Creutzmann/Mambo Photography/Getty Images News/Getty Images, (b)©Corbis

The Panama Canal opens for use across the Isthmus of Panama.

1914

Fidel Castro becomes the dictator of Cuba as a newly established communist state.

1959

Fidel Castro hands over power to his brother Raúl.

2008

➔**1900**

➔**2008**

1898

The United States engages in a war with Spain to liberate Cuba from Spanish rule. The Treaty of Paris of 1898 ends the war. Spain loses control over Cuba and cedes Guam and Puerto Rico to the United States.

The Mola: Colorful Kuna Culture

Created by Kuna women in Panama, molas are made from layers of cloth that are cut and stitched into brightly colored, intricate designs. Molas are worn by Kuna women as part of their everyday garments and are inspired by ancient and contemporary sources.

❝... [T]he relation between designs and images lies at the core of the Kuna theory of personhood. Kuna people consider persons as formed by elements that have their own independent life in the universe, which are then transformed and molded into human bodies by adult people. The Kuna theory of visual art is therefore identical to their ontology [nature of their existence]; it provides people with categories to know the world. Concepts such as image and design are embedded in social praxis [customs] and not just abstract ones. Images and designs are not categories of the aesthetic; they are category of praxis [customs].**❞**

—Paolo Fortis, *Kuna Art and Shamanism: An Ethnographic Approach*, 2012

Molas are an important cultural art form and serve as a source of livelihood for those who sell them in markets to tourists.

DBQ ▲ **CRITICAL THINKING**

1. *Interpreting Significance* Describe the importance of designs and images to the Kuna people.
2. *Analyzing Visuals* Describe some of the motif-designs seen in the photo. What are the clues that suggest the molas pictured are intended to be sold to tourists?

---

traditional size and structure of families in Central America and the Caribbean. Fertility rates have decreased. Women are delaying childbearing as well as increasing their participation in the labor force. Increases in the number of working women are **attributed** to advancing levels of education, as well as to social movements promoting gender equality. Still, households living below the poverty line tend to be larger and headed by single women, or to contain at least one elderly person living alone or in charge of the household.

attribute to explain by indicating a cause

The Arts

Indigenous artisans living before the arrival of Columbus in the pre-Columbian era produced the earliest art forms in the region. These included stonework, woodcarving, pottery, metalwork, and weaving. For example, the stone Hieroglyphic Stairway in Copán, Honduras, an ancient Maya city, was carved with over a thousand symbols. The work of contemporary artisans often reflects this ancient heritage. In Guatemala today, hand-woven textiles display ancient Maya symbols and weaving techniques as well as modern influences.

Cross-cultural influences also combine to produce musical styles and dances unique to particular regions of Central America and the Caribbean. Styles evolved from multiple musical traditions. These include indigenous wind and percussion instruments, European string instruments, and African drums, rhythms, and dances. Many countries have their own distinct dance or music styles. Rumba and salsa originated in Cuba, while merengue began in the Dominican Republic and Haiti. Latin jazz permeates the cultures of Cuba and the Spanish Caribbean.

✔ **READING PROGRESS CHECK**

Evaluating What dance and music styles are found in Central America and the Caribbean?

PHOTO: ©Bob Krist/Corbis; TEXT: Kuna Art and Shamanism: An Ethnographic Approach by Paolo Fortis. Copyright © 2003-2011 University of Texas Press. All rights reserved.

Economic Activities

GUIDING QUESTION *What economic factors have helped maintain an unequal distribution of wealth in Central America and the Caribbean?*

Countries in Central America and the Caribbean face the challenge of developing and diversifying economies in which wealth is unevenly distributed. In most Central American countries, agriculture employs more workers than other sectors. The legacy of the hacienda system hampers reform. Traditionally, these large farms were owned by wealthy families and worked by the poor. Such estates were known as *latifundia*, and the relationships between landowners and workers persist even today, with corporations owning the land and the poor providing the labor. These highly mechanized operations yield high returns for low investment in labor. As mechanization increases, rural farmers and workers, or campesinos (kam•puh•SEE•nohs), leave and look for work in cities. Some campesinos still farm *minifundia*, or small plots of land, to feed their families. However, workers rarely own the land and may be forced to work to pay off debts to landowners.

Some governments are passing laws to distribute farmland more fairly, and many campesinos have combined *minifundia* into large cooperative farms. But the legacy of economic inequality is difficult to overcome, and campesinos remain very poor. Large-scale growers farm cash crops, such as bananas, sugarcane, and coffee. Countries run **obvious** risks, however, if they depend on just one or two export products.

Many countries in the region are still considered less developed countries. In Central America, most industrial employment is in **cottage industries**, those which employ workers in their homes. Other industries include food, beverage, and tobacco processing and the production of textiles, clothing, shoes, furniture, and leather. In the Caribbean, manufacturing of textiles, electronics and clothing, oil refining, and offshore banking are important.

Service and technology industries have grown rapidly in recent years. In the Caribbean new types of tourism, such as **ecotourism**, provide new opportunities. The Dominican Republic, once dependent on sugar production, has capitalized on the telecommunications sector. Other new industries are offshore education, health services, and information technology. In 2005 the United States and six countries signed into law the Dominican Republic-Central American Free Trade Agreement (DR-CAFTA) to lower trade barriers.

latifundia in Latin America, large agricultural estates owned by families or corporations

minifundia in Latin America, small farms that produce food chiefly for family use

obvious easily discovered, seen, or understood

cottage industry a business that employs workers in their homes

ecotourism the practice and business of recreational travel based on concern for the environment

☑ **READING PROGRESS CHECK**

Describing What new industries are contributing to the Caribbean economy?

LESSON 2 REVIEW

Reviewing Vocabulary
1. *Inferring* Describe the relationship between *latifundia* and *minifundia*.

Using Your Notes
2. *Listing* Use your graphic organizer to list the major religions practiced in Central America.

Answering the Guiding Questions
3. *Evaluating* How did colonialism influence the history and government of Central America and the Caribbean?

4. *Finding the Main Idea* How have migration and rapid growth affected the population of Central America and the Caribbean?

5. *Identifying Cause and Effect* How have the traditions and beliefs of indigenous peoples, Africans, and Europeans shaped society and culture in Central America and the Caribbean?

6. *Identifying Central Issues* What economic factors have helped maintain an unequal distribution of wealth in Central America and the Caribbean?

Writing Activity
7. *Argument* Write a paragraph describing why the hacienda system was not beneficial to the people of Central America and Caribbean and how it is still practiced in a form today.

WHAT KIND OF DEVELOPMENT IS BEST FOR HAITI?

The Human Development Index ranks Haiti as the poorest country in the Western Hemisphere. Most of the population lives in absolute poverty, while a large percentage is unemployed or underemployed. Health care resources, including basic sanitation systems, are lacking. Although education is required for children between the ages of six and twelve, only a small percentage of children actually attend school because of a lack of facilities and staff.

In geography, the term *development* refers to improvements in the social and economic welfare of people as well as improvements in production and technology. The ways in which a country experiences development vary greatly—from NGO (nongovernmental organization) programs to government initiatives to foreign business arrangements. Funding in the form of foreign aid, investments, and loans makes development possible, but such support often includes special requirements or limits.

A devastating magnitude 7.0 earthquake on January 12, 2010, reduced much of Port-au-Prince, Haiti, to rubble. Southern areas of the country were also affected. Buildings of all kinds— from shantytowns to hospitals to national landmarks—were damaged or completely collapsed. Telephone service and electricity were knocked out. According to reports, more than 300,000 people lost their lives and over 600,000 were displaced. Estimates show the total cost of the disaster to be between $8 billion and $14 billion. In October 2010, after the earthquake, one of history's worst cholera outbreaks hit Haiti. Cholera is a waterborne disease that is spread through contaminated water. The outbreak was made worse by the earthquake's destruction and lack of public sewage systems. Such loss has only served to magnify Haiti's need for economic and social development.

Thony Belizaire/AFP/Getty Images

Create Market Opportunities

PRIMARY SOURCE

❝ Haiti has duty-free, quota-free access to the American market guaranteed for the next nine years, with generous rules of origin well-suited to the garment industry. . . .

Of course, market access is not enough: costs of production must be globally competitive. . . . In garments the largest single component of costs is labour. Due to its poverty and relatively unregulated labour market, Haiti has labour costs that are fully competitive with China, which is the global benchmark.

Market access and costs of production are not the only factors of importance: transport to market is also a fundamental consideration. . . . Haiti is on the doorstep of its market. Since it is the only low-wage economy in the region, it has a transport advantage over competing low-wage economies of several thousand miles. . . . ❞

—Paul Collier, *Haiti: From Natural Catastrophe to Economic Security,* A Report for the Secretary General of the United Nations, January 2009

Develop Local Agriculture

PRIMARY SOURCE

❝ Municipal governments should construct properly equipped marketplaces for the women who sell rural produce. The Haitian state should develop trade policies aimed at protecting the agricultural sector, and take the lead in fixing roads and ports, confronting deforestation and improving systems of water management. . . .

The return on the investment in the rural economy would be self-reliance, the alleviation of dangerous overcrowding in cities and, most important, a path toward ending Haiti's now chronic problems of malnutrition and food insecurity. . . . ❞

—Laurent Dubois and Deborah Jenson, "Haiti can be rich again," *New York Times,* January 8, 2012

What do you think? DBQ

1. **Drawing Conclusions** According to Collier, why is the garment industry an ideal market opportunity for Haiti?

2. **Identifying Central Issues** Why do Dubois and Jenson argue that protecting the agricultural sector is best for Haitian development?

3. **Evaluating** Who do you think makes the stronger argument, Collier or Dubois and Jenson? Explain.

netw⊙rks

There's More Online!

☑ **IMAGE** Deforestation in Central America

☑ **IMAGE** Urban Pollution in Central America

☑ **IMAGE** Pollution of Freshwater Sources

☑ **INTERACTIVE SELF-CHECK QUIZ**

☑ **VIDEO** People and Their Environment: Central America and the Caribbean

LESSON 3
People and Their Environment: Central America and the Caribbean

ESSENTIAL QUESTION · *How do physical systems and human systems shape a place?*

Reading HELPDESK

Academic Vocabulary

- **cooperation**
- **sustainable**

Content Vocabulary

- **sedimentation**
- **reforestation**

TAKING NOTES: *Key Ideas and Details*

IDENTIFYING Use a graphic organizer like the one below to take notes as you read about the issues relating to people and their environment in Central America and the Caribbean.

IT MATTERS BECAUSE

People and governments in Central America and the islands of the Caribbean struggle to develop modern economies without destroying the natural environment and the invaluable resources and biodiversity it provides. Balancing the needs of humans with sustainable environmental practices is both an important goal and a major challenge for countries of the subregion.

Managing Resources

GUIDING QUESTION *How do growing human needs affect resources and the environment in Central America and the Caribbean?*

Countries in Central America and the Caribbean face a daunting challenge: How can they preserve and manage their resources while developing their economies and meeting the increasing needs of a growing population? Water shortages, access to freshwater, and legal issues related to waterways are a big concern for several countries in the subregion. Urban populations in Central America and the Caribbean increase at a rate that strains their cities' ability to provide freshwater for personal and industrial use.

Water has been at the center of border disputes throughout Central America's history. The latest dispute is part of a conflict that is two centuries old. It involves a wetlands area that is part of a nature reserve owned by Costa Rica. The conflict erupted in October 2010 when Nicaragua began a controversial dredging project to redirect the San Juan River—a Nicaraguan-controlled waterway that forms part of the border—and posted Nicaraguan troops at the site. Costa Rica and other neighboring countries asked to have the troops withdrawn. The Nicaraguan government refused, saying it was reclaiming a natural resource. Costa Rica took the dispute to the United Nations. The United Nations ordered all troops to withdraw. This allowed Costa Rica to monitor potential damage to the wetlands environment, but did not stop Nicaragua from clearing sand from the river's channel.

(tl)©Danny Lehman/Corbis, (tcl)James Brunker/Alamy, (tcr)Orlando Sierra/AFP/Getty Images, (tr)John Coletti/Alamy

To increase the power supply necessary for industry, some countries in Central America and the Caribbean turn to the construction of hydroelectric power plants. Such plants are located within human-made dams that raise the level of water so that when intake channels are opened, water flows rapidly through the dams, turning water turbines. The water turbines capture the energy of fast-falling or flowing water and produce electricity from generators.

Costa Rica benefits from the power generated by hydroelectric plants. The plants supply energy to Costa Rica's industries, allowing them to increase production capabilities and become more competitive in international trade. In the early twenty-first century, about four-fifths of Costa Rica's electricity was produced from hydroelectric plants. Hydroelectric dams promote clean energy use by harnessing the power of water, a renewable resource.

However, hydroelectric systems threaten the natural environment of the areas from which they extract energy. The interaction of plant and animal species within an ecosystem is very sensitive and complex. Fish habitats, for example, are affected by factors such as water level, water velocity, and the availability of food and shelter from predators. The flooding created by hydroelectric dams dramatically alters the habitats and devastates native fish populations. Hydroelectric projects on El Salvador's Lempa River provide most of the country's power needs. These dams harm the natural environment, however, because of changing water levels and the construction of roads through otherwise unaffected landscapes. Similarly, the hydroelectric project on the Patuca River in Honduras blocks fish migrations, alters habitats, and threatens the survival of the region's indigenous population who depend on the environment.

Central America boasts a biodiversity hotspot with naturally fertile soil. Yet much of the subregion's timberland has been cleared by slash-and-burn farmers. In this process of cultivation, all plants are cut down and any trees are stripped of bark. After the plants and trees have dried out, they are set on fire. The ash from the fire adds nutrients to the soil, making agriculture more productive. Unfortunately, frequent rains leach away the beneficial nutrients and within a year or two the soil loses its fertility. Crop yields decline due to poor soils, and farmers move on to clear new parts of the forest. The spent land supports little growth of vegetation. Within just a few years, huge swaths of centuries-old rain forests have disappeared.

Deforestation is a major threat to biodiversity as shown here in Panama and throughout Central America and the Caribbean.

◀ CRITICAL THINKING

1. *Analyzing Visuals* What features of the land shown in the forest signify that deforestation has taken place?

2. *Predicting* How does the destruction of rain forests affect biodiversity?

Slash-and-burn cultivation is not the only activity that contributes to deforestation, or the clearing or destruction of forests. Commercial logging operations are key components of the region's economies. The logging companies harvest trees for timber and other products, which are sold as exports.

As Central American forests are depleted, habitats are lost, resources are threatened, and Earth's biodiversity dwindles. Deforestation severely alters biologically rich ecosystems of tropical rain forests and threatens various species of plants and organisms from which key medicines are derived. Scientists are trying to save species from extinction by creating corridors of vegetation that connect remaining areas of the forest to provide animals with access to habitats. Deforestation may also mean that less carbon dioxide is captured in plants. This would cause higher levels of carbon dioxide to remain in the atmosphere.

✔ **READING PROGRESS CHECK**

Gathering Information What are the effects of slash-and-burn farming?

Human Impact

GUIDING QUESTION *Why is soil erosion and soil decline such an issue in Central America and the Caribbean?*

As cities in Central America and the Caribbean experience rapid urbanization, population growth exceeds available resources like housing, running water, sewage systems, and electricity. Rural workers migrating to cities often cannot find jobs or adequate housing. Thousands are forced to live in slums or shantytowns on the edges of cities. Such communities often rest on dangerous slopes and near delicate wetlands. Mud slides, floods, and other natural disasters can wipe out entire communities. Because they lack running water and underground sewage systems, these areas are unsanitary. As a result, disease can spread rapidly. Cities that expand rapidly may also encroach on natural landscapes that have previously been untouched by human impact.

Runoff from chemical fertilizers and pesticides used on commercial farms pollutes freshwater sources in Central America and the Caribbean.

CRITICAL THINKING ▼

1. Assessing Explain how water pollution can affect areas far from the source of the pollution.

2. Hypothesizing How could pollution in waters cause sickness in humans even if it is not used for drinking?

The rapid growth of cities creates high pollution rates from overloaded sewage, electric, and water systems. Air pollution affects people in cities without adequate clean-air laws. Vehicles clog city streets and release massive amounts of exhaust (greenhouse gases) into the air. Industrial pollutants from factory smokestacks built by multinational firms under free-trade agreements also foul the air. Similarly, runoff from chemical fertilizers and pesticides used on commercial farms damages oceans and freshwater areas, including the Caribbean's coral reefs.

As cities in Central America and the Caribbean grow, the demand for food also increases. The effect of expanding agriculture on the natural environment can be devastating. To increase food supplies, Central American farmers and ranchers clear forested regions, which not only destroys forests but also causes the soil to erode and lose fertility over time. In the Caribbean, small islands colonized for intensive plantation agriculture became prone to soil erosion because of the elimination of natural vegetation. Furthermore, after more than 300 years of commercial agriculture, soil fertility has decreased to the point where large applications of fertilizer are necessary for plant production.

Another factor affecting vegetation loss and soil erosion is the construction of hotels and other structures to support the tourism industry. As rain falls, it washes soil eroded by construction and farming down hills and into the sea, resulting in the **sedimentation** of the reef system. Ultimately, sedimentation may kill coral beds in the water. When fertilizers also become part of the runoff into the sea, they damage coral further and threaten the hundreds of fish species, marine turtles, and sharks that live in the reefs. Communities that depend on the reefs for their livelihoods and food security are also affected.

City sewage is a major contributor to water pollution in Central America.

▲ **CRITICAL THINKING**

1. *Analyzing Visuals* What items in the photograph could pose a threat to sea animals?

2. *Describing* In what way could the pollution shown be an effect of rapid urbanization? Explain.

sedimentation the action or process of forming or depositing sediment

☑ **READING PROGRESS CHECK**

Assessing How has human activity influenced soil erosion and its threat to coral reefs?

Addressing the Issues

GUIDING QUESTION *Why is biodiversity protection so important in Central America?*

Central America and the Caribbean face many international challenges, including conflicts over natural resources and the need to prepare for wide-scale natural disasters. Regional **cooperation** in addressing issues that reach beyond national borders will help the subregion move forward. Government, private industries and grassroots efforts all contribute to protecting the environment.

If the issue of deforestation is not addressed, the rain forests of this subregion will be greatly reduced within 40 years. Although the threats to the world's rain forests are well known, the proposed strategies for preserving them are hotly debated. For instance, Costa Rica and other countries with rain forests listen to the advice of scientists and environmentalists, but they still face pressing social and economic realities. If Costa Rica were to ban the use of rain forest lands, for example, how would it provide for the people who would no longer have a way to support themselves? How would the country handle

cooperation a common effort

sustainable a method of harvesting or using a resource so that the resource is not depleted or permanently damaged

reforestation planting young trees or seeds on lands where trees have been cut or destroyed

Orlando Sierra/AFP/Getty Images

Indigenous Communities Protect Biodiversity

"José López Hernández, a member of the Oxlajuj No'j tribe, would chop down trees in Santa Maria de Jesus, Guatemala, before IDB (Inter-American Development Bank) started a project to restore local traditions and culture as a way to prevent further land degradation and to preserve the region's biodiversity.

Now Hernández is a leader of his indigenous community and is working, together with another 600 families, to plant 60,000 new trees by the end of the year using traditional organic methods."

—Silvia Lambiase, "Indigenous communities in Central America use traditions to protect biodiversity," Platform for Agrobiodiversity Research, May 2011

DBQ *HYPOTHESIZING*
Based on the passage above, what can be inferred about the traditional organic methods in regard to biodiversity?

population growth in coastal areas if vast parts of the interior were not open to settlement? Some of these answers lie in **sustainable** development—technological and economic growth that does not deplete the human and natural resources of a given area. Given time, of course, rain forests would regenerate on their own, but with a considerable loss of biodiversity. Laws requiring **reforestation**—the planting of young trees or the seeds of trees on the land that has been stripped—can help. Developing new methods of farming, mining, and logging and combining conservation with responsible tourism protect the forests and boost local economies.

Initiatives are also underway across Central America to develop "green businesses." The businesses include environmentally-friendly organic food production, renewable energy, and sustainable tourism. Large industries are looking at ways to increase energy efficiency and improve production practices in an effort to be more environmentally friendly.

Governments, international agencies, and grassroots groups are beginning to address the needs of the subregion's urban areas. Work is being done to create programs that limit rural-to-urban migration and to improve the infrastructure of cities. Programs that encourage cooperation with other countries to establish incentives for industries to reduce pollution have also been implemented. A United Nation's initiative in Central America called Reducing Emissions from Deforestation and Forest Degradation, or REDD+, is a program of environmental credits. PRODESEC (Economic Development Program for the Dry Region of Nicaragua) is a Nicaraguan plan aimed at providing new opportunities for rural families. Environmental laws and programs, however, have not reduced the risks of increased pollution associated with industrial growth.

Human activity is not the only source of potential problems for Central America and the Caribbean. The physical geography of the region leaves the area vulnerable to extreme weather-related disasters, such as hurricanes and floods, as well as devastating earthquakes. In order to increase the region's emergency preparedness, governments are cooperating in the use of sophisticated technology, such as satellite imaging and computer modeling, to forecast the direction and severity of natural disasters such as hurricanes.

Scientists are also gathering detailed information about volcanic eruptions in the Caribbean. Since a 1997 major volcanic eruption in Montserrat, scientists are closely monitoring the volcano, which continues to be active. Montserrat's volcano is similar to those on other continents, so the information and lessons learned will help produce more detailed forecasts and predictions around the world.

☑ **READING PROGRESS CHECK**

Exploring Issues Why is reforestation so important?

LESSON 3 REVIEW

Reviewing Vocabulary
1. *Determining Importance* What is the significance of sedimentation and reforestation?

Using Your Notes
2. *Describing* Use your graphic organizer to describe the threats to coral reefs from development and pollution.

Answering the Guiding Questions
3. *Assessing* How do growing human needs affect resources and the environment in Central America and the Caribbean?

4. *Explaining* Why is soil erosion and decline such an issue in Central America and the Caribbean?

5. *Applying* Why is biodiversity protection so important in Central America?

Writing Activity
6. *Informative/Explanatory* Think about the physical environment of the state in which you live. In two paragraphs, compare the ways urban populations in Central America and those in your state have modified their physical environments.

Directions: On a separate sheet of paper, answer the questions below. Make sure you read carefully and answer all parts of the questions.

Lesson Review

Lesson 1

1 *Evaluating* Which Central American waterways provide an important link among the region's countries? Give examples.

2 *Identifying* List some natural resources of the Central America and Caribbean subregion that are important to the global economy.

Lesson 2

3 *Summarizing* Describe the ways in which each of the following factors have influenced history and government in Central America and the Caribbean: indigenous cultures, colonialism, slavery, and struggles for freedom.

4 *Identifying Cause and Effect* What urban challenges have been caused by the migration of many people in Central America and the Caribbean to capital cities and major ports?

Lesson 3

5 *Interpreting* What does the phrase "sustainable development" mean? What is one example of sustainable development?

6 *Identifying Central Issues* Why is deforestation a major problem in Central America? Include reasons why many people cut down forests and reasons many others want to restore them.

Exploring the Essential Question

7 *Speculating* How do physical systems and human systems shape a place? Give specific examples showing why the highland regions have attracted human settlement and migration throughout Central America.

21st Century Skills

8 *Analyzing* Why is regional cooperation particularly important to address international challenges in Central America?

9 *Compare and Contrast* Create a chart of the countries in Central America, ranking them in order of population. Use the Internet to research the latest statistics.

10 *Making Connections* What circumstances might make environmental protection a low priority for some Central American people? List one government program aimed at meeting the needs of the people and protecting the environment.

College and Career Readiness

ETHNIC GROUPS IN THE CARIBBEAN

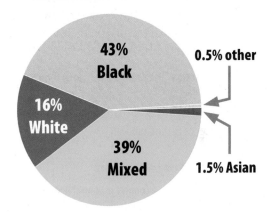

43% Black
0.5% other
16% White
39% Mixed
1.5% Asian

Source: Encyclopedia Britannica Almanac 2009

Use the circle graph to answer the questions below.

11 *Draw Conclusions* Based on what the circle graph shows about ethnic groups in the Caribbean, how do you think these ethnic groups are reflected in the culture of the subregion?

12 *Compare and Contrast* How does the percentage of White compare to the percentage of Mixed in the Caribbean?

Need Extra Help?

If You've Missed Question	1	2	3	4	5	6	7	8	9	10	11	12
Go to page	197	199	200	202	212	210	199	211	202	211	213	213

Directions: On a separate sheet of paper, answer the questions below. Make sure you read carefully and answer all parts of the questions.

Critical Thinking

13 *Exploring* Describe how physical features have kept some of Central America's people isolated.

14 *Drawing Conclusions* Write a paragraph describing the series of events that led to the exploration, conquest, and colonization of Central America and the Caribbean.

15 *Explaining* Describe how elevation can affect the climate and ecosystems more than distance from the Equator in some parts of tropical Central America. Provide specific examples.

Applying Map Skills

Use your Unit 3 Atlas to answer the following questions.

16 *Physical Systems* What is the predominant type of natural vegetation found in Central America?

17 *Places and Regions* What generalizations can you make about the population density in Honduras compared to the population density of El Salvador?

18 *Human Systems* Use your mental map of Central America to describe how the Panama Canal saves travel distance and time for ocean travel from the west coast of North America to the Caribbean.

Research and Presentation

19 *Explaining* Choose a country in Central America or the Caribbean. Write a paragraph describing the landforms within that country. Explain how those landforms affect trade both within the country and with other countries.

Writing About Geography

20 *Narrative* Use the Internet to research and write a three-paragraph account of a cultural event, such as a festival, holiday, or music or dance concert, important to people living in Central America or the Caribbean.

DBQ Analyzing Primary Sources

Read the excerpt and use it to answer the following questions.

At a summit on biodiversity in Japan, the World Future Council announced that Costa Rica had won the 2010 Future Policy award.

PRIMARY SOURCE

"*Costa Rica channels funds from a fuel tax, car stamp duty and energy fees to pay for nature reserve management and environmental services like clean air, fresh water and biodiversity protection.*

Landowners are paid to preserve old-growth forests and to plant new trees. As a result, forest cover has risen from 24% in 1985 to close to 46% today.

It has also established a national commission on biodiversity, comprising scientists, civil servants and indigenous representatives, which proposes policies to the government and promotes green education among the public.

'We are declaring peace with nature,' said Mario Fernández Silva, the ambassador of Costa Rica, referring also to his country's abolition of its army in 1958. 'We feel a strong sense of responsibility about looking after our wealth of biodiversity. Our attitude is not progressive; it is conservative. Our view is that until we know what we have, it is our duty to protect it.'"

—Jonathan Watts, "Costa Rica Recognized for Biodiversity Protection," The Guardian, October 2010

21 *Assessing* Why do you think Costa Rica decided to tax fuel, cars, and energy rather than other goods and services?

22 *Making Decisions* What choices do you think individuals can make to help prevent the loss of rain forests and what incentives can governments provide?

Need Extra Help?

If You've Missed Question	**13**	**14**	**15**	**16**	**17**	**18**	**19**	**20**	**21**	**22**
Go to page	196	200	199	170	172	168	199	204	214	214

Excerpted from "Forever Costa Rica," by Curtis Runyan © Nature Conservancy magazine, June 2011.

South America

networks

There's More Online about South America.

CHAPTER 9

ESSENTIAL QUESTION • *How do physical systems and human systems shape a place?*

Geography Matters...

South America is a land of beautiful natural wonders, including the vast Amazon rain forest. However, not all of South America is a lush jungle paradise. The region also includes snow-covered mountains and boasts the world's driest place— the Atacama Desert.

South America is diverse in culture as well as in landscape. The people of this subregion reflect a unique blend of indigenous and colonial heritage, the result of centuries of migration, trade, and conquest.

◄ Carnival in Brazil is held over five days preceding Lent, with parades featuring elaborate floats, drummers, and dancers.

Viviane Ponti/Lonely Planet Images/Getty Images

215

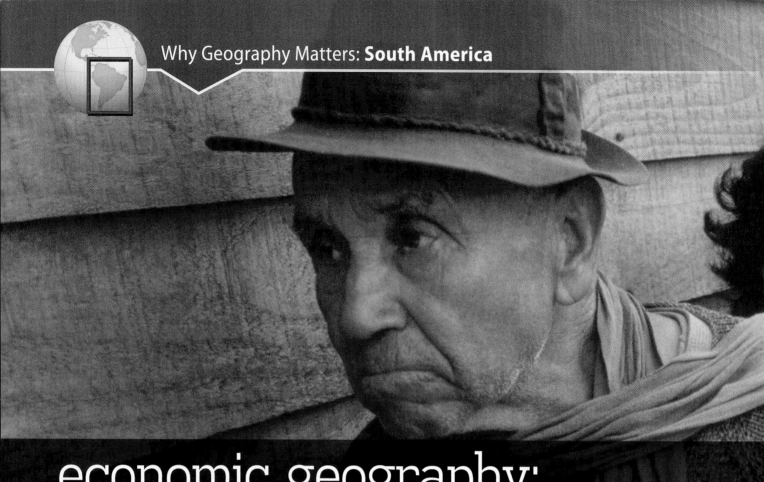

economic geography:
uneven development

As countries struggle to solve problems of urbanization by investing in infrastructures and incentives for investors, they may achieve economic growth. However, the issues of poverty and uneven development still remain as human challenges.

© Diego Goldberg/Sygma/Corbis

THERE'S MORE ONLINE

SEE how Brazil's urban population has grown • **VIEW** a diagram of the core-periphery theory

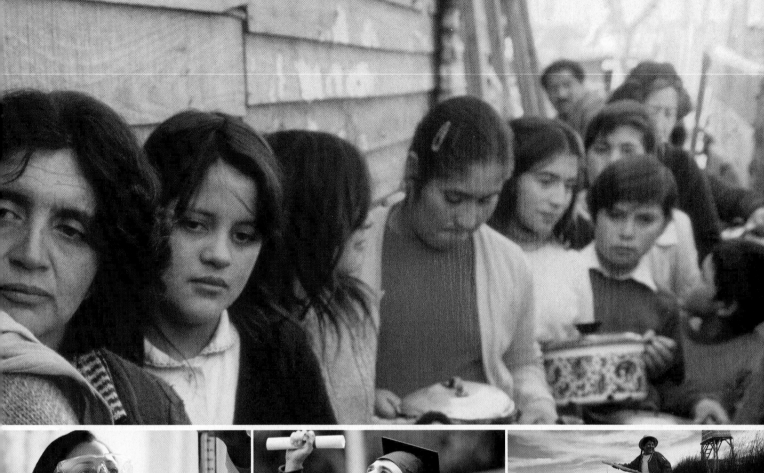

(t) ©Diego Goldberg/Sygma/Corbis, (c)Steve Cole/the Agency Collection/Getty Images, (c)Andres Rodriguez/Alamy, (cr) ©Michael S. Lewis/National Geographic Society/Corbis

How do core-periphery relations fuel uneven development?

Rural workers in search of better job opportunities tend to seek work in the core—cities, such as São Paulo, Brazil. Within a country, rural areas act as a periphery, contributing to burgeoning urban populations. The cities, or core, receiving the migrants often experience explosive growth. As the core grows, it begins to engulf surrounding smaller towns and cities. While the core offers job opportunities, it also presents challenges to governments, which struggle to maintain the infrastructure and services needed to support such a large, quickly growing urban population.

1. The World in Spatial Terms
Describe the migration of rural workers to urban areas in terms of both push factors and pull factors.

What are the effects of uneven patterns of development?

Many government programs focus on the metropolitan core areas and provide insufficient support to the periphery areas. This can leave the periphery with inadequate transportation, sanitation, education, and other essential services. Another result of uneven development is brain drain, or the process of educated and professional individuals leaving the periphery to go to the core for job opportunities. This deprives areas in the periphery of their most educated residents. For example, in the state of Ceará, home to some of the poorest Brazilians, there is a lack of incentives for establishing non-farming activities. This is a push factor for many residents to seek better opportunities in nearby core areas. Although the metropolitan areas have shown economic growth, the smaller urban areas and rural towns are still areas of widespread poverty.

2. Human Systems
How might the migration of workers from rural areas to large cities negatively impact rural areas?

How can the issue of uneven development be addressed?

Uneven patterns of development within countries are a result of unequal distribution of resources, capital, and infrastructure between the core and the periphery. Some governments work to create programs promoting non-farming activities in the periphery so there is more incentive for skilled workers to stay in these areas to develop business and industry.

3. Environment and Society
Imagine you are a candidate for mayor of a large city plagued by problems of overcrowding, poor sanitation, crime, and unregulated land use (squatting) as a result of rural workers flocking to your city for better jobs. Write a campaign speech detailing how you would address these problems if elected.

networks
There's More Online!

- ☑ **IMAGE** The Andes
- ☑ **DIAGRAM** South America: Elevation Profile
- ☑ **INFOGRAPHIC** El Niño
- ☑ **GRAPH** Oil Reserves of South America
- ☑ **INTERACTIVE SELF-CHECK QUIZ**
- ☑ **VIDEO** Physical Geography of South America

Reading **HELP**DESK

Academic Vocabulary

- **phenomenon**
- **predominant**

Content Vocabulary

- **cordillera**
- **altiplano**
- **escarpment**
- **llanos**
- **pampas**

TAKING NOTES: *Key Ideas and Details*

PARAPHRASING Use a graphic organizer like the one below to take notes on the biomes of South America, including any adaptations of plants and animals for survival in their biome.

South American Biomes

LESSON 1
Physical Geography of South America

ESSENTIAL QUESTION · *How do physical systems and human systems shape a place?*

IT MATTERS BECAUSE
The diverse landscapes of South America are very different from what we know in the United States. They have led to the people in South America developing very different lifestyles that are adapted to their physical environment.

Landforms

GUIDING QUESTION *How has South America's rugged landscape both attracted and isolated people?*

The Isthmus of Panama connects North America to South America. The subregion of South America spans 4,700 miles (7,564 km) from north to south, passing through the Equator near its widest point of 3,300 miles (5,311 km). Like Central America and Mexico, the most characteristic of South America's many landforms are its mountains. The Andes are the world's longest mountain chain. Some peaks in the Andes rise more than 20,000 feet (6,096 m) above sea level. The Andes consist of **cordilleras**, groups of several mountain ranges that run parallel to one another. Although known by different names, they are an extension of the Rocky Mountains that run from Canada south through the western United States and into Mexico and Central America. Because the cordilleras have established natural barriers between surrounding areas, many indigenous communities developed as isolated groups. As a result, some mountain villages exhibit centuries-old social customs.

The Andes encircle the **altiplano**, which means "high plain." The altiplano is an area that includes southeastern Peru and western Bolivia. It is the second-largest mountain plateau in the world.

In southern Argentina, hills and flatlands form the plateau of Patagonia. The presence of the Andes to the west produces a rain shadow that causes Patagonia to be dry, barren, and windy. The Patagonian region also extends across the Andes to southern Chile. Patagonia boasts dramatic valleys, glaciers, and fjords. The rugged Andes and Patagonia's landscape are a result of its location along the Ring of Fire.

(l)David Madison/Photodisc/Getty Images, (cr)Chlaus Lotscher/Peter Arnold/Getty Images

Heavy tectonic activity in the subregion changes and reshapes the landscape. But despite the threats of natural disasters, people have chosen to settle in the Andean highlands for thousands of years. The climates are cooler, the volcanic soil is good for agriculture, and natural resources are concentrated here.

In contrast to the high peaks of the western Andes, eastern South America is defined by broad plateaus and valleys. The Amazon Basin, located along the eastern base of the Andes, is the lowlands area drained by the Amazon River. Just south is the Mato Grosso Plateau, a sparsely populated plateau of forests and grasslands extending across Brazil, Bolivia, and Peru. Farther east are the Brazilian Highlands, a vast area spanning several climate and vegetation zones. Warm climates and open spaces make the Brazilian Highlands good for raising livestock. The Eastern Highlands plunge to the Atlantic Ocean, forming a steep slope called an **escarpment**. This escarpment presents obstacles for inland development. As a result, most of Brazil's population lives along the coast.

Narrow coastal lowlands hem the Atlantic and Pacific coasts of South America. South America's inland grasslands—the **llanos** (LAH•nohs) of Colombia and Venezuela and the **pampas** of Argentina and Uruguay—provide grazing for cattle. Ranchers on large estates employ cowhands, called *llaneros* or gauchos, to drive herds across the rolling plains. Known for its fertile soil, the pampas are one of the world's breadbaskets, producing wheat and corn.

cordillera parallel chains or ranges of mountains

altiplano Spanish for "high plain," a region in Peru and Bolivia encircled by the Andes

escarpment a steep cliff or slope between a higher and lower land surface

llanos fertile grasslands found in inland areas of Colombia and Venezuela

pampas grassy, treeless plains of southern South America

✔ **READING PROGRESS CHECK**

Explaining What features of South America's landscape have hindered or encouraged development?

The Andes extend along the western part of South America. They are the world's longest mountain chain and one of the highest.

David Madison/Photodisc/Getty Images

▲ **CRITICAL THINKING**

1. *Explaining* How do physical features affect human development?

2. *Describing* In what ways have the populations of the Andes region become dependent on their environment?

Water Systems

GUIDING QUESTION *How are South America's rivers important for economic development?*

Waterways are important for the subregion's economic development because they provide ways to transport goods and people within and between the countries of South America. As the Western Hemisphere's longest river and the world's second longest, the Amazon River flows about 4,000 miles (6,400 km) through the heart of South America. It begins in the headwaters of the Peruvian Andes, flows across the lowlands of the Amazon Basin in the interior of Brazil, and drains into the Atlantic Ocean. Hundreds of smaller rivers join the Amazon as it flows from the Andes to the Atlantic Ocean. Together these rivers form the Amazon Basin. The basin drains an area of more than 2 million square miles (5.2 million sq. km).

The Paraná, Paraguay, and Uruguay Rivers form the second-largest river system in Latin America, draining the rainy eastern half of South America. These rivers flow through the Pantanal, one of the world's largest tropical wetlands. After coursing through inland areas, the rivers flow into a broad estuary where the ocean tide meets a river current. This estuary, the Río de la Plata, or "River of Silver," flows into the Atlantic Ocean.

IMPACTS OF EL NIÑO
in Latin America

In an El Niño year, the normally low atmospheric pressure over the western Pacific rises, and the normally high pressure over the eastern Pacific drops. This reversal causes a change in wind patterns, which reverses the equatorial ocean currents and influences climates around the world.

El Niño's disruption of the ocean-atmosphere system in South America has a great impact on climate and economies in the region.

▲ **CRITICAL THINKING**

1. ***Sequencing*** Describe the order of events beginning with changes in the atmospheric pressure systems that lead to the impacts caused by El Niño.

2. ***Speculating*** Explain how the environmental impacts of El Niño could have negative economic effects on countries.

Population Patterns

GUIDING QUESTION *How has South America's physical geography influenced its population patterns?*

South America is the world's fourth-largest continent. The continent's 12 countries are home to nearly 400 million people. Like much of the rest of the developing world, population growth is steady, and so is the migration of people into large, urban areas.

South America's once high rate of population growth is beginning to slow. Urban populations now include fewer children as well as women and men with increased levels of education. Most people live on or near the coasts and along major rivers of the continent. These coastal regions offer favorable climates, fertile land, and access to transportation. The rain forests, deserts, and mountainous areas of South America's interior have discouraged human settlement.

South American countries tend to have low population densities. Ecuador, the most densely populated country in South America, has an average of 132 people per square mile (51 people per sq. km). Brazil has a population of nearly 200 million. However, because Brazil has about 3.2 million square miles (8.4 million sq. km), its average population density is about 60 people per square mile (23 people per sq. km). Despite population densities that are lower overall than other world regions, much of the economic and structural development is concentrated in major cities. Today about 80 percent of the subregion's population lives in urban areas.

In highly populated urban areas such as São Paulo, Buenos Aires, and Bogotá, finding employment and suitable living conditions is difficult for migrants arriving in the city. Rural-to-urban migrants seek higher wages, better living conditions, and sometimes an escape from the violence of drug cartels or criminal groups. Countries across the region are experiencing **brain drain** to North America and Europe as people search for a better life.

brain drain the loss of highly educated and skilled workers to other countries

✔ **READING PROGRESS CHECK**

Explaining Why do most South Americans live along the continent's coasts?

(l)©Bettmann/Corbis, (c)©John Van Hasselt/Sygma/Corbis, (r)Paul Zimmerman/WireImage/Getty Images

1946

Juan Perón is elected president of Argentina. He is also founder and leader of the Peronist movement, which combines populist and nationalistic policies.

1950

Perón is reelected president after 18 years in exile.

1973

Hugo Chávez is elected president of Venezuela. Key factors of his ideology (*chavismo*) are nationalism, a centralized economy, and a strong military.

1998

2000

2006

Michelle Bachelet is elected Chile's first female president; Cristina Kirchner is elected president of Argentina the following year.

2010

Dilma Rousseff becomes president of Brazil.

Dominant Ethnic Groups of South America

Caribbean Sea

VENEZUELA
GUYANA
SURINAME
FRENCH GUIANA
(Fr.)

COLOMBIA

EQUATOR

ECUADOR

Galápagos
Islands
(Ecuador)

PERU

BRAZIL

PACIFIC
OCEAN

BOLIVIA

PARAGUAY

TROPIC OF CAPRICORN

CHILE

ATLANTIC
OCEAN

URUGUAY

ARGENTINA

	African
	Mestizo
	European
	Indigenous

0 ___ 1,000 miles
0 ___ 1,000 kilometers
Lambert Azimuthal Equal-Area projection

Falkland
Islands (U.K.)

South
Georgia
Island (U.K.)

60°W   40°W   20°W

GEOGRAPHY CONNECTION

South America's population has been shaped by ethnic diversity, physical geography, migration, and urban growth.

1. **PLACES AND REGIONS** What landforms are found in the areas where the majority of indigenous people live?

2. **HUMAN SYSTEMS** What countries in South America have large African populations?

Society and Culture Today

GUIDING QUESTION *Why is South America one of the world's most culturally diverse areas?*

South America is home to an ethnically diverse population. Today many indigenous cultural groups inhabit the subregion, especially in rural or less populated areas. Most indigenous groups—of which there are more than 350—live in the Andes region of Ecuador, Peru, and Bolivia.

The Spanish and Portuguese were the first Europeans in South America. Enslaved Africans were later brought as laborers. After South American countries gained their independence, other European groups—French, Dutch, Italians, and Germans—moved to South America. In fact, Argentina's population is 97 percent European, as the majority of Argentines are descendants of Spanish and Italian immigrants. Towns in the lakes region of southern Chile exhibit architecture, cuisine, and traditions influenced by its German population.

Immigrants from Asia also arrived in South America. In Guyana, almost half of the population is of South Asian descent. People of Chinese descent have immigrated to Peru. Many people of Japanese descent live in Brazil, Argentina, and Peru. Spanish, Portuguese, Dutch, French, and English are each spoken in different parts of South America. In countries with South Asian populations, such as in Guyana and Suriname, people also speak Urdu, Javanese, and Caribbean Hindustani, a dialect of Hindi. As a result, people in South America are often bilingual. During the colonial period, some European languages blended with indigenous languages to form completely new languages.

The majority of South Americans are Roman Catholic. Carnival is celebrated in the week before the Roman Catholic observance of Lent, a 40-day period of fasting and prayer before Easter. People from around the world come to Rio de Janeiro to participate in Carnival celebrations. In addition, tens of millions of people practice syncretism, a combination of mixed religions, such as Macumba and Candomblé, which combine West African religions with Roman Catholicism. Other minority religions include Protestant Christianity, Hinduism, Buddhism, Shinto, Islam, Judaism, and Eastern Orthodox Christianity.

Education varies greatly throughout South America. Many countries support public education through high school, and literacy rates have risen steadily. Public universities provide higher education at little or no cost to students in many countries. Many children leave school before completion, however, to help support the family by selling goods in markets or engaging in household or farming duties.

In countries with stable economies and high standards of living, people have access to better health care and live longer, healthier lives. This results in a situation of **uneven development** among countries. The health of a country's people is linked to poverty, lack of sanitation, infectious diseases, and malnutrition. These conditions persist in rural areas and especially in the slums on the outskirts of cities where millions of people live in overcrowded conditions.

uneven development
condition in which some places do not benefit as much as others from social and economic advancement

Family and the Status of Women

In urban upper and middle classes, the family unit is likely to consist of a nuclear household—father, mother, and dependent children—rather than an extended family. Loyalty and responsibility toward the extended family, however, remain very strong. The *compadre* relationship, in which parents and godparents share in the upbringing of a child, is valued in parts of Latin America. However, changes brought about by urban society have diminished its overall importance.

The elevation of women's rights has grown as countries have established more stable governments and economies in the past few decades. The result is an increasing proportion of women entering the workforce. Work codes in Chile and Colombia provide benefits for pregnant employees, and women's earnings are increasing in comparison to men's wages. Although women can still face discrimination and mistreatment, there are signs of change. Some countries now provide shelters for abused women and enforce stricter penalties for offenders.

The Arts

Indigenous arts survive in many different forms. The massive buildings of the ancient Inca at Cuzco and Machu Picchu reveal a mastery of stone and engineering that are still studied today for their ingenuity. Traditional arts and crafts dating from before the arrival of the Europeans—such as weaving, ceramics, and metalworking—have been passed from generation to generation.

Music also has ancient ties. Panpipes are one of the most common pre-Columbian musical instruments from the Andean region. Musical traditions later mixed Native American, African, and European influences to create unique styles. The Brazilian samba, Chilean cumbia, and the Argentine tango complement the Cuban salsa and Dominican merengue to exemplify the diversity of music developed from a mixture of cultural and geographical roots.

✔ READING PROGRESS CHECK

Specifying How has South America's role as a cultural melting pot contributed to its unique cultural elements?

This funeral mask from the Moche culture of South America is pre-Columbian, meaning it dates from before the arrival of Europeans.

▼ CRITICAL THINKING

1. ***Theorizing*** Why might geographers use the term *pre-Columbian* when studying cultures? Think about the significance of the arrival of Europeans.

2. ***Making Connections*** What features do you recognize in the mask as being similar to a popular fad in youth cultures today?

Economic Activities

GUIDING QUESTION *How have South America's abundant natural resources contributed to its economic development?*

Several countries in the region have combined their abundance of natural resources with current changes in government and improved economic conditions. Argentina is not only rich in natural resources, but also has a highly literate population and a diversified economic base that have helped to fuel a strong expansion in its economy. Brazil, the largest country in South America, has been undergoing continuous growth and development since 1970. Taking advantage of its natural resources, Brazil has become a powerful country in economic terms. Chile has established

©Bowers Museum/Corbis

Agricultural Land Use in South America

Caribbean Sea

GUYANA
VENEZUELA   SURINAME
                FRENCH GUIANA
COLOMBIA   (Fr.)

EQUATOR
ECUADOR

Galápagos
Islands
(Ecuador)

PERU

PACIFIC
OCEAN

BOLIVIA

PARAGUAY

CHILE

BRAZIL

TROPIC OF CAPRICORN

ATLANTIC
OCEAN

URUGUAY

ARGENTINA

0   1,000 miles
0   1,000 kilometers
Lambert Azimuthal Equal-Area projection

Falkland
Islands (U.K.)

South
Georgia
Island (U.K.)

60°W   40°W   20°W

Legend:
- Scattered pastoralism
- Non-soybean grain crops
- Dairy products and flowers for export
- Mixed dry farming
- Soybeans
- Cattle
- Plantation
- Agroforestry and shifting cultivation
- Highland mixed subsistence-market
- Mixed
- Nonagricultural areas

GEOGRAPHY CONNECTION

South America's rich agricultural tradition presents some modern-day challenges to development.

1. **ENVIRONMENT AND SOCIETY** What physical feature in Argentina and Uruguay creates good areas for raising cattle?

2. **PLACES AND REGIONS** What generalization can you make about the location of agroforestry and shifting cultivation in South America?

widespread covering a wide area; prevalent

free-trade agreements with the United States, Turkey, Australia, and other countries to enhance its economic activities.

Resources and Industry

Land and water use in South America closely follow physical geography. Forestry prevails in the Amazon Basin. Ranching is **widespread** in the grasslands of the south. Herding llamas and alpacas occurs in the high Andean regions. Fishing occurs in major lakes, rivers, and along coastlines. Agriculture remains highly important in South America. More than 20 percent of the subregion's workforce is employed in the primary sector that includes farming, ranching, and fishing. As in other subregions of Latin America (Mexico and Central America and the Caribbean), the legacy of the hacienda system still exists in South America. Larger commercial and smaller subsistence agriculture exist side by side.

Agriculture dominates much of east-central Brazil and the nearby areas of Paraguay, Uruguay, and Argentina. South America's contribution to agricultural global trade includes grains, soybeans, coffee, cocoa, citrus, cattle, sugarcane, tobacco, and cotton. In fact, Brazil is the world's largest exporter of coffee. Brazil and Paraguay also cultivate today's fastest-growing crop in the global economy: soybeans. Paraguay is the sixth-largest producer of soybeans in the world.

Additionally, the coca plant thrives in the northwestern parts of South America. Coca use is popular among Bolivia and Peru's working class for its effects as a legal stimulant and appetite suppressant. Yet coca's derivative can also be used to make the illegal drug cocaine. Peru, Bolivia, and Colombia have nonetheless legalized coca farming. They did so because, sold in its legal form, it is a large source of profit for these countries.

Natural resources include timber, gold, silver, copper, iron ore, and tin. South America contains about one-fifth of the world's iron ore, which is used for steel making and machine building. Many countries are heavily dependent on exporting their natural resources. Energy resources include petroleum and natural gas. Venezuela, Ecuador, and Argentina are leading exporters.

Manufacturing is growing rapidly, but the region's geography varies greatly. Most manufacturing is concentrated in urban areas, especially the primate cities. Because the largest cities lie mainly along the coasts where transportation is the best, the vast interior of South America has few manufacturing plants.

The major road systems in South America include the Pan-American Highway that stretches through Chile as it links many cities north to south, and the Trans-Andean Highway that links cities in Chile and Argentina east and west. The Trans-Amazonian Highway was built by Brazil to access the Amazon rain forest for developing timber and mineral resources. The Transoceanic Highway was designed to link the Amazon River ports with Peru's ports on the Pacific to transport agricultural products to the global markets in Asia and Europe. Argentina and Brazil have well-developed rail systems, which are important modes of transportation along with the inland waterways. All South American capital cities and major cities have domestic and international airports.

Economic Integration

The increased global demand for raw natural resources and manufacturing has had an impact on the overall economic growth in the region's countries. The economic growth has affected Brazil, Chile, and Argentina more than other countries. Ecuador, Peru, Venezuela, and Bolivia have struggled to modernize their economies and improve standards of living. For example, Bolivia is one of South America's most impoverished and least developed countries. Political reforms in the 1990s stimulated economic growth, but Bolivia continues to struggle to improve conditions for its people.

The separatism that characterized South American countries in the past is giving way to new trade partnerships and cooperation on infrastructure that are mutually beneficial. In South America's current economy, investments flow more freely from one country to another. For example, Colombia has taken advantage of its more stable economy by promoting free-trade agreements with other countries. Colombia **implemented** the U.S.-Colombia Free Trade Agreement with the United States in 2012. Colombia is negotiating free-trade agreements with other Latin American countries, such as Chile and Mexico, as well as with countries outside South America. Additionally, Colombia has 59 bilateral or regional trade agreements—more than any other country.

implement to carry out or accomplish by concrete measures

Economic growth has been steady and strong across the region and has permitted countries to pay their foreign debt. If the economy does not grow, then the debt payments reduce the amount of money available to the government for essential services such as roads, water, flood control, and health care. Consistent, stable economic policies adopted by some South American countries in recent decades have contributed to steady economic growth and improved standards of living for people.

✓ **READING PROGRESS CHECK**

Inferring What is the benefit of cooperation between countries in South America?

LESSON 2 REVIEW

Reviewing Vocabulary

1. *Summarizing* Describe the concept of brain drain and how it relates to the countries of South America.

Using Your Notes

2. *Describing* Use your notes from the graphic organizer to describe South America's contributions to the arts. Include specific examples from the lesson.

Answering the Guiding Questions

3. *Evaluating* How have indigenous peoples and Europeans contributed to the creation of modern governments in South America?

4. *Making Connections* How has South America's physical geography influenced its population patterns?

5. *Explaining* Why is South America one of the world's most culturally diverse areas?

6. *Examining* How have South America's abundant natural resources contributed to its economic development?

Writing Activity

7. *Informative/Explanatory* Write a paragraph describing how countries in South America have established transnational and international relationships. Include examples of relations among the countries within South America and countries around the world.

AMAZON
IN THE BALANCE

The Amazon rain forest is the largest and most diverse rain forest on Earth, holding an abundance of plant and animal life. This rain forest and the life in it is under constant threat. Deforestation caused by logging, farming, and ranching threatens its depletion and the extinction of the plants and animals that call it home. In some ways the loss of the Amazon rain forest threatens us all.

FULL OF LIFE

In every 4 square miles of the Amazon rain forest, there are **1,500** flowering plant species, **750** tree species, and **900** tons of living plants. The Amazon is also home to **30 million** insect species, **175** lizard species, and **500** mammalian species. **One-third** of the world's birds also live in the Amazon.

CARBON CONTROL

The Amazon rain forest holds nearly **half** of the 247 billion tons of carbon contained in the world's tropical rain forests. Carbon is stored in the woods and roots, but when trees are **cut, burned, or decompose**, carbon is released into the atmosphere. This has **consequences** for the Earth's climate. According to NASA "...the average global temperature on Earth has **increased** by about 0.8^{0}Celsius (1.4^{0}Fahrenheit) since 1880. Two-thirds of the warming has occurred since 1975, at a rate of roughly 0.15-0.20^{0}C per decade."

LONG-TERM EFFECTS OF DEFORESTATION

"Forest clearing in Brazil has already claimed **casualties**, but the animals lost to date in the rainforest region are just **one-fifth** of those that will slowly die out as the full impact of the loss of habitat takes its toll. In parts of the eastern and southern Amazon, 30 years of **concerted deforestation** have shrunk viable living and breeding territories enough to condemn 38 species to regional **extinction** in coming years, including 10 mammal, 20 bird and 8 amphibian species, scientists found."

—Ian Sample, *The Guardian (UK)*, July 12, 2012

ON THE RANGE

Cattle ranching is the **leading** cause of deforestation in the Brazilian Amazon. It accounts for **60%** of forest clearing. Subsistence and commercial agriculture, development, and **logging** also drive deforestation.

Making Connections

1. ***Making Predictions*** What may happen to the rain forest if land clearing continues unchecked?

2. ***Drawing Conclusions*** How might the continued deforestation and clearing of the Amazon rain forest have a direct effect on your life?

3. ***Environment and Society*** What circumstances might make environmental protection of the Amazon rain forest a low priority for some Latin American countries?

*Interact with **Global Connections** Online*

net·works

There's More Online!

☑ **GRAPH** Per Capita CO_2 Emissions

☑ **IMAGE** EcoTaxi

☑ **IMAGE** Large-Scale Monoculture

☑ **INTERACTIVE SELF-CHECK QUIZ**

☑ **VIDEO** People and Their Environment: South America

LESSON 3
People and Their Environment: South America

(tl)©Paulo Fridman/Corbis, (tcl)Nilton Ricardo/Brazil Photos/Alamy, (tc)Mauricio Lima/AFP/Getty Images, (tcr)Alexandre Cappi/LatinContent/Getty Images, (tr)Steve Allen/Brand X Pictures/Getty Images

Reading HELPDESK

Academic Vocabulary

- **voluntary**
- **alternative**

Content Vocabulary

- **oxisol**
- **monoculture**

TAKING NOTES: *Key Ideas and Details*

IDENTIFYING Use a graphic organizer like the one below to take notes as you read about the issues that relate to people and their environment in South America.

	Impact on Environment
Deforestation	⟹
Large-scale agriculture	⟹
Grazing livestock	⟹
Urban Growth	⟹
Mining	⟹

ESSENTIAL QUESTION · *How do physical systems and human systems shape a place?*

IT MATTERS BECAUSE

The use of natural resources for economic advancement has benefited the livelihood and stability of many South American countries. It has also caused damage to the continent's natural environment and dramatic changes to its biodiversity.

Managing Resources

GUIDING QUESTION *How has the management of forest and agricultural resources impacted the environment in South America?*

South America is home to some of the largest reserves of forest and agricultural resources in the world. Countries are taking advantage of such resources to improve their economies and overall wealth. Extensive exploitation of resources comes at a cost, however. Issues such as deforestation, soil erosion, desertification, and pollution have been significant problems for all countries in South America.

Like the rain forests of Costa Rica and Panama, the rain forests of South America are threatened by intensive human activity. Rain forests harbor at least half of all animal and plant species on Earth. Deforestation is occurring at a rapid rate in the Amazon rain forest. This has reduced the diversity of plants and animals there. Brazil has the world's largest remaining expanses of tropical rain forest, but almost 20 percent of the Amazon rain forest has already been destroyed. The loss of biodiversity is also occurring in Brazil's lesser-known Atlantic Forest, one of Earth's richest and most threatened habitats. The Atlantic Forest now covers less than 10 percent of its original area.

Soil erosion in South America has diminished the ability of soils to produce food and vegetation. Intensive farming, construction, logging, fires, and overgrazing all increase the rate of soil erosion. Certain soil types in South America—such as the volcanic soils and

Europe

Sylvain Grandadam/Robert Harding World Imagery/Getty Images

UNIT **4**

1 Culture Europe's cultural heritage is as varied as its landscape. Centuries-old traditions, such as the running of the bulls in Pamplona, Spain, illustrate how the region's history is intertwined with its modern-day culture.

EXPLORE the REGION

EUROPE is a peninsula of peninsulas, with many pieces of land extending into the Atlantic Ocean and the Mediterranean Sea. Over the centuries, Europeans have taken advantage of their location, using the seas as a source of food and an avenue for trade and exploration. Europe is home to many different languages and culture groups. Throughout history, conflicts between competing nations have caused destruction in the region. In recent years, however, most of Europe has joined together in an economic and political union that has fostered peace and prosperity.

THERE'S MORE ONLINE

(tl)Westend61/the Agency Collection/Getty Images, (bl)Barry Wong/The Image Bank/Getty Images, (br)Jorg Greuel/Getty Images

3 Rivers   The Danube River flows from southern Germany to the Black Sea. Like many European rivers, it is an important commercial route as well as a scenic highlight.

4 Mountains   The Alps stretch 750 miles (1,200 km) through eight countries, separating Southern Europe from the north. The mountain range's imposing peaks are the source of many of the region's rivers.

2 Economy   Tourism is an important sector of the European economy. It comprises a wide variety of destinations and employs a large percentage of the European workforce. Venice, with its beautiful canals, is a top choice among travelers.

Europe
Physical

0   1,000 miles
0   1,000 kilometers
Two-Point Equidistant projection

100°W   120°W   140°W

NORTH AMERICA

80°W

Elevations

10,000 ft. (3,000 m)	
5,000 ft. (1,500 m)	
2,000 ft. (600 m)	
1,000 ft. (300 m)	
0 ft. (0 m)	
Below sea level	

—— National boundary

▲ Mountain peak

▼ Lowest point

60°W

PACIFIC OCEAN

180°

Bering Strait   **Bering Sea**

Chukchi Sea   Chukchi Peninsula

Wrangel I.

ARCTIC OCEAN

80°N

New Siberian Islands

East Siberian Sea

Klyuchevskaya Sopka 15,580 ft. ▲ (4,749 m)

Kamchata Peninsula

160°E

KOLYMA LOWLAND

Kolyma Mts.

Kolyma R.

Cherski Range

Sea of Okhotsk

Kuril Islands

40°W

ATLANTIC OCEAN

Iceland

ARCTIC CIRCLE

60°N

Svalbard

Franz Josef Land

Severnaya Zemlya

Kara Sea

Taymyr Peninsula

Laptev Sea

Verkhoyanski Mts.

Lena R.

Stanovoy Range

Sakhalin

140°E

20°W

Faeroe Islands

Norwegian Sea

Shetland Islands

Hebrides

Ireland

Celtic Sea

Great Britain

North Sea

Scandinavia

Gulf of Bothnia

Kola Peninsula

Barents Sea

Novaya Zemlya

Yamal Peninsula

SIBERIA

CENTRAL SIBERIAN PLATEAU

Lower Tunguska R.

Yenisey R.

Yablonovyy Range

Lake Baikal

120°E

Bay of Biscay

Mont Blanc 15,771 ft. (4,807 m)

Meseta

Pyrenees

ALPS

IBERIAN PENINSULA

Corsica

Apennines

Balearic Islands

Sardinia

BALKAN PENINSULA

Sicily

Rhine R.

Baltic Sea

NORTHERN EUROPEAN PLAIN

Carpathian Mts.

Dnieper R.

Dniester R.

Northern Dvina R.

WEST SIBERIAN PLAIN

Ob' R.

URAL MOUNTAINS

Irtysh R.

Tobol R.

Angara R.

Sayan Mts.

ASIA

N E W S (compass)

0°

40°N

Tagus R.

Don R.

Volga R.

Ural R.

Danube R.

Black Sea

Caucasus Mts.

Mt. Elbrus 18,510 ft. (5,642 m)

Caspian Sea shoreline -92 ft. (-28 m)

Caspian Sea

Crete

Cyprus

Mediterranean Sea

TROPIC OF CANCER

20°N

AFRICA

0°

20°N

EQUATOR

40°E

UNIT 4
REGIONAL ATLAS

MAP STUDY

1. **Physical Systems** What are the major peninsulas in Europe?

2. **Human Systems** Which countries in Europe are landlocked?

(Inset map)

▲ Ben Nevis 4,406 ft. (1,343 m)

North Sea

Skagerrak

Kattegat

Gotland

Jutland

Great Britain

Prins Alexander Polder -23 ft. (-7 m)

Zealand

Baltic Sea

NORTHERN EUROPEAN PLAIN

20°E

Thames R.

English Channel

Elbe R.

Oder R.

Vistula R.

Dnieper R.

Seine R.

Rhine R.

Dniester R.

Loire R.

Mont Blanc 15,771 ft. (4,807 m)

ALPS

Danube R.

Massif Central

Rhône R.

Po R.

HUNGARIAN PLAIN

Carpathian Mts.

Pyrenees

Riviera

Adriatic Sea

Danube R.

BALKAN PENINSULA

Balkan Mts.

Black Sea

Corsica

Apennines

Sardinia

Balearic Islands

Tyrrhenian Sea

Sicily

▲ Etna 10,902 ft. (3,323 m)

Ionian Sea

Aegean Sea

40°N

0   200 miles
0   200 kilometers
Lambert Azimuthal Equal-Area projection

Mediterranean Sea

Crete

Europe
Political

○ National capital

Abbreviations

AUST.	Austria
B.&H.	Bosnia & Herzegovina
BELG.	Belgium
CZECH REP.	Czech Republic
EST.	Estonia
HUNG.	Hungary
KOS.	Kosovo
LIECH.	Liechtenstein
LITH.	Lithuania
LUX.	Luxembourg
MAC.	Macedonia
MONT.	Montenegro
NETH.	Netherlands
SERB.	Serbia
SLOV.	Slovenia
SWITZ.	Switzerland

NORTH AMERICA

ATLANTIC OCEAN

ARCTIC OCEAN

PACIFIC OCEAN

Greenland (Den.)

Bering Strait

Bering Sea

Chukchi Sea

Wrangel I.

East Siberian Sea

New Siberian Islands

Laptev Sea

Severnaya Zemlya

Kara Sea

Novaya Zemlya

Franz Josef Land

Barents Sea

Svalbard

Norwegian Sea

Faeroe Islands (Den.)

Reykjavik
ICELAND

ARCTIC CIRCLE

RUSSIA

Lena R.
Aldan R.
Lower Tunguska R.
Yenisey R.
Ob' R.
Angara R.
Lake Baikal
Kolyma R.
Northern Dvina R.
Volga R.
Don R.
Ural R.
Irtysh R.
Tobol R.

Sea of Okhotsk

Kuril Islands

Sakhalin

ASIA

Caspian Sea

Moscow

Kyiv (Kiev)

IRELAND
Dublin
UNITED KINGDOM
London
NORTH America
North Sea
Celtic Sea
PORTUGAL
Bay of Biscay
Lisbon
SPAIN
Madrid
ANDORRA
Corsica
FRANCE
Paris
BELG.
LUX.
NETH.
DENMARK
GERMANY
Berlin
SWITZ.
AUST.
ITALY
Rome
Sardinia
Balearic Islands
MALTA
Sicily
NORWAY
Oslo
SWEDEN
Stockholm
FINLAND
Helsinki
Gulf of Bothnia
Baltic Sea
EST.
LATVIA
LITH.
POLAND
CZECH REP.
SLOVAKIA
HUNG.
SLOV.
CROATIA
B.&H.
MONT.
SERB.
KOS.
MAC.
ALB.
BUL.
GREECE
Athens
Crete
BELARUS
UKRAINE
MOLDOVA
ROM.
Black Sea
Mediterranean Sea
CYPRUS
Nicosia

AFRICA

TROPIC OF CANCER

EQUATOR

20°N

0°

N W E S

0   1,000 miles
0   1,000 kilometers
Two-Point Equidistant projection

100°W   120°W   140°W   80°W   60°W   40°W   20°W   60°N   80°N   180°   160°E   140°E   120°E   100°E   80°E   60°E   40°N   20°E   40°E

Inset map

Stockholm
SWEDEN
North Sea
DENMARK
Baltic Sea
Copenhagen
ESTONIA
Tallinn
LATVIA
Riga
LITH.
Vilnius
UNITED KINGDOM
London
Thames R.
English Channel
NETH.
Amsterdam
BELG.
Brussels
Elbe R.
Berlin
GERMANY
Rhine R.
Seine R.
Loire R.
Paris
FRANCE
LUX.
Luxembourg
Prague
CZECH REP.
POLAND
Warsaw
Vistula R.
BELARUS
Minsk
Dniester R.
UKRAINE
Kyiv (Kiev)
SLOVAKIA
Danube R.
LIECH.
Vaduz
Vienna
AUST.
Bratislava
Budapest
HUNG.
MOLDOVA
Chişinău
ROMANIA
Bern
SWITZ.
SLOV.
Ljubljana
Zagreb
CROATIA
B.&H.
Sarajevo
Belgrade
SERB.
Bucharest
Rhône R.
Po R.
Andorra la Vella
ANDORRA
MONACO
SAN MARINO
Rome
ITALY
Corsica (France)
Sardinia (Italy)
Adriatic Sea
MONT.
Podgorica
KOS.
Pristina
BULGARIA
Sofia
Skopje
MAC.
Tirane
ALBANIA
VATICAN CITY (within Rome)
Balearic Islands (Spain)
Sicily (Italy)
MALTA
Valletta
GREECE
Athens
Aegean Sea
Crete (Greece)

0   200 miles
0   200 kilometers
Lambert Azimuthal Equal-Area projection

Europe
Climate and Vegetation

0 ___ 1,000 miles
0 ___ 1,000 kilometers
Two-Point Equidistant projection

160°W

Climate
- Semi-arid (steppe)
- Humid subtropical
- Marine west coast
- Mediterranean
- Humid continental
- Subarctic
- Tundra and high altitude
- Ice cap

NORTH AMERICA

100°W   120°W   140°W

180°

Bering
Sea

PACIFIC
OCEAN

80°W

Chukchi
Sea

East
Siberian
Sea

Bering Strait

160°E

60°W

ARCTIC
OCEAN

Laptev
Sea

Sea of
Okhotsk

140°E

Reykjavík ★

ATLANTIC
OCEAN

40°W

ARCTIC CIRCLE

Kara
Sea

Barents
Sea

Vladivostok ●

20°W

Norwegian
Sea

Oslo ★

North
Sea

Gulf of Bothnia

Helsinki ★

Stockholm ★

Irkutsk ●

120°E

Celtic
Sea

London ★

Baltic Sea

Berlin ★

Moscow ●

Bay of
Biscay

Paris ★

Kyiv (Kiev) ●

N
W · E
S

40°N

Madrid ★

Adriatic Sea

Rome ★

Bucharest ★

Black Sea

Caspian Sea

ASIA

60°E   80°E   100°E

Athens ★

Mediterranean Sea

0°

TROPIC OF CANCER

20°N

AFRICA

Vegetation
- Desert scrub and desert waste
- Temperate grassland
- Mediterranean scrub
- Deciduous forest
- Coniferous forest
- Mixed forest (deciduous and coniferous)
- Tundra
- Ice cap
- Highland (vegetation varies with elevation)

80°W   100°W   120°W   140°W   160°W   60°N   180°

60°W

40°W

20°W

ARCTIC CIRCLE

140°E

40°N

120°E

0°

EQUATOR

0°   20°E   40°E   60°E   80°E   100°E

20°E

40°E

Europe
Economic Activitiy

NORTH AMERICA

Land Use
- Commercial farming
- Livestock raising
- Nomadic herding
- Primarily forest
- ■ Manufacturing and trade
- Commercial fishing
- Little or no activity

Resources
- Coal
- Petroleum
- Natural gas
- Iron ore
- Tin
- Zinc
- Bauxite
- Cobalt
- Uranium
- Ⓝ Nickel
- Copper
- Lead
- Manganese
- Gold
- Silver
- Platinum
- Gems

100°W 120°W 140°W 160°W 180°

80°W 60°W 40°W 20°W

80°N 60°N 40°N

ARCTIC CIRCLE

ATLANTIC OCEAN

ARCTIC OCEAN

PACIFIC OCEAN

Bering Strait Bering Sea Chukchi Sea East Siberian Sea Laptev Sea Kara Sea Barents Sea Sea of Okhotsk

Norwegian Sea Celtic Sea North Sea Gulf of Bothnia Baltic Sea Bay of Biscay

Northern Dvina R. Ob' R. Irtysh R. Tobol R. Ural Volga R. Don R. Dnieper R. Danube R.

Yenisey R. Lower Tunguska Lena R. Aldan R. Angara R. Kolyma R. Lake Baikal

160°E 140°E 120°E

Black Sea Caspian Sea Mediterranean Sea

ASIA

AFRICA

TROPIC OF CANCER

0° 20°N

20°E 40°E 60°E 80°E

EQUATOR

INDIAN OCEAN

0 1,000 miles
0 1,000 kilometers
Two-Point Equidistant projection

UNIT 4
REGIONAL ATLAS

MAP STUDY

1. **Environment and Society** Describe the relationship between land use and climate.

2. **Human Systems** In which parts of Europe is nomadic herding practiced?

Europe
Population Density

NORTH AMERICA

160°W

120°W

100°W

140°W

80°W

180°

60°W

Bering
Sea

PACIFIC
OCEAN

Chukchi
Sea

160°E

40°W

ARCTIC
OCEAN

East
Siberian
Sea

Sea of
Okhotsk

ATLANTIC
OCEAN

ARCTIC CIRCLE

Laptev
Sea

140°E

Kara
Sea

120°E

Barents
Sea

Norwegian
Sea

Novosibirsk

Glasgow
Dublin
Birmingham Manchester
London Amsterdam Stockholm Helsinki
The Hague Hamburg Copenhagen St. Petersburg
Paris Berlin Nizhniy Yekaterinburg Omsk
Frankfurt Warsaw Novgorod
Madrid Lyon Prague Minsk Moscow Kazan' Chelyabinsk
Lisbon Zurich Vienna Kyiv Ufa
Barcelona Milan Budapest Kiev Samara
Rome Belgrade Dnipropetrovs'k
Naples Bucharest Rostov
Sofia
Athens Black Sea Caspian Sea

Bay of
Biscay

North
Sea

Gulf of Bothnia

Adriatic Sea

Mediterranean Sea

ASIA

N
W   E
S

0   1,000 miles
0   1,000 kilometers
Two-Point Equidistant projection

100°E

120°E

EQUATOR

INDIAN
OCEAN

20°E

40°E

60°E

80°E

UNIT 4
REGIONAL ATLAS

MAP STUDY

1. *Human Systems* Which European countries have the highest population densities?

2. *Places and Regions* How does Russia's overall population density compare to that of the rest of Europe?

Cities
(Statistics reflect metropolitan areas.)
- ■ Over 5,000,000
- □ 2,000,000–5,000,000
- ⊙ 1,000,000–2,000,000

POPULATION

Per sq. mi.	Per sq. km
1,250 and over	500 and over
250–1,249	100–499
63–249	25–99
25–62	10–24
2.5–24	1–9
Less than 2.5	Less than 1

Northern Europe

ESSENTIAL QUESTION · *How do physical systems and human systems shape a place?*

Geography Matters...

The countries of Northern Europe rank among the highest in the world in happiness surveys. Maybe it is the beautiful landscapes, high standard of living, cutting-edge cities, and eco-friendly planning that lead their citizens to this contentment. The region is also a destination for visitors to view the amazing phenomenon of the northern lights—a spectacular array of colorful lights appearing in the sky that can be seen in the northern reaches of Europe, attracting people from all over the world.

◀ This hockey player proudly wears Sweden's flag painted on his face.

Tomasz Tomaszewski/National Geographic Stock

volcanic eruption
in Iceland

The island of Iceland is a geologically young country located within the Mid-Atlantic Ridge, a 10,000-mile- (16,093-km-) long undersea mountain range. This ridge sits at the juncture of two tectonic plates, resulting in frequent seismic activity, including earthquakes, volcanic eruptions, and geothermal venting. Iceland experiences all of these as a result of its location along the Mid-Atlantic Ridge.

What was the physical process that occurred?

In 2010 the long-dormant Eyjafjallajökull, one of Iceland's largest volcanoes, became increasingly active. The world watched as several small earthquakes were followed by intense and frequent quakes. On March 21, lava erupted from a 0.3-mile- (500-m-) long vent in the Fimmvörduháls Pass. The volcano, often called Eyja (EYE·YAH) for short, spewed fountains of lava up to 328 feet (100 m) high and released lava flows up to 66 feet (20 m) thick. The glacier above the lava melted, sending mud and ice into the rivers. This caused flooding that damaged roads and farmland. Huge plumes of smoke and ash were sent nearly seven miles (11 km) into the atmosphere, where they were carried by wind toward continental Europe.

1. **Physical Systems** How did the eruption of the volcano lead to massive flooding?

How did this affect transportation systems in Europe?

The airborne ash posed a dangerous threat to jet engines. Volcanic ash consists of tiny pieces of glass. When those pieces get sucked into a jet engine, they melt and cause the engine to seize up. Within two days of the 2010 eruption, air travel authorities across northern and central Europe were forced to cancel thousands of commercial flights in the largest disruption of peacetime air travel in history. Millions of passengers were stranded or delayed as the ripple effects of the eruption spread across air travel systems worldwide. Many Europeans were stranded abroad. Other transportation systems, such as trains and ferry boat operations, benefited as passengers who could not travel by air made other arrangements.

2. **Human Systems** Imagine you were a traveler stranded in Northern Europe by the eruption of Eyja. How would you cope with the situation? Write a paragraph detailing your experience.

What were the resulting economic effects?

Airlines lost hundreds of millions of dollars because of canceled flights, and Europe's ability to respond effectively to a crisis was called into question. Tourism in Europe was also affected. Many tourists ended up spending their money elsewhere. Countries that relied heavily on air travelers to sustain their tourist sectors were negatively affected by the sudden disruption. Exports and imports were impeded as well. Countries with perishable exports that depend on air travel suffered extra losses. Iceland eventually experienced an increase in tourism, however, as visitors sought out views of the now-famous volcano.

3. **Environment and Society** How could individual governments across a region cooperate to minimize the economic disruption caused by events like the eruption of Eyja?

(tl)Páll Jökull - www.flickr.com/photos/palljokull/Flickr/Getty Images, (tc)Pascal Le Segretain/Getty Images, (tr)Ulrich Baumgarten/Getty Images

EYJAFJALLAJÖKULL

EYJAFJALLAJÖKULL

At its peak the ash cloud covered most of Western Europe and parts of Eastern Europe and Russia.

ENGINE TROUBLES

Erodes fan blades

Blocks air filters

Clogs fuel nozzles

Ash particles

DUBLIN

OSLO GARDERMOEN

AMSTERDAM SCHIPHOL

HELSINKI

LONDON HEATHROW

STOCKHOLM ARLANDA

COPENHAGEN

PARIS-CHARLES de GAULLE

DOMODEDOVO MOSCOW

THE TIME LINE

DAY 1	DAY 2	DAY 3	DAY 4	DAY 5	DAY 6
APRIL 14, 2010	APRIL 15, 2010	APRIL 16, 2010	APRIL 17, 2010	APRIL 18, 2010	APRIL 19, 2010
THE ERUPTION	**SPREADING ASH**	**GROWING CONCERN**	**AT A STANDSTILL**	**STILL STRANDED**	**TRAVEL RESUMES**
Eyjafjallajökull erupted on April 14, 2010, releasing a 4-mile high- (6.4 km-) plume of smoke into the sky.	The drifting ash cloud forced northern European countries to close their airspace; 6,000 flights were cancelled.	France, Lithuania, and Hungary closed their airspace. Iceland's airlines were unaffected.	All of Europe's major airports were closed; 17,000 flights were cancelled.	80 percent of Europe's airline traffic was still grounded. Millions of passengers worldwide were stranded.	Selected zones of European airspace reopened to flight traffic.

GLOBAL GDP LOSS

$-2.6 BILLION

$-957 MILLION

$-591 MILLION

$-517 MILLION

EUROPE   **AMERICA**   **MEAF***   **ASIA**

Airlines lost an estimated **$200 million** per day.

6.8 million passengers were stranded worldwide.

*****MIDDLE EAST AND AFRICA**

networks

There's More Online!

☑ **DIAGRAM** Gulf Stream Effects on Northern Europe

☑ **IMAGE** Fjords of Northern Europe

☑ **IMAGE** Geyser in Northern Europe

☑ **INTERACTIVE SELF-CHECK QUIZ**

☑ **VIDEO** Physical Geography of Northern Europe

LESSON 1
Physical Geography of Northern Europe

ESSENTIAL QUESTION • *How do physical systems and human systems shape a place?*

Reading **HELP**DESK

Academic Vocabulary

• **emerge**
• **migrate**

Content Vocabulary

• **glaciation**
• **fjord**
• **geothermal energy**
• **hot spring**
• **geyser**

TAKING NOTES: *Key Ideas and Details*

LISTING Use a graphic organizer like the one below to list the landforms, water systems, climate regions, and resources of Northern Europe.

Landforms	
Water Systems	
Climate Regions	
Resources	

IT MATTERS BECAUSE

Europe is a large peninsula made up of numerous peninsulas, such as the Scandinavian Peninsula and the Jutland Peninsula found in Northern Europe. The unique physical geography, shaped by glaciers and plate tectonics, and the cold northern climate have influenced the lives of people in this subregion.

Landforms

GUIDING QUESTION *How did the last ice age impact the landforms of Northern Europe?*

Glaciation has been the primary process by which the landforms of Northern Europe came to be as they are today. During the last ice age, the process of glaciation scoured the land and shaped the landforms. Ice filled the valleys and carved out long, narrow, steep-sided **fjords** (fee•AWRDS) that are now filled with seawater. Plains were scraped flat by the glaciers that covered the land, while the mountains in the region were made steeper and more rugged.

The ice that covered Northern Europe during the last ice age was over one mile (1.6 km) thick. It was so heavy that it pressed the land down into the Earth's mantle. Over time, as the ice melted and lessened the weight on the land beneath, the land began to rise in a process called continental rebound. The entire land surface continues to rise today. When the ice sheet melted about 10,000 years ago, it also gouged the surface of the land and left in its wake innumerable islands, rivers, and streams as well as countless lakes.

Northern Europe is made up of five countries. Norway and Sweden are found on the scenic Scandinavian Peninsula. The Jutland Peninsula forms the mainland part of Denmark and extends into the North Sea. Although not situated on the Scandinavian Peninsula, Denmark is considered part of the cultural region called Scandinavia. Finland lies in the eastern part of the region, and the island country of Iceland is located in the North Atlantic Ocean.

(tl)Jose Maria Mellado/Photonica/Getty Images, (tc)©Doug Pearson/JAI/Corbis, (tr)Guy Edwardes/age fotostock

Most of Norway and northern Sweden are mountainous, but in southern Sweden lowlands slope gently to the Baltic Sea. Glaciers from the last ice age left behind thousands of sparkling lakes in these two countries as well as in Finland. Many deep fjords lie on the Atlantic coastline of the Scandinavian Peninsula.

Svalbard is an archipelago in the Arctic Ocean that constitutes the northernmost part of Norway. This group of islands is located about 400 miles (644 km) north of the mainland, midway between mainland Norway and the North Pole. Today glaciers and snowfields cover more than 50 percent of the islands' land. Glaciers cut the former plateau into fjords and valleys, and the geological processes of folding and faulting cause the mountainous landscape to **emerge**. The landforms of Svalbard were created through repeated ice ages and the folding and faulting associated with continental drift and plate tectonics. Norway's strongest earthquake, measuring 6.5 on the Richter scale, occurred in Svalbard on March 6, 2009.

Finland is mostly flat with a few hills and mountains. Over 10 percent of its area is covered with inland waters such as lakes and rivers. Its rugged coastline is deeply indented with bays and inlets, and the offshore region is dotted with thousands of islands.

Formerly a possession of Denmark, Iceland is an island country in the North Atlantic. Iceland is located 186 miles (300 km) east of Greenland and 621 miles (1,000 km) west of Norway. It is situated on a geological hot spot along the Mid-Atlantic Ridge. The island is very geologically active. With about 200 volcanoes, volcanic activity is frequent. Earthquakes are also frequent but rarely result in serious damage.

Although considered to be a European country, Iceland sits partly on ocean crust shared with the North American continent, as it straddles the Mid-Atlantic Ridge that marks the boundary between the Eurasian and North American tectonic plates. The tectonic activity caused by these plates' separating is the source of the abundant **geothermal energy** in the region. Iceland's many rivers and waterfalls are also harnessed to produce hydropower. These two natural resources provide Iceland with sustainable and inexpensive sources of energy.

Additional physical features include Iceland's numerous mountains, countless **hot springs**, rivers, small lakes, waterfalls, glaciers, and **geysers**. Glaciers cover roughly 11 percent of the island. The largest, Vatnajökull, is nearly 1,300 feet (400 m) thick and covers about 8 percent of the island. It is by far the largest glacier in Europe. The word *geyser* is derived from a geyser in Iceland named Geysir.

During the last ice age, glaciers deposited sand and gravel on the Jutland Peninsula's flat western side and carved fjords on the coastline of the east. Flat plains make up most of the Jutland Peninsula's interior in Denmark. The Kingdom of Denmark also includes the Faeroe Islands and Greenland in the North Atlantic.

glaciation a process by which glaciers form and spread

fjord a long, steep-sided glacial valley now filled by seawater

emerge to rise from an obscure position or condition; to become visible

geothermal energy a form of energy conversion that captures heat energy from within Earth

hot spring a spring whose water issues at a temperature higher than that of its surroundings

geyser a spring that throws forth intermittent jets of heated water and steam

Fjords and geysers are both signature features of the Northern European landscape.

▼ **CRITICAL THINKING**

1. *Comparing* How are geysers similar to volcanoes?

2. *Hypothesizing* How might fjords have influenced the history of Northern Europe?

During the last ice age, glaciers scraped across the land, making the mountains steeper and more rugged.

▲ **CRITICAL THINKING**

1. *Analyzing Visuals* Describe the landforms shown in the picture.

2. *Hypothesizing* How might the same landscape have looked during the last ice age?

At 839,399 square miles (2.1 million sq. km), Greenland is the world's largest island. The Faeroe Islands, an island group and archipelago, are located about halfway between Iceland and Norway. They are made up of volcanic rocks with high and rugged cliffs.

✔ **READING PROGRESS CHECK**

Identifying What landforms were created by glaciation during the last ice age in Northern Europe?

Water Systems

GUIDING QUESTION *Why is the landscape of Northern Europe dotted with so many lakes?*

Continental glaciers covered much of Northern Europe during the last ice age. The scouring action of these glaciers created a landscape dotted with hundreds of thousands of lakes. After the glaciers melted, the debris left behind on a flat landscape blocked rivers and trapped water like dams.

The landscape of Iceland is geologically young. It is characterized by impressive waterfalls and an abundance of small lakes and numerous, swift-moving rivers that are filled by glacier meltwater and heavy rainfall. The majority of rivers in Iceland consist of meltwater from glaciers. Thus, they contain large amounts of glacial debris that makes the water cloudy. The longest river is called Thjórsá. Located in the southern region of Iceland, it extends for 143 miles (230 km).

Many of the rivers of the Scandinavian Peninsula are short and do not provide easy connections between cities. Norway's chief rivers stem from the mountains of Norrland. These rivers mostly flow toward the southeast with many falls and rapids, eventually emptying into the Gulf of Bothnia or the Baltic Sea. The country's longest river is the Klar-Göta. It flows 447 miles (719 km) until it reaches Lake Vänern. The Glåma River drains an area of 16,236 square miles (42,051 sq. km), running almost the entire length of Norway from north to south.

The Kemi River in Finland is harnessed for hydroelectric power. It rises near the Russian border and flows generally southwest for about 300 miles (483 km)

Guy Edwardes/age fotostock

to the Gulf of Bothnia in the town of Kemi. The Muonio and Torne Rivers flow along the border of Finland and Sweden. Sweden has many small hydroelectric power plants that harness the power of the country's rivers.

☑ **READING PROGRESS CHECK**

Explaining From what landforms do most rivers originate?

Climate, Biomes, and Resources

GUIDING QUESTION *How does Northern Europe's location affect its climate and vegetation?*

Latitude, mountain barriers, wind patterns, and distance from large bodies of water influence Northern Europe's climate patterns. Climate regions include marine west coast, humid continental, subarctic, and tundra. Location influences vegetation patterns. Natural vegetation varies from forests and grasslands to tundra plants. In Iceland, the Gulf Stream creates a mild climate even though the country is located in higher latitudes.

Climate Regions and Biomes

Strong interrelationships exist between climate and plant and animal life in Northern Europe. The arctic tundra regions lie in the extreme northern parts of Scandinavia and Iceland. Due to dry conditions, poor soil quality, extremely cold temperatures, and frozen ground, vegetation in this climate is limited. Arctic tundra plants must adapt to the short, cold growing seasons. The frozen ground prevents plants with deep roots, like trees, from growing. Animals in the alpine zone **migrate** to lower elevations in the winter to escape the cold and find food. South of the tundra biome is the subarctic climate region. This covers most of the northern half of Scandinavia. It has long, very cold winters and short, cold to mild summers. The vegetation is limited to only the few species that can tolerate the cold conditions. Along the Atlantic coast and in southern Sweden, the climate is the marine west coast type that has milder

migrate to move from one place to another

The Gulf Stream is a warm ocean current that originates in the Gulf of Mexico. As it crosses the Atlantic Ocean, the Gulf Stream splits into two currents, the North Atlantic Drift and the Canary Current. The North Atlantic Drift warms the climates of Northern Europe.

▲ **CRITICAL THINKING**

1. *Analyzing Visuals* In what direction is the Gulf Stream's North Atlantic Drift moving as it approaches Europe?

2. *Identifying* Over what other part of Europe does the Gulf Stream flow before it reaches Scandinavia?

Exploiting the energy stored under Earth's surface has a long human history. There is evidence that ancient Romans, Chinese, and Native American cultures used hot mineral springs for bathing, cooking, and eating. Geothermal power plants today use steam from underground hot spots to power electricity-generating turbines. They are usually located in zones of recent volcanic activity or where tectonic plates meet. Iceland has both of these criteria and is the location of many state-of-the-art geothermal power plants.

IDENTIFYING How is Iceland's location favorable for generating energy from geothermal plants?

winters. This area supports a vast coniferous forest. The rest of the Scandinavian Peninsula and Finland has a humid continental climate with cold, wet winters. The temperatures of most of the region are warmer than most other regions of similar latitudes due to the influence of the Baltic Sea, inland waters, and airflows from the Atlantic that are warmed by the Gulf Stream.

Natural Resources

Finland's peat deposits cover nearly one-third of the country. Peat is vegetable matter found in swamps. It is dug up, chopped into blocks, and dried so it can be burned. Hydroelectric power and geothermal energy are significant renewable resources in Finland. The diversity of minerals in the country includes modest amounts of iron ore, nickel, zinc, cobalt, copper, and chromium. Trees, however, are Finland's most important natural resource. In 2012 Finland exported over 10 percent of the paper and paperboard traded on the global market.

Deposits of iron ore lie in Sweden near Kiruna. Other natural resources in the area include gold, copper, lead, and zinc. Forest products are a major source of revenue. In 2010 Sweden exported 10 percent of the world's sawnwood, a wood used to make lumber. Hydroelectric power is a renewable resource in Sweden as it is in most Northern European countries.

Norway is one of the world's leading producers of hydroelectric power. One important use of this hydroelectric power is the production of aluminum, a process that requires large amounts of electricity. Although Norway lacks the mineral bauxite from which aluminum is made, it imports bauxite and utilizes its hydroelectric power to produce aluminum. Magnesium is an important natural resource in Norway. Norway is also Europe's largest oil producer and the world's second-largest natural gas exporter. Other European countries rely on Norway as an important supplier of both sources of fuel. In 2009 Norway exported 2.184 million barrels per day on the world market.

Denmark also has oil and natural gas, but renewable wind energy is its most important energy source. Denmark ranks number one in the world for electricity generated from renewable sources. This makes it a model that other countries can follow to transition to renewable energy.

Iceland has vast energy resources. Only a small fraction of the hydroelectric power of the country's rivers has been tapped, however. Geothermal energy provides heat for the entire capital city of Reykjavík and several other communities. It also provides steam for industrial energy and is used in commercial vegetable farming in greenhouses.

☑ **READING PROGRESS CHECK**

Describing How have animals in the arctic and alpine tundra biomes adapted to survive in the harsh climate?

LESSON 1 REVIEW

Reviewing Vocabulary
1. *Explaining* Explain the significance of glaciation, fjords, and geysers to the landscape of Northern Europe.

Using Your Notes
2. *Listing* List the natural resources found in Northern Europe by using your graphic organizer.

Answering the Guiding Questions
3. *Assessing* How did the last ice age impact the landforms of Northern Europe?

4. *Explaining* Why is the landscape of Northern Europe dotted with so many lakes?

5. *Making Generalizations* How does Northern Europe's location affect its climate and vegetation?

Writing Activity
6. *Informative/Explanatory* Write an essay that describes the geographic factors that contribute to the water systems of Iceland.

networks

There's More Online!

- ☑ **DIAGRAM** Population Pyramid of Finland
- ☑ **IMAGE** Community Art in Norway
- ☑ **TIME LINE** Northern Europe: Democracy and Independence
- ☑ **INTERACTIVE SELF-CHECK QUIZ**
- ☑ **VIDEO** Human Geography of Northern Europe

(tl)INTERFOTO/Personalities/Alamy, (tcl)K Salminen/age fotostock, (tcl)Arctic-Images/The Image Bank/Getty Images, (tcr)Sean Gallup/Getty Images News/Getty Images, (trl)John Borthwick/Lonely Planet Images/Getty Images

Reading **HELP**DESK

Academic Vocabulary

- **integrate**
- **achieve**

Content Vocabulary

- **continental**
- **entrepôt**
- **break-of-bulk**
- **welfare state**

TAKING NOTES: *Key Ideas and Details*

OUTLINING Use a graphic organizer like the one below to take notes about society and culture in Northern Europe.

Elements of Culture	
Family and the Status of Women	
The Arts	

LESSON 2
Human Geography of Northern Europe

ESSENTIAL QUESTION · *How do physical systems and human systems shape a place?*

IT MATTERS BECAUSE

The countries of Northern Europe, also called the Nordic countries, have fascinating histories and cultures. Scandinavia (Denmark, Norway, and Sweden) and Finland are lands of castles, Viking museums, and progressive contemporary cities. Iceland is the least densely populated country in Europe and has blended strong cultural traditions with a modern capital city.

History and Government

GUIDING QUESTION *What influenced the creation of new social classes and peaceful democracies in Northern Europe?*

Geographically, the Scandinavian Peninsula includes mainland Sweden, mainland Norway, and part of Finland. The Jutland Peninsula includes mainland Denmark and a small part of Germany. The history of Northern Europe has been shaped by thousands of years of migrations, invasions, and trade. The Nordic countries have similar structures of societies and cultural traits, which collectively differentiate them from mainland Europe. From about A.D. 793 to 1050—a period of European history known as the Viking Age—Scandinavian Viking raiding parties in powerful warships roamed the coastal waters of Europe. Their territories expanded over time, and the numerous Viking kingdoms emerged as Denmark, Sweden, and Norway.

The Rise of Northern Europe

In Norway the coasts, mountainous terrain, and fjords formed strong natural boundaries. Communities there remained independent of each other. By A.D. 800, some 30 small kingdoms existed in Norway. The sea was the easiest way of communication between the Norwegian kingdoms and the outside world. During the Viking Age, Norsemen—also known as Vikings—built ships of war and sent them on raiding expeditions into western and eastern Europe. Their language, Old Norse, became the basis for present-day Nordic languages.

Northern Europe **255**

Norway, Sweden, Denmark, and Iceland shared similar cultures, languages, and religions. A Christian mission was established in Denmark in the 800s, but conversion to Christianity progressed slowly in Northern Europe. From the 1100s, what is now Finland began sharing common developments as it became economically **integrated** with Sweden. In the 1300s, Denmark, Norway (with Iceland), and Sweden (with Finland) were united under one regent in the Kalmar Union, which Denmark dominated. The increased trade during this time connected the countries to mainland Europe, and Nordic society became increasingly **continental**. In 1523 Sweden became a separate kingdom. Denmark's domination over Norway lasted until 1814, when the king was forced to cede Norway to the king of Sweden. Iceland, Greenland, and the Faeroe Islands remained under the control of Denmark.

Industrialization, Democracy, and Independence

The nineteenth century brought many changes to Europe and the Nordic region. Industrialization required a larger population as more people were needed for jobs. This influx brought new social classes as well as different socioeconomic levels. The political culture changed as these new classes of people brought new ideas and, ultimately, democracy and independence to the countries of Northern Europe.

In the period following World War II, the Nordic countries developed democracies. Norway, Sweden, Denmark, Finland, and Iceland are now governed by democratically elected parliaments, although Norway, Sweden, and Denmark are constitutional monarchies. They also share similar traits in the policies implemented during the postwar period. For example, all Nordic countries have large, tax-funded public welfare sectors and extensive social-democratic legislation. These programs support health care, education for elementary through college students, and retirement income for senior citizens.

✅ **READING PROGRESS CHECK**

Inferring Why do the Nordic countries share so many similar characteristics today?

integrate to blend into a functioning whole

continental relating to or characteristic of a continent

INTERFOTO/Personalities/Alamy

TIME LINE ⌄

DEMOCRACY
and Independence ➡

The countries of Northern Europe have a shared history of wars, kingdoms, independence, and cooperation.

CRITICAL THINKING ▶

1. *Describing* Discuss the results of the disintegration of the Kalmar Union, including the subsequent status of the countries.

2. *Identifying* As of 1995, which Northern European countries were members of the European Union?

Denmark's King Frederick VI cedes Norway to Sweden.

1814

1300s ➡

1389 Denmark, Norway, and Sweden are united under the rule of Margaret of Denmark, forming the Kalmar Union.

1523 The Kalmar Union is dissolved when Sweden rebels and becomes independent under King Gustav I Vasa. Norway loses its status and becomes a Danish province (1536).

256

Directions: On a separate sheet of paper, answer the questions below. Make sure you read carefully and answer all parts of the questions.

Lesson Review

Lesson 1

① *Identifying Cause and Effect* Describe the role of glaciation in the formation of the fjords of Northern Europe.

② *Explaining* Explain the location of the Faeroe Islands. Why are they not considered part of Scandinavia?

③ *Drawing Conclusions* What factors influence the climate in Northern Europe? Give examples.

Lesson 2

④ *Evaluating* Describe how the welfare states benefit the people of Northern Europe. How are they funded?

⑤ *Analyzing* How has the history of Northern Europe been shaped by migrations, invasions, and trade?

⑥ *Summarizing* Explain how industrialization affected the societies of Northern Europe in the nineteenth century.

Lesson 3

⑦ *Identifying Central Issues* Describe how human activities have created water and air pollution in Northern Europe.

⑧ *Explaining* Explain why Northern Europe's waters and forests are in need of protection.

⑨ *Comparing and Contrasting* How are private and public groups in Northern Europe addressing environmental issues?

21st Century Skills

Use the following chart to answer the questions below.

Norway's Gross Domestic Product by Economic Sector	
Agriculture	2.6%
Industry	39.7%
Services	57.7%

Source: CIA World Factbook, 2011

⑩ *Using Graphs, Charts, Diagrams, and Tables* What sector of Norway's economy in the chart constitutes the smallest percentage of the country's total GDP?

⑪ *Decision Making* What types of imports do you think Norway must rely on due to the percentages shown in the chart? Explain your answer.

⑫ *Economics* Describe the types of jobs that may be part of Norway's service industry.

College and Career Readiness

⑬ *Explaining Continuity and Change* As a sociologist, you have been hired by the United Nations Population Fund to list and describe the new family structure in Northern Europe and explain why the change has occurred. Conduct the necessary research, and then write a one-page report evaluating the causes of the change.

Need Extra Help?

If You've Missed Question	❶	❷	❸	❹	❺	❻	❼	❽	❾	❿	⓫	⓬	⓭
Go to page	250	251	253	259	255	256	261	262	264	265	265	265	259

Directions: On a separate sheet of paper, answer the questions below. Make sure you read carefully and answer all parts of the questions.

Critical Thinking

14 *Making Generalizations* How does tectonic activity affect the islands of Northern Europe?

15 *Identifying Central Issues* What is the relationship between glaciation and hydroelectric power in Northern Europe?

16 *Evaluating* Write a paragraph discussing the economic characteristics of the Nordic model and why it has been considered successful.

17 *Drawing Conclusions* How has Northern Europe's culture been affected by its physical environment?

Applying Map Skills

Use the Unit 4 Atlas to answer the following questions.

18 *Place and Regions* Using the political map of Europe, which Northern European countries lie partially within the Arctic Circle?

19 *Physical Systems* Describe the elevation of Finland compared to that of Norway as shown on the physical map of Europe.

20 *Environment and Society* Use your mental map to list the seas that border the countries of Northern Europe.

Exploring the Essential Question

21 *Identifying Cause and Effect* Create a flow chart that shows the geologic events that led to the eruption of the Eyjafjallajökull volcano in Iceland. How did it affect the economy and transportation? How is this event an example of how physical systems and human systems shape a place?

DBQ Analyzing Primary Sources

Use the document to answer the following questions.

In Denmark, some families have decided to generate their own electricity as a way of protecting the environment.

PRIMARY SOURCE

"*We like using computers, and watching the television. And we like taking hot showers like everybody else. By having a windmill we can do that with a good conscience, because we know that the power we consume doesn't expose the environment. Now we produce electricity without producing carbon dioxide. It feels right to take our little part of the responsibility of getting a better environment.***"**

—Hans Chresten Jeppesen, *National Geographic News,* November 4, 2009

22 *Analyzing* Describe what the Jeppesen family is doing to help limit their impact on global warming.

23 *Speculating* How could the governments of countries in Northern Europe and elsewhere encourage more people to take action like the Jeppesen family?

Research and Presentation

24 *Research Skills* Use the Internet and other resources to gather information about religions in Northern Europe and how they have changed throughout the centuries. Create a multimedia presentation to share your findings.

Writing About Geography

25 *Informative/Explanatory* Use standard grammar, spelling, sentence structure, and punctuation to write a one-page essay discussing current population trends in Northern Europe, including birthrates, life expectancy, and international migration.

Need Extra Help?

If You've Missed Question	**14**	**15**	**16**	**17**	**18**	**19**	**20**	**21**	**22**	**23**	**24**	**25**
Go to page	251	251	260	255	243	242	242	248	266	266	258	259

Northwestern Europe

ESSENTIAL QUESTION · *How do physical systems and human systems shape a place?*

Why Geography Matters
Suburban Growth and Transportation

Lesson 1
Physical Geography of Northwestern Europe

Lesson 2
Human Geography of Northwestern Europe

Lesson 3
People and Their Environment: Northwestern Europe

Geography Matters...

The countries of Northwestern Europe have long been at the crossroads of many cultures. There you will see a colorful mix of the old and the new. People in the subregion enjoy a higher standard of living than many other subregions. This is reflected in high levels of education, long life expectancies, and industrialized economies. The people of Northwestern Europe include an ethnic mix. Some countries are made up of two or more ethnicities that have blended over the centuries. Switzerland, for example, has three official languages.

◀ A young girl from Ireland is dressed for traditional folk dancing.

Vidler Steve/age fotostock

suburban growth
and transportation

With a population approaching 11 million, Paris is a megacity. Since 1921 its growth has been primarily in the suburbs, while the urban core's population has dropped. Providing effective transportation for surburban populations is an ongoing challenge. Important to meeting this challenge is understanding the city's geography.

THERE'S MORE **ONLINE**

READ a news article about suburbanization in Paris • *SEE* a map of Paris's population density

What is fueling suburban growth in Paris?

Various factors contribute to suburban growth in Paris. Housing prices have risen dramatically in the urban core. This is largely a result of gentrification, a process in which older areas of a city are revived by wealthier residents. This has caused many people to look for more affordable housing in locations outside the city. Deindustrialization has also occurred in the urban core, a process in which older manufacturing industries shut down. Also, many housing units have been converted into office space. These two processes have led residents to migrate to suburbs to find new jobs and places to live.

1. **Human Systems** What trends have contributed to the population growth in the Paris suburbs while the urban core population has declined?

How does this affect transportation needs in Paris?

The Paris Metro subway system has served the city since 1900 and is considered one of the world's finest. The Metro trains are both fast and frequent. However, as population growth has shifted from the urban core to the suburbs, the geographic pattern of the Metro has presented problems. Rail lines have been built to connect the city to the suburbs in a "hub and spoke" arrangement. This makes travel between suburbs and city easier, but rail travel across the region typically requires a trip into the urban core to change trains. This takes extra time, and residents increasingly use automobiles to commute, causing significant congestion. Official campaigns to discourage daily automobile use have not had much success because, even with reduced traffic congestion, it can still be more efficient to drive than to take the train.

2. **Places and Regions** How has this population shift affected the transportation needs of Paris's residents?

How is Paris responding to these needs?

The Paris Metro system is slated for a significant upgrade aimed at serving the growing suburban population and alleviating traffic problems mainly caused by commuters. A highlight of the planned system expansion (the Metro Grand Paris Plan) is a 96-mile (154.5-km) automated rail line linking suburban areas and connecting to existing Metro lines. Transportation planners think that the suburb-to-suburb expansion of the Metro system will eliminate needless trips to the city center. They also predict that the fast travel times on the new lines will encourage travelers to use the Metro system rather than automobiles.

3. **Physical Systems** How are transportation planners responding to regional travelers' needs?

(l)Egon Bömsch/Alamy, (c)Bernard Rouffignac/Photononstop/Getty Images, (c)Grant Faint/Photographer's Choice/Getty Images, (cr)©William Ryall 2007

netw🌐rks

There's More Online!

☑ **IMAGE** The Alps of Northwestern Europe

☑ **INFOGRAPHIC** Netherlands' Reclaimed Lands: Polders

☑ **MAP** Gulf Stream Effects on Climates in High-Latitude Northern Europe

☑ **INTERACTIVE SELF-CHECK QUIZ**

☑ **VIDEO** Physical Geography of Northwestern Europe

LESSON 1
Physical Geography of Northwestern Europe

ESSENTIAL QUESTION · *How do physical systems and human systems shape a place?*

Reading HELPDESK

Academic Vocabulary

- **consist**
- **significant**

Content Vocabulary

- **loess**
- **dike**
- **polder**
- **mistral**
- **foehn**
- **avalanche**

TAKING NOTES: *Key Ideas and Details*

PARAPHRASING Use a graphic organizer like the one below to take notes on the climate regions of Northwestern Europe.

Climate	Influences on Climate

IT MATTERS BECAUSE

Together, the United Kingdom, Ireland, France, the Netherlands, Belgium, Switzerland, Germany, Austria, Liechtenstein, Monaco, and Luxembourg make up the subregion known as Northwestern Europe. Within these countries lie some of Europe's most iconic landscapes, from the banks of the Seine River that runs through Paris to the jagged, awe-inspiring Alps.

Landforms

GUIDING QUESTION *How did the Northern European Plain affect the development of Europe?*

Northwestern Europe's landscape **consists** of plains interrupted by mountains. Scoured by Ice-Age glaciers, the Northern European Plain, or Great European Plain, is an area of relatively flat and low-lying land. It stretches from southeastern England and western France to central France and across Germany. The plain's fertile soil and wealth of rivers originally drew farmers to the area. The southern edge is especially fertile because it is covered by deposits of **loess**, a fine, rich, wind-borne sediment left by glaciers.

In contrast, the Alps are a high and jagged mountain range that lies to the south of the Northern European Plain. Created by the folding of the Earth's crust and shaped by glaciation, the Alps mountain system forms a crescent that runs from southern France through Switzerland and Austria to the Balkan Peninsula. Mont Blanc, the highest peak in the Alps, stands in France on the border with Italy at a height of 15,771 feet (4,807 m).

The Central Uplands lie between the Alps and the Northern European Plain—in parts of eastern France, southern Belgium, and southern Germany. This landform is made up of low rounded mountains, hills, and high plateaus with scattered forests. The Central Uplands are rich in natural resources.

The British Isles lie northwest of the mainland. They consist of the two large islands—Great Britain and Ireland—and thousands of smaller islands. The rugged coastline of the British Isles features rocky cliffs that

(tl)Mark Harris/Photodisc/Getty Images, (tcl)Henni Conodul/age fotostock, (tcr)Monty Rakusen/Culutra/Getty Images, (tr)Allan Baxter/The Image Bank/Getty Images drop to deep bays. Mountains, plateaus, and valleys make up most of northern and western Great Britain. Low hills and rolling plains dominate in the south and in Ireland.

✓ READING PROGRESS CHECK

Explaining What makes the Northern European Plain good land for agriculture?

Water Systems

GUIDING QUESTION *How have the rivers in Europe's heartland contributed to the region's development?*

Water plays a crucial role in the lives and economic activities of many people who live in Northwestern Europe. Most of Northwestern Europe lies within 300 miles (483 km) of a sea or ocean coast, so ocean transportation is important. In addition, people here depend on the many rivers that flow across the subregion for transportation, trade, and recreational activities.

The Alps are the location of major water sources. Eleven **significant** European lakes surround the Alps. Most are in valleys that were formed during the geological uplift of the Alps. They are long, narrow, deep lakes and have provided good places for people to settle because of water power for industry and convenient water routes for transportation. The spectacular scenery also makes the lake areas popular as tourist attractions.

In the Netherlands, water can be friend or foe. Approximately 25 percent of the country lies below sea level. There are extensive coastal dunes, but they have not always been helpful in keeping out the North Sea waters.

consist to be composed of or made up of

loess fine, yellowish, brownish topsoil made up of particles of silt and clay, carried and deposited by the wind

significant important

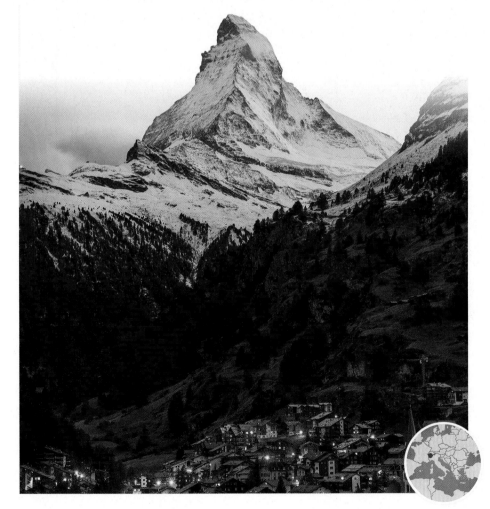

The Alps are considered relatively young mountains in geologic terms.

◀ CRITICAL THINKING

1. *Identifying Central Issues* How do the Alps benefit the people of Northwestern Europe?

2. *Speculating* What features show that the Alps are relatively young?

Mark Harris/Photodisc/Getty Images

dike large bank of earth and stone that holds back water

polder low-lying area from which seawater has been drained to create new land

Since the Middle Ages, the Dutch have built **dikes**, or large banks of earth and stone, to hold back the water. With the dikes as protection, they have reclaimed land from the sea. These reclaimed lands, called **polders**, once were drained and kept dry by the use of windmills. Today, other power sources run pumps to remove water. Polders provide hundreds of thousands of acres of land for farming and settlement. Still, stormy seas have breached the dikes, creating devastating floods.

The rivers of Northwestern Europe have differing characteristics. Although relatively short, England's Thames River allows oceangoing ships to reach the port of London. On the European mainland, however, the relatively long rivers provide links between inland areas as well as to the sea. The Rhine River, the most important river in Northwestern Europe, flows from the Swiss Alps through France and Germany and into the Netherlands. It connects many industrial cities of the interior to the port of Rotterdam on the North Sea.

BUILDING
Dikes and Polders

Polders act as a regulator of sea water to areas that are below sea level. Dikes and polders have allowed cities like Amsterdam and Rotterdam to grow and prosper.

Sea level · Dike · Windmill · Mainland · Pump station · Canal · Polder · Excess water is drained out to sea. · Water pump · Area reclaimed from the sea

ROTTERDAM: CITY BELOW THE SEA
One of the most important junctions of goods flows in the world.

350 million consumers in the European market are served by the port of Rotterdam.

430 tons of cargo and passengers passed through the port of Rotterdam in 2011.

Ranked #1 port in Europe
Ranked #5 port in the world

AMSTERDAM · ROTTERDAM

Reclaimed land from the sea

Polders provide hundreds of thousands of acres for farming and settlement in the Netherlands, but the dikes built to protect them are no guarantee against flooding.

▲ **CRITICAL THINKING**

1. *Synthesizing* What natural resources were needed to create polders?

2. *Drawing Conclusions* What geographic factors pushed the Dutch—and enabled them—to reclaim so much land from the sea?

Henri Conodu/age fotostock

The Seine River, whose source is in central France, flows northwest through Paris and empties into the English Channel. It carries most of France's inland waterway traffic. The Loire River is the longest river in France. From its source in southern France, it flows north and then west before emptying into the Atlantic Ocean. Silting, shallowness, and fluctuations in volume throughout its course limit the use of the Loire for navigation. Canals connect the river with both the Seine and Rhône river systems. Important to conservationists and the environment, these systems form unique habitats for migratory birds.

The Rhône River begins in the Swiss Alps and flows through Switzerland and France. It is the only major river in the subregion that flows directly to the Mediterranean Sea. Important for hydroelectric power, the Rhône is 505 miles (813 km) long and has a drainage basin of 37,750 square miles (97,775 sq. km).

Another important European river that runs through the subregion is the Danube River. Both historically and economically important, the Danube River runs from southern Germany's Black Forest through Austria and into Eastern Europe. After flowing some 1,770 miles (2,850 km), it empties into the Black Sea.

✔ **READING PROGRESS CHECK**

Listing For what activities are Europe's waterways used?

Climate, Biomes, and Resources

GUIDING QUESTION *Why does most of Northwestern Europe have a generally mild climate?*

Several factors affect climate in Northwestern Europe. These factors include the presence of the Alps and the location of the subregion near or along large bodies of water. The winter storms that originate over the North Atlantic Ocean also affect Northwestern Europe's climate.

Climate Regions and Biomes

Northwestern Europe generally has a mild climate compared with other regions located at the same latitude. This results from the North Atlantic Current, a powerful warm-ocean current. It is warm because it is a continuation of the Gulf Stream that emerges from the tropical waters of the Caribbean. This warm water flows along the coast of Northwestern Europe and carries warm, maritime air that bathes the coasts and also blows far inland across the Northern European Plain.

When the moist Atlantic winds reach the Alps, the winds rise up the slopes and the temperature of the air cools. This cooling air produces snow that covers the mountains in winter. Local winds in a region sometimes cause changes in the normal weather pattern. For example, the **mistral**, a strong north wind from the Alps, can send gusts of bitterly cold air into southern France. At other times, dry winter winds called **foehns** (FUHRNS) blow down from the mountains into valleys and plains. Foehns can trigger **avalanches**, which are destructive masses of ice, snow, and rock sliding down mountainsides. Avalanches represent a serious natural hazard in the Alps. They destroy everything in their paths, threatening skiers, hikers, and villages.

As is the case anywhere, climate influences the distribution of biomes in Northwestern Europe. Most of the subregion has a marine west coast climate with mild winters, cool summers, and abundant rainfall. This type of climate produces soils that are often rich in humus—a material formed from decaying leaves and other organic matter that makes soil extremely fertile. A mild Mediterranean climate is found in Monaco, which enjoys an average annual temperature of 61°F (16°C) and receives about 60 days a year of rain.

Natural vegetation in Northwestern Europe includes varieties of deciduous and coniferous trees. Deciduous trees, or those that lose their leaves seasonally,

Analyzing
PRIMARY SOURCES

The Alps

"... I learned to ski in ... [t]he Alps, the great upper story of Europe. From the mountains, in all directions, flow mighty rivers, the Rhine, Rhône, Po, and Danube; a necklace of immortal cities lies in the surrounding foothills and plains.... You are simultaneously in the center of civilization and the most majestic, thrilling wilderness.... The timberline is relatively low; above the trees typically are meadows that for centuries have been used for summer grazing. There are few wild animals.... These are [t]he Alps. High up, near the sky."

—James Salter, "The Alps," *National Geographic Traveler*, October 1999

DBQ *ANALYZING PRIMARY SOURCES* Why does Salter describe the Alps as being in the center of civilization and yet a wilderness?

mistral a strong northerly wind from the Alps that can bring cold air to southern France

foehn a dry wind that blows from the leeward sides of mountains, sometimes melting snow and causing avalanches; term used mainly in Europe

avalanche a large mass of ice, snow, and rock that slides down a mountainside

Although coal creates pollution, it still has the advantages of being abundant and easy to ship and burn.

▲ CRITICAL THINKING

1. *Drawing Inferences* Why do you think Europeans are using less coal, even if it is relatively plentiful?

2. *Analyzing Visuals* Based on this photograph and prior knowledge, what effects might coal mining have on the environment?

such as ash, beech, and oak, thrive in the subregion's marine west coast climate. Coniferous trees—cone-bearing fir, pine, and spruce—are found in cooler, alpine mountain areas up to the timberline, the elevation above which trees cannot grow. The wildlife in this region includes deer, brown bears, badgers, squirrels, and numerous songbirds.

Natural Resources

Northwestern Europe's abundant supply of coal and iron ore fueled the development of modern industry in the 1700s. Today, people in the subregion still rely on coal, but it is being replaced by oil, natural gas, nuclear, and hydroelectric energy sources. Vast oil and natural gas deposits under the North Sea contribute greatly to Europe's energy needs. France, which lacks large oil and gas reserves, has invested heavily in nuclear power. People in the Netherlands rely on natural gas for a majority of their energy needs, but continue to use some wind power. Mountainous Switzerland and Austria get most of their electricity from renewable sources, such as hydroelectric plants. They also have substantial timber resources. The peat bogs of Ireland serve as a source of fuel, especially in the rural countryside where it is used to heat homes. In contrast to much of the subregion, Germany has relatively few natural resources and imports more than half of its energy needs.

✔ READING PROGRESS CHECK

Drawing Conclusions Why does Northwestern Europe have a generally mild climate compared with other places at the same latitude?

Monty Rakusen/Culutra/Getty Images

LESSON 1 REVIEW

Reviewing Vocabulary

1. *Explaining* Explain the relationship between foehns and avalanches.

Using Your Notes

2. *Expressing* Use your notes to describe the factors that influence the climate regions of Northwestern Europe.

Answering the Guiding Questions

3. *Identifying* How did the Northern European Plain affect the development of Europe?

4. *Describing* How have the rivers in Europe's heartland contributed to the region's development?

5. *Explaining* Why does most of Northwestern Europe have a generally mild climate?

Writing Activity

6. *Informative/Explanatory* Write an essay explaining which geographic factors contribute to climate differences between the highlands area of the Alps and the Northern European Plain.

274

when people settle in places near cities. Middle-class and working-class populations may find more affordable housing in suburbs.

Many cities emerged as key trading centers along navigable rivers, such as the Thames and the Seine. As populations in these areas increased, people moved to these growing urban centers. Today this part of Europe is home to some of the world's most famous cities: Paris, London, Brussels, and Amsterdam.

☑ **READING PROGRESS CHECK**

Making Connections How has migration shaped the population of Northwestern Europe?

Society and Culture Today

GUIDING QUESTION *How has its location as a crossroads of Europe influenced society and culture in Northwestern Europe?*

Many countries in Northwestern Europe have one or more official languages. Most people in the region speak languages that are part of the Indo-European language family, including German and French. A small number of Basque people speak Euskera, one of few languages that are unrelated to any other.

Christianity is the primary religion in the subregion, consisting mainly of Roman Catholics and members of various Protestant sects. Nevertheless, most people in the subregion today are only nominally members of a religion. They may state that they have a religion but do not attend church or regularly practice their faith.

Compulsory education for children and **comprehensive** health care and social services are valued in Northwestern Europe. The population is 98 to 100 percent literate, and laws in each country provide for the education of children through secondary school. Europeans have been innovators in education and developing programs to improve quality of life. This subregion values caring for their people through education, health care, and other social programs.

The Status of Women

Northwestern Europe has made great strides regarding equality and political empowerment for women. Countries support equal work opportunities for women and promote public policies to balance work and family life. Roughly an equal number of women and men receive degrees in higher education. Women also find opportunities in political participation. They were generally granted suffrage (the right to vote) before women in the United States and other areas of the world. Margaret Thatcher became the first female prime minister of Great Britain in 1979, Gro Harlem Brundtland became prime minister of Norway in 1981, and Angela Merkel became the chancellor of Germany in 2005.

The Arts

Northwestern Europe has long been a leading force in literature, architecture, music, and the visual arts. Influential artistic movements began here. In the early 1800s, Romanticism was embodied in the works of Lord Byron and Ludwig van Beethoven. Realism, which **focused** on the realistic depiction of everyday life, became prominent in the mid-1800s. This was a change from the emotional style of Romanticism. Later in the century, another group of French painters, the impressionists, moved outdoors to capture immediate experiences, or impressions, of the natural world.

☑ **READING PROGRESS CHECK**

Identifying How have countries of Northwestern Europe worked for equality for women?

comprehensive covering completely or broadly; inclusive

focus to concentrate attention or effort

The Paris fashion scene influences design and fashion trends throughout the world.

▼ **CRITICAL THINKING**

1. *Analyzing Visuals* Describe what you think is about to take place in the picture.

2. *Classifying* How is fashion a part of culture? Explain your answer.

Martin Bureau/AFP/Getty Images

Original members, 1993
Members joining in 1995
Members joining in 2004
Members joining in 2007
Members joining in 2013

ARCTIC CIRCLE

FINLAND
SWEDEN

ESTONIA
LATVIA
LITHUANIA

North Sea
DENMARK

IRELAND UNITED KINGDOM

NETHERLANDS
BELGIUM GERMANY POLAND

LUXEMBOURG
CZECH REPUBLIC
FRANCE
SLOVAKIA
AUSTRIA HUNGARY
SLOVENIA
CROATIA
ROMANIA
ITALY
BULGARIA

Baltic Sea

ATLANTIC OCEAN

PORTUGAL SPAIN

Mediterranean Sea

MALTA

GREECE

CYPRUS

Black Sea

0   400 miles
0   400 kilometers
Lambert Azimuthal
Equal-Area projection

50°N   60°N   40°N   20°W   30°N   10°E   20°E   30°E   40°E   50°E   60°E

GEOGRAPHY CONNECTION

The European Union was made up of 27 countries in 2013.

1. **PLACES AND REGIONS** Which Northwestern European countries are members of the EU?

2. **HUMAN SYSTEMS** Why do you think original members of the EU were from Northwestern Europe?

Economic Activities

GUIDING QUESTION *Which economic activities form the core of Northwestern Europe's economy?*

Northwestern European economies have developed various combinations of agriculture, manufacturing, services, and technology (which includes banking, tourism, and high-tech industries). As in all parts of the world, economic activities in Northwestern Europe are related to the natural resources, the people, and the culture of the area.

Resources, Power, and Industry

The Industrial Revolution made Europe the birthplace of modern industry. The development of industry is often linked to the availability of raw materials. In the 1800s, Europe's large deposits of coal and iron ore, especially in the Ruhr and the Middle Rhine districts in Germany and the Lorraine-Saar district in France, sparked the growth of heavy industry—the manufacture of machinery and industrial equipment. Northwestern Europe's leading industrial centers remain in these areas. Today, the region produces everything from computers to transportation equipment. However, service industries provide a large percentage of countries' GDPs.

More than 70 percent of workers in Northwestern Europe are employed in service industries. Among the top industries are international banking and insurance. Switzerland and the United Kingdom are the leaders in these fields. Tourism is another large service industry, especially in Great Britain, the Netherlands, and France.

The economies of the subregion are supported by transportation and communication systems that are of a high quality everywhere and state-of-the art in many places. Transportation infrastructure in Northwestern Europe includes highways, high-speed railroads, waterways, and airline routes. Modern communications systems also link most parts of Europe to one another and to the rest of the world.

Although largely industrialized, Northwestern Europe also has fertile farmland. Agricultural crops vary with the geography of the region. Olives, citrus fruits, and grapes grow in warm Mediterranean areas of southern France. Farther north, in the cooler plains region, farmers grow wheat, rye, and other grains as well as raise livestock. Northern countries, such as the Netherlands, are major producers of dairy products. Farmers use advanced technology to make the best use of limited agricultural space and climate differences.

In the twentieth century, people began flocking to cities. This left a shrinking population of farmers who struggled to meet the demand for food. Eventually, agricultural companies arose and created **agribusiness**, an industry engaged in large-scale, corporate farming operations.

agribusiness an industry engaged in agriculture on a large scale, sometimes including the manufacture and distribution of farm supplies

Many Northwestern European countries have made the switch from coal to natural gas, wind, and solar collections for energy production. The subregion is a global leader in the development and use of wind energy. The European Union has developed long-term plans that focus on biofuels, solar, wind, and other renewable energy resources. This is part of an effort to preserve the environment and to decrease reliance on imported fuels, such as oil.

The European Union

Trade, both within Europe and between Europe and the rest of the world, is changing as the result of the European Union (EU). The EU, which unites much of Europe into one trading community, or bloc, enjoys one of the largest volumes of trade in the world. The EU accounts for about one-sixth of the world's trade in goods.

The EU has achieved its large trade activity by eliminating tariffs and other trade barriers among member nations. The EU has also formed favorable trade relationships with other countries, including the United States and China. Its educated and technologically advanced workforce, its locations with access to water transportation, and its adaptability to changes in global trade give the EU trade advantages over other regions of the world.

✓ **READING PROGRESS CHECK**

Specifying What service industries rank among the most important in Northwestern Europe?

LESSON 2 REVIEW

Reviewing Vocabulary

1. *Explaining* Describe what is meant by the *Cold War,* and explain how it affected Germany's geography.

Using Your Notes

2. *Describing* Use your graphic organizer to describe society and culture of today in Northwestern Europe.

Answering the Guiding Questions

3. *Identifying* How have new ideas influenced the development of governments and economies in Northwestern Europe?

4. *Describing* How have physical geography, migration, and world events shaped the population of Northwestern Europe?

5. *Explaining* How has its location as a crossroads of Europe influenced society and culture in Northwestern Europe?

6. *Identifying* Which economic activities form the core of Northwestern Europe's economy?

Writing Activity

7. *Informative/Explanatory* Write an essay about the development of current population patterns of Northwestern Europe.

How beneficial is THE EUROPEAN UNION?

In November 2012, the European Union (EU) was awarded the Nobel Peace Prize. The world's reaction was decidedly mixed. The prize was awarded, not to an individual or an organization, but to this entire group of 27 countries that, at the time, was in the midst of an economic crisis. A nagging question was raised once again: Just how beneficial is the European Union?

The EU was officially created in November 1993, with the goal of establishing a more unified and economically and politically stable Europe. EU supporters argue that its single market has provided companies a strong business arena. Businesses do not have to worry about currency exchanges and complex international tax laws. Competition, innovation, and productivity have increased, leading to lower costs and a wider variety of consumer products. Moreover, students can study abroad in any of the EU member countries. Most importantly, after two major wars in the last century, Europe has enjoyed a long era of peace.

However, EU detractors point out that the single market has always been inherently unstable. As evidence, they point to the financial crisis Europe experienced beginning in 2010. The EU became entangled in a multibillion euro crisis, and EU members hotly debated the merits and ramifications of bailouts. Since 2008, the EU has scrambled to support failing economies in Greece, Italy, and Spain. In 2010 and 2011, for example, the EU made loans to Greece to rescue its faltering economy. The loans were conditional on the adoption of strict austerity measures by the government of Greece, which led to a worsening recession and increased social unrest. Such social unrest in countries suffering from failing economies has pitted the poorer southern European countries against the generally more affluent northern European countries. EU bureaucracy also slowed down solutions. Some worry that the EU has gone beyond the goal of economic unity, and that larger countries dominate at the expense of smaller ones.

Jonathan Kitchen/Photographer's Choice/Getty Images

PHOTO: (t)Jonathan Lavick/Stockbyte/Getty Images; (c)Frederick Florin/AFP/Getty Images; (l)Adrian Peacock/Photodisc/Getty Images; TEXT: The Nobel Peace Prize for 2012 by the Norwegian Nobel Committee. Copyright © Nobel Media AB 2013. All rights reserved.

The EU Is a Stabilizing Force

PRIMARY SOURCE

❝ The union and its forerunners have for over six decades contributed to the advancement of peace and reconciliation, democracy and human rights in Europe.... The dreadful suffering in World War II demonstrated the need for a new Europe. Over a seventy-year period, Germany and France had fought three wars. Today war between Germany and France is unthinkable. This shows how, through well-aimed efforts and by building up mutual confidence, historical enemies can become close partners.... The stabilizing part played by the EU has helped to transform most of Europe from a continent of war to a continent of peace. ❞

—Thorbjorn Jagland, the chairman of the Norwegian Nobel Committee, October 2012

The EU Is Driving Europe into Recession

PRIMARY SOURCE

❝ You only have to open your eyes to see the increasing violence and division within the EU which is caused by the Euro project. Spain is on the verge of a bail-out, with senior military figures warning that the Army may have to intervene in Catalonia. In Greece people are starving and abandoning their children through desperate poverty and never a week goes by that we don't see riots and protests in capital cities against the troika [a committee composed of the European Commission, the European Central Bank, and the International Monetary Fund that organized loans to rescue the economies of faltering EU countries] and the economic prison they have imposed.... The last attempt in Europe to impose a new flag, currency and nationality on separate states was called Yugoslavia. The EU is repeating the same tragic mistake.... Rather than bring peace and harmony, the EU will cause insurgency and violence. ❞

—Nigel Farage, UK Independence Party leader, October 2012

What do you think? **DBQ**

1. *Finding the Main Idea* According to Jagland, how has the EU benefited Europe?

2. *Identifying Central Issues* What arguments does Farage give to support his opinion that the EU has failed?

3. *Evaluating* Who do you think makes the stronger argument, Jagland or Farage? Explain.

netw⊙rks

There's More Online!

☑ **IMAGE** Fishing Vessel in Northwestern Europe

☑ **IMAGE** Environmental Activism

☑ **GRAPH** Overfishing in Northwestern Europe

☑ **MAP** Oxygen Depletion in Coastal Marine Ecosystems

☑ **INTERACTIVE SELF-CHECK QUIZ**

☑ **VIDEO** People and Their Environment: Northwestern Europe

LESSON 3
People and Their Environment: Northwestern Europe

ESSENTIAL QUESTION · *How do physical systems and human systems shape a place?*

Reading HELPDESK

Academic Vocabulary

- **isolate**
- **ensure**

Content Vocabulary

- **acid deposition**
- **Kyoto Protocol**
- **cap-and-trade**

TAKING NOTES: *Key Ideas and Details*

SUMMARIZING Use a graphic organizer like the one below to take notes about the human impact on the environment in Northwestern Europe and how people are addressing the issues.

Humans and Their Environment

Human Impact	Addressing the Issues

IT MATTERS BECAUSE

Humans have been changing the environment of Northwestern Europe for thousands of years. Today, very little of the region has escaped substantial alteration from its natural condition. Nevertheless, a healthy and beautiful environment is important to the region's people. They have been successful in finding ways to reverse the environmental damage. But water quality and pollution are still major issues facing Northwestern Europe.

Managing Resources

GUIDING QUESTION *How has modern development resulted in challenges to the management of resources in Northwestern Europe?*

With its highly developed and industrialized economy, Northwestern Europe consumes large amounts of natural resources and generates considerable waste products. While many natural resources are imported from other parts of the world, some of them are extracted from mines, quarries, forests, and other operations within the subregion. Such activities cause pollution and land-use changes that require careful management.

Pollution contaminates marine and animal life and creates health hazards for humans. France and other countries that border the Mediterranean Sea use the sea for waste disposal. In the past, bacteria broke down most of the waste. However, growing populations and tourism along the coasts have increased the environmental problems. Small tides and weak currents tend to keep pollution where it is discharged.

Overfishing has been a problem in the subregion as well. Global fishing levels are estimated to be four times greater than the amount of fish left to catch. Adding to the problem, native species of seaweed and shellfish often compete with foreign species carried into local waterways by ships. Unsustainable fishing remains a major issue for the European Union (EU), where 75 percent of stocks are overfished and catches are only a fraction of what they were 15 to 20 years ago. However, in many fishing grounds that

(l)Voorspoels Kurt/Alamy, (r)Boris Horvat/AFP/Getty Images

affect Great Britain and Germany, progress has been made. Reports show that overfishing in the northeast Atlantic, the North Sea, and the Baltic Sea declined from 72 percent in 2010 to 47 percent in 2012.

As new roads and railways intersect Northwestern Europe, the fragmentation of the landscape limits the migration of wildlife and causes some animal populations to become **isolated**. The highest levels of fragmentation in the subregion are found in the Benelux countries (Belgium, the Netherlands, and Luxembourg), followed by Germany and France. Fragmentation also reduces the development of healthy ecosystems upon which human and animal life rely.

isolate to place or keep by itself

✔ READING PROGRESS CHECK

Evaluating How has industry affected Northwestern Europe's environment?

Human Impact

GUIDING QUESTION *In what ways have human activities impacted the environment in Northwestern Europe?*

The degradation of marine and coastal ecosystems is occurring in Northwestern Europe. This trend has intensified due largely to overfishing, agriculture, pollution, tourism, industrial chemicals, and coastal development.

Air pollution is a problem throughout Europe due to manufacturing industries and the heavy use of vehicles. Industrial fumes and vehicle exhaust cause eye irritation, asthma, and respiratory infections in people who live in industrial areas of Northwestern Europe. Factories built during the Communist era in East Germany, for example, still give off soot, sulfur, and carbon dioxide into the air. Some former Communist countries are closing polluting factories. Yet they are putting more cars on the road, emitting sulfur, nitrogen oxides, and carbon.

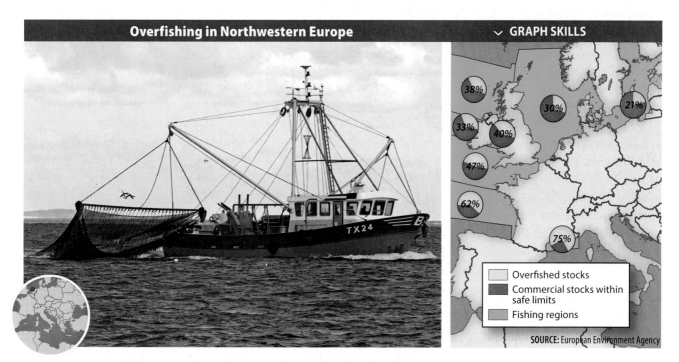

Overfishing in Northwestern Europe

⌄ GRAPH SKILLS

38%
30%
21%
33%
40%
47%
62%
75%

Overfished stocks
Commercial stocks within safe limits
Fishing regions

SOURCE: European Environment Agency

Overfishing in the coastal waters of Northwestern Europe has led to reduced catches for fishers of the subregion.

▲ CRITICAL THINKING

1. *Analyzing Visuals* What body of water suffers the most from overfishing?

2. *Making Predictions* What might be the consequences if overfishing in this subregion continues?

Voorspoels Kurt/Alamy

Oxygen Depletion in Coastal Marine Ecosystems

- ○ Eutrophic
- ● Hypoxic
- ● Systems in recovery

In *eutrophic areas*, the water is overenriched with nutrients, often from agriculture, resulting in algal "blooms" harmful to aquatic life. In *hypoxic areas*, the water has been depleted of oxygen, creating "dead" zones.

North Sea

IRELAND
NETHERLANDS
UNITED KINGDOM
GERMANY
BELGIUM
ATLANTIC OCEAN
LUXEMBOURG
AUSTRIA
SWITZERLAND
LIECHTENSTEIN
Bay of Biscay
FRANCE
MONACO
Adriatic Sea
Mediterranean Sea

0   200 miles
0   200 kilometers
Lambert Azimuthal
Equal-Area projection

SOURCE: World Resources Institute

GEOGRAPHY CONNECTION

Marine coastal ecosystems in Northwestern Europe have been polluted, resulting in areas of over-enriched nutrients that are dangerous to aquatic life and areas where aquatic life cannot survive.

1. *HUMAN SYSTEM* Which type of oxygen depletion is the result of the agricultural pollutants?

2. *PLACES AND REGIONS* How does France compare to the United Kingdom in terms of areas in recovery?

acid deposition wet or dry airborne acids that fall to the ground

In the 1970s and 1980s, industries in Northwestern Europe built taller smokestacks to carry pollution away from their communities. This worked locally, but pollution drifted across national borders. This pollution, containing acidic chemicals, combines with moisture in the air and falls as acid rain. Polluted clouds drift from the industrial belt of the subregion, and **acid deposition**—wet or dry acid pollution that falls to the ground—withers forests and damages rivers in other areas. Acid pollution even damages buildings, especially those made of limestone. During winter, snow carries the industrial pollution to the ground. During spring, meltwater—the result of melting snow and ice—carries the acid into lakes and rivers. As acid concentrations build, fish and other aquatic life die. Many lakes in Northwestern Europe have a declining fish population or even no fish at all.

☑ READING PROGRESS CHECK

Identifying Central Issues What human activities have led to withered forests and damaged rivers and buildings?

Addressing the Issues

GUIDING QUESTION *What actions have been taken to address environmental issues in Northwestern Europe?*

In recent decades, Europeans have made concerted efforts to clean up the environment. Countries in the EU can face legal action if they do not respect environmental protection laws. Individual countries are also addressing the consequences of pollution. For example, cities in Northwestern Europe now protect buildings and statues with acid-resistant coatings. In 2012 the mayor of Paris proposed a ban on older vehicles that produce higher emissions from the city and surrounding suburbs.

The EU and others continue to develop ways to protect the environment. Many power plants now burn natural gas instead of lignite coal. Natural gas burns cleaner and results in less pollution. Some countries are also developing alternative fuel sources, such as solar and wind.

Global climate change is another process that has implications for Northwestern Europe. Scientists have not identified the exact cause of climate change, but it appears clear that emissions of greenhouse gases from human activities are partly responsible for it.

In response, all the countries in the Northwestern Europe have ratified the **Kyoto Protocol**—an amendment to the international treaty on climate change designed to reduce the amount of greenhouse gases emitted by specific countries. The Kyoto Protocol sets emissions targets for participating countries and establishes a system of **cap-and-trade**.

Under a cap-and-trade system, a limit is set on the amount of air pollution that can be emitted. Businesses that produce amounts lower than the cap receive credits for the difference between their actual emissions and the cap.

Kyoto Protocol an amendment to the international treaty on climate change designed to reduce the amount of greenhouse gases emitted by specific countries

cap-and-trade a method for managing pollution in which a limit is placed on emissions and businesses or countries can buy and sell emissions allowances

Boris Horvat/AFP/Getty Images

Many Northwestern Europeans, such as these protesters, are deeply concerned about damage to the environment in their subregion.

▲ **CRITICAL THINKING**

1. *Speculating* What are some environmental issues that concern citizens in Northwestern Europe that may be the cause of this protest?

2. *Identifying Cause and Effect* Describe how protests like the one pictured can lead to change in government policies.

Cap-and-Trade Systems

In 2005 the European Union began a cap-and-trade system aimed at reducing the impact of pollutants that are linked to the burning of fossil fuels. The program sets a maximum overall limit (cap) on greenhouse gas emissions. Sources of the emissions receive permits to emit (emissions allowances), which can be traded. Greenhouse gas producers in the program are responsible for monitoring emissions and using permits to pay for emission privileges. The program is designed to provide both flexibility and an incentive to reduce emissions. Between 2005 and 2010, greenhouse-gas emissions dropped 13 percent.

CONSIDERING ADVANTAGES AND DISADVANTAGES What are some potential pros and cons of a cap-and-trade system? Discuss which industries may have objections to a strict cap-and-trade program and provide effective counter-arguments.

ensure to make sure or certain

Then the businesses can "trade" their pollution credits, so that a business that produces pollution at levels higher than the cap may purchase credits from businesses who earned credits. This creates an economic incentive to reduce the emissions, and the cap can be lowered over time, so that emissions continue to decline.

Protecting water resources in the subregion has also become a high priority. By the end of 2015, as required by the EU Water Framework Directive, several rivers in Europe will be required to have major restoration to meet the new regulations. Progress has been made, but the rivers in the region still suffer from floodplain drainage, industrial discharge, and the heavy use of fertilizers.

PRIMARY SOURCE

❝It will require a substantial investment but the costs of reviving Europe's rivers will be more than repaid by long-term savings in flood damage, water treatment and public health. WWF will be keeping a close eye on EU countries to make sure they live up to these commitments.❞

—Julian Scola, "Europe's Rivers Ready for Revival," World Wide Fund for Nature, April 20, 2001.

In 1992 the Convention on Biological Diversity (CBC) was signed by 159 governments at the United Nation's Earth Summit in Rio de Janeiro, Brazil, creating the first treaty to provide a legal framework for biodiversity conservation. Following the convention, the United Kingdom created the UK BAP (Biodiversity Action Plan) a national strategy to conserve, protect and enhance biological diversity. It describes the United Kingdom's biological resources, including species and habitats, with a detailed plan for protecting their resources.

To address air pollution and improve air quality, the government of the United Kingdom allows community governments to forward fines imposed by the European Union to the local offenders. Local authorities can remove the most polluting vehicles from the roads. Governments of the countries in Northwestern Europe are not only working on solutions to environmental problems together, but with nongovernmental organizations (NGOs) as well. For example, the European Union's common fisheries policy is designed to **ensure** fisheries are more sustainably managed to prevent overfishing.

Three NGOs that are working to protect, restore, and improve the marine and coastal ecosystems are Greenpeace, Oceana, and Seas At Risk. The mission of all three of these international organizations is to limit further degradation of marine areas and loss of species, and to provide awareness, education, and concrete steps to arrest further damage. Seas At Risk, in particular, focuses its efforts on the North Atlantic and the Irish and North Seas.

✓ **READING PROGRESS CHECK**

Explaining How is the EU encouraging countries to develop new ways to protect the environment?

LESSON 3 REVIEW

Reviewing Vocabulary
1. *Explaining* Explain the Kyoto Protocol.

Using Your Notes
2. *Describing* Use your graphic organizer to describe how Northwestern Europe is addressing the issues of pollution in the region.

Answering the Guiding Questions
3. *Identifying* How has modern development resulted in challenges to the management of resources in Northwestern Europe?

4. *Describing* In what ways have human activities impacted the environment in Northwestern Europe?

5. *Explaining* What actions have been taken to address environmental issues in Northwestern Europe?

Writing Activity
6. *Argument* Imagine that you live in a polluted area of Northwestern Europe. Write a letter to the editor of a local newspaper advocating steps to halt environmental damage.

Directions: On a separate sheet of paper, answer the questions below. Make sure you read carefully and answer all parts of the questions.

Lesson Review

Lesson 1

1 *Describing* Describe the geography of the Northern European Plain and its appeal to early farmers.

2 *Explaining* Explain how water plays an important role in the lives of Northwestern Europeans.

3 *Drawing Conclusions* What are some factors affecting climate in Northwestern Europe? Why are they significant?

Lesson 2

4 *Evaluating* How did the Industrial Revolution change the human geography of Northwestern Europe?

5 *Discussing* Describe the impact of the aging population on population patterns in Northwestern Europe.

6 *Summarizing* What are the main service industries that fuel Northwestern Europe's economy?

Lesson 3

7 *Identifying Central Issues* Describe the problem of overfishing in Northwestern Europe and provide some viable solutions to reduce its impact.

8 *Explaining* What is the main reason for poor air quality and air pollution in parts of Northwestern Europe?

9 *Comparing and Contrasting* Explain cap-and-trade and the Kyoto Protocol. How are they related?

Exploring the Essential Question

10 *Making Connections* Write a paragraph that explains the effect of the Seine River on the location of cities in France.

21st Century Skills

Use the following graph to answer the questions below.

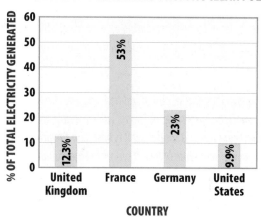

ELECTRICITY GENERATED FROM NUCLEAR FUEL

% OF TOTAL ELECTRICITY GENERATED vs COUNTRY

United Kingdom: 12.3%
France: 53%
Germany: 23%
United States: 9.9%

Source: CIA World Factbook

11 *Compare and Contrast* How does France compare to Germany and United Kingdom in regard to nuclear fuels generated?

12 *Identifying Cause and Effect* What factors may contribute to the higher percentage of nuclear fuels generated in Northwestern Europe, as compared to the amount generated in the United States?

College and Career Readiness

13 *Economics* As a consultant for a global conglomerate, you have been hired to research the importance of transportation and communications systems to the economy of Northwestern Europe. What are the strengths and weaknesses of the current systems, and how can they be improved?

Need Extra Help?

If You've Missed Question	❶	❷	❸	❹	❺	❻	❼	❽	❾	❿	⓫	⓬	⓭
Go to page	270	271	273	276	278	280	284	286	287	273	289	289	280

Directions: On a separate sheet of paper, answer the questions below. Make sure you read carefully and answer all parts of the questions.

DBQ Analyzing Primary Sources

Use the document to answer the questions that follow.

> PRIMARY SOURCE

"*The peoples of Europe, in creating an ever closer union among them, are resolved to share a peaceful future based on common values.*

Conscious of its spiritual and moral heritage, the Union is founded on the indivisible, universal values of human dignity, freedom, equality and solidarity; it is based on the principles of democracy and the rule of law. It places the individual at the heart of its activities, by establishing the citizenship of the Union and by creating an area of freedom, security and justice."

—Charter of Fundamental Rights of the European Union, 2000

14 *Identifying* According to the excerpt, what core values are held by the European Union?

15 *Paraphrasing* In your own words, summarize the meaning of the Charter of Fundamental Rights of the European Union.

Research and Presentation

16 *Research Skills* Use the Internet and other resources to gather information about one of Northwestern Europe's most famous cities, such as Paris, London, Brussels, or Amsterdam. What is the city most known for? How is it alike and different from other cities in the subregion? Create a multimedia presentation to share your findings.

Applying Map Skills

Use your Unit 4 Atlas to answer the following questions.

17 *Places and Regions* Describe the location and extent of the Alps in relation to the Northern European Plain.

18 *Physical Systems* Use the physical map of Europe from the unit atlas to list the bodies of water that border Great Britain.

19 *The World in Spatial Terms* Use your mental map of Europe to describe the location of Ireland and Great Britain in relation to continental Europe.

Critical Thinking

20 *Identifying Central Issues* How have urbanization and immigration shaped the population patterns of Northwestern Europe?

21 *Making Generalizations* What are the characteristics of internal migration and suburbanization, and how did they affect population patterns?

22 *Drawing Conclusions* Discuss the ways in which human activities have affected marine and coastal ecosystems in Northwestern Europe.

Writing About Geography

23 *Informative/Explanatory* Use standard grammar, spelling, sentence structure, and punctuation to write a one-page essay about Paris's rise to megacity status. Use library and Internet resources to gather information to discuss the city's population and how it has changed over the decades. Describe the factors regarding the decline in out-migration and a higher natural population rate increase.

Need Extra Help?

If You've Missed Question	14	15	16	17	18	19	20	21	22	23
Go to page	290	290	278	242	242	242	278	278	286	269

Southern Europe

ESSENTIAL QUESTION · *How do physical systems and human systems shape a place?*

Geography Matters...

Have you ever heard of the Seven Wonders of the World? Europe is the continent where that idea was born. There have been many lists, so no one knows for certain what all of the Seven Wonders were. The cultures of Southern Europe have created architectural wonders. Among them are the Roman Colosseum, the Parthenon of ancient Greece, and the elaborate Spanish palace called the Alhambra. The natural wonders of Southern Europe are also abundant, from the Mediterranean Sea to Basque Country—home of the largest cavern in Europe.

◀ Greece's many islands and long coastlines have always supported a strong tradition of fishing.

Steve Outram/Photographer's Choice/Getty Images

labor migration
from North Africa

The migration of people from place to place is an important theme in the study of geography. Throughout history people have migrated in search of better opportunities. One modern example is the migration of North African workers to the countries of Southern Europe. This has had important consequences both for the North African migrants and the countries of Southern Europe.

Migration: North Africa to Southern Europe

Why do people from North Africa migrate to Southern Europe?

Several factors cause North African workers to migrate to Southern Europe in search of employment. The economic opportunities in Southern Europe represent major pull factors. Migrants are drawn to the Southern European countries in search of employment. Demographic changes have made this trend greater. The population of Southern Europe has aged. Young workers are needed to fill the spaces left as older people retire and stop working, thus increasing the demand for migrants. Other factors push people out of North Africa. As countries in North Africa face increasingly high unemployment and wages decline for workers, many choose to leave their home countries—a push factor. In addition, political instability and conflict arising from the "Arab Spring" uprisings and the Libyan civil war have forced many to flee to Southern Europe.

1. **Human Systems** Describe the causes of North African migration to Southern Europe. Include both push and pull factors.

How is Southern Europe affected by this migration?

An economic boom has resulted in increased demand for low-skilled labor in several Southern European countries. Italy and Spain, with their long Mediterranean coastlines, have become popular entry points for undocumented migrants as they fill the need for service industry jobs. During the prosperous economic times, the labor of these migrants helped fuel the engine of economic development by providing sources of cheap labor. However, besides providing labor, North African migrants (whether documented or not) have influenced the cultural development of the receiving countries of Southern Europe. The migrants have enriched the cultural landscape of their new countries. They have brought their religious beliefs, social customs, arts, cuisine, and dress to Southern Europe. In some places in Southern Europe the change has come at the expense of the migrants. The arrival of migrants from North Africa has led to racism and to a backlash against them as they try to establish their new lives in Southern Europe.

2. **Human Systems** How does North African migration to Southern European countries benefit the receiving countries?

How is North Africa affected by this migration?

North Africa is affected both negatively and positively by the migration to Southern Europe. Many North African migrants send some of their pay home to help support their families. The money sent by a migrant to his or her home country is called a remittance. These remittances have reduced poverty in their countries of origin. The remittances also help the country even when the money goes to families that are relatively wealthy since it contributes to the economy of the North African home country. Other positive effects include more funds being available for North Africans to invest in children's health and education and a reduction in child labor. Not all of the effects are positive, as the migrations can cause labor shortages in the home country.

3. **Environment and Society** What are some of the potential long-term effects of the remittances by North African migrants? Write a paragraph discussing the possible impacts on North African society and on the economies of the home countries.

(l)©David Turnley/Corbis, (c)Directphoto/age fotostock/SuperStock, (r)Andrew Woodley/Alamy

THERE'S MORE ONLINE

SEE a graph of remittances from Southern European countries • *WATCH* a video about migrants

There's More Online!

- ☑ **IMAGE** Mining in Spain
- ☑ **INFOGRAPHIC** Rivers of Europe
- ☑ **IMAGE** Massifs in Southern Europe
- ☑ **INTERACTIVE SELF-CHECK QUIZ**
- ☑ **VIDEO** Physical Geography of Southern Europe

LESSON 1
Physical Geography of Southern Europe

ESSENTIAL QUESTION · *How do physical systems and human systems shape a place?*

Reading HELPDESK

Academic Vocabulary

- **access**
- **resource**

Content Vocabulary

- **massif**
- **tungsten**

TAKING NOTES: *Key Ideas and Details*

LISTING Use a fishbone graphic organizer like the one below to list the landforms and rivers of Southern Europe.

IT MATTERS BECAUSE

The seaports of the Iberian Peninsula, the mountains and rivers of the Italian Peninsula, and the rocky islands of Greece have played important roles in the history of Southern Europe. Its geographic location has made this subregion—which includes the countries of Italy, Spain, Andorra, Greece, Portugal, Vatican City, Malta, Cyprus, and San Marino—important for trade and agriculture stretching back more than 3,000 years.

Landforms

GUIDING QUESTION *What two types of physical features dominate Southern Europe's physical geography?*

Geographically, Europe is a continent made up of peninsulas. The southern part of the region includes three major peninsulas: the Iberian Peninsula, the Italian Peninsula, and the Balkan Peninsula.

Extending off southwestern Europe, the Iberian Peninsula is the location of Spain and Portugal. This landmass separates the Atlantic Ocean from the Mediterranean Sea, leaving only the 20-mile-(32-km-) wide Strait of Gibraltar to connect them. Coastal plains give way to the Meseta, a large plateau that makes up most of the interior of the peninsula. To the north, the Iberian Peninsula is separated from the rest of Europe by the Pyrenees, mountains that have isolated the peninsula's residents for centuries. The independent principality of Andorra is located high in the Pyrenees between modern-day Spain and France. Andorra owes its political autonomy to its isolated location in the mountains.

The most southwestern of Europe's mountain ranges, the Pyrenees stretch from the Bay of Biscay on the Atlantic side of Spain to the Mediterranean Sea. At its widest the Pyrenees range is a daunting 80 miles (128 km) across. The Pyrenees are characterized by flat-topped **massifs**—a body of mountain ranges formed by fault-line activity. The forces of plate tectonics are responsible for the rise of these massifs, and earthquakes occur as the mountains are built up.

(tl)Alessandra Sarti/Alamy, (tc)Luis Castaneda/age fotostock, (tr)Hoberman Collection/Universal Images Group/Getty Images

Italy occupies the Italian Peninsula, which extends from the south of Europe into the heart of the Mediterranean Sea. Plains cover only about one-third of the Italian Peninsula. The largest is the plain of Lombardy along the Po River in the north. The coastline of Italy varies from high, rocky cliffs to long, sandy beaches, and has several well-sheltered ports to support trade. The Apennine Mountains run down the spine of the peninsula all the way through the center of the large island of Sicily off the southwestern tip of Italy. The range is about 1,245 miles (2,000 km) long.

To the north of the Italian Peninsula lie the majestic Alps, the most recognizable range of mountains on the European continent. They loom over Southern Europe and form a natural barrier between the Italian Peninsula and Northern Europe. Because the Alps are the highest mountain range in Europe, they are also the source of Europe's largest and most important rivers. These rivers flow north into France and Germany or south where they empty into the Mediterranean, Adriatic, and Black Seas.

In southeastern Europe, the Balkan Peninsula is bounded by the Adriatic and Ionian Seas to the west and the Aegean and Black Seas to the east. Greece is the southernmost country on the Balkan Peninsula. The numerous mountains on this peninsula have limited the area's potential for communication and development. However, this has been offset by the region's easy **access** to the sea. Greece is known for the large numbers of islands—nearly 2,000—that spread out from its coastline in the Aegean Sea.

The islands that lie south of the mainland of the Iberian, Italian, and Balkan Peninsulas are geographically and politically important to Spain, Italy, and Greece. They serve as trading posts in the Mediterranean. Rugged mountains form the larger islands of Sicily, Sardinia, Corsica, Crete, and Cyprus. Tectonic activity is characteristic of this region. Sicily, the largest of these islands, is dominated by Mount Etna. At 10,700 feet (3,261 m), Mount Etna is Europe's tallest active volcano. Smaller island groups include Spain's Balearic Islands in the Mediterranean, Italy's Lipari Islands in the Tyrrhenian Sea, and the three islands of Malta in the Mediterranean.

☑ **READING PROGRESS CHECK**

Identifying What seas surround the Balkan Peninsula?

massif a body of mountain ranges formed by fault-line activity

access a way to approach or enter

Massifs, such as these in the Dolomite range of the Alps in northeast Italy, developed as a result of tectonic activity.

▼ **CRITICAL THINKING**

1. *Identifying Cause and Effect* Why do massifs form?

2. *Drawing Inferences* Other than the Alps, where else can massifs be found in Southern Europe?

Water Systems

GUIDING QUESTION *How do Southern Europe's rivers compare to those of Northwestern Europe?*

The two major rivers on the Iberian Peninsula are the Tagus and the Ebro. Both rivers play crucial roles in the economy and ecology of the region. However, their roles are limited because they, like all rivers on the Iberian Peninsula, are generally too shallow for large ships.

The Tagus River begins near the eastern edge of Spain and travels westward for 626 miles (1,007 km) through Portugal to the Atlantic Ocean. In northern Spain, nearly 200 tributaries, mostly from the rainy Pyrenees, feed the Ebro River, Spain's longest river. The steep gorges and rocky terrain that this river flows through make it inaccessible to boats. However, the Ebro has been dammed to provide a significant portion of Spain's hydroelectric power. In addition, the

RIVERS of Europe

RHINE — 77,692 ft³ (2200 m³) — 820 MILES (1319 km)

ELBE — 11,200 ft³ (317 m³) — 724 MILES (1165 km)

SEINE — 10,000 ft³ (280 m³) — 482 MILES (776 km)

THAMES — 1,856 ft³ (53 m³) — 210 MILES (338 km)

NORTHWESTERN

SOUTHERN

VOLUME OF RIVER WATER PER SECOND — 9,182 ft³ (260 m³) — 252 MILES (405 km)

RIVER SHAPE — LENGTH OF RIVER

ft³ = cubic feet   c³ = cubic meters

TIBER — 9,182 ft³ (260 m³) — 252 MILES (405 km)

PO — 405 MILES (652 km) — 48,400 ft³ (1370 m³)

EBRO — 14,126 ft³ (400 m³) — 565 MILES (909 km)

TAGUS — 14,126 ft³ (2200 m³) — 626 MILES (1007 km)

Although critically important to the ecology of the region, the rivers of Southern Europe have less importance as transportation and trade routes than the rivers of Northwestern Europe.

▲ CRITICAL THINKING

1. *Contrasting* How are the rivers of Southern Europe different from the rivers of Northwestern Europe in regard to length?

2. *Evaluating* Is there a correlation between length of rivers in Europe and the volume of water flow? Provide an example to support your answer.

resulting reservoirs provide water to an impressive network of irrigation canals that support the agriculture of Spain.

The Apennines form a mountain range that runs down the center of the long and narrow Italian Peninsula. This has created rivers that are steep, short, and relatively narrow and shallow, and not suitable for transportation by boat. In the north, however, the Po River runs through the plain of Lombardy. Although it is Italy's longest and most significant river, it is still only 405 miles (652 km) long.

Venice is located at the mouth of the Po on the Adriatic Sea. The city has built a complicated system of dikes and canals to help control the river's outflow. Efforts to claim marshy areas of the Po Delta for small farms failed due to floods, especially in the 1950s and 1960s. Nevertheless, the drainage basin of the Po River forms Italy's largest and most fertile agricultural plain, covering 27,062 square miles (70,091 sq. km).

The Tiber River is Italy's second-longest river and has great historical significance. A mere 252 miles (405 km) long, this short river is nevertheless very important to Italy's economic history. It is the primary water source for the capital, Rome. Civitavecchia, on the lower part of the river, is a significant port and naval harbor for Rome. The Tiber River empties into the Tyrrhenian Sea.

Greece, on the southern tip of the Balkan Peninsula, has a mountainous terrain. Its rivers are short, unsuitable for navigation, and unusable for irrigation. In their upper courses, near their sources, the rivers flow in broad, gently sloping valleys. However, in their middle courses they plunge through basins into narrow gorges. In their lower courses, as they near their mouths, they meander across coastal plains into marshy deltas. Northeastern Greece is home to the Maritsa River, located in a low valley full of marshes. The Maritsa marks Greece's border with Turkey. In northeastern Greece, the two main rivers are the Vardar and the Aliákmon.

Glacial movement in the last ice age did not reach Southern Europe's peninsulas. As a result, the landforms of these countries lack the natural lakes or reservoirs found in Northern Europe and Northwestern Europe. The climate of Southern Europe is also much drier than farther north, another reason why there are fewer rivers and lakes in this subregion.

☑ **READING PROGRESS CHECK**

Describing Describe the importance of the Po River.

Climate, Biomes, and Resources

GUIDING QUESTION *Why does Southern Europe's climate make it popular with tourists and ideal for agricultural activities?*

Southern Europe's location on the Mediterranean Sea influences the climate and biomes of the subregion. The climate also makes the subregion a popular vacation destination. The subregion is particularly suited for growing grapes, olives, and shrub herbs and raising goats and other livestock.

Climate Regions and Biomes

The Alps separate two major climate zones: the marine west coast climate to the north and the warm Mediterranean climate of Italy and the Balkans to the south. The Alps block most Atlantic winds from the north, causing less precipitation to fall in Southern Europe. Generally, Southern Europe experiences the warm, dry summers and the mild, rainy winters characteristic of the Mediterranean climate.

The Mediterranean climate of Southern Europe results, in part, from the warm waters of this sea. Average yearly rainfall across Southern Europe is less than 30 inches (76 cm), and most of the yearly rainfall occurs in the winter months.

Analyzing
PRIMARY SOURCES

Olive Oil Production in Italy

"From a geographical point of view, the prime area of Italian olive production is located between 30 and 45 °N, which in general is the geographical distribution of suitable areas for olive-tree production in the northern hemisphere. In the Mediterranean area, Italy represents the central point of olive production because of its history and environmental conditions. . . . In terms of olive-oil production, Italy ranks second in the world (after Spain), producing an average oil quantity over the last four years of 550,000 tonnes."

—G. Fontanazza, "Importance of olive-oil production in Italy," *Integrated Soil and Water Management for Orchard Development*, 2005

DBQ **DRAWING CONCLUSIONS**
What aspect of Italy's geography makes the country a suitable location for growing olives?

Spain has more mineral resources than Italy due to their different geological histories.

1. *Making Generalizations* Why does Italy have fewer mineral resources than Spain?

2. *Analyzing Visuals* Based on the photo, what might be a drawback to plentiful mineral resources?

tungsten an extremely rare heavy-metal element essential in high-tech industry

resource a usable stock or supply

The plants and animals native to the subregion are well-suited for less water and the long summer dry period. The coastal areas are covered in chaparral, or shrubs and shrub trees that are drought resistant. Most agriculture in Southern Europe takes place on coastal plains that receive more rainfall and develop a thicker topsoil as a result of runoff and sediment left by rivers.

The ecosystems in the Mediterranean are diverse and ecologically sensitive to climate change. Gorges channel water away from the land, leaving much of the region warm, dry, and covered in scrub plants. The coastal plains have rich sedimentary soil and a high diversity of plant life, and support most of the regional agriculture. However, they are prone to flooding.

Natural Resources

Italy has few mineral resources. Portugal, however, has large deposits of copper. Northern Spain, along the Pyrenees and the Atlantic Ocean, is rich in coal, tin, and **tungsten.** Tungsten is an extremely rare heavy-metal element which is essential in high-tech industries. Spain's mining operations in search of this valuable natural element are unlikely to be reduced.

Both Italy and Spain have benefited from the production of hydroelectricity. Greece has many rivers suitable for producing hydroelectricity. However, Greece has not fully developed these **resources.**

☑ **READING PROGRESS CHECK**

Describing Describe the ecosystem of the Mediterranean climate.

Luis Castaneda/age fotostock

LESSON 1 REVIEW

Reviewing Vocabulary
1. *Explaining* Explain what massifs are and how they are formed.

Using Your Notes
2. *Listing* Use your graphic organizer to list the major landforms in Southern Europe.

Answering the Guiding Questions
3. *Identifying* What two types of physical features dominate Southern Europe's physical geography?

4. *Comparing* How do Southern Europe's rivers compare to those of Northwestern Europe?

5. *Expressing* Why does Southern Europe's climate make it popular with tourists and ideal for agricultural activities?

Writing Activity
6. *Informative/Explanatory* Write a paragraph describing how the rivers of Southern Europe can be an asset for the countries in the subregion.

(tl)©Bettmann/Corbis, (tc)Fine Art Images/SuperStock/Getty Images, (tcr)Photodisc/Getty Images, (tcr)©Bettmann/Corbis, (tr)Angel Navarrete/Bloomberg/Getty Images

netw⌀rks

There's More Online!

- ☑ **IMAGE** Classic Greek Architecture
- ☑ **IMAGE** Classic Roman Architecture
- ☑ **TIME LINE** Foundations of Western Civilization
- ☑ **INTERACTIVE SELF-CHECK QUIZ**
- ☑ **VIDEO** Human Geography of Southern Europe

Reading **HELP**DESK

Academic Vocabulary

- **issue**
- **accurate**

Content Vocabulary

- **city-state**
- **Renaissance**
- **complementarity**

TAKING NOTES: *Key Ideas and Details*

IDENTIFYING Use a graphic organizer like the one below to take notes about the human geography of Southern Europe.

Southern Europe

History and Government ▶ ▢

Population Patterns ▶ ▢

Society and Culture Today ▶ ▢

LESSON 2
Human Geography of Southern Europe

ESSENTIAL QUESTION • *How do physical systems and human systems shape a place?*

IT MATTERS BECAUSE

The European countries in the region of the Mediterranean Sea have a rich cultural heritage. Greece was the birthplace of classical civilization, the Roman Empire was born in Italy, and Spain and Portugal became leaders in the Age of Exploration. Today the region is highly developed and plays an important role in geopolitics and the world economy and as a cultural center.

History and Government

GUIDING QUESTION *What characteristics of early civilizations are evident in Southern Europe today?*

Evidence of the cultural inheritance of Southern Europe is visible in the ruins of ancient civic architecture, such as the Parthenon. Those majestic ruins are a reminder of the lasting impact that past civilizations have had on this region and around the world. The civilizations of ancient Greece and Rome laid the foundations for European—and Western—civilization.

Early History

Ancient Greece was a collection of **city-states,** each independent with its own form of government and society. They were linked only by their common language and shared cultural identity. The classical period of Greek history reached its height in the 400s B.C. The city-state of Athens introduced the concept of democracy to the world. The city-state of Sparta was built on the glory of war. Sparta's generals created battle strategies that are still studied and used today. The mythology of the Greeks and Romans has influenced and inspired the art and culture of Western civilization for over 2,000 years.

The Roman Republic, founded in Italy, was established on the rule of law and the balance of power. It was the foundation of the largest empire of the ancient world. The empire reached the height of its power in 27 B.C. and experienced a resurgence around 200 years later, in A.D. 180.

city-state an independently governed community consisting of a city and the surrounding lands, notably present in ancient Greece

Renaissance rebirth; the period in European civilization characterized by a surge of interest in classical learning and values

issue a vital or unsettled matter

The Path to Today

The period that followed the fall of the Roman Empire in the West is known as the Middle Ages. Beginning in Italy in the 1300s, a new era called the **Renaissance** (from the French word for "rebirth") signaled a major cultural revival. Much of Europe experienced a renewed interest in the arts, politics, science, and philosophy.

This blossoming was possible because of the level of wealth and stability that the Italian city-states achieved. Wealth and stability made it possible for people to devote themselves to study and innovative thought. They could also afford to be patrons for artists, writers, and thinkers. This cultural revival eventually spread across Europe and was later carried by explorers to other regions of the world.

At the beginning of the Renaissance, Europe was lagging behind the Chinese and Islamic empires in innovation and enterprise. However, by the 1400s Southern Europe, led by Portugal and Spain, was exploring new opportunities beyond the Mediterranean. By the early 1500s, both countries had established footholds in the Americas where they would build empires. Portugal also established colonies in India. New trade routes and colonization created enormous wealth.

The 1800s and 1900s brought dramatic changes to all of Europe. In the Mediterranean, small kingdoms, principalities, and city-states that shared a linguistic and cultural identity began to unify under the concept known as nationalism. Others pushed for independence from older multiethnic regimes. As a result, the modern countries of Southern Europe emerged.

The road to political stability in Southern Europe was not easy due to the history of invasions from North Africa, Asia, and other parts of Europe. Today, Southern European countries have democratically elected governments. Portugal, Spain, Italy, and Greece are members of the European Union (EU). The subregion was mainly agricultural until the 1980s when industrialization arrived. Economic **issues** have been a problem in recent decades, however. Spain, Italy, and Greece have experienced high unemployment rates, face credit problems, and are undergoing the transition to an industrial and, more recently, a service economy.

✓ **READING PROGRESS CHECK**

Making Connections What was the impact of nationalism on the development of modern Southern Europe?

(bl)©Bettmann/Corbis, (br)Fine Art Images/SuperStock/Getty Images

TIME LINE ⌄

FOUNDATIONS
of Western Civilization ➡

The origin of modern democratic principles can be traced to the ancient civilizations of Greece and Rome.

▶ **CRITICAL THINKING**

1. Identifying What Allied victory during World War II led to the invasion of Italy?

2. Describing What is the significance of the Hellenic Period?

The Early Roman Republic flourishes.

509–27 B.C.

The Hellenic Period marks the high point of Athenian culture and the birth of democracy.

479–323 B.C.

1000 ➡

The Renaissance, which begins in Italy, sparks a renewal of Western civilization.

1300–1600

1400–1700

The Age of Exploration spawns the foundation of the Spanish Empire.

Population Patterns

GUIDING QUESTION *How have migration and aging populations affected Southern Europe's population patterns?*

Europe is one of the most densely populated regions of the world. It is a relatively small region with a large population, which is mostly urbanized. Opportunities for education, employment, and social services during the past 50 years have been greater in urban areas. Vatican City, just 0.17 square miles (0.44 sq. km) in area, is the most densely populated state in Southern Europe. In contrast, Greece has about 222 people per square mile (86 people per sq. km). It is still the most agrarian (based on farming) and rural of all Southern European countries.

As in most European countries, the populations of Italy, Spain, and Greece are aging. The birthrates are not high enough to replace the current generation. In the coming decades there will be a greater number of older people than young people. As older people die, the population will decline.

General political and economic stability in the last 50 years has meant an increase in the quality of life and the alternatives open to individuals. One result is an increase in income. A consequence of the increase is smaller families. Many people are choosing not to have children since the economic cost of raising a child is great. Another reason is the full inclusion of women into society. Women are postponing having children to pursue careers and, as a consequence, are having fewer children.

The care of the aging is another issue in Southern Europe. Traditionally, younger generations cared for older family members. This practice is declining in Europe where older people do not have adult children to care for them. The aging population is beginning to strain the social welfare systems of countries, and governments must often provide the costs of caring for an older society.

Demands for necessary labor are being met by migrants within the European Union. Newly admitted EU countries in Eastern Europe have larger proportions of their populations living in rural areas. They are attracted to work in urban centers throughout the EU for wages. Often they are unskilled, but have a strong work ethic and make good employees. Economic migration throughout the EU has resulted in a growing diversity of cultures and ethnic groups.

(bl)©Bettmann/Corbis, (br)Angel Navarrete/Bloomberg/Getty Images

Connecting Geography to SCIENCE

Demography

Demography, the statistical study of human populations, tracks changes in populations and seeks to identify trends, their causes, and possible consequences. Tracking fertility (birth) and mortality (death) rates is an important part of the work of a demographer, but patterns of immigration, civil or regional armed conflict, and the effects of health care and disease on populations are also important factors. The work of demographers can influence government decisions, including immigration policy, economic planning, and decisions about the distribution of services within the country.

IDENTIFYING Describe the work of a demographer and provide an example of how it might affect government policy.

1915
Italy enters World War I by declaring war on Austria-Hungary.

➔ 1900

1943
During World War II, Allies launch invasion of Italy after defeating Axis forces in North Africa. Italy surrenders, but Germany continues to fight.

1950–2000
Political and economic stability is established throughout Southern Europe.

High unemployment in Spain and Italy and near economic collapse in Greece lead to protests.

➔ 2000

2012

The largest cities in Southern Europe are economic and cultural hubs founded on ancient trade routes. Within the city of Rome, the modern capital of Italy, is Vatican City. It is an independent country that is home to the Roman Catholic Church. Vatican City was formally established as an independent city-state in 1929. This country is unique in its purpose because it relies on a **complementarity,** or an interdependence, with Rome for its goods and services. Greece has its modern capital in the ancient city of Athens. Spain's major cities, such as Madrid and Barcelona, were once the capitals of kingdoms. Today regions of Spain maintain separate cultural, and even linguistic, identities. Andalusia and the Basque regions have greater autonomy since their ethnic backgrounds are different from the rest of Spain. The importance of trade is apparent with Portugal, since its capital, Lisbon, is also its largest port.

☑ **READING PROGRESS CHECK**

Summarizing What factors are resulting in a decreasing population in the region?

Society and Culture Today

GUIDING QUESTION *How have religion, the arts, and Southern Europe's rich intellectual traditions shaped society and culture today?*

Southern Europe is culturally dominated by two legacies. The first is its location as a geographic crossroad for the cultures surrounding the Mediterranean Sea. The second is the historic events that occurred as a place where civil societies first formed representative governments elected by the people. As a result of empires and territorial control, the countries share roots in language, religion, and the arts. The people of Italy, Spain, and Portugal all speak Romance languages that evolved from Latin, the language of the Romans. Many inhabitants of these countries are also heavily influenced by the Roman Catholic Church.

Throughout the region, literacy is very high (about 95 percent). The high literacy rate reflects the strong government support of education. Education is compulsory for all children, and university education is increasingly common.

Southern European governments have developed as providers of health care and other social welfare services that are funded by taxes. Spain and Italy have well-established universal health care systems that cover the basic health care needs of the people. Spain has more doctors per person than many other countries in Europe. Greece is more rural and has faced greater challenges in meeting the needs of health care among a less dense, more geographically distributed population.

complementarity
relationship between two places in which one produces something the other needs, resulting in an exchange

The Greek Parthenon (left) and the Roman Colosseum (right) are two of the greatest architectural achievements of ancient Southern Europe. Both have provided inspiration for later builders and architects up to the present day.

▼ **CRITICAL THINKING**

1. Speculating Why do you think that these structures have been so inspirational?

2. Making Connections Name two modern buildings that have features like the Parthenon or Colosseum.

Family and Status of Women

Italian women have achieved a high level of gender equality. They enjoy access to higher education and success in business. However, integration of women into Italian politics suffers from a long period of male dominance and poor attitudes toward the contributions of qualified women.

Like Italy, Greece and Spain have a family-centered culture. In Greece, the family is the primary social support. Despite the rise of a large middle class since World War II, it is still common for members of extended families to live together. In the 1980s, laws were enacted that changed traditional family life but also created more freedoms. Spain enjoys similar social structures to Italy and Greece with smaller families and gender equality in higher education and business.

The Arts

The ancient Greeks and Romans developed many important and basic elements of architecture, including columns, arches, and domes. The classical architectural style of these two cultures remains popular to this day, especially in civic buildings. The ancient Greeks valued a type of idealized realism in sculpture; the human figure is portrayed **accurately,** or perfectly, in Greek art. The Romans embraced Greek realism and carried it further, depicting people much as they really were.

accurate free from error

The Renaissance saw a revival of the classical style and its subjects in the arts. Artwork during the Renaissance also reflected a trend towards realism through new techniques such as perspective. Italian Renaissance artists, such as Leonardo da Vinci (1452–1519) and Michelangelo (1475–1564), created many of the era's masterpieces.

Spain has given the world some of the best examples of modern art. Artists in this genre include Pablo Picasso (1881–1973) and architect Antoni Gaudí (1852–1926). Salvador Dalí (1904–1989) led the surrealist movement. Surrealist works and ideas changed the definition of art in the twentieth century.

☑ **READING PROGRESS CHECK**

Identifying Describe the major contributions of the Romans and Greeks to the arts.

Economic Activities

GUIDING QUESTION *What are the characteristics of Southern Europe's economy today?*

The global economic downturn at the beginning of the twenty-first century had a major impact on the subregion. Greece's fragile economy, based on small, family-owned businesses and agriculture, collapsed under a real-estate crash caused largely by a lack of regulation. In 2012 the general unemployment rate topped 25 percent, but youth unemployment was over 50 percent. Recovery was supported by other EU countries.

Resources, Power, and Industry

For all Southern European countries, access to the Mediterranean Sea and the Atlantic Ocean has been essential to economic development and stability. Like many developed countries, Southern European countries have seen steady decreases in their manufacturing sectors in the last 30 years. Instead, the service sectors have become more important, with an increase in industries such as banking, retail and wholesale sales, tourism, health care, and telecommunications. The countries in this subregion have relied on their warm, attractive climate, proximity to the sea, and rich food culture to attract tourists. Tourism is a major part of the service sector of the economy.

Since the 1990s, Spain's economy has grown as a result of lower costs of operating factories and a large supply of workers. Membership in the EU provided a market for Spanish products. Although it was severely affected by the global recession, Spain has continued its important economic role in the region.

Italy went through an industrial reconstruction after World War II. Areas along the Po River valley became the leading industrial region of the country. The EU opened other countries for the import of Italian Fiat cars and trucks, clothing, and home furnishings. Today Italy continues to thrive on high-tech engineering and metallurgical manufacturing. Certain portions of Italy's agricultural market play an important part in the export market, namely olive oil and wine.

Greece's economy is one of the least developed in Southern Europe. Natural resources are limited and industrialization has been slow to develop. A large sector of the economy is tourism. EU membership has been beneficial to agriculture and industrialization, since EU investments and subsidies compensate for low productivity. Today Greece faces major challenges to reduce its public spending. Generous social programs, tax evasion, and persistently high unemployment continue to be issues facing Greece.

Southern Europe faces many challenges in meeting its energy needs. The EU in general is the world's largest energy importer and is focusing more on natural gas due to its plans to reduce carbon dioxide emissions. While regional resources are limited, just south across the Mediterranean are rich oil- and natural gas-producing countries in North Africa. Pipelines on the seafloor deliver important energy supplies.

Future Prospects

Prior to 2009, membership in the EU was good for countries' economies. Member countries were doing well, and trade among them was growing annually. Easier migration within the EU across the borders between countries was filling the labor shortage. Living standards increased steadily.

In the twenty-first century the EU faces its first great test of economic stability within the subregion. The Southern European members are dealing with economic issues such as too much national debt, too high unemployment, and too much governmental spending. Those problems have stretched resources and goodwill very thin across all of Europe. Tensions have grown as the EU has provided special assistance to the Southern European members through loans and subsidies.

☑ **READING PROGRESS CHECK**

Summarizing Why is the Greek economy one of the least developed in Southern Europe?

LESSON 2 REVIEW

Reviewing Vocabulary

1. *Drawing Conclusions* How did city-states influence the political history of Italy and Greece?

Using Your Notes

2. *Summarizing* Use your lesson graphic organizer to describe society and culture today in Southern Europe.

Answering the Guiding Questions

3. *Identifying* What characteristics of early civilizations are evident in Southern Europe today?

4. *Expressing* How have migration and aging populations affected Southern Europe's population patterns?

5. *Discussing* How have religion, the arts, and Southern Europe's rich intellectual traditions shaped society and culture today?

6. *Describing* What are the characteristics of Southern Europe's economy today?

Writing Activity

7. *Informative/Explanatory* Write a paragraph describing the contributions Southern Europe has made to the arts, both classical and modern art. Include specific time periods and forms of art and architecture.

netw⊙rks

There's More Online!

☑ **IMAGE** Southern Europe Addressing the Issues

☑ **IMAGE** Pollution Around the Mediterranean Sea

☑ **MAP** Soil Erosion in Southern Europe

☑ **INTERACTIVE SELF-CHECK QUIZ**

☑ **VIDEO** People and Their Environment: Southern Europe

LESSON 3
People and Their Environment: Southern Europe

ESSENTIAL QUESTION · *How do physical systems and human systems shape a place?*

(tl)SuperStock/age fotostock, (tc)Marco Secchi/Getty Images News/Getty Images, (tr)Cyril Bana/Author's Image/PunchStock

Reading HELPDESK

Academic Vocabulary

- **promote**
- **factor**

Content Vocabulary

- **pollution hot spot**

TAKING NOTES: *Key Ideas and Details*

PARAPHRASING Use a web diagram like the one below to take notes about the human impact on the environment of Southern Europe.

People and Their Environment: Southern Europe

Managing Resources — Human Impact

IT MATTERS BECAUSE

The Mediterranean Sea is essential to the culture and economy of Southern Europe. The human population growth of the region is slowing after 100 years of rapid expansion. Maintaining the balance between human needs and the environment is often challenging with such a large population.

Managing Resources

GUIDING QUESTION *What are the threats that require closer management of resources in Southern Europe?*

Southern Europe's landscape is full of natural wonders—beautiful rivers, seas, and forests. Across most of the region, the climate is suitable for outdoor activity year-round, encouraging tourists and residents to spend a good deal of time exploring the outdoors. Human settlement has greatly increased over the last few decades, which has resulted in a number of concerns for the area's resources and environment.

One major environmental concern in the region is the presence of large algae blooms in the Adriatic Sea. This sea forms a smaller part of the Mediterranean Sea between Italy and Greece. The algae blooms are evidence of the effects of human activity on delicate marine biomes. These visible clusters of organisms appear when there is an imbalance in the ecological structure. Warmer water, chemical fertilizer runoff from agriculture, and human settlements are all possible causes for an algae bloom. The bloom is potentially harmful because it can use up the dissolved oxygen in the water, killing fish and other marine life. Toxins produced by particularly harmful algae blooms can kill marine life and humans who consume affected marine life.

The Mediterranean climate is known for its long, dry summers and wet, mild winters. However, changes in the global climate have made summers unpredictable. Some summers turn into droughts and others produce unseasonable rain. Both scenarios can result in excessive soil promote to help something grow or develop

factor something that actively contributes to the production of a result

erosion. Loss of topsoil and soil erosion can threaten the natural balance of an ecosystem by removing potential footholds for vegetation. When bare rock is exposed, the land becomes vulnerable to other environmental threats, such as fire or desertification. When there is no plant life to hold dry soil to steep hillsides, rain can wash away already thin topsoil on the rocky terrain of much of the subregion. Too much rain can also **promote** the growth of vegetation that can become fuel for a fire.

Dry season fires are a natural part of Mediterranean climate ecosystems. They are necessary to maintain the native vegetation. However, fire is a contributing **factor** in the deforestation of human-modified areas. Climate change is also a significant contributing factor to deforestation. The Mediterranean region is experiencing longer and hotter summers with less humidity and more wind. Both low humidity and more wind provide conditions that increase the incidence of forest fires.

Higher altitude areas are also affected by climate change. Parts of the Pyrenees and the Alps that normally experience sufficient snowfall and cold temperatures to maintain glaciers and forests of coniferous trees are changing. The glaciers are shrinking, while the forests are being attacked by beetles and other insects that flourish in the warmer temperatures. Awareness of these effects of climate change is needed to protect the environments that support life in the Mediterranean areas.

✓ READING PROGRESS CHECK

Making Connections How is climate change related to soil erosion in Southern Europe?

Soil Erosion in Southern Europe

GEOGRAPHY CONNECTION

Soil erosion threatens the ecosystems of Southern Europe.

1. *ENVIRONMENT AND SOCIETY* Which area of Southern Europe is most affected by soil erosion?

2. *THE USES OF GEOGRAPHY* For whom might this map be a valuable tool? Provide an example of how it might be used.

ANNUAL SOIL EROSION RISK

Very low risk	Very high risk
Low risk	Bare land
Medium risk	No data
High risk	

Human Impact

GUIDING QUESTION *What are the environmental concerns regarding tourism in Southern Europe?*

Southern Europe is well known for its food culture. In the coastal areas, this means fish and other seafood are common in the local diets. As a result, overfishing has occurred. Fish stocks are not being replenished as quickly as they are being depleted. As the fish stocks have declined, the overall health of the marine biome has declined as well.

Two sectors of the economy in Southern Europe that have seen tremendous growth in the last 50 years are industrial manufacturing and tourism. Industrial pollution is a major threat to the surrounding seas and to the quality of water and agricultural resources. Development associated with tourism causes major damage to coastal ecosystems, loss of natural habitat, overuse of freshwater resources, as well as pollution and waste.

Tourism inundates regions that are normally sparsely populated with huge numbers of people. More people make a greater impact on the local environment. Waste management issues are the primary concern. One popular form of vacationing is via cruise ship. Oil spills and oil pollutants from ships of all kinds are a growing concern in the subregion.

Pollutants and other human impacts in the water systems have given rise to **pollution hot spots** in the Mediterranean Sea. These hot spots are locations where such impacts have led to the degradation, and even the death, of the local ecosystem. These spots do not support marine life. The water is warmer and more vulnerable to the development of algae and other microorganisms, some of which are harmful to humans.

☑ **READING PROGRESS CHECK**

Summarizing How are manufacturing and tourism affecting the environment?

Industrialization and development along coastal Southern Europe has led to numerous problems in the marine biomes of the region.

▲ **CRITICAL THINKING**

1. *Analyzing Visuals* How might the human activity in this photo affect the nearby waterway?

2. *Formulating Questions* If you were an ecologist, what are two questions you would ask to evaluate the health of this waterway?

pollution hot spot a location where pollution and other human activities have led to the degradation, or even death, of an ecosystem

Addressing the Issues

GUIDING QUESTION *How are groups, governments, and others addressing environmental issues in Southern Europe?*

As early as 1975, European countries in the Mediterranean region recognized that rapid population expansion, industrialization, and a general increase in human activity were threatening the region's natural beauty and resources. The Mediterranean Action Plan was created at that time to help plan a way to curb damage to the environment. Each participating country set goals for reducing its environmental impact on the region, and governments in the region have been diligent in addressing the issues. Legislation has been enacted in all participating countries to try to regulate the effects of industrialization and population growth.

In 1990 the European Union (EU) created the European Environmental Agency. Its purpose is to help evaluate threats to the region and create plans to effectively deal with environmental issues. This agency makes independent reports to the European Union, which can then work with member countries to enact action plans to address environmental concerns.

SuperStock/age fotostock

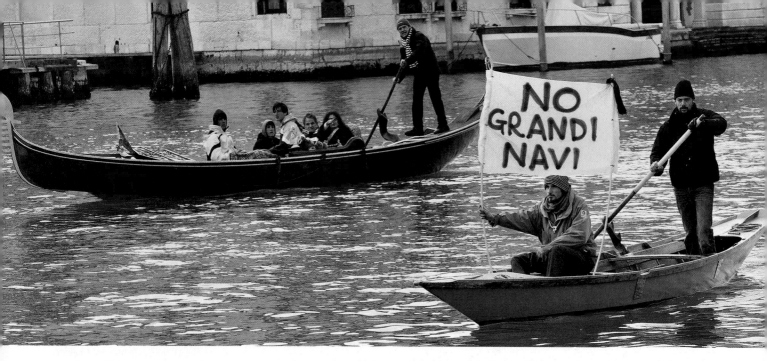

In Venice, Italy, demonstrators protest against the docking of huge cruise ships—*grandi navi*—in the city's bay, or lagoon.

▲ **CRITICAL THINKING**

1. *Identifying Central Issues* What types of damage might very large cruise ships cause?

2. *Speculating* Why might some people in Venice want cruise ships to keep docking in the lagoon, even if they cause problems?

Many nongovernmental organizations (NGOs) are dedicated to addressing issues that affect the region and that may have a lasting impact on the welfare of the global environment. These include the World Wildlife Fund (WWF), Earthwatch, and the Nature Conservancy. These organizations work to solve the problems, including industrial pollution, algae blooms, and soil erosion, which are prominent in Southern Europe. There are many ways for individuals to get involved with organizations that protect the regions of the world that are essential to a region's cultural and natural heritage.

Deforestation is also a growing concern in Europe. The Forest Stewardship Council (FSC), an international NGO, is leading a grassroots campaign in Europe to reforest or protect forest biomes. Specifically, they are working to reduce soil erosion and the risk of forest fires and fighting causes of global warming. Certification is one way to ensure that forest resources are being used responsibly. When resource extractors obtain certification, consumers can be assured that they support products and services that avoid exploiting forest resources. A public awareness campaign has helped to some extent, although it is hard to measure by how much. Teaching the public about these issues helps people make decisions that support responsible stewardship of the environment.

☑ **READING PROGRESS CHECK**

Identifying What is the role of the European Environmental Agency?

Marco Secchi/Getty Images News/Getty Images

LESSON 3 REVIEW

Reviewing Vocabulary

1. *Making Predictions* How might pollution hot spots in the waters of Southern Europe affect future economic activity in the region?

Using Your Notes

2. *Explaining* Use your web diagram to describe the major impacts humans are having on the environment in Southern Europe.

Answering the Guiding Questions

3. *Evaluating* What are the threats that require closer management of resources in Southern Europe?

4. *Explaining* What are the environmental concerns regarding tourism in Southern Europe?

5. *Discussing* How are groups, governments, and others addressing environmental issues in Southern Europe?

Writing Activity

6. *Informative/Explanatory* In a paragraph, explain how soil erosion poses a threat to the ecosystems of the subregion. What can be done to lessen the negative effects of climate change and the problem of soil erosion in Southern Europe?

Directions: On a separate sheet of paper, answer the questions below. Make sure you read carefully and answer all parts of the questions.

Lesson Review

Lesson 1

❶ *Analyzing* Several of the rivers of Southern Europe are not navigable because they are not wide or deep enough for large ships to travel on. Why are rivers still important to the region?

❷ *Explaining* What factors of the Mediterranean climate of Southern Europe make it ideal for growing grapes and olives?

❸ *Describing* Describe the geographic locations of the three major Southern European peninsulas.

Lesson 2

❹ *Evaluating* How has Southern Europe contributed to the arts? Provide specific examples.

❺ *Identifying Central Issues* How have changes in population growth affected demographics in Italy?

❻ *Understanding Relationships* What effect does tourism have on the economies of Southern Europe?

Lesson 3

❼ *Identifying Cause and Effect* What are the causes and consequences of deforestation in the region?

❽ *Summarizing* What key factors have contributed to pollution in the Mediterranean Sea?

❾ *Analyzing* In what ways have the governments of Southern Europe taken steps to reduce pollution?

Exploring the Essential Question

❿ *Making Connections* Create a poster illustrating one of the three peninsulas of Southern Europe. Show how the peninsula has been influenced by human systems, including history, art, politics, and economics. Use photos, graphs, charts, and maps.

21st Century Skills

Use the chart to answer the questions below.

The Labor Force in Southern Europe			
	% of Labor Force in Agriculture	% of Labor Force in Industry	% of Labor Force in Services
Greece	12.4	22.4	65.1
Italy	3.9	28.3	67.8
Portugal	11.7	28.5	59.8
Spain	4.3	24	71.7

Source: CIA World Factbook

⓫ *Economics* In which two countries is agriculture still a relatively significant contributor to the economy?

⓬ *Compare & Contrast* What generalization can you make about Spain and its level of economic development compared to the other countries listed?

Critical Thinking

⓭ *Making Generalizations* What long-term effects will Italy's aging population have on the country?

⓮ *Comparing and Contrasting* Compare the levels of development and industrialization in northern Italy to those in Greece. What factors help explain the differences?

⓯ *Predicting* Why should Southern European countries be concerned about high rates of unemployment among educated youth?

Need Extra Help?

If You've Missed Question	❶	❷	❸	❹	❺	❻	❼	❽	❾	❿	⓫	⓬	⓭	⓮	⓯
Go to page	296	297	294	303	301	304	306	307	307	294	309	309	301	303	303

Directions: On a separate sheet of paper, answer the questions below. Make sure you read carefully and answer all parts of the questions.

DBQ Analyzing Primary Sources

Use the document to answer the following questions.

Federico García Lorca (1898–1936) was a Spanish poet and playwright who was part of a group of avant-garde Spanish artists that also included Salvador Dalí.

PRIMARY SOURCE

Rider's Song

Córdoba
Far away and alone.

Black pony, big moon,
and olives in my saddle-bag.
Although I know the roads
I'll never reach Córdoba.

Through the plain, through the wind,
black pony, red moon.
Death is looking at me
from the towers of Córdoba.

Ay! How long the road!
Ay! My valiant pony!
Ay! That death should wait me
before I reach Córdoba.

Córdoba.
Far away and alone.

—Federico García Lorca, *The Selected Poems of Federico García Lorca*, 2006

16 *Analyzing* The North African poets of the Middle Ages used literary forms of allusion and repetition similar to those used by the author of this poem. Why would there be similarities between Spanish and North African poetry?

17 *Interpreting* What aspects of Spain's physical geography can be identified in this poem?

Applying Map Skills

Use your Unit 4 Atlas to answer the following questions.

18 *Places and Regions* What climate and vegetation types are found in Greece?

19 *Human Systems* What generalizations can you make about the population density of Spain in regard to its physical geography? What city is the exception?

20 *Physical Systems* Use your mental map of Southern Europe to describe the location of the major mountain ranges in Italy.

College and Career Readiness

21 *Reaching Conclusions* Imagine you are an economic geographer working for a nongovernmental organization (NGO). Write a recommendation for the Greek government explaining the benefits of investing in locally produced power. What natural resources can the Greek government use?

Research and Presentation

22 *Research Skills* Use Internet and library resources to gather information about the country of Andorra. Focus specifically on the unique government structure of this small country. Create a presentation explaining the history of this country's government. How did it become a dual principality and how does it maintain the system today?

Writing About Geography

23 *Informative/Explanatory* Use standard grammar, spelling, sentence structure, and punctuation to write a one-page essay suggesting possible solutions to the problem of pollution in the Mediterranean Sea. Who will need to participate for the solution to be successfully implemented?

"Canción de Jinete by Federico García Lorca copyright © Herederos de Federico García Lorca, from Obras Completas (Galaxia/Gutenberg, 1996 edition). English-language Translation, Rider's Song, by Stephen Spender and J. L. Gili copyright © Stephen Spender, J. L. Gili and Herederos de Federico García Lorca. All rights reserved. For information regarding rights and permissions of Lorca's works in Spanish and in any other language, please contact lorca@artslaw.co.uk or William Peter Kosmas, Esq., 8 Franklin Square, London W14 9UU, England."

Need Extra Help?

If You've Missed Question	**16**	**17**	**18**	**19**	**20**	**21**	**22**	**23**
Go to page	310	310	244	246	242	298	294	307

Eastern Europe

ESSENTIAL QUESTION · *How do physical systems and human systems shape a place?*

Geography Matters...

Borders in Eastern Europe have quite a history. Political changes have made the subregion's borders a source of division and conflict. Physical features also play a role. The Danube River divides countries physically, but it also helps connect Eastern and Western Europe economically. The Eastern European countries of today were once part of large powerful empires. Most modern countries of the subregion did not gain independence until after World War I. The subregion today is a unique land with a turbulent past, but with richness in literature, art, and culture.

◀ A young Roma woman in Slovakia holds her child.

Matt Cardy/Alamy

the breakup of Yugoslavia

The former Yugoslavia was historically a crossroads. Because the countries of Eastern Europe have a long history of belonging to different empires at different times, they have a complex mix of ethnic and religious influences. Conflict was common, and the region became known as the "powder keg" of Europe.

The Former Yugoslavia

Ethnic and religious diversity characterized the former Yugoslavia and contributed to its breakup by 1995. The map shows the countries today that were once part of Yugoslavia.

Ethnic and religious tensions resulted in bloody conflict between Serbs and Bosnians in the 1990s.

Nationalism was also a strong force at work. In 2008 Kosovo was the last country to declare its independence.

AUSTRIA

HUNGARY

Danube R.

SLOVENIA
Ljubljana

Zagreb

CROATIA

ROMANIA

BOSNIA AND HERZEGOVINA

Sarajevo

Belgrade

SERBIA

Adriatic Sea

ITALY

MONTENEGRO

Priština

KOSOVO

Podgorica

Skopje

BULGARIA

MACEDONIA

ALBANIA

GREECE

—— Former boundary of Yugoslavia

0 — 100 miles
0 — 100 kilometers
Lambert Azimuthal Equal-Area projection

15°E   20°E   25°E
45°N
40°N

The Road to a New Era

From the 1950s to the 1980s, revolts against communist rule periodically swept Eastern Europe. In 1989 large-scale public demonstrations contributed to the fall of the communist governments in the east. By 1991 the nationalist protests and financial crisis led to new, ethnic-based countries within Yugoslavia. Slovenia, Croatia, Bosnia-Herzegovina, and Macedonia all seceded from Yugoslavia, leaving just Serbia and Montenegro. Power struggles by ethnic and political groups resulted in civil war. What followed was a horrific practice called **ethnic cleansing**. Ethnic cleansing was used mainly by military units, but civilians did participate. The "cleansing" resulted in certain ethnic groups being killed or expelled from the disputed territories where they lived. Ethnic cleansing affected all groups, but Bosnian Croats and Bosnian Muslims were affected the most. Military aircraft from NATO countries bombed Serbian military and civilian targets to stop Serbia's support of the ethnic cleansing. Eventually international peacekeeping efforts were able to dislodge Serb leaders and end the conflict. Balkanization continued, and Montenegro declared independence from Serbia in 2006, with Kosovo declaring independence in 2008. In less than 20 years, the country of Yugoslavia was divided into eight smaller independent countries.

ethnic cleansing the expelling from a country or genocide of an ethnic group

✓ **READING PROGRESS CHECK**

Expressing What did the public demonstrations of 1989 in Yugoslavia lead to?

Population Patterns

GUIDING QUESTION *How have wars, migrations, and changing political borders influenced the population patterns of Eastern Europe?*

Most Eastern Europeans are ethnically Slavic. Slavs are descended from Indo-European peoples who migrated from Asia and settled in the region. Eastern Europe contains west Slavs, including Poles, Czechs, and Slovaks, and south

©Yevgeny Khaldei/Corbis

- The United States and its allies in Western Europe form the North Atlantic Treaty Organization (NATO).

1949

- The UN recognizes independent Slovenia, Croatia, and Bosnia as member states.

1992

- Macedonia is admitted to the UN. Czechoslovakia splits into the Czech Republic and Slovakia. The European Union is created.

1993

- The EU admits the Czech Republic, Hungary, Latvia, Lithuania, Poland, Slovakia, and Slovenia.

2004

- Kosovo declares independence from Serbia.

2008

➡ 1950 ············· **➡ 2000**

1989
- World War II ends and the Cold War begins.

1945
- Public demonstrations lead to the fall of the communist governments in Eastern Europe. Yugoslavia descends into civil war.

2006
- Montenegro declares itself independent from Serbia.

2007
- Bulgaria and Romania join the European Union.

Ethnic Groups in Eastern Europe

The Roma are the largest minority group in Europe. Thought to have migrated west from India into Europe, many live today in Bulgaria, Romania, Hungary, Slovakia, and the Czech Republic. The Roma continue to live nomadic lives, clashing with European governments as they move farther west.

Some of the earliest Slavs migrated from Asia thousands of years ago to settle in Poland.

The central portion of Eastern Europe includes countries that once formed the country of Yugoslavia. Ethnic and cultural groups living here include southern Slavic peoples such as Eastern Orthodox Serbs, Roman Catholic Croats, and Bosnian Muslims.

The Pomaks are thought to be descendants of native Bulgarians who converted to Islam during the period of Ottoman rule. Today they live in parts of Turkey, Macedonia, Kosovo, Greece, and Albania as well as Bulgaria.

Slavic
- Poles
- Czechs
- Slovaks
- Slovenes
- Croats
- Muslims
- Bulgars
- Russians
- Macedonians
- Serbs
- Montenegrins

Non-Slavic
- Magyars
- Albanians
- Romanians
- Turks
- Pomaks
- Latvians
- Lithuanians
- No group over 50%

0 400 miles
0 400 kilometers
Lambert Azimuthal Equal-Area projection

GEOGRAPHY CONNECTION

Historical migrations to Eastern Europe resulted in a subregion that is very ethnically diverse.

1. **PLACES AND REGIONS** Which Eastern European countries have the largest populations of Roma today?

2. **HUMAN SYSTEMS** What major religions do the southern Slavic people practice?

Slavs, including Serbs, Croats, Slovenes, and Macedonians. Slavs are an ethnic group defined mainly by their language. They speak languages today that have words and grammar that vary by country, but are traced to the same Indo-European root languages.

Another ethnic group, the Roma, traces its language to Indo-European origins. Roma are thought to have migrated from northern India to Europe centuries ago. Today there are several million Roma who live in Europe. They generally have less education, poorer health care, and shorter life expectancies than other groups.

Population density and distribution in Eastern Europe are influenced by geographic factors. For example, Poland has fertile soil and ample water resources that support large populations. Montenegro, by contrast, is the least populous country in Eastern Europe. It is located along the Dalmatian coast with a jagged coastline of narrow coastal plains backed by rugged limestone mountains and plateaus. The capital, Podgorica, has the main population concentration. Located along the Adriatic Sea, it provides access to the Mediterranean and therefore global transportation routes.

Industrialization throughout the 1900s led to urbanization in Eastern Europe. The majority of Eastern Europe's population today lives in and around large towns and cities. However, during much of the twentieth century, about half the population lived in rural areas. The cities that developed in the subregion were located near navigable water, such as the Danube and Vistula Rivers. Water transportation is not as expensive as other forms of transport and provided the cities with advantages as industrial and trade centers. During the Cold War, Eastern Europe was the major trading partner for the Soviet Union. New trade connections were necessary after 1991, and cities have new economic objectives based on tourism and services. Joining the European Union (EU) has been a major advantage to those Eastern European countries that met the requirements for membership.

Eastern European countries faced many difficult economic and political circumstances during and following World War II. Following the war there were large-scale internal and external migrations. Eastern European populations experienced heavy declines. Many Jewish people emigrated to Israel and to countries on other continents. People left in great numbers to escape communism. With the fall of communism, Eastern European countries underwent additional changes in economics, politics, and social conditions that continue to affect them today.

✔ READING PROGRESS CHECK

Describing How did industrialization affect population distribution in Eastern Europe?

Society and Culture Today

GUIDING QUESTION *How has conflict affected society and culture in Eastern Europe?*

Since almost all of Eastern Europe was a part of the Soviet bloc, the region has a recent history of free education for the population. As a result, literacy rates are improving throughout the region. However, in the transition to democratic governments, some countries have faced economic challenges. A lack of funds has affected the education systems in some countries. The health care system, known as "cradle to grave," was affected by the change to the market economy and by the lower birthrates and higher numbers of senior citizens. Despite this, most Eastern Europeans continue to have access to quality health care provided by national governments. The European outlook on health care makes it a basic human right.

Religious and ethnic differences were at the heart of conflict in the Balkan Peninsula in the 1990s. Tensions between Serbs and other ethnic groups such as Croats and Bosnians erupted into armed violence. Layered on top of such tensions were also religious divisions between people practicing Eastern Orthodox Christianity and Islam. Roman Catholicism and Judaism are also practiced in the region.

Family and Status of Women

In traditional Eastern European culture, the family is the basic social unit and serves to reinforce social values. However, twentieth-century industrialization, World War I, and new political movements introduced greater mobility and attracted rural populations to urban areas to find work. Since the beginning of industrialization and urbanization in Europe, families have tended to live in small houses or apartments. Eastern European families today have fewer children, and extended family members often live in other places. Women have made gains in education and, as a result, are employed in professional jobs.

The Arts

Many traditional forms of art and music exist throughout Eastern Europe. Folk and classical music are particularly important among Czech, Hungarian, Slovak, and Slovene peoples. In larger cities,

A crucifix stands against a bullet-riddled wall in the former Yugoslavia.

▼ **CRITICAL THINKING**

1. *Describing* In what ways has war shaped Eastern European culture and politics?

2. *Explaining* How has religion played a role in the conflicts of Eastern Europe?

©Ron Haviv/VII/Corbis

contemporary music from Western Europe and the United States is also popular. Some areas have thriving nightclub scenes. Some of the large countries also have a long musical history and continue to perform the works of famous composers in historic concert halls and opera houses.

Literature is a valued art form in Eastern Europe. The region has produced world-famous writers, such as Czech-born Franz Kafka (1883–1924), known for his novel *The Metamorphosis*. The most common themes in Eastern European literature reflect the region's history. Many authors wrote about the despair and darkness that fell over the region due to war, ethnic conflict, and economic depressions.

☑ **READING PROGRESS CHECK**

Explaining What are the common themes in Eastern European literature?

Economic Activities

GUIDING QUESTION *How have Eastern European economies developed since the fall of communism?*

Eastern European countries have made the transition to market economic systems over the past 20 years. The change to a market economy has presented both problems of production and consumption—of providing items that others want to buy and managing the changes in ownership from government to private corporations. The countries located nearest to Western Europe benefited early from the market system and received investments from countries such as Germany, Sweden, and the United Kingdom. Slovenia, for example, prospered after the fall of communism because it developed economic relationships with Germany. Political instability in the 1990s resulted in countries such as Macedonia having less success attracting investment than countries neighboring on Western Europe.

Agriculture and Industry

While Eastern Europe was largely industrialized under communist rule, many parts maintained their agricultural roots. The countries with Mediterranean climates such as Bulgaria, Serbia, Croatia, Macedonia, Bosnia and Herzegovina,

Growing crops such as these potatoes in Romania reflects the agricultural roots of Eastern Europe.

▼ **CRITICAL THINKING**

1. *Analyzing Visuals* How would you describe the type of farming you see in the picture?

2. *Contrasting* How was Eastern European society changed in terms of industrialization during the era of communist rule?

Diana Mayfield/Lonely Planet Images/Getty Images

The Russian Core

ESSENTIAL QUESTION · *How do physical systems and human systems shape a place?*

Geography Matters...

People rely on natural resources to provide for their everyday needs. The harsh geography of the Russian Core has stimulated empire building. For centuries, the people of the subregion have sought out waterways that connect to other lands and resources. This subregion has a violent political history, as well as a rich cultural heritage. After the fall of the Communist regime, the Russian Core today is claiming a new identity in the global economy.

◀ Russian girl wearing a colorful head scarf during a traditional religious festival

Russ Images/Alamy

Russia's shrinking population

The breakup of the Union of Soviet Socialist Republics (USSR) in 1991 signaled the beginning of a decline in Russia's population. Government officials and demographers (scientists who study population data) expressed deep concern over reduced population size and its threat to the country's society and economy. Leaders hope that improvements in the economic circumstances will help reverse the decline.

What happened to Russia's population after 1991?

When Vladimir Putin became Russia's president and prime minister in 1999, he inherited a demographic crisis with potentially dire outcomes for the country. Population declined by five million between 1992 and 2002 and has continued to decline, dropping from 145.2 million in 2002 to 142.9 million in 2011. Life expectancy plummeted to its lowest peacetime level ever from a high of 70.1 years in 1986–1987 to only 59 years for men and 72 years for women in 1992. Demographers voiced concerns about the country's ability to maintain its army, industry, and agricultural production.

1. **Human Systems** What problems can a steep population decline cause for a country? Explain the effects on labor, government services, prosperity, and the environment.

Why did Russia's population decline?

An important factor contributing to Russia's negative population growth rate during the 1990s was a steep decline in the birth (fertility) rate. At the same time, there was an increase in the death (mortality) rate and a sharp drop in life expectancy. In 1991, for the first time in the postwar history of Russia, the number of deaths exceeded that of births. The total population decreased by 30,900 in 1992 and by 307,600 in 1993. The life expectancy decreased largely due to poor nutrition and health care. Poverty, stress, alcoholism, heavy smoking, suicide, diseases such as HIV/AIDS, strict immigration policies, and homicide all made the trend more severe. The trend was especially strong among working-age men.

2. **Human Systems** Why did Russia's population begin to decline after the collapse of the Soviet Union? Summarize the factors contributing to the decline.

What is being done to address the issue?

Russian president Putin directed his government to halt the drop in population, offering subsidies and financial incentives for women to have more children. He has also proposed new initiatives to battle alcoholism and to provide better education and housing opportunities for all Russians. Another proposal is an attempt to use the immigration policy to lure Russians living abroad to return and to attract skilled foreign workers. This last component of Putin's proposal is one that Russian leaders have historically avoided. They feared a loss of national identity if foreigners were allowed to enter Russian society.

3. **Places and Regions** Why have Russian immigration officials historically resisted allowing non-Russians to immigrate?

(l)Vladimir Godnik/Getty Images, (c)Maxim Marmur/AFP/Getty Images, (r)Alexander Petrow/Epsilon/Getty Images News/Getty Images

2.5 million
1.5 million
500,000
0
500,000
1.5 million
2.5 million

⊕ BIRTHS

December 1991
USSR Falls

⊖ DEATHS

2006
Government makes
efforts to stem tide

USSR Era	Russian Federation Era
1971	1991	2011

CAUSES OF THE DECLINE

AGING POPULATION
Russians overall are getting older. By 2015 20% will be over age 60.

MOSCOW CALLING
Despite an increase in incentives, Russia has difficulty attracting immigrants.

ALCOHOLISM
Since the fall of the USSR, alcoholism has increased. It killed 98,000 Russians in 2011 alone.

HIV/AIDS
Uncommon in the USSR, HIV boomed after its fall and continues to spread. The number of Russians living with HIV doubled from 2006 to 2011.

BRAIN DRAIN
Due to a lack of employment, many educated Russians leave the country. The vast majority who do are under 35 years old.

FEWER BIRTHS
A low birthrate of 1.61 births per woman and a high infant mortality rate of 7.2 deaths per 1,000 live births contribute to Russia's shrinking natural increase.

Tim E White/Photolibrary/Getty Images

netw⊙rks

There's More Online!

☑ **IMAGE** Russia's Shipping Industry

☑ **IMAGE** The Ural Mountains

☑ **INFOGRAPHIC** The Volga River Basin—Heart of Russia

☑ **INTERACTIVE SELF-CHECK QUIZ**

☑ **VIDEO** Physical Geography of the Russian Core

Reading **HELP**DESK

Academic Vocabulary

- **corresponding**
- **challenge**

Content Vocabulary

- **chernozem**
- **permafrost**
- **continentality**

TAKING NOTES: *Key Ideas and Details*

PARAPHRASING Use a chart like the one below to take notes about the landforms, waterways, climates, biomes, and resources of the Russian Core.

Physical Geography of the Russian Core		
Landforms	Water Systems	Climate, Biomes, and Resources

LESSON 1
Physical Geography of the Russian Core

ESSENTIAL QUESTION · *How do physical systems and human systems shape a place?*

IT MATTERS BECAUSE

The lowlands, plains, and plateaus of the Russian Core—Russia, Ukraine, and Belarus—helped shape human activity for centuries. In some cases, the harsh climate of the subregion made it a forbidding environment for people to settle and live in. Some were drawn to the few mountainous areas found in this region for their beauty and abundant natural resources despite difficult winters. Certainly the diverse topography of the Russian Core makes for some of the most beautiful geography in the world.

Landforms

GUIDING QUESTION *How do interconnected mountain ranges and plains shape human activities in the Russian Core?*

Russia has two notable mountain ranges, the Ural Mountains and the Caucasus Mountains. The other distinct landforms that dominate Russia are plains, hills, and plateaus. The Ural Mountains form a natural barrier between European Russia and Siberian Russia. Rich in iron ore and mineral fuels, they run in a north-to-south direction and span 1,230 miles (2,000 km) from the Arctic Ocean almost to the Caspian Sea. The Caucasus Mountains run east to west along the southwestern portion of the country, forming a natural barrier between Russia and countries to the south. The highest peak of the Caucasus Mountains is Mount Elbrus, which soars to 18,510 feet (5,642 m). It is the highest mountain in Russia and Europe. Mount Elbrus is also a popular tourist destination for skiing and mountain climbing.

To the west of the Ural Mountains is the Northern European Plain. The southern part of the plain has navigable waterways and rich, black soil called **chernozem** that supports agriculture. As a result, the majority of Russia's population lives here. This vast plain is an extension of the plain that begins in France and stretches across Northwestern Europe. This eastern part of the plain is sometimes called the Russian Plain

(tl)©Evgeny Prokofyev/Alamy, (tr)©Pashkov Andrey/Alamy

because as it reaches the Russian Core it broadens and becomes very expansive.

East of the Ural Mountains and extending to the Pacific Ocean are the vast stretches of plains and plateaus that make up much of Siberia. The West Siberian Plain covers about one-third of Siberia and is one of the largest low-lying flatlands in the world. It is known for its harsh continental climate and some of the world's largest swamps and wetlands.

Ukraine is the second-largest country in Europe, occupying the southwest portion of the Russian Plain. Its two main landforms are vast plains and plateaus, with mountains found only in small areas in the west and south, accounting for less than five percent of the country's landmass. In the south, the Isthmus of Perekop connects the Crimean Peninsula to the mainland.

Belarus is the smallest of the three Slavic republics that were once part of the Soviet Union. Belarus is a landlocked country and lies entirely on the Northern European Plain. Glacier scarring accounts for the flat terrain and some 11,000 lakes found in Belarus. There are also numerous swamps and rivers.

✓ READING PROGRESS CHECK

Naming Which mountains form a natural boundary between European Russia and Siberian Russia?

Water Systems

GUIDING QUESTION *What role do rivers of the Russian Core play in the economic activities of the region?*

The combined waterways of the Russian Core have played important roles in its social and economic development from early times. The rivers and lakes of Russia, Ukraine, and Belarus are key to the subregion's growth, expansion, and success. Included in the water system is the Volga River, one of the world's greatest rivers.

The Volga River and its many tributaries make up the Volga River system. Draining most of western Russia, the Volga travels 2,293 miles (3,690 km), making it the longest river in Europe. It starts in the Valdai Hills west of Moscow and travels across much of southern Russia before emptying into the Caspian Sea. The Volga River system is an important commercial, transportation, and hydroelectric resource for millions of Russians.

In Ukraine, the Dnieper River is the longest in the country, stretching 609 miles (908 km) beginning in the northwestern plains and traversing southeast to drain into the Black Sea. Other major waterways in Ukraine include the Southern Bug, Dniester, and even a small portion of the Danube River. Like the Volga in Russia, the Dnieper provides hydroelectric power and transportation as well as enabling commerce. But the most important function of these rivers collectively is to supply water to Ukrainians. This is done through an intricately constructed series of canals.

The Ural Mountains extend for well over a thousand miles, cutting through many varied landscapes.

▲ CRITICAL THINKING

1. **Classifying** Why are the Ural Mountains considered a natural boundary?

2. **Speculating** How might the Ural Mountains have influenced settlement patterns in the areas they span?

chernozem rich, black topsoil found in the Northern European Plain, especially in Russia and Ukraine

Evgeny Prokofyev/Alamy

THE VOLGA RIVER BASIN—Heart of Russia

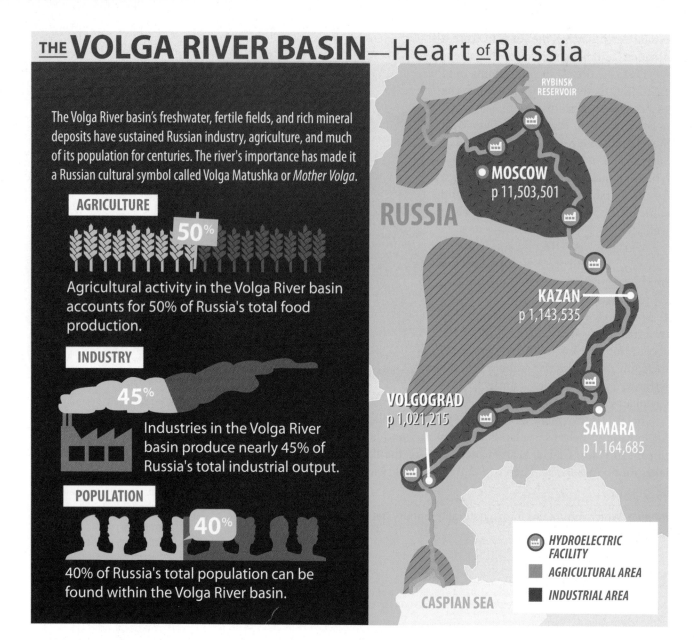

The Volga River basin's freshwater, fertile fields, and rich mineral deposits have sustained Russian industry, agriculture, and much of its population for centuries. The river's importance has made it a Russian cultural symbol called Volga Matushka or *Mother Volga*.

AGRICULTURE

50%

Agricultural activity in the Volga River basin accounts for 50% of Russia's total food production.

INDUSTRY

45%

Industries in the Volga River basin produce nearly 45% of Russia's total industrial output.

POPULATION

40%

40% of Russia's total population can be found within the Volga River basin.

RYBINSK RESERVOIR

RUSSIA

MOSCOW p 11,503,501

KAZAN p 1,143,535

VOLGOGRAD p 1,021,215

SAMARA p 1,164,685

Legend:
- HYDROELECTRIC FACILITY
- AGRICULTURAL AREA
- INDUSTRIAL AREA

CASPIAN SEA

Western Russia's Volga River connects Moscow to the Caspian Sea. Called *Matushka Volga*, or "Mother Volga," the river is vital to Russia.

▲ **CRITICAL THINKING**

1. Assessing Based on the infographic, describe the importance of the Volga River to Russia in regard to hydropower generation.

2. Identifying Central Issues Why is the Volga River called *Mother Volga*? Provide at least two reasons.

There are few freshwater lakes across Ukraine. Small saltwater lakes can be found in the Black Sea Lowland and Crimea, while *limans*, or large saline lakes, are found along the coast. In recent years Ukraine has created some artificial lakes and reservoirs.

Belarus was once an important passage route for inland navigation between the Baltic Sea and the Black Sea. Each with its own tributaries, the largest rivers in the country include the Dnieper River, Berezina River, Pripyat' River, Neman River, Bug River, and Western Dvina. Narach is the country's largest lake, measuring 31 square miles (79.6 sq. km).

Waterways are also a prominent feature of Siberia. Rivers such as the Ob', Irtysh, Yenisey, and Lena begin in the south of Siberia and flow northward, emptying into the Arctic Ocean. Ice in the Arctic Ocean blocks the rivers from reaching the ocean for much of the year. This is problematic in the springtime when the ice blockage prevents the surging rivers from reaching the ocean. Instead, the water floods out across the low-lying plains, causing extensive areas of swamps and floodlands.

One of the most renowned Siberian lakes is Lake Baikal. Located in southeast Russia, Lake Baikal is the oldest lake in the world (25 million years). It is also the deepest at 5,715 feet (1,742 m). It holds one-fifth of all unfrozen freshwater found on the planet. Known as the "Galapagos of Russia" because of its age and isolation, Lake Baikal has many unusual freshwater marine species, which are of exceptional value to scientists who study how species evolve.

☑ **READING PROGRESS CHECK**

Identifying Which river provides western Russia with hydroelectric power?

Climate, Biomes, and Resources

GUIDING QUESTION *What are the general climate conditions in much of the Russian Core?*

Russia's vast expanse of land extends from east to west, and all of the country lies in the same high latitude range. The result is that the dominant characteristic of the subregion is cold, snowy winters. There is, however, variation in how cold and how long the winters last, and the warmth and length of the summers. These variations in climate have shaped settlement patterns from the earliest times.

Climate Regions and Biomes

The tundra occupies the parts of the subregion that are farthest north and covers about 10 percent of Russia. Here, the sky stays dark for many weeks before and after the winter solstice that occurs around December 22 each year. Then, for several weeks during the summer, there is continuous sunlight. Its short growing season and thin, acidic soil lying just above the **permafrost** limit the kinds of plants that can grow there. Only mosses, lichens, algae, and dwarf shrubs thrive in the tundra.

permafrost a permanently frozen layer of soil beneath the surface of the ground

South of the tundra lies the subarctic climate zone. While not as severe as the tundra climate, this zone only has four months of the year when the temperature rises above 50°F (10°C). The rest of the year is cold. The biome here is boreal forest, or taiga, which consists of broad expanses of coniferous evergreen trees.

Russia's midlatitude climates, found in the western region, are not as severe and have milder winters and warmer summers. Although still relatively cold, these climates are where most Russians live and where much of Russia's agricultural production takes place. The natural biome here is deciduous forest, although much of it has been cleared for agriculture and construction.

An area between the Black and Caspian Seas north of the Caucasus Mountains and along Russia's border with Kazakhstan make up Russia's steppe climate region. The steppe is a broad, open grassland in which seas of grass stretch to the horizon in every direction. The region's chernozem soil supports the production of wheat, barley, rye, oats, and other crops. Sunflowers, mint, and beans also flourish here.

The southern part of Ukraine lies in a humid continental climate zone where warmer, humid air from the Atlantic Ocean makes the climate milder than farther north. The greatest amount of precipitation falls in the warmer summer season, with the maximum precipitation occurring in late June and July.

Belarus has a humid continental climate moderated by maritime influences from the Atlantic Ocean. Average January temperatures range from the mid-20°sF (about −4°C) in the southwest to about 17°−19° F (about −8°C) in the

Russia's shipping industry includes both river- and sea-going vessels.

▲ **CRITICAL THINKING**

1. *Speculating* Why is Russia's coal, although abundant, difficult to produce?

2. *Describing* Where does Russia rank in the production of nickel, aluminum, and platinum-group metals?

corresponding showing a direct connection between two things

continentality effect of extreme variation in temperature and very little precipitation within the interior portions of a landmass

challenge to arouse or stimulate especially by presenting with difficulties

northeast, but thaw days are frequent. **Corresponding** to these temperature differences, the frost-free period decreases from more than 170 days in the southwest to 130 days in the northeast. Maximum temperatures in July are generally in the mid-60°sF (about 18°C). Rainfall is moderate, although higher than over most of the vast Russian Plain.

Warmer air from the Atlantic Ocean moderates temperatures in western Russia. Most of Russia, however, lies well within the Eurasian landmass, far away from any moderating ocean influences. As a result, much of the country's interior has more extreme variations in temperature and little precipitation. This climatic effect within the interior areas of a landmass is called **continentality**.

Natural Resources

Russia's physical geography benefits but also **challenges** its people. The country has an abundance of natural resources, including one-fifth of the world's forests. Much of this wealth, however, lies in remote and climatically unfavorable areas and is difficult to tap or utilize. For example, Russia holds large petroleum deposits and 16 percent of the world's coal reserves; however, the country's biggest coalfields lie in remote areas of eastern Siberia. Russia is also a leading producer of natural gas, but much of it is located in northern Siberia. It also leads the world in nickel production and ranks among the top three producers of aluminum, gemstones, and platinum-group metals. Russia's rivers make it a leading producer of hydroelectric power.

There are considerable resources of iron and other ores in parts of Ukraine. Other resources such as mountain wax, granite, and graphite are among the country's most abundant. Ukraine also produces various salts and has a rich base for metallurgical, porcelain, and chemical industries.

Natural resources in Belarus include fuel sources such as peat, oil, and natural gas deposits. The country also contains small quantities of a variety of rocks and minerals, such as chalk, sand, clay, gravel, granite, and limestone.

☑ **READING PROGRESS CHECK**

Classifying How does Russia's location in the high latitudes affect its climate?

Pashkov Andrey/Alamy

LESSON 1 REVIEW

Reviewing Vocabulary

1. *Explaining* Explain the significance of permafrost and continentality.

Using Your Notes

2. *Listing* Use your chart to list and describe the locations of the major landforms of the Russian Core.

Answering the Guiding Questions

3. *Identifying* How do interconnected mountain ranges and plains shape human activities in the Russian Core?

4. *Describing* What role do rivers of the Russian Core play in the economic activities of the region?

5. *Explaining* What are the general climate conditions in much of the Russian Core?

Writing Activity

6. *Informative/Explanatory* Think about the locations of Russia's seas. Write a paragraph describing how the locations of these seas affect Russia's economy.

netw⊙rks

There's More Online!

- ☑ **GRAPHS** Ethnic Composition of the Russian Core
- ☑ **GRAPH** GDP Per Capita in the Russian Core
- ☑ **TIME LINE** Soviet Era of the Russian Core
- ☑ **INTERACTIVE SELF-CHECK QUIZ**
- ☑ **VIDEO** Human Geography of the Russian Core

Reading **HELP**DESK

Academic Vocabulary

- **acquire**
- **trigger**
- **decline**

Content Vocabulary

- **czar**
- **Russification**
- **satellite**
- **perestroika**
- **glasnost**
- **black market**
- **privatization**

TAKING NOTES: *Key Ideas and Details*

IDENTIFYING Use a graphic organizer like the one below to take notes on society and culture today in the Russian Core.

Human Geography of the Russian Core: Society and Culture Today

Lasting effects of Soviet government policies ⟶ ☐

Family and Status of Women ⟶ ☐

The Arts ⟶ ☐

(tl)©Swim Ink/Corbis, (tcl)©The Dmitri Baltermants Collection/Corbis, (trl)©Peter Turnley/Corbis

---

LESSON 2
Human Geography of the Russian Core

ESSENTIAL QUESTION · *How do physical systems and human systems shape a place?*

IT MATTERS BECAUSE

The Russian Revolution, the Soviet era, and the collapse of the Soviet Union have had profound effects on the Russian Core's economies, politics, and people. This subregion's countries have been forced to redefine who they are and to restructure their economies based on new ideas about political and economic systems.

History and Government

GUIDING QUESTION How have the Russian Core's historical roots and modern ideas influenced the history and government of the region?

For most of the last century, Russia was part of the Soviet Union. Ruled by a Communist government, it challenged the United States and other democracies for global influence. Then the Soviet Union collapsed, and Russia, Belarus, and Ukraine emerged as independent countries.

Early History

Russia's historical roots go back to the A.D. 600s when Slav farmers, hunters, and fishers settled near the waterways of the Northern European Plain. The Slav communities were once organized into a loose union of city-states known as Kievan Rus. Ruled by princes, the leading city-state, Kiev, controlled a prosperous trading route, using Russia's western rivers to link seaports and trade centers on the Baltic and Black Seas.

In the early 1200s, Mongols from Central Asia invaded Kiev and many of the Slav territories. Although the Mongols allowed the Slavs self-rule, they continued to control the area militarily for more than 200 years.

In the late 1600s, **Czar** Peter I—known as Peter the Great—came to power, determined to modernize Russia. Under his rule, Russia enlarged its territory, built a strong military, and developed trade with Western Europe. To **acquire** seaports, Russia gained land along the Baltic Sea from Sweden. St. Petersburg became the new capital and a major port city. It was carved out of the wilderness along the Gulf of Finland, providing access to the Baltic Sea and giving Russia "a window on the West."

czar ruler of Russia until the 1917 revolution; originally from Latin word *Caesar,* title of Roman emperors

acquire to gain possession

Russification in nineteenth and twentieth century Russia and the Soviet Union, a government program that required everyone in the empire to speak Russian and to become a Christian; assignment of some Russian-speaking people to non-Russian ethnic regions

trigger to set off

satellite a country controlled by another country, notably Eastern European countries controlled by the Soviet Union by the end of World War II

During the late 1700s, Empress Catherine the Great continued to expand Russia's empire and gained a long-desired warm-water port on the Black Sea. A later royal family of Russia, the Romanovs, continued expansion, resulting in acquisitions in both Eastern and Central Europe. These new territories and their people brought many non-Russians under Russian rule. A program known as **Russification** was begun to make those peoples more "Russian."

Czar Alexander II's limited reforms caused many former serfs to move to cities. There they faced the poor conditions and low wages of factory work. Non-Russian peoples from newly colonized regions of the Russian Empire faced prejudice. The government also insisted colonized people become like Russians. Russification became government policy, and people were required to speak Russian and follow Eastern Orthodox Christianity to receive jobs and benefits.

Beginning in 1891, under Czar Alexander III, Russia expanded into Siberia with the construction of the Trans-Siberian Railroad. Nearly 6,000 miles (9,700 km) long, it connected Moscow to Vladivostok. Once completed in 1916, the railroad opened Russia's Asian eastern region to settlement.

Revolution and Change

One of the biggest proponents for greater economic equality was the German philosopher Karl Marx, the founder of modern communism. He advocated two principles: the public ownership of all land and means of production, and a classless society with an equal sharing of wealth. During World War I, which began in 1914, numerous strikes and demonstrations were organized by Russian workers who suffered hardships because of the war. They protested, demanding "bread and freedom." This unrest **triggered,** or set off, the Russian Revolution of 1917. Czar Nicholas and his family were murdered, signaling the demise of Europe's last absolute monarchy. What emerged was the Communist-controlled Union of Soviet Socialist Republics (USSR), or the Soviet Union.

The Soviet Union played a pivotal role in the Allied victory over Germany during World War II. Following the war, the Soviet Union occupied much of Eastern Europe. Several countries in the region were controlled as **satellites** under Communist rule. Following World War II, the Communist Soviet Union was engaged in a political and ideological war with the West, particularly the United

©Swim Ink/Corbis

SOVIET ERA
of the Russian Core ➡

Discontent with inequality in Russian society led to revolution and freedom from generations of czarist rule in the early 1900s. This was followed by new ideas about political and economic systems that led to changes in Russia in the 1990s.

▶ **CRITICAL THINKING**

1. *Explaining* What did the Soviet Union gain from the Nazi-Soviet Nonaggression Pact?

2. *Describing* How did post-World War II agreements change Russia's influence in Europe?

1910 ➡

Lenin dies. Joseph Stalin emerges as new leader.

1924

German-Soviet Nonaggression Pact gives Soviet Union a sphere of influence in Estonia, Latvia, Lithuania, and eastern Poland.

1917

Revolution forces Czar Nicholas II to abdicate the throne. Vladimir Lenin becomes leader of Russia.

1922

The Union of Soviet Socialist Republics (USSR) is established.

1939

Germany invades western Poland, starting World War II. The Soviets occupy eastern Poland.

342

States, in the Cold War. Tensions during this time brought the world to the brink of nuclear war and shaped modern economic and political policies.

The enormous costs of the Soviet Union's military and inefficient economic policies weakened its role in the world. In 1985 Mikhail Gorbachev, a reform-minded official, became the leader of the Soviet Union. He instituted a policy of economic restructuring called **perestroika** (PEHR•uh•STROY•kuh) and a policy of greater political openness called **glasnost** (GLAZ•nohst). Political reform was set in motion. Satellites controlled by the Soviet Union began to replace their Communist governments in 1989. In 1991 a failed coup led to the collapse of the Soviet government and regions of the country began declaring their independence. Today there are 15 independent countries, including Russia, Ukraine, and Belarus.

The newly independent countries emerged from a long period of both imperial Russian and Soviet domination. Belarus is a good example. Although Belarusians share an ethnic identity and language, they had never enjoyed unity and political sovereignty, except during a brief period in 1918. Belarusian history is a study of regional conflicts and territorial claims by neighbors to the east and west.

Since the beginning of Russian occupation, Chechnya has unsuccessfully sought independence. Several oil and gas pipelines vital to the Russian economy run through Chechen territory. In May 2000, Russian president Vladimir Putin established direct rule of Chechnya to try to stop the rebels. In 2003 a new constitution was passed that gave Chechnya a significant amount of autonomy. The new Chechen government is now struggling to recover from the violence waged by the Chechen rebels and to bring stability and peace to Chechnya.

✔ **READING PROGRESS CHECK**

Explaining How was St. Petersburg important to the expansion of the Russian Empire?

Population Patterns

GUIDING QUESTION *What factors have shaped population patterns in the Russian Core?*

About 80 percent of all Russians live west of the Ural Mountains. This is due in part to the rich soils, waterways, and a milder climate than that of eastern Russia, also known as Siberia. Densely settled western Russia includes the country's

perestroika in Russian, "restructuring"; part of Gorbachev's plan for reforming the Soviet economy and government

glasnost Russian term for new openness in areas of politics, social issues, and media; part of Gorbachev's reform plans

(t)©The Dmitri Baltermants Collection/Corbis, (b)©Peter Turnley/Corbis

→ 1940

1941 Germany attacks the Soviet Union, drawing it into World War II.

1943 Germany fails to take Stalingrad. Soviet troops launch a counter-offensive. Berlin is eventually captured.

1945 Germany is defeated in World War II. USSR gains influence over Eastern Europe.

1953 Joseph Stalin dies. Nikita Khrushchev becomes Communist party leader.

1962 Cuban missile crisis erupts over presence of Soviet missiles in Cuba.

1985 Mikhail Gorbachev comes to power and proposes perestroika.

1991 The USSR is dissolved.

→ 1990

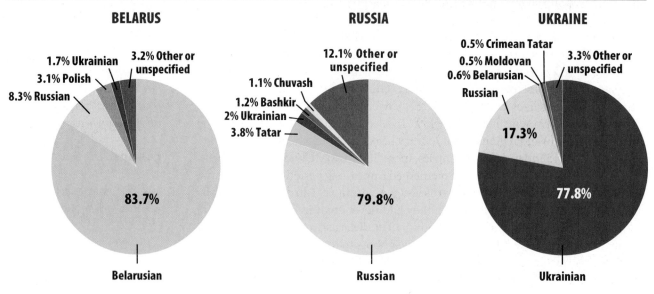

BELARUS

1.7% Ukrainian
3.1% Polish
8.3% Russian
3.2% Other or unspecified
83.7%
Belarusian

RUSSIA

12.1% Other or unspecified
1.1% Chuvash
1.2% Bashkir
2% Ukrainian
3.8% Tatar
79.8%
Russian

UKRAINE

0.5% Crimean Tatar
0.5% Moldovan
0.6% Belarusian
Russian
3.3% Other or unspecified
17.3%
77.8%
Ukrainian

Source: CIA World Factbook

The diversity of people in Russia has led to many ethnic groups demanding greater self-rule or independence. In some places, like Chechnya, some groups have resorted to violent methods, such as terrorism.

▲ **CRITICAL THINKING**

1. **Analyzing**   Which groups make up over 10 % of the Ukrainian population?

2. **Categorizing**   Which country shown has the largest population of "Other" ethnic groups? What factors may account for the higher percentage?

---

decline to become less in amount or number

industrialized cities. The major industrial city is Moscow, Russia's capital. Following the reestablishment of Russia as a country in 1991, the controls over where people could live were removed. People migrated where there were jobs and other opportunities. As a result, Russian cities experienced increases in population. By 2000 the population of ethnic Russians began to **decline** due to a greatly reduced birthrate. The decline has continued into the twenty-first century.

A shrinking population is a major problem for Russia. To offset falling birthrates, there are government programs that pay monthly allowances to parents with babies and children. The birthrate of 1.4 babies per woman in 2010 is far below the 2.1 needed to naturally maintain the population. Reducing mortality rates is also necessary. Death rates, particularly among males between the ages of 25 and 45, have increased to levels that outpace new births. As a result of having fewer younger people, Russia's overall population has been simultaneously decreasing and growing older.

✓ **READING PROGRESS CHECK**

Making Connections   Why do most Russians live west of the Ural Mountains?

Society and Culture Today

GUIDING QUESTION   *How have past policies affected the Russian Core today?*

The policies of the past have had lasting effects on Russia's culture. The Soviet government severely discouraged religious practices and discriminated against different ethnic groups. It actively promoted atheism, or the belief that God or another supreme being does not exist. In the late 1980s, however, the government relaxed its restrictions on religion. Since then, millions of Russians are rediscovering their religious faiths and traditions. Similarly, more than 100 languages are spoken in the country today, but Russian is the official language.

During the Soviet era, education was free and mandatory. The emphasis was on math, science, and engineering rather than on language, history, and literature. The curriculum changed dramatically after Russian independence. Schools began to emphasize a more balanced approach to learning, including language, history, and literature. Today, students have a choice of different types of schools, but the country's economy has limited funding for schools.

Family and Status of Women

Living conditions in Russia affect family life. Due to a housing shortage, most people live in large apartment blocks. While space in the traditional apartments is tight, new housing developments since 1991 offer space and living conditions similar to those in Western Europe and the United States.

The status of women in Russian society is a combination of the past and the present. Women have always had a very important role working in industry. During World War II, women not only fought in every branch of the Soviet military but also took jobs in the wartime industrial workforce. That trend continued into the 1970s. In the 1990s, there were increasing financial pressures and shrinking government programs in Russia. This meant more women entered the workforce out of necessity. Full-time employment of women contributed to Russia's declining birthrate. Higher education, better-paying jobs, and feminist groups and social organizations have resulted in greater rights for women. As a consequence, many women marry later in life, many have careers, and they expect to be treated as equals in a society that has traditionally been dominated by men.

The Arts

Russian arts are characterized by a list of well-known artists. Painters such as Viktor Vasnetsov and composers such as Pyotr (Peter) Tchaikovsky contributed to the richness of Russian culture. The works of poets Aleksandr Pushkin, Boris Pasternak, and Anna Akhmatova and novelists Lev (Leo) Tolstoy and Fyodor Dostoyevsky have made Russian literature famous. Russian ballet and theater also have an international reputation.

The period of communism limited individual artistic expression and directed artists to glorify the government's achievements in their works, an approach known as socialist realism. Artists who did not follow these guidelines were punished. Accounts of imprisonment and punishment among artists during the period are shown in works such as Aleksandr Solzhenitsyn's *The Gulag Archipelago.*

Beginning in the mid-1980s, activity in the arts renewed as loosening government controls allowed the printing of previously unpublished works and new materials. During the height of Communist repression, books were smuggled from Russia and printed in other languages, becoming best sellers in other countries. Many were not available to Russian readers until after 1991.

☑ **READING PROGRESS CHECK**

Analyzing What contributed to a resurgence of the arts in the 1980s?

Economic Activities

GUIDING QUESTION *How has the economy of the Russian Core changed since 1991?*

Under Communist leaders, the Soviet Union operated as a command economy in which the government made key economic decisions. The government owned banks, factories, farms, mines, and transportation systems. The government decided what and how much to produce, where and how to produce it, and who would benefit from the profits. It also controlled the pricing of most goods and decided where they would be sold.

Analyzing
PRIMARY SOURCES

Poetry in Contemporary Russia

"Since the collapse of the Soviet empire in 1991, Russian poetry has begun to resemble American poetry in ways that are both fascinating and sad. What's fascinating is how talented, and how different from one another, Russia's young poets are. What's sad is how little they are read, and how little they matter. Whatever reach contemporary poetry had in Russian society has vanished like wood smoke."

—Dwight Garner, "When Poets Rocked Russia's Stadiums," *New York Times*, June 3, 2010

DBQ *SPECULATING* Why do you think poetry has lost its importance in Russian society today?

black market illegal trade of scarce or illegal goods, usually sold at high prices

privatization a change to private ownership of state-owned companies and industries

Unemployment was nearly nonexistent, but wages were low. Some people could not afford consumer goods, or goods needed for everyday life. Even when people had enough money, such goods were hard to find. Scarce products could be bought on the **black market**, illegal trade in which scarce or illegal goods are sold at high prices. Most workers, however, could not afford such high prices.

When Mikhail Gorbachev came to power in 1985, the Soviet command economy was under strain. He began to move the country toward a market economy, in which businesses became owned by private individuals and companies. Gorbachev reduced government controls, allowed people to start small businesses, and encouraged foreign investment. Boris Yeltsin, Gorbachev's successor, expanded this process.

Russia's economy continued to change after 1991. Yeltsin removed governmental price controls and encouraged **privatization**—a shift to private ownership—of state-owned companies. This favored wealthy people or businesses with money to purchase large companies. Those who made money sometimes invested their profits outside the country in Europe and the United States. Most Russians were not able to benefit from this new wealth at the time.

The Russian economy experienced many successes throughout the 1990s. More consumer goods were available to meet pent-up demands. Greater demands resulted in prices that soared, and many people could not afford to buy the goods. Between 1990 and 1995, Russia experienced an economic depression as its GDP fell by 50 percent. Following a 1998 financial crisis, the ruble—Russia's currency—lost 71 percent of its value. The international community issued loans and credit to help the country.

Russia has experienced steady economic growth since 1998, due in part to increases in productivity, wages, consumption, and a growing middle class. The anchor in the economic success has been its vast supplies of natural resources that are marketed to industrialized countries.

Resources, Power, and Industry

The fertile triangle is the most productive agricultural area of western Russia, extending into Ukraine. With its tip at St. Petersburg and its base stretching from Novosibirsk to Odessa, the triangle has rich soils and a favorable climate. Ukraine's crop production is highly developed. It is referred to as a breadbasket because of the grain crops it grows. Yields of grain and potatoes are among the highest in Europe.

The Russian manufacturing and service sectors are expanding, while the oil and gas sector has grown rapidly. Russia's most important industry is petroleum extraction and processing, and the country is one of the world's largest producers of crude oil. While energy resources dominate Russian exports, minerals also provide important export income. Russian forests produce one-fifth of the world's softwood, and Russian supertrawlers, or fish-factory ships, process catches from the Atlantic and Pacific Oceans for sale on the global market.

Belarus is the one country that has remained closely aligned with Russia since 1991. The two countries remain industrially integrated as a carryover from the past. After 1991 Belarus experienced less privatization of industry and land. Instead, Belarus looked to Russia to be a strong trading partner in exchanging its agricultural products for oil and natural gas.

The gross domestic products (GDPs) of the Russian Core countries have fluctuated since the breakup of the Soviet Union.

▼ **CRITICAL THINKING**

1. **Analyzing Visuals** Which Russian Core country has consistently had the highest GDP since 1990?

2. **Assessing** In approximately what year did the GDP of Belarus surpass the GDP of Ukraine?

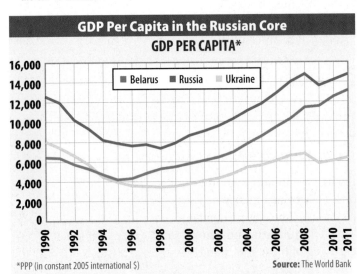

GDP Per Capita in the Russian Core

GDP PER CAPITA*

■ Belarus ■ Russia ▫ Ukraine

16,000
14,000
12,000
10,000
8,000
6,000
4,000
2,000
0

1990 1992 1994 1996 1998 2000 2002 2004 2006 2008 2010 2011

*PPP (in constant 2005 international $)   **Source:** The World Bank

Trade and Interdependence

Russia has focused on becoming a full partner in the global community by expanding trade and building international relationships. The country is a major source of oil and natural gas. These resources account for more than 50 percent of Russia's exports.

Relationships with neighboring countries are important to Russia. Ukraine imports petroleum, petroleum products, and natural gas, as well as many other products. Ukraine has the major seaport on the Black Sea and exports grain, sugar, iron ore, coal, and manganese. In the twenty-first century, Russia and Ukraine remain dependent upon one another for trade. However, Ukraine is wary of past Russian dominance of the Ukrainian people. Therefore, Ukraine's government has discussed making application to join the European Union (EU), which would switch its closest relationships from Russia to Europe. However, acceptance to the EU is dependent on economic, political, and social criteria on which Ukraine must make progress. In the meantime, Ukraine is developing its economy and its national identity within the region.

A major highway system links Moscow with other major cities in Russia, but winters are severe on roadways and repairs are both necessary and expensive. Because of its size and climate extremes, Russia depends on railroads and waterways for much of its transportation needs. Major cities are found where the Trans-Siberian Railroad and other railroads cross large rivers. Millions of tons of goods travel along thousands of miles of navigable inland waterways, which include seaports and inland cities. Pipelines are effective in transporting petroleum products, although constructing and maintaining them is challenging in large regions, such as Siberia, with a harsh climate.

Russia, Ukraine, and Belarus experience the effects of pressure for rapid modernization of their economies. Some of the world's best software companies are operated by people in Ukraine. Belarus is a crossroads for trade from Western Europe to Moscow and the rest of Russia. It has had to improve highways, communications, and repair services for long-haul trucks and railroads that transport goods both to and from Russia. All three countries are in the process of identifying their future roles in Europe and the world. All have traditions and cultures going back for centuries. However, they are each new countries, having become independent after 1991, and face political, economic, and social changes.

☑ **READING PROGRESS CHECK**

Identifying What economic transition has Russia been making since the mid-1980s?

Connecting Geography to ECONOMICS

Privatization

After 1991, private ownership of companies in Russia was allowed. The government established a stock exchange and gave vouchers to citizens enabling them to purchase shares in private companies, but these were often sold for cash instead. In 2001 the sale of land was legalized. These changes were a challenge for Russians who were accustomed to government control of nearly every aspect of their lives. It was complicated by a poor economy, inflation, and a drop in personal income. Corruption and organized crime have become issues in Russia, and they hinder investment by international businesses.

SPECULATING Why has the transition from a command to a market economy after 1991 affected Russia in both positive and negative ways?

LESSON 2 REVIEW

Reviewing Vocabulary

1. *Explaining* Explain the significance of black market and privatization.

Using Your Notes

2. *Describing* Use your graphic organizer to describe how the policies of the Soviet era affect society and culture in the Russian Core today.

Answering the Guiding Questions

3. *Identifying Cause and Effect* How have the Russian Core's historical roots and modern ideas influenced the history and government of the region?

4. *Identifying* What factors have shaped population patterns in the Russian Core?

5. *Describing* How have past policies affected the Russian Core today?

6. *Explaining* How has the economy of the Russian Core changed since 1991?

Writing Activity

7. *Informative/Explanatory* Write a paragraph describing the conditions in Russian society that led to the Revolution of 1917. Be sure to include the influence of German philosopher Karl Marx in your answer.

Arctic Oil Frontiers

Some of Earth's largest oil and natural gas fields lie under the icy waters of the Arctic Ocean. As extraction technology has improved and as demand for these resources has increased, states that border the Arctic Ocean have staked their claims and energy companies have begun to plan for drilling operations. But are energy companies prepared for the hazards involved in Arctic drilling? Are the rewards worth the risk?

Legend:
- Int'l Waters
- Agreed Borders
- Danish Claim
- Russian Claim
- Canadian Claim
- US Claim
- Future Oil Field
- Future Gas Field

HOW MUCH OIL IS THERE ?

160 BILLION BARRELS

It could power the world for the next **5 years**.

GREENLAND

NORWAY

HOW DOES IT COMPARE?

256 BILLION BARRELS

160 BILLION BARRELS

60 BILLION BARRELS

20 BILLION BARRELS

SAUDI ARABIA	ARCTIC OIL	RUSSIA	UNITED STATES

CANADA

UNITED STATES

ARCTIC OCEAN

RUSSIA

IS IT SAFE TO EXTRACT?

Extensive construction could alter the ecologically pristine environment.

Sound vibrations could disturb marine habitats.

Narwhal

Oil spills could harm marine and coastal wildlife.

Making Connections

1. *Speculating* Besides harm to the environment, what other risks with Arctic drilling can you foresee?

2. *Human Systems* What patterns do you notice about the claims made on the Arctic Ocean by bordering states? Explain why you think these claims are fair or unfair.

3. *Considering Advantages and Disadvantages* Using what you know about the risks and rewards of Arctic drilling, write a one-page summary statement explaining why you think plans for drilling in the Arctic Ocean are positive or negative.

*Interact with **Global Connections** Online*

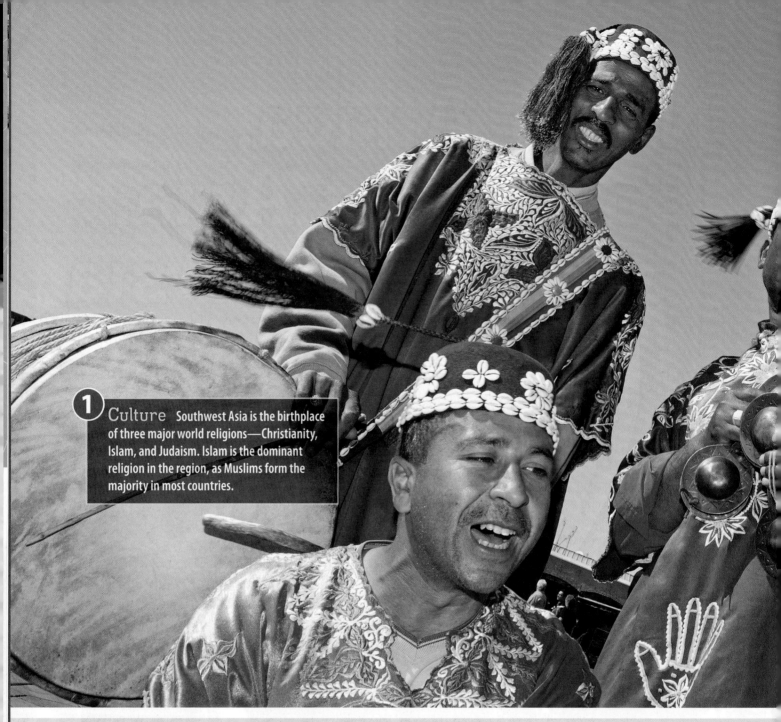

1 Culture Southwest Asia is the birthplace of three major world religions—Christianity, Islam, and Judaism. Islam is the dominant religion in the region, as Muslims form the majority in most countries.

EXPLORE the REGION

Stretching from the Atlantic Ocean in the west to the borders of China in the east, **NORTH AFRICA, SOUTHWEST ASIA,** and **CENTRAL ASIA** are three distinct but similar subregions. This is a vast area of contrasts, with plentiful oil and scarce water, bustling cities and large expanses of uninhabited land, soaring mountains and flat plains, ancient landmarks and modern technology, extreme wealth and struggling poverty.

Frans Lemmens/Alamy

THERE'S MORE ONLINE

③ Deserts The Rub' al-Khali covers more than 250,000 square miles (650,000 sq. km) and is one of the major deserts found in the region.

④ History Home to ancient kingdoms and empires that rose and fell thousands of years ago, the region has left a lasting mark on history as the source of such common practices as farming wheat and writing with an alphabet.

② Oil Much of the energy used around the world comes from this region, which includes six of the top ten countries with the most oil reserves and five of the top ten oil producers.

(l)Mustafa Ozer/AFP/Getty Images,
(t)Berno Blüchel/age fotostock,
(b)Tim Barker/Lonely Planet Images/Getty Images

North Africa, Southwest Asia, and Central Asia

Physical

140°E
120°E
100°E
80°E

ATLANTIC OCEAN

ARCTIC CIRCLE

0   1,000 miles
0   1,000 kilometers
Lambert Azimuthal Equal-Area projection

40°W
40°N
60°N

EUROPE

THE STEPPES

Irtysh R. (Ertis R.)

Lake Balkash

Caspian Depression

Aral Sea

Caspian Sea

Turan Lowland

Syr Dar'ya

Kyzyl Kum

TIAN SHAN

80°E

20°W

Black Sea

Caucasus Mts.

Pontic Mts.

ANATOLIA

Taurus Mts.

Ararat 16,945ft. (5,165 m)

Kara-Kum

Amu Dar'ya

Ismoili Somoni 24,590 ft. (7,495 m)

PAMIRS

HINDU KUSH

Khyber Pass

SOUTH ASIA

Strait of Gibraltar

ATLAS MOUNTAINS

Mediterranean Sea

Suez Canal

Syrian Desert

Mesopotamia

Euphrates R.

Tigris R.

Elburz Mts.

Damavand 18,934ft. (5,771 m)

PLATEAU OF IRAN

Zagros Mts.

20°N

S   A   H   A   R   A

Ahaggar Mountains

TROPIC OF CANCER

Sinai Peninsula

Dead Sea shoreline -1,312 ft (-400 m)

Persian Gulf (Arabian Gulf)

Strait of Hormuz

Gulf of Oman

Niger R.

Aïr

Tibesti Mts.

Nile R.

Aswān High Dam

Lake Nasser

Hejaz

Red Sea

Asir

Rub' al-Khali (Empty Quarter)

Arabian Sea

Boundary claimed by Sudan

Gulf of Aden

0°
EQUATOR

The red dashed line represents the northern boundary of a region known as the Transition Zone, an area of increasing Islamic influence.

Elevations

10,000 ft. (3,000 m)
5,000 ft. (1,500 m)
2,000 ft. (600 m)
1,000 ft. (300 m)
0 ft. (0 m)
Below sea level

— National boundary
▲ Mountain peak
▼ Lowest point
■ Mountain pass

AFRICA SOUTH OF THE SAHARA

20°S

Mediterranean Sea

Euphrates R.

Sea of Galilee

Jordan R.

Syrian Desert

Gulf of Suez

Dead Sea shoreline -1,312 ft (-400 m)

Sinai Pen.

Gulf of Aqaba

30°N

40°E

INDIAN OCEAN

TROPIC OF CAPRICORN

20°E
40°E
60°E

North Africa, Southwest Asia, and Central Asia

Political

Legend:
- ✪ Capital city
- • City

0 ——— 1,000 miles
0 ——— 1,000 kilometers
Lambert Azimuthal Equal-Area projection

ARCTIC CIRCLE

60°W

40°W

20°W

ATLANTIC
OCEAN

EUROPE

100°E

N
W   E
S

Astana ✪

Lake
Balkash

Irtysh R. (Ertis R.)

KAZAKHSTAN

Almaty •

Aral
Sea

Syr Dar'ya

Bishkek ✪
KYRGYZSTAN

Tashkent ✪
UZBEKISTAN

80°E

Amu Dar'ya

TAJIKISTAN
Dushanbe ✪

GEORGIA
Tbilisi ✪
Istanbul •
Yerevan ✪
ARMENIA

Caspian Sea

TURKMENISTAN
Ashkhabad ✪

Kabul ✪

Strait of
Gibraltar

Rabat ✪
Casablanca •
Oran •
Algiers ✪

Tunis •

Black Sea

TURKEY
Ankara ✪

Baku ✪
AZERBAIJAN

Mashhad •

AFGHANISTAN

MOROCCO

TUNISIA

Tripoli ✪

Mediterranean Sea

Tigris R.
Tehran ✪

IRAN

SOUTH
ASIA

20°W

LAAYOUNE
WESTERN
SAHARA
(Morocco)

ALGERIA

LIBYA

Benghazi •

Baghdad ✪

IRAQ

Shīrāz •

UNITED
ARAB
EMIRATES

20°N

MAURITANIA
Nouakchott ✪

MALI

TROPIC OF CANCER

Alexandria •
Cairo ✪
EGYPT

Luxor •

KUWAIT
Kuwait ✪

Manama ✪
BAHRAIN

Riyadh ✪

Doha ✪
QATAR

Persian Gulf (Arabian Gulf)

Abu
Dhabi ✪

Gulf of
Oman

Masqat ✪

Tombouctou
(Timbuktu) •

Red Sea

Jidda •

SAUDI
ARABIA

OMAN

Arabian
Sea

NIGER

BURKINA
FASO

CHAD

Boundary
claimed
by Sudan

ERITREA

YEMEN

Sanaa ✪

Aden •

Gulf of Aden

SUDAN

0°   EQUATOR

The red dashed line represents
the northern boundary of a region
known as the Transition Zone, an
area of increasing Islamic influence.

AFRICA SOUTH OF
THE SAHARA

Inset map

TURKEY

Aleppo •

Euphrates R.

SYRIA
Homs •

Tripoli •
Beirut ✪
LEBANON
Damascus ✪

IRAQ

Tel Aviv-
Jaffa •
Jerusalem ✪

Amman ✪
WEST
BANK

GAZA
STRIP
ISRAEL

JORDAN

EGYPT

SAUDI
ARABIA

30°N

40°E

20°S

0°

20°E

40°E

UNIT 5
REGIONAL ATLAS

MAP STUDY

1. **Environment and Society** How would you describe the elevation of the region? What impact would it have on human settlement patterns?

2. **Human Systems** Which countries have areas that are part of the African Transition Zone?

North Africa, Southwest Asia, and Central Asia
Climate and Vegetation

Climate
- Semi-arid (steppe)
- Arid (desert)
- Humid subtropical
- Marine west coast
- Mediterranean
- Humid continental
- Subarctic
- Tundra and high altitude

Vegetation
- Tropical rain forest
- Tropical grassland (savanna)
- Desert scrub and desert waste
- Temperate grassland
- Mediterranean scrub
- Deciduous forest
- Coniferous forest
- Mixed forest (deciduous and coniferous)
- Highland (vegetation varies with elevation)

ARCTIC CIRCLE

ATLANTIC OCEAN

EUROPE

0   1,000 miles
0   1,000 kilometers
Lambert Azimuthal Equal-Area projection

N
W   E
S

Astana

Aral Sea

Bishkek

Caspian Sea

Tashkent

80°E

Dushanbe

Black Sea

Tbilisi

Ankara

Ashkhabad

Kabul

Mediterranean Sea

Damascus

Tehran

SOUTH ASIA

Strait of Gibraltar

Rabat

Algiers

Tunis

Tripoli

Baghdad

Amman

Cairo

Persian Gulf (Arabian Gulf)

Gulf of Oman

Nouakchott

TROPIC OF CANCER

Riyadh

Red Sea

Arabian Sea

Boundary claimed by Sudan

Sanaa

Gulf of Aden

The red dashed line represents the northern boundary of a region known as the Transition Zone, an area of increasing Islamic influence.

INDIAN OCEAN

EQUATOR

TROPIC OF CANCER

20°N

Boundary claimed by Sudan

The red dashed line represents the northern boundary of a region known as the Transition Zone, an area of increasing Islamic influence.

EQUATOR

20°E   40°E   60°E

TROPIC OF CAPRICORN

40°E   60°E

Land Use

- Commercial farming
- Livestock raising
- Nomadic herding
- Primarily forest
- Manufacturing and trade
- Commercial fishing
- Little or no activity

Resources

- Coal
- Petroleum
- Natural gas
- Iron ore
- Zinc
- Bauxite
- Cobalt
- N Nickel
- Copper
- Lead
- Manganese
- Gold
- Silver
- Uranium

EUROPE

Strait of Gibraltar

Black Sea

Caspian Sea

80°E

100°E

120°E

ARCTIC CIRCLE

60°W

40°W

40°N

20°W

20°N

Mediterranean Sea

TROPIC OF CANCER

Red Sea

Persian Gulf (Arabian Gulf)

Gulf of Oman

SOUTH ASIA

Arabian Sea

Gulf of Aden

INDIAN OCEAN

The red dashed line represents the northern boundary of a region known as the Transition Zone, an area of increasing Islamic influence.

ATLANTIC OCEAN

0° EQUATOR

AFRICA SOUTH OF THE SAHARA

20°S

20°E

40°E

0°

0	1,000 miles
0	1,000 kilometers

Lambert Azimuthal Equal-Area projection

UNIT 5
REGIONAL ATLAS

MAP STUDY

1. **Physical Systems** Which types of climate cover most of the region? Why will these climates create challenges for people?

2. **Environment and Society** Compare the vegetation map with the economic activity map. Which type of vegetation supports livestock raising?

North Africa, Southwest Asia, and Central Asia
Population Density

MAP STUDY

1. *Places and Regions* Which countries in the region are the most densely populated?

2. *The World in Spatial Terms* What are North Africa's largest cities? Where are they located?

EUROPE

Strait of Gibraltar

Rabat
Casablanca
Fès
Algiers
Tunis

Black Sea
Istanbul
Bursa
Ankara
İzmir
Adana
Aleppo
Mosul
Beirut
Damascus
Tel Aviv-Jaffa
Jerusalem
Amman

Tbilisi
Yerevan
Baku
Tabriz

Caspian Sea

Mashhad
Tehran
Eşfahān
Baghdad
Shīrāz

Almaty
Tashkent
Andizhan
Dushanbe
Kabul

SOUTH
ASIA

Tripoli
Benghazi

Mediterranean Sea

Alexandria
Cairo
Asyūṭ

Kuwait

Persian Gulf (Arabian Gulf)

Gulf of Oman

Dubai

TROPIC OF CANCER

Madinah (Medina)
Riyadh

Jidda
Makkah (Mecca)

Red Sea

Arabian Sea

Sanaa

The red dashed line represents the northern boundary of a region known as the Transition Zone, an area of increasing Islamic influence.

ATLANTIC OCEAN

EQUATOR

AFRICA SOUTH OF THE SAHARA

INDIAN OCEAN

TROPIC OF CAPRICORN

Cities
(Statistics reflect metropolitan areas.)

■ Over 5,000,000
□ 2,000,000–5,000,000
⊙ 1,000,000–2,000,000

POPULATION

Per sq. mi.		Per sq. km
1,250 and over		500 and over
250–1,249		100–499
63–249		25–99
25–62		10–24
2.5–24		1–9
Less than 2.5		Less than 1

0   1,000 miles
0   1,000 kilometers
Lambert Azimuthal Equal-Area projection

North Africa

ESSENTIAL QUESTION · *How do physical systems and human systems shape a place?*

Why Geography Matters
Choke Point: Suez Canal

Lesson 1
*Physical Geography
of North Africa*

Lesson 2
*Human Geography
of North Africa*

Lesson 3
*People and Their Environment:
North Africa*

Geography Matters...

North Africa is home to the Sahara, the world's largest hot desert, and the Nile Delta, the outflow of the world's longest river. North Africa also has access to thousands of miles of coastline and many favorable trade routes. There is a long history of trade across the subregion, with various ethnic groups moving in and out of North Africa. Today the culture of North Africa has developed as an eclectic mix of the indigenous Berbers and those of their trading partners throughout Europe and Asia.

◀ This young man is a Berber, one of the oldest ethnic groups in the region. Many Berbers live in the Atlas Mountains and in the Sahara.

Kimberley Coole/Lonely Planet Images/Getty Images

choke point:
Suez Canal

*For thousands of years, countries have sought shorter trade routes to boost imports and exports. Seafaring traders for all countries have looked for shortcuts called choke points. The **Suez Canal,** linking the Mediterranean with the Red Sea, is a good example. If the Suez Canal were blocked, economies worldwide would be affected.*

DeA/A. VERGANI/Alamy

THERE'S MORE ONLINE

SEE a map of sea routes before and after the construction of the canal   •   *READ* a quote about the Suez Canal (t)DEA/A. VERGANI/Alamy, (cl)Spacephotos.com/age fotostock, (c)Jack Guez/AFP/Getty Images, (cr)Hazlan Abdul Hakim/E+/Getty Images

What is a choke point?

Where is the Suez Canal?

Why is the Suez Canal important?

A choke point is a narrow channel or canal that connects two larger bodies of water. Choke points are valuable trading routes that save vessels from traveling thousands of miles to reach a destination. Historically, countries have used choke points to create a trade monopoly by restricting trade through the narrow passage.

Today, choke points are a part of major trade networks for oil and petroleum products. In an energy-dependent world, these integral passageways power our daily lives. Roughly half of the world's oil production is shipped and moved through a choke point. Any disruption in shipment would add thousands of miles to shipping distance, raise energy prices worldwide, and possibly increase political tensions.

1. Environment and Society How have choke points influenced modern maritime trade routes?

The Suez Canal is in northern Egypt and connects the Red Sea to the Mediterranean Sea. The canal is 101 miles (163 km) long but is only 179 feet (55 m) wide at its narrowest point. Its depth has been increased over time to accommodate larger vessels. Before the completion of the canal in 1869, maritime traders were forced to sail around the tip of Africa, known as the Cape of Good Hope, which added some 4,000 miles (6,437 km) to their journey.

The canal was constructed as a shorter route between Europe and Asia. Today the Suez Canal's importance has increased due to the rise of the oil trade in the region. It now serves as a major passage for ships carrying oil and petroleum products from North Africa and Southwest Asia to Europe, North America, and the rest of the world. The canal is also a major source of revenue for Egypt. Ships pay a toll for passing through the strategic waterway. Each year the Suez Canal generates about $5 billion in foreign currency for Egypt.

2. Human Systems How has the purpose of the Suez Canal changed since its construction?

The Suez Canal is one of the world's most important passageways for oil and petroleum products. Northbound transit through the canal supplies part of Europe's and North America's energy needs. In recent years, southbound transit has delivered more of the energy resources needed by Asia's growing economies.

Egypt controls the Suez Canal. Thus, the internal political stability of Egypt affects the world's energy security. Conflict in Egypt could jeopardize the security of the canal or shippers' confidence in the canal's safety. This could force vessels to use alternate routes to ensure that cargo reaches intended locations. Increased shipping times and costs would have a direct impact on oil and petroleum prices worldwide.

3. The World in Spatial Terms How could conflict in one country affect an uninvolved country far away?

networks

There's More Online!

☑ **IMAGE** Nile River Valley

☑ **MAP** Tectonic Plates

☑ **INTERACTIVE SELF-CHECK QUIZ**

☑ **VIDEO** Physical Geography of North Africa

LESSON 1
Physical Geography of North Africa

(tl)©Natalie Tepper/Arcaid/Corbis, (tr)NASA/Alamy

Reading HELPDESK

Academic Vocabulary

- **derive**
- **pose**

Content Vocabulary

- **alluvial soil**
- **wadi**
- **phosphate**

TAKING NOTES: *Key Ideas and Details*

ORGANIZING Use a graphic organizer like the one below to take notes on the water systems of North Africa.

North Africa: Water Systems

Nile ⟶ ☐

The Aswān High Dam ⟶ ☐

ESSENTIAL QUESTION · *How do physical systems and human systems shape a place?*

IT MATTERS BECAUSE

The physical geography of an area directly determines what natural resources are available to the people who live there. Physical features such as mountain ranges act as natural barriers to the rainfall and water that support human settlement. Rivers and oases provide water for irrigation and farming. Variations in climate influence local living conditions, which can affect the rate of economic and population change.

Landforms

GUIDING QUESTION *What physical process is responsible for shaping the landscape in North Africa?*

The landscapes of North Africa have been shaped by the action of four tectonic plates: the African, Arabian, Anatolian, and Eurasian plates. The movements and interactions among these plates build mountains, shift landmasses, and cause earthquakes. Africa's longest mountain range, the Atlas Mountains, was the result of the interaction of the African and Eurasian plates. The Atlas Mountains extend from southwest Morocco to northeast Tunisia in a region known as the Maghreb. The range consists of a northern section, the Tell Atlas, and a southern section, the Saharan Atlas. Both sections of the Atlas Mountains have deep gorges that cut through the sediment as a result of runoff **derived** from annual rainfall.

Western Sahara and Mauritania occupy a large region in northwest Africa on the edge of the Sahara. The picturesque, sandy, dune-covered areas in the desert are called *ergs*. However, the majority of the desert is not covered by sand. Two desert features are *regs*—stony plains covered with rocky gravel also known as "desert pavement"—and *hamadas*—flat, sandstone plateaus.

The Ahaggar is a large, mountainous plateau region in the north-central Sahara. The highest mountain in the Ahaggar is Mount Tahat at 9,573 feet (2,918 m) in southeastern Algeria. Parts of the plateau also extend into Niger in the south and Libya in the northeast. Black volcanic landforms and a pink granite massif are found in the rocky desert of the Ahaggar plateau. Volcanic necks formed as magma cooled when the volcanoes were active, creating impressive landforms in the Ahaggar.

The Tibesti Mountains in northeastern Niger and southern Libya rise steeply above the surrounding plains. The mountain range is about 300 miles (480 km) long and about 175 miles (280 km) at its widest point. The Tibesti Mountains are the location of the highest point in the Sahara—the volcanic summit of Emi Koussi in northern Chad. It has an elevation of 11,204 feet (3,415 m).

✔ **READING PROGRESS CHECK**

Identifying What major landform has been created by the interaction of the Eurasian and African plates?

Water Systems

GUIDING QUESTION *How has the Nile River been the lifeblood of North Africa?*

Rivers are the lifeblood of North Africa. They support the growth of plant and animal life. Rivers also promote human settlement and development. At 4,160 miles (6,693 km) in length, the Nile River is the world's longest river. It originates in the highlands of Ethiopia and Uganda and flows north through the desert.

The fertile land along the river's banks and the Nile Delta gave birth to some of the world's earliest civilizations. Today more than 90 percent of Egypt's people live in the Nile Delta or along the course of the river even though the area covers only 3 percent of Egypt's land.

Before the 1950s, the people in Egypt anticipated and depended on the annual flooding of the Nile River. The floodwaters carried sediments that were deposited along the river banks and formed rich **alluvial soil** that made for productive agricultural lands.

Beginning in the 1950s, the Aswān High Dam and other dams were built farther upstream along the Nile River. The dams were built to control the river's flow and reduce flooding. Dams were also built to provide year-round irrigation and hydroelectric power in Egypt. However, without the annual deposition of new alluvial material, soil fertility diminishes and requires the use of fertilizers. Away from the Nile, in deserts, infrequent but intense rainstorms can create **wadis** (WAH•dees), or streambeds that remain dry until a heavy rain. The rainstorms often produce flash flooding. During a flash flood, wadis fill with so much sediment that they can rapidly become mudflows, or moving masses of wet soil. The mudflows **pose** a danger to humans and animals.

Bodies of water also help define boundaries in the subregion. For example, edging the coast of North Africa, the Mediterranean Sea separates Africa and Europe. To the region's east, Egypt's Sinai Peninsula, the Red Sea, and the Gulf of Aden separate North Africa from the Arabian Peninsula.

✔ **READING PROGRESS CHECK**

Naming Which river in North Africa is the world's longest river?

derive to receive or obtain from a specified source

alluvial soil rich soil made up of sand and mud deposited by running water

wadi in the desert, a streambed that is dry except during a heavy rain

pose to present

The meeting of four tectonic plates shapes the landscape of North Africa.

▼ **CRITICAL THINKING**

1. **Identifying Cause and Effect** What landforms were created as a result of tectonic activity in North Africa?

2. **Explaining** Why does North Africa experience such a significant amount of tectonic activity?

©Natalie Tepper/Arcaid/Corbis

This true-color image shows the Nile River valley, the Nile Delta, and the strategic location of the Suez Canal.

CRITICAL THINKING ▲

1. Identifying Cause and Effect
What factors influence the green vegetation in the region shown in the image?

2. Categorizing Describe the landscapes that appear as tan and brown areas in the image.

phosphate a natural mineral containing chemical compounds often used in fertilizers

Climate, Biomes, and Resources

GUIDING QUESTION *How does water scarcity define the climate of North Africa?*

North Africa has a varied climate because it is the meeting place of humid and cold air masses that come from the north and hot, tropical air masses that come from the south. Despite the region's ample amount of rainfall in some parts during the rainy season, desert is the dominant feature of the North African environment.

Climate Regions and Biomes

Water scarcity defines the subregion's variations in climate. Today's North African landscape is most commonly associated with vast stretches of sand and the occasional watering hole. Meanwhile, ancient cave paintings show that this part of the African continent was once wet and green. These opposing views of the landscape indicate changing climates over time because of physical processes. Moreover, changes in climates across the region have affected natural vegetation and human activities in the past and continue to influence the region.

The entire area covered by the Sahara is classified as desert because its annual precipitation averages 10 inches (25 cm) or less. The Sahara, the largest hot desert in the world, is about 3.5 million square miles (9.1 million sq. km) and covers almost 75 percent of North Africa. In recent decades, droughts have expanded the extent of the Sahara.

The desert regions of North Africa experience extreme changes in climate. The Sahara mostly consists of two types of climates: an arid (desert) and a semi-arid (steppe) climate. Northern parts of the Sahara have winters with freezing temperatures, while the southern Sahara is characterized by milder, dry winters and summer rains. The harsh desert climate makes vegetation growth sparse.

Despite their arid conditions, the North African deserts can support some vegetation, such as cacti and drought-resistant shrubs. Small-scale farming is possible in oases, places in the desert where underground water surfaces. Oases are very fertile due to the ready availability of water. Because they are a source of water and can support the growth of trees— such as olive trees and date palms— many oases in the Sahara are also the sites of villages, towns, and cities.

The Sahel is a flat plain that stretches about 3,300 miles (5,400 km) across the subregion from Senegal to Sudan south of the Sahara. The narrow band of semi-arid land forms a transition zone between the Sahara and the savannas to the south. The most common animal in the Sahel is the Senegal gerbil. Baobab and acacia trees are also found in the transition zone. Cattle feed on the Sahel's prickly cram-cram grass.

A long belt of grassland that spans nearly 5,000 miles (8,046 km) runs east to west, south of the Sahara and the Sahel. Precipitation in this semi-arid climate region usually averages less than 14 inches (36 cm) annually. This amount of rain is enough to support the growth of shrubs and some trees, as well as short grasses that feed sheep, goats, and camels.

NASA/Alamy

Some parts of North Africa enjoy the benefits of a mild Mediterranean climate. Occurring in parts of Morocco, Tunisia, and Libya, this climate consists of cool, rainy winters and hot, dry summers. Countries with Mediterranean climates boost their economies by exporting citrus fruits, olives, and grapes. They are mostly exported to Europe and North America. Some of these countries also benefit from tourism. People from colder climates seek the sunshine and warmth of the North African Mediterranean climate.

Coastal and highland areas near mountain ranges usually receive the most rainfall. Warm, moist air driven off the sea by prevailing winds condenses as it rises over the mountains and returns to Earth as rainfall. The North African coast near the Atlas Mountains, for example, averages more than 30 inches (76 cm) of rain each year. The rainfall is enough to support agriculture and forests.

In areas where more than 14 inches (36 cm) of rain falls yearly, farmers can raise food grains such as barley, oats, and wheat. These crops can be grown without irrigation. Areas with this much rainfall have an abundance of trees. Unfortunately, deforestation for agriculture has reduced tree cover in the Atlas ranges. The reduced tree cover has had negative effects on animal life.

Natural Resources

North Africa boasts an abundance of natural resources. The region's most significant resources are **phosphates**, natural gas, and petroleum. Morocco is one of the leading exporters of phosphates. Algeria has the tenth-largest natural gas reserves in the world. Additionally, it ranks sixteenth in oil reserves and sixth among the world's gas exporters.

Petroleum exports have enriched some of the countries of North Africa, particularly Algeria and Libya. Heavy reliance on petroleum exports can be risky, however. When oil prices go up and down on world markets, as they did in the late 1990s and after the Arab Spring in 2011, the region's economies suffer.

Some countries in North Africa are trying to make their economies more diverse as a way to decrease their reliance on oil and mineral exports. Algeria has tried to attract increased investment in business and industry. Libya is investing in infrastructure, agriculture, and fisheries.

Minerals also provide revenue for North African countries. Deposits of chromium, gold, lead, manganese, and zinc are sprinkled across the region. Discoveries of iron ore and copper deposits indicate that the region may contain up to 10 percent of the world's iron ore reserves.

✔ **READING PROGRESS CHECK**

Specifying What natural resources are most prevalent in North Africa?

Analyzing PRIMARY SOURCES

Saving Mediterranean Habitats Worldwide

"Mediterranean climate regions cover only 2.2 percent of Earth's land surface, yet they account for 20 percent of all known plant species. Only the tropical rainforests of the western hemisphere and southeast Asia have a greater density of plant species. Mediterranean climates are found in only five places on Earth: California and northern Baja California, the basin of the Mediterranean Sea, southwestern Australia, the western cape of South Africa and the central coast of Chile.

They all share similar landscapes. Mediterranean climates are characterized by hot, dry summers and mild, wet winters. . . ."

—Rebecca Shaw, Nature Conservancy, *Saving Mediterranean Habitats Worldwide*, February 25, 2011

DBQ *IDENTIFYING CENTRAL ISSUES* Why is it important to protect Mediterranean biomes?

LESSON 1 REVIEW

Reviewing Vocabulary

1. *Interpreting* What physical forces cause the creation of wadis?

Using Your Notes

2. *Describing* Use your graphic organizer to describe the Nile Delta.

Answering the Guiding Questions

3. *Summarizing* What physical process is responsible for shaping the landscape in North Africa?

4. *Evaluating* How has the Nile River been the lifeblood of North Africa?

5. *Paraphrasing* How does water scarcity define the climate of North Africa?

Writing Activity

6. *Informative/Explanatory* Identify a climate in the North African subregion that has supported agricultural production and human growth. Identify another climate in this subregion that has restricted the establishment of settlement or agriculture. What physical features and specific natural resources differentiate these two types of climates?

netw🌐rks

There's More Online!

☑ **IMAGE** Traditional Crafts of North Africa

☑ **MAP** North Africa: Invasions and Migrations

☑ **TIME LINE** North Africa: Path to Independence

☑ **INTERACTIVE SELF-CHECK QUIZ**

☑ **VIDEO** Human Geography of North Africa

LESSON 2
Human Geography of North Africa

ESSENTIAL QUESTION · *How do physical systems and human systems shape a place?*

Reading **HELP**DESK

Academic Vocabulary

- **principal**
- **partner**

Content Vocabulary

- **domesticate**
- **hieroglyphics**
- **geometric boundary**
- **nationalism**
- **nomad**
- **bedouin**

TAKING NOTES: *Key Ideas and Details*

IDENTIFYING Use a graphic organizer like the one below to take notes on the indigenous peoples and independence movements of North African countries.

Human Geography of North Africa

Early Peoples and Civilizations	Invasions and Independence

IT MATTERS BECAUSE

For centuries, North Africa has been the birthplace and meeting place of cultures, from the Maghreb area in the west to the Nile Valley in the east. Today the subregion exhibits a cultural vibrancy that rests on this deep historical foundation. Ethnic groups, religious changes, and political movements such as the Arab Spring make this a dynamic and important part of the world.

History and Government

GUIDING QUESTION *How have the Sahara, the Nile River valley, and multiple invasions influenced the history and government of North Africa?*

North Africa has a long history. Some of the oldest human civilizations began in the fertile Nile River valley. People from nearby places, including Greece, Rome, Arabia, and Europe, have all influenced North Africa.

Early Peoples and Civilizations

Hunters and gatherers settled throughout North Africa about 10,000 years ago. The subregion's farmers were among the first to **domesticate** plants and animals. Under this process, plants and animals once wild became raised by people for food, clothing, and transportation.

The Egyptian civilization developed in the fertile Nile River valley about 6,000 years ago. Annual floods from the Nile deposited rich soil on the floodplain. During dry seasons, Egyptians used sophisticated irrigation systems to water crops. This enabled farmers to grow two crops each year. The Egyptians also developed a calendar with a 365-day year and a system of writing known as **hieroglyphics** (HY•ruh•GLIH•fihks). The large pyramids the Egyptians built were used as tombs for their rulers and demonstrate their abilities in mathematics and engineering.

In the A.D. 600s, invasions of Arab armies moved westward from the Arabian Peninsula. These invasions heavily influenced the cultures of North Africa. As Arab rule spread across North Africa, so did the Muslim religion. Muslim and Jewish exiles fleeing persecution by the Spanish Inquisition

(tl)Transcendental Graphics/Hulton Archive/Getty Images, (tc)(l)Keystone-France/Gamma-Keystone/Getty Images, (tc)(r)STR/AFP/Getty Images, (tr)Stefan Kiefer/Alamy infused Morocco with Spanish culture in the 1400s. The Arabs retained control over much of North Africa until the early 1500s, when the Ottoman Empire began expanding westward from what is now Turkey. The Ottomans were Muslims, or followers of the Islamic faith. With time, the Ottomans were expelled from North Africa as ethnic groups took up military efforts to end their rule. Resistance to Ottoman rule was finally ended with World War I in 1918.

Location enabled other cultural influences on the subregion. Europe was geographically nearby and established colonial control as the Ottomans left. Algeria was invaded by the French in the mid-1800s. French influence was imprinted on the country, but unrest simmered beneath the colonial surface. Italy colonized Libya, and France occupied additional territory south of the Atlas Mountains. There were disagreements over who would rule North Africa. As a result, the European colonial powers drew **geometric boundaries** that followed straight lines and did not account for natural and cultural features. The boundaries often created conflict between groups of people. Tensions occurred because the local methods of governance were not the same as European methods of governance.

Independence and Power Struggles

In the 1800s, an educated, urban middle class developed in North Africa. This new middle class came in contact with worldwide anticolonial thought and adopted European ideas of **nationalism**. This development stirred demands for self-rule that provided the basis for the modern countries that emerged.

Egypt gained independence from the United Kingdom in 1922. Algeria won independence from the France in 1962 when a nationalist movement led a violent revolt. Since independence, Algeria has developed its resources and raised its living standards. Libya won independence from Italy in 1951 and was then ruled by a Western-friendly monarchy until 1969 when a coup led by Colonel Muammar

domesticate to adapt plants and animals from the wild for human use

hieroglyphics an ancient writing system used in Egypt in which pictures and symbols represent words or sounds

geometric boundary a fixed limit or extent that typically follows straight lines

nationalism a belief in the right of a nation to be an independent state

GEOGRAPHY CONNECTION

North Africa's location near Europe and Southwest Asia has made it vulnerable to numerous migrations and invasions over the centuries.

1. **PLACES AND REGIONS** Which cities were affected by Islamic invasions until the 750s?

2. **HUMAN SYSTEMS** Which later migration route is similar to that of Islamic invasions?

North Africa: Invasions and Migrations

Legend:
- Islamic invasions, until 750s
- Jewish exiles from Christian Spain, late 1400s
- Ottoman Empire, 1500s
- Muslim exiles from Christian Spain, early 1600s
- France, early 1800s
- U.S. and Britain, World War II
- Present-day boundary

0 — 500 miles
0 — 500 kilometers
Lambert Azimuthal Equal-Area projection

al-Qaddafi overthrew the monarchy. Tunisia, one of the major fronts during World War II, separated from France in 1956. Its history is similar to Libya, with autocratic regimes ruling until 2011. Morocco also won independence from France in 1956. Today Morocco is a constitutional monarchy. Though most countries adopted constitutions, their elected presidents often ruled as dictators once they were in office. Their rule often stifled human rights and free speech.

Political and ethnic tensions have risen often in North Africa. A protest by a Tunisian citizen that resulted in his death in 2010 led to riots. The act was to protest the lack of economic opportunities. News of the event resulted in revolts against oppressive, nondemocratic governments in Egypt, Libya, and Tunisia. The movement became known as the Arab Spring. Citizen protests in Morocco did not overthrow the government, but the king was forced to concede political and social reform. Beginning in 2011, new governments were elected in Egypt, Libya, and Tunisia.

☑ **READING PROGRESS CHECK**

Examining What movement led the demand for self-rule in North African countries?

Population Patterns

GUIDING QUESTION *How have indigenous ethnic groups, migrations, and climate shaped population patterns in North Africa?*

While European connections continue in the coastal regions of North Africa, a widespread influence throughout the subregion is a mix of indigenous and Arab cultures. The Berbers were the indigenous people of North Africa before the Arab invasions. They maintain their traditional non-Arabic language. Most of the 15 million Berbers live today as farmers. They had previously been pastoral **nomads**, groups of people who move from place to place with herds of animals depending on the season and the availability of grass for grazing and water. The other **principal** ethnic group in North Africa is the Arab people. United by language, Arabs first migrated from the Arabian Peninsula to North Africa in the

nomad a member of a wandering pastoral people

principal most important, consequential, or influential

Transcendental Graphics/Hulton Archive/Getty Images

TIME LINE ⌄

NORTH AFRICA
Path to Independence ➡

The countries of North Africa share a common history of colonialism, independence, and journeys toward democracy.

CRITICAL THINKING ▶

1. *Describing* Discuss the effects of colonization on the recent history of North Africa.

2. *Identifying* Which countries changed governments as a result of the Arab Spring?

1800s France and Great Britain take control over areas of North Africa.

1800 ➡ .... **1950 ➡** ....

1922 Great Britain grants Egypt independence, but maintains control of the Suez Canal.

1952 Military coup topples government of Egypt; military leader Gamal Abdel Nasser later becomes president of Egypt.

1954 Algerian nationalist group, the National Liberation Front, launches guerrilla war against France.

374

A.D. 600s. Some of the Arabic-speaking peoples are nomadic **bedouins** (BEH•duh•wuhns) who migrated to North Africa from deserts in Southwest Asia.

Geographic factors, especially the availability of water, have influenced settlement in the subregion. Because water is scarce, people have for centuries settled along seacoasts and rivers. The Mediterranean and Atlantic coasts and the Nile River valley hold most of the subregion's people. The Nile River valley is one of the world's most densely populated areas. Other population centers in North Africa include Casablanca, Tunis, Tripoli, and Cairo. Egypt's most populous city and its capital, Cairo, dominates the country's social and cultural life. Cities have grown rapidly as people migrate from rural areas in search of a better life. Cities face problems providing services because growth has occurred quickly.

Emigration to regions outside North Africa is high due to higher availability of employment opportunities in other regions. The region's close proximity and historical ties to European countries such as Spain and France pull migrants from Morocco, Algeria, and Tunisia. Migrants from Egypt, an ex-British colony, migrate to the United Kingdom and other English-speaking countries such as the United States, Canada, and Australia.

☑ **READING PROGRESS CHECK**

Naming Who are the people indigenous to North Africa?

Society and Culture Today

GUIDING QUESTION *How have Islam and the Arabic language helped define much of the culture and society of North Africa?*

When the Arabs invaded North Africa, they brought Islam. Calls to worship occur five times each day in countries with Muslim populations. A muezzin, or crier, delivers the call to prayer for each local mosque. Following the movements of the imam, or prayer leader, individuals bow and kneel, touching their foreheads to the ground in the direction of the holy city of Makkah (Mecca) in Saudi Arabia.

bedouin member of the nomadic desert peoples of North Africa and Southwest Asia

(bl)©Hulton-Deutsch Collection/Corbis, (bc)Keystone-France/Gamma-Keystone/Getty Images, (br)STR/AFP/Getty Images

1956 March—France recognizes independence of Morocco and Tunisia.

1962 Brutal Algerian war of independence with France ends; Algeria becomes an independent country.

1969 Libyan army overthrows king. Muammar al-Qaddafi becomes head of state.

2011 Arab Spring leads to regime change in Tunisia, Libya, and Egypt.

2010 ➜

1956 July—President Nasser of Egypt nationalizes Suez Canal, leading to Suez Crisis.

2010 Young Tunisians launch protests that ultimately forces president to flee.

Islamic patterns are found in weaving and embroidery throughout the subregion.

▲ **CRITICAL THINKING**

1. *Analyzing Visuals* Describe the designs shown in the picture above.

2. *Making Connections* Why are there no animals or people in the designs of traditional art in North Africa?

As Islam spread, so did the Arabic language. Non-Arab Muslims learned Arabic in order to read the Quran, Islam's holy book. Moroccans speak an Arabic dialect that is a result of Berber influence. Egyptians speak Arabic with a different accent than do Moroccans, Tunisians, and Algerians. People in Algeria and Morocco also speak French as a result of colonization and influence on education.

Family and Status of Women

Since achieving independence in the 1950s and 1960s, North African countries have been redefining their political, social, and cultural systems. Class status influences family size. Upper-class families have fewer children than lower-class families. Historically, extended family members have lived nearby, so children usually grow up with many cousins, aunts, and uncles. Today, families are living more spread out and do not necessarily live close to one another.

Decades of urbanization have led nomadic peoples to settle in villages and cities. Very few groups still roam the desert as they did centuries ago. They have access to schools, transportation, and health care near where they live. Families have plans that educated children will earn enough money to support the older members of the family.

Although women participated in winning independence, women's issues were not priorities in the years immediately following independence. Government rule often stifled human rights. This affected women's issues as well, and many women are not permitted to accept jobs even when they have an education level higher than men.

The Arts

From the earliest times, the peoples of North Africa have expressed themselves through the arts. The ancient Egyptians built huge pyramids that still stand as marvels of construction. One of Algeria's most popular forms of music, *raï*, involves various instruments and poetic lyrics. Contemporary Egyptian music blends traditional Arabic and Western styles. Egypt has experienced a revival of folkloric dance and traditional crafts. The arts of weaving, embroidery, and metalworking are heavily influenced by Islamic patterns found in local architecture. Elaborate geometric designs are featured in these works because of the traditional Islamic prohibition against the representation of living beings.

☑ **READING PROGRESS CHECK**

Assessing What has been the primary influence of language in North Africa?

Economic Activities

GUIDING QUESTION *What factors have helped many countries in North Africa become rising middle-income countries?*

North Africa, unlike other regions of the continent, has historically engaged in a range of manufacturing processes. By the end of the 1800s, much of the region, however, was regarded by European countries as a source of raw materials. Although manufacturing increased during World War II, output accelerated quickly only after the decline of European control. Despite expansion since the 1950s, manufacturing output among the countries varies greatly.

Only a small part of the subregion's land is arable, or suitable for farming. Agriculture plays a smaller role in countries with desert climates. It is also less important in countries with economies based on oil, such as Libya and Algeria.

Stefan Kiefer/Alamy

Areas of North Africa that have a Mediterranean climate are best suited for growing a variety of crops. Barley, wheat, citrus fruits, grapes, vegetables, olives, figs, dates, and almonds are grown. When rainfall is below normal, however, harvests of major crops such as wheat and barley seldom meet local needs, which increases dependence on imported food. Seafood is an important food source, and fishing vessels catch sardines and mackerel from the Atlantic Ocean.

Wealth from oil, natural gas, and mining has helped develop economies in the subregion. Petroleum and oil products are North Africa's main export commodities, or economic goods. Natural gas is also a major export. Coal and copper mining and cement production are important as well.

Tourism plays a significant role in the subregion's economies. Certain areas serve as popular travel destinations because of their historical importance, blend of religious history, or pleasant climates. The Valley of the Kings and the pyramids draw tourists to Egypt. Port el-Kantaoui near Sousse, Tunisia, is a vacation beach destination. Carthage, an ancient city, is rich in historical artifacts.

Advancements in transportation and communications have brought improvements to the subregion. However, the physical environment and government control have limited some development here. Water transportation is very important to North Africa. Ships load and unload cargo and passengers at ports on the Mediterranean Sea, Red Sea, and Atlantic Ocean, including oil, natural gas, minerals, and tourists. Because of its accessibility to Europe and low-cost labor, Morocco has become an appealing trading **partner**—especially for Spain, France, and Italy. Other North African countries also have significant trade relations with countries outside the region.

partner one associated with another, especially in an action

Despite economic progress within North Africa, its countries suffer from high rates of unemployment, poverty, wealth disparity, and political upheaval. Key economic challenges include reducing socioeconomic disparities in government spending, increasing available jobs for the young, and establishing more diverse industries. The Arab Spring has resulted in a number of stressful aftereffects in North Africa as well. These include high food and fuel prices that have strained many of the resources and budgets of the subregion's governments. Political instability, public unrest regarding dishonest elections, and the protection of human rights are concerns expressed in the new Arab Spring democracies.

✓ READING PROGRESS CHECK

Explaining How have North Africa's natural resources influenced economic growth?

LESSON 2 REVIEW

Reviewing Vocabulary

1. *Locating* Where did the bedouin people originate?

2. *Defining* What political concept introduced by North Africa's middle class is a response to the region's history of colonialism?

Using Your Notes

3. *Summarizing* Use your graphic organizer to describe the early peoples of North Africa.

Answering the Guiding Questions

4. *Gathering Information* How have the Sahara, the Nile River valley, and multiple invasions influenced the history and government of North Africa?

5. *Explaining* How have indigenous ethnic groups, migrations, and climate shaped population patterns in North Africa?

6. *Summarizing* How have Islam and the Arabic language helped define much of the culture and society of North Africa?

7. *Describing* What factors have helped many countries in North Africa become rising middle-income countries?

Writing Activity

8. *Informative/Explanatory* Identify two natural resources in North Africa crucial to economic activity and trade. In a paragraph, provide specific examples of how these resources increased trade and interdependence between North Africa and the world.

The Arab Spring

The Arab Spring was a series of protests and revolutions that began to spread across the Arab world in late 2010. All protesters shared the same general goals: less corruption in their country's leadership and more openly democratic forms of government. The outcomes of the Arab Spring have differed between countries. In some places, like Egypt, the protests have resulted in major changes in government. In others, like Syria, protests and government repression have broken down into full-scale civil wars. The lasting outcomes of the Arab Spring, positive or negative, are yet to be fully realized. Here we look at the beginnings of the Arab Spring.

MAJOR PROTESTS   **MINOR PROTESTS**   **PROTESTS SUPPRESSED**

CHANGE IN GOVERNMENT   **LEADER KILLED**   **CIVIL WAR**

TUNISIA

MOROCCO   **ALGERIA**

LIBYA   **EGYPT**

1

3

4

TUNISIA

December 17, 2010–January 4, 2011 — As a protest against the government, Mohammed Bouazizi sets himself on fire in the city of Sidi Bouzid. Shortly afterward, protests against the authoritarian rule of President Zine El Adidine Ben Ali begin in the capital, Tunis. Ben Ali flees the country in mid-January.

LIBYA

February 4, 2011 – February 15, 2011 — Calls for protest on the Internet result in demonstrations in the cities of Benghazi, Tripoli, and across the country. Clashes between protesters and the government turn into a civil war which ends on October 23, 2011, with a rebel victory.

EGYPT

January 25, 2011 – February 11, 2011 — Protests in the cities of Cairo, Alexandria, and across the nation erupt against Egyptian President Hosni Mubarak's regime in mid-January. Mubarak steps down from power shortly after the protests begin, allowing Egypt to hold its first ever democratic elections in late 2011.

SYRIA

February 14, 2011 – March 18, 2011 – **Angry** with corruption in their government, Syrian protesters organize a "Day of Rage" for mid-February on Facebook and Twitter. Sporadic protests in the capital of Damascus and across the country pop up in the following days. Government suppression turns the protests into a civil war.

– 5.

 ## BAHRAIN

February 14, 2011 – March 18, 2011 – During the Bahrain "Day of Rage" protests on February 14th, a young protester is killed by police. This causes the situation to escalate. The government of Bahrain, however, has been largely able to suppress protest efforts.

YEMEN

January 14, 2011 – March 11, 2011 – Inspired by events in Tunisia and Egypt, Yemeni people protest against the corrupt government of President Ali Abdullah Saleh. As police fire on the protesters, Yemen erupts in civil war. However, Saleh bows to outside pressure and steps down in late 2011.

2

Making Connections

1. Places and Regions What were the two most common outcomes of states experiencing major protests?

2. Human Systems Many protest groups used public sites like Facebook and Twitter to organize and communicate. How might this practice have inspired and encouraged protests elsewhere?

3. The Uses of Geography Create a time line of events for one of the states above, detailing some of the protests, changes in government, or conflicts. Then, write a short summary explaining the overall positive or negative effects of the movement on that state.

*Interact with **Global Connections** Online*

netw⊙rks

There's More Online!

- ☑ **IMAGE** Managing the Great Man-Made River project
- ☑ **IMAGE** Oasis in Tinghir, Morocco
- ☑ **INFOGRAPHIC** Aswān High Dam
- ☑ **INTERACTIVE SELF-CHECK QUIZ**
- ☑ **VIDEO** People and Their Environment: North Africa

LESSON 3
People and Their Environment: North Africa

ESSENTIAL QUESTION · *How do physical systems and human systems shape a place?*

Reading **HELP**DESK

Academic Vocabulary

- **subsequent**
- **relevant**

Content Vocabulary

- **aquifer**

TAKING NOTES: *Key Ideas and Details*

Summarizing Use a graphic organizer like the one below to summarize ways that North Africa is managing its resources.

Managing Resources

IT MATTERS BECAUSE

The relationship of humans with their environment is constantly changing. While humans have achieved impressive advancements in energy and power production, such innovation comes at a cost. For example, the need for freshwater and the process of oil extraction have often negatively affected the North African environment. The subregion's water scarcity and oil wealth make trade an essential activity for many North African countries.

Managing Resources

GUIDING QUESTION *Why are water resources in such demand in North Africa?*

As populations and economies grow, so does demand for water resources. Clean drinking water ensures the continued livelihood and development of human populations. Much water in North Africa is used for irrigation, which allows for continued food production. Places with fewer water resources, such as Western Sahara and Libya, are more dependent on trade to import food. This not only leaves populations dependent on outside food sources, but also creates a vulnerability to changes in market prices for food. Agricultural land loss due to urbanization and windblown sands is also an issue affecting food production.

Much of the freshwater in North Africa comes from rivers, oases, and **aquifers**, or underground sources of water. Aquifers can be problematic for countries with coastal access. As water is removed from the coastal aquifers, it is replaced by salty seawater. Water contaminated by increased salinity in the aquifers becomes unusable for agriculture. The contaminated water is also unfit to drink.

The source of water for consumption and agriculture in North Africa varies by country. Egypt, for example, gathers surface water from large rivers, particularly the Nile. Other countries, such as Algeria and Libya, have more limited water resources. As another form of water extraction, these countries use a process called desalination. Desalination is a

(tl)©Walter G. Allgower/imagebroker/Corbis, (tr)ZUMA Press/Alamy process in which the salt is removed from salt water. Building and maintaining desalination plants, however, can be expensive for most countries.

Libya's Great Man-Made River is an ambitious effort to supply freshwater to this country that consists of 95 percent desert and relies on food imports to feed its population. Nearly 40,000 years ago, when North Africa had a more temperate climate, rainwater collected in underground reservoirs beneath Libya. It was not until the mid-1950s that oil exploration in the southern Libyan Desert exposed these vast quantities of fresh, clean groundwater. The Great Man-Made River is a pumping and pipeline system that attempts to utilize this extensive source of groundwater.

The Great Man-Made River provides 70 percent of the Libyan population with water for drinking and irrigation. It was funded by profits from Libya's nationalized oil sector. The project uses two pipelines to pump water from large aquifers beneath the Sahara to farms along Libya's coastal region. The first two phases of the project were completed in 1993 and 1996. The third phase of the project, completed in 2009, links the pipelines of the first two phases. Two **subsequent** phases will extend the distribution network to the northeast coast and unite the two transporting systems into a single network.

However, environmental challenges exist for Libya's Great Man-Made River. Scientists fear that the pipelines could drain aquifers in Libya and neighboring countries. They also fear that pumping aquifers near the Mediterranean could draw in salt water from the sea. In most areas, the rate of water extraction far exceeds the rate at which water is replaced. Such rates of extraction cannot be sustained in the coming decades. Although providing water for the current population is important, an equally **relevant** concern is having water for future generations.

✔ **READING PROGRESS CHECK**

Specifying What is a project that has provided water to the region's inhabitants?

aquifer underground water-bearing layers of porous rock, sand, or gravel

subsequent later; following after another event

relevant having to do with the matter at hand

This dense palm oasis in Tinghir, Morocco, is irrigated by a network of pipes and irrigation canals.

CRITICAL THINKING ▼

1. **Analyzing Visuals** What environmental challenges does irrigation pose in North Africa?

2. **Hypothesizing** What type of water source is likely to be used to irrigate the oasis shown?

Human Impact

GUIDING QUESTION *How have modern economic activities impacted North Africa's environment?*

North Africa faces many challenges related to economic activity and its effects on the environment. Oil and fishing are two very profitable economic sectors for the subregion. High-volume production for these economic sectors, however, can alter the physical environment. The plants and animals in the area also suffer when habitats are lost or damaged.

Mass tourism is another main cause behind ecological loss in the subregion. Areas along the coast that receive the most visits from tourists have suffered the greatest losses. Some locations that were once pristine are now severely damaged beyond repair. Increased use of water in hotels, swimming pools, and golf courses—especially during the summer—is also a concern since the subregion suffers from water shortages.

North Africa's location puts it at risk for oil spills since tankers pass the subregion's coasts. Oil spills can occur on land or on the seas. Following an oil spill at sea, winds can sweep the oil slicks toward coastlines, affecting the ecosystems of the land as well as the marine environments. Past oil spills have

EGYPT'S Aswān High Dam

🖒 **BENEFIT**   🖓 **DRAWBACK**

NILE DELTA

Egypt's Aswān High Dam, located near the Egypt-Sudan border, brings both benefits and drawbacks for the human geography of the area.

🖒 **IRRIGATION FOR FARMING**

The dam helps provide irrigation to thousands of new acres of farmland.

🖓 **FLOODING AND FARMS**

When the Nile River flooded, it deposited fertile silt on agricultural lands. The Aswān High Dam, by preventing flooding, has resulted in a decrease in agricultural productivity.

🖒 **POWERING COMMUNITIES**

The dam provides hydroelectric power to Egypt, about 10 billion kilowatt-hours. This is enough to power nearly 870,000 average U.S. homes.

🖓 **POPULATION RELOCATION**

About 90,000 Egyptians and Sudanese had to be relocated because of the dam.

🖒 **FLOOD CONTROL AND SAFETY**

The Aswān High Dam controls flooding along the Nile River. Before construction of the dam, flooding could be unpredictable and dangerous. The dam makes it safer to live along the Nile.

🖓 **THREAT TO CULTURE**

The flooding caused by building the dam would have submerged the ancient temple of Abu Simbel. The temple was dismantled by hand and moved to higher ground.

ASWĀN HIGH DAM

The Aswān High Dam and other modern dams constructed farther up the Nile control the river's flow, resulting in both positive and negative effects for the people of the subregion.

▲ CRITICAL THINKING

1. **Assessing** How does the Aswān High Dam benefit the people of North Africa?

2. **Posing Questions** What questions would you ask the government agency in charge of the dam about its effects on the environment of North Africa?

DigitalGlobe/Getty Images

exposed the lack of safety precautions related to oil extraction and the necessity of establishing a faster response to oil spills. The need for greater cooperation between countries to prevent future spills is crucial to limit the disastrous effects of oil spills.

The combination of oil by-products, sewage, and industrial waste threatens the North African coast. Marine biodiversity is threatened by the pollutants. In Libya and Algeria, waste from petroleum refineries and raw sewage seeps into the rivers and coastal waters, damaging ecosystems.

The European Union's 2010 bilateral trade agreement with Morocco allows for greater freedom in trade with Europe. The drawback of increased trade is that fish populations and other marine life may be endangered. Many fish populations have been in rapid decline in the Mediterranean during the past decade. For example, bluefin tuna have been overfished for decades and are at serious risk of extinction if overfishing and unsustainable fishing practices in the region are not stopped. Marine turtles have also been negatively affected by the destruction and disturbance of nesting sites.

In 1970 Egypt completed the Aswān High Dam, located about 600 miles (966 km) south of Cairo, Egypt. The dam controls the Nile's floods. It also created Lake Nasser and provides irrigation for about 3 million acres (1.2 million ha) of land while also increasing the Egyptian fishing industry. Electricity for Egypt is generated by the dam. The completion of the dam marked the first time in history that the Nile's annual flood could be controlled by humans.

Despite the economic boosts of the Aswān High Dam and Lake Nasser, these construction projects have had negative consequences. These negative impacts affect the health and livelihood of Egypt's people and their livestock. For instance, when the dam was built, it changed the ecosystem. Some plants and animals could adapt, but others could not. Ultimately, the new ecosystem provided the perfect environment for an increase in waterborne diseases such as malaria and dysentery. In addition, fertile silt has accumulated behind the dam, soil that formerly went downstream and revitalized farmland.

There are three basic concerns in much of North Africa that relate to human impact on the environment: population growth, agricultural performance, and environmental degradation. Population growth—coupled with longer life spans and the ability to feed the population—is increasingly a challenge to the subregion. Making the matter worse is the low agricultural productivity faced by many farmers in North Africa.

Land degradation from desertification is an environmental issue in all of the North African countries and is human induced. Soil erosion as a result of farming, overgrazing, and destruction of vegetation contributes to desertification. Increasingly, evidence shows that land degradation is a driver of climate change. The other causes of land degradation include drought, population pressure, and local agricultural and land use policies.

✓ READING PROGRESS CHECK

Identifying Cause and Effect How has tourism in North Africa negatively affected the environment?

The Great Man-Made River project uses pipelines to pump water from large aquifers beneath the Sahara to farms along Libya's coastal region.

▲ CRITICAL THINKING

1. *Assessing* How was the Great Man-Made project funded?

2. *Evaluating* What is the risk caused by pumping water from the underground aquifers?

ZUMA Press/Alamy

Geopolitics

Regional cooperation and international aid is important during times of regional unrest and government changes. Fuel scarcities due to the Libyan conflict in 2011 affected the amount of energy supplied to the Great Man-Made River and desalination plants. During the conflict, water shortages in Tripoli resulted from power outages, attacks on the staff managing the water network, and a reservoir held hostage by pro–al-Qaddafi fighters. United Nations agencies, the International Committee of the Red Cross, and neighboring countries such as Greece and Malta delivered millions of liters of water to Tripoli. Government resistance to change and the dependence of large populations on projects such as the Great Man-Made River exemplify how vulnerable Libya and other North African countries are to political and environmental crises.

EXPLAINING How can water supplies be vulnerable during times of political unrest and conflict?

Addressing the Issues

GUIDING QUESTION *How are environmental issues in North Africa being addressed?*

In North Africa, the environmental impacts of economic activities and growing populations is of increasing concern. Natural resources are important to economic activity, but resource extraction and harvesting are also major contributors to pollution and degradation in the region. The Middle East and North African countries (MENA) hold approximately 61 percent of the world's oil reserves. About 45 percent of the world's natural gas reserves are also located in these countries. Because the world relies on access to oil, global markets are affected by changes in the government policies of these countries. Therefore, cooperation among MENA countries can settle issues of fluctuating oil supplies to better meet the demand. Cooperation among countries can also help establish preventative measures related to oil spills and cleanup. The Partnership for African Fisheries (PAF) is working to strengthen the fishing industry. One goal of the partnership is to enforce stricter regulations to prevent overharvesting of certain species.

Determining water rights between countries is also a concern in North Africa. For example, Libya's Great Man-Made River receives water supplies from the Nubian Sandstone Aquifer System and other aquifers in the region. The aquifers are located beneath multiple political borders in the region. As a result, a joint authority for the management of the aquifer system was established in 1992. The joint authority encourages cooperation between Egypt and Libya in managing the Nubian Sandstone Aquifer System.

At the international level, the World Bank is developing plans with several countries to invest in modern irrigation practices to address the needs of agriculture. The plans call for increasing water supply sources in North Africa. The World Bank is also working to develop water policies and institutions that will improve water quality in the region. Groundwater management and wastewater collection and treatment are central to these plans. Projects in Egypt and Morocco hope to increase irrigation efficiency and implement the reuse of treated wastewater. Improving the way farmers' associations are managed is another goal of projects in North Africa. In Algeria, the World Bank is promoting the creation of independent public companies that offer water supply delivery service to small cities and towns. The World Bank also supports organizations such as the Arab Water Council as a means to gather and share information related to water supply issues.

☑ **READING PROGRESS CHECK**

Classifying Name two industries in North Africa that have required stricter regulation and enforcement to preserve natural resources.

LESSON 3 REVIEW

Reviewing Vocabulary
1. *Defining* What is an aquifer, and why is it important to the region?

Using Your Notes
2. *Summarizing* Use your graphic organizer to describe two human-made structures built to manage North Africa's resources.

Answering the Guiding Questions
3. *Evaluating* Why are water resources in such demand in North Africa?

4. *Identifying Cause and Effect* How have modern economic activities impacted North Africa's environment?

5. *Synthesizing* How are environmental issues in North Africa being addressed?

Writing Activity
6. *Argument* Using examples of economic activities or human-made projects, identify three ways in which natural resources in North Africa have been managed to profit and sustain the subregion's economies.

Directions: On a separate sheet of paper, answer the questions below. Make sure that you read carefully and answer all parts of the questions.

Lesson Review

Lesson 1

1 *Identifying Cause and Effect* Briefly explain how tectonic activity has shaped the North African landscape. Provide an example of a land formation that illustrates its effects.

2 *Explaining* Why does such a great percentage of Egypt's population make a home on such a small percentage of the country's land? In your response, identify the area where this large population concentration lives.

3 *Analyzing* Explain whether this statement is accurate: "Sand covers most of the Sahara, and the Sahara and other deserts in North Africa are unable to support any substantial vegetation."

Lesson 2

4 *Exploring Issues* Provide a summary of why it has been difficult to maintain peace in North Africa.

5 *Making Connections* Discuss influences that have led to a change in lifestyle for nomadic peoples in North Africa.

6 *Identifying Cause and Effect* Identify at least three types of crops grown in North Africa, and name the type of climate that provides desirable conditions for crop growth. Explain why harvests of major crops sometimes fail to meet people's needs.

Lesson 3

7 *Explaining* Discuss how the location of North Africa makes the countries vulnerable to oil spills and what steps can be taken to prepare for such events.

8 *Considering Advantages and Disadvantages* Explain why petroleum refining and the Aswān High Dam can both be assessed as providing benefits and detriments.

9 *Diagramming* Create a graphic organizer to compare and contrast these organizations: Middle Eastern and North African countries (MENA) and Partnership for African Fisheries (PAF). Include the reasons for the formation of each organization as well as each organization's actions.

21st Century Skills

Use the graph to answer the questions that follow.

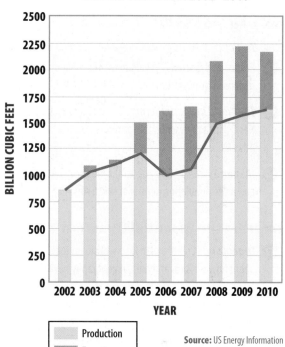

EGYPT: NATIONAL GAS PRODUCTION, CONSUMPTION AND EXPORTS FROM 2002–2010

Production
Exports
Consumption

Source: US Energy Information Administration, *International Energy Statistics*

10 *Creating and Using Graphs, Charts, Diagrams, and Tables* Explain the relationship between Egypt's natural gas production and consumption during 2002.

11 *Compare and Contrast* Explain Egypt's natural gas consumption between 2002 and 2010.

12 *Identifying and Explaining Continuity and Change* Did Egypt's natural gas consumption rise, fall, or remain the same from 2009 to 2010? What impact, if any, did this have on exports of natural gas? What is a logical reason for this impact?

Need Extra Help?

If You've Missed Question	1	2	3	4	5	6	7	8	9	10	11	12
Go to page	368	369	370	372	374	377	382	382	384	385	385	385

Directions: On a separate sheet of paper, answer the questions below. Make sure that you read carefully and answer all parts of the questions.

Critical Thinking

13 *Interpreting Significance* Discuss the impact of groundwater availability on exports and the economy in general in North Africa.

Applying Map Skills

Refer to the Unit 5 Atlas to answer the following questions.

14 *Places and Regions* What are the three major climate regions in North Africa?

15 *The World in Spatial Terms* Use your mental map of North Africa to describe the location of the Atlas Mountains.

16 *Human Systems* What generalizations can be made about the population density of North Africa?

Exploring the Essential Question

17 *Organizing* Create a plan for a website to teach students about how physical and human geography have shaped North Africa. Your goal for this project is to provide general categories with hyperlinks. Create a document that describes each page category for your site and a design for the home page. Include at least three links for each page of the website. Include the following aspects of North Africa: landforms, water systems, climates, biomes, history and government, language and religion, economic activities and trade, and environmental issues.

College and Career Readiness

18 *Reaching Conclusions* Imagine you are an intern working for a company that is developing a report to analyze the pros and cons of desalination in Algeria and Libya. Write an overview for the company, briefly explaining the process of desalination and why it is important in Algeria and Libya. Your overview should include a chart listing the pros and cons of the process for these countries. Based on the overview, write a paragraph explaining whether you think desalination should be pursued in Algeria and Libya.

DBQ Analyzing Primary Sources

Use the quote to answer the following questions.

PRIMARY SOURCE

"*Egypt is a large, complex, very important country . . .*"

—Hillary Rodham Clinton, U.S. Secretary of State, CNN News interview,
January 30, 2011

19 *Evaluating* Secretary of State Hillary Rodham Clinton made this statement in 2011. Explain whether this statement would have been equally accurate during earlier periods of history.

20 *Identifying Central Issues* What events in Egypt in 2011 would have given rise to the focus of Secretary of State Clinton and the world?

Research and Presentation

21 *Problem Solving* Utilize Internet and library sources to research information regarding the Great Man-Made River and create a multimedia presentation. Your presentation should explain the following: how and why the Great Man-Made River began, the accomplishments of the first three phases, the accomplishments of phases completed or in progress after the first three phases, and suggestions that could effectively be incorporated into a current or future phase. Within your presentation, provide an analysis as to whether the Great Man-Made River has provided effective solutions for the people of Libya. Include maps and images of places, as well as charts, diagrams, or graphs in your presentation.

Writing About Geography

22 *Argument* Use standard grammar, spelling, sentence structure, and punctuation to write a paragraph explaining why Algeria should diversify its sources of revenue.

Need Extra Help?

If You've Missed Question	**13**	**14**	**15**	**16**	**17**	**18**	**19**	**20**	**21**	**22**
Go to page	380	362	360	364	365	380	372	374	381	371

The Eastern Mediterranean

ESSENTIAL QUESTION · *How do physical systems and human systems shape a place?*

Geography Matters...

Because its fertile land supported the development of agriculture, the Eastern Mediterranean became one of the birthplaces of civilization. Many of its cities have been continuously inhabited for thousands of years, and several of the world's most widespread religions have roots here. This subregion is marked by a deep, complex history of political and religious conflict. Modern tensions have developed many distinct ideologies that reside in close proximity.

◄ The Eastern Mediterranean is home to many ethnicities and religions.

©Blaine Harrington III/Corbis

Israeli-Palestinian conflict

Conflict between Israelis and Palestinians has not evolved from one single factor such as a dispute over religion or land. Rather, the conflict has resulted from years of differences over cultural, religious, economic, and geographic issues. At the heart of the conflict lie the West Bank, the Gaza Strip, and East Jerusalem.

What is the conflict?

In the 1880s and through the 1940s many European Jews immigrated to Palestine. This immigration took place in the context of the Zionist movement, which aimed to return Jews to what they saw as their homeland. Prior to World War I the Ottoman Empire controlled Palestine, and the British controlled it after. In 1948 the British withdrew and the State of Israel was established. Israel was then attacked by Arab states in the first of several wars. In 1967 Israel captured the West Bank, Gaza, and East Jerusalem, and these areas have become the focus of demands for the creation of a Palestinian state. Agreements like the Oslo Accords of the mid-1990s have attempted to resolve the conflict.

1. **Human Systems** How could the Zionist movement have led to the creation of the State of Israel?

What is the Israeli perspective?

Palestine, more properly "The Land of Israel," is the historic home of the Jewish people, from which they were expelled by force. Even after that expulsion, some Jews continued to live there, and the yearning for return was never abandoned. After centuries of persecution culminating in the Holocaust, this hope was finally realized. The Jews accepted a 1947 U.N.-sponsored plan to partition Palestine into two states; the Arabs did not accept this and instead attacked Israel. The conquests of 1967, which include East Jerusalem and other areas that are the heartland of historic Israel, resulted from an essentially defensive war. The fact that reasonable Israeli peace initiatives have been regularly rejected reinforces other evidence that many Palestinian leaders continue to strive not for peace but for the end of Israel as a Jewish state.

2. **Places and Regions** What issues make the conflict between the Israelis and the Palestinians so complex?

What is the Palestinian perspective?

Palestinian Arabs are the indigenous people of the area. Many were driven out during the war that followed the establishment of Israel, and those that remained are treated as second-class citizens and have been denied the right to self-determination. Israeli expansionism in the wake of the 1967 war has placed even larger sections of the Palestinian homeland under foreign rule. Settlement building in the occupied territories violates international law and makes it more and more difficult to bring about a two-state solution. This activity, along with Israel's insistence on retaining all of Jerusalem, demonstrates that Israel's assertions that it seeks peace and is willing to accept a Palestinian state are insincere.

3. **Human Systems** How has conflict resulted in changing borders and demographics?

(l)Nir Kafri/AFP/Getty Images, (c)Uriel Sinai/Getty Images News/Getty Images, (r)Abbas Momani/AFP/GettyImages

The Oslo Accords of 1993 and 1995 provided for different levels of authority in the Gaza Strip and the West Bank. Though the accords are still in effect they are unpopular with both the Palestinian and Israeli public.

WEST BANK

GAZA STRIP

Double border fences and buffer zones

■ Palestinian Civil and Security Control

■ Palestinian Civil Control, Israeli Security Control

■ Israeli Civil and Security Control

■ Israeli Settlements

┅ Security Barrier

┅ Planned Security Barrier

⚠ Border Crossing

JERUSALEM

netw⊙rks

There's More Online!

☑ **IMAGE** Fish Farm in Israel

☑ **IMAGE** Jordan River Valley

☑ **MAP** Eastern Mediterranean Precipitation

☑ **INTERACTIVE SELF-CHECK QUIZ**

☑ **VIDEO** Physical Geography of the Eastern Mediterranean

Reading **HELP**DESK

Academic Vocabulary

- **range**
- **transform**

Content Vocabulary

- **rift valley**
- **kibbutz**
- **moshav**

TAKING NOTES: *Key Ideas and Details*

SUMMARIZING Use a graphic organizer like the one below to write about the water systems of the Eastern Mediterranean.

Water Systems of the Eastern Mediterranean		
The Jordan River	The Sea of Galilee	The Gulf of Aqaba

LESSON 1
Physical Geography of the Eastern Mediterranean

ESSENTIAL QUESTION · *How do physical systems and human systems shape a place?*

IT MATTERS BECAUSE

Landforms, waterways, and climate affect settlement and economic activities in any region. These factors interact in the Eastern Mediterranean, which has been home to people for centuries. It has a diverse and changing landscape. While the subregion is generally temperate, water scarcity greatly affects agricultural production and other aspects of life.

Landforms

GUIDING QUESTION *How have physical features affected the human geography of the Eastern Mediterranean?*

The Eastern Mediterranean subregion includes the countries of Syria, Jordan, Lebanon, and Israel and the Palestinian territories. This area is also known as the Levant, French for "rising," referring to the sun rising in the East. Some of the most prominent landforms in the area are the Anti-Lebanon Mountains, the Syrian Desert, the Jordan Rift Valley, and the Negev Desert.

Syria, the northernmost country in the Eastern Mediterranean, is bordered by Turkey on the north, Iraq on the east, Jordan on the south, and Lebanon and Israel on the southwest. To the southwest of Syria is a territory called the Golan Heights. This territory consists of a rocky plateau that is officially part of Syria, but most of it has been occupied by Israel since 1967.

The Anti-Lebanon mountain **range** runs along the border between Syria and Lebanon. The mountains have low population density since they have thin soil that is not useful for agriculture. They are frequented mainly by nomadic herders. The highest point in the Anti-Lebanon Mountains is Mount Hermon. At 9,232 feet (2,814 m), it is also the highest point on the land surrounding the Mediterranean Sea. The southern and western slopes of Mount Hermon, which are located in the Golan Heights, have been developed for tourist activities. A favorite tourist activity is skiing. The summit of the mountain is located in both Lebanon and Syria.

(tl)Spacephotos/age fotostock, (tr)Börje Olsson/Alamy

To the east of the Anti-Lebanon mountain range is the Syrian Desert. It covers parts of Saudi Arabia, Jordan, and Iraq. The desert is composed of gravel, not sand. Sparsely populated by nomadic tribes, the desert is used by the general public as a roadway. Southern Syria also includes high lava plains. Some of Syria's great cities, such as Aleppo, lie on the steppes, a semi-arid region east of the coastal country. Another feature is a narrow plain on the eastern edge of the Mediterranean Sea that stretches from the Turkish to the Lebanese borders.

The Syrian Desert takes up four-fifths of Jordan's total territory. The Jordan Rift Valley is located on the western border of the country. A **rift valley** is a valley formed by the separation of tectonic plates. The Jordan Rift Valley is deep, reaching 1,312 feet (400 m) below sea level on the coast of the Dead Sea. This is the lowest point on the Earth's land surface. Between the Syrian Desert and the Jordan Rift Valley is a large plateau where most cities in Jordan are located.

Lebanon is mountainous. The Anti-Lebanon Mountains, which run from north to south, cover the eastern part of the country. The Lebanon Mountains lie in the western part of the country. They also run north to south and cover almost the entire length of the country. Both the Anti-Lebanon and Lebanon Mountains run parallel to the eastern coast of the Mediterranean Sea.

The northern and central parts of Israel, which border Lebanon, are hilly regions. The beaches of western Lebanon that border the eastern Mediterranean Sea continue south into Israel. The Galilee Mountains in the north of Israel are low in comparison with the Anti-Lebanon Mountains. The Galilee area consists of rocky terrain that is unsuitable for agriculture. The Negev Desert occupies most of southern Israel. The western side of the desert connects Israel with the Sinai Peninsula. The Negev Desert lies on top of fault lines and is geographically unique because of the erosion craters that dot the landscape.

✓ READING PROGRESS CHECK

Identifying What factors have limited the settlement of the Anti-Lebanon Mountains?

Water Systems

GUIDING QUESTION *What role has water played in the human systems of the Eastern Mediterranean?*

There are several water bodies in the Eastern Mediterranean subregion. These include the Euphrates and Jordan Rivers, the Sea of Galilee, the Dead Sea, the Gulf of Aqaba, and the Mediterranean Sea itself. The Euphrates River, which originates in Turkey, is the most important river in Syria and provides the entire country with water. A complex irrigation network has watered the valley and supported farming there for some 7,000 years. The Euphrates helps irrigate Syria today. The river was dammed in the 1970s. Lake al-Assad was formed behind the dam. It was named after Syria's dictator.

The Jordan River, which has a tributary in Syria, flows through all of the countries in the Eastern Mediterranean. The river flows along

range a series of things in a row; a series of mountains

rift valley a valley formed by the separation of tectonic plates

Located in the Jordan River valley, the Dead Sea is fed by the Jordan River north of the sea as well as other smaller streams and springs.

▼ CRITICAL THINKING

1. *Analyzing Visuals* How can you tell that the Dead Sea is located in the Jordan Rift Valley?

2. *Classifying* What type of border does the Jordan River form between Jordan, Israel, and the West Bank?

Spacephotos/age fotostock

This fish farm in the Gulf of Aqaba in Israel once used cages to raise fish.

Börje Olsson/Alamy

▲ **CRITICAL THINKING**

1. *Analyzing Visuals* What other activities can be seen in the photo?

2. *Assessing* How might the fish in the cages have been contaminated given their location?

the Syrian-Lebanese border, through the Anti-Lebanon Mountains and near Mount Hermon. It then flows south into the Sea of Galilee. There the river becomes a natural border between the Israeli-occupied West Bank and Jordan before it empties into the Dead Sea. The river is important for irrigation and agriculture. Farmers grow oranges, bananas, beets, and other vegetables on the river's eastern banks.

Many of the streams in the desert regions of the Eastern Mediterranean subregion, mainly in inland areas, flow only intermittently, appearing suddenly and disappearing just as quickly. In the subregion's deserts, runoff from infrequent rainstorms creates wadis (WAH•dees), or streambeds that remain dry until a heavy rain. Irregular rainstorms often produce flash flooding. During a flash flood, wadis fill with so much sediment that they can rapidly become mudflows, or moving masses of wet soil. The mudflows pose dangers to humans and animals and can destroy agriculture.

The Sea of Galilee in western Israel supports some marine life, even though it is relatively salty for a freshwater body. The sea has supported communities of people for millennia due to its low elevation and moderate climate. The Dead Sea, which lies more than 1,300 feet (400 m) below sea level, is the lowest body of water on Earth. The salty waters make objects so buoyant that people can remain afloat in the sea without swimming.

The Gulf of Aqaba, located south of Israel and Jordan, connects the Eastern Mediterranean countries with the Red Sea and the Indian Ocean. The Mediterranean Sea allows people in the Eastern Mediterranean countries to navigate to Turkey and North Africa, as well as a number of countries in Western and Eastern Europe. Syria has a relatively short coastline that stretches between Turkey in the north and Lebanon in the south, while Lebanon and Israel have longer Mediterranean coastlines. The coastal areas are more densely populated than the interior areas of Syria and Lebanon.

Lake al-Assad is a large source of fish. It is also used for irrigation and agriculture in the surrounding lands and provides drinking water for the city of Aleppo. The southern part of the lake is home to trees such as the Aleppo pine. The lake is also a resting point for migrating birds.

✔ **READING PROGRESS CHECK**

Identifying What bodies of water does the Gulf of Aqaba connect?

Climate, Biomes, and Resources

GUIDING QUESTION *What defines the climate of the Eastern Mediterranean?*

Rainfall is limited throughout the Eastern Mediterranean subregion. Much of the area contains semi-arid (steppe) and arid (desert) climates. Areas with slightly greater moisture include the Mediterranean, highland, and humid subtropical climates.

Climate Regions and Biomes

The coastal regions of the Eastern Mediterranean have a Mediterranean climate with hot, dry summers and mild, rainy winters. Mediterranean climates support the growth of citrus fruits, olives, apples, apricots, and grapes. Natural vegetation is varied and includes oaks, evergreen conifers, and many types of wildflowers. Thickets of woody bushes and short trees known as Mediterranean shrub cover many parts of the region. Wildcats, wild boars, gazelles, hyenas, hares, badgers, and tiger weasels are found in the Eastern Mediterranean. The coastal areas are also important stopping places for more than 400 species of migratory birds.

Interior areas of the subregion have much less rainfall. As one moves inland, the climate becomes humid subtropical and then **transforms**, or changes, into semi-arid steppe and arid desert. The semi-arid climate zones usually receive about 14 inches (36 cm) of rain each year. Areas with less than 10 inches (25 cm) of rain per year are classified as deserts. Animal life in the deserts of the subregion includes geckos, lizards, and vipers. Plants in the deserts have adapted to the dry conditions and include a few scattered tree species, scrubs, and herbs. Overall, the landscape in Israel and Lebanon is extremely varied and can change within short distances. Syria and Jordan are dominated by semi-arid and arid climates.

Soils vary across the subregion just as climates do. Eastern Mediterranean coastal soils are alluvial, making them richer than the inland soils. East of the coastal mountain ranges, in the more arid inland parts of Syria and Jordan, topsoil is often lost to strong winds. These dry areas need to be irrigated in order to produce agricultural products because of the lack of precipitation. The Syrian Desert, for example, receives less than 5 inches (13 cm) of rainfall each year. Similarly, the coast of the Dead Sea is desert and cannot support many types of plant life aside from halophytes, or plants that can grow in salty soil.

transform to change in form or appearance

kibbutz a communal farm or settlement in Israel

moshav a cooperative settlement of small individual farms in Israel

GEOGRAPHY CONNECTION

Differences in annual precipitation across the Eastern Mediterranean have affected and continue to affect vegetation and human activities.

1. *PLACES AND REGIONS* How does the annual precipitation in Israel compare to that of Jordan?

2. *THE WORLD IN SPATIAL TERMS* Describe how the physical geography of Israel and Lebanon influence precipitation in the two countries.

Eastern Mediterranean Precipitation

Annual Precipitation (cm/year)
- More than 125
- 75–125
- 50–75
- 25–50
- 10–25
- 0–10

Lambert Azimuthal Equal-Area projection

Analyzing
PRIMARY SOURCES

**Lebanon's Forests:
Going up in Smoke**

"In 2002 alone, there were over 15,000 fires that became out of control in this small country. The yearly fires are a huge threat to Lebanon's already vulnerable forests.

Most forest fires in Lebanon start when a fire set by a farmer to clear his orchards and fields of grasses and stubble gets out of control. The problem has become worse over the last decade because of changing land use.

In the past, villagers used scrubby bushes from the forests as fuel for cooking, which kept the dry undergrowth thin. Now there's electricity, so the undergrowth stays thick—perfect fuel for fires."

—World Wildlife Fund,
"Lebanon's Forests: Facing
New Threats"

DBQ *ANALYZING PRIMARY SOURCES* How has technological progress resulted in threats to the forests?

Natural Resources

Many countries in the Eastern Mediterranean were once heavily wooded in the past, but have since cut down most of their trees. Cedar trees have been used by people in the region dating back to around 2500 B.C. The cedar trees are prized natural resources due to their fragrance and commercial value as a source of lumber. Most are found in the western part of the Anti-Lebanon Mountains. The cedar appears on the flag of Lebanon and is esteemed as a symbol of happiness and prosperity.

Minerals are also important to the economies of the Eastern Mediterranean subregion. Minerals—such as bromine and magnesium, gypsum from the Negev, and marble in the Galilee area—are produced. In addition, Syria produces chrome and manganese ores, asphalt, iron ore, and rock salt. Jordan has fewer mineral resources.

Recently, large oil and natural gas reserves were discovered below the waters of the Mediterranean Sea in this region, including offshore parts of Israel. There have been other finds in the Palestinian territories. These finds are still being explored, and currently, production of oil and gas is of minor importance to these countries. While petroleum exports could enrich the region, they also would make the Eastern Mediterranean economies fluctuate, or rise and fall, in the global market. Some countries in the subregion seek to diversify their economies so that they are not reliant on single commodities. Many countries are also turning to tourism as a source of revenue.

All Eastern Mediterranean countries have relied on agriculture for centuries. However, the Israeli government has always made a specialized effort to create farming communities. They supply these communities with farming equipment and irrigation education. The two types of communities are the **kibbutz** and the **moshav**. The communities have supported the cultivation of crops such as peanuts, sugar beets, and cotton in addition to dairy farming. Extensive farm mechanization has recently contributed to the increase in the value of Israel's agricultural production. Through these efforts, Israel produces a major portion of its food supply and imports the remainder.

The biggest challenge for agriculture in the Eastern Mediterranean is the lack of water. Many farmers have begun to use drip irrigation as an alternative to more costly and wasteful flood-field or spraying irrigation methods. Thin soil makes cultivation of certain crops very difficult in certain parts of the subregion. Underground springs created by cracked limestone make irrigation of the lower slopes of Lebanon and Israel very successful, however.

☑ **READING PROGRESS CHECK**

Naming What is the dominant climate of the Eastern Mediterranean?

LESSON 1 REVIEW

Reviewing Vocabulary
1. *Specifying* What geographical characteristics define the Negev Desert?

Using Your Notes
2. *Summarizing* Use your graphic organizer to discuss the water systems of the Eastern Mediterranean subregion.

Answering the Guiding Questions
3. *Making Generalizations* How have physical features affected the human geography of the Eastern Mediterranean?

4. *Evaluating* What role has water played in the human systems of the Eastern Mediterranean?

5. *Synthesizing* What defines the climate of the Eastern Mediterranean?

Writing Activity
6. *Informative/Explanatory* Write a paragraph describing plants found in the Eastern Mediterranean subregion, including the climate zones and biomes where they are found.

netw⊕rks
There's More Online!

- ☑ **IMAGE** Family and Religion in the Eastern Mediterranean
- ☑ **GRAPHS** Ethnic Groups of the Eastern Mediterranean
- ☑ **GRAPH** GDP per Capita
- ☑ **INTERACTIVE SELF-CHECK QUIZ**
- ☑ **TIME LINE** Israeli-Palestinian Conflict: An Elusive Peace Process
- ☑ **VIDEO** Human Geography of the Eastern Mediterranean

Reading HELPDESK

Academic Vocabulary

- **rely**
- **output**

Content Vocabulary

- **monotheism**
- **prophet**
- **mosque**
- **stateless nation**

TAKING NOTES: *Key Ideas and Details*

SUMMARIZNG Use a graphic organizer like the one below to record details about how the settlement of Israel affected population patterns in the Eastern Mediterranean.

Jewish Immigration · Coastal Plains → Population Patterns → Ethnicities and Religious Sects

---

LESSON 2
Human Geography of the Eastern Mediterranean

ESSENTIAL QUESTION · *How do physical systems and human systems shape a place?*

IT MATTERS BECAUSE

The Eastern Mediterranean is the birthplace of Judaism, Christianity, and Islam. Throughout history, including more recent times, the subregion has been characterized by political instability and conflict based on a combination of ethnic, religious, cultural, economic, and political factors.

History and Government

GUIDING QUESTION *How have Judaism, Christianity, and Islam shaped the politics and culture of the Eastern Mediterranean?*

As a bridge between Europe, Africa, and Asia, the Eastern Mediterranean has been influenced by cultural groups from each of these continents throughout history. The rise of influential religions has profoundly affected its history, geography, and culture.

Civilizations and Religion

The subregion has been under the control of powerful cultures and empires over the centuries. The capital city of Syria, Damascus, is one of the oldest continuously settled urban centers in the world. Judaism, Christianity, and Islam share the same territory, and they also share many beliefs, especially **monotheism**, or belief that there is only one God.

Judaism is the oldest of these monotheistic religions. Jews trace their origin to the ancient Israelites who created the kingdom of Israel along the Eastern Mediterranean coast. The kingdom, which included what is now Israel and the West Bank, had Jerusalem as its capital and religious center. Judaism teaches obedience to God's laws as the means for obtaining God's favor. These laws are described in the Hebrew Bible, which contains the Torah, the books of the prophets, and the sacred writings.

A Jewish teacher named Jesus began preaching throughout Israel. His central teachings emphasized faith in God's grace as the means for obtaining God's favor and entry into heaven after death. After Jesus' death, his teachings became the basis of a new religion—Christianity.

(tl)Ahmad Gharabli/AFP/Getty Images, (tcl)Stan Honda/AFP/Getty Images, (tcc)Marwan Naamani/AFP/Getty Images, (tcr)Alex Brandon/AFP/Getty Images, (tr)©David Turnley/Corbis

monotheism belief in one God

prophet person believed to be a messenger from God

mosque in Islam, a house of worship

The Christian Scriptures included the Old Testament, or Hebrew Bible, and the New Testament, a collection of works about the life of Jesus and the earliest Christian communities.

Islamic tradition holds that in A.D. 610 revelations from God were received by Muhammad. He was a merchant in the city of Makkah (Mecca) in the Arabian Peninsula. Muhammad began preaching that people should turn away from sin and worship the one true God. His followers claimed that he was the last in a line of **prophets**, or messengers, that included Abraham and Jesus. Muslim beliefs were recorded in the Quran, the sacred text of Islam. By the A.D. 700s, Islam had spread through the subregion and into Europe and the eastern horn of Africa. Islam had profound influences in these areas. One of the new features in the land use of cities and villages was the **mosque**, or place of worship.

Independence and Conflict

Over the centuries, Islamic empires in the Eastern Mediterranean rose and fell. Some were highly advanced civilizations that had great achievements in science, literature, and art. By the late 1800s, Western European powers conquered large areas of the subregion. These colonies generally provided few resources for the European powers until oil was discovered.

Following the defeat of the Ottomans in World War I, occupation of the Eastern Mediterranean was divided between Britain and France for protection from invasion, as well as for economic development, health and education. Lebanon and Syria gained independence from France in 1943 and 1946, respectively. Jordan gained independence from the British in 1946. The remaining disputed territory was Palestine.

Soon after World War II, armed conflicts broke out among Jewish and Arab ethnic groups in Palestine. The Jews wanted an internationally recognized homeland in a part of Palestine. The Palestinians wanted all of Palestine. The Jews accepted a 1947 United Nations (UN) plan to divide Palestine into a Jewish

©David Turnley/Corbis

TIME LINE ⌄

ISRAELI-PALESTINIAN CONFLICT
An Elusive Peace Process ➜

For decades, Israelis and Palestinians have been in conflict with one another.

▶ **CRITICAL THINKING**

1. Analyzing What are the key issues that have created conflict between Israelis and Palestinians?

2. Explaining How have other world organizations tried to support the peace process?

Oslo Accords: Palestinians recognize Israel's right to exist. Israel agrees to give considerable control of some occupied territories to Palestinians.

1993

1990 ➜

1994

Yasir Arafat becomes leader of the Palestinian territories' self-ruling government.

Israeli prime minister Yitzhak Rabin is assassinated. Shimon Peres becomes the new prime minister.

1995

July—An agreement for peace is made at a Camp David summit. Arafat cannot bring himself to accept it.

2000

➜ **2000**

2000

September—After the Camp David summit ends, a Palestinian intifada, or uprising, begins.

state and an Arab state. The Arabs rejected it, thus setting up a struggle for control of the land. Both groups had a deep religious and ethnic attachment to the land. The British withdrew from the disputed Palestinian territory, and the Jews proclaimed the independent state of Israel in May 1948.

Between 1948 and 2013, tensions over Arab opposition to Israel and Israel's concern for its security led to seven military conflicts. They involved Israel in conflict with Egypt, Jordan, Syria, Lebanon, and Hezbollah, a paramilitary group dedicated to destroying Israel. In the 1967 Israeli-Palestinian conflict, Israel gained control of the West Bank, Gaza Strip, and Golan Heights.

The wars following the birth of Israel forced many Arabs living in Palestine and Jews living in various Southwest Asian and North African countries to become refugees. The status of Arab refugees is an ongoing issue in the Israeli-Palestinian dispute. In addition, the Palestinians want an independent country of their own in the West Bank. The Palestinian people have become a **stateless nation**, an ethnic group without a formal country. Although Palestinians are religiously diverse, today most are Muslim with a significant number of Christians living near Jerusalem.

In 2006 Hamas, an Islamic political party, won elections in the West Bank and Gaza Strip. Hamas supported terrorist attacks on Israel. In 2012, in response to rocket attacks from Hamas-governed Gaza, the Israeli military launched air strikes followed by ground forces. Many targeted areas suffered extensive damage.

Independence in the 1940s was not smooth for Lebanon, Syria, or Jordan. By 1970 Lebanon had descended into a civil war based on religious politics between Christians and Muslims. Following the 1983 Israeli invasion, a group of Lebanese Shia clerics formed Hezbollah. Their goal was to drive Israel from Lebanon. The Syrian government intervened in the conflict, although Lebanese of all religions were opposed to Syrian interference. The civil war ended in 1989 with a new government that divided all administrative positions based on equal ratios of Christians and Muslims.

stateless nation an ethnic group without a formal country

Ahmad Gharabli/AFP/Getty Images

2002

June—In a speech, U.S. president George W. Bush outlines plans to revive the Israeli-Palestinian peace process.

2002 — November—The UN and the EU join with the United States and Russia to create a road map for peace.

2005

Israel disengages from the Gaza Strip, relocating all Israeli settlers.

2008

2006–2008 — Hamas rocket attacks from Gaza continue despite Israeli disengagement. Israel invades Gaza in response.

2012

After renewed fighting between Israel and Hamas, Egypt brokers a cease-fire between the two parties.

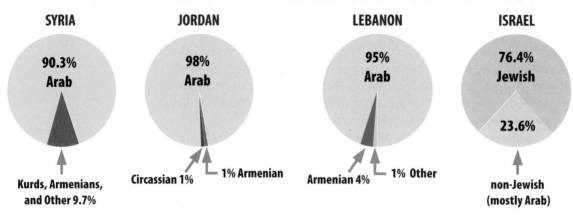

Source: CIA World Factbook

The Eastern Mediterranean is inhabited by millions of Arab and Jewish people. Arabs share similar customs and speak Arabic. Most Jewish people in the region live in Israel, where the official language is Hebrew.

▲ **CRITICAL THINKING**

1. **Contrasting** How is Israel's ethnic and religious composition different from that of other countries in the Eastern Mediterranean?

2. **Comparing** Other than large Arab populations, what do Syria, Lebanon, and Jordan have in common in terms of ethnic makeup?

While Syria was technically declared an independent country in 1946, it required a cultural convergence of ethnic groups. Territories with different religions and ethnicities that had been ruled separately by the French were joined under one government. Mergers of territory were disputed, and opposition to national independence persisted. Syria went through a series of military coups influenced by Egyptian and Iraqi politics. In 1971 Hafez al-Assad was elected president and remained in this position until 2000. His son, Bashar al-Assad, succeeded him.

In 2011, following a series of uprisings against dictators, prodemocracy Syrians rebelled against Bashar al-Assad's rule. This revolt grew out of the larger Arab Spring movement, but al-Assad's regime used lethal force against protesters and did not quickly collapse. In 2013 al-Assad's government was still resisting the revolutionary fighters.

Jordan is a constitutional monarchy. In 2012 people protested for lower food and gas prices and were successful. Jordan did not experience an "Arab Spring."

☑ **READING PROGRESS CHECK**

Listing What are the three monotheistic faiths rooted in the Eastern Mediterranean?

Population Patterns

GUIDING QUESTION *How have migrations, claims to ancestral homes, and boundary disputes affected population patterns in the Eastern Mediterranean?*

The dry, desert climate causes the majority of people in the Eastern Mediterranean to live along coastal plains. For example, over 2 million of Lebanon's 4 million inhabitants live in its capital city, Beirut, on the Mediterranean coast. The two smallest countries in the subregion, Lebanon and Israel, are more densely populated. While Syria and Jordan have lower population densities, the majority of people live in the western parts of those countries. Few people live in the deserts of the east.

The population of the subregion includes many different ethnicities and religious sects. In Lebanon, Christians of various sects make up about 40 percent of the total population. Christians make up smaller percentages in Jordan and Syria. There is also a population of Kurds in Syria. A minority of Christians live among the Muslim majority in Gaza and the West Bank. Among the major Muslim groups, the largest number adhere to the Sunni branch of the religion. In Syria and Lebanon, there are Muslims who adhere to the teachings of the Shia branch.

The demographics of the Eastern Mediterranean have changed drastically in the past century. After 1990 nearly 900,000 immigrants moved to Israel from Russia. It was the largest immigration to Israel since independence in 1948. Today over 67 percent of Israel's Jews are native born. Due to military conflicts and resettlement programs, many Palestinians were displaced beginning with the war following Israel's declaration of independence. They relocated to refugee settlements in neighboring Arab countries. In 2012 the UN reported that there were 4,797,723 registered Palestinian refugees living in 58 camps in Syria, Lebanon, and Jordan. These include several generations of descendants of the original refugees. This has been a serious humanitarian issue that hampers peace talks between Israel and the Palestinians.

☑ **READING PROGRESS CHECK**

Discussing How have demographics in the Eastern Mediterranean changed?

Society and Culture Today

GUIDING QUESTION *How has history influenced the society and culture of the Eastern Mediterranean today?*

Hebrew and Arabic are both official languages in Israel. The primary language in all other areas of the subregion is Arabic. English is also spoken in some areas.

Public schools in the subregion are generally free. Private schools, which often offer a better standard of education, require fees. Education is mandatory, but the quality differs from country to country. Literacy rates vary from 80 percent in Syria to 97 percent in Israel. Most countries require schooling until the age of 15.

Family life in the Eastern Mediterranean often includes the extended family.

▼ **CRITICAL THINKING**

1. *Classifying* Would you characterize the family shown as a nuclear family or as an extended family?

2. *Making Connections* In what ways is the family in the picture similar to yours?

Development

Development projects sponsored by non-governmental organizations (NGOs) and private donors are introducing changes to the region. Two-thirds of Syrian homes have a satellite dish providing access to foreign television broadcasts. The role of the Internet and cellular phones in mobilizing people during the Arab Spring emphasized the importance of technology. The bloggers and social media users were informed instantly of developments in the revolts in Egypt, Libya, and later Syria. The Internet also improves the quality of life for individuals. Cell phones give access to information about health care, local food markets, and transportation. Investments in the communications infrastructures contribute to a more informed, connected society.

EXPLAINING How did access to Internet-based information affect the protest movements of the Arab Spring?

rely to be dependent

output something produced; a mineral, agricultural, or industrial production

Affordable health care is something each country strives to achieve. Hospitals are often government-owned. However, such hospitals are crowded. As a result, Lebanon, Jordan, and Syria have private hospitals that provide equal or better health care at a higher cost. Both public and private health care facilities are generally located in urban areas. Rural dispensaries are available to treat minor health issues. In Israel medical insurance is mandatory for all citizens, partially subsidized by the government, and of high quality.

Family and Status of Women

Family life usually includes the extended family and often involves religious worship. Marriage, property, and in some cases divorce are a combination of civil and religious law. Religious laws differ among ethnic groups and guide arranged marriages and the division of property upon divorce.

In Israel, women generally enjoy equal rights. Women's rights in the Arab world have been changing drastically in the past few decades. In some countries women drive cars, own property, obtain advanced education, and pursue high-level jobs in government and industry. In other places, girls and women have few opportunities outside the home. They have restricted social interactions and attend schools set aside for girls. While government regulations provide legal rights for girls and women, the exercise of those rights may vary greatly from place to place. The greatest changes have been made in urban places where opportunities are more numerous. Women participated in large numbers in the Arab Spring in 2011. This participation was a clear indication of their increased influence on political and social issues.

The Arts

For thousands of years, the peoples of the Eastern Mediterranean have expressed themselves through the arts and architecture. Artists and writers have also found inspiration in religion. Muslim scholars wrote about Islamic achievements and translated Greek writings into Arabic. These works later added to European knowledge about the ancient world. Syrian, Jordanian, and Lebanese poetry, which addressed themes such as female beauty, spirituality, and love, rivaled those of the Europeans. The poetry of the early people of the subregion also influenced Roman culture and thought. Arabs have been forging new methods for creativity and expression for centuries.

Paintings by Jordanian and Syrian artists have come to center on political themes. New expressions of romantic literature have removed the past and focused on the social realism of the present. Palestinian refugees have contributed to Syrian, Jordanian, and Lebanese social realism. The art forms of Israel reflect the insecurity and tensions of a country that occupies a relatively small area.

☑ **READING PROGRESS CHECK**

Describing In what ways has art changed in Eastern Mediterranean countries?

Economic Activities

GUIDING QUESTION *What has spurred economic growth in the Eastern Mediterranean?*

Large percentages of the world's known oil and natural gas reserves lie in North Africa, Central Asia, and Southwest Asia. However, oil resources are not distributed evenly. The Eastern Mediterranean is not well endowed with oil. Economic growth has been dependent upon agriculture, relatively small deposits of minerals, and manufacturing.

Resources, Power, and Industry

In recent years Israel has been exploring deposits of natural gas offshore in the Eastern Mediterranean. In addition to its investments in natural gas, Israel maintains a diversified economy with support from the international market. Lebanon and Jordan do not have significant natural resources to export. As a result, their economies **rely** on other exports, largely agricultural products, with the help of foreign investments.

Service industries play significant roles in the subregion's economies. For example, Israel's growing high-tech sector accounts for the largest share of the country's overall manufacturing **output**. Tourism is typically another profitable industry. Ancient monuments, sacred religious sites, sunny Mediterranean beaches, and archaeological sites attract visitors. However, conflicts and instability in Syria, Israel, the Palestinian territories, and Lebanon have negatively affected tourism.

Trade and Transport

Road systems are unevenly distributed across the subregion. In some countries, mountains and deserts make road building difficult and costly. In recent years, however, economic development and the growing number of vehicles have resulted in the construction of more roads. Since World War II, the growth of the air travel industry has benefited the Eastern Mediterranean. All major cities are transportation centers for goods and people.

Water transportation is vital to the subregion. Ships load and unload cargo at ports along the Mediterranean Sea. The Strait of Tirān—which links the Gulf of Aqaba and the Red Sea—is of strategic and economic importance. It is the gateway to Jordan's only port city, Aqaba. A pipeline transports oil from Iraq to Aqaba. Canned and processed fruits and fish products are exported to many places.

☑ **READING PROGRESS CHECK**

Describing How have Eastern Mediterranean countries diversified their economies?

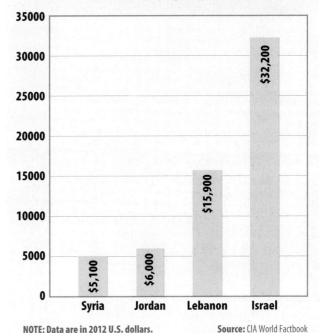

⌄ GRAPH SKILLS

GDP PER CAPITA (PPP), 2011–2012

- Syria: $5,100
- Jordan: $6,000
- Lebanon: $15,900
- Israel: $32,200

NOTE: Data are in 2012 U.S. dollars. Source: CIA World Factbook

The economies of the Eastern Mediterranean countries are based on the resources available in each country.

▲ CRITICAL THINKING

1. *Contrasting* How does Israel's GDP compare to that of the other Eastern Mediterranean countries?

2. *Assessing* How can countries that lack natural resources maintain a high GDP?

LESSON 2 REVIEW

Reviewing Vocabulary

1. *Labeling* Define the following terms and use each one appropriately in a sentence: monotheism, prophet, and mosque.

Using Your Notes

2. *Summarizing* Use your graphic organizer to describe how the settlement of Israel has affected the population patterns of the Eastern Mediterranean.

Answering the Guiding Questions

3. *Stating* How have Judaism, Christianity, and Islam shaped the politics and culture of the Eastern Mediterranean?

4. *Classifying* How have migrations, claims to ancestral homes, and boundary disputes affected population patterns in the Eastern Mediterranean?

5. *Making Connections* How has history influenced the society and culture of the Eastern Mediterranean today?

6. *Synthesizing* What has spurred economic growth in the Eastern Mediterranean?

Writing Activity

7. *Informative/Explanatory* Choose one Eastern Mediterranean country. Research to write a paragraph describing how the country's political history is linked to its economic activities today.

networks
There's More Online!

- ☑ **IMAGE** High-Tech Greenhouse in Israel
- ☑ **IMAGE** Landfill in Lebanon
- ☑ **IMAGE** Oases
- ☑ **INFOGRAPHIC** Managing Israel's Water Resources
- ☑ **INTERACTIVE SELF-CHECK QUIZ**
- ☑ **VIDEO** People and Their Environment: The Eastern Mediterranean

CONSERVATION

RAINFALL is scarce, leaving little fresh drinkable water.

DRIP IRRIGATION helps conserve water and is used in more than 90% of Israel's agriculture.

Reading **HELP**DESK

Academic Vocabulary

- **preliminary**
- **adequate**

Content Vocabulary

- **fertilizer**
- **pesticide**
- **desertification**
- **overgrazing**

TAKING NOTES: *Key Ideas and Details*

SUMMARIZING Use a graphic organizer like the one below to take notes on the ways agricultural pollutants affect the Eastern Mediterranean.

Eastern Mediterranean: People and Their Environment	
Agricultural Pollutants	Process of Desertification

LESSON 3
People and Their Environment: The Eastern Mediterranean

ESSENTIAL QUESTION · *How do physical systems and human systems shape a place?*

IT MATTERS BECAUSE
The Eastern Mediterranean subregion has a long and important history of human habitation and human impact on the environment. That impact has become an increasingly serious problem that some people and governments are currently addressing.

Managing Resources

GUIDING QUESTION *What resources are at risk in the Eastern Mediterranean?*

The Mediterranean Sea is at risk for a number of reasons. Overfishing, contamination, a rise in sea surface temperatures due to climate change, and the introduction of invasive species have all caused environmental damage. The Suez Canal connects the Red Sea to the Mediterranean Sea, allowing invasive species to enter. Invasive fish from the Red Sea have left Eastern Mediterranean reefs almost bare of native species and have greatly changed the ecology of the region. Also, the waters of the Eastern Mediterranean have been nearly depleted of fish due to overfishing. Another problem caused by human activity is water pollution. Sewage, oil spills, and chemical **fertilizers** and **pesticides** used for agriculture contaminate the Mediterranean Sea and have limited its use as a water source. The pollutants have also contributed to the vast depletion of marine life. Water pollution affects the limited number of freshwater supplies in the area and adds to the region's problem of water scarcity.

Several tectonic plates meet in the Eastern Mediterranean Sea. This contact zone is an interesting deep-sea terrain with unique landforms and aquatic life. Scientists have only recently begun to study the ecosystems of these deep waters. They still have much to learn. However, because the fish resources of the continental shelf (the areas between the shoreline and the deeper waters) are seriously depleted, Mediterranean commercial fishers are turning to these deep-sea habitats as a new source of fish. This is putting additional strain on the natural resources of the Eastern Mediterranean subregion.

(tl)Paul Gapper/Alamy, (tc)©Tomasz Grzyb/Demotix/Corbis, (tr)©Pallava Bagla/Corbis

Air pollution is also a serious problem in the Eastern Mediterranean. High levels of urbanization and immigration to the area in the past few decades have increased air pollution. Unplanned urbanization not only strains city services but also amplifies the air pollution problem produced largely by using old buses and taxis for transportation in densely populated urban communities. The unregulated burning of fossil fuels and wood for heating also contributes to air pollution. Air pollution problems have been difficult to address, mainly due to long-term unrest in the subregion and a resulting lack of environmental policies and services.

Heat waves in the subregion compound the problems of air pollution in these cities. Air pollutants have also been linked to health hazards. In the past few years, an increasing number of patients have been admitted to hospitals for respiratory diseases. Many cases of respiratory diseases are a result of poor air quality.

The wetlands of the Eastern Mediterranean are another resource at risk. They serve as an important stopping place for over 250 species of migratory birds. In Lebanon, the Aammiq Wetland is the most significant remaining freshwater wetland in the country. The area covered by the Aammiq today is only a remnant of the marshes and lakes that once existed. Bird populations in the region also face serious threat from hunters. The government of Lebanon banned hunting in 1994, but environmental groups report that the hunting laws are ignored. Soaring aquatic birds and raptors are most at risk. This is a worldwide issue since these birds migrate to and from breeding grounds outside the subregion.

☑ READING PROGRESS CHECK

Identifying What risks to human health are posed by pollution in the Eastern Mediterranean?

fertilizer a chemical or natural substance added to soil or land to increase its fertility

pesticide a chemical used to kill insects, rodents, and other pests

Although desalination is expensive, more countries are using it as alternative water sources become scarce.

▼ CRITICAL THINKING

1. *Assessing* What is Israel's annual desalination output? How much of the country's drinking water is obtained from desalination?

2. *Evaluating* Describe the progress Israel has made in regard to water conservation.

MANAGING ISRAEL'S
Water Resources

DESALINATION

50% OF DRINKING WATER COMES FROM SEAWATER

159 billion gallons = ISRAEL'S ANNUAL DESALINATION **OUTPUT**

CONSERVATION

RAINFALL is scarce, leaving little fresh drinkable water.

CONSUMPTION for personal use has decreased by 18.5% since 2008

DRIP IRRIGATION helps conserve water and is used in more than 90% of Israel's agriculture.

EDUCATION about responsible water usage has helped reduce waste.

RECYCLING

ISRAEL IS THE WORLD LEADER in water recycling, according to a United Nations report.

70% of its wastewater is recycled for use in agriculture

LEBANON · SYRIA · MEDITERRANEAN SEA · HAIFA · HADERA · SEA OF GALILEE · TEL AVIV · ASHDOD · ASHQELON · Gaza Strip · West Bank · JORDAN RIVER · DEAD SEA · JORDAN · ISRAEL · EGYPT

⬛ DESALINATION PLANT

Human Impact

GUIDING QUESTION *What human activities have affected the physical environment of the Eastern Mediterranean?*

desertification process in which arable land becomes desert

overgrazing grazing so heavily that the vegetation is damaged and the ground erodes

Long-term deforestation and desertification are other problems in the Eastern Mediterranean. **Desertification** refers to a process in which arable land, or land that is suitable for growing crops, turns into desert. Desertification is caused by a number of factors, such as **overgrazing** and deforestation. Overgrazing in certain areas of Syria and Jordan has led to the desertification of grasslands and deforestation of the mountain ranges. This change has contributed to climate change in the subregion. Overgrazing is also a significant cause of soil erosion.

Lebanon's legendary forests have a long history of exploitation. During the Middle Ages the forests were cleared for farmland, and trees were used for fuel and construction. In the early 1900s, the Ottomans controlled much of the Eastern Mediterranean. They pursued a policy of very aggressive deforestation in order to fuel their railways and keep up with industrialized Europe. The forests continue to face threats from overgrazing, unregulated tourism, and a high occurrence of forest fires. Lebanon had once been almost completely covered in forests, but by 2012 just 5 percent of the land was covered with forest. The Lebanese cedar—long a prized symbol of the country—survives today only in a small number of patches, although there have been efforts to conserve

ANALYZING PRIMARY SOURCES

Desertification in Syria

The UN has stated that 80 percent of Syria's land is at risk for desertification. The causes include both human-made and natural factors. Abdulla Tahir Bin Yehia, head of the Food and Agriculture Organization (FAO) in Syria, discusses the causes of desertification.

❝'Traditionally, communities had methods to avoid desertification, such as rotation or leaving an area unused. This allowed the vegetation to grow back,' said Bin Yehia. 'But modernization and centralization takes the decision out of their hands.'

He said rising demand for meat from a growing and increasingly affluent population was also contributing to land degradation.

'Syria's estimated livestock stands at 14–16 million. But it is only that low because many died during the drought. Prior to this the national herd stood at around 21 million. We need to study how much livestock the land can take,' said Bin Yehia.

Desertification can be irreversible, such as when an aquifer dries out and the land sinks in on itself, destroying the structure. Flora and fauna species that lose their natural habitat can become extinct. ❞

—United Nations, "Syria: Act Now to Stop Desertification, Says FAO," *Humanitarian News and Analysis*, June 15, 2010

Modern farming practices in Syria do not allow fields to lie fallow. This puts them at risk for desertification when droughts occur.

DBQ ▲ **CRITICAL THINKING**

1. Assessing How did traditional methods protect the land from desertification?

2. Describing How can desertification become irreversible?

Paul Gapper/Alamy

404

and protect these trees. This deforestation destroys animal habitats. As a result, large mammals such as wolves and wild boars—as well as migratory birds that stop in the forests on their way from Europe to Africa—are endangered.

Agriculture is one of Israel's most highly developed economic activities. It remains successful despite the lack of freshwater resources in the country. Israeli lands are not naturally suited to large-scale agricultural production. Only about 15 percent of Israel's land is arable. One major source of water for irrigation is from Lake Tiberias (the Sea of Galilee). It is an important freshwater source for the subregion.

In Israel, water is carried through canals in the National Water Carrier project built in the 1960s. Desalination plants and water recycling programs also provide water for agricultural irrigation. Israel will need to find new sources of freshwater as its agricultural sector grows. New agricultural methods to increase crop yields will also become important as land is exhausted.

This landfill in the coastal city of Saida, Lebanon, has become a major environmental hazard.

▲ CRITICAL THINKING
1. *Hypothesizing* Why might a city locate a landfill along the coast?
2. *Sequencing* How do landfills cause the pollution of waterways?

✔ READING PROGRESS CHECK

Explaining Why has Lebanon experienced so much deforestation?

Addressing the Issues

GUIDING QUESTION *How are environmental issues being addressed in the Eastern Mediterranean?*

In recent years, Eastern Mediterranean governments have made real attempts to address serious environmental issues. In 2011 the Israeli Clean Air Law came into effect. This legislation provides a framework for reducing air pollution by imposing responsibilities on national and local government authorities and the country's industrial plants. The law aims to improve air quality and protect biodiversity. Also in 2011, the Ministry of the Environment in Lebanon, working with the United Nations Environmental Program, launched a project whose goal is sustainable management of marine and coastal biodiversity.

The Jordanian government has instituted an environmental education program in schools to address that country's severe water shortage. The importance of maintaining biodiversity in the country is also taught as part of the program. The Royal Society for the Conservation of Nature (RSCN) is an organization in Jordan with broad authority to manage the natural resources of the country. It has outlawed semiautomatic hunting weapons in Jordan. The government has determined how many and what types of animals can be hunted in any given season. The RSCN has also addressed biodiversity by planning a system of wildlife reserves. One success story is that of the Arabian oryx, a type of antelope with large horns. The Arabian oryx was extinct in the

©Tomasz Grzyb/Demotix/Corbis

New agricultural methods, such as those in this high-tech greenhouse in Israel, are being developed to increase crop yields.

©Pallava Bagla/Corbis

▲CRITICAL THINKING

1. Assessing Why is it important for governments to collaborate with international organizations in developing new agricultural methods?

2. Speculating How could research help increase agricultural yields?

preliminary something done in preparation for something more important

adequate satisfactory or acceptable

wild but was reintroduced to the wild in the Azraq wetlands in the deserts of eastern Jordan in 1978. The populations have since thrived under legal protection from hunters and habitat loss.

A number of international organizations have offices throughout the Eastern Mediterranean. These include the United Nations Development Program (UNDP), the World Wildlife Fund (WWF), and the International Union for the Conservation of Nature (IUCN). Their urgent projects address desertification, unsustainable water extraction and use, biodiversity and habitat loss, and threats to sensitive marine ecosystems. The **preliminary** plan for most of these organizations is to address environmental issues by introducing legislation. They have been unsuccessful in getting most governments to take environmental projects seriously because most projects that are put on legislative agendas are severely underfunded. Local nongovernmental organizations, such as the Society for the Protection of Nature in Lebanon, also address environmental issues. However, these organizations are generally smaller and less effective than international environmental organizations.

In 2003 the WWF began a campaign to protect the cedar trees of Lebanon from fires and destruction caused by overgrazing. Fires threaten about 28 percent of Lebanon's forests. The project, which has met with some success, utilizes information and communication technologies (ICTs) to achieve its goal of protecting the endangered forests.

All governments in the subregion could more **adequately** address air and water pollution. Investing in energy-efficient public transportation systems would help alleviate the air pollution caused by urbanization. Furthermore, marine life could gradually increase in the Mediterranean Sea if countries would employ sustainable fishing methods and legislate to protect the sea from agricultural runoff and other pollution.

✓ READING PROGRESS CHECK

Describing How has the Jordanian government addressed the issue of water scarcity in the country?

LESSON 3 REVIEW

Reviewing Vocabulary

1. Naming Define overgrazing and desertification and write two sentences describing the connection between them.

Using Your Notes

2. Explaining Use your notes from your graphic organizer to describe the impact of agricultural pollutants in the Eastern Mediterranean.

Answering the Guiding Questions

3. Stating What resources are at risk in the Eastern Mediterranean?

4. Categorizing What human activities have affected the physical environment of the Eastern Mediterranean?

5. Differentiating How are environmental issues being addressed in the Eastern Mediterranean?

Writing Activity

6. Informative/Explanatory Pick one of the international organizations discussed in the lesson. Using the Internet to search its website, pick one of its Eastern Mediterranean projects. Write a paragraph identifying the main issue that the project seeks to address, why the issue is important, and what the organization plans to do about the issue.

Directions: On a separate sheet of paper, answer the questions below. Make sure you read carefully and answer all parts of the questions.

Lesson Review

Lesson 1

1 *Making Generalizations* How do water systems and water scarcity affect economic activities in the Eastern Mediterranean?

2 *Describing* How does the climate change as one travels inland, away from coastal regions, in the Eastern Mediterranean?

3 *Explaining* Why do some economies in the Eastern Mediterranean subregion often fluctuate, or rise and fall, in the global market?

Lesson 2

4 *Exploring Issues* How have political activities affected settlement and population patterns in the Eastern Mediterranean?

5 *Analyzing* What activities have encouraged economic growth in the economies of the Eastern Mediterranean?

6 *Making Connections* Why is the Eastern Mediterranean an important connection between Judaism, Christianity, and Islam?

Lesson 3

7 *Finding the Main Idea* How have human activities affected the Eastern Mediterranean environment?

8 *Making Connections* What efforts have been made to preserve natural resources in the Eastern Mediterranean?

9 *Evaluating* How successful have international organizations been at addressing environmental issues in the Eastern Mediterranean?

21st Century Skills

Use the graph to answer the following questions.

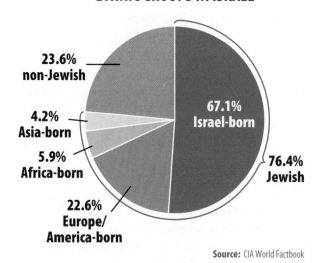

ETHNIC GROUPS IN ISRAEL

23.6% non-Jewish

4.2% Asia-born

5.9% Africa-born

67.1% Israel-born

76.4% Jewish

22.6% Europe/ America-born

Source: CIA World Factbook

10 *Creating and Using Graphs, Charts, Diagrams, and Tables* From what part of the world have most non-native Israeli Jews emigrated?

11 *Time, Chronology, and Sequencing* What factors might be responsible for attracting non-native Jews to Israel?

Exploring the Essential Question

12 *Drawing Conclusions* Using a map of the Eastern Mediterranean, describe the locations of the physical features that exist in those countries—such as lakes, mountains, and rivers—and the major capital cities. Based on this information, write a paragraph describing how geographic features and urbanization in those countries are related.

College and Career Readiness

13 *Problem Solving* Think about the issue of water scarcity in the Eastern Mediterranean. As an environmental geographer, where would you recommend desalination plants be built? Consider population centers, energy needs, and water sources. Explain your reasoning.

Need Extra Help?

If You've Missed Question	**1**	**2**	**3**	**4**	**5**	**6**	**7**	**8**	**9**	**10**	**11**	**12**	**13**
Go to page	391	393	394	398	400	395	402	405	406	407	407	390	391

Directions: On a separate sheet of paper, answer the questions below. Make sure you read carefully and answer all parts of the questions.

DBQ **Analyzing Primary Sources**

Use the document to answer the following questions.

In 1917, when Great Britain ruled parts of the Eastern Mediterranean subregion, the government issued the Balfour Declaration. This letter promised that the British government would support the Zionist effort to secure a homeland for the Jewish people. In time, the Declaration helped to lay the foundation for the establishment of the present-day state of Israel.

PRIMARY SOURCE

"*Foreign Office*
November 2nd, 1917
Dear Lord Rothschild,
I have much pleasure in conveying to you, on behalf of His Majesty's [the British] Government, the following declaration of sympathy with Jewish Zionist aspirations which has been submitted to, and approved by, the Cabinet:

'His Majesty's Government view with favour the establishment in Palestine of a national home for the Jewish people, and will use their best endeavours to facilitate the achievement of this object, it being clearly understood that nothing shall be done which may prejudice the civil and religious rights of existing non-Jewish communities in Palestine, or the rights and political status enjoyed by Jews in any other country.'

I should be grateful if you would bring this declaration to the knowledge of the Zionist Federation.

Yours sincerely
Arthur James Balfour"

—The Balfour Declaration,
November 2, 1917

14 *Evaluating* How did the Balfour Declaration contribute to the formation of a Jewish state?

15 *Identifying Central Issues* What did the British say about the rights of the people who already lived in this area?

Critical Thinking

16 *Identifying Cause and Effect* What factors have created the need for Palestinian refugee camps in the subregion?

17 *Evaluating* How have demographics in the subregion changed since 1947? Be sure to use examples from the chapter.

18 *Drawing Inferences* What might result if the Israeli agriculture industry does not focus on immediate environmental issues?

Applying Map Skills

Refer to the Unit 5 Atlas to answer the following questions.

19 *Human Systems* Which capital cities of the Eastern Mediterranean are located near major waterways?

20 *Physical Systems* Name the major mountain ranges in the Eastern Mediterranean.

21 *Environment and Society* Think about your mental map of Israel. Using this mental map, describe the location of major areas of conflict.

Research and Presentation

22 *Evaluating* Use Internet and library resources to gather information about the factors that have shaped Lebanon's current political system. Create a multimedia presentation outlining the country's current political structure, its interaction with the global economy, and the most pressing environmental issues the country is facing. The presentation should include facts, figures, and images regarding ethnic and religious diversity in the country.

Writing About Geography

23 *Informative/Explanatory* Use standard grammar, spelling, sentence structure, and punctuation to write a one-page essay describing the origins of the Israeli-Palestinian conflict. What role might international organizations play in resolving this conflict?

Need Extra Help?

If You've Missed Question	**14**	**15**	**16**	**17**	**18**	**19**	**20**	**21**	**22**	**23**
Go to page	408	408	399	399	405	361	360	361	397	396

The Northeast

ESSENTIAL QUESTION · *How do physical systems and human systems shape a place?*

◀ Turkish youth in an outdoor market in İstanbul, Turkey

Rebecca Erol/Alamy

Geography Matters...

In the Northeast, cultures are fundamentally tied to religion. Islam is the most widely practiced of them all. For hundreds of years, it has defined people's lives in the subregion. The Northeast is not homogenous, however. Many different ethnic groups live in the Northeast, including Arabs, Persians (Iranians), Turks, and Kurds.

a stateless nation:
the Kurds

A nation is a group of people who share a common language, religion, culture, and common institutions. When a national group has a state or country of its own, it is called a nation-state. Albania, Iceland, and Japan are examples of nation-states. Other nations, such as the Kurds of the Northeast, do not have their own country. They are considered "stateless nations."

The Distribution of Kurdish People

The name *Kurdistan* ("Land of the Kurds") can represent different things. Iran terms its northwestern province "Kordestān," while Iraq has an area referred to as a "Kurdish autonomous region."

The Kurds are an indigenous ethnic minority of 25–30 million living in mountainous areas in Armenia, Iran, Iraq, Syria, and Turkey.

Most Kurds live in areas of Iran, Iraq, and Turkey that form a loose geographic region known as Kurdistan. The Kurds are sometimes referred to as a stateless nation: a large ethnic group with its own culture and language that has no political state of its own.

Areas with large Kurdish populations

0 — 400 miles
0 — 400 kilometers
Lambert Azimuthal Equal-Area projection

What is a stateless nation?

The term *stateless nation* refers to a group that identifies itself as a nation based on common ethnic, linguistic, and religious identity, but that lacks majority status in any nation-state. Some stateless nations are native minority populations within a larger state. Examples of such minority populations include the Uyghur people in China and the Catalan people in Spain. Others are dispersed across several countries in numbers insufficient to form a majority in any one state. The Yoruba people of Nigeria, Benin, and Togo are an example of this latter category. Another such group is the Kurdish people. The Kurds are an Islamic pastoral people who have lived in parts of the Northeast for at least three thousand years, but have never achieved the status of a nation-state. Today they constitute a stateless nation of an estimated 25 to 30 million people. The Kurds are the largest ethnic group in the world that does not have its own country.

1. Human Systems Why are the Kurds considered a stateless nation?

Who are the Kurds?

The Kurds are the fourth-largest ethnic group in the Northeast, after Arabs, Persians, and Turks. The majority are Sunni Muslims, although some are Shia Muslims, Christians, or followers of other faiths. Kurds speak dialects of the Indo-European language family that are related to Persian. They live in an area that straddles mountain and plateau regions of eastern Turkey, northeastern Iran, northern Iraq, northwestern Syria, southern Armenia, and eastern Azerbaijan. There are also Kurdish communities in Georgia, Kazakhstan, Lebanon, and parts of Europe. For centuries, their lands were part of the vast multiethnic Ottoman and Persian Empires. After World War I, an independent Kurdish state was proposed, but that plan was abandoned. The Kurds found their lands divided among several new nation-states. Since then, Kurdish nationalist movements have led to conflicts with those countries. During the Iran-Iraq War (1980–1988), the Iraqi regime of Saddam Hussein embarked on a campaign to crush Kurdish resistance. This included the use of chemical weapons against Iraqi Kurdish civilians. Land mines planted during the war continue to take the lives of Kurds who live along the Iran-Iraq border.

2. Places and Regions Why do the Kurds want their own state?

What is the nation-state of Kurdistan?

Kurdistan literally means "Land of the Kurds." Calls for an independent Kurdish state date back to at least the 1800s. In 1880 a Kurdish sheikh in Turkey wrote to an American missionary, explaining the need for an independent Kurdish state: "We are . . . a nation apart. We want our affairs to be in our own hands." In the twenty-first century, however, there is still no Kurdish nation-state. Some Kurds (particularly educated urban residents) have formed a nationalist movement aimed at gaining autonomous (self-governing) rule. After years of brutal suppression and the fall of the regime of Saddam Hussein, Iraqi Kurds were able to gain autonomous civil authority over a section of northern Iraq, and in 2005 a Kurdish parliament was formed. Kurdish populations in Syria, Turkey, and Iran—encouraged by the success of the Iraqi Kurds—continue to push for more political autonomy.

3. Human Systems What progress have Kurds made toward achieving the goal of an autonomous nation-state?

(l)Ahmad Al-Rubaye/AFP/Getty Images, (c)©Peter Turnley/Corbis, (r)Safin Hamed/AFP/Getty Images

THERE'S MORE ONLINE

READ a primary source quote about Kurds in Iraq   •   **VIEW** an image of the Kurdish people

networks

There's More Online!

☑ **IMAGE** A River Flows Through Coastal Turkey

☑ **MAP** Physical Map: Turkey, Iran, and Iraq

☑ **INTERACTIVE SELF-CHECK QUIZ**

☑ **VIDEO** Physical Geography of the Northeast

LESSON 1
Physical Geography of the Northeast

Reading HELPDESK

Academic Vocabulary

- **exceed**
- **sustain**

Content Vocabulary

- **pastoralism**

TAKING NOTES: *Key Ideas and Details*

IDENTIFYING As you read about the physical geography of the Northeast, use a graphic organizer like the one below to record the major features of the subregion.

Landforms of the Northeast

ESSENTIAL QUESTION · *How do physical systems and human systems shape a place?*

IT MATTERS BECAUSE

The mountains that dominate the Northeast have shaped and continue to shape human history. Some of the earliest civilizations sprang from this ancient land. Today, the Northeast subregion is home to the countries of Iran, Iraq, and Turkey.

Landforms

GUIDING QUESTION *What features dominate the physical geography of the Northeast?*

The Northeast is a largely mountainous area, but it also features a range of other significant landforms and environments, particularly the broad plain of the Tigris and Euphrates Rivers in Iraq. The subregion occupies a part of the Earth where several tectonic plates converge. The movement of these plates has produced the many mountain ranges that dominate the Northeast. Movement along the fault lines produces earthquakes.

The North Anatolian Fault extends much of the distance across Turkey. At about 750 miles (1,200 km) long, it rivals the San Andreas Fault of California in length. As the plates slide past each other in an east-west direction, they can rupture and cause destructive earthquakes. Over the last seven decades, the North Anatolian Fault has been among the most active faults in the world, with nearly a dozen quakes **exceeding** 6.7 on the Richter scale of earthquake magnitude. Recently, the East Anatolian Fault in eastern Turkey has also become more active, causing several large quakes in the twenty-first century. Earthquakes are also frequent in Iran. In recent years, several devastating quakes have taken tens of thousands of lives.

Ancient volcanic activity also contributed to the mountainous terrain of the Northeast. Even today, a number of volcanoes dot the region. Turkey alone has over a dozen volcanoes within its borders.

The Pontic Mountains and the Taurus Mountains rise from the Turkish landscape. The highest peak in the Pontic range rises to 12,900 feet (3,932 m). The Taurus Mountains have many peaks which

(tl)Dorling Kindersley/Alamy, (tr)Atta Kenare/AFP/Getty Images exceed 10,000 feet (3,048 m). Between these ranges lies the Anatolian Plateau, or central massif. It stands 2,000 to 5,000 feet (610 to 1,524 m) above sea level. The plateau also has extensive sections of relatively flat terrain. East of the Pontic range, camel-backed Mount Ararat—at a height of 16,945 feet (5,165 m)—overlooks the Turkish-Iranian border.

To the east of Turkey, in Iran, lie the mountains of the Elburz and Zagros ranges. The Elburz Mountains cover the northern portion of Iran with a series of peaks that often exceed 9,000 feet (2,743 m). The tallest of these mountains is Iran's highest peak—the volcanic Mount Damāvand, which rises to 18,934 feet (5,771 m). The Zagros Mountains were formed by the collision of the Eurasian and Arabian plates. This mountain system stretches for 932 miles (1,500 km) from southwestern Iran and extending into Iraq at its northwestern end. The tallest peak in this range, Zard Kūh in Iran, is 14,921 feet (4,548 m) high.

In the western part of the subregion lies the Anatolian Peninsula, the location where the continents of Europe and Asia meet. This great peninsula is surrounded to the north by the Black Sea, to the east and south by the eastern Taurus Mountains and the Mediterranean Sea, and to the west by the Aegean Sea. It was historically referred to as Asia Minor. Today, it constitutes the Asian part of Turkey.

Over thousands of years, rivers created an alluvial plain that covers the central and southern third of Iraq. This alluvial plain is a low-lying area that is frequently marshy and covered with lakes and swamps. It is this land that produced and **sustained** the earliest civilizations in human history.

exceed to be greater than

sustain to give support to

✅ **READING PROGRESS CHECK**

Describing How has tectonic activity shaped the Northeast?

GEOGRAPHY CONNECTION

The Northeast is a largely mountainous area that includes some of the most important waterways in the world.

1. *PHYSICAL SYSTEMS* Name the two major mountain ranges in Turkey and describe their locations.

2. *ENVIRONMENT AND SOCIETY* On the map, find the Dardanelles, Sea of Marmara, and Bosporus. How are these features important to trade?

Physical Map: Turkey, Iran, and Iraq

Elevations
- 10,000 ft. (3,000 m)
- 5,000 ft. (1,500 m)
- 2,000 ft. (600 m)
- 1,000 ft. (300 m)
- 0 ft. (0 m)
- Below sea level
- — National boundary
- ▲ Mountain peak

EUROPE, Sea of Marmara, Bosporus, Black Sea, Dardanelles, TURKEY, Anatolia, Taurus Mts., Pontic Mts., Mt. Ararat ▲ 16,945 ft. (5,165 m), Caspian Sea, ASIA, Elburz Mts., Mts. Damāvand 18,934 ft. (5,771 m), Plateau of Iran, Mediterranean Sea, Tigris R., Mesopotamia, Zagros, Euphrates R., IRAQ, Kārun R., Mts., IRAN, 30°N, AFRICA, Red Sea, Persian Gulf (Arabian Gulf), Strait of Hormuz, Gulf of Oman, TROPIC OF CANCER, 30°E, 40°E, 50°E, 60°E

0 400 miles
0 400 kilometers
Lambert Azimuthal Equal-Area projection

A river flows through the Turkish plains near the Mediterranean coast.

▲ **CRITICAL THINKING**

1. *Analyzing Visuals* Based on the photo, what human purposes might this river serve?

2. *Interpreting Significance* What is the significance of this river's close proximity to the sea?

Water Systems

GUIDING QUESTION *How is the Tigris-Euphrates river system important to the human geography of the Northeast?*

The waterways of the Northeast have supported the birth of great civilizations and played vital roles in the fates of nations and empires. Linking the Aegean Sea and the Black Sea is a key waterway known as the Turkish Straits. It marks the border between Asia and Europe and consists of three smaller waterways called, from west to east, the Dardanelles, the Sea of Marmara, and the Bosporus. Through history, control of this waterway has produced military and commercial advantages for those who possessed it. The ability to stop or profit from traffic between the Mediterranean Sea and the Black Sea has been fought over repeatedly.

The Dardanelles is a narrow strait. At one point, it is only three-fourths of a mile (1 km) in width and at no point along its 38 miles (61 km) is it greater than 4 miles (6 km) wide. It opens on its eastern end to the small but deep Sea of Marmara. To the east of that point begins the Bosporus, which is another long, narrow strait. The Bosporus runs for 19 miles (31 km) before emptying into the Black Sea. The Black Sea encompasses a vast area, covering about 180,000 square miles (466,200 sq. km).

In the mountains of far eastern Turkey, within 50 miles (80.5 km) of each other, two great rivers begin their descent to the sea. These are the Tigris and the Euphrates. The rivers bound a region known since ancient times as Mesopotamia, which means "the land between the rivers." It is in this region that some of the world's earliest advanced civilizations arose. From its source, the Euphrates flows through the Taurus Mountains and southwest toward the Mediterranean coast in southern Turkey. But before reaching the Mediterranean, it turns southeast and continues in this direction for most of its 2,235-mile (3,596 km) journey across Syria and Iraq.

The shorter Tigris, at 1,180 miles (1,900 km) in length, flows from a mountain lake. It moves steadily southeast into Iraq. There, the Tigris and the Euphrates gradually converge. They almost touch near the Iraqi capital, Baghdad. They then separate and converge again. The two rivers eventually meet to form the Shatt al Arab. This river then flows the final 120 miles (193 km) to the Persian Gulf.

The dominant water feature in northern Iran is the Caspian Sea, which is slowly shrinking because of evaporation and a reduction of water flowing into the sea from the Volga River. This great inland sea forms the northern border of the country, where its waters are salty. The Caspian Sea serves as an important transportation link between Iran and other Asian countries. The Caspian also yields a number of valuable resources, notably oil and natural gas. Iran also has an active fishery in the Caspian Sea. Southern Iran is bounded by the Persian Gulf (or Arabian Gulf) and the Gulf of Oman. Between these lies the narrow Strait of Hormuz, a vital and strategic outlet from the Persian Gulf to the Indian Ocean.

✔ **READING PROGRESS CHECK**

Explaining What is the significance of the Dardanelles, the Bosporus, and the Sea of Marmara?

Dorling Kindersley/Alamy

Climate, Biomes, and Resources

GUIDING QUESTION *How do mountains influence climate in Turkey and Iran?*

The climate of the Northeast is heavily influenced by two major factors: mountain ranges and proximity to major bodies of water. Coastal and highland areas near mountain ranges usually receive the most rainfall. This is because moist, warm air is driven off the sea by the prevailing winds.

Climate Regions and Biomes

In Turkey, the areas along the coast enjoy a Mediterranean climate. Temperatures are subtropical, and typical Mediterranean scrub plants are native to the subregion. The summers are warm and dry, and the winters are mild and rainy. A similar climate prevails in the northern part of Iran along the Caspian Sea. The most valuable soils in Turkey are the alluvial soils found in the valleys, deltas, and basins of the lowland areas.

The interior of Turkey features a semi-arid steppe climate, which means it is generally dry with mild temperatures. Grasslands are the dominant vegetation type in the interior areas. This climate supports **pastoralism**—the raising of livestock—which is widely practiced in these areas. Western Iran also has an area of semi-arid steppe, although the temperatures are generally higher than in Turkey.

The high mountains of the subregion form barriers that block moisture from the major bodies of water from penetrating far inland. Thus, apart from the coasts, much of the Northeast experiences arid or semi-arid climates. Deserts cover western and southern Iraq and eastern and southern Iran.

Natural Resources

The most significant resources of the Northeast today are fossil fuels and the vital waterways that transport them. Iraq and Iran are among the world's leading producers of petroleum and natural gas. Iraq has many untapped mineral resources. Known reserves include sulfur and phosphates. Iran's major natural resources include chromium, copper, iron ore, lead, manganese, zinc, and sulfur. In addition to hydroelectric power, Turkey's natural resources include coal, iron ore, copper, mercury, and gold.

Turkey's geographic location is also an asset. In 2006 a pipeline that carries oil from an oil field in the Caspian Sea to the Mediterranean opened. Turkey benefits substantially from the transit fees. Other pipeline projects are planned to transport oil to Europe through Turkey from other countries.

☑ **READING PROGRESS CHECK**

Identifying What are the key natural resources of the Northeast subregion?

Analyzing
PRIMARY SOURCES
Iraq's Energy Resources

"Among all oil exporters, Iraq in 20 years' time will vault past Russia from third to second place, behind only Saudi Arabia, says IEA. The agency expects Iraq single-handedly to provide 45 percent of global production growth between now and 2035.

Iraq is 'very, very rich' in energy assets, said Fatih Birol, IEA chief economist and the report's main author. Iraq benefits from having some of the lowest production costs in the world and 'easy geography,' he said."

—Thomas K. Grose, "Iraq Poised to Lead World Oil Supply Growth, but Obstacles Loom," *National Geographic Daily News*, October 9, 2012

DBQ *DRAWING CONCLUSIONS*
Why would the IEA chief economist say that Iraq has "easy geography"?

pastoralism the raising of animals for food and other products

---

LESSON 1 REVIEW

Reviewing Vocabulary

1. *Identifying* Write a sentence or two explaining pastoralism and how the geography of Turkey supports it.

Using Your Notes

2. *Listing* Use your graphic organizer to describe the significance of the North Anatolian Fault.

Answering the Guiding Questions

3. *Describing* What features dominate the physical geography of the Northeast?

4. *Evaluating* How is the Tigris-Euphrates river system important to the human geography of the Northeast?

5. *Making Connections* How do mountains influence climate in Turkey and Iran?

Writing Activity

6. *Informative/Explanatory* Write a paragraph explaining why the alluvial soils of the Northeast have been so central to the development of the human geography of the subregion.

networks

There's More Online!

☑ **IMAGE** Women Wearing Burkas

☑ **MAP** Civilizations and Empires of the Northeast

☑ **TIME LINE** Independence and Turmoil

☑ **INTERACTIVE SELF-CHECK QUIZ**

☑ **VIDEO** Human Geography of the Northeast

LESSON 2
Human Geography of the Northeast

ESSENTIAL QUESTION · *How do physical systems and human systems shape a place?*

(tl)Thomas Hartwell/TIME & LIFE Images/Getty Images, (tcl)Tom Stoddart Archive/Hulton Archive/Getty Images, (tcr)Janet Wishnetsky/Alamy, (tr)Ali Al-Saadi/AFP/Getty Images

Reading HELPDESK

Academic Vocabulary

- **assume**
- **participate**

Content Vocabulary

- **natural boundary**
- **culture hearth**
- **cuneiform**
- ***qanat***
- **ziggurat**
- **embargo**

TAKING NOTES: *Key Ideas and Details*

IDENTIFYING As you read about the human geography of the Northeast, use a graphic organizer like the one below to list the subregion's historical civilizations and empires and record their characteristics.

The Northeast

Ancient Civilizations	Characteristics

IT MATTERS BECAUSE
In the Northeast, cultures are fundamentally tied to religion. Islam is the most widely practiced religion in the subregion. For hundreds of years, it has defined people's lives. Today, Islam is the fastest-growing religion in the world.

History and Government

GUIDING QUESTION *How have ancient civilizations and the discovery of oil impacted the Northeast?*

The Northeast saw the rise of several great civilizations. The earliest was Sumer, followed by the Babylonian, Persian, and Ottoman civilizations. More recently, the subregion's rich oil resources have helped make it the focus of the international quest to obtain and control energy supplies.

Civilizations and Empires

The Tigris and Euphrates Rivers formed a **natural boundary** around a historical region known as Mesopotamia. Mesopotamia was one of the world's first **culture hearths**. A culture hearth is a center from which cultures develop and then spread to other places.

The rich alluvial soils of Mesopotamia supported the development of agriculture. As humans learned to raise their own food, they were able to settle in permanent villages. Farmers could produce a surplus, which supported growing populations. In larger, settled communities, people now had time to develop new tools and organize and govern communities. In this way, civilizations emerged.

Mesopotamia was home to the Sumerian civilization that developed some 5,000 years ago. The Sumerians grew crops year-round and used canals to irrigate their fields. To keep records, they developed an early system of writing based on **cuneiform** which consists of wedge-shaped symbols pressed into clay tablets. In time the Sumerians created a system of mathematics and a code of law. About 1900 B.C., the Babylonian civilization emerged and dominated Mesopotamia.

The Persian Empire arose to the east of Mesopotamia, in what is now Iran. By 500 B.C. the Persians had conquered nearly all of the present-day territory of Iraq, Jordan, Israel, and Turkey. One of their great engineering achievements was the building of **qanats**, or underground canals. *Qanats* reduced the evaporation of water as it flowed from the mountains to farmlands.

The Ottoman Empire has its roots on the Anatolian Plateau. At its peak, the empire extended into North Africa, western Asia, and southeastern Europe. The defeat of the Ottomans and Germans in World War I ended some 600 years of rule.

Oil and the Modern Era

From the remnants of the Ottoman Empire, the modern country of Turkey was established. Turkey is aligned militarily with Western Europe as a member of the North Atlantic Treaty Organization (NATO). It has applied for EU membership.

Once part of the Ottoman Empire, Iraq became an independent state in 1932. It has experienced considerable turmoil since. This includes a war with Iran (1980–1988) and two wars against the United States and coalition forces. The most recent occurred between 2003 and 2011. It resulted in the arrest, conviction, and execution of dictator Saddam Hussein and the establishment of a parliamentary democracy.

The last Persian Empire lost most of its territory to the Ottoman Empire, but the core of Persia—modern-day Iran—remained an independent country. In 1979 a revolution overthrew the secular government headed by the shah, and an Islamic Republic was formed. Islamic scholars, known as mullahs, came to power. They continue to dominate Iranian politics. Recently, international tensions have risen as Iran is suspected of developing nuclear weapons.

Fossil fuels are a major natural resource in the subregion. Oil was discovered near the Persian Gulf in 1908. By 1911 the first commercial oil well was operating in Iran. Over time, the wells produced large quantities of oil and came under the control of countries and national oil companies, benefiting their economies.

natural boundary a boundary created by a physical feature, such as a mountain, river, or strait

culture hearth a center in which cultures develop and from which they are spread

cuneiform a system of writing using wedge-shaped symbols that were pressed into clay tablets

qanat an underground canal first built by the ancient Persians

GEOGRAPHY CONNECTION

The Northeast produced several great civilizations and empires.

1. *PLACES AND REGIONS* Which present-day country in the Northeast was not part of the Ottoman Empire?

2. *PHYSICAL SYSTEMS* Which area benefited from its physical geography, which included the Tigris and Euphrates Rivers?

Civilizations and Empires of the Northeast

Nile Valley, 3000 B.C.
Mesopotamia, 3000 B.C.
Umayyad Empire, A.D. 750
Abbasid Empire, A.D. 800
Ottoman Empire, 1640

0   1,000 miles
0   1,000 kilometers
Albers Equal-Area Conic projection

In 1960 Iran and Iraq joined with several oil-producing neighbors to form the Organization of Petroleum Exporting Countries (OPEC). This group agreed to regulate oil production to keep oil prices high. As the global demand for oil grew, OPEC **assumed**, or gained, more power over global oil prices.

assume to gain or acquire

☑ **READING PROGRESS CHECK**

Explaining How did agriculture contribute to the formation of the Sumerian civilization?

Population Patterns

GUIDING QUESTION *How have ethnic diversity and Islam shaped the population patterns of the Northeast?*

The majority of people in the Northeast are Muslims. Their religion, Islam, has a significant role in their cultures. While Islam is the primary religion, the population is ethnically diverse. Major ethnic groups include Turks, Iranians, Arabs, and Kurds.

This ethnic diversity persists because over the past 8,000 years many peoples have occupied the subregion. Each group added its own customs and beliefs to the local culture. Turkic peoples migrated to the western Anatolian Plateau from Central Asia in the A.D. 1000s. The name *Iran* means "land of the Aryans" (AR•ee•uhnz), from the ancient Indo-European people who settled in Iran.

Today the majority of people living in Iraq are Arabs, with ethnic ties to the Arabian Peninsula. The Kurds are an ethnic group that has lived for thousands of years in the mountainous border areas of Turkey, Iraq, and Iran. Kurds have no country of their own, though they call the land they live in Kurdistan. Some Kurds **participate** in efforts to achieve independence and have their own country.

participate to take part in

Turkey and Iran each have populations of about 80 million people, while about 31 million people live in Iraq. Population density ranges from about 246 people per square mile (95 people per sq. km) in Turkey to about 117 people per square mile (45 people per sq. km) in Iran. All three countries are increasingly urban.

Thomas Hartwell/TIME & LIFE Images/Getty Images

TIME LINE ∨

INDEPENDENCE
and Turmoil ➔

Turkey, Iran, and Iraq have experienced decades of struggles for political and military power in the Northeast.

▶ **CRITICAL THINKING**

1. *Analyzing* How did the Iranian Revolution affect Iran and its neighbors?

2. *Describing* What role has the United States played in events in this region?

1979

January—Shah of Iran forced to flee due to Iranian Revolution. Ayatollah Ruhollah Khomeini takes over as Iran's supreme leader.

1979 ➔

1979

November—Iranian militants seize U.S. embassy in Tehran to protest U.S. decision allowing the shah to seek medical treatment in the United States. Dozens of embassy workers are held hostage for more than a year.

July—Saddam Hussein becomes president of Iraq and begins work to secure his power as a dictator.

1979

1980

Fearing the spread of the Iranian Revolution, Iraqi forces invade Iran. The Iraq-Iran War lasts eight years.

The cities of Istanbul, Turkey; Tehran, Iran; and Baghdad, Iraq, dominate social and cultural life in their respective countries. These cities have become overcrowded due to the rapid influx of villagers seeking opportunities available in the urban centers. Iran has tried to address this problem by moving some of its government offices to towns away from the capital, Tehran.

✓ READING PROGRESS CHECK

Identifying What are the major ethnic groups in the Northeast?

Society and Culture Today

GUIDING QUESTION *How do religion and language influence life in the Northeast today?*

The legacies of the ancient civilizations of the Northeast can be seen today. Languages and customs differ depending on ancestral homes. Arabic is the most commonly spoken language in Iraq. Most Iraqi Arabs are Shia Muslims, but about 35 percent are Sunni Muslims. A small percentage is Christian or another faith. Iranians speak Farsi, also called Persian, and the majority are Shia. Turks speak Turkish, and most practice Islam. Modern Turkish culture blends Turkish, Islamic, and Western elements. Most Kurds are Sunni Muslims and speak Kurdish, a language distinct from, but related to, Farsi.

Turkey, Iraq, and Iran all support systems of free public education. The amount of education a child receives in a lifetime is 13 years in Iran, 12 years in Turkey, and 10 years in Iraq. Iran has made great strides in educating its population since the Islamic Revolution. Before the revolution, its literacy rate was under 50 percent. Today, the literacy rate is 77 percent in Iran, more than 78 percent in Iraq, and over 87 percent in Turkey.

Health care varies in the subregion. Iraq is still struggling to rebuild hospitals after years of war. In other countries, hospitals are government-owned and may suffer from doctor shortages, especially in rural areas. In recent years, private

(tr)Atta Kenare/AFP/Getty Images, (bl)Tom Stoddart Archive/Hulton Archive/Getty Images, (br)Ali Al-Saadi/AFP/Getty Images

1984
Kurdish nationalist organization, Kurdistan Workers' Party (PKK), launches armed campaign against Turkey. Actions lead to reprisals by the Turkish government and further PKK attacks.

1985
Iraq invades Kuwait. Persian Gulf War begins in early 1991 with air strikes by a U.S.-led coalition. The war ends six weeks later in an Iraqi defeat.

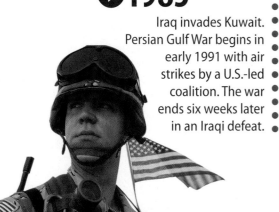

1990s
International fears grow over Iran's nuclear program and development of nuclear, chemical, and biological weapons.

1990
2003
U.S.-led coalition invades Iraq, overthrows Saddam Hussein, and turns over power to an interim Iraqi government.

2011
February—Street demonstrations erupt in Iran during the "Arab Spring."

2011
December—United States completes its withdrawal of troops from Iraq.

Some Muslim women wear a chador—a long, loose garment that covers the body—in public.

▲ **CRITICAL THINKING**

1. *Analyzing Visuals* How would wearing the chador affect women's interactions in public?

2. *Inferring* What does the chador reveal about the status of women?

ziggurat a large temple built by the Sumerians

hospitals in Turkey have begun competing with state hospitals. The competition has led to improvements in state-owned hospitals.

Family and Status of Women

Family life usually includes the extended family. In Turkey, the average household consists of a nuclear family with 3.8 persons, but the extended family remains important. The average household size in Iran is similar, with 3.6 persons per household. In recent years, family size in Iran has decreased due to a government-supported family planning program and the economic effects of embargoes intended to stop Iran's nuclear program. The average household in Iraq is much larger than in Turkey and Iran, with an average of 6.7 persons.

The conservative Islamic culture of the Northeast defines the roles and opportunities available to women. Government policies in some countries have improved the status of women. In Turkey, a 2002 law gave women equal rights with men for marriage, divorce, and property. In Iran, however, women's rights were curtailed after the establishment of the Islamic Republic in 1979. Many public places were segregated by sex, and female government workers were required to observe an Islamic dress code. In the more secular state of Iraq, women and men have equal rights under the law. However, lack of women's rights and gender equity are issues that hinder development in much of the Northeast.

The Arts

The early civilizations of the Northeast created sculptures, fine metalwork, and large buildings. In Mesopotamia the Sumerians built large, mud-brick temples called **ziggurats**. These temples were shaped like pyramids and rose high above the flat landscape. Sumerian architecture was also characterized by the construction of arches and domes.

Literature in the Northeast is based on strong oral tradition, epics, and poetry. The *Rubáiyát* by the Persian poet Omar Khayyám, who lived some 900 years ago, is the most famous example. *Osman's Dream* is an Ottoman epic dating to the 1200s that foretells of a great future empire.

The subregion also has a long tradition of fine carpet weaving that dates back to ancient Persia. Iran is still a center for the production of Persian rugs. These spectacular rugs are works of art and considered the finest made in the world today.

☑ **READING PROGRESS CHECK**

Differentiating In what ways does Iran differ from Iraq and Turkey in terms of society and culture?

Economic Activities

GUIDING QUESTION *On what natural resources do the economies of the Northeast depend?*

Oil and natural gas are the major economic resources in the Northeast. There are other valuable resources and economic activities in the subregion, however. For example, the Caspian Sea supports Iran's fishing industry. Agriculture produces many fruits and nuts, such as figs and pistachios, of very high quality and value.

Janet Wishnetsky/Alamy

Resources, Power, and Industry

Both Iran and Iraq produce far more oil than they need for domestic consumption, and so are leading exporters of oil. Revenues from oil exports are enormously important to their economies. As members of OPEC, Iran and Iraq have considerable influence in global affairs. For example, OPEC raised oil prices during the 1970s and placed an **embargo** on oil shipments to the United States and other industrialized countries. OPEC actions can also have positive effects. OPEC restored stability to the oil market and supported economic recovery during the global recession in 2009. Events unrelated to OPEC also show the influence that Iraq and Iran can have in global affairs. For example, in 2006 and 2008, oil prices rose sharply due to the ongoing conflict in Iraq and tensions over Iran's nuclear ambitions.

embargo a ban on trade

Turkey does not have the oil or natural gas resources that its regional neighbors do, but it does produce hydroelectric power. Other key industries in Turkey include textiles, food processing, and manufacturing autos and electronics. Turkey's service sector also forms a significant part of the economy.

Trade and Interdependence

The countries of the Northeast and the rest of the world depend on one another. Industrialized countries need oil from the region; the region needs industrial products for its markets. Iran operates oil-refining facilities, although Iraq's oil-refining facilities were closed or greatly restricted during the 2003–2011 U.S.-led war. Natural gas and oil pipelines are examples of the advanced transportation systems that crisscross this subregion. Much of this network is designed to move oil and gas from the fields to various ports. Since Turkey has access to the Mediterranean and the rest of Europe, many pipelines from oil fields end in Turkish ports. Other lines deliver oil to ports on the Persian Gulf and the Gulf of Oman. Much of the oil shipped from the Persian Gulf must pass through the Strait of Hormuz. The narrow strait is a choke point for shipping. Military action, a shipping accident, or political decisions could possibly close the strait. During most years, there are about 17 supertankers passing through the strait each day.

Television and radio broadcasting is expanding in the subregion, though government control of the media in many places limits programming. Advances in satellite technology are improving communication services. Wireless service and solar-powered radiophones are bringing telephone service to more people. Cell phones are now common, and many people have computer and Internet access.

☑ **READING PROGRESS CHECK**

Identifying What is the most significant natural resource in the Northeast?

LESSON 2 REVIEW

Reviewing Vocabulary

1. *Applying* Write a paragraph about the history of the Northeast that includes the words *culture hearth, cuneiform, qanat, natural boundary, embargo,* and *ziggurat.*

Using Your Notes

2. *Summarizing* Use your notes from the graphic organizer to write two paragraphs summarizing the history of the Northeast.

Answering the Guiding Questions

3. *Explaining* How have ancient civilizations and the discovery of oil impacted the Northeast?

4. *Describing* How have ethnic diversity and Islam shaped the population patterns of the Northeast?

5. *Making Connections* How do religion and language influence life in the Northeast today?

6. *Identifying* On what natural resources do the economies of the Northeast depend?

Writing Activity

7. *Informative/Explanatory* Write a paragraph comparing and contrasting modern Turkey, Iran, and Iraq. Include aspects of their economies, cultures and population patterns.

HOW HAVE SUNNI AND SHIA BELIEFS LED TO CONFLICT?

Within Islam, there are two main branches—Sunni and Shia. The divisions between these groups go back to the early years of the religion. The death of Muhammad in A.D. 632 led to a dispute over the succession of leadership in Islam. What we today call the Shia insisted that the Muslim leader, or caliph, be a descendant of Ali, the son-in-law and cousin of Muhammad, and his wife Fatima, Muhammad's daughter. Conversely, the Sunni believed that the leader should be selected by the Islamic community on the basis of personal qualities and abilities.

The split between the two groups has persisted over the centuries. Today Sunnis are the majority of Muslims. It is estimated that between 85 and 90 percent of all Muslims worldwide are Sunni. Only in Iran, Iraq, Bahrain, and Azerbaijan do Shias make up the majority of Muslims.

Iran is a country dominated by Shia Muslims. With this dominance has come some conflict—and not all of it having to do purely with matters of deeply held religious faith. In many places, Sunnis and Shias have vied for political power and sought to advance their people's interests at the expense of others.

After the Iranian Revolution, Islamic law was reintroduced. Although the new government was a republic with elements of a parliamentary democracy, it was also a theocracy. Shia clerics have final approval of legislation, with the intention of ensuring compliance with Islamic law. Iran does permit and have religious minorities. These include a small but significant Sunni Muslim population (many of whom are Kurds), Zoroastrians, Christians, and Baha'is. However, relations between the government and its Sunni minority have been complicated, and members of religious minorities are treated as second-class citizens.

Ata Kenare/AFP/Getty Images

Charges of Sunni Mistreatment in Iran

PRIMARY SOURCE

❝ Iran's Sunni Muslims . . . face widespread discrimination by Iranian authorities. The United States Commission on International Religious Freedom 2012 Report states that 'Sunni leaders have reported . . . abuses and restrictions on their religious practice, including detentions and abuse of Sunni clerics . . . and bans on Sunni teachings in public schools and Sunni religious literature, even in predominantly Sunni areas'. Sunnis are not allowed to build mosques in large cities and have been banned from conducting separate Eid prayers.

Attacks in Sunni-populated areas are not unusual. In May, one person was killed and two injured after police forces opened fire on protestors in Sistan-Baluchistan province. They were protesting against the recent arrest of local Sunni clerics. Last year, at least 12 people believed to be Sunni protestors were killed in Ahwaz, a southwestern city in the Khuzestan province, in clashes with security forces. Several Sunni mosques have been destroyed—the Abu Hanifa Mosque in Sistan-Baluchistan was bulldozed in 2008—or converted into parks. ❞

—Mona Moussavi, *ISS Voices*, October 10, 2012

Charges of Sunni Terrorism in Iran

PRIMARY SOURCE

❝ Jundallah [a pro-Sunni group] was designated as a Foreign Terrorist Organization on November 4, 2010. Since its inception in 2003, Jundallah, a violent extremist organization that operates primarily in the province of Sistan va Balochistan of Iran, has engaged in numerous attacks resulting in the death and maiming of scores of Iranian civilians and government officials. Jundallah's stated goals are to secure recognition of Balochi cultural, economic, and political rights from the government of Iran and to spread awareness of the plight of the Balochi situation through violent and nonviolent means. In October 2007, Amnesty International reported that Jundallah has by its own admission, carried out gross abuses such as hostage-taking, the killing of hostages, and attacks against non-military targets. ❞

—The United States Institute of Peace, *U.S. Terrorism Report: MEK and Jundallah*, August 23, 2011

What do you think? DBQ

1. *Drawing Conclusions* How might religious differences affect how countries interact?

2. *Identifying Cause and Effect* How might the actions of a group like Jundallah promote the denial of rights to Sunni groups in Iran?

3. *Identifying Cause and Effect* How might the actions of the Iranian government promote the activities of groups like Jundallah?

(c)©Ayman Oghanna/Corbis, (r)STR/AP Images

net**w**rks

There's More Online!

☑ **IMAGE** Drained Marshland

☑ **INFOGRAPHIC** Water Pollution in the Northeast

☑ **MAP** Tigris-Euphrates River Basin

☑ **INTERACTIVE SELF-CHECK QUIZ**

☑ **VIDEO** People and Their Environment: The Northeast

Reading **HELP**DESK

Academic Vocabulary

- link
- monitor

Content Vocabulary

- **feeder stream**
- **marsh**

TAKING NOTES: *Key Ideas and Details*

IDENTIFYING As you read about people and their environment in the Northeast, use a graphic organizer like the one below to record the environmental dangers facing the subregion.

LESSON 3
People and Their Environment: The Northeast

ESSENTIAL QUESTION · *How do physical systems and human systems shape a place?*

IT MATTERS BECAUSE

The Northeast contains valuable natural resources. Human utilization of those resources, however, can sometimes place other aspects of the environment at risk. Countries in the subregion attain variable levels of success in maximizing the benefits from their resources while minimizing the costs associated with obtaining them.

Managing Resources

GUIDING QUESTION *How and why are water resources at risk in the Northeast?*

Turkey has long utilized its many rivers for hydroelectric power and other purposes. In recent years, Turkey has been engaged in a regional development project called the Southeast Anatolia Project. The project is known as GAP in Turkey. GAP involves the construction of nearly two dozen dams along the Tigris and Euphrates Rivers and a number of their **feeder streams**, or tributaries. The plan also calls for a large number of hydroelectric stations capable of generating huge amounts of energy. The project will divert water to irrigate about 4.2 million acres (1.7 million ha) of land. Finally, GAP is also an economic development program that provides a number of benefits to a poor region of Turkey.

Such massive projects threaten existing water systems in a number of ways. The dams change the course and ecology of rivers. They also inundate lands that were once dry and divert water that would otherwise flow downstream. The effect on water quality and quantity is significant. Many ecosystems and habitats are disturbed or even destroyed.

Southern Iraq is characterized by **marshes**. The lower portions of the Tigris and Euphrates feature the Mesopotamian Marsh, which covers 15,000 square miles (38,850 sq. km). It is the largest wetlands area in the Northeast. In recent decades the Iraqi government has undertaken the draining of many of these marshlands. Marsh drainage was used to create farmland and divert water for agricultural purposes.

Essam Al-Sudani/AFP/Getty Images

Some marshlands were also drained for political reasons. The people who historically lived on these lands—the Marsh Arabs—had developed a culture and economy based upon the marsh. The Marsh Arabs became involved in uprisings against the regime of Saddam Hussein in the early 1990s. During the uprisings they took refuge in the tall reeds. The Iraqi government retaliated and drained large portions of the marshlands. The Marsh Arabs, whose cultural **link** to the land went back thousands of years, were forced to relocate.

More recently, another threat to the marshes has become a concern: drought. The problem is made worse by the heavy diversion of water for irrigation projects for agriculture in places like Turkey. These factors have lowered water levels in many wetlands areas to dangerous levels. Drought has also led to wetlands destruction in Iran.

feeder stream a tributary that feeds a larger river

marsh a wetland typically covered with grasses

link a connecting structure

✔ **READING PROGRESS CHECK**

Explaining How did politics play a part in the destruction of the Mesopotamian Marsh in the 1990s?

Human Impact

GUIDING QUESTION *What human activities result in air, land, and water pollution in the Northeast?*

Human activity impacts the environment in many ways. A major threat is the pollution of air, land, and water. Human activity can also lead to the loss or degradation of natural resources, such as soils and forests. War has also had

GEOGRAPHY CONNECTION

The Tigris and Euphrates Rivers define the boundaries of the historical region known as Mesopotamia.

1. *PLACES AND REGIONS* In what country is the Southeast Anatolia Project located?

2. *THE WORLD IN SPATIAL TERMS* What region of the Tigris-Euphrates Basin has the most irrigated land?

Tigris-Euphrates River Basin

Legend:
- Dam
- Intermittent river
- Southeast Anatolia project
- Irrigated land or planned irrigation
- Tigris-Euphrates drainage basin

TURKEY, SYRIA, IRAQ, IRAN, JORDAN, KUWAIT, SAUDI ARABIA

Mediterranean Sea, Caspian Sea, Red Sea, Persian Gulf (Arabian Gulf)

Tigris R., Euphrates R.

0 200 miles / 0 200 kilometers
Lambert Azimuthal Equal-Area projection

negative effects on the environment. During the Persian Gulf War in 1991, Iraqi troops retreating from Kuwait set fire to more than 700 oil wells. Huge black clouds of smoke polluted the area. Iraqi troops also dumped about 250 million gallons (946 million l) of oil into the Persian Gulf. Thousands of fish and other marine life died when the oil spill spread along the coastal areas of the Persian Gulf. Smoke from the oil well fires threatened millions of birds. Oil pollution from routine shipping also adversely affects the environment of the Persian Gulf.

Water pollution in the Caspian Sea is an issue. The Caspian Sea is the world's largest inland body of water, measuring 750 miles (1,200 km) from north to south with an average width of 200 miles (320 km). It has historically served as a key fishery, notable for its sturgeon and the valuable caviar. Today, however, pollution threatens these fish populations.

There are several sources of biological and chemical pollutants in the Caspian Sea. Untreated waste from the Volga River flows into the sea. The river also receives chemicals that run off from industries located along its shores, which then flow into the Caspian. In Iran it has been common practice to dump garbage from seaside cities and towns directly into the Caspian Sea. Runoff from farms—carrying pesticides, fertilizers, and detergents—also flows into the sea, adding to what is a major environmental problem.

The Caspian Sea has experienced a boom in oil and natural gas finds in recent decades. The processes used to extract these resources, however, have contributed to the increasing levels of pollution in the Caspian Sea.

▲ **CRITICAL THINKING**

1. *Evaluating* How might the pollution of water sources limit their value to people?

2. *Problem Solving* What steps could governments or individuals take to reduce water pollution in the Northeast?

Perhaps the biggest threat to the Caspian's waters comes from the large oil-extraction operations that take place throughout the sea. Much of this work is carried on by countries other than Iran that rim the Caspian. The effects of pollution from oil and gas operations have been devastating. Each year, thousands upon thousands of tons of hydrocarbons and heavy metals are spilled into the Caspian. This pollution is likely to get worse as Iran and other countries increase their exploitation of the large reserves of oil and gas beneath the Caspian Sea.

The Black Sea, which forms Turkey's northern border, is also plagued by severe pollution. The contributors are agricultural runoff and industrial pollution related to the shipping of oil. Much of this pollution comes from countries other than Turkey. Of particular concern to Turkey is the danger of an oil spill in the Bosporus. This major shipping channel sees a great deal of traffic by oil tankers. With a population of more than 10 million, Turkey's largest city—İstanbul—straddles the Bosporus. Thus, the potential for environmental disaster is high. In addition, Turkey's ability to regulate traffic in the strait is limited by international agreements designed to ensure the free flow of shipping.

In Iraq, the Tigris and the Euphrates have also suffered a serious decline in water quality. Some of this comes from agricultural runoff, as water used for irrigation flows back into the rivers. Pollution from industrial sources and sewage also threatens the rivers.

Air pollution is a problem in all of the countries of the Northeast, especially in urban areas. The population of these areas has been growing. With that increase have come concerns about air quality. Streets and roadways are frequently clogged with cars, many of which are older or poorly maintained models that produce heavy exhaust. Iran's capital, Tehran, has especially suffered from severe air pollution. In recent years there have been many instances of school and public office closures due to high air pollution levels. The Iranian government has blamed air pollution for thousands of deaths in the city.

In Iraq problems with air pollution raised the concern of U.S. military officials, who worried about the effects on American soldiers stationed there. Iraq's frequent dust storms produce large amounts of airborne particulates, many contaminated with substances such as lead, a toxin still used in Iraqi gasoline. Iraq has traditionally done little to ensure good air quality.

The health of forests and soils is threatened in several parts of the Northeast. In Iraq, for example, soils in many areas have become increasingly saline—salty—to the point that many areas have been abandoned for agricultural use. High salinity results from heavy irrigation in a place where evaporation happens quickly. Iraq has also had major problems with overgrazing and deforestation, which have left the soil exposed to seasonal rains. Many tons of soil have been carried down the Tigris and the Euphrates and been deposited in the Persian Gulf.

Iran has sizable forests in the northern part of the country. These, however, have been heavily exploited, especially in recent decades. Loss of forest resources is a major concern in the country. In 1963 Iran nationalized its forests, but that failed to stop deforestation.

Turkey also has seen many of its historical forest resources destroyed. Areas once covered with forests are now open plains. The loss of forests has brought a loss of soil. Erosion in hillier areas is a serious problem for a country that depends heavily on agriculture.

☑ **READING PROGRESS CHECK**

Describing In what ways does oil extraction pose an environmental threat to the Northeast?

Connecting Geography to **SCIENCE**

Soil Erosion in Iraq

Iraq is an ancient land that has supported agriculture and pastoralism longer perhaps than any other land on Earth. The removal of trees and the cultivation of soil began thousands of years ago. This has exposed soils in the mountain and foothill regions of northern Iraq to erosion. Massive amounts of soil have flowed down these mountains and hills over the centuries. These soils have been deposited at the mouth of the Shatt al Arab, the river formed where the Tigris and Euphrates meet and an area long contested by Iraq and Iran. Over the years, the river's delta has extended many miles into the Persian Gulf. The Iranian city of Ābādān, for example, was located on the Persian Gulf 1,000 years ago. Today it is some 30 miles (50 km) inland, situated on an island in the Shatt al Arab.

IDENTIFYING CAUSE AND EFFECT
What led to the extension of the alluvial deposits at the mouth of the Shatt al Arab?

Iraqi Marsh Arabs examine their canoes in what had once been marshland.

▲ **CRITICAL THINKING**

1. *Analyzing Visuals* How would you describe the land shown in the photo that was once covered with marshes?

2. *Making Connections* How might the changed landscape affect the lives and livelihoods of the Marsh Arabs?

monitor to watch closely, evaluate

Addressing the Issues

GUIDING QUESTION *How are environmental issues being addressed in the Northeast?*

In Turkey there are many active groups working to address environmental issues. Opposition to GAP has been strong. In Turkish universities, scientists have questioned some of the claimed benefits of the project. Activists have also sought to prevent the destruction of archaeological treasures that would occur if land is flooded for GAP. Supporters of GAP point out that hydropower is a relatively clean way to address rising energy needs. Environmentalists have placed a heavy emphasis on reducing Turkey's historical dependence on coal, which is highly polluting. There is lively debate in Turkey about the best ways to balance competing interests.

Political forces from outside Turkey are also having an impact. Turkey continues to make its case for entry into the European Union (EU). In an effort to attain this goal, it has strengthened certain environmental laws to meet EU standards.

The loss of wetlands in Iraq has prompted action from Iraqis as well as from people around the world. Organizations such as Wetlands International and the International Wetlands Conference are working on solutions to the threats. Iran has also been acting to protect its many valuable wetlands. In 1971 Iran hosted the Ramsar Convention. This meeting produced an international treaty, the Ramsar Convention on Wetlands, to **monitor**—or closely watch—and preserve wetlands resources around the world. Today, this convention brings together governments and nongovernmental organizations (NGOs) from around the world to protect wetlands of all types.

The Iranian government has also been vocal in raising concerns about pollution in the Caspian Sea. Many of these complaints, however, have been aimed at other nations who contribute to the problem. In 2013, for example, Iran threatened to sue the government of Azerbaijan over pollution coming from its oil platforms in the Caspian.

☑ **READING PROGRESS CHECK**

Evaluating What factors must Turkey balance when evaluating the benefits of its GAP project?

Essam Al-Sudani/AFP/Getty Images

LESSON 3 REVIEW

Reviewing Vocabulary

1. *Applying* Using the words *feeder stream* and *marsh*, explain the environmental challenges facing the Northeast.

Using Your Notes

2. *Describing* Use your graphic organizer to write a paragraph describing the major causes of pollution in the Northeast.

Answering the Guiding Questions

3. *Analyzing* How and why are water resources at risk in the Northeast?

4. *Describing* What human activities result in air, land, and water pollution in the Northeast?

5. *Problem Solving* How are environmental issues being addressed in the Northeast?

Writing Activity

6. *Argument* Write an argument either for or against the development of the Southeast Anatolia Project, or GAP, in Turkey.

Directions: On a separate sheet of paper, answer the questions below. Make sure that you read carefully and answer all parts of the questions.

Lesson Review

Lesson 1

1 *Analyzing* What is the meaning of the word *Mesopotamia*? Based on the location of Mesopotamia, explain whether this name is logical for the region.

2 *Interpreting Significance* Identify the dominant water feature in northern Iran. Explain its significance to the area.

3 *Making Decisions* Imagine you are considering the possibility of pursuing agricultural interests in the Northeast subregion. Would you select the steep slopes or the river basin? Explain.

Lesson 2

4 *Identifying Cause and Effect* In Mesopotamia, humans learned to raise their own food. How and why did this change population and settlement patterns?

5 *Describing* When did Turkic peoples migrate to the Anatolian Peninsula? Provide an overview of the cultural blend that constitutes modern Turkish culture and identify the language spoken and religion practiced by most Turks.

6 *Identifying Perspectives* Explain why other countries in the Northeast could benefit by following Turkey's example of an alternative energy source. Within your response, identify the specific alternative energy source used in Turkey and explain at least one likely reason that Turkey developed it.

Lesson 3

7 *Identifying Central Issues* Discuss one human factor and one factor of the physical environment that has posed threats to marshlands in Iraq.

8 *Summarizing* Summarize the threats to the Caspian Sea, explaining which is likely the greatest threat and why.

9 *Making Connections* Identify the causes and consequences of soil erosion in the Northeast.

21st Century Skills

Use the graph below to answer the following questions.

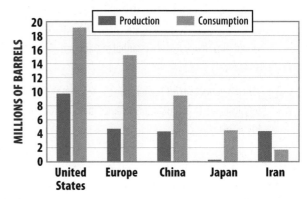

CRUDE OIL PRODUCTION AND CONSUMPTION PER DAY, 2010

Source: U.S. Department of Commerce Economics & Statistics Administration, http://www.esa.doc.gov/Blog/2012/02/22/crude-oil-and-gasoline-prices-2012

10 *Using Graphs, Charts, Diagrams, and Tables* Which country shown on the graph produced the most barrels of crude oil per day in 2010? How did this relate to its consumption?

11 *Compare & Contrast* Other than Japan and China, which country produced the fewest barrels of crude oil per day in 2010? How did this relate to its consumption? Discuss the relationships among production, consumption, and likelihood of exports. Then tie this to the likely imports of the country that produced the most barrels of crude oil per day.

Critical Thinking

12 *Exploring Issues* Write a one-page essay explaining this statement: "The modern-day Northeast is a product of early historical events in the region, though more recent events threaten the human and physical geography of the area."

Need Extra Help?

If You've Missed Question	**1**	**2**	**3**	**4**	**5**	**6**	**7**	**8**	**9**	**10**	**11**	**12**
Go to page	414	414	415	416	419	421	424	426	427	429	429	419

Directions: On a separate sheet of paper, answer the questions below. Make sure that you read carefully and answer all parts of the questions.

Applying Map Skills

Refer to the Unit 5 Atlas to answer the following questions.

13 *The World in Spatial Terms* Use your mental map of Turkey and Iraq to describe the course of the Euphrates River.

14 *Environment and Society* Use the Unit Atlas to explain where the greatest threat of pollution of the Black Sea would most likely exist—İstanbul or Baghdad—and why.

15 *The World in Spatial Terms* Use the scale on the physical map to estimate the distance in miles from Tehran to Ankara.

Exploring the Essential Question

16 *Making Connections* You are a television broadcast journalist who has been assigned a broadcast segment focusing on the following statement: "The ongoing conflicts between Shia Muslims and Sunni Muslims can be viewed through the perspectives of the physical and human geography of the region." Your broadcast report must open with a broad overview of the interaction between physical and human systems of the subregion. Write your report, and deliver it orally to the class. Include visuals, such as maps and photos.

College and Career Readiness

17 *Examining Information* You are working for a firm that has interests in Turkey. Your boss is concerned about the possibility of oil spills in the Bosporus Strait, and you have been asked to write a memo to summarize the following: the likelihood of oil spills in the Bosporus; the reasons for this likelihood; issues with cleaning up after an oil spill in the Bosporus. Within your memo, you have been asked to reference at least two other oil spills that have occurred in the world—with an explanation as to the causes and the mitigation after the spills. Prepare the memo for your boss, clearly and logically supporting your statements.

DBQ Analyzing Primary Sources

Use the excerpt to answer the following questions.

PRIMARY SOURCE

"*Iraq's energy sector holds the key to the country's future prosperity and can make a major contribution to the stability and security of global energy markets. Iraq is already the world's third-largest oil exporter and has the resources and plans to increase rapidly its oil and natural gas production as it recovers from three decades punctuated by conflict and instability. Success in developing Iraq's hydrocarbon potential and effective management of the resulting revenues can fuel Iraq's social and economic development. Failure will hinder Iraq's recovery and put global energy markets on course for troubled waters.*"

—International Energy Agency, "Executive Summary," *Iraq Energy Outlook*, 2012

18 *Interpreting Significance* Reread the following statement from the quote: "Failure will hinder Iraq's recovery and put global energy markets on course for troubled waters." What did the writer most likely mean by this statement?

19 *Drawing Conclusions* Why would oil exports alone fail to hold the key to Iraq's future prosperity?

Research and Presentation

20 *Gathering Information* Use Internet and library sources to research the Southeast Anatolia Project, known as GAP, in Turkey. Identify the purpose of GAP and note the reasons for opposition to GAP within Turkey.

Writing About Geography

21 *Narrative* Use standard grammar, spelling, sentence structure, and punctuation to write a paragraph that describes what you might see around you if you were to fly in an airplane from Ankara to an area near the summit of Mount Ararat.

Need Extra Help?

If You've Missed Question	**13**	**14**	**15**	**16**	**17**	**18**	**19**	**20**	**21**
Go to page	360	360	360	409	427	430	430	424	413

The Arabian Peninsula

ESSENTIAL QUESTION · *How do physical systems and human systems shape a place?*

netw⊙rks

There's More Online about the Arabian Peninsula's geography.

CHAPTER 18

Geography Matters...

Once a land of nomadic herders, the Arabian Peninsula entered the modern world during the twentieth century with a sudden burst of wealth and power created by the vast deposits of oil beneath its land. It is a region of contrasts. Its teeming cities perch on the edge of the sea or on inland oases, surrounded by seemingly endless deserts. The oil beneath those deserts has created almost boundless wealth. However, it is not the most precious resource on the peninsula. The most sought-after resource of all is water.

◄ Herding sheep and goats is an ancient tradition on the Arabian Peninsula.

Morales/age fotostock

431

migrant workers *in the* Arabian Peninsula

The Arabian Peninsula is flanked by the Red Sea and the Persian Gulf, far from the island countries of Indonesia and the Philippines. Yet this peninsula has drawn thousands of migrants from these and other countries who migrate here in search of work. This "guest" population, however, faces several other difficulties in addition to living thousands of miles from their homes and families.

Karim Sahib/AFP/Getty Images

Why are migrant workers coming to the Arabian Peninsula?

Migrant workers are pushed out of their homelands because of increasing poverty and unemployment. This was especially true after the global financial crisis that broke out in 2007–2008. Many migrant workers flock to the Arabian Peninsula because of the promise of wages that are much higher than they could ever earn in their native countries. Even Yemen, which has high levels of poverty, sends many workers to Saudi Arabia.

In 2010 Saudi Arabia was the world's second-largest source of jobs for migrant workers, followed by the United States. Another reason Muslim migrants in particular are attracted to the Arabian Peninsula is that they will be able to fulfill one of the pillars of Islam—the pilgrimage to Makkah (Mecca).

1. Human Systems What are the push and pull factors influencing migration to the Arabian Peninsula? How do similar religious beliefs affect the cultural geography of the subregion?

Who makes up the majority of migrant workers, and what challenges do they face?

In Indonesia, about 90 percent of the total population is Muslim, and nearly half of all migrant workers in Saudi Arabia are from Indonesia. The millions of migrant workers include women who travel each year in search of domestic work. They come from such Asian countries as the Philippines, Sri Lanka, and Indonesia, and from some African countries including Ethiopia, Egypt, and Madagascar.

Many are products of the *kafala* system, in which an employer sponsors a migrant worker by paying travel expenses and providing room and board. In some cases, however, the system is abused, and the sponsor exercises tight control over the migrant worker. Sometimes sponsors confiscate, or take control of, workers' passports. Some of the challenges migrant workers face include unsanitary living conditions, hunger, low wages, a lack of health care, and discrimination. All too often, they become the victims of human rights violations. Some critics say the system closely resembles slavery; female migrant workers, especially, are treated like property.

2. Human Systems What challenges do female migrant workers face, and what influence does the *kafala* system have on their rights?

What actions can be taken to help migrant workers?

Migrant workers have few rights and are not protected under most labor laws. Many work nearly 100 hours per week with little or no compensation for overtime work; few are given leisure time. Under the *kafala* system, they are indebted to their sponsors, who prevent them from leaving to visit family in their home countries. The workers are also prevented from seeking other employment.

International groups such as Amnesty International and other organizations are pressuring governments and organizations in the region to take action to crack down on human rights abuses. The United Nations is also investigating court rulings that have favored and empowered employers in labor disputes that involve migrant workers.

3. Human Systems Write a paragraph explaining how government and community action could improve the lives of female migrant workers in the Arabian Peninsula.

(l)Rabi Karim Photography/Flickr RF/Getty Images, (c)Ghaith Abdul-Ahad/Getty Images News/Getty Images, (r)Munir Uz Zaman/AFP/Getty Images

THERE'S MORE ONLINE

VIEW a map of migrant worker home countries • **READ** about issues of migrant workers' rights in Saudi Arabia

networks

There's More Online!

☑ **IMAGE** Red Sea Coral Reef

☑ **MAP** Physical Geography: The Arabian Peninsula

☑ **INTERACTIVE SELF-CHECK QUIZ**

☑ **VIDEO** Physical Geography of the Arabian Peninsula

LESSON 1
Physical Geography of the Arabian Peninsula

Academic Vocabulary

- **nonetheless**
- **estimate**

Content Vocabulary

- **peninsula**
- **arid**
- **dune**
- **monsoon**
- **simooms**
- *shamal*

TAKING NOTES: *Key Ideas and Details*

IDENTIFYING As you read about the landforms of the Arabian Peninsula, use a graphic organizer like the one below to identify the characteristics of each major area.

Arabian Peninsula

Area	Characteristics

ESSENTIAL QUESTION · *How do physical systems and human systems shape a place?*

IT MATTERS BECAUSE

The physical systems of the Arabian Peninsula affect the distribution of its two greatest resources, oil and water. The location and types of landforms contribute to extremes of climate. Human activities shape the peninsula through the movement of people, the creation of settlements, and the use of natural resources.

Landforms

GUIDING QUESTION What are the major physical characteristics of the Arabian Peninsula?

About 20 to 30 million years ago, Africa and Arabia formed a single tectonic plate. Over the millennia, tectonic forces created a rift valley which separated the plate into two landmasses. The valley filled with water, becoming the Red Sea. Positioned almost equally north and south of the Tropic of Cancer, this area east of the Red Sea is now the Arabian Peninsula.

The Red Sea became the western and southwestern boundaries of the newly formed **peninsula**. A peninsula is an area of land almost entirely surrounded by water. The waters of the Gulf of Aden, the Arabian Sea, the Gulf of Oman, and the Persian Gulf form natural boundaries on the southern and eastern parts of the peninsula. The southern borders of Israel, Jordan, and Iraq make up its northern boundary.

On the west, the Arabian Shield runs from north to south along the Red Sea coast. A shield is a landform composed of hard, ancient rocks. Here the Shield forms a tall mountain range. Along the western side of the Shield there is a steep escarpment, or cliff, that looms above the narrow strip of coastal plain that runs the length of the Red Sea. Extinct volcanoes cover the surface of the Shield. The lava beds produced by once-active volcanoes create the wide black bands typical of the western Arabian landscape.

To the east of the peaks of the Arabian Shield lies the Najd, the central plateau of the Arabian Peninsula. Unlike most plateaus, it does not end abruptly with a steep cliff. Instead, it slopes gently from west to east and ends at sea level on the Persian Gulf coast.

(tl)Franco Banfi/WaterFrame/Getty Images, (tr)Michele Falzone/AWL Images/Getty Images

The Arabian Peninsula is a region dominated by the **arid**, or very dry, land of its interior. The 1.2 million square miles (3 million sq. km) of the peninsula are mostly unsuitable for human settlement. In much of the peninsula, silt and dust left by winds and sediment deposited by ancient seas have created soil suitable for agriculture. **Nonetheless**, a lack of water has meant that less than two percent of the peninsula is used for agriculture.

The Arabian desert has two areas with distinct characteristics. The northern desert, An Nafūd, consists of reddish, crescent-shaped **dunes**—ridges of sand formed by wind. This desert stretches across an area of 25,000 square miles (65,000 sq. km). Far to the south of An Nafūd is the Rub' al-Khali. Known as "the Empty Quarter," the Rub' al-Khali has ten times the area of the northern desert. It is the world's largest uninterrupted area of sand. It is also one of the driest regions on Earth. These four areas—the central plateau, the northern desert, the Rub' al-Khali, and the Arabian Shield—make up most of the Arabian Peninsula.

✔ **READING PROGRESS CHECK**

Describing How did tectonic activity form the Red Sea?

Water Systems

GUIDING QUESTION *How do the waters surrounding the Arabian Peninsula affect life there?*

Thousands of miles of mostly tropical seas surround three sides of the Arabian Peninsula. The Red Sea runs south along the western edge of the Arabian Shield and then to the southeast, where it meets the waters of the Gulf of Aden. The Arabian Sea lies to the southeast of the peninsula. On the northeast lie the Gulf of Oman and the Persian Gulf. These bodies of water isolate the peninsula from

---

peninsula a portion of land nearly surrounded by water

arid excessively dry

nonetheless in spite of; nevertheless

dune a mound or ridge of sand formed by the wind

GEOGRAPHY CONNECTION

The Arabian Plate broke free of Africa millions of years ago. It is still moving northeast at a rate of approximately .039 inches (4 mm) per year.

1. PHYSICAL SYSTEMS What physical characteristics might the Arabian Peninsula share with the African continent?

2. PHYSICAL SYSTEMS What will eventually happen to the Persian Gulf as the Arabian Plate moves northeast over time?

Physical Geography: The Arabian Peninsula

Elevations

10,000 ft. (3,000 m)
5,000 ft. (1,500 m)
2,000 ft. (600 m)
1,000 ft. (300 m)
0 ft. (0 m)
Below sea level

— National boundary
➡ Direction of plate movement
〰 Tectonic plate boundary

EURASIAN PLATE

An Nafūd

Persian Gulf (Arabian Gulf)

plate movement

ARABIAN SHIELD

Red Sea

NAJD

ARABIAN PLATE

Gulf of Oman

TROPIC OF CANCER

opening ocean

Rub' al-Khali

20°N

Arabian Sea

INDIAN PLATE

AFRICAN PLATE

0   500 miles
0   500 kilometers
Lambert Azimuthal Equal-Area projection

Gulf of Aden

SOMALI PLATE

10°N

40°E   50°E   60°E   70°E

In 1902 an American named Willis Carrier invented air-conditioning. At first it was used commercially in the United States in public buildings like movie theaters and hotels. It was a boon to productivity in the summertime. Air-conditioning allowed cities in the southern United States to attract industries from other parts of the country. On the Arabian Peninsula, air-conditioning made it possible for people from temperate climates to work during the summers in the hot, steamy coastal cities or on the oil pipelines that cross the deserts. In recent years, the demand for air-conditioning on the Arabian Peninsula has increased as industries and tourism have expanded.

ANALYZING How does air-conditioning affect productivity in hot weather?

monsoon a seasonal wind that brings warm, moist air from the oceans in summer and cooler, dry air from inland in winter

simoom a hot, dry, suffocating wind that blows from time to time in the Arabian Peninsula

shamal a northwesterly wind in the Persian Gulf area

its neighbors to the east and west. However, they are vital to the life of the peninsula. The Red Sea and the Persian Gulf, in particular, are necessary to the economies of the region.

The Red Sea, on the western side of the Arabian Peninsula, is the northernmost tropical sea in the world. It is also an important shipping route. Ships entering from the Suez Canal in the north carry goods through the Red Sea to the Gulf of Aden. Most of the products transported between Europe and Asia pass through the long and narrow Red Sea.

On the east, the Persian Gulf serves the international shipping needs of all the oil-producing countries of the peninsula. The warm waters of the Persian Gulf are crowded with traffic. Huge tankers bring oil from Saudi Arabia, Kuwait, Qatar, Bahrain, and the United Arab Emirates to the rest of the world.

Although oil is plentiful on the Arabian Peninsula, freshwater is scarce. The subregion receives little rainfall. Rapidly growing populations are quickly using up groundwater, the water in the earth that supplies wells and springs. Dotted across the Arabian landscape are small oases, which are gradually going dry as groundwater is depleted. Small, temporary rivers form in wadis when there is rain. Wadis are streambeds that remain dry when no rain falls. They are not reliable sources of water because rainfall is unpredictable.

☑ **READING PROGRESS CHECK**

Locating Where are the principal bodies of water that border the Arabian Peninsula?

Climate, Biomes, and Resources

GUIDING QUESTION *How does the climate of the Arabian Peninsula affect its biomes?*

For the most part, the Arabian Peninsula is a desert with an extremely dry climate. In much of the peninsula, the average rainfall is less than 4 inches (10 cm) per year. The temperature in some places has been recorded as high as 129°F (54°C). The noontime sun is directly overhead at least once a year for land on or south of the Tropic of Cancer. The heat of the sun combined with the lack of moisture for cloud cover makes the peninsula one of the hottest places on Earth.

Climate Regions and Biomes

A hot and dry desert biome like the Rub' al-Khali has little rainfall. There is no surface water. The little rain that falls often evaporates before it reaches the ground. The Rub' al-Khali has few forms of animal or plant life. Some of the desert areas in the northern part of the peninsula have a greater abundance of plants and animals. These areas are similar to the Mojave and Sonoran Deserts in the United States.

The dry heat of the desert areas in summer contrasts with the extreme humidity of some coastal areas, which are affected by their closeness to water. Dew and fog add to the humidity. Seasonal **monsoon** winds bring heavy rains in the summer.

Wind is a significant factor in the climate of the Arabian Peninsula. The heat of the deserts in the summer sometimes generates hot, suffocating winds called **simooms**. The simooms last only about 20 minutes, but they destroy plant and animal life. These winds are so dangerous, in fact, that the word *simoom* actually means "poison" in Arabic. In areas near the Persian Gulf, sand- and dust-laden winds called **shamals** occur in midwinter and early summer. The word *shamal* means "north," and the winds blow from the north or northwest. These winds carry millions of tons of sand and silt into the Rub' al-Khali.

The Red Sea is an aquatic biome. The coral reefs shelter about 1,200 different species of fish and many species of plants. About 10 percent of the species found in the Red Sea are not found in any other marine environment. This aquatic biome is unique because its corals can endure extreme heat and salinity. The salinity is high because of high rates of evaporation and a lack of rain over the area. Violent, dust-laden winds churn the waters in ways that would prove fatal to other reefs. In spite of their hardiness, the reefs are subject to the same threats as other aquatic biomes: heavy pollution and overuse.

Natural Resources

At one time, the Earth was a huge aquatic biome. Animal and plant life teemed in ancient seas. Over time, the remains of these organisms were buried under the sedimentary rock. Millions of years of intense heat and pressure eventually transformed the remains of these organisms into crude oil. This fossil fuel is one of the world's most sought-after resources.

The Arabian deserts lie over a vast deposit of crude oil. This deposit is **estimated** to be at least 25 percent of the world's proven reserves, and oil is the region's principal export. Discovered in Saudi Arabia in 1938, rapid expansion of the oil industry did not occur until after World War II. Of the seven countries of the Arabian Peninsula, only Bahrain does not have significant petroleum reserves.

The Arabian Peninsula has few other significant resources. Some of these resources include fish, pearls, and salt and gypsum produced from saline flats. Mineral resources include iron ore, gold, copper, limestone, and some marble.

The resource that is most crucial to the peninsula, however, is freshwater. Drinkable water is rapidly being depleted because the population has grown dramatically. Aquifers beneath the desert sands cannot be replenished at the rate they are being used. The countries of the area are looking for ways to increase their supplies of this critical resource.

✔ READING PROGRESS CHECK

Classifying What characteristics of the Rub' al-Khali make it a desert biome?

The Red Sea is known for its extensive coral reefs.

▲ CRITICAL THINKING

1. *Hypothesizing* Why would it be important to protect the Red Sea reefs?

2. *Describing* What resources does the Red Sea provide?

estimate to judge approximately; to determine roughly

LESSON 1 REVIEW

Reviewing Vocabulary

1. *Comparing* What do *shamals* and simooms have in common?

2. *Contrasting* How do monsoons differ from *shamals* and simooms?

Using Your Notes

3. *Analyzing* Use your graphic organizer on the Arabian Peninsula's landforms to write a paragraph analyzing the characteristics that make some areas of the peninsula more suitable for some activities than others.

Answering the Guiding Questions

4. *Differentiating* What are the major physical characteristics of the Arabian Peninsula?

5. *Evaluating* How do the waters surrounding the Arabian Peninsula affect life there?

6. *Making Connections* How does the climate of the Arabian Peninsula affect its biomes?

Writing Activity

7. *Explanatory* In a paragraph, discuss why the Red Sea and the Persian Gulf are vital to the economy of the Arabian Peninsula.

Franco Banfi/WaterFrame/Getty Images

networks

There's More Online!

- ☑ **TIME LINE** Islam and Saudi Arabia
- ☑ **CHART** Arabian Peninsula Country Statistics
- ☑ **IMAGE** Sanaa, Yemen
- ☑ **INTERACTIVE SELF-CHECK QUIZ**
- ☑ **VIDEO** Human Geography of the Arabian Peninsula

(tl)Hassan Ammar/AFP/Getty Images, (tc)Popperfoto/Getty Images, (tr)Roslan Rahman/AFP/Getty Images

Reading **HELP**DESK

Academic Vocabulary

- **document**
- **strategy**

Content Vocabulary

- **sheikdoms**
- **Sunni**
- **Shia**
- **Ibadhism**
- **shari'ah**
- **hajj**
- **choke point**

TAKING NOTES: *Key Ideas and Details*

ORGANIZING As you read the lesson, take notes on how the people of the Arabian Peninsula have been governed in the past and how they are governed today.

Arabian Peninsula: Government

Past	Today

LESSON 2
Human Geography of the Arabian Peninsula

ESSENTIAL QUESTION · *How do physical systems and human systems shape a place?*

IT MATTERS BECAUSE

History and physical geography have had profound effects on the Arabian Peninsula. A similar history, religion, and culture have shaped the lives of its people. The discovery of vast oil and natural gas resources in the twentieth century produced widespread changes and major challenges to the human population, as well as effects on the physical environment.

History and Government

GUIDING QUESTION *How did history, culture, and geography help influence the types of governments on the Arabian Peninsula?*

The people of the Arabian Peninsula are overwhelmingly ethnic Arabs and religious Muslims. Religion, ethnicity, and a shared culture unite and define the subregion. From the earliest times, the people of the peninsula lived in tribes—social groups based on family relationships. Individual tribes controlled specific areas of the region, called **sheikdoms**. Some tribes were sedentary, and others were nomadic.

The bedouin nomads of Saudi Arabia had a tribal structure but were not confined to a single place. Bedouins occupied most of the Arabian Peninsula. Primarily herders of camels, goats, and sheep, the bedouin tribes moved throughout the Arabian Peninsula and beyond. Different tribes of nomads became powerful from time to time and dominated much of the area. Nevertheless, there was no formal government. Each tribal leader had absolute power, and there was no authority above him.

Similarly, powerful families controlled the settled areas along the eastern and southern coasts. These lands were more vulnerable to foreign control than the vast area of the interior. Throughout history, foreigners invaded these areas. The ancient Greeks established a trade center in the Persian Gulf near what is now Kuwait. The Ottoman Empire was a strong presence on the Arabian Peninsula for several hundred years until the

Saud family finally drove them from the northern part of the peninsula in the early twentieth century. Despite the interference of foreigners, however, the power of the families remained.

sheikdom territory ruled by an Arab tribal leader

Today, the majority of governments in the Arabian Peninsula are monarchies. Saudi Arabia, Oman, and Qatar are absolute monarchies. Kuwait and Bahrain are constitutional monarchies. The United Arab Emirates is a federation of emirates, formed from seven smaller states that were headed by emirs, or Arab rulers. In 1990 North Yemen and South Yemen were united to form a republic. However, southern secessionists and northern insurgents have continued to oppose the union. By 2012, civil unrest and violence had destabilized the government. The activities of al-Qaeda militants increased in number in recent years, and ocean piracy along the coasts remains a problem.

✔ **READING PROGRESS CHECK**

Drawing Conclusions How has the role of nomadic tribes affected the way that most people are governed in the Arabian Peninsula?

Population Patterns

GUIDING QUESTION *How have climate and history helped determine the population patterns of the Arabian Peninsula?*

The harsh climate of the Arabian Peninsula has had a strong influence on human settlement. Most of the subregion's population lives along the coasts of the Persian Gulf and the Red Sea. Bedouin herders once moved from oasis to oasis in the arid interior of the peninsula. Today, most bedouins have settled in towns and cities.

The population of the Arabian Peninsula was once almost exclusively Arab. Today, as the table shows, some countries in the subregion are more heterogeneous. The makeup of the population began to change with the development of the petroleum industry. The oil fields and construction jobs attracted guest workers from South Asia, mainly from India and Bangladesh. Poverty and unemployment in South Asia pushed these workers toward the oil-wealthy Arab countries. The prospect of high wages continues to pull guest workers to the subregion despite the long work hours and crowded housing conditions.

Arabian Peninsula Country Statistics				
Country	**Population**	**Median Age**	**Principal Ethnic Groups**	**GDP Billions $**
Bahrain	1,248,349	30	Bahrainis, Asians	31.5
Kuwait	2,646,314	28	Kuwaiti, other Arabs, South Asians, Iranians	155.5
Oman	3,090,150	25	Arabs, Baluchis, South Asians, Africans	82.8
Qatar	1,951,591	32	Arabs, South Asians, Iranians	184.3
Saudi Arabia	26,534,504	26	Arabs, Asians, Africans	691.5
United Arab Emirates	5,314,317	30	South Asians, other Arabs and Iranians, Emiratis	262.1
Yemen	24,771,809	17	predominantly Arabs	58.7

Source: World Almanac and Book of Facts 2013

⌄ **CHART SKILLS**

This chart shows some population statistics for countries of the Arabian Peninsula. It also gives each country's GDP, or gross domestic product. GDP is a measure of a country's wealth.

◀ **CRITICAL THINKING**

1. ***Differentiating*** In what ways does Yemen's population differ from that of other Arabian Peninsula countries?

2. ***Hypothesizing*** Why might few immigrants be attracted to Yemen?

Before the discovery of oil in 1932, the cities of the Arabian Peninsula were already important economic centers. Their populations were small, however, compared to the numbers of people living in rural areas and in desert towns and villages. In addition, there were smaller clans of nomads grazing sheep, camels, and goats where there was vegetation for the animals.

Today about 80 percent of the population in the Arabian Peninsula lives in cities. Most of these cities, like Jidda (Saudi Arabia), Kuwait (Kuwait), and Doha (Qatar), are located on either the Red Sea or the Persian Gulf. They thrive because they are important centers of their countries' oil industries.

One major city is located in the interior, away from the coast. Riyadh, a center of oil refining, is also the capital of Saudi Arabia and home to the royal palace of the king. Other important cities are near the coast. Makkah (Mecca) is the holiest city of Islam. Millions of religious pilgrims visit the city every year. Lying 50 miles (80 km) from the Red Sea, it houses the Grand Mosque. Madinah (Medina), which is farther inland, is several hundred miles north of Makkah. The tomb of Muhammad is in Madinah. The city is considered a holy city second in importance only to Makkah.

☑ **READING PROGRESS CHECK**

Drawing Conclusions Why does most of the population live along the coast?

Society and Culture Today

GUIDING QUESTION *How does Islam affect life on the Arabian Peninsula?*

Arabic is the language common to all Arab people of the Arabian Peninsula. Many people also speak a second language, often English. Immigration to the peninsula has added to the number of languages spoken, but most immigrants learn the Arabic necessary to function in their jobs and everyday life.

Hassan Ammar/AFP/Getty Images

TIME LINE ⌄

ISLAM and
Saudi Arabia ➡

In the seventh century, Islam originated in Makkah, a city on the Arabian Peninsula. Islam had a profound influence over the culture and governance of the peoples of the Arabian Peninsula.

▶ **CRITICAL THINKING**

1. *Describing* Why did Islam split into two branches?

2. *Analyzing* Why do millions of Muslims travel to Saudi Arabia each year?

500 ➡

610 According to Islamic teachings, an angel tells Muhammad that he is chosen to be Allah's prophet.

570 Muhammad is born in the city of Makkah.

622 To escape persecution, Muhammad and his followers leave Makkah and travel to Madinah.

630 Muhammad leads his followers from Madinah to Makkah. Muslims dedicate the Kaaba to worship of Allah.

632 Muhammad dies. Much of the Arabian Peninsula is united under Islam.

The majority of immigrants are Muslims. When Muhammad founded Islam, the new religion spread rapidly across the peninsula. Arab armies conquered Persia, much of the Byzantine Empire, North Africa, and Spain. Their military successes eventually ended, but Islam continued to diffuse across Asia and Africa.

The unifying fervor of the early years of Islam was broken by a growing conflict within the religion. After Muhammad's death, arguments arose over who would succeed him. As a result, Islam broke into two branches, the **Sunni** and the **Shia**. The conservative sect of **Ibadhism** broke off from the main branches of Islam in A.D. 657. Ibadhites became the major religious group in Oman.

Islamic law, or **shari'ah**, governs every aspect of a Muslim's life. Shari'ah specifies what a person owes to Allah, or God, and what he or she owes to other human beings. A Muslim's duty to Allah is satisfied by observing such practices as making a religious journey to Makkah, called the **hajj**, at least once in one's lifetime. Shari'ah also shapes criminal and civil laws that govern the responsibility of an individual to other people.

With the exception of Yemen, the countries of the Arabian Peninsula have universal health care. However, medical practice varies greatly. Qatar spends large amounts on health care, and its citizens are the most satisfied with their health care system. People in Yemen, which does not offer universal health care, receive little or no preventive care. Overall, in Yemen it is estimated that 48 percent of women and 37 percent of men have the necessary access to health care.

Education for males is compulsory in all of the countries of the subregion. Boys and girls attend separate schools, and higher education is not as available to women as it is to men. For example, in Saudi Arabia there were no schools for girls until the 1960s, a situation that changed by the twenty-first century.

Saudis must attend school from the ages of six to eleven. Saudi Arabia has a literacy rate of 86.6 percent. Qatar, with a literacy rate of 96 percent, requires children to attend school until they are seventeen. In Yemen, a much poorer

Sunni a branch of Islam that regards the first four successors of Muhammad as his rightful successors

Shia a branch of Islam that regards Muhammad's son-in-law Ali and the imams as his rightful successors

Ibadhism a conservative form of Islam distinct from Sunni and Shia sects

shari'ah Islamic law derived from the Quran and the teachings of Muhammad

hajj in Islam, the yearly pilgrimage to Makkah that Muslims must make at least once in a lifetime

(t)Popperfoto/Getty Images, (b)Roslan Rahman/AFP/Getty Images

Islam divides into two main branches— Sunni and Shia.

661

1000 →

1744

Muhammad bin Saud establishes the first Saudi state.

Abdul Aziz Al-Saud captures the Hejaz, unites warring tribes.

1925

1932

Modern Kingdom of Saudi Arabia is established as an Islamic state with the Quran as its constitution.

2000 →

2012

More than 3 million people travel to Makkah to perform the annual hajj.

country without oil resources, students are supposed to stay in school until they are age fourteen. Only about 64 percent of Yemen's population is literate.

Family and the Status of Women

Most marriages in the subregion are arranged between families. Although not required, the arranged marriages of cousins have been a tradition. Once they are married, most women are expected to stay at home to take care of the children and provide a comfortable home for their husbands. In most countries, a married woman has domestic help, usually female Asian guest workers. The male head of household is the ultimate authority in the home.

Throughout the peninsula, men and women socialize separately. They are also segregated in public, particularly in Saudi Arabia. Usually women wear an *abaya*, a long black robe, over their clothing. In most of the countries, women cover their hair but are permitted to show their faces. In most of Saudi Arabia, however, women are considered immodest if their faces are not covered by a veil.

There has been political progress for women in several countries in the subregion. Most countries guarantee women equal rights, although equality is not enforced. Women in Saudi Arabia are denied voting rights, but King Abdullah recently granted women the right to vote in municipal elections. Kuwaiti women secured the right to vote in 2005. Elsewhere in the region, women's suffrage is permitted, but women have little political power.

Sanaa, Yemen, is a UNESCO World Heritage City and famous for its ancient architecture.

▲ CRITICAL THINKING

1. *Analyzing Visuals* What clues reveal that this is an old city?

2. *Analyzing* Sanaa's population has increased dramatically and is growing almost twice as fast as the country as a whole. Why do you think this is so?

The Arts

The Arabian Peninsula has produced many art forms. There is a strong oral tradition, and the culture is rich in poetry and folk art. The founding of Islam was accompanied by the flowering of religious art. In order to glorify the words of Allah, artists produced rich decoration called arabesques, often created in intricate geometric designs. Inlaid tile designs were used on buildings as far away as Spain and in other places in the Mediterranean.

Calligraphy is another characteristic Islamic art form. The word *calligraphy* means "beautiful writing." Arab calligraphers have used their script in the pages of the Quran, as well as to decorate textiles, walls, walkways, and building facades.

Islamic architecture is uniquely beautiful. Yemen's architecture features tall buildings, sometimes tinted in various colors, that have stood for over 2,000 years. The capital city, Sanaa, and several Yemeni towns have been designated as World Heritage sites because of their extraordinary architecture.

☑ READING PROGRESS CHECK

Identifying How does Islam affect the daily life of people on the Arabian Peninsula?

Economic Activities

GUIDING QUESTION *What impact does the oil industry have on life on the Arabian Peninsula?*

Oil was discovered on the Arabian Peninsula in 1932, but it was not until 1938 that the largest Saudi Arabian oil fields were discovered. U.S. and British oil companies became the major producers of Saudi oil. With time, Saudi Arabian

Michele Falzone/AWL Images/Getty Images

engineers were trained, often in the United States and the United Kingdom, so they could fill management positions in the oil fields. By the end of the twentieth century, the Saudi Arabian government was in control of the production and export of its oil. In 1960 Saudi Arabia joined with Kuwait and other oil-producing countries to form the Organization of the Petroleum Exporting Countries (OPEC). Later Qatar and the United Arab Emirates and other countries outside the peninsula joined OPEC. The members of OPEC meet regularly to set production quotas. Every five years, OPEC **documents** its **strategy** for dealing with the long-term demands for oil.

The Arabian Peninsula is one of the most strategically important regions in the world for oil. Maintaining good relations with its neighbors is important. This is because two strategic waterways, the Strait of Hormuz on the Persian Gulf and the Bab el Mandeb, a strait on the Red Sea, are two of the world's leading oil-shipping **choke points**. Choke points are strategically located narrow passages between two bodies of water. Blocking either of these straits would prevent the export by shipping of the crude oil and petroleum products from the Arabian Peninsula.

Oil has produced great wealth for some countries on the peninsula. Some countries have used this wealth wisely. Saudi Arabia, with its enormous oil income, has made large investments in infrastructure and education. Bahrain, with relatively small reserves of oil, has successfully diversified its economy into banking and finance. Oman and the United Arab Emirates have invested their oil revenue. The investments help to soften the effects of fluctuations in oil prices in the international markets. Despite its oil wealth, Kuwait has not invested in its infrastructure. Yemen's oil production is small compared to other countries in the subregion. It has an unstable government and poorly developed economy. It has also been affected by internal division and conflict between Sunni and Shia groups, each wanting to control the country.

Although petroleum is the major product of the peninsula, there are other important industries. Cement manufacturing is a leading industry, as is ship repair. Bahrain and Oman have fishing industries, and commercial aquaculture is being developed. Tourism is important to both Bahrain and the United Arab Emirates. The Red Sea coastline of Saudi Arabia is a growing magnet for tourists because of its white sand beaches and its magnificent coral reefs. Yemen has a small textile industry.

Although arable land is scarce and precipitation is limited, a variety of fruits, vegetables, and some grains are grown. Agriculture consumes most of the water in the subregion. Food crop production is being developed that uses less water and has a high yield. Water is the subregion's most important natural resource issue.

document to put in writing in order to have as a record

strategy a plan or method

choke point a strategic, narrow waterway between two larger bodies of water

✔ **READING PROGRESS CHECK**

Explaining How has Saudi Arabia used its oil wealth to improve life in the country?

LESSON 2 REVIEW

Reviewing Vocabulary
1. *Determining Importance* What is the significance of shari'ah?

Using Your Notes
2. *Explaining* Use your graphic organizer on the governments of the Arabian Peninsula to explain how the governments of the subregion have or have not changed throughout history.

Answering the Guiding Questions
3. *Inferring* How did history, culture, and geography help influence the types of governments on the Arabian Peninsula?

4. *Evaluating* How have climate and history helped determine the population patterns of the Arabian Peninsula?

5. *Explaining* How does Islam affect life on the Arabian Peninsula?

6. *Drawing Conclusions* What impact does the oil industry have on life on the Arabian Peninsula?

Writing Activity
7. *Argument* In a paragraph, discuss the advantages and disadvantages of monarchy as a form of government for a country in the Arabian Peninsula. Conclude with your opinion supported by reasons.

netw⊙rks
There's More Online!

☑ **IMAGE** Dubai's Air Pollution

☑ **IMAGE** Golf Course in the Desert

☑ **IMAGE** Drilling Equipment

☑ **INTERACTIVE SELF-CHECK QUIZ**

☑ **VIDEO** People and Their Environment: The Arabian Peninsula

LESSON 3
People and Their Environment: The Arabian Peninsula

Reading **HELP**DESK

Academic Vocabulary

- **furthermore**
- **coincide**

Content Vocabulary

- **geopolitics**
- **desalination**

TAKING NOTES: *Key Ideas and Details*

IDENTIFYING As you read about environmental concerns of the Arabian Peninsula, use a graphic organizer like the one below to identify the causes and effects of the environmental problems.

Environmental Concerns

Cause	Effect

ESSENTIAL QUESTION • *How do physical systems and human systems shape a place?*

IT MATTERS BECAUSE

The petroleum industry has had an enormous impact on the environment of the Arabian Peninsula. The growth of the population, due partly to the arrival of millions of foreign workers, is putting a serious strain on the subregion's water supplies. In addition to the scarcity of water, environmental concerns such as pollution and climate change are an ongoing challenge to the people and governments of the peninsula.

Managing Resources

GUIDING QUESTION *How has the extraction of petroleum resources impacted the environment of the Arabian Peninsula?*

Oil is the most widely used energy source in the modern world. For industrialized nations, oil is one of the most sought-after natural resources. Petroleum products are used for the generation of power and the operation of vehicles. Oil is a fossil fuel, created over millions of years, and thus is a nonrenewable resource. The relative scarcity of this resource makes it an important factor in the economies of the Arabian Peninsula. The peninsula's vast petroleum reserves also make it a focus of international interest.

Events in the oil-producing countries generate concern throughout the world. Unrest, such as that generated by the series of uprisings known as the Arab Spring, is magnified in importance by the effect it might have on oil production or transportation.

Physical, human, and economic geography can influence government policy. Such **geopolitics** determines how world powers respond to shifts in power on the peninsula. The 1991 Persian Gulf War is an example of such a response. In late 1990, Iraq invaded Kuwait, a neighbor of Saudi Arabia. The presence of Iraqi forces in Kuwait was not only a violation of international law, but it was also a threat to oil-rich Saudi Arabia. The United States, heading a large multinational force, pushed Iraq's forces back and restored the control of Kuwait to the Kuwaitis.

(tl)Sorge/Alamy, (tc)Barry Iverson/arabianEye/Getty Images, (tr)©Bryan Denton/Corbis

Oil extraction comes with a price. Oil pollution, whether from extraction, processing, or spills, releases hundreds of chemical gases into the air. These pollutants linger in the air until rain carries them to the ground, where they contaminate the groundwater and the soil.

One of the most dramatic examples of oil pollution occurred during the Persian Gulf War. In a successful attempt to sabotage Kuwait's oil industry, Iraqi forces set fire to oil wells and released 250 million gallons of oil into the Persian Gulf. Hundreds of miles of coastline were soaked with oil. The fishing industry, a vital part of the Persian Gulf economy, was dealt a serious blow.

The extraction and processing of petroleum has other consequences. Climate studies show that the average temperature of the peninsula has increased over the past three decades. All of the countries of the subregion have been affected by climate change.

Rainfall has decreased over much of Saudi Arabia. Persian Gulf countries like Oman are experiencing flash floods that erode the topsoil. Drought has affected the fertility of the soil and altered times for planting crops. Farmers report that increased temperatures have led to an increase of insect pests and plant diseases. Higher temperatures also increase the evaporation of water used for irrigation.

Kuwait has experienced both a sharp drop in rainfall and a dramatic rise in average temperature. The country has experienced tornado winds and an increase in sandstorms and flooding. Offshore, seawater temperatures have risen at three times the global rate.

Climate change is also causing a rise in sea level. The archipelago country of Bahrain is particularly vulnerable to rising sea levels. A significant rise would be a disaster for the small country, whose population lives mainly on the coasts, as well as for its marine ecology.

Most scientists think that these changes can be attributed, at least in part, to human activities. The burning of fossil fuels releases carbon dioxide into the air. These carbon emissions are called greenhouse gases because they contribute to

geopolitics government policy as it is influenced by physical, human, and economic geography

Air pollution creates a dense blanket over Dubai, United Arab Emirates.

▼ **CRITICAL THINKING**

1. *Drawing Inferences* How does air pollution affect the lives of people?

2. *Explaining* Why is air pollution such a problem on the Arabian Peninsula?

"You have thousands of tankers entering the Gulf and washing their tanks illegally. Between the tankers, pollution from urban centres and the brine disposed of from desalination plants, the Gulf is almost dead."

—Dr. Shawki Barghouti, director-general of the International Centre for Biosaline Agriculture, from "Desalination Threat to the Growing Gulf," *The National*, August 31, 2009

DBQ *ANALYZING* What does the speaker mean by "the Gulf is almost dead"?

furthermore besides; in addition

A golf course has been carved out of the barren desert of the Arabian Peninsula.

▶ **CRITICAL THINKING**

1. *Assessing* Is maintaining a golf course in the middle of a desert a good use of water? Why or why not?

2. *Analyzing Visuals* How can you tell that this golf course is not suited to the environment?

the warming climate. Heavy use of fossil fuels for desalination, air-conditioning, and oil processing add greenhouse gases to the atmosphere. Carbon emissions were the topic of a special meeting of a United Nations convention on climate change in December 2012. The meeting was held in Qatar, which has the highest rate of carbon dioxide emissions per capita of any country in the world.

☑ **READING PROGRESS CHECK**

Identifying How have the advantages of the Arabian Peninsula's resources created a problem for its environment?

Human Impact

GUIDING QUESTION *How has a growing population created a set of related environmental challenges in the subregion?*

In 2012 the Arabian Peninsula was home to approximately 68 million people, and the population is expected to reach 120 million by 2050. Saudi Arabia's population alone is projected to grow by 333 percent over the next 34 years. The population of Kuwait is increasing at an annual rate of 1.88 percent, and Oman's is increasing at a rate of 4.93 percent. Yemen's population is growing by about 700,000 people each year. As the number of people increases, so does the human impact on the environment.

The need for water is critical. Surface water created by rainfall exists in only a few places, and the evaporation rate is high because of the hot, arid climate. **Furthermore**, climate change is reducing the amount of rainfall in some areas, while increasing it in others. As weather patterns change, severe flooding can occur. These increases in rainfall overwhelm the existing drainage system, eroding the soil and endangering the human population.

Every country in the peninsula extracts increasingly higher annual percentages of water from its aquifers every year. This increase is not sustainable. The aquifers contain fossil water—water trapped in the underground rocks millions of years ago. Only minute amounts of surface water reach these depths

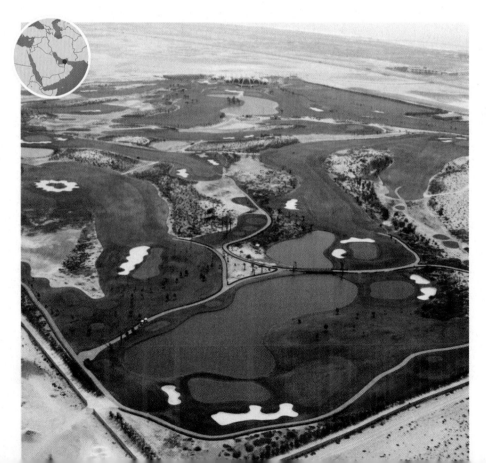

Barry Iverson/arabianEye/Getty Images

over time. Not only is the water virtually a nonrenewable resource, removing it from the aquifers draws in seawater. The seawater contaminates the aquifer and makes it unfit for drinking. The impact of the rising population on water resources has reached a crisis stage in many countries. For example, Sanaa, the capital city of Yemen, is estimated to have only a three-year supply of water left.

Yemen shows that an increasing population and the effects of climate change can contribute to political instability. Warring factions, coupled with widespread corruption and a weak government, prevent Yemen from enacting solutions to its problems. A 40 percent unemployment rate and widespread poverty are made worse by decreased rainfall and a resulting water crisis. Without governmental leadership, individuals and groups are extracting water as quickly as they can in order to make easy short-term profits. Meanwhile, the large rural population, which depends on agriculture, is starved for water. Yemen is at the point of collapse.

The Arabian Peninsula is facing growing desertification. There are many factors that **coincide** to turn arable land into desert. Reduced rainfall can gradually extend a desert into a once fertile area. Mismanagement of the land is another factor. As the population of farmers has increased, so have the herds of sheep and goats. Domestic animals take a heavy toll on vegetation that anchors topsoil. Increasing temperatures reduce the fertility of the soil, and unusually violent rainstorms wash it away. All the countries of the peninsula are affected to some degree by these conditions, but the four Persian Gulf States—Bahrain, Kuwait, Qatar, and the United Arab Emirates—are the most desertified countries on the Arabian Peninsula.

Drilling equipment sits at a farm in Yemen. Yemen faces a water crisis due to overconsumption and mismanagement. Illegal drilling in aquifers is rampant.

▲ **CRITICAL THINKING**

1. *Analyzing* What factors have led to Yemen's water shortage?

2. *Making Predictions* What are some ways Yemen might address water shortages?

coincide to happen in the same place and at the same time

✔ **READING PROGRESS CHECK**

Identifying Which challenges to the environment are related to population?

Addressing the Issues

GUIDING QUESTION *How have people and governments on the Arabian Peninsula addressed the environmental challenges they face?*

Water use in most countries of the Arabian Peninsula is similar to that of more developed countries with far greater supplies of freshwater. Saudi Arabia uses the same amount of water per capita as other more developed countries. Qatar uses almost twice that amount.

One response to the need for water has been to build **desalination** plants. Desalination removes salt from seawater, as well as minerals from undrinkable groundwater. Current methods of desalination require enormous amounts of energy. For example, desalination accounts for 25 to 30 percent of the energy used in Saudi Arabia.

As groundwater has been depleted, the countries of the subregion have greatly increased their dependence on desalination. Qatar is an extreme example of this dependence. It relies on desalination for 99 percent of its water needs. Yemen, on the other hand, has just recently begun to use desalination.

desalination the removal of salt from seawater or from brackish groundwater to make it usable for drinking and irrigation

©Bryan Denton/Corbis

Water Management

Yemen, like the other countries of the Arabian Peninsula, has a limited water supply. Farmers use 90 percent of that water every year, irrigating their crops as they have done for centuries. Plentiful groundwater supplies exist in eastern Yemen, but the cost of transporting it to other parts of the country is too high.

Yemen's water problems stem in part from the lack of a stable government. In other parts of the peninsula, governments reach out to farmers, teaching them how to use water effectively. Yemen does not have an agricultural service, nor does the government have the ability to regulate consumption of water. Without a stable government, there can be no countrywide solutions to the water problem.

ASSESSING What should be government's role in managing a country's resources?

Desalination is not a perfect solution. Desalinated water must be blended with water from aquifers to make it drinkable. Even then, many people prefer the taste of imported bottled water to that of their local tap water.

Perhaps more important, desalination has environmental costs. Disposal of the brine that results from the desalination process presents problems. If it is collected in pools, it can seep into the groundwater and contaminate the aquifer. If it is piped into the sea, it increases the salinity of the water, interfering with the ecology of the coastal waters.

Historically, all of the governments of the Arabian Peninsula have subsidized the cost of water, and cheap water has led to overconsumption. To cater to wealthy citizens and attract tourists, the Persian Gulf countries have built resorts and spas that feature swimming pools and waterfalls, among other water-intensive attractions. Even the less wealthy are accustomed to using water freely to wash their cars and water their gardens.

By far the greatest overconsumption, however, has been in the practice of agriculture. During the 1970s, the demand for meat and dairy products increased as the urban population grew richer. An effort was made to supply meat from local ranches, but within 10 years most meat was imported. Many governments became alarmed by increasing prices on the world market and concerned about their countries' dependence on foreign food. In response, the governments of Kuwait, Bahrain, the United Arab Emirates, and Saudi Arabia joined in an effort to encourage agriculture. However, about 85 percent of the region's annual water supply is used for agriculture, and much of it is wasted due to inefficient irrigation. Yemen provides another example of impractical water use. Much of its groundwater goes toward khat cultivation, the leaves of which are chewed as a mild stimulant.

Today, many countries on the Arabian Peninsula, particularly Saudi Arabia, are promoting a variety of changes in how agriculture is practiced. Greenhouse agriculture, which uses much less water than other methods of growing food, is being encouraged. Scientists are developing salt-tolerant plants that can make use of water that is too salty for drinking. Several countries are also looking abroad to find new solutions to their food problems. They are investing in land overseas. They plan to use the natural resources and labor of the host countries to produce food for import to the peninsula. Qatar, for example, is purchasing farmland in such distant places as Sudan, Australia, Kenya, Brazil, Argentina, Turkey, and Ukraine.

☑ **READING PROGRESS CHECK**

Explaining What are the advantages and disadvantages of desalination as a solution for water scarcity issues in the subregion?

LESSON 3 REVIEW

Reviewing Vocabulary
1. *Explaining* Write a paragraph explaining how geopolitics can create conflict between countries.

Using Your Notes
2. *Summarizing* Use your graphic organizer on the environmental concerns in the Arabian Peninsula to write a paragraph summarizing the environmental situation in the subregion.

Answering the Guiding Questions
3. *Evaluating* How has the extraction of petroleum resources impacted the environment of the Arabian Peninsula?

4. *Making Connections* How has a growing population created a set of related environmental challenges in the subregion?

5. *Explaining* How have people and governments on the Arabian Peninsula addressed the environmental challenges they face?

Writing Activity
6. *Argument* In a paragraph, take the position of a water resource manager of a city in the subregion, and explain to citizens why it is necessary to conserve water.

Directions: On a separate sheet of paper, answer the questions below. Make sure you read carefully and answer all parts of the questions.

Lesson Review

Lesson 1

1 *Describing* Describe how the Red Sea interacted with a single tectonic plate to create the Arabian Peninsula.

2 *Explaining* Why is wind such a significant factor in the climate of the Arabian Peninsula? How do simooms and *shamals* affect plant and animal life there?

3 *Contrasting* The two greatest resources in the Arabian Peninsula are oil and water. Why is freshwater even more crucial than oil to countries in the subregion?

Lesson 2

4 *Identifying Cause and Effect* How did the development of the petroleum industry affect the makeup of the population on the Arabian Peninsula?

5 *Evaluating* Why is the Arabian Peninsula considered one of the most strategically important regions in the world?

6 *Identifying* What countries in the Arabian Peninsula belong to OPEC? What is the role of that organization?

Lesson 3

7 *Making Connections* How does the extraction of oil cause damage to air, land, and water throughout the Arabian Peninsula?

8 *Summarizing* What are some of the drawbacks associated with current methods of desalination?

9 *Explaining* Why is the 1991 Persian Gulf War an example of the importance of oil production to geopolitics?

Critical Thinking

10 *Hypothesizing* Oil is a nonrenewable resource. How might the economies of the Arabian Peninsula be different if oil were renewable?

11 *Drawing Conclusions* Why do you think millions of immigrants have been drawn to the Arabian Peninsula even though their employers may treat them unfairly?

21st Century Skills

Use the graph to answer the following questions.

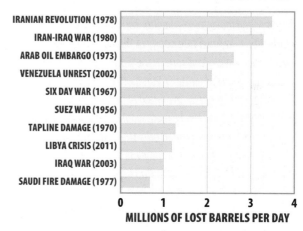

HUMAN EFFECTS ON OIL SUPPLY

	MILLIONS OF LOST BARRELS PER DAY
IRANIAN REVOLUTION (1978)	
IRAN-IRAQ WAR (1980)	
ARAB OIL EMBARGO (1973)	
VENEZUELA UNREST (2002)	
SIX DAY WAR (1967)	
SUEZ WAR (1956)	
TAPLINE DAMAGE (1970)	
LIBYA CRISIS (2011)	
IRAQ WAR (2003)	
SAUDI FIRE DAMAGE (1977)	

Sources: Energy Information Administration (EIA), Bank of America Merrill Lynch Global Commodities Research, Money Morning staff research

12 *Understanding Relationships Among Events* According to the graph, how do war, revolution, and social unrest affect global oil production?

13 *Drawing Conclusions* What other factors may also have affected daily oil production during these events?

College and Career Readiness

14 *Multicultural Societies* Tourism has become an important way to diversify the economy in Bahrain and the United Arab Emirates. Choose one of these countries. Imagine that you have been hired by a travel agency to design a three-day tour of that country. Research the major tourist destinations, cultural highlights, and other activities. Write a travel itinerary for each day of the trip.

Need Extra Help?

If You've Missed Question	1	2	3	4	5	6	7	8	9	10	11	12	13	14
Go to page	434	436	437	439	443	443	445	447	444	444	439	449	444	443

Directions: On a separate sheet of paper, answer the questions below. Make sure you read carefully and answer all parts of the questions.

DBQ Analyzing Primary Sources

Use the document to answer the following questions.

Strict interpretations of Islamic law and tradition on the Arabian Peninsula have created many restrictions for women. This is particularly the case in Yemen, the country with the least developed infrastructure.

PRIMARY SOURCE

"*By almost all indicators—health, education, and economic opportunity—women fare poorly in Yemen. It has one of the highest rates of child marriage in the world, and 60 percent of Yemeni women are illiterate. Infant mortality rates are also among the world's worst, attributed to the lack of prenatal and postnatal health care. Unlike men, women cannot easily get divorced, and they have limited property and inheritance rights. The country ranks dead last among 135 countries in the World Economic Forum's Global Gender Gap Report.*

Al Abdeli is an assistant professor of accounting at a university in Taizz, and she enjoys more freedoms than most of her peers. She credits this to growing up in Taizz and having an open-minded father who 'did not go to university but is knowledgeable about the world.' She is also a poet who for years openly expressed her loathing for Ali Abdullah Saleh's regime. 'I put some of my dreams in my poetry and implanted them in the minds of my students,' she said."

—Joshua Hammer, "Days of Reckoning," *National Geographic,* September 2012

15 *Analyzing* How do traditions of marriage and divorce contribute to the difficulties women face in Yemen?

16 *Problem Solving* How might greater access to education improve the condition of women in Yemen?

17 *Interpreting* What dreams do you think Al Abdeli implanted in the minds of her students?

Applying Map Skills

Use the Unit 5 Atlas to answer the following questions:

18 *Physical Systems* Use your mental map of the Arabian Peninsula to describe the location of the Red Sea, the Arabian Sea, and the Persian Gulf.

19 *Places and Regions* Describe the overall population density pattern of the Arabian Peninsula.

20 *Environment and Society* What physical and human features help explain why the Bab el Mandeb and the Strait of Hormuz are called choke points?

Exploring the Essential Question

21 *Diagramming* Design two diagrams showing how vast deposits of crude oil under the sands of the Arabian deserts have affected the human and physical systems of the Arabian Peninsula. In each diagram, label the factors involved, and explain how the human and physical systems interact.

Research and Presentation

22 *Research Skills* Using the Internet and other resources, research the history and culture of nomadic bedouin tribes in the Arabian Peninsula. Describe how bedouin culture evolved to meet the demands of life in the enormous central deserts of the interior. How do bedouins live today? Create a multimedia presentation to share your findings.

Writing About Geography

23 *Argument* Use standard grammar, spelling, sentence structure, and punctuation to write an essay arguing the benefits of a diverse economy for countries in the Arabian Peninsula. Give examples of how countries in the subregion have successfully diversified their economies, and discuss the potential consequences of relying too heavily on oil.

Need Extra Help?

If You've Missed Question	**15**	**16**	**17**	**18**	**19**	**20**	**21**	**22**	**23**
Go to page	442	441	442	360	364	360	439	438	443

Central Asia

ESSENTIAL QUESTION · *How do physical systems and human systems shape a place?*

Geography Matters...

Central Asia can be a challenging area in which to live. Much of it is arid. Farmers rely on snowmelt from the mountains to provide water to farm land in the valleys. Rocky highlands prevent farming in many areas but provide pasture for sheep. Forbidding deserts stretch over large expanses.

However, much of Central Asia is quite beautiful. Snow-covered mountains tower over deserts of stark beauty. Dazzling tiles adorn buildings. In some cities, centuries-old mosques stand alongside gleaming glass skyscrapers.

◄ This man in Kabul, Afghanistan, faces traditional and modern ways.

©Alison Wright/Corbis

Afghanistan's troubled history

Located on trade routes that link areas from East Asia, South Asia, Southwest Asia, and Europe, Afghanistan has been "the crossroads of Central Asia" for thousands of years. Becoming an independent country in 1747, Afghanistan has had a troubled history. In recent decades, Afghanistan has faced the external forces of invasion, as well as the internal forces of ethnic and tribal diversity. Both threaten the country's stability.

Ethnic Groups of Afghanistan

0   200 miles
0   200 kilometers
Lambert Azimuthal Equal-Area projection

UZBEKISTAN

TAJIKISTAN

CHINA

TURKMENISTAN

The Hazaras are mostly Shia Muslims, which represent a minority sect of Islam in Afghanistan. They have been persecuted for their beliefs.

• Mazār-e Sharīf

Tajiks comprise the second most prevalent group and are also primarily Sunni Muslims. Unlike other ethnic groups, they are spread out across the country.

35°N

• Herāt

AFGHANISTAN

Kabul

INDIA

IRAN

Kandahār

N
W   E
S

PAKISTAN

Pashtuns are the largest ethnic group and historically have been the dominant political group in Afghanistan. They are Sunni Muslims. The Taliban is a Pashtun group.

30°N

Ethnic Groups

Aimak	Pashtun
Baluchi	Tajik
Hazara	Turkmen
Kirghiz	Uzbek
Nuristan	Other
Pamir	

60°E   65°E   70°E   75°E

Who are the people of Afghanistan?

Who has invaded and tried to rule Afghanistan?

Why does Afghanistan continue to suffer?

Afghanistan has a complex human geography. The country is home to dozens of different ethnic groups and subgroups. The map shows only the largest. While Pashtuns form the largest group, even they do not make up a majority of the country's people. Afghanistan's people speak about 50 languages. The two most widely spoken are Pashto and Dari. Pashto is the language of the Pashtun. Dari is a Farsi, or Persian, dialect. Dari often serves as a lingua franca (common language) in Afghanistan.

Ethnic differences are not necessarily strong divisions. People from different ethnic groups marry one another and live in the same areas. Perhaps the strongest tie Afghanistan's people feel is to their local tribes. For many people, their bond to the tribe is stronger than their feeling of national identity, which has made it difficult to organize them into a single modern country.

1. Human Systems How might tribal loyalties affect a leader's ability to govern Afghanistan?

For thousands of years, Afghanistan has been the target of foreign invaders. Persians ruled the area until Alexander the Great conquered it around 330 B.C. Alexander was followed by the Scythians, then the Huns, and then the Turks. Later, Arabs conquered the area. Persians ruled again, followed by the Turks once more. Then two different Mongol conquerors invaded Afghanistan.

The Soviet Union invaded Afghanistan in 1979. However, the Soviets withdrew after ten years of costly fighting. In the 1990s, a fundamentalist Islamic group took control of Afghanistan. They began to shelter Islamic terrorists. The terrorists established training camps for their fighters, some of whom attacked the United States on September 11, 2001. When the government refused to close the terrorist camps and even refused to hand over the terrorist leader, the United States and other countries sent troops to fight them. The government was quickly defeated and an elected government eventually took office. However, rebels have continued to fight against the new regime.

2. Places and Regions Why do you think Afghanistan has been invaded by outside forces so frequently?

Afghanistan is among the poorest countries in the world. More than three-fourths of its workforce makes its living by farming using traditional methods. As a result, incomes are low. For example, more than a third of the people live in poverty. Just over a third of the workforce is unemployed.

Afghanistan's economy has begun to recover since the fundamentalists were ousted, and the economy has the potential for development. The country has reserves of natural gas and petroleum, precious metals such as gold and silver, and gems. However, conflict has hampered efforts to develop the country's natural resources. For example, continuing conflict has interfered with infrastructure development. Government corruption has also been a significant problem in trying to develop the economy and a stable society. As a result, Afghanistan still depends heavily on aid from the international community.

3. Environment and Society What opportunities and what challenges does Afghanistan face today?

(l)Musadeq Sadeq/AP Images, (c)©Bernd Obermann/Corbis, (r)©Ahmad Masood/X01429/Reuters/Corbis

THERE'S MORE ONLINE

READ about Afghanistan's place in the world • **VIEW** an image of the Soviet invasion of Afghanistan

netw‍rks

There's More Online!

- ☑ **IMAGE** Mountains in Tajikistan
- ☑ **IMAGE** Resort on Caspian Sea
- ☑ **IMAGE** Herding in Armenia
- ☑ **IMAGE** Farming in Afghanistan
- ☑ **INTERACTIVE SELF-CHECK QUIZ**
- ☑ **VIDEO** Physical Geography of Central Asia

Reading **HELP**DESK

Academic Vocabulary

- **unique**
- **reverse**

Content Vocabulary

- **steppe**
- **cereal**

TAKING NOTES: *Key Ideas and Details*

IDENTIFYING As you read about the physical geography of Central Asia, use a graphic organizer like the one below to identify major features of the subregion.

LESSON 1

Physical Geography of Central Asia

ESSENTIAL QUESTION · *How do physical systems and human systems shape a place?*

IT MATTERS BECAUSE

Central Asia has long been home to varied peoples. It has also served as the location of trade routes connecting Asia and Europe. For thousands of years, the people of this subregion have carved a life from a rugged land of mountains, deserts, and grasslands.

Landforms

GUIDING QUESTION *What are the major landforms of Central Asia?*

Central Asia stretches from the Black Sea to China. It is bounded on the north by Russia, on the east by China and Pakistan, and on the south by Iran and Turkey. It is a distinct subregion defined by its physical and human geography. The land is challenging, marked mainly by highlands, a dry climate, and vast expanses of desert. All the countries of the subregion, except Georgia, are landlocked, meaning they have no access to the sea. The many cultures of the subregion share some **unique**, or distinctive, characteristics. They have been formed by the ethnic traditions and ways of life of the people and largely by Islam. All the countries in Central Asia, except Afghanistan, have something of a common history as well. They were part of the Soviet Union, gaining their independence in the 1990s.

Central Asia can be divided into two sections. The western section, sometimes called the Caucasus, lies between the Black Sea and the Caspian Sea. It consists of the three small countries of Georgia, Armenia, and Azerbaijan. The eastern section, reaching from the eastern shores of the Caspian Sea to China and Pakistan, includes the six "stans": Kazakhstan, Uzbekistan, Turkmenistan, Kyrgyzstan, Tajikistan, and Afghanistan. The suffix -*stan* is an ancient Persian word that means "land of" or "place of." In most of these countries, the first part of the name identifies the major ethnic group living there.

The western section gets its name from the Caucasus Mountains, which straddle the area between the Black Sea and Caspian Sea. The mountains actually consist of two ranges. The higher peaks, called the

(tl)©Nevada Wier/Corbis, (tc)Michael Runkel/Alamy, (tcl)©Kim Karpeles/Alamy, (tc)©Alex Treadway/National Geographic Society/Corbis, (tr)©Andrew Peacock/Lonely Planet Images/Getty Images

Greater Caucasus, are to the north. They include Mount Shkhara, in Georgia. At 17,063 feet (5,201 m), it is the third-highest peak in the Caucasus. The lower mountain range, the Lesser Caucasus, is to the south.

The Central Asian countries have several high mountain ranges. Like the Caucasus, they were formed and continue to rise because of the collision of tectonic plates. The eastern ranges include the high Tian Shan of Kazakhstan and Kyrgyzstan; the Altay Shan, which run through those countries and Tajikistan; the Pamirs of Tajikistan; and the Hindu Kush of Afghanistan. Some of the peaks in these chains, like Nowshak in the Hindu Kush, soar over 24,000 feet (7,315 m) above sea level. These mountains dominate the eastern part of the subregion. They are separated by high plateaus and basins. Some of these basins, like the Fergana Valley of eastern Uzbekistan, have water and fertile soil. The Fergana, which covers an area about the size of Massachusetts, is thickly settled and marked by farms. Other basins are vast, dune-covered *kums* (KOOMZ), or deserts.

While the eastern part of the subregion has mountains that soar miles high, the western part slopes downward. The Turan Plain, around the Aral Sea, is only 200 to 300 feet (61 to 91 m) above sea level. The Caspian Depression, on the north shore of the Caspian Sea in southwestern Kazakhstan, is below sea level.

The Kara-Kum, or black-sand desert, covers most of Turkmenistan. The Kyzyl Kum, or red-sand desert, blankets the western half of Uzbekistan. These deserts are each about the size of Arizona. The deserts are sparsely settled. Afghanistan has a fertile plateau to the north of the Hindu Kush and a dry plateau to the south.

The Caucasus section is dominated by the Caucasus Mountains. Small patches of lowland plains lie along the shores of the Black and Caspian Seas. The mountains are flanked north and south by foothills and divided by valleys.

Tectonic activity built the mountains of this subregion. As is typical along plate boundaries, earthquakes frequently strike here. Afghanistan, surrounded by active plate boundaries, has an estimated 500 significant earthquakes a year. Tajikistan averages a major destructive quake every 10 or 15 years. A 1948 earthquake in Turkmenistan destroyed the capital city of Ashkhabad and killed about 176,000 people. Forty years later, a quake shook Armenia and caused such damage that its manufacturing capacity fell by 25 percent.

✔ **READING PROGRESS CHECK**

Explaining How do the landforms of Central Asia affect where people live?

Analyzing
PRIMARY SOURCES

Ancient Geography Text

"In proceeding from the [Caspian] Sea towards the east, on the right hand are the mountains . . . extending as far as India. They . . . stretch to this part from the west in a continuous line, bearing different names in different places."

—Strabo from *Geography,* Book XI, Chapter 8, C.A.D. 21

DBQ *MAKING CONNECTIONS*
Through what present-day countries would you pass if following the route through Central Asia as described by this ancient geographer?

unique distinctive

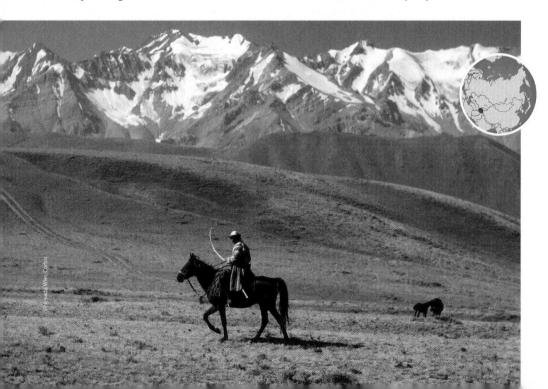

Fertile areas of Central Asia are dwarfed by the arid deserts and high, rocky mountains.

◄ **CRITICAL THINKING**

1. *Analyzing Visuals* Do you think this is the Kara-Kum or the Kyzyl Kum? Why do you think so?

2. *Hypothesizing* What challenges might this landscape pose to the people of the subregion?

Resorts, such as this one in Baku, Azerbaijan, dot some parts of the coast of the Caspian Sea, where people go to vacation.

▲ **CRITICAL THINKING**

1. Identifying Which countries in the subregion border on the Caspian Sea?

2. Summarizing What economic benefits does the Caspian Sea provide?

reverse the opposite

Water Systems

GUIDING QUESTION *What water features are important to the people of Central Asia?*

Central Asia is home to a large inland sea, the Caspian Sea, and a large lake, Lake Balkhash. The Caspian is the world's largest inland sea. Thousands of years ago it was connected to the Black Sea and, through it, to the Mediterranean. When the climate is warmer and precipitation is low, the Caspian Sea loses more water from evaporation than it gains from rivers emptying into it. In cooler, rainier times, the **reverse**, or opposite, is true.

Three countries from the subregion, plus Russia and Iran, border the Caspian Sea. Sturgeon fishing has long been an important industry. The sturgeon's eggs are eaten as caviar, an expensive delicacy. In recent years, oil and natural gas found under the Caspian Sea have become very important. Along some coastal areas of the sea are beaches and resort hotels.

East of the Caspian Sea is the Aral Sea. Until the 1960s, it was one of the world's largest inland seas. Today the sea is several small, shallow, disconnected bodies of water separated by dry land. This drying of the Aral Sea began in the 1960s when water from the main tributary was diverted to irrigate fields. This diminished the water supply to the Aral Sea. Also, the sea sits in a hot, dry region where high temperatures evaporate the water faster than rainfall replaces it.

Similar problems have hurt Lake Balkhash, in eastern Kazakhstan. The Ile River provided most of the lake's water. When the river was dammed, the lake's water level began to drop. The water also grew more saline, or salty.

The subregion's main rivers are the Amu Dar'ya and Syr Dar'ya. Both form in the subregion's center, fed by mountain rivers. Both rivers flow into the Aral Sea. Turkmenistan's Karakum Canal connects the Amu Dar'ya to the Caspian Sea.

✓ **READING PROGRESS CHECK**

Identifying Cause and Effect How have humans affected the water systems of Central Asia?

Michael Runkel/Alamy

Africa South of the Sahara
Economic Activity

Land Use

- Commercial farming
- Subsistence farming
- Livestock raising
- Nomadic herding
- Primarily forest
- ■ Manufacturing and trade
- Commercial fishing
- Little or no activity

EUROPE

NORTH AFRICA

Mediterranean Sea

Red Sea

Gulf of Aden

TROPIC OF CANCER

Gulf of Guinea

EQUATOR

The red dashed lines represent the northern and southern boundaries of a region known as the Transition Zone, an area of increasing Islamic influence.

ATLANTIC OCEAN

Mozambique Channel

INDIAN OCEAN

TROPIC OF CAPRICORN

Resources

Coal		Nickel	(N)
Petroleum		Copper	
Natural gas		Lead	
Iron ore		Manganese	
Tin		Gold	
Zinc		Platinum	
Bauxite		Gems	
Cobalt		Chromite	
Uranium			

0   1,000 miles

0   1,000 kilometers

Lambert Azimuthal Equal-Area Conic projection

UNIT 6
REGIONAL ATLAS

MAP STUDY

1. **Physical Systems** Describe the sequence of climate regions as one moves north and south of the Equator.

2. **Environment and Society** How can you explain the presence of nomadic herding as well as the lack of agriculture in parts of some countries in the African Transition Zone?

Africa South of the Sahara
Population Density

EUROPE

CENTRAL ASIA

Cities
(Statistics reflect metropolitan areas.)
- ■ Over 5,000,000
- □ 2,000,000–5,000,000
- ⊙ 1,000,000–2,000,000

POPULATION

Per sq. mi.	Per sq. km
1,250 and over	500 and over
250–1,249	100–499
63–249	25–99
25–62	10–24
2.5–24	1–9
Less than 2.5	Less than 1

Mediterranean Sea

NORTH AFRICA

SOUTHWEST ASIA

TROPIC OF CANCER

20°N

Red Sea

Gulf of Aden

ATLANTIC OCEAN

Dakar
Bamako
Ouagadougou
Kano
Kaduna
Abuja
Ibadan
Conakry
Monrovia
Abidjan
Lomé
Benin City
Accra
Lagos
Port Harcourt
Douala
Yaoundé
Gulf of Guinea

Khartoum

Addis Ababa

Mogadishu

EQUATOR

The red dashed lines represent the northern and southern boundaries of a region known as the Transition Zone, an area of increasing Islamic influence.

Brazzaville
Kinshasa
Mbuji-Mayi
Luanda
Lubumbashi

Kampala
Nairobi

Dar es Salaam

INDIAN OCEAN

Lusaka

Harare

Antananarivo

Mozambique Channel

TROPIC OF CAPRICORN

Tshwane (Pretoria)
Maputo
Johannesburg
Durban

Cape Town
Port Elizabeth

UNIT 6
REGIONAL ATLAS

MAP STUDY

1. **The World in Spatial Terms** Which part of the region has the highest overall population density?

2. **Human Systems** Which southern African country has a low overall population density? What physical feature accounts for this?

0		1,000 miles
0	1,000 kilometers	

Lambert Azimuthal Equal-Area Conic projection

The Transition Zone

ESSENTIAL QUESTION · *How do physical systems and human systems shape a place?*

Geography Matters...

The Transition Zone is a challenging land because of transitions in both the human and physical geography. It is a mix of Muslim, Christian, and animist cultures. It is also a transition from the Sahara to the Tropics. Climate and geography, as well as the legacy of colonialism, have led to turmoil in the subregion.

In recent decades, the Transition Zone has experienced extended periods of drought. Overpopulation has contributed to the problem. People are trying to find new ways to cope with or to reverse the damage to the land and face the environmental challenge.

◄ A Muslim woman in Lomé, Togo

Godong/age fotostock

diffusion: Muslim and non-Muslim cultures

Africa is a continent with great diversity in its physical, cultural, and economic geography. The Transition Zone is a region that lies south of the Sahara and has been increasingly influenced by Islam. At the same time, Christian and traditional religions are also practiced. Some countries in this zone are predominantly Muslim, while others are more religiously mixed.

Sunday Alamba/AP Images

Why do Muslim and non-Muslim cultures exist in Africa?

The Sahel, which means "shore" in Arabic, serves as a cultural and ecological gateway. To the north are primarily Muslim countries and more arid land. To the south are mostly non-Muslim countries with less arid to tropical environments. This cultural gateway exists because of the continent's long history of interaction with other cultures.

When trade flourished in the West African kingdoms and East African kingdoms between 800 and the 1500s, the region encountered new cultures. Arab and Berber traders exchanged trade goods with the people of these kingdoms. They also shared ideas, including their Islamic faith. Christianity also spread in different parts of the region through trade with merchants from the Mediterranean. In Axum (Ethiopia), an East African trading state, Christianity became the official religion in the fourth century.

1. The World in Spatial Terms How did the location of Africa's kingdoms and trading states influence its cultural and religious diversity?

What happens when these distinct cultures encounter each other?

The slave trade, European colonialism, and independence also contributed to cultural diffusion in Africa. All of the following countries make up the Transition Zone: Somalia, Djibouti, Guinea, Guinea-Bissau, Gambia, and Senegal, as well as parts of Eritrea, Ethiopia, Sudan, Chad, Cameroon, Niger, Nigeria, Benin, Togo, Ghana, Burkina Faso, Côte d'Ivoire, Liberia, Sierra Leone, Mali, and Mauritania. Ethnic conflicts occurred in post-colonial Africa because Europeans drew these political borders with no regard for ethnic borders.

As a result, the Transition Zone has seen cultural and ethnic conflicts in many countries. In Nigeria, a civil war rooted in ethnic tensions erupted not long after independence. The "Biafran War" resulted in the deaths of about a million Ibo people, mainly from starvation. In Senegal, a rebellion spurred by religious and cultural differences began in 1984. The area's residents were mainly Christian or followers of traditional religions—unlike the majority of Senegalese, who were Muslim. The conflict ended in 1998.

2. Human Systems Has cultural diffusion in Africa through interaction with other cultures been positive or negative for the Transition Zone? Explain.

How are distinct cultures encountering one another today?

The Transition Zone is still a hotbed for conflict. Sudan, for example, has experienced ongoing conflicts since its independence in 1956. The government has oppressed its Christian population, which mainly lives in the south. Even though South Sudan became independent from Sudan in 2011, conflict continues.

In the Darfur region of Sudan, fighting broke out in 2003 between Muslim villagers, who identified themselves as African, and government-backed militias. The militias were made up of Muslim nomads who identified themselves as Arab. Since then, nearly 3 million people have been displaced and hundreds of thousands killed. Many of the Darfur refugees have fled to the neighboring country of Chad, but refugee camps there have swelled beyond capacity. Peace deals have been signed and a peacekeeping force tries to maintain peace. Critics predict, however, that change will be slow.

3. Environment and Society Write a paragraph explaining how the environment of Darfur might be contributing to this conflict.

(l)Lisa Harrison, (c)McPHOTO/age fotostock, (r)Scott Nelson/Getty Images News/Getty Images

THERE'S MORE ONLINE

VIEW an interactive image of Lake Chad • *EXPLORE* a map of religions in the Transition Zone

netw✷rks

There's More Online!

☑ **IMAGE** Lake Chad in 1973

☑ **IMAGE** Lake Chad in 2001

☑ **MAP** The African Transition Zone

☑ **INTERACTIVE SELF-CHECK QUIZ**

☑ **VIDEO** Physical Geography of the Transition Zone

Reading **HELP**DESK

Academic Vocabulary

- **area**
- **benefit**

Content Vocabulary

- **transition zone**
- **Sahel**
- **delta**
- **harmattan**

TAKING NOTES: *Key Ideas and Details*

IDENTIFYING As you read about the geography of the African Transition Zone, use a graphic organizer like the one below to note the characteristics of the land, water, and climate of the area.

The African Transition Zone		
Land	Water	Climate

LESSON 1
Physical Geography of the Transition Zone

ESSENTIAL QUESTION · *How do physical systems and human systems shape a place?*

IT MATTERS BECAUSE

The African Transition Zone lies south of the Sahara, almost entirely within tropical latitudes, and marks the shift from desert to tropical forests, as well as from Muslim to Christian and animist cultures. The subregion is known for its remarkable wildlife and extraordinary physical geography. It also has the world's fastest-growing population and hundreds of ethnic groups.

Landforms

GUIDING QUESTION *What is the Transition Zone?*

Africa immediately south of the Sahara is an **area** of land that transitions between the desert climate of North Africa and the tropical savanna of Equatorial Africa. This subregion, with a total area of 1,178,800 square miles (3,053,200 sq. km), in the widest part of Africa is known as the African Transition Zone. A **transition zone** is a physical area in which the land undergoes a radical change such as from arid to tropical.

The African Transition Zone also represents a transition between the Islamic cultures of North Africa and the Christian and animist cultures to its south. Extending east to west from Senegal to Somalia, this region is home to a very diverse population. Hundreds of ethnic groups coexist in an area influenced by the African cultures that originated in the region and by the Arab and European cultures that came later.

Part of the Transition Zone is a geographical area called the **Sahel**, which means "shore" or "coast" in Arabic. The word describes the appearance of vegetation in the Sahel as a "coastline" marking the border of the sands of the Sahara. The topography is mainly flat with a series of plateaus that range in elevation from about 650 feet (200 m) to about 1,300 feet (400 m). This steppe region runs in a band across Africa from the Atlantic Ocean in the west to the Red Sea in the east. It spans more than 3,000 miles (4,800 km) and ranges in width from about 125 miles (200 km) to 250 miles (400 km). East of the Sahel is a region known as the Horn of Africa for its shape.

Image courtesy NASA/Goddard Space Flight Center Scientific Visualization Studio

The Sahel receives little rainfall. On average, 4 to 8 inches (10 to 20 cm) of rain falls annually, mostly in June, July, and August when the sun is high in the sky and the temperatues are hot. The rest of the year is a little cooler but very dry, so only low-growing grasses, shrubs, and acacia trees can grow there. Most people in the Transition Zone have traditionally herded livestock. The short grasses of the Sahel provide one of the largest pastoral and herding zones in the world. Overgrazing, however, has stripped the land bare in some places.

☑ **READING PROGRESS CHECK**

Identifying Describe the human and geographic features that characterize the African Transition Zone.

Water Systems

GUIDING QUESTION *What rivers flow through the Transition Zone?*

The African Transition Zone experiences dryness, infrequent rainfall, and often drought. A drought in the 1970s caused a famine that killed thousands of people. Over the past century, annual precipitation has decreased. This, in turn, is shrinking some of the subregion's natural lakes. However, the Sahel still has a number of lakes, rivers, and wetlands. These water resources, in addition to rain, are important to the livelihood of many people.

At the southern edge of the Sahara, Lake Chad is bordered by Nigeria, Niger, Chad, and Cameroon. It was once the second-largest wetland in Africa and supported a rich diversity of animal and plant life. It has been important to

area a geographical region; the amount of space that the surface of a place covers

transition zone an area in which the properties of the land undergo a radical change

Sahel steppe region extending from Senegal to Somalia that receives little rainfall

GEOGRAPHY CONNECTION

The African Transition Zone is both a physical transition between desert and savanna and a cultural transition from Muslim cultures to Christian and animist cultures in the south.

1. *THE WORLD IN SPATIAL TERMS*
   What three bodies of water does the Transition Zone meet in the east?

2. *HUMAN SYSTEMS* What countries in the Transition Zone may be more influenced by Muslim culture than by Christian culture? Explain.

The African Transition Zone

Legend:
- Southern Africa
- East Africa
- Equatorial Africa
- West Africa
- North Africa
- Transition Zone
- Boundary of Transition Zone

0–1,000 miles
0–1,000 kilometers
Lambert Conformal Conic projection

the livelihoods of more than 20 million people but is threatened with extinction. Although fed by three rivers—the Chari, the Logone, and the Yobe—the lake is shrinking. The drought of the 1970s completely dried up the northern portion of the lake. Even during years of normal rainfall, the water level remains low. Because of the dry climate, much of the lake's water evaporates or becomes increasingly saline. The land is left dry and unable to support life. As Lake Chad shrinks, the desert expands.

Lake Volta in Ghana is one of the largest human-made lakes in the world. It is about 250 miles (400 km) long and covers 3,275 square miles (8,482 sq. km). The lake itself lies entirely in Ghana. However, six countries—Benin, Burkina Faso, Côte d'Ivoire, Ghana, Mali, and Togo—share the Volta River system. Lake Volta was created in the 1960s by damming the Volta River south of Ajena in Ghana. The new lake flooded more than 700 villages, forcing more than 70,000 people to find new homes. It was originally created as a reservoir to store water for generating hydroelectric power. People **benefit** from the lake because it supplies irrigation for farming and has several fisheries. It also serves as a transportation route linking different parts of the country. The hydroelectric plant also generates electricity used throughout Ghana.

Unfortunately, there are negative consequences of this reservoir. These include reduced agricultural productivity because annual floods no longer occur and bring new silt, and because the land higher up on the current lake shore is not as fertile as the land now on the lake bottom. Also, the ecology of the river is completely different because running water and annual floods were replaced by standing water. This has resulted in an increase in disease.

The Niger (NY•juhr) River is known by many names along its course. All of them have roughly the same meaning—"great river." The Niger is the main artery in western Africa, extending about 2,600 miles (4,183 km) in length. Originating in the highlands of Guinea, the river forms an arc flowing northeast, then curving southeast to its mouth on the Nigerian coast. In addition to being important to agriculture, the Niger River is a major means of transportation.

This great river does not flow as one well-defined stream all the way along its course to the Atlantic Ocean. In central Mali, the river spreads out across a

benefit to gain

delta an often triangular-shaped section of land formed as the waters of a river slow down and split into many channels as they deposit sand and silt that has been carried downriver

Once nearly the size of Lake Erie, the surface area of Lake Chad has declined dramatically in the last 50 years. The photo on the left was taken in 1973 and the one on the right in 2001.

▶ **CRITICAL THINKING**

1. Assessing By about what percentage does the surface area of Lake Chad appear to have shrunk?

2. Making Connections How do you suppose the lake's shrinkage has affected the surrounding land?

broad plateau and inland **delta**, an area where a river slows down and spreads out into many smaller channels. These small river channels create a large wetland that supports wildlife. This inland delta also provides local farmers with water. Further downstream, at Aboh in southern Nigeria, the Niger splits into a delta that stretches 150 miles (241 km) north to south and extends to a width of about 200 miles (322 km) where the river empties into the Gulf of Guinea.

The Senegal River is 1,015 miles (1,633 km) long in West Africa and forms the border between Senegal and Mauritania. The sources of the river are in Guinea and the wetter, southwestern part of Mali. Its drainage basin encompasses some 174,000 square miles (450,000 sq. km). Roughly 3.5 million people live near the river. In 1972 Mali, Mauritania, and Senegal founded an organization to manage the river basin. Guinea joined in 2005. Two dams have helped the inhabitants of the area make better use of the river. The Manantali Dam, built in 1986, helps prevent flooding during the rainy season and provides freshwater during the dry season. In 2002 the dam began generating hydroelectric power. The Diama Dam was built near the mouth of the river in 1988. It helps prevent the intrusion of salt water during the dry season. The dam also creates reserves for irrigation.

Two major tributaries of the Nile River pass through the Transition Zone. These are the Blue Nile and the White Nile. The Blue Nile, the main source of the fertile soil along the banks of the Nile River, stretches 850 miles (1,368 km) and originates in the mountains of Ethiopia. In 2012 Ethiopia began construction of a 6000-megawatt hydroelectric dam on the river. Expected to be completed in 2015, the Grand Millennium Dam will be the largest hydroelectric power plant in Africa. It will also create Ethiopia's largest artificial lake.

The White Nile River, which is longer than the Blue Nile River, originates in the mountains to the west of Lake Victoria in Burundi. The Blue Nile and the White Nile meet at the city of Khartoum in Sudan. From there, the famous Nile River flows northward into Egypt on a winding course across the desert to the Mediterranean Sea.

☑ **READING PROGRESS CHECK**

Summarizing How is Lake Volta used by the residents of the region?

Connecting Geography to MATH

Geometry

The Sahel is a deceptively huge area—1,178,800 square miles (3,053,200 sq. km). Consider the area of Texas, Tennessee, and the contiguous United States in relation to the area of the Transition Zone.

Texas:
268,800 square miles (696,200 sq. km)

Tennessee:
42,180 square miles (109,200 sq. km)

United States:
3,718,710 square miles (9,631,420 sq. km)

Compare Tennessee and Texas. How many times larger is Texas than Tennessee? Now answer the questions below.

ANALYZING How many times larger is the Sahel than Texas? How many times smaller is the Sahel than the United States?

Image courtesy NASA/Goddard Space Flight Center Scientific Visualization Studio

Climates, Biomes, and Resources

GUIDING QUESTION *How does the climate affect life and resources in the Transition Zone?*

There are two hot seasons in the African Transition Zone. One occurs between February and April and the other between September and October. The rainy season occurs between May and August, with most of the rain falling in the southern part of the subregion. Some years, this results in a fair amount of rain. However, the rain may be isolated to one area or be so intense that it damages crops. Most of the rain occurs during the summer months. During December and January, a hot, dry wind known as the **harmattan** blows from the north. It often carries dust and sand from the Sahara. In western Africa, the harmattan is sometimes referred to as "the doctor" because its dryness—in contrast to the typical humid air of the subregion—is considered invigorating.

harmattan a hot, dry wind that blows from the northeast or east in the western Sahara

In most countries of the Transition Zone, water is such a precious resource that rain and life are one and the same. Rain helps determine climate, and in turn vegetation, in most parts of the subregion. In many areas, rainfall is the only water source. Over the past century, annual precipitation has decreased. For example, parts of interior Eritrea get almost no rainfall. Sadly, a vicious cycle of soil erosion, insufficient water, deforestation, and drought has plagued the area for decades. Long periods of drought and overuse of the land destroyed vegetation, making the land unusable. Droughts have caused crops to fail, killed livestock, and led to famines.

This vast biome is also an important home to a diverse population of flora and fauna. For much of the year, the African Transition Zone is a vast expanse of dry soil. During the rainy season, however, the Sahel comes alive with plant life. In addition to grasslands, the Sahel is home to the baobab tree and the jujube, whose fruit is used to feed herds. Although the summer months provide abundant food for animal life, at other times of the year wildlife must remain constantly on the move, foraging for water and vegetation. The diverse population of animals includes migratory birds, which use the wetlands as a rest and feeding area. Wild dogs, cheetahs, lions, elephants, giraffes, warthogs, and gerbils are some of the many different animals that roam the African Transition Zone.

Farming and nomadic herding remain the important traditional economies of the subregion. Unfortunately, deforestation and drought are making these lifestyles more difficult. However, new discoveries and better use of oil, natural gas, and coal, as well as the mining of uranium, gold, and iron deposits may help the people of the Transition Zone develop new economies.

☑ **READING PROGRESS CHECK**

Determining Importance Why is it important that the people of the Transition Zone manage their resources in the most efficient way?

LESSON 1 REVIEW

Reviewing Vocabulary
1. *Making Connections* Why is the Sahel called a transition zone?

Using Your Notes
2. *Summarizing* Use your graphic organizer on the geography of the African Transition Zone to write a paragraph summarizing how the land supports the people, flora, and fauna of the subregion.

Answering the Guiding Questions
3. *Finding the Main Idea* What is the Transition Zone?

4. *Classifying* What rivers flow through the Transition Zone?

5. *Identifying Central Issues* How does the climate affect life and resources in the Transition Zone?

Writing Activity
6. *Informative/Explanatory* In a paragraph, discuss how the climate affects the people and the flora and fauna of the Transition Zone.

netw⊙rks

There's More Online!

☑ **IMAGE** Dogon Mask

☑ **IMAGE** Mine Workers

☑ **MAP** Kingdoms and Empires of the Transition Zone

☑ **INTERACTIVE MAP** Crisis in Darfur

☑ **TIME LINE** Droughts and Hunger

☑ **INTERACTIVE SELF-CHECK QUIZ**

☑ **VIDEO** Human Geography of the Transition Zone

Reading **HELP**DESK

Academic Vocabulary

- **enhance**
- **potential**

Content Vocabulary

- **domesticate**
- **animist**
- **sanitation**
- **patriarchal**
- **clan**
- **nuclear family**
- **oral tradition**
- **subsistence farming**

TAKING NOTES: *Key Ideas and Details*

IDENTIFYING As you read about the human geography of the African Transition Zone, use a graphic organizer like the one below to identify examples of how history, culture, and economics have worked together to create the Transition Zone of today.

(tl)©Stapleton Collection/Corbis, (tcl)©Alain Nogues/Sygma/Corbis, (tc)Ton Koene/Alamy, (tcr)Daphne Ouwersloot/Alamy, (tr)Eric Feferberg/AFP/Getty Images

LESSON 2
Human Geography of the Transition Zone

ESSENTIAL QUESTION · *How do physical systems and human systems shape a place?*

IT MATTERS BECAUSE

The Transition Zone has been home to many peoples, empires, and kingdoms for millennia. European colonization in the late 1800s and early 1900s had lasting effects on the subregion. The peoples of the Sahel maintain their distinct cultures and traditions but face many similar challenges.

History and Government

GUIDING QUESTION *What factors influenced the formation of the countries of the Transition Zone?*

Between about 9000 B.C. and 4000 B.C., the northern half of Africa received greater amounts of rain than today. At that time, what is now the Sahara was a savanna of extensive grasslands, lakes, and rivers. Nomadic people who once hunted for food settled in places, adopted agriculture, and **domesticated**, or tamed, animals. Around 4000 B.C., however, a climate shift began to occur. It became hotter and drier. Some scientists think that this shift happened quite abruptly, perhaps in as few as 400 years beginning in about 2000 B.C. As a result of the changes, the people in the farming communities migrated south.

Although much of the area became desert, the Nile Valley remained well watered, giving rise to Egypt and its civilization. Between about 2000 B.C. and 1000 B.C., the Egyptians pushed south, bringing various peoples along the Nile under their control. When Egypt's authority weakened, these peoples rose to power.

One of the civilizations Egyptians encountered as they pushed south was Kush. Around 2000 B.C., the Kush river civilization became a powerful kingdom in what is now Sudan. From about 2000 B.C. to about 1500 B.C., the Kushites controlled the Egyptian territory to the north. After retreating from Egypt, the Kushites pushed south and built a civilization around a new capital, Meroë. Kush flourished until about A.D. 300, when Meroë was attacked by Axum, a trading empire in Ethiopia. Axum was a great trading power from about A.D. 100 to A.D. 700.

Kingdoms and Empires of the Transition Zone

Axum, c. A.D. 350
Kush, c. 590 B.C.–A.D. 350
Mali, c. 1200–1500
· · · Trade route
Songhai, c. 1450–1590
Ghana, c. 700–1205
Hausa States, c. 1000–1800

GEOGRAPHY CONNECTION

Throughout recorded history, major civilizations have risen in Africa, particularly in the Transition Zone.

1. **ENVIRONMENT AND SOCIETY** How did location affect the rise of these early kingdoms?

2. **HUMAN SYSTEMS** Which kingdom overtook the Kushites?

domesticate to adapt plants and animals from the wild for human use

enhance to improve or increase

potential possible or likely

Empires of the Transition Zone

In addition to the kingdoms of Kush and Axum in the eastern part of the Transition Zone, new trading empires gained strength in the western part. The present-day country of Mali is named after one of them, the Mali Empire. One of Mali's early kings, Sundiata Keita, helped Mali flourish by expanding trade routes for gold and salt. He also conquered many of the surrounding territories. The most famous ruler of the Mali Empire was Mansa Musa. A Muslim, Musa **enhanced** the prestige and power of Mali through a famous pilgrimage to Makkah (Mecca). Accompanied by a huge entourage, Musa apparently dispensed so much gold on his journey to Makkah that the price of gold fell drastically in places he visited.

The Mali Empire became legendary in the Islamic world and Europe, and helped Islam spread. Mali, which extended west to the Atlantic, had as its center the wealthy city of Timbuktu. Another empire, Songhai (SAWNG•hy), grew rich from the gold-for-salt trade started in the western empire of Ghana. Songhai eventually took over Mali and expanded east. It prospered until about A.D. 1600, when the Moroccans overran it.

Word of the wealth of Africa's kingdoms reached Europe. As early as the 1200s, Timbuktu became an important center of trade in gold and salt between Africa and Europe. By the 1600s and 1700s, Europeans were trading extensively with Africa. They traded for African gold and other goods and for enslaved people.

Colonization and Independence

In the 1800s, European powers regarded the region as a source of raw materials and as a **potential** market for finished goods. European countries also laid claim to African territory. A conference, known as the Berlin Conference, was held between 1884 and 1885 to regulate European colonization. All of the Sahel was under European control by 1914.

European rulers knew little about Africa's political and social systems, and no Africans participated at the Berlin Conference. As a result, Europeans created colonial boundaries that often cut across cultural, religious, or traditional boundaries, merging **animist** cultures with Muslim societies. Animist cultures believe all elements of nature, such as animals and mountains, have spirits. The introduction of Christianity added to the tension. Religious friction set African peoples against one another and strengthened European rule.

In the mid-1900s, Africans began to demand a share in government. Educated Africans launched independence movements. In the second half of the century, the colonies became independent of European rule.

These new countries faced difficult challenges, often the result of their colonial legacy. European powers, for example, set up colonial economies that met European, rather than African, needs. In addition, colonial governments did not involve Africans much in government, nor did they give Africans models for democracy. At independence, many of the new countries kept the political boundaries set by the colonial powers. Within the new countries, rival ethnic groups struggled for power. Civil wars erupted.

Conflicts also arose between countries in the Sahel. Border disputes have lasted for years between Somalia and Ethiopia. The collapse of governments as a result of warring factions, drought, and famine further weakened the newly independent countries. While some countries in the African Transition Zone continue to struggle, new opportunities from the discoveries of oil, gas, and uranium deposits may help to bring stability if governments can manage these resources properly.

Conflict in Sudan

The countries in the Transition Zone are teeming with cultural differences. At times, these differences have led to conflicts such as between Sudan's north and south. Arabic-speaking Muslims live mostly in the northern cities and favor Islamic-oriented governments. People in the south live mostly in rural areas, are focused on a subsistence economy, and prefer a secular government. These differences led to a conflict in which nearly 300,000 people died and an estimated 2.7 million people were displaced between 1983 and 2005.

A peace agreement was finally signed in 2005, which provided considerable independence for Sudan's southern provinces. Although the agreement ended the conflict between the north and the south, it did not address the conflict in the western Darfur region of the country. The civil war in Darfur occurred because non-Arab Sudanese accused the government of favoring Sudanese Arabs. Finally, in 2011, the southern provinces of Sudan held a referendum and voted for independence. South Sudan is now an independent country, with Juba as its capital.

More recently, in January 2011, South Sudan shut down all of its oil fields after a disagreement about the fees Sudan demanded to transport the oil. In May 2011, Sudan seized control of Abyei, a disputed oil-rich border region, after three days of clashes with South Sudanese forces. On September 27, 2011, the presidents of Sudan and South Sudan signed agreements of cooperation. The status of Abyei, however, was not addressed. The future of the region is still uncertain.

☑ **READING PROGRESS CHECK**

Identifying Cause and Effect In what ways has colonialism affected the countries of the Transition Zone?

Analyzing
PRIMARY SOURCES
The Conflict in Sudan

"We left because of war. For the last one and a half years we have been bombed by planes every day. We lived in the forest; there was no chance for school for the children, no healthcare or medicine. We got food from the ground, but not corn. We would collect water in the early mornings. This has happened all seasons."

—a 36-year-old mother of nine at the Jamam refugee camp in South Sudan, from Médecins Sans Frontières, January 18, 2013

DBQ **IDENTIFYING CAUSE AND EFFECT** In what ways did the people of Sudan suffer from the conflict between the north and the south?

animist pertaining to traditional religious beliefs in which nature and objects, such as animals and mountains, are thought to have spirits

Population Patterns

GUIDING QUESTION *What are the population patterns of the Transition Zone?*

The African Transition Zone is more than just a geographical transition—it is also a cultural transition between Muslim North Africa and the animist and Christian south. Many different cultural groups inhabit the changing environment of the Transition Zone.

There are not only differences in religious beliefs, but also differences in ways of life. Herders, farmers, nomads, and city dwellers make up the people of the Transition Zone. Despite rapid population growth, population density currently remains low. The average population density of the subregion is about 58 people per square mile (23 people per sq. km). Only a small portion of the total land area, however, is suitable for agriculture. Thus, population throughout the subregion is not evenly distributed.

According to some studies, the population of the countries in the Transition Zone will reach 100 million by 2020 and 200 million by 2050. More than half of these people are expected to live in Burkina Faso, Mali, and Niger. It is difficult to imagine how the environment can support such a large population.

While there are relatively few cities, there are several important ones. The ancient city of Timbuktu, now known as Tombouctou, is still standing. Nouakchott, the capital of Mauritania, has a population of about 710,000 people. Sudan has two important cities as well. Khartoum, its capital, has a population of over 5 million. The second, Port Sudan, handles the bulk of Sudan's external trade.

Cities in the Sahel will continue to grow in the years to come. Increasing drought and environmental damage have spurred people who once lived in rural areas to migrate to cities such as Niamey in Niger and Bamako in Mali.

☑ **READING PROGRESS CHECK**

Gathering Information Which religions can be found in the Transition Zone?

©Stapleton Collection/Corbis

TIME LINE ⌄

DROUGHTS and HUNGER →

For centuries, the peoples of the Sahel have experienced famine brought about by droughts. As the droughts have become more severe in recent years, climate scientists have collected evidence to determine if they are, in part, caused or intensified by human activities.

▶ **CRITICAL THINKING**

1. *Identifying Cause and Effect*
What have been some of the effects of the droughts in the Sahel?

2. *Analyzing* What efforts have been made to deal with the effects of drought?

1600 →
European travelers observe a major drought in the Sahel and create the first modern record.

1640

1740s–1750s
The Great Famine in northern Nigeria, Niger, and Mali leads to massive migration and disruption of trade.

1800 →

1820s–1830s
Major droughts lead to famine from Senegal to Chad and contribute to decline of Bornu Empire

Human Impact

GUIDING QUESTION *What causes desertification and water shortages?*

Desertification puts pressure on the people that live in the Transition Zone. Agriculture and livestock suffer, as does the natural biodiversity of the area. In addition to infrequent rainfall, various human activities also cause desertification. One example of harmful human management is overgrazing.

Overgrazing is caused by an excessive number of livestock feeding too long in one area. Overgrazing kills plant roots. Too many animals in one area also compact the soil, thus reducing its capacity to hold water. Other detrimental human activities include poor agricultural practices and deforestation, which include the stripping of trees for firewood and clearing of land for farming. The main trigger of these harmful human activities is overpopulation. When more people live on the land, more livestock and crops are needed for food and more trees are needed for use as fuel. Just as drought and poor land use practices led to the Dust Bowl in the Great Plains of the United States, these factors have hastened desertification in the Sahel. A growing population dependent on the land in an already arid region puts further stress on that land. This stress triggers even more harm to the environment, in turn leading to more desertification.

Human activity in the Transition Zone has exacerbated desertification and created water management problems. For example, the availability of water in the Lake Chad Basin has decreased not only because of climate change but also because of over-demand. Planners have recognized that because of very high evaporation rates, it will not be enough to efficiently manage the water supply. The quality of the water that remains is also a concern. Commercial cotton and rice farmers are using agricultural chemicals that are polluting the water in Lake Chad. In other areas, poorly planned irrigation projects have pumped out too much groundwater, which causes wells to go dry. After the drought of the late

desertification the destruction of land in arid and semi-arid climates

circumstance an event or fact that accompanies or determines another

stress pressure or strain

GEOGRAPHY CONNECTION

The risk for desertification in the Sahel varies but overall is increasing.

1. **THE WORLD IN SPATIAL TERMS** Which biomes border the fragile zone?

2. **ENVIRONMENT AND SOCIETY** Why would it be important for people and governments to know which areas were in the high risk fragile zone?

The Sahel's Vulnerable Zone

50 days Length of rainy season
Limit of the fragile zone
Desert region (nomadic herding)
High risk fragile zone
Average risk fragile zone
Capital city

1960s—which lasted until the 1980s—humanitarian aid poured into the region in an effort to create water resources for the people. Though well-intentioned, the results were not beneficial. The irrigation systems that were built attracted disease-carrying insects and could not withstand the frequent droughts. The systems also interrupted the natural flow patterns of water. Boreholes and wells that were dug also added to the problem. Many boreholes and wells were dug in areas unsuitable for livestock and agriculture. This led to larger herds occupying smaller areas and resulted in further overgrazing and desertification.

☑ **READING PROGRESS CHECK**

Summarizing How has human activity affected the Transition Zone?

Addressing the Issues

GUIDING QUESTION *How are environmental problems in the Sahel being addressed?*

Once an ecosystem is damaged, it is difficult to reverse the effects. However, by improving water management and land use practices, some remediation may be possible. Water management improvements include better crop varieties and new water conservation technologies. Many of the efforts to stave off desertification are headed by international groups such as the International Atomic Energy Association (IAEA). In June 2012, the IAEA began a project with 13 countries in the region. The project aims to enhance knowledge and understanding of the five large aquifers in the Sahel, the source of most freshwater in the subregion.

The United Nations Convention to Combat Desertification (UNCCD) also works to stop desertification of the Sahel. The UNCCD helps to reduce the effects of drought and desertification and to restore land productivity in dry regions. Within the UNCCD, groups such as the Committee on Science and Technology and a Roster of Experts create action plans or programs to combat specific aspects of desertification.

This aerial view of a town in the Darfur region of Sudan shows the barren landscape produced by desertification.

▼ **CRITICAL THINKING**

1. Analyzing Visuals As shown in this photograph, describe the landscape of areas affected by desertification.

2. Drawing Conclusions Why is it important to reverse desertification in the Transition Zone?

©Richard Baker/In Pictures/Corbis

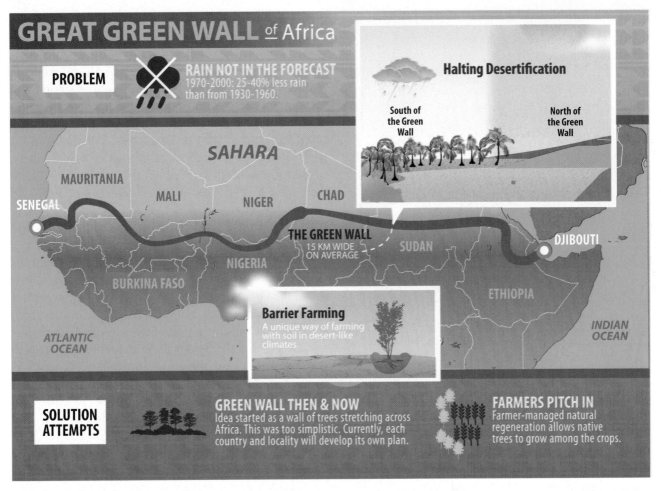

GREAT GREEN WALL of Africa

PROBLEM

RAIN NOT IN THE FORECAST
1970-2000: 25-40% less rain than from 1930-1960.

Halting Desertification

South of the Green Wall

North of the Green Wall

SAHARA

MAURITANIA

MALI

NIGER

CHAD

SENEGAL

THE GREEN WALL
15 KM WIDE ON AVERAGE

SUDAN

DJIBOUTI

NIGERIA

BURKINA FASO

ETHIOPIA

ATLANTIC OCEAN

INDIAN OCEAN

Barrier Farming
A unique way of farming with soil in desert-like climates.

SOLUTION ATTEMPTS

GREEN WALL THEN & NOW
Idea started as a wall of trees stretching across Africa. This was too simplistic. Currently, each country and locality will develop its own plan.

FARMERS PITCH IN
Farmer-managed natural regeneration allows native trees to grow among the crops.

The Great Green Wall is an ambitious project on the part of 11 countries to combat desertification. It will stretch across the continent. Expectations are high. One way the wall is expected to help is by blocking desert winds, which dry the soil.

▲ **CRITICAL THINKING**

1. *Analyzing Visuals* How will the Great Green Wall combat desertification?

2. *Drawing Conclusions* Why is it important for farmers to support the project?

Another group is the Permanent Interstate Committee for Drought Control in the Sahel (CILSS). The CILSS was established in 1973 following the severe drought that struck the region. Current member states include Gambia, Guinea-Bissau, Mauritania, Senegal, Burkina Faso, Mali, Niger, Chad, and Cape Verde. The CILSS works to improve food security and living conditions, water and land management, and natural resources management in the Sahel.

Small-scale efforts that local farmers can employ have been the most successful changes to date. Exchange of information on best farming and water management practices can be particularly helpful. Simple water-saving irrigation techniques, such as drip irrigation, have also proved useful. However, the challenge to adapt new technologies to local circumstances remains.

On a small scale, local people are also discovering ways to improve their land. After the drought in the 1980s, farmers in Burkina Faso started to experiment with traditional planting pits to reclaim severely dry land. These farmers increased the depth and width of the pits and began adding manure at the bottom before planting. These changes improved soil fertility and production. The techniques allowed farmers to increase their yield, prevent further soil erosion, and improve the soil quality. This technique has spread throughout the surrounding countries and has helped to rehabilitate between 500,000 and 750,000 acres (200,000 and 300,000 ha).

Various projects are underway to combat desertification. This man in Tombouctou, Mali, is irrigating a field.

▲ CRITICAL THINKING

1. *Analyzing Visuals* What purpose do you suppose the shallow trenches serve?

2. *Classifying* How would you describe the environment shown in the photograph?

There are differing techniques and views on how to combat desertification, and the problem is not confined to the Sahel. In some areas, such as Australia, scientists are monitoring when and where desertification occurs to better understand why it happens. They hope this information will aid in finding an appropriate solution that can be applied no matter where desertification is a problem. Others are thinking about radical solutions such as adding bacteria to sand dunes that would calcify, or harden, the sand. The hope is that the hardened sand would create a wall between the deserts and semi-arid land on the other side.

Others think that applying better water and land use management can reverse the problem. This has led to a unique initiative by many of the African countries within the Sahel to stop the spread of the Sahara. It is called the Great Green Wall, and it is a $2 billion project backed by the United Nations. The name comes from the plan that calls for a wall of trees to divide the semi-arid Sahel from the Sahara. The goal is for the trees to stop the southward spread of the Sahara. When it is completed, the Great Green Wall will stretch 4,970 miles (8,000 km) from Senegal to Djibouti in a band roughly 9 miles (15 km) wide. Planners expect that the barrier will be wide enough to block desert winds and will help hold moisture in the soil and also help to reduce soil erosion.

Senegal was one of the first countries to support the Great Green Wall project, and it has already benefited from the initiative. Senegal hopes that the project will help increase food security and prevent poverty. The planners hope to support local community efforts to sustainably manage the trees that are planted. To help build local support, seven varieties of acacia tree—including one that produces a fruit that can be used as animal food—were planted. People in local communities have harvested vegetables from gardens planted near the trees and fenced to keep out livestock. People have improved their nutrition, and they been able to sustain themselves throughout the year without having to leave for work in towns.

☑ READING PROGRESS CHECK

Making Connections How is the Permanent Interstate Committee for Drought Control in the Sahel working to combat desertification in the Transition Zone?

frans lemmens/Alamy

LESSON 3 REVIEW

Reviewing Vocabulary

1. *Determining Importance* Write a paragraph discussing some of the causes of desertification.

Using Your Notes

2. *Applying* Use your graphic organizer on the environment in the Transition Zone to write a paragraph evaluating the problems and some possible solutions.

Answering the Guiding Questions

3. *Identifying Central Issues* What water challenges do people of the Sahel face?

4. *Identifying Cause and Effect* What causes desertification and water shortages?

5. *Finding the Main Idea* How are environmental problems in the Sahel being addressed?

Writing Activity

6. *Argument* In a paragraph, explain what may happen if the environmental problems of the Transition Zone are not resolved.

Directions: On a separate sheet of paper, answer the questions below. Make sure you read carefully and answer all parts of the questions.

Lesson Review

Lesson 1

1 *Defining* Why is the region immediately south of the Sahara called the Transition Zone?

2 *Identifying Cause and Effect* How has the climate of the Sahel affected Lake Chad?

3 *Problem Solving* What are some ways that countries in the Transition Zone manage water resources?

Lesson 2

4 *Describing* Describe how culture changes as one moves from north to south in the Transition Zone.

5 *Making Predictions* Why are urban populations in the Sahel likely to grow in the years to come?

6 *Identifying Cause and Effect* What are some ways that poverty affects people in the Transition Zone?

Lesson 3

7 *Explaining* How has human activity in the Transition Zone contributed to desertification?

8 *Evaluating* After the drought that lasted from the late 1960s to the 1980s, irrigation systems were built to create water resources for people. Why were these systems ineffective?

9 *Analyzing* What steps have international organizations taken to reduce desertification in the Sahel?

Critical Thinking

10 *Explaining* In the mid-1900s, colonies in the Sahel gained independence from European powers. How did the legacy of colonialism create problems for the new countries?

11 *Comparing and Contrasting* Compare the Great Green Wall project to the proposal to create a wall of hardened sand. Which project do you think is more likely to stop the Sahara from moving south?

21st Century Skills

Use the chart to answer the questions that follow.

Country	Imports	Exports
Burkina Faso	Capital goods, foodstuffs, petroleum	Gold, cotton, livestock
Chad	Machinery & transport equipment, industrial goods, foodstuffs, textiles	Oil, cattle, cotton
Senegal	Food & beverages, capital goods, fuels	Fish, groundnuts, petroleum products, phosphates, cotton
Sudan	Foodstuffs, manufactured goods, refinery & transport equipment, medicines & chemicals, textiles, wheat	Gold, oil & petroleum products, cotton, sesame, livestock, groundnuts, sugar

Source: CIA World Factbook

12 *Economics* What general statement can you make about the types of commodities these four countries export?

13 *Compare and Contrast* How do Chad's imports differ from Senegal's imports? How are Senegal's imports similar to those of Burkina Faso?

Need Extra Help?

If You've Missed Question	1	2	3	4	5	6	7	8	9	10	11	12	13
Go to page	484	486	486	491	492	493	497	498	498	491	500	501	501

Directions: On a separate sheet of paper, answer the questions below. Make sure you read carefully and answer all parts of the questions.

College and Career Readiness

14 **Problem Solving** As a researcher for the United Nations Convention to Combat Desertification, you have been asked to report on how farmers in Burkina Faso used traditional planting pits to reclaim severely dry land. Write a two-page report describing their techniques. Do you recommend that these techniques be used more widely? Why or why not?

DBQ Analyzing Primary Sources

Use the document to answer the following questions.

Long regarded as a failed state, Somalia has had persistent problems dealing with hunger.

PRIMARY SOURCE

"*Food is power in Somalia. Militia groups have routinely descended on the arable lands of central Somalia during harvest and claimed the crops for themselves. Pirates on the Indian Ocean have waylaid dozens of foreign vessels bearing food aid. Food prices were high here even before last year's worldwide spike, thanks to drought, militia roadblocks, and a devalued currency. The result is that millions now depend on food aid. The fresh fighting is pushing the country toward an unprecedented humanitarian crisis.*"

—Robert Draper, "Shattered Somalia," *National Geographic Magazine*, September 2009

15 **Interpreting** Why does the author say that in Somalia "food is power"?

16 **Speculating** How might geography have affected the availability of food in Somalia?

17 **Evaluating** Is food aid an effective way to address the problem of hunger? Why or why not?

Applying Map Skills

Use the Unit 6 Atlas to answer the following questions.

18 **Physical Systems** Use your mental map of Africa to describe the location of two or more major rivers in the Transition Zone. What countries do they connect?

19 **Environment and Society** What regions in the Transition Zone are most likely to be affected by desertification in the coming years?

20 **Human Systems** How do bodies of water affect population density in the Transition Zone?

Exploring the Essential Question

21 **Making Connections** Recall what you have learned about the problem of water shortage in the Transition Zone. Create a poster showing how drought, human activity, and desertification interact as part of a cycle.

Research and Presentation

22 **Research Skills** Use Internet and other resources to gather information and write a report about recent developments in the conflict in Sudan. What has happened since the presidents of Sudan and South Sudan signed agreements of cooperation? Has the conflict in Darfur been resolved?

Writing About Geography

23 **Informative/Explanatory** Use standard grammar, spelling, sentence structure, and punctuation to write an essay describing how drought has affected population patterns in the Transition Zone. How do these patterns reflect changes in traditional rural life and in the extended family?

Need Extra Help?

If You've Missed Question	14	15	16	17	18	19	20	21	22	23
Go to page	499	502	502	502	476	476	480	497	491	492

East Africa

ESSENTIAL QUESTION · *How do physical systems and human systems shape a place?*

◀ Many young East Africans enjoy both tradition and modern ways of life.

Roy Toft/National Geographic Stock

Geography Matters...

East Africa is a land of magnificent landscapes, from the volcanic peak of Kilimanjaro to the lakes of the Great Rift Valley. It is also, however, a land of desert and drought.

East Africa has attracted many countries over the centuries. Its prime location on the Red Sea and Indian Ocean drew the interest of people looking for trade outposts and entrances into the continent. Outsiders colonized almost all the land. Today, East Africa is working toward improving its economies, environments, and standards of living in the face of challenges such as population growth, corruption, and environmental degradation.

export crops
and East Africa

About 80 percent of East Africa's population relies on subsistence farming. However, the subregion's per capita agricultural income and production is one of the lowest in the world. Still, agriculture is of vital importance and makes up about 40 percent of the subregion's gross domestic product (GDP). Even when local farmers try to produce agricultural commodities, the global market is highly competitive, which makes it difficult for them to access.

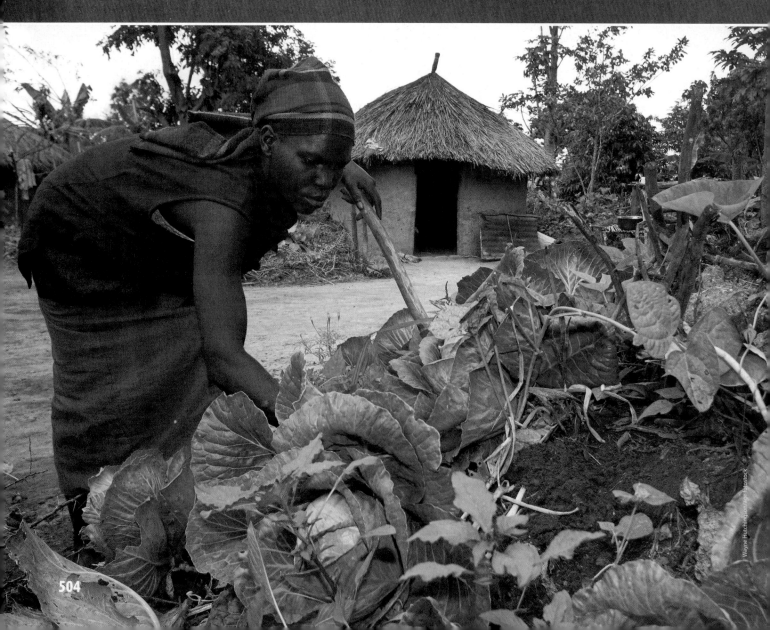

Wayne Hutchinson/age fotostock

Directions: On a separate sheet of paper, answer the questions below. Make sure you read carefully and answer all parts of the questions.

Lesson Review

Lesson 1

1 *Defining* Why is the region immediately south of the Sahara called the Transition Zone?

2 *Identifying Cause and Effect* How has the climate of the Sahel affected Lake Chad?

3 *Problem Solving* What are some ways that countries in the Transition Zone manage water resources?

Lesson 2

4 *Describing* Describe how culture changes as one moves from north to south in the Transition Zone.

5 *Making Predictions* Why are urban populations in the Sahel likely to grow in the years to come?

6 *Identifying Cause and Effect* What are some ways that poverty affects people in the Transition Zone?

Lesson 3

7 *Explaining* How has human activity in the Transition Zone contributed to desertification?

8 *Evaluating* After the drought that lasted from the late 1960s to the 1980s, irrigation systems were built to create water resources for people. Why were these systems ineffective?

9 *Analyzing* What steps have international organizations taken to reduce desertification in the Sahel?

Critical Thinking

10 *Explaining* In the mid-1900s, colonies in the Sahel gained independence from European powers. How did the legacy of colonialism create problems for the new countries?

11 *Comparing and Contrasting* Compare the Great Green Wall project to the proposal to create a wall of hardened sand. Which project do you think is more likely to stop the Sahara from moving south?

21st Century Skills

Use the chart to answer the questions that follow.

Country	Imports	Exports
Burkina Faso	Capital goods, foodstuffs, petroleum	Gold, cotton, livestock
Chad	Machinery & transport equipment, industrial goods, foodstuffs, textiles	Oil, cattle, cotton
Senegal	Food & beverages, capital goods, fuels	Fish, groundnuts, petroleum products, phosphates, cotton
Sudan	Foodstuffs, manufactured goods, refinery & transport equipment, medicines & chemicals, textiles, wheat	Gold, oil & petroleum products, cotton, sesame, livestock, groundnuts, sugar

Source: CIA World Factbook

12 *Economics* What general statement can you make about the types of commodities these four countries export?

13 *Compare and Contrast* How do Chad's imports differ from Senegal's imports? How are Senegal's imports similar to those of Burkina Faso?

Need Extra Help?

If You've Missed Question	**1**	**2**	**3**	**4**	**5**	**6**	**7**	**8**	**9**	**10**	**11**	**12**	**13**
Go to page	484	486	486	491	492	493	497	498	498	491	500	501	501

export crops
and East Africa

About 80 percent of East Africa's population relies on subsistence farming. However, the subregion's per capita agricultural income and production is one of the lowest in the world. Still, agriculture is of vital importance and makes up about 40 percent of the subregion's gross domestic product (GDP). Even when local farmers try to produce agricultural commodities, the global market is highly competitive, which makes it difficult for them to access.

Why do farmers in East Africa grow certain crops?

Many of the crops that are grown in East Africa are the legacy of colonialism. Colonial powers made the economic decisions about what crops would be grown for consumption and for export. They based these decisions on profit for themselves, not for the benefit of the African people. Many colonial governments set up plantations that each produced a single crop for export because export crops brought in the most income for the colonial government.

In East Africa, export crops included tea, coffee, tobacco, cotton, cashews, copra, and sisal. After decolonization, it was difficult or even impossible for the newly independent governments to modify this system. The result is the current situation, in which these countries are dependent on agricultural exports and there is little support for small-scale subsistence farmers.

1. **Human Systems** How has East Africa's colonial legacy influenced what crops are produced?

How has growing single crops for export affected the people in East Africa?

Reliance on the success of a single export crop can be risky. For example, if climate conditions are extreme—too hot or cold, or too dry or wet—a crop might be wiped out or limited for a season. Sometimes the single crop grown for export requires more labor or resources such as water and fertilizer to cultivate and harvest. Essentially, the crop requires a large investment in terms of human and capital resources. These factors can affect the return on investment.

Another factor is the fluctuation in market demand and price. If demand for a product is low, then the price at which a farmer can sell will be low as well. The decision to produce one single crop over another may result in the loss of resources that could have been used elsewhere. If the government decides that the best land should be devoted to growing one export crop, that land might be unavailable for growing food. Because many farmers in East Africa rely on subsistence farming for their livelihood, they are most affected by agricultural decisions.

2. **The World in Spatial Terms** How have agricultural decisions about export crops affected subsistence farmers in East Africa?

What steps have been taken to help farmers?

To achieve better crop yields in East Africa, some organizations are working on solutions. For example, the goal of the Smallholder Cash and Export Crops Development Project is to help small growers of coffee in Rwanda by setting up cooperatives. The hope is that by working together, they can produce high-quality coffee that will fetch a high price. In another project, public land is being made available to small growers of tea.

Also, East African governments have encouraged the use of modern farm technology, such as genetically engineered seeds, soil conservation, and fertilizers. However, these actions have created new marketing problems. Some countries in Europe have boycotted certain crops from East Africa to protest the use of some of these techniques.

3. **Environment and Society** Write a paragraph evaluating the actions taken to improve crop yields in East Africa. Propose an idea that governments or other organizations can initiate to help farmers in the subregion.

(l)©Tuul/Robert Harding World Imagery/Corbis, (c)Peter Bowater/Alamy, (r)Chicago Tribune/McClatchy-Tribune/Getty Images

THERE'S MORE ONLINE

READ about cash crops in East Africa • **EXPLORE** a map of East African natural resources

European Colonization

Vasco da Gama's voyage around the Cape of Good Hope in the late 1400s began the Portuguese exploration of the East African coast. The Portuguese established control over the Indian Ocean and brought Roman Catholicism to Ethiopia. Portuguese dominance led to the decline of Arab influence there, as well as to hostility toward foreign Christians and Europeans that lasted into the 1900s.

Others also made their way to East Africa. Railroads were built inland from the coast of the Indian Ocean. New medicines were discovered to treat tropical diseases and reduce health risks. Interior Africa gradually opened to immigrants from Europe and Asia who settled the region in the 1800s. A famous European explorer to the subregion was a British doctor and missionary David Livingstone. One of his goals was to end the slave trade, but he also called for Europeans to spread commerce, Christianity, and their civilization throughout Africa.

European powers—mainly Britain, France, Portugal, and Germany—took advantage of new opportunities. They competed fiercely to expand their empires and to protect their trade routes. In less than 40 years following 1880, Africa was divided into more than 40 colonial divisions with borders that ignored traditional tribal territories. In East Africa, present-day Eritrea came under Italy's control. With the opening of the Suez Canal, Eritrea's Red Sea location made it a valuable prize. Italy tried to establish itself in Ethiopia but was never fully successful. The British took control of Kenya and Uganda, while the Germans took control of Tanzania, Rwanda, and Burundi. Europeans also disrupted African village life by replacing traditional farming with plantation agriculture. New colonial economies focused on cash crops, such as coffee and tea, for export to world markets and ignored the crops relied upon by the people.

Colonialism created enormous problems for Africa's peoples. Some Africans attempted to resist colonial rule. Most, however, failed. Independence movements began to change the geopolitics of Africa in the 1950s. The colonies began winning self-government. Under increasing pressure, European powers granted

Paul Joynson Hicks/age fotostock

TIME LINE ⌄

A Cultural
CROSSROADS →

For centuries, traders came to East Africa from Europe, East Asia, and Southwest Asia. This trade resulted in significant cultural exchanges, as traders from these far-flung places worked with African traders.

▶ **CRITICAL THINKING**

1. *Describing* From which parts of the world did merchants come to trade in East Africa?

2. *Analyzing* How did the location of East Africa lead to it becoming a center of trade and cultural exchange?

1 A.D. →

1–100 Greeks have an established trade network along the East African coast.

600s Arabs develop trading stations along the coast of East Africa and begin spreading Islam.

late 900s Settlers from Arabia and Persia found the trading city of Kilwa on an island off the coast of Tanzania.

c. 1420s Zheng He travels to East Africa, bringing back exotic animals for the Chinese emperor.

a large number of colonies their independence in the 1960s. Independence, however, was not a simple cure for long-standing issues. These newly independent countries often faced internal strife and leadership issues. In Uganda, a dictator ruled for eight years in the 1970s, causing social disintegration, human rights violations, and economic decline.

Conflict between **indigenous** peoples resulted in extraordinary dislocations. In Rwanda and Burundi, the colonial powers had favored the Tutsi people and had given them positions of power over the majority Hutu. In 1959, the resentment among the Hutus sparked a violent rebellion. Conflict between the two peoples continued for decades. The country erupted in violence over the discrimination against Hutus in 1994. In this violent outbreak, called the Rwanda genocide, the Hutus killed hundreds of thousands of Tutsi people.

Conflict also arose between several African countries. Major issues include the location of national borders and control of natural resources. Governments collapsed and periods of military rule followed. Famine and drought further weakened some of the newly independent countries. However, people continued to work toward nation building and improving the quality of life for citizens.

indigenous native to a place

✔ **READING PROGRESS CHECK**

Describing Describe some effects of European colonization on East Africa.

Population Patterns

GUIDING QUESTION *How is the population of the subregion distributed?*

With many different indigenous peoples and so many peoples coming to East Africa over the centuries, the population of the subregion is as diverse as the terrain. East Africa's population today reflects the many languages, belief systems, and ways of life that originated here and that have been introduced into this subregion. For example, European and Arab influences are everywhere.

(l)Leemage/Universal Images Group/Getty Images, (r)Cordier Sylvain/hemis.fr/Getty Images

Portuguese explorer Vasco da Gama reaches Tanzania.

1498

The Portuguese control most of the East African coast.

1506

1450 ➔

1527

Portuguese cartographer Diego Ribero is the first to depict the Comoros Islands on a map.

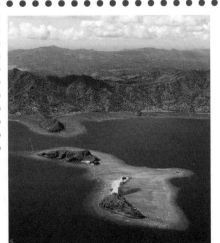

Muslim traders from the Indian Ocean coast exchange weapons, cloth, and beads for ivory and slaves in Uganda.

1840s

1800 ➔

1699

The Portuguese are driven out of Zanzibar by Omani Arabs.

East Africa **513**

East Africa's population can be found in the cities as well as in the countryside.

(l)©Torleif Svensson/Corbis, (r)Tom Cockrem/Lonely Planet Images/Getty Images

▲ **CRITICAL THINKING**

1. *Contrasting* What are some differences between the nomad and the city dweller?

2. *Drawing Conclusions* Why do you think that most people of European descent in East Africa live in the cities?

approximately about or nearly

lingua franca a common language used among people with different native languages

incorporate to blend or thoroughly combine

Despite the mingling of cultures, in many East African countries, one ethnic group forms the majority of the population. For example, the Bantu are in the majority in Uganda and Tanzania. In Rwanda and Burundi, the Hutu are the majority. In contrast, Ethiopia has more than ten distinct ethnic groups.

East Africa is home to city dwellers, nomads, and farmers. Geography has played an important role in population distribution. Due to widely diverse land and climate characteristics, settlement is varied. Most urban centers lie along the ocean coasts or along major rivers. However, some inland cities—such as Nairobi in Kenya and Addis Ababa in Ethiopia—have developed along trade routes. Urbanization is low in East Africa, varying from **approximately** 26 percent of the population in Tanzania to 11 percent in Burundi. Fresh water is the magnet for people in East Africa. Population densities are high on the fertile lands near water since crop irrigation increases yields.

☑ **READING PROGRESS CHECK**

Understanding Relationships What are some ethnic groups who still influence the subregion today?

Society and Culture Today

GUIDING QUESTION *What is life like for the people of East Africa?*

With the movement of many ethnic groups through East Africa over the centuries, languages, religions, and ways of life are a mosaic of influences. However, many indigenous traditions and languages, differing religious practices, and vibrant forms of art remain in the culture of today.

There are hundreds of indigenous languages spoken in East Africa. Because of European colonization, however, English and French often serve as a **lingua franca**, or common language. They are spoken throughout the subregion. In addition to their religion, Islam, Arab settlers brought their language, Arabic, to East Africa, and it is a secondary language in the region. Some indigenous languages have **incorporated** European, Arabic, and other elements and are widely spoken. An example is Swahili in Kenya and Tanzania. Swahili combines elements from local Bantu languages, Arabic, Persian, Portuguese, Hindi, and English. It has tens of millions of speakers in East Africa and other areas as well.

Because of the Arab and European influences of the past, most people in the subregion are either Muslim or Christian. But traditional animist religions are (l)Kimberley Coole/Lonely Planet Images/Getty Images, (c)Christian Aslund/Lonely Planet Images/Getty Images, (r)©Godong/Robert Harding World Imagery/Corbis

What challenges do women in West Africa face?

Women in West Africa are equal to men under the law. In reality, however, they face many hardships. Many are forced into marrying much older men at a young age. When women marry early, they have few opportunities for education. Some of these women are in polygamous marriages. Polygamy is the practice of having more than one spouse at a time. While illegal in some parts of the world, polygamy is a cultural and religious tradition in many others. It is legal in parts of West Africa. Polygamy is particularly common in agricultural societies where women do most of the farmwork. Multiple wives and their children provide the labor necessary for a man to expand his agricultural holdings. Polygamous marriages offer little financial security to women.

Studies show that when women are able to earn income or receive benefit payments, they spend more money on education and health care for their children. This promotes economic growth and development.

1. **Human Systems** How does tradition in West Africa affect women's rights?

Why do these conditions exist?

Despite efforts to improve the status of women, discrimination and other challenges persist. Many gender inequalities are deeply rooted in cultural and religious beliefs and practices across West Africa. These traditions support limiting opportunities for women and girls in education, the economy, and political life. Rural areas of West Africa are more likely to hold onto traditional ways of life. Additionally, it is more difficult to monitor and enforce laws in rural areas. Fewer opportunities for education and health care are available in the countryside, and transportation to cities is often difficult and expensive. Isolated, the women of rural West Africa have few options.

These deeply held traditions and beliefs have prevented some countries from signing the Maputo Protocol (also known as the Protocol to the African Charter on Human and Peoples' Rights on the Rights of Women in Africa), which took effect in 2005. The goal of the protocol is to end discrimination and violence against women while also promoting women's rights in Africa. However, some countries have refused to sign or ratify the protocol, citing issues with its stances on polygamy, contraception, and abortion.

2. **Human Systems** Why have some West Africans been resistant to supporting the equality of women?

What steps have been taken to address these issues?

Organizations such as the Office of the United Nations High Commissioner for Refugees are working to change the nationality laws for states that do not provide equal rights to women. Gender inequality in nationality laws can prevent women from passing on their citizenship to their children. Children without citizenship face many problems, including little or no access to education and health care. Sierra Leone is one West African country that has made partial reform in this area.

In 2012 the New Partnership for Africa's Development (NEPAD) and the Economic Community for West African States (ECOWAS) signed an agreement for the empowerment of West African women who live in rural areas. This grant of one million euros will offer funding for an organization that supports African female entrepreneurs in agriculture and related endeavors.

3. **Places and Regions** How have other regions around the world worked to promote change and boost economic development for West African women?

THERE'S MORE ONLINE

VIEW a West African gender population map • *READ* about the changing lives of West African women

There's More Online!

- ☑ **IMAGE** The Akosombo Dam
- ☑ **IMAGE** The Senegal River
- ☑ **IMAGE** Oases
- ☑ **INFOGRAPHIC** Elevation Profile of West Africa
- ☑ **INTERACTIVE SELF-CHECK QUIZ**
- ☑ **VIDEO** Physical Geography of West Africa

LESSON 1
Physical Geography of West Africa

ESSENTIAL QUESTION · *How do physical systems and human systems shape a place?*

Reading HELPDESK

Academic Vocabulary

- **parallel**
- **environment**

Content Vocabulary

- **reservoir**
- **river plain**
- **conflict diamonds**

TAKING NOTES: *Key Ideas and Details*

IDENTIFYING As you read about the physical geography of West Africa, use a graphic organizer like the one below to note the characteristics of the land, water, and biomes of the subregion.

Physical Geography of West Africa		
Land	Water	Biomes

IT MATTERS BECAUSE

West Africa has a rich and diverse landscape, which includes shoreline, lakes, rivers, low plains, and highlands. These landscapes and the subregion's resources are important to the people for many reasons. West Africa is rich in resources and minerals such as oil, natural gas, uranium, gold, and diamonds.

Landforms

GUIDING QUESTION *What landforms dominate the West African landscape?*

West Africa includes the southern half of the bulge of the African continent that extends west into the Atlantic Ocean. West African countries include Benin, Cape Verde, Côte d'Ivoire, Gambia, Ghana, Guinea, Guinea-Bissau, Liberia, Nigeria, Senegal, Sierra Leone, and Togo. The African Transition Zone cuts through the northern part of this subregion. The land in the African Transition Zone changes from deserts to a tropical savanna. All of the countries of West Africa, except Cape Verde, have part of their territory in the Transition Zone. The northern border of West Africa is the Sahara.

West Africa is a large area. Its size contributes to its considerable diversity, as does its location in the Transition Zone. West Africa is largely a tropical region with a landscape that includes desert, shoreline, low plains, highlands, and rain forests.

West Africa has a long coastline. The coastal region is made up of sandy beaches, thick mangrove swamps, lagoons, and broad coastal plains. Lagoons are shallow bodies of water that are separated from the ocean by islands, which lie **parallel** to the shoreline. The Ebrié Lagoon is the largest lagoon in West Africa with a surface area of 218 square miles (566 sq. km). Rain forests once covered much of the West African coast. However, rain forests have been drastically reduced as land was cleared for agriculture and logging of prized hardwood trees such as African mahogany and iroko, also known as African teak.

530

(tl)©iStockphoto.com/szymanskim, (tc)Bob Burch/Photoshot, (tr)Author's Image/PunchStock

Mountains rise up behind the broad coastal plains of the subregion. In the south-central part of West Africa are the Guinea Highlands. This mountainous plateau rises several thousand feet above the coastal plains in southeastern Guinea, northern Sierra Leone, Liberia, and northwestern Côte d'Ivoire. The plateau contains the Nimba Range, the Loma Mountains, and the Tingi Mountains. These highlands are covered with savanna and rain forest.

In the southeast are the Cameroon Highlands, which lie between Nigeria and Cameroon. Forest and grasslands with a rich biodiversity cover these highlands. They form part of a chain of former volcanoes that stretch inland from the sea. The largest and only active one of these volcanoes is Mount Cameroon, which rises to a height of 13,353 feet (4,070 m).

Low plains, sandy soil, and grasslands cover much of inland West Africa. The Sahara, the largest desert in Africa, is in the north. It is a vast and nearly uninhabited region. The landforms here are shaped mostly by winds and infrequent rainstorms. They include sand dunes, salt flats, gravel plains, stone plateaus, and dry valleys.

parallel lying in the same direction with an equal distance between

✔ **READING PROGRESS CHECK**

Identifying What types of landscapes are found in coastal regions?

Water Systems

GUIDING QUESTION *How are saltwater and freshwater resources important to the people of West Africa?*

The water systems of West Africa are not only important to wildlife but to people as well. The lagoons and mangrove swamps are important parts of the coastal ecosystem of the subregion. They provide food and shelter for fish, shellfish, mollusks, wildfowl, and marine mammals. Flounder and bluefish use the

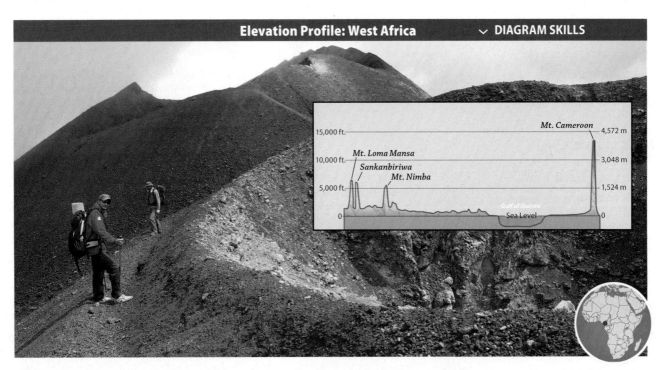

Elevation Profile: West Africa ⌄ **DIAGRAM SKILLS**

Mount Cameroon is the highest point in West Africa and an active volcano. It is much higher than the peaks in the Guinea Highlands.

▲ **CRITICAL THINKING**

1. *Classifying* What is the second-highest peak shown in the diagram?

2. *Analyzing Visuals* What appears to coat the surface of the mountain on which the hikers are standing?

©iStockphoto.com/szymanskim

marshes as nurseries and winter habitats. Ducks, geese, and other wild birds stop over during migration. Unfortunately, due to environmental disasters such as oil spills, global warming, and clearing of the mangrove for the farming of shrimp, these swamps are slowly disappearing.

The mangrove swamps are also very important to coastal people. They provide a source of food such as crabs, clams, oysters, and fish. Even the mangrove fruits themselves can be eaten. The mangrove trees are also useful. They are often collected for firewood and can be used for construction.

The Atlantic Ocean provides an important resource for the people of West Africa—fish. Fishing not only provides food for the people of the subregion, but is also a key source of revenue. Commercial fishing is particularly important in the Gulf of Guinea, which is part of the eastern tropical Atlantic Ocean system off the coast of western Africa. In 2006 a fishery committee was established to manage fishing among the countries of Liberia, Côte d'Ivoire, Ghana, Togo, Benin, and Nigeria. All these countries share fish stocks in the Gulf of Guinea.

Lake Chad, in west-central Africa, is bordered by Nigeria, Niger, Chad, and Cameroon. Lake Chad was once the second-largest wetland in Africa, supporting a great diversity of animal and plant life. Although Lake Chad has since shrunk by about 90 percent, it is still vital to the region for irrigation of farmland and fishing.

Lake Volta is a human-made lake and is the fourth-largest **reservoir** in the world. A reservoir is an artificial or natural lake where water is stored and used to supply farms, homes, and businesses in the area with freshwater. The lake was made by damming the Volta River, and its creation flooded more than 700 villages, forcing over 70,000 people to find new homes. The trade-off is that it has provided a consistent source of freshwater and electricity. Recently a project has begun to try to harvest the trees that were submerged when the lake was created. This project will provide additional revenue and work for the people of the region. Another new industry in the region is tourism. The town of Akosombo, which grew around the dam, is the starting point for many fishing excursions and for water sports. Fishing for Volta perch, African tiger fish, Nile tilapia, and several varieties of catfish is a favorite of charter expeditions.

reservoir a natural or artificial lake used as a source of water

Lake Volta, an artificial lake, was formed by the Akosombo Dam. Today the dam provides electricity for most of Ghana's needs.

▼ **CRITICAL THINKING**

1. *Analyzing Visuals* How do you suppose the landscape shown in the photo contrasts with its appearance before the construction of the Akosombo Dam?

2. *Explaining* Why is hydroelectricity an important resource in Ghana?

The number of fish in the Senegal River has increased since the Senegal River Basin Multi-Purpose Water Resources Development project began managing irrigation and water resources in the area.

◀ **CRITICAL THINKING**

1. *Analyzing* How can the health of waterways affect surrounding communities?

2. *Making Inferences* In what other ways can water resource management benefit surrounding communities?

While the Atlantic Ocean, Lake Chad, and Lake Volta are important to the economy and people of West Africa, the subregion also has a number of significant rivers. The Volta River, the main river system in Ghana, runs 1,000 miles (1,600 km) long. The northern four-fifths of the Volta River valley are now covered by Lake Volta, which was formed by the Akosombo Dam. The Volta River Basin covers an estimated area of 154,440 square miles (400,000 sq. km).

The Senegal River flows north from the Guinea Highlands through four countries: Guinea, Mali, Mauritania, and Senegal. The best agricultural land along the river is the **river plain** between the towns of Bakel and Dagana in Senegal. A river plain, also known as an alluvial plain, is a plain formed by the deposit of sediment over a long period of time by one or more rivers. After the yearly floods have retreated, crops such as millet, rice, and vegetables are sown. While agriculture is the largest economic activity along the river, fishing is the second largest. Additionally, two dams along the river, the Manantali and the Diama, provide electricity to the subregion.

The Niger (NY•juhr) River, which runs about 2,600 miles (4,183 km), is the third-longest river in Africa. The Niger is the main river of western Africa. Commercial shipping takes place on about 80 percent of the river. Fishing is also an important industry to the people who live along the Niger. The Benue River, the longest tributary of the Niger River, is about 673 miles (1,083 km) long. A considerable amount of trade moves along the Benue River, including petroleum, cotton, and peanuts.

The Niger River has two deltas, a vast inland delta and a delta where the river empties into the Atlantic Ocean. The inland delta is in Mali, where the river spreads out across the plains, creating a large wetland and agricultural area. The delta at the river's mouth is in Nigeria. At one time, fishing and the production of palm oil were strong industries in the region. However, the discovery of oil in the 1950s in the Niger Delta region has changed that. Oil production and spills have polluted the land and water. This pollution has destroyed the livelihoods of farmers and fishers of the Niger Delta.

river plain a plain formed by the deposit of sediment over a long period of time by one or more rivers

✔ **READING PROGRESS CHECK**

Understanding Relationships How do the water systems of West Africa benefit the people of the subregion?

Author's Image/PunchStock

Climates, Biomes, and Resources

GUIDING QUESTION *How do the tropical climates of West Africa support different biomes?*

environment natural surroundings

The **environment** of West Africa changes dramatically along a north-to-south latitudinal climate pattern. In the north it begins with the desert, which transitions into the semi-arid steppe, savanna grassland, and finally tropical forest. These environments have different climates, biomes, and resources.

In the north, where some countries of West Africa border the Sahara, the climate is hot and dry with very little rainfall. Parts of the African Transition Zone are steppe, with low-growing grasses, shrubs, and acacia trees. This area receives about 4 to 8 inches (10 to 20 cm) of rainfall a year.

As one moves farther south in West Africa, the land slowly becomes savanna. During the rainy season, the average rainfall here is about 15 to 25 inches (38 cm to 64 cm) a month. The rainy season begins in May and ends in November as warm, moist air is drawn from the Gulf of Guinea. At other times of the year, when the wind reverses and blows from the Sahara, rainfall averages about 4 inches (10 cm) a month. The land is covered with grasses and trees, such as the acacia and baobab. Many types of mammals and birds—such as giraffe, gerbils, foxes, elephants, and pygmy hippos—live in the savanna.

Still farther south, West Africa has tropical forests. Its tropical wet and dry forests, also known as tropical seasonal forests, receive rainfall during the rainy season followed by the dry and hot months. These tropical dry forests have less densely growing trees than the tropical wet forests farther south. The tropical wet forests receive plentiful rain and are abundant with semideciduous trees, evergreens, or semievergreens. They are also rich in spices, nuts, and legumes. These forests are at risk for deforestation as the human population increases.

conflict diamonds
diamonds that are mined in war-torn areas and are used to finance wars

Oil, natural gas, coal, gold, and uranium deposits are just some of the natural resources found in West Africa. Another major resource is diamonds. In recent years, diamonds from Africa have been referred to as **conflict diamonds**, or blood diamonds. Conflict diamonds are diamonds that are mined in a war zone. Often the money from the sale of diamonds in these areas is used to finance war. For example, during the conflict in Sierra Leone in the 1990s, rebels used forced labor to mine diamonds. These diamonds were then sold in neighboring countries and then in European markets. The money earned from these sales was used to finance the civil war in Sierra Leone. As people around the world became aware of conflict diamonds, efforts were made to prevent their purchase.

✓ **READING PROGRESS CHECK**

Summarizing What are some of the natural resources in West Africa?

LESSON 1 REVIEW

Reviewing Vocabulary
1. *Describing* Write a paragraph describing how reservoirs are used in West Africa.

Using Your Notes
2. *Summarizing* Use your graphic organizer on the physical geography of West Africa to write a paragraph summarizing the characteristics of the land, water, and biomes of the subregion.

Answering the Guiding Questions
3. *Categorizing* What landforms dominate the West African landscape?

4. *Identifying Central Issues* How are saltwater and freshwater resources important to the people of West Africa?

5. *Analyzing* How do the tropical climates of West Africa support different biomes?

Writing Activity
6. *Informative/Explanatory* In a paragraph, discuss how the land of West Africa affects life in the subregion.

networks

There's More Online!

- ☑ **IMAGE** Lagos, Nigeria
- ☑ **INFOGRAPHIC** Today's Nigerian Family
- ☑ **TIME LINE** Struggle for Power
- ☑ **INTERACTIVE SELF-CHECK QUIZ**
- ☑ **VIDEO** Human Geography of West Africa

LESSON 2
Human Geography of West Africa

ESSENTIAL QUESTION · *How do physical systems and human systems shape a place?*

Reading HELPDESK

Academic Vocabulary

- **convert**
- **scope**

Content Vocabulary

- **infrastructure**
- **griot**
- **e-commerce**

TAKING NOTES: *Key Ideas and Details*

IDENTIFYING As you read about the human geography of West Africa, use a graphic organizer like the one below to identify examples of how history, a growing population, and economic activities affect life today.

IT MATTERS BECAUSE

West Africa has seen many peoples, empires, and kingdoms flourish. The peoples of West Africa are incredibly diverse. Colonial rule brought changes and challenges to West Africa, and the cultures of the subregion modified. Some aspects of traditional cultures were destroyed, while others remained intact, often incorporating elements of European culture, such as a language.

History and Government

GUIDING QUESTION *How does the legacy of colonialism affect governments in West Africa today?*

Thousands of years ago, the Sahara was a much wetter area that supported considerable plant and animal life and, in turn, human populations. As the climate of the Sahara changed, however, people adapted by moving south to greener and more habitable regions. These seminomadic peoples herded animals and developed a system of agriculture. As populations grew, settlements grew into cities and empires.

Empires of West Africa

The first empire to emerge in West Africa was the Ghana Empire. It became one of the richest trading civilizations of West Africa. Ghana profited from its location midway between the salt mines in the Sahara and the gold mines farther south. Archaeologists believe the empire began around A.D. 300. It lasted until about the thirteenth century. During its time as a trading empire, the kingdom prospered by imposing taxes on trade goods. Muslim traders from North Africa sent caravans of goods and salt across the Sahara to Ghana. Gold from Ghana was traded for salt. Muslim traders also brought Islamic beliefs and customs to the kingdom. Eventually, many Ghanaians **converted** to Islam.

The Mali Empire developed after the small state of Kangaba broke away from Ghana. Sundiata Keita, one of Mali's early kings, helped Mali flourish. He took over Timbuktu and made it an important center of trade and scholarship. By the fourteenth century, the power of the Mali Empire began to wane and another empire, Songhai (SAWNG•hy), arose.

(tl)©William Campbell/Sygma/Corbis, (tc)©Patrick Robert/Corbis, (tr)©George Esiri/Reuters/Corbis

convert to change from one system, use, or method to another that is quite different

Songhai broke from Mali after the death of Mali's most well-known king, Mansa Musa. Sunni Ali Ber conquered the cities of Timbuktu and Djenné, expanding his empire to include most of the West African savanna. The empire prospered until about A.D. 1600, when it fell to the Moroccans.

The Hausa city-states were located between the Niger River in what is now northern Nigeria and ended at Lake Chad in the east. These independent states, the first of which emerged around A.D. 1000, formed loose alliances. The Hausa city-states remained independent for the most part, until they were conquered in the early 1800s by the Fulani. Each of the states served a different role within the alliance. Some provided goods, others soldiers or access to trade. As a result, the states became important to international trade.

The kingdom of Benin, which occupied the area of present-day Nigeria, developed into an important power from the thirteenth to the nineteenth century. Benin became an extremely organized state by the late 1400s. Between the fifteenth and eighteenth centuries, the kingdom grew in wealth and power through trade with the Portuguese and Dutch. By the nineteenth century, however, infighting weakened the kingdom. At the same time, suppression of the slave trade, from which Benin had greatly profited, led to the kingdom's decline.

The Colonization of Africa

As contact with Europe increased, word of the riches to be found in Africa spread. Europeans soon began to view Africa as a source of resources and opportunities. As European countries laid claim to territories in Africa, disputes over territory arose. At the Berlin Conference between 1884 and 1885, 14 European countries met in an effort to sort out how territory would be divided.

With no African input at the Berlin Conference, colonial boundaries were drawn with little regard for African ethnic boundaries. One example of this is in Nigeria, where boundaries were drawn that merged Muslim societies with animist cultures. As Christianity was introduced, additional tensions emerged. These and other issues set African peoples against one another and strengthened European rule.

©Patrick Robert/Corbis

TIME LINE ⌄

Struggle for POWER ➔

The oldest republic in Africa, Liberia has experienced a lengthy period of warfare as warlords vie for power.

▶ **CRITICAL THINKING**

1. Explaining How was Liberia influenced by the United States?

2. Analyzing How would you characterize the political situation in Liberia since 1980?

Portuguese traders first arrive in Liberia — **1461**

1400 ➔

Portuguese and British traders transport Africans to the Americas as part of the transatlantic slave trade. — **1500s**

Liberia becomes independent with a constitution modeled after that of the United States. — **1847**

1900 ➔

American Colonization Society (ACS) is formed in the United States to send free blacks to Africa. — **1816**

The one country Europeans never colonized is Liberia. African Americans freed from slavery established it on land purchased for them by the American Colonization Society. The African Americans who relocated to Liberia had a different culture from the indigenous people who already lived in the area. Over time, the differences widened, causing conflicts between the people of Liberia. In 1847 Liberia declared independence. Soon after, a constitution was written.

Decolonization and Difficulties

Europeans maintained their colonies in West Africa until the mid-1900s. But even after attaining independence, these new countries faced daunting challenges. Years of strife afflicted many of the countries. In Nigeria, for example, the ill-conceived boundaries drawn by the colonial powers led to increasing tensions between Muslims and Christians. In 2009 the Nigerian government began fighting against a militant Islamist group that sought to establish an Islamist state in Nigeria.

Another problem is corruption. In Nigeria, for example, the discovery of oil and gas made the country particularly vulnerable. Nigerian leaders used their offices to steal billions of dollars paid by foreign companies for oil extraction rights. To give an idea of the **scope**, or extent, of the problem, some estimate that close to $400 billion was stashed in foreign bank accounts between 1960 and 1999. In the early 2000s, Nigerians and new leaders waged a war against corruption.

scope the extent of an activity or influence

Changes in infrastructure and culture produced by European colonization persisted even after independence. The countries' economies were still dependent on providing raw materials to European countries. They were largely unable to revive their traditional cultural knowledge and practices that had been lost during colonization. As a result, they had to adopt the market economy and consumer culture of the European powers. These lasting changes made it impossible for the new countries to chart an independent course that would be more beneficial to their own interests.

☑ **READING PROGRESS CHECK**

Evaluating In what ways has colonialism affected the countries of West Africa?

(tl)©William Campbell/Sygma/Corbis, (tr)Jerry Lampen/AFP/Getty Images, (b)Pascal Guyot/AFP/Getty Images

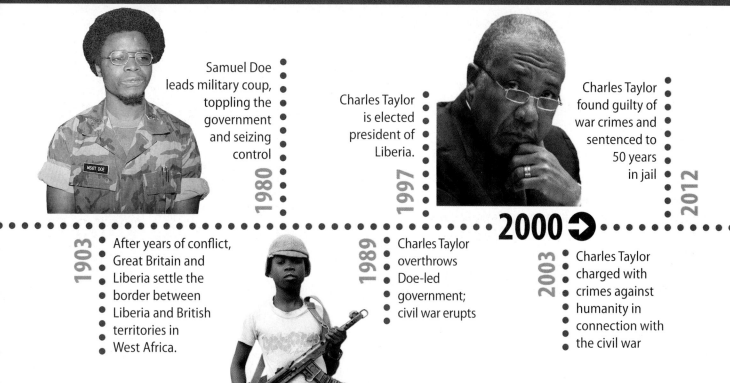

Samuel Doe leads military coup, toppling the government and seizing control

1980

Charles Taylor is elected president of Liberia.

1997

Charles Taylor found guilty of war crimes and sentenced to 50 years in jail

2012

2000 ➔

1903 After years of conflict, Great Britain and Liberia settle the border between Liberia and British territories in West Africa.

1989 Charles Taylor overthrows Doe-led government; civil war erupts

2003 Charles Taylor charged with crimes against humanity in connection with the civil war

Population Patterns

GUIDING QUESTION *How is a booming population affecting life in West Africa?*

Africa's population is increasing rapidly. Most people in West Africa live along the coast and river plains. About half of the people in West Africa live in crowded urban locations. West Africa is one of the most populous regions of Africa, and the country of Nigeria has the largest population of any country in Africa. Lagos, the commercial center of Nigeria, is Africa's largest city with an estimated population of more than 20 million. In 2012, about 170 million people lived in Nigeria. By 2025, the country's population is projected to exceed 190 million.

Demands created by climate change, deforestation, increasing population, and decreasing food supplies have prompted many Nigerians to move to urban areas. This pattern is also occurring in other parts of West Africa. People are moving to urban areas in hopes of finding better job opportunities, health care, public services, and education.

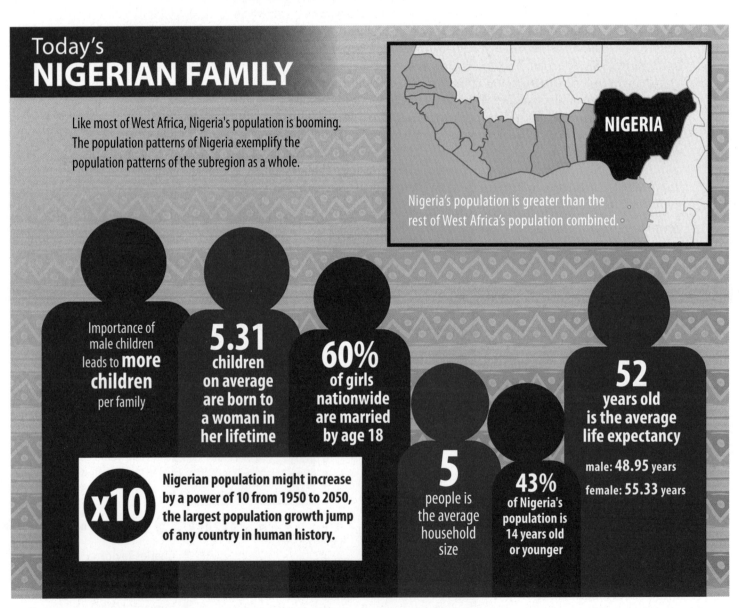

Today's NIGERIAN FAMILY

Like most of West Africa, Nigeria's population is booming. The population patterns of Nigeria exemplify the population patterns of the subregion as a whole.

NIGERIA

Nigeria's population is greater than the rest of West Africa's population combined.

Importance of male children leads to **more children** per family

5.31 children on average are born to a woman in her lifetime

60% of girls nationwide are married by age 18

x10 Nigerian population might increase by a power of 10 from 1950 to 2050, the largest population growth jump of any country in human history.

5 people is the average household size

43% of Nigeria's population is 14 years old or younger

52 years old is the average life expectancy

male: 48.95 years
female: 55.33 years

Nigeria is Africa's most populous nation and continues to grow rapidly. The pace of population growth threatens to undermine efforts to improve the standard of living for Nigerians.

▲ CRITICAL THINKING

1. Analyzing What are the consequences of having a very young population?

2. Problem Solving What steps could Nigeria's government take to slow population growth?

©George Esiri/Reuters/Corbis

The Nigerian city of Lagos is the most populous city in Africa and one of the fastest growing.

◀ **CRITICAL THINKING**

1. *Analyzing Visuals* How does the photograph illustrate the effects of rapid urbanization?

2. *Drawing Inferences* How does rapid urbanization affect the daily lives of a city's residents?

West Africa's rate of urbanization, or movement of people from rural areas to cities, is among the world's fastest. New population growth has caused cities to spread out into the countryside. This expansion of population in urban areas in West African countries has placed a strain on existing **infrastructure**—which includes electricity, water, roads, and information and communications technology. Limitations on infrastructure reduce productivity and limit economic development. However, with the help of international investors, West African countries are attempting to improve infrastructure to meet the demands of their ever-increasing populations.

infrastructure the set of systems that affect how well a place or organization operates, such as telephone or transportation systems, within a country

☑ **READING PROGRESS CHECK**

Evaluating How has West Africa been affected by population growth?

Society and Culture Today

GUIDING QUESTION *How is life in West Africa a mix of the ancient and the modern?*

West Africa is home to a very large number of ethnic groups—about 500 in Nigeria alone. This immense diversity includes peoples that have lived in West Africa for centuries. It also includes groups that have settled there more recently. Some ethnic groups that live in West Africa have been divided into different countries by political boundaries originally established by European colonial powers.

Of the ethnic groups in West Africa, the five largest are the Yoruba, Hausa, Fulani, Ibo, and Akan. Estimated at more than 30 million, the Yoruba live in areas of Nigeria, Benin, Togo, and Ghana. Some 30 million Hausa live in the West African countries of Nigeria, Ghana, and Côte d'Ivoire as well as in Chad, Niger, and Cameroon.

Hundreds of languages are spoken in West Africa. Arabic is common in the northern areas of the subregion. Yoruba and its many dialects, which are part of the Congo-Kordofanian language group, are widely spoken. In the southern region of West Africa, Yoruba is taught in primary and secondary schools, at universities, and in television and radio broadcasting schools. It is also commonly used in printed materials such as books, newspapers, and pamphlets. Part of the legacy of colonial rule was the introduction of European languages. Thus, English and French are widely spoken. In Nigeria, English is the official language. Using English facilitates communication in a country where more than 500 languages are spoken.

Sociology is the study of the development, structure, and function of humans in society. As more West Africans move from rural to urban areas in search of better health care, sanitation, and work, social norms are changing. West Africans have had to adapt to new ways of life in an urban setting. One example of this is the shift from living with extended families to living in nuclear families. This often leads to cultural changes and lost traditions.

ANALYZING What examples of change in social norms do you see in your own society?

griots traditional oral historians, storytellers, singers, and musicians of West Africa

Regardless of ethnic group or language, the diverse people of West Africa value religion and strong family ties. The dominant religions practiced in the subregion are Islam, Christianity, and the traditional animist religions of Africa. Although followers of different religions often coexist peacefully, conflict sometimes occurs.

Education in West Africa is inconsistent. Literacy rates in West Africa are one of the lowest in the world. As of 2010, for example, the literacy rate in Sierra Leone was only 42 percent. In Ghana, however, education has been a priority, and government spending on education has steadily increased since the 1960s. As of 2010, the literacy rate in Ghana was 67 percent, the highest in the subregion.

Poverty is a key factor in determining access to health care in West Africa. Health care is uneven and limited. As a result, infant mortality rates are high throughout the subregion. In Sierra Leone, for example, 185 children out of every 1,000 die before they reach the age of five. High death rates are caused by poor sanitation, health conditions, and inadequate nutrition.

Family and the Status of Women

Many people in West Africa still live in extended families made up of parents, children, grandparents, and sometimes aunts, uncles, and cousins. Generally in these traditional societies, men and women have different roles. Traditionally, women look after children and the home while men earn a living to support the family. In the cities, however, the nuclear family—made up of husband, wife, and children—is rapidly replacing the extended family. The role of women is also beginning to change.

Women in West Africa play a vital role in the family. In recent years, men have begun to move away from home to find jobs, leaving women and children behind. Yet, in traditional societies, there are few jobs available for women that would provide income to care for a family. Things are beginning to change, however. Women are beginning to enter into professions and establish small businesses like selling vegetables in local markets or opening a local beauty salon. These small businesses give women control over their own lives and pave the way for participation in other aspects of their society such as politics. For example, the Women Peace Security Network Africa (WIPSEN-Africa) was established in 2006 in Ghana to promote women's strategic participation and leadership in peace and security governance in Africa.

The Arts

Much of the art in West African culture expresses religious beliefs. Music and dance are a part of everyday life, and entire communities participate. Dancers wear masks honoring specific deities, the spirits of their ancestors, or to honor special occasions, such as a birth. West African music has become popular around the world and has influenced contemporary Western music. Kanye West and Sting are only two of the popular Western musicians who have borrowed from African music. In fact, the entire blues and jazz tradition of the United States has its roots in the music enslaved Africans brought with them. **Griots** are an ancient tradition and an important part of the history of art in West Africa. They are oral historians, storytellers, singers, and musicians. Their songs, music, and stories help the people remember their cultural heritage.

Another art form is weaving. The Ashanti of Ghana are expert weavers, known for kente cloth. Kente is a brightly colored cloth consisting of bands of fabric sewn together. The Ashanti king and royal family have worn this cloth for centuries. Today people around the world also wear kente. Kente has become a symbol of Africa for many African Americans.

☑ **READING PROGRESS CHECK**

Evaluating What ancient tradition can be found in West Africa today?

Economic Activities

GUIDING QUESTION *What natural resources are available in West Africa for economic development?*

People in West Africa earn their living in many different ways. Some West Africans run small businesses selling locally made products such as baskets, art, and jewelry. Using **e-commerce**, or buying and selling on the Internet, people sell their products to customers around the world. The Internet broadens the market for locally made products. Agriculture, however, is the main economic activity of more than half of the people in West Africa.

e-commerce buying and selling on the Internet

Many of the people in West Africa live by subsistence farming in which people focus on growing food to feed themselves and their families. These people are largely self-sufficient, growing their own food and building their own houses. They are generally not involved in a wage-earning economy.

A small percentage of the population works at commercial farming. Commercial farmers produce crops on a large scale. These cash crops are grown and sold for profit. One important cash crop in West Africa is cacao, which is used to make cocoa powder and chocolate. Côte d'Ivoire is the world's largest exporter of cacao, as well as a major exporter of palm oil and coffee beans. Cacao trees are native to Central and South America, and the crop was introduced to West Africa during the colonial period. Small family-run farms have the greatest success in growing cacao. A very small percentage of people along the coast also work in commercial fishing.

Although difficult and risky, mining is also an important economic activity in the region. Coal, gold, uranium, and natural gas are some of the natural resources found in West Africa. Oil has also been discovered in some countries. Immense oil reserves make Nigeria the region's only member of the Organization of Petroleum Exporting Countries (OPEC). Nigeria's economy has been heavily dependent on this single resource, but Nigerian leaders have tried to diversify the economy in recent years. Diamonds, another important resource for West Africa, have a sordid history. Conflict diamonds have financed wars in Sierra Leone, Liberia, and Côte D'Ivoire.

Trade groups have helped increase trade within the region. Various countries have formed regional trade associations, such as the Economic Community of West African States (ECOWAS). This group of 15 countries was founded in 1975 to promote economic integration across the region. ECOWAS is expanding trade within the region and outside of Africa. In 2012 ECOWAS signed an agreement with China for cooperation in infrastructure development, trade, and investment.

☑ **READING PROGRESS CHECK**

Evaluating What economic activities can be found in West Africa?

LESSON 2 REVIEW

Reviewing Vocabulary
1. *Describing* Write a paragraph describing how griots contribute to the culture of West Africa.

Using Your Notes
2. *Analyzing* Use your graphic organizer about the human geography of West Africa to write a paragraph about the complexity and difficulties the subregion faces.

Answering the Guiding Questions
3. *Drawing Conclusions* How does the legacy of colonialism affect governments in West Africa today?

4. *Evaluating* How is a booming population affecting life in West Africa?

5. *Making Connections* How is life in West Africa a mix of the ancient and the modern?

6. *Listing* What natural resources are available in West Africa for economic development?

Writing Activity
7. *Informative/Explanatory* In a paragraph, discuss how the people of West Africa, both past and present, have shaped the subregion.

Conflict
DIAMONDS

Diamonds are one of Africa's most important natural resources. Today, about $8.5 billion annually worth of rough diamonds are from Africa. Groups have found ways to exploit this natural resource for their own gain. Conflict diamonds, also known as blood diamonds, are ones that are mined in rebel-controlled areas. The miners, and those who lived near the mines, have often been threatened, harmed, or even killed.

1. Funding War with Diamonds

Workers use shovels, picks, and their own hands to dig diamonds out of tunnels, open pits, and mines. In Angola, Democratic Republic of the Congo, Côte d'Ivoire, Liberia, Sierra Leone, and other African countries, rebel groups use force to steal these diamonds or the mines themselves. These same groups then sell these rough, uncut diamonds and use the profits to purchase weapons from Eastern European states like Ukraine and Bulgaria, or to fund civil wars and government takeovers.

2. Selling Conflict Diamonds

Some conflict diamonds are smuggled into neighboring African countries. These diamonds are then mixed with diamonds hat have been mined legally, making the illegal gems undetectable. Buyers in London and other large diamond markets then purchase the diamonds unaware of their source.

3. Cutting Diamonds

Diamonds are then cut and polished and sold to consumers throughout the world. Some of these diamonds are used in jewelry while others are used for industrial purposes.

Establishing the Kimberley Process

The United Nations established the Kimberley Process to reduce the amount of conflict diamonds. In this process a diamond must be certified that it is "conflict-free" before it can be sold.

Making Connections

1. **Places and Regions** How does the demand for diamonds in other parts of the world fuel civil war and conflict in Africa?

2. **Human Systems** Why has it been difficult to stop the flow of conflict diamonds?

3. **The Uses of Geography** Using what you know about the trade in conflict diamonds, write a one-page letter to the United Nations suggesting additional steps that could be taken to address the issue of conflict diamonds.
   *Interact with **Global Connections** Online*

networks

There's More Online!

☑ **IMAGE** Deforestation in Ghana

☑ **IMAGE** Sanitation Problems

☑ **INTERACTIVE SELF-CHECK QUIZ**

☑ **VIDEO** People and Their Environment: West Africa

LESSON 3
People and Their Environment: West Africa

(tl)Max Milligan/AWL Images/Getty Images, (tcl)Jean-Marc Giboux/Getty Images News/Getty Images, (tcr)Kimberley Coole/Lonely Planet Images/Getty Images, (tr)©Godong/Robert Harding World Imagery/Corbis

Reading **HELP**DESK

Academic Vocabulary

- **goal**
- **demonstrate**

Content Vocabulary

- **carrying capacity**
- **erosion**
- **fishery**

TAKING NOTES: *Key Ideas and Details*

IDENTIFYING As you read about the people and their environment in West Africa, use a graphic organizer like the one below to identify problems and the different ways in which groups are trying to address these problems.

People and Their Environment: West Africa

Problems | Solutions

ESSENTIAL QUESTION · *How do physical systems and human systems shape a place?*

IT MATTERS BECAUSE

West Africa is faced with multiple challenges. These include drought, increasing population, deforestation, and the degradation of water resources. These challenges have contributed to poverty, hunger, conflict, and war. Finding solutions to these problems may help to reverse the losses and end the crises that afflict the countries of West Africa.

Managing Resources

GUIDING QUESTION *What challenges resource management in West Africa?*

The poor health and wellness conditions of many West Africans makes survival a chief **goal**, or aim—well ahead of environmental concerns. And yet, many of the environmental issues contribute to poor health and wellness. The ongoing water and sanitation crisis kills large numbers of people and limits economic development. The poor, especially women and children in rural areas, are most at risk. Water and sanitation problems also threaten the growing informal settlements near cities where the poor have migrated seeking work.

Sanitation problems include the lack of toilets and polluted water supplies. These problems result in illness, premature death, and loss of productivity while people are sick or looking for health care. Some of the illnesses that result from poor sanitation are infant diarrhea, cholera, malaria, and respiratory infections. Poor sanitation also costs countries money through such things as health care and lost hours of work. For example, poor sanitation costs Benin $104 million per year. Ghana loses $209 million per year, and the cost to Nigeria is $3 billion per year. Tackling these problems by providing better health care, education, and sanitation would help improve the quality of life for everyone in the subregion.

One obstacle to change and economic efficiency in some countries of West Africa is corruption. Corruption has a negative impact on social and economic development as well as on the environment and resource management. Corruption in West African countries has often led to violent political conflict. These conflicts are financed by the unlawful sale of weapons or the illegal extraction and sale of natural resources such as diamonds, gold, and timber. The use of money and natural resources to fund these conflicts hinders economic development. Fortunately, national and international groups are making efforts in West African countries to address corruption, health, sanitation, and access to education and health care.

These are not the only problems West Africans face, however. Difficulty in meeting the needs of the people also leads to the mismanagement of natural resources. Such mismanagement is part of a legacy of colonialism, during which massive numbers of people were forced to migrate from mainly rural areas to other rural areas or even urban areas. This resulted in the abandonment of long-held practices learned over generations that were no longer relevant in the new areas and which ultimately caused environmental harm in some places.

Many of the countries in West Africa depend on natural resources for economic development. However, the failure to properly manage soils, forests, lakes, rivers, and wildlife will only lead to further problems. There is great concern that the mismanagement of resources will lead to food insecurity for large numbers of West Africans.

✓ **READING PROGRESS CHECK**

Explaining How does poor sanitation affect the people and countries of West Africa?

Human Impact

GUIDING QUESTION *What environmental challenges does West Africa face today?*

The population of countries south of the Sahara is increasing at a rate of 3.2 percent a year. While that may not sound high, it means—if the rate stays the same—that the population will double in about 22 years. This rate of population

goal the aim of an activity

Ghana has the highest rate of deforestation in Africa.

▼ **CRITICAL THINKING**

1. Analyzing Why is it important to stop deforestation?

2. Identifying Cause and Effect How has the issue of management of resources affected deforestation?

Local Fishermen Near Lake Chad

"There are constant arguments over territory between fishermen.... It's difficult to determine boundaries on water, yet the gendarmes [police from Cameroon and Chad] always come after us and seize our fishing nets and traps and we have to pay heavily to get them back."

—Muhammad Sanusi, a fisher in Dogon Fili, "Lake Chad fishermen pack up their nets," BBC News, January 15, 2007

DBQ **IDENTIFYING CAUSE AND EFFECT** How has drought and desertification affected fishers in some areas?

carrying capacity the maximum population of any given species that an environment can sustain

erosion the wearing away of soil

fishery a place for catching fish; the fishing industry

increase is one of the highest in the world. Such rapid growth has placed tremendous pressure on the **carrying capacity** of the land. Carrying capacity is the maximum population of a species the environment can sustain without depleting or degrading the resources available. The growth in population has already led to an environmental crisis in many parts of Africa. One hopeful idea is that humans can increase the carrying capacity of the land by employing more sustainable technologies. Examples include solar cooking technology that reduces the need to burn wood and small-scale irrigation pumps that work like bicycle pedals and allow subsistence farmers to irrigate their fields.

In addition to the heavy load on the environment caused by the increase in population, the breakdown of traditional systems of agriculture is also causing environmental deterioration on a massive scale. This deterioration, in turn, contributes to widespread poverty, malnutrition, and famine. In some countries, the breakdown in farming is contributing to political instability and civil war.

Overgrazed and overworked soils make farming difficult and can damage the long-term health of the soil. The use of heavy farm machinery, frequent tilling, and the clearing of forests have caused **erosion** and desertification. Together with the mismanagement of the land, drought and infrequent rainfall in West Africa also cause desertification. Once desertification begins, it puts pressure on the people that live in the area. Agriculture and livestock suffer, as well as the area's natural biodiversity. All these problems result in food production that has fallen short of the needs of the West African population.

Another problem facing West Africa is deforestation. Logging creates serious consequences, but the lumber industry maintains a relatively small output. Coastal countries with rain forests, such as Ghana and Côte d'Ivoire, export significant amounts of valuable hardwoods, such as teak, ebony, African walnut, and rosewood. Forests are also cut down to make room for farming, mining, and settlement. Unfortunately, most of this logging is not done sustainably, which threatens the natural biodiversity of the area and also contributes to desertification and erosion.

Fishing is an important industry in West Africa. **Fisheries** in the West African Marine Ecoregion (WAMER) generate about $400 million a year, making them a key source of revenue. In Senegal alone, over 600,000 people depend directly on fishing and related industries. Here, individual fishers catch over 80 percent of the fish while industrial fishing fleets catch about 20 percent. However, in Senegal and other West African countries, unsustainable fishing practices are leading to dwindling stocks in the ocean fisheries. More and more boats are searching for fewer fish. The competition is leading to even more destructive methods of fishing. These methods include the use of chemicals, dynamite, and bottom trawling. Pollution has also negatively affected fishing in countries such as Nigeria. Oil spills in the Niger Delta, where the local people are fishers and farmers, has led to dwindling fish stocks and reduced soil quality.

Lake Chad is another resource that has been affected by human activity. Drought, desertification, and climate change have caused Lake Chad to shrink. But there are other causes for the changes to the size of the lake as well. The need for water to irrigate farmland as well as for consumption has led to the damming of the rivers that supply Lake Chad. This has also contributed to the reduction in the size of the lake. The shrinking lake has led to smaller fish stocks and, in turn, to arguments over territory between fishers. Conflicts also arise between farmers and fishers over the use of the water from Lake Chad for irrigation. At times, these arguments have led to violence.

✓ **READING PROGRESS CHECK**

Summarizing How does human activity impact the carrying capacity of the land?

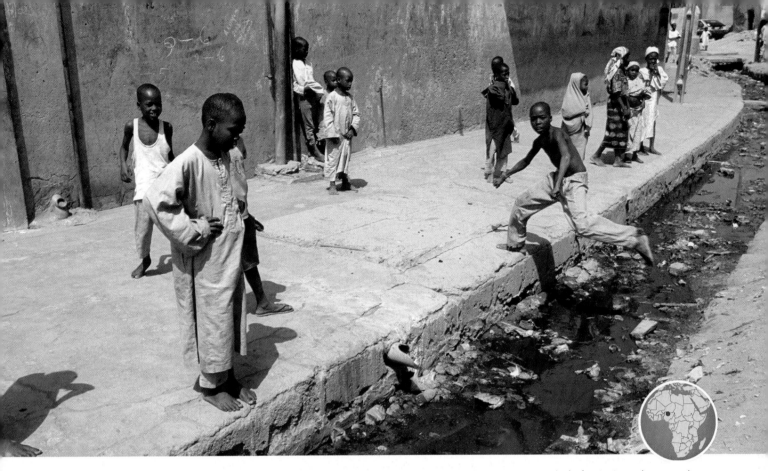

Jean-Marc Giboux/Getty Images News/Getty Images

Addressing the Issues

GUIDING QUESTION *What steps are being taken to combat these environmental challenges?*

West African countries face many environmental problems. These include the degradation of land, desertification, drought, pollution, and the reduction of fish stocks and water resources that reduce the availability of food for both humans and animals. These problems increase the likelihood for conflicts to arise over competition for natural resources. Many individuals, cities, countries, and national and international organizations are working to combat the problems the subregion is facing. Some farmers, for example, have started to practice conservation farming, a land-management technique that helps protect farmland. By planting different crops where they will grow best, farmers protect farmland. Better fertilizers, seeds, and irrigation practices have increased crop yields and production.

Some farmers have begun to revive traditional land-management practices to reverse desertification and have **demonstrated** that these practices can be effective. Such practices include using planting pits to reclaim severely dry land. Additionally, help has come from Oxfam International, a nongovernmental organization (NGO). Oxfam International is a confederation of 17 organizations working together in more than 90 countries around the world to end poverty. Together with Oxfam, farmers in the Yatenga Province of Burkina Faso are building stone contour bunds, or embankments, to control rainwater runoff. This technique has rehabilitated about 740,000 acres (300,000 ha) of land. The additional food produced on the land helps feed about 500,000 people. Other farmers in the region have started to plant native plants in and around crop fields. These food-producing, drought-resistant trees and shrubs help improve soil fertility, retain moisture, and reduce erosion.

Lack of sanitation and access to clean water are major problems for some areas in West Africa.

▲ **CRITICAL THINKING**

1. *Analyzing* How can poor sanitation affect a community?

2. *Making Connections* How would people benefit from better management of water resources?

demonstrate to clearly show

In Nigeria, President Goodluck Jonathan is attempting to address water shortages in the state of Benue through the Greater Makurdi Water Works project. The water treatment plant is designed to bring clean water daily to the people of Benue. However, corruption is hampering this and other projects within the region. For example, in the case of the Makurdi Water Works project, the government council agreed to a water treatment plant with a capacity of 26 million gallons (98 million l). However, the governor of Benue state signed a contract for a treatment plant of half the capacity at the same price of full capacity, raising questions about corruption. Additionally, the plant and the state of Benue do not have the pipe structure necessary to supply the water. The people of Benue will continue to suffer with insufficient clean drinking water until more contracts are signed and the government spends additional money.

On a larger scale, the Global Water Initiative (GWI) works in various regions around the world to improve access to clean water and sanitation. In West Africa, GWI works with Burkina Faso, Ghana, and Senegal as well as Mali and Niger. One of the main objectives of the GWI in West Africa is to improve access to clean water and sanitation for vulnerable populations in the region. The group has six projects in West Africa that focus on these objectives in the countries surrounding the Gambia, Volta, and Niger River basins. In villages in Ghana where the GWI initiated projects, the rate of access to safe water has increased by an estimated 50–80 percent.

The United Nations (UN) is working on the use and availability of water in the subregion through the Integrated Water Resources Management (IWRM) project. The IWRM recognizes that water is key to economic and social development in West Africa and elsewhere. The project promotes the coordinated development and management of water, land, and related resources in order to maximize the economic and social welfare of people in the region without harming the environment.

Other projects around the region are working to improve access to clean water, thereby improving health and sanitation. Efforts by farmers and larger organizations to improve soil quality will help increase crops, thus reducing hunger. These improvements will encourage governments to make greater efforts to reduce corruption and to increase access to health care and education. These changes and others will help bring stability and security to the people of the region and reduce the incidences of conflict.

☑ **READING PROGRESS CHECK**

Evaluating In what ways are West Africans trying to combat their environmental problems?

LESSON 3 REVIEW

Reviewing Vocabulary

1. ***Describing*** In what ways has the human population put pressure on the land?

Using Your Notes

2. ***Summarizing*** Use your graphic organizer on the people and their environment in West Africa to write a summary of the problems the subregion faces and to propose possible solutions.

Answering the Guiding Questions

3. ***Synthesizing*** What challenges resource management in West Africa?

4. ***Evaluating*** What environmental challenges does West Africa face today?

5. ***Classifying*** What steps are being taken to combat these environmental challenges?

Writing Activity

6. ***Informative/Explanatory*** In a paragraph, write about the cause-and-effect relationship between people and the environment in West Africa.

Directions: On a separate sheet of paper, answer the questions below. Make sure you read carefully and answer all parts of the questions.

Lesson Review

Lesson 1

1 *Explaining* Why are mangrove swamps an important part of the ecosystem of West Africa?

2 *Evaluating* How successful was the damming of the Volta River at providing benefits for people in the area?

3 *Comparing and Contrasting* Describe how biomes and climates change as one moves from north to south in West Africa.

Lesson 2

4 *Identifying Cause and Effect* How did the colonial boundaries drawn by European countries at the Berlin Conference affect people in the region?

5 *Assessing* Why has the discovery of oil and gas reserves in Nigeria made the country more vulnerable?

6 *Summarizing* West Africa has one of the fastest rates of urbanization in the world. What are some of the reasons for this population trend?

Lesson 3

7 *Describing* Describe how the ongoing sanitation and water crisis affects the economy and health of people in the subregion.

8 *Making Connections* How does the clearing of forests contribute to desertification in West Africa? What are some of the economic and social consequences of this process?

9 *Problem Solving* What steps are some West African farmers taking to manage resources more effectively?

Critical Thinking

10 *Hypothesizing* How might diamonds purchased in Europe fuel political and social instability in West Africa?

11 *Drawing Conclusions* Why do you think e-commerce has become a popular way for West Africans to market locally produced goods?

21st Century Skills

Use the chart to answer the questions that follow.

People Vulnerable to Food Insecurity, March 2012	
Country	**Number of People Food Insecure**
Burkina Faso	2,852,280 food insecure
Chad	3,622,200 food insecure, of which 1,180,300 severe
Gambia	713,433 in areas at risk
Mauritania	700,000 food insecure, of which 290,000 severe
Niger	6,112,089 food insecure, of which 1,916,855 severe
Senegal	850,000 food insecure

Source: www.oxfam.org

12 *Geography Skills* Which country in West Africa has the highest number of people vulnerable to severe food insecurity? What is the number?

13 *Creating and Using Graphs, Charts, Diagrams, and Tables* How does food insecurity in Senegal compare to Burkina Faso? Why does the table describe Gambia as at risk?

College and Career Readiness

14 *Problem Solving* Imagine that you have been hired by Oxfam to help identify potential food shortages in West Africa. Choose a country from the region. Use the Internet to research food production and population growth in that country. How fast is the population growing? Does the production of food meet the needs of the growing population? Write a brief response assessing whether or not the country faces a food shortage.

Need Extra Help?

If You've Missed Question	**1**	**2**	**3**	**4**	**5**	**6**	**7**	**8**	**9**	**10**	**11**	**12**	**13**	**14**
Go to page	532	532	534	536	537	538	544	546	547	534	541	549	549	547

Directions: On a separate sheet of paper, answer the questions below. Make sure you read carefully and answer all parts of the questions.

DBQ Analyzing Primary Sources

Use the document to answer the following questions.

The ongoing food crisis in West Africa has many causes and consequences. Below are the observations of a journalist who has written about Oxfam's work in West Africa.

PRIMARY SOURCE

"*Some of my most vivid images of the Sahel food crisis are of hands and feet.*

When I traveled to Senegal recently to document Oxfam's work on the crisis, I met with women farmers who lost their last harvest to erratic rains. Several times I noticed an injury to a woman's foot or finger—usually something simple that without medical care had become so serious that it was disabling or worse: from the look of it, some would require amputation. But none of the women had any food stocks left and, with prices on the rise, none could afford to purchase enough for her family to eat, so a visit to the doctor was out of the question."

—Elizabeth Stevens, "Sahel Food Crisis: The Cost of Climate Change," October 2012

15 *Interpreting* How does the author's observation of women farmers in Senegal help you understand the title of the article?

16 *Making Connections* Describe how climate change, inadequate medical care, and food shortages combine to create problems in productivity.

17 *Analyzing Primary Sources* Why do you think the author begins her description of the food crisis with the image of hands and feet?

Applying Map Skills

Use the Unit 6 Atlas to answer the following questions.

18 *Places and Regions* Use your mental map of West Africa to describe the length and location of the Niger River and the Niger Delta.

19 *Human Systems* How has rapid population growth caused cities to spread out into the countryside in West Africa? Give examples.

20 *Environment and Society* What are some of the freshwater resources in West Africa? How important are they to the economies of the subregion?

Exploring the Essential Question

21 *Analyzing* Use the Internet and library resources to research the colonial history of a country in West Africa. Who colonized the region? Did colonial boundaries cut across ethnic lines? What happened when colonial rule ended? Create a time line showing how colonial rule affected people in the country you chose.

Research and Presentation

22 *Research Skills* The Niger Delta is one of the world's most important ecosystems. However, the discovery of oil in the 1950s has created problems. Use the Internet to research the environmental and social impact oil has had on the Niger Delta. How severe is the problem of pollution? How have oil companies and the Nigerian government responded? Write your conclusions in paragraph form and share them with the class.

Writing About Geography

23 *Informative/Explanatory* Use the Internet to research challenges facing a city in West Africa. Using standard grammar, spelling, sentence structure, and punctuation, write a two-page essay describing the ways rapid urbanization creates pressure on infrastructure.

Need Extra Help?

If You've Missed Question	**15**	**16**	**17**	**18**	**19**	**20**	**21**	**22**	**23**
Go to page	550	550	550	476	476	476	537	533	538

Equatorial Africa

ESSENTIAL QUESTION · *How do physical systems and human systems shape a place?*

CHAPTER 23

Geography Matters...

Equatorial Africa lies in the heart of Africa. Great expanses of rain forest teem with wildlife. Through the forest flows the mighty Congo River—one of the longest in the world—and its many tributaries. These waterways form a vast transportation network that the people of Equatorial Africa have used for centuries.

Along these streams are many villages. Most of the people of these villages make their living as farmers. Others work in the region's many mines, extracting gold, diamonds, and other minerals. Still others live and work in huge port cities, where the rivers of the heart of Africa flow into the Atlantic Ocean.

◄ A chief from Ndian and Koupe-Manengouba in Cameroon

Thomas Imo/Photothek/Getty Images

South Sudan: independence *and* conflict

South Sudan declared its independence from Sudan in 2011, after centuries of ethnic and religious conflict. The celebration of newly won freedom did not last long, however. This break only fueled tensions, as both countries continue to dispute borders, which affects in part the economic survival of South Sudan.

Sudan and South Sudan Conflicts

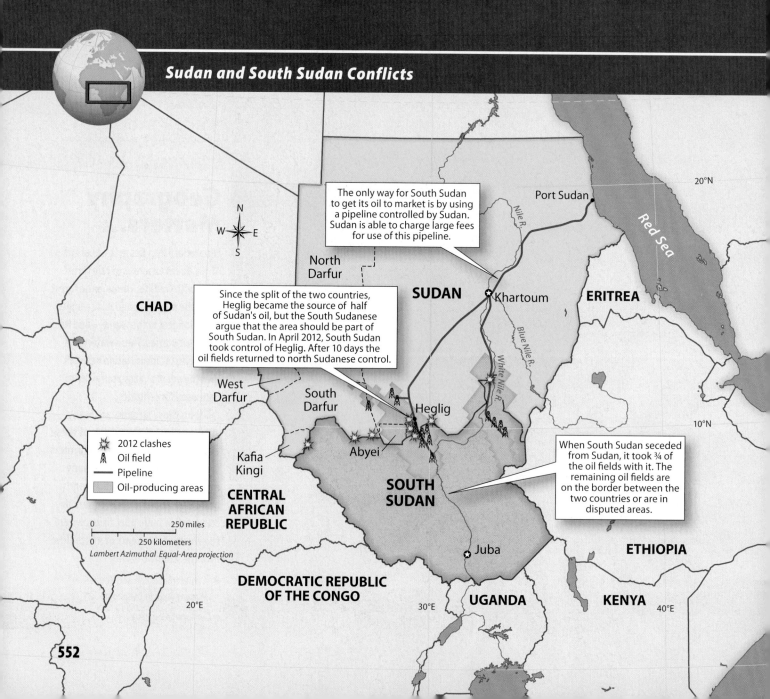

The only way for South Sudan to get its oil to market is by using a pipeline controlled by Sudan. Sudan is able to charge large fees for use of this pipeline.

Since the split of the two countries, Heglig became the source of half of Sudan's oil, but the South Sudanese argue that the area should be part of South Sudan. In April 2012, South Sudan took control of Heglig. After 10 days the oil fields returned to north Sudanese control.

When South Sudan seceded from Sudan, it took ¾ of the oil fields with it. The remaining oil fields are on the border between the two countries or are in disputed areas.

Legend:
- 2012 clashes
- Oil field
- Pipeline
- Oil-producing areas

0   250 miles
0   250 kilometers
Lambert Azimuthal Equal-Area projection

Map labels: CHAD, North Darfur, West Darfur, South Darfur, Kafia Kingi, CENTRAL AFRICAN REPUBLIC, DEMOCRATIC REPUBLIC OF THE CONGO, SUDAN, Khartoum, ERITREA, Port Sudan, Red Sea, Nile R., Blue Nile R., White Nile R., Heglig, Abyei, SOUTH SUDAN, Juba, UGANDA, KENYA, ETHIOPIA, 20°N, 10°N, 20°E, 30°E, 40°E

552

Climates, Biomes, and Resources

GUIDING QUESTION *How is the rain forest structured and how does this affect plant and animal life?*

Equatorial Africa is called "equatorial" because it is located on and near the Equator. It is a region of high temperatures and tropical climates. Most of the subregion has a tropical rain forest climate. There is a smaller area that has a tropical wet/dry (savanna) climate. The highland areas that surround most of the Congo Basin experience highland or **montane** climates. The word *montane* comes from the Latin word for "mountain," and it simply means "of mountainous country."

montane referring to a mountainous area

Warm temperatures prevail in the tropical wet climate zone of Equatorial Africa. More than 60 inches (150 cm) of rainfall per year soak the land. Rainfall varies seasonally, but there is not a truly dry season. Daily, rain falls on an amazing number and variety of life-forms. The tropical rain forests of Equatorial Africa are among the biggest—covering about half of the subregion—and densest in the world.

ANATOMY
of the Rain Forest

Emergent Layer reaches 240 ft. (73 m)
- Stretches from 100 to 240 feet
- Umbrella-shaped canopies
- Smooth, straight trunks
- Very few branches

Canopy reaches 130 ft. (40 m)
- Stretches from 60 to 130 feet
- Most animal life found in this layer
- Leaves have "drip spouts" that allow rain to run off trees

Understory reaches 60 ft. (18 m)
- Stretches from 15 to 60 feet
- Contains trunks of canopy trees, shrubs, plants and small trees
- Little air movement, high humidity, constant shade

Undergrowth reaches 15 ft. (5 m)
- Usually completely shaded

The rain forest is not a simple forest, but a complex web of life. Distinct forms of vegetation grow at different layers.

▲ **CRITICAL THINKING**

1. Hypothesizing Why might there be little growth on the rain forest floor?

2. Differentiating What is the difference between the understory and the canopy?

Rain forests are complex biomes. They are home to a wide variety of animal life, including birds, bats, rodents, monkeys, baboons, chimpanzees, elephants, wild boar, and okapi. A huge number and variety of insects also live there. Some of these, such as the tsetse fly and mosquito, carry diseases.

Rain forests have four basic layers. The topmost layer of the rain forest is the emergent layer, where trees stick out of, or emerge from, the canopy—the next highest layer. The **canopy** is a layer of trees and leaves with a maximum height of 130 feet (39.6 m). Orchids, ferns, and mosses grow among the branches of the canopy, which provide most of the plant food for rain forest animals. Below the canopy is the layer called the **understory**. Shrubs, ferns, and mosses grow together at this level of the rain forest, which rises 15 to 60 feet (4.6 to 18 m). The lowest layer is the undergrowth, where little sunlight penetrates and consequently few shrubs and herbs can grow. The soils in the tropical rain forest biome are typically not very fertile because the heavy rains leach, or dissolve and carry away, nutrients from the soil. Here animals feed mostly on insects.

Tropical grassland with scattered trees—known as savanna—covers much of southern Equatorial Africa. Rainfall is seasonal in this climate zone, with alternating wet and dry seasons. In the wettest areas, which are closest to the Equator, six months of almost daily rain is followed by a six-month dry season. Average annual rainfall is about 35 to 45 inches (90 to 115 cm).

Mineral resources are **abundant** through much of Equatorial Africa. In the Democratic Republic of the Congo alone, the following are mined: cadmium, cobalt, copper, gold, manganese, petroleum, silver, tin, and zinc. The value of these mineral resources is estimated to be about $25 trillion. The Congo Basin also holds major diamond deposits. Gabon, the Republic of the Congo, Cameroon, and Equatorial Guinea have oil and natural gas reserves.

Water is a major natural resource in the subregion. Equatorial Africa receives abundant rainfall. Yet controlling water for practical uses, such as irrigation and hydroelectric power, is difficult because the amounts of rainfall are irregular and unpredictable. These challenges, combined with a lack of financial support, result in untapped hydroelectric power potential in parts of the subregion.

Solar power is another renewable energy source that has been harnessed in Equatorial Africa. Rural electrification programs involve installing small-scale solar power systems. These continue to expand in some parts of the subregion.

The rain and the sun together give rise to the massive tropical rain forests, which may be Equatorial Africa's most important natural resource. Tropical rain forests are a rich source of food, medicine, fibers, oils, rubber, and, wood. Deforestation, however, is threatening many of the forests in the subregion.

☑ **READING PROGRESS CHECK**

Differentiating What is the difference between the tropical rain forest and tropical wet/dry areas of Equatorial Africa?

canopy top layer of a rain forest, where the tops of tall trees form a continuous layer of leaves

understory a lower layer of the rain forest

abundant present in large amounts; plentiful

LESSON 1 REVIEW

Reviewing Vocabulary
1. *Describing* Describe the layers of a rain forest using the words *canopy* and *understory*.

Using Your Notes
2. *Identifying* Use your graphic organizer to write a paragraph describing the major land and water features of Equatorial Africa.

Answering the Guiding Questions
3. *Identifying* What huge landform covers most of Equatorial Africa?

4. *Describing* How is the Congo River central to life in Equatorial Africa?

5. *Explaining* How is the rain forest structured and how does this affect plant and animal life?

Writing Activity
6. *Informative/Explanatory* Think about the physical features of Equatorial Africa. Write a paragraph describing how specific landforms affect the course of the subregion's rivers.

Reading **HELP**DESK

Academic Vocabulary

- **intermittent**
- **hub**

Content Vocabulary

- **cash crop**
- **plantation**

TAKING NOTES: *Key Ideas and Details*

IDENTIFYING Use a graphic organizer like the one below to take notes about the human geography of Equatorial Africa.

Human Geography of Equatorial Africa

History and Government → ☐

Population Patterns → ☐

Society Today → ☐

(tl)©Jacques Pavlovsky/Sygma/Corbis, (tc)Brennan Linsley/AP Images, (tc)Themba Hadebe/AP Images, (tc)Giovanni Mereghetti/age fotostock, (tr)Lionel Healing/AFP/Getty Images

LESSON 2
Human Geography of Equatorial Africa

ESSENTIAL QUESTION · *How do physical systems and human systems shape a place?*

IT MATTERS BECAUSE

Equatorial Africa reflects a wide variety of cultures. The arrival of Europeans fundamentally affected life for the Africans and led to a long struggle for freedom. The legacy of European colonization is still felt in the subregion. Today, Equatorial Africa is a rich mix of African and European influences, ancient and modern lifestyles, and challenges and hope.

History and Government

GUIDING QUESTION *What challenges does democracy face in Equatorial Africa?*

Thousands of years ago, people lived nomadic lifestyles across Africa. They eventually settled to grow food and livestock. By the time Europeans began to arrive in the 1400s, peoples of the region had established enduring cultures.

The first inhabitants of the Congo Basin are thought to be the Mbuti, once known as Pygmies, whose descendants still live in the region today. They were already living there when the great migration of Bantu-speaking peoples arrived in the region. The Bantu-speaking peoples, who were farmers and herders, had established settlements by A.D. 800. Although the origins of the Bantu and their migration routes are debated, historians believe that they spread across one-third of the continent. The Bantu founded the kingdoms of Kongo (Congo), Luba, and Lunda.

Slavery and European Colonization

When European explorers landed along Equatorial Africa's coasts in the late 1400s, they were interested in trade. They sought gold, ivory, textiles, and enslaved Africans. European ships carried away enslaved people from areas that are now the Republic of the Congo, Gabon, and Cameroon. Gabon later became a center of the slave trade.

Millions of people from the African interior were sold into slavery. Nzinga Mbemba, the king of Kongo, deplored the actions of some African rulers. He also complained to the king of Portugal about Portuguese slave merchants:

The Atlantic Slave Trade

NORTH AMERICA

American Colonies

30°N

West Indies

Cuba
Jamaica
Haiti

ATLANTIC OCEAN

EUROPE

ASIA

TROPIC OF CANCER

AFRICA

Senegal
Sierra Leone

Slave Coast

EQUATOR

PACIFIC OCEAN

SOUTH AMERICA

Brazil

Congo

Angola

Mozambique

INDIAN OCEAN

30°S

TROPIC OF CAPRICORN

Cape Colony

Main source of enslaved Africans
Slave trade route
Sugar and tobacco
Cheap European manufactured goods

N
W   E
S

0   2,000 miles
0   2,000 kilometers
Miller projection

30°W   0°   30°E   60°E

GEOGRAPHY CONNECTION

The Atlantic slave trade began in the 1500s and continued into the 1800s.

1. **HUMAN SYSTEMS** What were the major destination points for enslaved people leaving Africa?

2. **THE USES OF GEOGRAPHY** How can maps help historians understand the past?

cash crop a farm crop grown to be sold or traded rather than used by the farm family

PRIMARY SOURCE

66 [They] seize upon our subjects, sons of the land and sons of the noblemen, and cause them to be sold; and so great, Sir, is their corruption and licentiousness that our country is being utterly depopulated. 99
—Nzinga Mbemba, the king of Kongo, quoted in *East Along the Equator*, 1987

Once captured and sold, enslaved Africans faced a terrible trip across the Atlantic Ocean as human cargo in a ship's hold. The passage from Africa claimed millions of lives. The loss of so many young people to the slave trade did great harm to the societies they left behind.

Large areas of Equatorial Africa were not colonized until the 1800s. Various obstacles had previously prevented colonization, including malaria and the Congo River cataracts. By the 1900s, however, European powers held colonies throughout the subregion. Colonizers were Spain, Germany, France, Belgium, Britain, and Portugal.

France gained control of what is now the Republic of the Congo in the late 1800s by engineering treaties with local rulers who wanted protection against Belgium, France's main rival in the region. The French changed the local economy into one based on income from resource extraction and growing **cash crops** for export. In 1878 King Leopold II of Belgium began establishing trade posts along the Congo River. He soon convinced other European powers to grant him control of what was then termed the Congo Free State and what is today called the Democratic Republic of the Congo. This territory and its people were considered the king's personal property. He treated the people terribly.

European rulers, traders, and missionaries promoted European culture, weakened traditional African cultures, and often treated Africans harshly. African village life was also disrupted by the replacing of locally centered agriculture with large commercial farms, called **plantations,** that grew cash crops. This harsh treatment of native peoples by European colonists led to the many struggles for independence across Africa during the twentieth century.

Independence

Resistance to colonial rule in Africa was growing by the mid-1900s. In one year alone, 1960, all of the countries of Equatorial Africa achieved independence. Thus, 1960 is remembered as "the year of independence."

People in most of the new countries experienced periods of ethnic strife, harsh rule, and human rights abuses after independence. In the Democratic Republic of the Congo, serious instability led to the reign of the dictator Mobutu Sese Seko from the late 1960s until the 1990s. Although the countries of the subregion were technically democracies, this period saw human rights abuses, one-party rule, and **intermittent** civil war.

Today the governments of Equatorial Africa have varying degrees of stability. Gabon is a stable republic. It has been developing its economy through the creation of a national park system (to promote ecotourism), developing industry, and educating its residents to work in financial services, education, and health care. In contrast, the Central African Republic has experienced military coups and periods of instability. Although today it is nominally a democracy, rebel groups, not the government, control land and people in the countryside. Abundant natural resources in the subregion have helped some countries achieve relative stability, but problems persist. Oil revenues helped Equatorial Guinea achieve greater stability, but most residents have a low standard of living, and the government has a history of human rights abuses.

☑ **READING PROGRESS CHECK**

Theorizing Why might the countries of Equatorial Africa have suffered harsh rule after independence?

Population Patterns

GUIDING QUESTION *How does geography influence where the people of Equatorial Africa live?*

Life for people in Equatorial Africa is influenced by the subregion's tropical climates, thick tropical forests, and tropical grasslands. Equatorial Africa is mostly rural, despite having prominent urban centers. The dense growth of natural vegetation in this equatorial region makes large-scale intensive plantation agriculture difficult. Most people exist by subsistence agriculture or raising cattle. In some places, coffee, rubber, and cacao are grown for export.

As a whole, Equatorial Africa is one of the least densely populated regions on the continent. The northern part of the Democratic Republic of the Congo, for example, is tropical forest. Large tracts of government forest reserves discourage cities, although there are scattered villages along the Congo and its tributaries. The southern part of the country, however, has millions of people. Kinshasa, the capital city, is the political, cultural, and economic **hub** of the region. It is home to some 8 million people. Just across the Congo River is Brazzaville, the capital of the Republic of the Congo, which is home to 1.2 million. In contrast, the largest city in São Tomé and Príncipe is the capital, São Tomé, with only 60,000 residents. Overall, Gabon is the most urbanized country in Equatorial Africa, with 86 percent of its population living in urban areas. At the other extreme is the newly independent South Sudan, where only 22 percent of the population lives in urban areas.

☑ **READING PROGRESS CHECK**

Explaining Why is much of Equatorial Africa thinly populated?

Connecting Geography to HISTORY

Slave Forts

European colonizers built a series of forts along coastal Africa. These were fortified trading posts from which they conducted their trade in gold and enslaved Africans. Built by the Portuguese in 1482, Elmina Castle was the first such fort in Africa south of the Sahara. Africans captured in the interior of the continent were held in the fortress until ships arrived to transport them across the Atlantic in chains. This trip, known as the "Middle Passage," could last from three to six weeks. Located in present-day Ghana, Elmina Castle is a national museum and has been declared a World Heritage Monument. Over 100,000 tourists visit the site annually.

UNDERSTANDING HISTORICAL INTERPRETATION How would touring Elmina Castle help a visitor understand the history of Equatorial Africa?

plantation a large commercial farm growing crops for export

intermittent occurring at irregular intervals; occasional

hub the center of an activity or region

Society and Culture Today

GUIDING QUESTION *How does conflict shape life in Equatorial Africa?*

Equatorial Africa contains hundreds of ethnic groups. The Democratic Republic of the Congo alone has more than 200 groups. The largest ethnic groups in that country are Bantu, peoples who speak one of the hundreds of Bantu languages. One indigenous group, the Mbuti, lives in the rain forest of eastern Democratic Republic of the Congo and is relatively unchanged by outside influences. In the northwestern part of the country, Fang and Bantu ethnic groups are the dominant ethnic groups.

The approximately 200,000 people of the islands of São Tomé and Príncipe are from many different ethnic groups. Many of the people are descendants of Portuguese settlers and people freed from slavery in Africa. These Portuguese Africans are known as *forros* and compose the country's elite. Much of the population is descended from groups that migrated to the islands in the late 1400s.

Along with hundreds of ethnic groups come hundreds of languages. About 700 local languages are spoken in the Democratic Republic of the Congo alone. French is widely spoken throughout most of Equatorial Africa, reflecting its colonial heritage. It is the official language of the Democratic Republic of the Congo and the Republic of the Congo. People from different sublanguage groups often communicate in pidgin, a simplified speech used among people who speak different languages.

Christianity, Islam, and indigenous religions are practiced in Equatorial Africa. Traditional indigenous religions are numerous and diverse, but they have many common elements. For example, most traditional religions profess a belief in the existence of a supreme being and a ranked order of deities. Animism, or the belief that natural or inanimate objects possess spirits, is also widespread.

European colonialism greatly influenced Equatorial Africa's religious practices. In many areas today, a majority of people observe different forms of Christianity. Roman Catholicism, brought by Belgian and French colonists and

©Jacques Pavlovsky/Sygma/Corbis

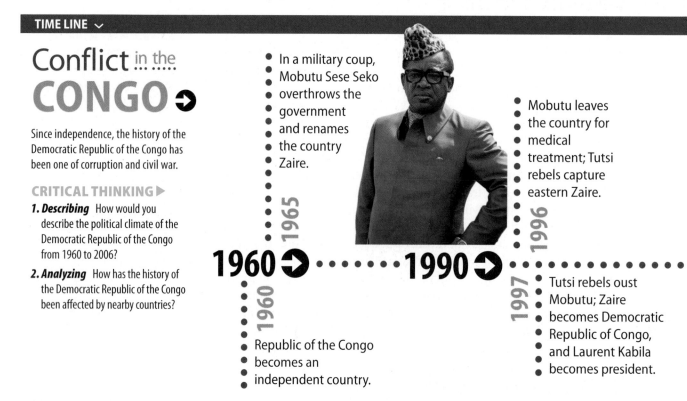

TIME LINE ⌄

Conflict in the
CONGO ➡

Since independence, the history of the Democratic Republic of the Congo has been one of corruption and civil war.

CRITICAL THINKING ▶

1. *Describing* How would you describe the political climate of the Democratic Republic of the Congo from 1960 to 2006?

2. *Analyzing* How has the history of the Democratic Republic of the Congo been affected by nearby countries?

1960 ➡ .... 1990 ➡ ....

1965 — In a military coup, Mobutu Sese Seko overthrows the government and renames the country Zaire.

1960 — Republic of the Congo becomes an independent country.

1996 — Mobutu leaves the country for medical treatment; Tutsi rebels capture eastern Zaire.

1997 — Tutsi rebels oust Mobutu; Zaire becomes Democratic Republic of Congo, and Laurent Kabila becomes president.

missionaries, is common. About half of the population of the Democratic Republic of the Congo is Roman Catholic. The subregion is home to an increasing number of Muslims, as well.

Religion and family life are interwoven in Equatorial Africa. Some ethnic groups in the Democratic Republic of the Congo, for example, live according to a mix of indigenous and Christian beliefs. Many observe strict divisions of labor between men and women, depending on whether they farm or raise livestock. Several generations—and even people from different families—might share the same dwelling.

During the colonial era, religion dominated education systems in the subregion. After countries gained their independence, efforts were made to modernize education, but many rural areas still lacked sufficient support. Today, literacy ranges from about 56 percent in the Central African Republic to about 94 percent in Equatorial Guinea. Literacy among women is commonly lower than that of men.

Under Spanish rule, Equatorial Guinea achieved a high literacy rate as well as a decent health-care system. In general, however, Equatorial Africa's countries lack the resources to halt preventable diseases that Western countries have been able to stop. The lack of safe drinking water, the shortage of vaccines for curable diseases, and the rising rate of HIV infection and number of AIDS victims remain primary health concerns.

Ethnic conflicts continue to plague parts of Equatorial Africa. In the Democratic Republic of the Congo (once known as Zaire) the long dictatorship of Mobutu Sese Seko was followed by ethnic strife and one of the bloodiest conflicts since World War II. Ethnic conflict spiked after 1994, when the country received a heavy influx of refugees who were fleeing the fighting in the neighboring countries of Rwanda and Burundi.

✓ **READING PROGRESS CHECK**

Hypothesizing How might Equatorial Africa's ethnic and linguistic diversity be an obstacle to political stability?

(c)Themba Hadebe/AP Images, (bl)Brennan Linsley/AP Images, (br)Lionel Healing/AFP/Getty Images

Analyzing
PRIMARY SOURCES
Crime in Kinshasa

"Most reported criminal incidents in Kinshasa involve crimes of opportunity, which include pick-pocketing and petty theft. The majority of the crimes are committed by 'sheggehs,' who are generally homeless street children. Travel in certain areas of Kinshasa, Kisangani, Lubumbashi and most other major cities is generally safe during daylight hours, but travelers are urged to be vigilant against criminal activity that targets non-Congolese."

—U.S. Department of State, International Travel for U.S. Citizens

DBQ ***ANALYZING PRIMARY SOURCES*** Who are "sheggehs"?

Cease-fire agreement signed by rebel groups; sporadic fighting continues

President Laurent Kabila assassinated; his son Joseph Kabila becomes president

1999

1998

2001

2002

2006

2000 ➜

Kabila's government challenged by a rebellion backed by Rwanda and Uganda

Agreement signed to establish a government of national unity

First free presidential election in four decades takes place, returning Joseph Kabila to office.

DEMOCRATIC REPUBLIC OF THE CONGO 2008 GDP

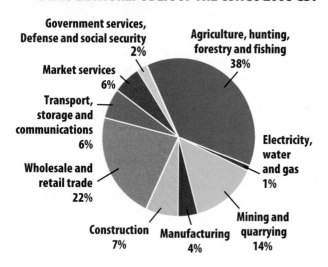

Government services,
Defense and social security
2%

Market services
6%

Transport,
storage and
communications
6%

Wholesale and
retail trade
22%

Construction
7%

Manufacturing
4%

Mining and
quarrying
14%

Electricity,
water
and gas
1%

Agriculture, hunting,
forestry and fishing
38%

Source: http://focusafrica.gov.in/Sector_Profile_DRC.html

Agriculture is the most important economic activity in the Democratic Republic of the Congo, accounting for more than one-third of the country's GDP.

▲ CRITICAL THINKING

1. *Analyzing Visuals* What is the country's smallest economic sector?

2. *Drawing Conclusions* How does the economy of the Democratic Republic of the Congo exemplify Equatorial Africa's colonial legacy?

Economic Activities

GUIDING QUESTION *Why is the economic activity of Equatorial Africa so varied?*

The lives of people throughout Equatorial Africa are changing. The region is becoming more closely tied to the global economy. Like the countries in which they live, individuals face challenges, yet changing economic activities also offer new opportunities.

Farming is the main economic activity in Equatorial Africa. Some countries still depend on single-crop economies; others produce a variety of agricultural goods. Most farmers are subsistence farmers, providing primarily for the needs of their family or village. After they meet their own needs, farmers often sell or trade any extra harvest or livestock at a local market. Common crops are bananas, cacao, cassava (a root), corn, sweet potatoes, and yams.

A smaller percentage of the population works in commercial farming, in which farms produce crops on a large scale. These cash crops are grown and exported to global markets. Most commercial farms are large, foreign-owned plantations. They produce palm oil, peanuts, cacao, and sisal (a vegetable fiber used in rope, drywall and car interiors).

The colonial economic systems played an important role in the growth of commercial crops in Equatorial Africa. Today these same crops are the subregion's main agricultural exports. Most cash crops leave Africa to be processed elsewhere, just as during the colonial period. In the Democratic Republic of the Congo, crops and other goods are exported from Matadi, a port on the Congo River.

Logging has serious consequences, but the lumber industry maintains a relatively small output. Coastal countries with rain forests, such as Gabon, export significant amounts of ebony, mahogany, and natural rubber. Wood from the Okoume tree—found mainly in Gabon, the Republic of the Congo, and Equatorial Guinea—is famous as a good raw material for plywood.

Fishing also represents a portion of the subregion's economic activity. The Congo and its tributaries are also teeming with fish. At Boyoma Falls, the local fishers use conical baskets to trap fish as they pass through the rapids. The shallow ocean area near the coast are excellent fishing waters. Along the southwestern coast, commercial fishing vessels catch herring, sardines, and tuna for export. Few countries, however, build and support large commercial fishing fleets.

The subregion is rich in mineral resources. Equatorial Guinea, the Republic of the Congo, and the Democratic Republic of the Congo are all major oil producers. Mining of copper and other minerals has been important in the Democratic Republic of the Congo since the colonial period. In spite of rich mineral resources, however, many people in these countries do not benefit directly. Governments have often mismanaged the income from mineral wealth. In the Democratic Republic of the Congo, for example, Mobutu Sese Seko appropriated a huge fortune from the state-owned mining company, Gécamines, for his own private fortune. In other cases, foreign mine owners send their profits abroad.

Most of Equatorial Africa never developed manufacturing industries and lacks the infrastructure to process natural resources. Today many countries are receiving foreign loans to industrialize. Progress, however, is slow and unemployment is a major problem. In 2010, for example, the unemployment rate in the Democratic Republic of the Congo was a staggering 61 percent.

Demand for manufactured goods has increased, and locally produced goods have replaced some imported items. Today the subregion's industrial workers process food or produce textiles, paper goods, leather products, and cement. Some assemble electric motors, tractors, electronics, and automobiles.

Equatorial Africa faces many obstacles to industrialization. Educational systems are still developing, and more people must gain new skills. Hydroelectric resources are plentiful but untapped, and electrical outages occur. Political conflicts interrupt economic planning and divert resources from development projects.

Well-developed transportation and communications systems are essential to industry and trade. However, creating and maintaining such systems is difficult. Roads and railways must cross vast distances and varied terrain in the region. Many countries lack modern roads and transport systems except in urban areas.

Some Africans run small businesses selling local products such as baskets, art, and jewelry. Using e-commerce, or selling and buying on the Internet, people sell their products. Thus, the Internet broadens the market for local products.

The subregion has long relied on radio, with state-run stations providing broadcast programming. Television reaches fewer people because the land-relay systems for transmitting TV signals are costly and less available outside urban areas. Satellite technology is helping to expand the reach of television.

Land-line telephone service is also limited, especially in rural areas. However, cell phone usage is increasing. According to a United Nations report, Africa has the highest growth rate of cell phone subscriptions. Satellite and wireless technology have broadened access to phone service and the Internet. Today, more than half of the people in large cities in the region have access to the Internet. Because it is less expensive to set up cell phone towers than to connect telephone wire over vast distances, some areas now have cell phone service where they never before had access to a telephone. Increasingly, these cell phones are smart phones with Internet access which opens new opportunities.

The use of cell phones and smart phones has grown dramatically in Equatorial Africa.

▲ CRITICAL THINKING

1. *Considering Advantages and Disadvantages* What advantages do you think cell phones offer over land lines in this region? Might there be any disadvantages to their use?

2. *Making Predictions* Would you expect smart phone use to increase or decrease in the future? Why?

 READING PROGRESS CHECK

Identifying Cause and Effect Why has it been difficult to develop transportation and communications networks in Equatorial Africa?

Giovanni Mereghetti/age fotostock

LESSON 2 REVIEW

Reviewing Vocabulary
1. *Explaining* What effect did the introduction of colonial cash crops have in Equatorial Africa?

Using Your Notes
2. *Summarizing* For each section of the graphic organizer, write one sentence that summarizes your notes.

Answering the Guiding Questions
3. *Expressing* What challenges does democracy face in Equatorial Africa?

4. *Identifying Cause and Effect* How does geography influence where the people of Equatorial Africa live?

5. *Identifying Central Issues* How does conflict shape life in Equatorial Africa?

6. *Assessing* Why is the economic activity of Equatorial Africa so varied?

Writing Activity
7. *Informative/Explanatory* Write a paragraph describing how technology is changing how people in Equatorial Africa communicate.

networks

There's More Online!

☑ **IMAGE** Cooking With Biofuels

☑ **IMAGE** Refugee Camp

☑ **INFOGRAPHIC** Poaching of the African Elephant

☑ **INTERACTIVE SELF-CHECK QUIZ**

☑ **VIDEO** People and Their Environment: Equatorial Africa

Reading **HELP**DESK

Academic Vocabulary

- **priority**
- **hamper**

Content Vocabulary

- **internally displaced person**
- **biofuel**

TAKING NOTES: *Key Ideas and Details*

IDENTIFYING Use a graphic organizer like the one below to take notes about threats to natural resources in Equatorial Africa.

Equatorial Africa: Managing Resources and Human Impact	
Natural Feature	Threat

LESSON 3
People and Their Environment: Equatorial Africa

ESSENTIAL QUESTION · *How do physical systems and human systems shape a place?*

IT MATTERS BECAUSE

The dense rain forest of Equatorial Africa is an area of abundant resources. Banana, orange, and mango trees grow wild. The rain forest is alive with the sounds of wildlife, and a vast treasure in mineral wealth lies beneath the ground. Such resources can provide a bright future to this often-troubled region, but managing them in a sustainable way poses challenges.

Managing Resources

GUIDING QUESTION *What makes managing resources in Equatorial Africa especially challenging?*

The people of Equatorial Africa face tremendous difficulties in achieving a better life. Many environmental challenges threaten the subregion's supply of food, the health of its people, and its plant and animal life. Yet its people, like their neighbors around the globe, look to the future with hope.

Today many of the people in Equatorial Africa must focus on survival. While there is wealth in the subregion, it tends to be concentrated in the hands of the few. Poverty and hunger are bitter enemies for most of the population. When the focus is on survival, management of natural resources can be less of a **priority**.

And yet, Equatorial Africa has a great number of natural resources. The Democratic Republic of the Congo, for example, has tremendous amounts of fertile land. Rapidly moving streams are resources that could be harnessed to generate electricity. The country also has many mineral resources, including large reserves of copper, cobalt, and industrial diamonds. Although mining is extensive, the full economic benefits that could be gained from mining have not been realized.

Extensive conflict in the subregion is a major cause of misery. Fighting and looting severely **hamper** food distribution and normal economic activity. In the Democratic Republic of the Congo, rival

(tl)Jack Guez/AFP/Getty Images, (tc)Heiner Heine/age fotostock, (tr)Marcel Pepeira/age fotostock factions fight for control of regions and wealth. In South Sudan, ethnic groups bent on eliminating other groups terrorize the countryside. In the Central African Republic, pockets of the countryside are lawless.

The Office of the United Nations High Commissioner for Refugees estimates the "population of concern" in Equatorial Africa is about four million people. These are refugees and **internally displaced persons**, individuals who have been forced from their homes but remain in their own country. Humanitarian organizations such as Doctors Without Borders (Médecins Sans Frontières) and the International Red Cross have helped by sending medical teams and relief workers, but the persistence of the conflicts limits their effectiveness.

Along with conflict, poverty takes an enormous toll on the people of the subregion. Africa south of the Sahara is the poorest region in the world, and some of the countries in Equatorial Africa are the poorest of the poor. With a per capita gross domestic product (GDP) of $300, the Democratic Republic of the Congo is arguably the poorest country in the world. For comparison, the United States has a per capita GDP of about $48,000. The reality of the region's deep and widespread poverty demands that attention be focused on meeting immediate human needs rather than on managing natural resources.

A further challenge to the management of natural resources is widespread corruption in the governments of the region. Government officials who might work to manage natural resources are subject to bribery by businesses who want to maximize profits. Enforcement of environmental regulations is also hampered by corruption.

✓ **READING PROGRESS CHECK**

Identifying Cause and Effect Why has effective resource management been given a low priority in Equatorial Africa?

Human Impact

GUIDING QUESTION *How have human activities impacted the environment in Equatorial Africa?*

Equatorial Africa is a beautiful region of forests, rivers, mountains, plains, and wildlife. The natural resources of this land are vast. Unfortunately, human activities have placed much of this land in jeopardy. Water pollution, for example, is a major issue in much of the region. This is partly due to the dumping of raw sewage. In the Central African Republic, Equatorial Guinea, and the Republic of the Congo, tap water is not potable, or suitable for drinking.

Soil erosion has become a major ecological problem in Equatorial Africa. In the Democratic Republic of the Congo, for example, soil erosion affects agricultural productivity and thus causes food shortages. Soil erosion also leads to landslides and mud flows. These can have devastating consequences when they happen near rapidly growing cities, such as Bamenda City in Cameroon. Mudslides can destroy buildings that were hastily and poorly constructed.

These children live in a refugee camp in the Democratic Republic of the Congo. The country has endured decades of civil war and ethnic strife.

▲ **CRITICAL THINKING**

1. ***Analyzing Visuals*** Based on the photograph, what hardships are experienced by refugees from ethnic conflicts? What other basic needs might these children lack?

2. ***Making Connections*** How does ethnic strife contribute to the region's poverty?

priority something given or meriting attention before competing alternatives

hamper to make difficult; to impede

internally displaced person a refugee within his or her own country

Jack Guez/AFP/Getty Images

Equatorial Africa **567**

The Congo Basin is one of the great forest regions of the world. Forests cover close to half of the basin's 2 million square miles (5.2 million sq. km). These forests, however, are under threat. Bananas, pineapples, cocoa, tea, coffee, and cotton are grown as cash crops on large commercial plantations. As farmers clear more land to grow these crops, the rain forest is seriously denuded, or stripped of trees. Commercial loggers also diminish the rain forest, as do local people who clear the land for small farming plots. Millions of refugees and internally displaced persons are forced to live off the land. They take wood for heating and cooking fires. About 3,500 square miles (9,100 sq. km) of the forest are lost each year.

Deforestation does not mean simply the loss of trees. It means less wood for the lumber industry, a loss of biodiversity, and accelerating climate change. It also leads to desertification, which in this case refers not to the advancing of a desert, but to the degradation of a formerly forested area. Many areas of Equatorial Africa that used to be forestland are now considered part of the savanna. Deforestation also contributes to soil erosion.

Air pollution, both inside and outdoors, is also a growing issue. In many parts of Equatorial Africa, household cooking is still done by burning **biofuels**— fuels such as charcoal or wood that are made from living matter. These pose a health threat as the smoke they produce pollutes household air. Household air pollution is much higher in countries that burn biofuels than in industrialized countries. Women, because they do most of the cooking, are especially at risk for the respiratory illnesses that pollution from burning biofuels can cause. Outdoor air pollution is also a problem. Vehicle emissions in growing, crowded cities, such as Kinshasa, pose a health threat to residents.

The Ivory Trade

African elephants once roamed in great numbers across the continent. Biologists estimate that in 1930, Africa was home to between 5 and 10 million elephants. During the last century, however, elephants were slaughtered by the tens of thousands for meat, for sport, and especially for their ivory tusks. Both male and female African elephants grow tusks—the world's main source of ivory. Ivory is a form of dentin that is excellent for carving and valued for its beauty. The trade in ivory dates back thousands of years. In the 1800s and 1900s, however, the demand for ivory from India, Europe, and the United States grew dramatically. By the mid-1900s, elephant populations had declined sharply. When the price of ivory soared in the 1970s, elephant tusks increased in value. Poachers began illegally killing elephants for their tusks. As many as 80,000 elephants a year were shot.

The ivory trade has had devastating consequences for the African elephant. During the 1980s, approximately half of Africa's elephant population was wiped out by poachers and only about 600,000 elephants remained. In 1989 African elephants were placed on the endangered species list by the

biofuel fuel created from living matter, such as trees

Millions of people in Equatorial Africa, such as these Mbororo women in Cameroon, rely on wood to make fires for cooking and heating. This accelerates deforestation and causes indoor air pollution.

▼ **CRITICAL THINKING**

1. *Analyzing Visuals* Based on the photo, what dangers are posed by indoor cooking with biofuels?

2. *Problem Solving* What technologies might be used to combat this problem?

Heiner Heine/age fotostock

ELEPHANT POACHING IN 2011

PERCENTAGE OF DEATHS DUE TO POACHING

Central Africa | Eastern Africa | Southern Africa | Western Africa

NOTE: Central Africa is sometimes referred to as Equatorial Africa.

Sources: http://www.cntraveler.com/2010; http://ngm.nationalgeographic.com/2012

Convention on International Trade in Endangered Species (CITES). Trade in ivory was banned worldwide. Organizations were established to coordinate the ban on the ivory trade and help African countries stop the poaching of elephants. The ban has only been partly successful, however. In recent years, the demand for ivory has surged. This is due in part to the huge demand for ivory ornaments in China.

Most elephants in Equatorial Africa are found in Cameroon. Armed poachers travel to Cameroon on horseback from Sudan and Chad across the Central African Republic, where almost all the elephants have already been wiped out. Even in protected areas, such as Cameroon's Bouba N'Djida National Park, more than half of the park's 400 elephants have been killed by poachers.

Hunting

In Equatorial Africa, local people do not hunt wild animals for sport as much as they hunt them for food. Warfare in the subregion has led to a widespread availability of firearms. It has also displaced millions of people who are extremely poor. This has proved a deadly combination for the subregion's wildlife. People hunt to feed themselves, and a growing number of people hunt to sell the meat to others.

The term *bushmeat* refers to the meat of animals hunted in "the bush," or forest. Among these animals are endangered or threatened species. Hunted animals include monkeys, antelopes, gorillas, and bonobos. In the Democratic Republic of the Congo, it is estimated that more than one million pounds of bushmeat are consumed each year.

☑ **READING PROGRESS CHECK**

Identifying Cause and Effect How has conflict in Equatorial Africa put pressure on the subregion's resources?

An area's elephant population will decline if more than 50 percent of the deaths are due to poaching. In 2011 all subregions in Africa south of the Sahara exceeded 50 percent. Some experts believe this rise in poaching could lead to the extinction of elephants by 2025. The loss of elephants could lead to ecosystem collapse in some areas.

▲ **CRITICAL THINKING**

1. ***Exploring Issues*** Do you think that the poachers or the people who buy ivory from poachers should be targeted by officials trying to stop the ivory trade? Explain.

2. ***Hypothesizing*** In what region are the most elephants killed by poachers? Why do you suppose more elephants are killed there than in other parts of Africa?

Marcel Pepeira/age fotostock

Addressing the Issues

GUIDING QUESTION *What steps are being taken to combat these environmental challenges?*

Increasingly, the protection of tropical rain forests is a priority in the subregion. Various countries have created forest reserves to protect tropical forests. Logging companies are also getting involved, using scientific tree farming and replanting projects to protect and renew forests. In one program, more than 10 million trees were planted on a tree plantation outside Virunga National Park in the Democratic Republic of the Congo. The Central African Forest Commission (COMIFAC) is an international organization coordinating the conservation and sustainable management of the forest ecosystem in the region. COMIFAC works in conjunction with a network of member countries, international institutions, and nongovernmental organizations (NGOs).

Some NGOs and other charitable organizations, along with Western governments and entrepreneurs, are working with local African communities to solve the indoor air pollution problem. New technologies may offer a solution to this serious health issue. Some organizations are inventing ways to make biofuel cookstoves cleaner and more efficient. Others have developed solar-powered cookstoves. Some are developing clean technologies to provide lighting. One of the most recent innovations is the development of gravity-powered lighting. This type of light uses gravity to power a small light for a short period of time. It has the potential to offer cheap lighting that is safe for both people and the environment.

To counter the ivory trade, governments in Equatorial Africa have deployed troops and special police forces to combat poachers. Unfortunately, they are often too few in number and poorly trained. Private groups like the World Wildlife Fund support local people's efforts to oppose poachers and pressure governments in the region to devote resources to antipoaching efforts. Nongovernmental organizations and conservationists have, however, come under criticism for allegedly disregarding the poverty and human suffering that exist in the region. The widespread economic hardship in Equatorial Africa creates a situation in which poaching is a tempting way of making a living. This makes the antipoaching campaign even more of a challenge.

Similarly, decreasing the trade in bushmeat is a difficult task given the ongoing political and economic struggles in the region. Humanitarian and development efforts aim to increase people's standard of living in hopes that this will in turn decrease the demand for hunted wildlife.

☑ **READING PROGRESS CHECK**

Identifying What actions are being taken to protect Africa's tropical forests?

LESSON 3 REVIEW

Reviewing Vocabulary
1. *Expressing* What problem does the use of biofuels create?

Using Your Notes
2. *Expressing* Use the notes in your graphic organizer to write a paragraph about environmental issues in Equatorial Africa.

Answering the Guiding Questions
3. *Identifying Central Issues* What makes managing resources in Equatorial Africa especially challenging?

4. *Identifying Cause and Effect* How have human activities impacted the environment in Equatorial Africa?

5. *Listing* What steps are being taken to combat these environmental challenges?

Writing Activity
6. *Argument* Write an e-mail to a United Nations official arguing for or against the creation of more protected wildlife areas in Equatorial Africa.

Directions: On a separate sheet of paper, answer the questions below. Make sure you read carefully and answer all parts of the questions.

Lesson Review

Lesson 1

1 ***Explaining*** What physical feature dominates Equatorial Africa? What functions does it serve?

2 ***Making Connections*** What is the economic value of the subregion's rain forests?

3 ***Identifying Cause and Effect*** Why do Equatorial Africa's water resources remain underdeveloped, despite abundant rainfall?

Lesson 2

4 ***Sequencing*** Create a graphic organizer showing the rise and fall of colonial powers in what is now the Republic of the Congo.

5 ***Explaining*** What role did colonialism play in the development of commercial agriculture?

6 ***Identifying Central Issues*** Describe the quantity of mineral resources in Equatorial Africa. Explain why the people of the area see few benefits from these resources.

Lesson 3

7 ***Summarizing*** Why is Africa south of the Sahara the poorest region in the world, despite its many resources?

8 ***Identifying Cause and Effect*** Develop a diagram with labels to show the causes and effects of soil erosion in Equatorial Africa. Write a paragraph to explain the information in your diagram.

9 ***Making Connections*** Discuss the connection between deforestation and desertification in Equatorial Africa.

Critical Thinking

10 ***Organizing*** After studying Equatorial Africa, you have decided you would like to write a book about the people, land, and climate of this subregion. Write a one-page outline for your book to show the major topics you will include.

21st Century Skills

Use the cartoon below to answer the questions that follow.

" 'Born in conservation,' if you don't mind. 'Captivity' has negative connotations."

11 ***Evaluating Primary Sources*** Explain why you think elephants were chosen as the topic of this cartoon. What conservation efforts are taking place in Equatorial Africa?

12 ***Understanding Relationships Among Events*** What relationship does poverty have to conservation in Equatorial Africa?

Applying Map Skills

Use the Unit 6 Atlas to answer the following questions.

13 ***The World in Spatial Terms*** Use your mental map of Equatorial Africa to describe the spatial relationship among the Congo Basin, the Katanga Plateau, and the Republic of the Congo.

14 ***Places and Regions*** What is the largest country in Equatorial Africa? Provide the approximate size of this country in square miles.

15 ***Environment and Society*** Discuss the connections between the population of Kinshasa, Democratic Republic of the Congo, and the pollution levels from vehicle emissions.

Need Extra Help?

If You've Missed Question	**1**	**2**	**3**	**4**	**5**	**6**	**7**	**8**	**9**	**10**	**11**	**12**	**13**	**14**	**15**
Go to page	554	558	558	560	564	564	567	567	568	551	571	571	476	476	476

J.B. Handelsman/The New Yorker Collection/www.cartoonbank.com

Directions: On a separate sheet of paper, answer the questions below. Make sure you read carefully and answer all parts of the questions.

DBQ Analyzing Primary Sources

Conflict is a major contributor to poverty in Equatorial Africa. Read the news report below to see how conflict in the Central African Republic affects access to and production of food.

PRIMARY SOURCE

"*The U.N. World Food Program [WFP] expressed concern about a potential food crisis in CAR [the Central African Republic]. It said conflict that erupted during a rebellion . . . had sparked the situation.*

The WFP said trade has been interrupted in parts of the country controlled by the rebel group. Basic food prices for areas under government control, meanwhile, have increased 40 percent since December.

WFP called on the rebel group to let humanitarian workers into its parts of CAR. An estimated 800,000 people live in areas under rebel control.

'We are very concerned about prospect for the 2013 growing season, which is due to start in just a few weeks,' Rockaya Fall, the Food and Agriculture Organization's country representative, said in a statement. 'Land preparation, which should have begun, is behind schedule in many places, due to insecurity."

—"Food Crisis Plagues War-Torn CAR," United Press International, February 18, 2013

16 *Identifying Cause and Effect* In what ways is conflict contributing to a food crisis in the Central African Republic?

17 *Differentiating* Distinguish between the roles played by the rebels and the government in the food crisis.

Exploring the Essential Question

18 *Synthesizing* Create a short guide in the form of a brochure to provide an overview of the ways physical systems and human systems shape Equatorial Africa. Within your brochure, include the following headings: *History, Politics, Population, Society, Culture,* and *Economics.*

College and Career Readiness

19 *Clear Communication* Imagine that you are applying for a position as an intern with the Office of the United Nations High Commissioner for Refugees (UNHCR). The focus of this internship is on helping refugees and internally displaced persons in Equatorial Africa. Write a one-page essay to explain why you should be chosen to serve in this position. Within your essay, provide the following: your knowledge of the relevant problems in the subregion, including the conditions that have led to the problems and the relevance of the work of Doctors Without Borders (Médicins Sans Frontières) and the International Red Cross in the subregion.

Research and Presentation

20 *Gathering Information* With a partner, conduct research to learn more about conflict diamonds, also known as blood diamonds, mined from Africa. Develop a multimedia presentation to answer the following questions: What are conflict diamonds? Why are they known as conflict diamonds? What are some issues related to conflict diamonds? What steps are being taken to address these issues? Include the following in your multimedia presentation: audio and video of relevant news reports, maps, diagrams, and excerpts from printed news reports.

Writing About Geography

21 *Narrative* Use standard grammar, spelling, sentence structure, and punctuation to write a paragraph that describes a visit to the equatorial rain forest. In your paragraph, include details that relate to all five senses and to the wildlife, plant life, and resources of the rain forest.

Need Extra Help?

If You've Missed Question	**16**	**17**	**18**	**19**	**20**	**21**
Go to page	572	572	551	567	566	558

Southern Africa

ESSENTIAL QUESTION · *How do physical systems and human systems shape a place?*

Why Geography Matters
Southern Africa and HIV/AIDS

Lesson 1
Physical Geography of Southern Africa

Lesson 2
Human Geography of Southern Africa

Lesson 3
People and Their Environment: Southern Africa

Geography Matters...

Southern Africa is a difficult land to categorize. Most of it is located on a continent, but it also consists of a large island and several small islands. The continental part is surrounded on three sides by ocean, but most of its landforms are plateaus and highlands.

The people of Southern Africa have a history that goes back tens of thousands of years. Yet their culture has been heavily influenced by Europeans who arrived just a few centuries ago. Some people in Southern Africa live in densely packed cities. Others live in the desert much the same way as their ancestors did. Christianity is widespread, but many traditional religions are still practiced as well.

◀ Jeffreys Bay in South Africa is one of the world's top destinations for surfers.

Kelly Cestari/ASP/Getty Images

Southern Africa
and HIV/AIDS

In Africa, HIV/AIDS has reached epidemic proportions. Southern Africa has
the highest number of people infected with HIV/AIDS on the continent.
Nearly one in five people in the region is suffering from the disease. New
incidences of HIV infections have declined somewhat, but the disease is still
devastating the economic and social health of Southern Africa.

World Distribution of HIV 2010

Worldwide, about 34 million people
were living with HIV by the end of 2011.
In the same year, more than 2.5 million
people became newly infected with HIV.
More than two-thirds of these were
living in Africa south of the Sahara.

ARCTIC
OCEAN

NORTH
AMERICA

PACIFIC
OCEAN

ATLANTIC
OCEAN

EUROPE

ASIA

AFRICA

PACIFIC
OCEAN

SOUTH
AMERICA

INDIAN
OCEAN

AUSTRALIA

More than half of
those living with HIV
in Africa south of the
Sahara are women.

Parts of Southern Africa—including Botswana,
Lesotho, and Swaziland—are among those
hardest hit by the disease. Many factors affect
the spread of HIV such as education, access to
medical care, and level of government funding
for programs for prevention and treatment.

**Percentage of
Adult Population
(ages 15-49)
Infected by HIV, 2010**

Less than 0.1%

0.1% – < 0.5%

0.5% – < 1.0%

1.0% – < 5.0%

5.0% – <15.0%

15.0% – 28.0%

No data

SOURCE: UNAIDS

0   4,000 miles

0   4,000 kilometers

Mercator projection

60°N

30°N

0°

30°S

60°S

150°W   120°W   90°W   60°W   30°W   0°   30°E   60°E   90°E   120°E   150°E

What are HIV and AIDS?

HIV (human immunodeficiency virus) is an infection that can be spread from one person to another. Over time, the disease attacks a person's immune system and can advance to AIDS (acquired immunodeficiency syndrome). The disease can be spread through blood or other bodily fluids, as well as from a pregnant mother to her unborn child. In a few instances, the disease has spread through organ transplants. Symptoms of HIV/AIDS range from flulike symptoms to none at all. In some cases, people who are infected with HIV show no symptoms for years, so without annual medical exams it can be difficult to diagnose in its early stages.

Southern African countries have the highest number of people living with HIV/AIDS on the continent. It is believed to have emerged in the Democratic Republic of the Congo in the 1970s. However, the disease did not become an epidemic in Africa until the following decade. It first spread east and then south within the continent. By the late 1980s, the virus had become a regional epidemic.

___

1. Human Systems Why is HIV/AIDS so dangerous to humankind?

Why have the people of Southern Africa been so badly affected by HIV/AIDS?

Much of the population of Southern Africa lives in poverty. Literacy is low. Medical care is limited or too expensive for the majority of the population. These factors weigh heavily into the high rates of HIV/AIDS in the subregion. In some countries of Southern Africa, about 40 percent of the adult population is living with HIV/AIDS. Percentages are over 40 percent in some urban areas. More than a million Southern Africans die each year because of HIV/AIDS, resulting in lower life expectancy and a rise in the number of orphans. Some scientists believe that differences between HIV subtypes have an effect on transmission rates. There is also some evidence that genetic factors and parasitic worm infections, which are common in Africa south of the Sahara, may contribute to the epidemic in the region.

There are other factors that may contribute to the high rate of HIV/AIDS cases in Southern Africa. These include social instability, rapid urbanization, labor migration, gender inequality, and sexual violence. Additionally, ineffective political leadership failed to address the crisis in the 1990s when the disease became an epidemic in the region.

___

2. Human Systems Why are the people of Southern Africa so vulnerable to HIV/AIDS?

What is being done to combat HIV/AIDS in Southern Africa?

Some governments in Southern Africa, as well as organizations such as the World Health Organization (WHO), have been working to educate people on how the disease is contracted and passed to others. This includes counseling programs, controlling incidences of other sexually transmitted diseases, which can cause the spread of HIV, and caring for people living with HIV/AIDS. Efforts have also been made to improve the safety of blood supplies.

Early testing for HIV has helped identify cases that can then be treated with antiretroviral drugs (ARVs). These drugs delay the progression of HIV to AIDS. WHO began a program in 2003 to treat 5 million people in the region with ARVs. Even though WHO failed to reach its goal, it set a precedent for countries in Southern Africa such as Botswana, Namibia, and South Africa to make ARVs accessible to those living with HIV.

___

3. Human Systems How are people working to combat HIV/AIDS, and how successful are their efforts?

(l)Heiner Heine/Alamy, (c)Dave Hogan/Getty Images Entertainment/Getty Images, (r)BSIP/Universal Images Group/Getty Images

THERE'S MORE ONLINE

SEE a map of HIV infection rates in Africa • *READ* a quote about efforts to fight HIV/AIDS in Southern Africa

- ☑ **CHART** Wildlife Reserves in Southern Africa
- ☑ **IMAGE** Baobab Tree
- ☑ **IMAGE** The Long Wall of Namib
- ☑ **IMAGE** Moremi Game Reserve
- ☑ **INTERACTIVE SELF-CHECK QUIZ**
- ☑ **VIDEO** Physical Geography of Southern Africa

LESSON 1
Physical Geography of Southern Africa

Reading **HELP**DESK

Academic Vocabulary

- **altitude**
- **source**

Content Vocabulary

- **escarpment**
- **delta**
- **Mediterranean climate**
- **rain shadow**

TAKING NOTES: *Key Ideas and Details*

IDENTIFYING As you read about the physical geography of Southern Africa, use a graphic organizer like the one below to keep track of its physical features.

ESSENTIAL QUESTION · *How do physical systems and human systems shape a place?*

IT MATTERS BECAUSE
The terrain of Southern Africa produces a wide variety of biomes—tropical, desert, and temperate regions. Life and culture are very different in each of these areas. In addition, Southern Africa's rivers are essential for both the natural environment and human activity in the subregion.

Landforms

GUIDING QUESTION *What is the most dominant physical feature of Southern Africa?*

Southern Africa consists of 14 countries, 10 of which are on the mainland. The other four are island countries. Imagine a line drawn roughly along the northern borders of Angola, Zambia, Malawi, and Mozambique. Everything south of that line is part of Southern Africa. Extend the line out into the Indian Ocean, and it runs just above Comoros and Madagascar. Mauritius is east of Madagascar, while Seychelles is northeast of Madagascar.

If Angola, Zambia, Malawi, and Mozambique form the "top row" of Southern Africa, then Namibia, Botswana, Zimbabwe, and the southern half of Mozambique form the "second row." At the southern tip of the continent is South Africa. The country of Lesotho (luh•SOO•too) lies entirely within South Africa. Swaziland is on the South Africa-Mozambique border.

The mainland of Southern Africa is surrounded on three sides by ocean. Most of the land sits at a high **altitude**, or height above sea level, of over 2,000 feet (610 m). Along the coasts of Angola, Namibia, and South Africa lies the coastal plain, a narrow strip of land that varies from a few miles wide to a few dozen miles wide. Looming up behind this coastal plain are high steep cliffs known as the Great **Escarpment**. These cliffs form an almost unbroken U-shape. They run from the west coast south to the Cape of Good Hope and then curve northeast to the South Africa-Mozambique border. The Drakensberg Range is part of the Great

(tl)Juan Carlos Muñoz/age fotostock, (tc)G Fischer/age fotostock, (tr)PhotoStock-Israel/Flickr/Getty Images

Escarpment. The range rises to more than 11,000 feet (3,353 m) and runs along the southern edge of South Africa. As the escarpment reaches Mozambique, it turns north to follow along Mozambique's western border and on through Malawi. Because most of Mozambique lies outside of the Great Escarpment, its landscape is mainly made up of soft, rolling hills. Only in the northwestern interior of the country does the land rise above 600 feet (183 m).

The land inside the Great Escarpment is mostly hills and plateaus. A plateau, or "tableland," is a stretch of flat land that is higher than its surroundings. The most striking part of the Southern African landscape is a plateau that is over a mile high. The Highveld, as it is called, sits more than 50 miles (80 km) from the coast. However, given its 6,000 foot (1,829 m) altitude, it is visible from a long distance.

The highest parts of Namibia and Angola are near the coast, along the Great Escarpment. Zimbabwe's highest terrain is in the center of the country. Zambia's high points are in the northeast, where they merge with Mozambique's highlands along the Great Escarpment. In contrast, most of Botswana sits in a great basin. While still much higher than the coastal lands, it is lower than the surrounding landforms.

The landforms surrounding Botswana on the west into Namibia or northwest into Angola are once again hills and plateaus. This undulating terrain continues westward until it rises up over the Great Escarpment. It then drops down onto the narrow Atlantic coastal plain.

In a sense, the terrain of Madagascar is a smaller version of that on the mainland. The middle of the island, like the mainland, is a series of plateaus surrounded by an escarpment. Unlike the mainland, groups and masses of volcanoes are strewn among the Madagascar plateaus. Tsiafajavona, one of these volcanic peaks, is 8,671 feet (2,643 m) high.

In the north, the plateaus give way to volcanoes that slope down to the sea. The southern edges of the plateaus tower above the Indian Ocean. To the east and west are escarpments. The escarpments, some of which are impassable, separate the highlands from the lower areas on either side. The eastern part of the island is a narrow coastal strip, rather like the Atlantic coast of Southern Africa. The western side is more varied, with low plateaus and rolling hills.

☑ **READING PROGRESS CHECK**

Describing Where is the highest terrain in Southern Africa found?

Water Systems

GUIDING QUESTION *How do people harness the power of the subregion's rivers?*

The Okavango River runs southeast from central Angola to northern Botswana. It starts on a plateau, and for a while it flows along the border between Angola and Namibia. Then it turns south and heads into Botswana. Most rivers empty into a lake or ocean, but the Okavango ends inland. The river just spreads wider and wider until it forms an inland **delta** and swamps.

The Orange River flows nearly from one side of Southern Africa to the other. It starts in Lesotho, on the Highveld, just over 100 miles (161 km) from the coast. It flows west across South Africa, then forms part of the border between South Africa and Namibia before it empties into the Atlantic Ocean.

altitude height above sea level

escarpment a long cliff that separates lands at two different altitudes

delta an often triangular-shaped section of land formed as the waters of a river slow down and split into many channels as they deposit silt and sediment

The Long Wall of Namib is a section of the Namib Desert that runs along the coast of southern Namibia.

▼ **CRITICAL THINKING**

1. *Making Connections* Less than 100 miles (161 km) inland from this spot is fertile land. Why does that fertile land not extend to the ocean?

2. *Analyzing Visuals* Why is the water in this picture not being used to irrigate the arid land?

Juan Carlos Muñoz/age fotostock

Wildlife abounds in the Moremi Game Reserve in Botswana.

▲ CRITICAL THINKING

1. *Constructing a Thesis* How can an area that is as wet as a swamp be part of a desert?

2. *Identifying Cause and Effect* Why do the waters of the Okavango River collect in northern Botswana and not flow into some other area?

The Zambezi River comes into contact with several countries along its course to the sea. Starting in eastern Angola, it flows south through western Zambia. It forms the border between Zambia and the countries of Namibia, Botswana, and Zimbabwe. Along the southern stretch of the border with Zimbabwe, the river suddenly drops straight down, forming Victoria Falls. Once it has cleared the northern tip of Zimbabwe, the Zambezi heads east through Mozambique, where it flows near the southern tip of Malawi before emptying into the Indian Ocean.

The Limpopo River begins in northern South Africa and flows north for a distance before it turns eastward and forms the border between South Africa and Botswana and the border between South Africa and Zimbabwe. Then it flows across Mozambique to the ocean.

None of these rivers are navigable, except for short stretches across the coastal plain. They originate in highlands and as they flow through the steep, terrain, rapids and waterfalls mark their courses. These conditions are not suitable for transportation, but they are useful for generating electricity. Two of the largest hydroelectric dams in Africa—the Kariba Dam and the Cahora Bassa Dam—are located on the Zambezi River.

Aquatic wildlife such as hippopotamuses and crocodiles thrive in and along these rivers. The northeastern part of the Okavango Delta is set aside as the Moremi Game Reserve. Hundreds of species including lions, cheetahs, hippopotamuses, buffalo, wild dogs, and many types of birds and fish live there. Two wildlife areas on either side of the Limpopo, in South Africa and in Mozambique, recently joined with each other and with several sanctuaries in Zimbabwe to form the Great Limpopo Transfrontier Park. This extended preserve provides a safe haven for lions, leopards, hippopotamuses, elephants, giraffes, and many other species.

✓ READING PROGRESS CHECK

Explaining What makes a river suitable for generating electricity?

Climates, Biomes, and Resources

GUIDING QUESTION *How do the climates of Southern Africa affect its biomes?*

Southern Africa is large with many different landforms. The result is a great variety of climates. These different climates and landforms create a number of different biomes with distinct characteristics.

G Fischer/age fotostock

Climate Regions and Biomes

The coastal areas of Southern Africa have marine climates, which means they are greatly affected by weather conditions and systems that blow in from the open ocean. The Cape of Good Hope and the area to its immediate northeast have a **Mediterranean climate** similar to that of Greece and Italy. It is not typical of Southern Africa, however. Moving up the eastern coast to Mozambique, the climate becomes tropical wet/dry. The winters are warm and the summer rainy season stretches from November through March. The western coast up through Namibia and Angola has an arid, or desert, climate.

The interior of the subregion is generally hot, although temperatures can dip below freezing in the higher elevations. The eastern areas experience a fair amount of precipitation, but toward the west the rainfall drops off. The deserts of Botswana and Namibia receive very little rain.

The climate of Madagascar is determined by its central plateau and the warm, wet winds off the open ocean. The eastern coast of the island has a tropical wet climate, while the interior plateau has a highland climate. The western side is in a **rain shadow**, and much of it is desert.

Given the number of different climates in the subregion, Southern Africa has many different biomes. Much of the coastal plain along the Atlantic coast is a desert biome. On the Indian Ocean side, grassland biomes give way to forest as one moves north. In the interior of the continent, on the highlands inside the Great Escarpment, the vast majority of the land falls into one of two types, savanna or desert.

The north and the east are covered with savanna—vast grasslands dotted with small stands of trees. Many of the most recognized African mammals, such as giraffes, zebra, and jackals, live on the savanna. This biome is also home to animals known as the Big Five: lions, leopards, elephants, Cape buffalo, and rhinoceroses. These animals became known as the Big Five because they were so dangerous and difficult to hunt. However, they are now the five species most tourists want to see while on safari.

The south and the west of the inland area are mostly desert biome. The Kalahari Desert occupies much of the Botswana basin. It stretches southwest to where Botswana meets Namibia and South Africa. Here, it blends into the Namib Desert. The Namib continues down to the coast, then runs north between the ocean and the Great Escarpment through all of Namibia and into southern Angola.

Most of the Namib Desert is quite arid. The southern parts are covered with seemingly endless sand dunes—brick red inland and yellow along the coast—some of which can be 800 feet (244 m) tall and 20 miles (32 km) long. Inland, bushes and tall grasses have adapted to grow in the sand dunes. Antelope and ostriches live here as well. Farther north in the interior desert, rivers can be found, and with them elephants, rhinoceroses, hyenas, and more.

The arm of the Namib Desert that runs along the coast is quite different. It is almost completely arid, with little or no plant life. Some reptiles and insects have adapted to this biome, but no larger animals live here. As in the interior portion, though, the most northern reaches have more water. Succulents grow here, and it is home to various marine birds, such as pelicans, flamingos, and even penguins.

In the southern part of the Kalahari, rain is scarce. Drought-tolerant grasses and scrub are all that can take root. Herds of antelope such as wildebeest and springbok roam the area. The central part of the Kalahari gets some rain, and shrubs can grow there. Acacia trees provide homes for birds, rodents, and insects. The northern Kalahari is hardly desertlike at all because of the rivers that flow through it. Many smaller animals, such as wild dogs, foxes, anteaters, and porcupines, live here. Plants such as pond lilies and reeds thrive here as well.

Connecting Geography to MATH

Output of the Kariba Hydroelectric Dam

The Kariba Dam, located on the Zambezi River between the countries of Zambia and Zimbabwe, produces 6.7×10^8 kilowatt-hours (kWh) of electricity per year. Some of that electricity goes to Zambia and some to Zimbabwe. Average per capita electricity use in Zimbabwe is approximately 850 kWh.

APPLYING If half the output of Kariba Dam were sent to Zimbabwe, how many people's electricity needs could be met? Round your answer to the nearest thousand.

Mediterranean climate a climate marked by warm, dry summers and cool, rainy winters

rain shadow a condition created when winds blow in primarily one direction over mountains or other elevated terrain: the altitude change causes clouds to drop their precipitation on the near side of the mountain, leaving the land on the far side, where winds descend, with little rain or snow

The fruit of the baobab tree is eaten or used to make a soft drink. The bark is used to make rope, fabric, and strings for instruments. The leaves are eaten or used to make medicines.

▲ **CRITICAL THINKING**

1. **Constructing a Thesis** Why might Africans like to use the baobab tree as a symbol for the entire continent?

2. **Drawing Conclusions** Why might someone living in or near the desert like to have a baobab tree growing nearby?

source a point of origin

The most striking tree found in the northern Kalahari is the baobab. The baobab tree is often used as a symbol for all of Africa. Its thick trunk stretches high before splitting into a tangle of skinny branches, making the tree look as if it is upside-down with its roots in the air. Baobabs can live for hundreds of years and reach a diameter of 30 feet (9 m). The trunks can be hollowed out and used to collect rainwater or even as a shelter.

Natural Resources

Southern Africa's most important resource is its vast mineral wealth. Gold and copper are mined today, just as they were by ancient peoples. Coal, nickel, iron, cobalt, manganese, and uranium are found in abundance. Deposits of gemstones, especially diamonds, have drawn miners for several centuries. Additionally, Southern Africa's mineral wealth made it attractive to countries for colonization.

Unfortunately, Southern Africa's valuable resources are a **source** of controversy. In most cases, foreign companies own the mines that extract minerals and gemstones. They hire local workers and pay them very little money to work in dangerous conditions. The profits the foreign companies make are taken out of Southern Africa and do little to benefit the subregion.

One major exception is found in Botswana. In 1978 the government formed a partnership with a multinational company called De Beers. The partnership, known as Debswana, mines and sells Botswana's diamonds. The profits are split between the country and the company. This has given Botswana one of the healthiest, fastest-growing economies in all of Southern Africa.

The history of diamond mining can be seen in Kimberley, South Africa, at a site called the Big Hole. It was a hill until diamonds were discovered. From 1871 to 1914, 22.5 million tons (20.4 million t) of dirt and rock were removed, with picks and shovels. Today, the 700-foot- (213-m-) deep hole is a tourist destination.

☑ **READING PROGRESS CHECK**

Evaluating What types of biomes are found in Southern Africa?

PhotoStock-Israel/Flickr/Getty Images

LESSON 1 REVIEW

Reviewing Vocabulary

1. **Applying** Write a paragraph describing the altitudes of the various areas in Southern Africa. Be sure to mention the Great Escarpment.

Using Your Notes

2. **Synthesizing** Use your graphic organizer on the features of Southern Africa's physical geography to write a paragraph describing the four major rivers of the subregion and their courses.

Answering the Guiding Questions

3. **Identifying** What is the most dominant physical feature of Southern Africa?

4. **Explaining** How do people harness the power of the subregion's rivers?

5. **Applying** How do the climates of Southern Africa affect its biomes?

Writing Activity

6. **Informative/Explanatory** In a paragraph, describe the major biomes of Southern Africa.

networks

There's More Online!

- ☑ **GRAPH** Gold Production
- ☑ **IMAGE** Cecil Rhodes
- ☑ **IMAGE** Johannesburg
- ☑ **IMAGE** Shaka Zulu
- ☑ **MAP** European Colonization of Africa
- ☑ **TIME LINE** Apartheid and Its Legacy
- ☑ **INTERACTIVE SELF-CHECK QUIZ**
- ☑ **VIDEO** Human Geography of Southern Africa

LESSON 2

Human Geography of Southern Africa

ESSENTIAL QUESTION · *How do physical systems and human systems shape a place?*

Reading HELPDESK

Academic Vocabulary

- **policy**
- **prohibit**

Content Vocabulary

- **urbanization**
- **sanitation**
- **subsistence farming**
- **commercial farming**

TAKING NOTES: *Integration of Knowledge and Ideas*

IDENTIFYING As you read about the human geography of Southern Africa, use a graphic organizer like the one below to identify examples of how history, population, culture, and economics have worked together to create the subregion as it is today.

Southern Africa

| History Culture | Population Economics |

IT MATTERS BECAUSE

Southern Africa's human geography blends components from the very earliest humans, traditional African cultures, and European influences. Southern Africa today is full of contrasts—white and black, traditional and modern, wealth and poverty, urban and rural.

History and Government

GUIDING QUESTION *What cultures have influenced life in Southern Africa?*

Fossils that predate those of modern humans can be found in Southern Africa. Fossils of the first true humans are found north of the subregion in the valleys of Tanzania and Kenya. The first people to live in Southern Africa were the San, who arrived more than 20,000 years ago. Today, their descendants live in Botswana, Namibia, and Angola.

The Bantu peoples originated in central Africa, but began spreading across the continent some 3,500 years ago. The term *Bantu* refers to a group of about 500 related languages and to the various peoples who speak them. By about A.D. 300, Bantu peoples had migrated to Southern Africa.

One of the Bantu peoples, the Shona, established a city called Great Zimbabwe. By A.D. 1000, Great Zimbabwe had a population of between 12,000 and 20,000 people. The inhabitants farmed, raised cattle, and mined and traded gold. For 400 years, Great Zimbabwe was the center of a huge trading empire. Sometime in the 1400s, however, the city was abandoned. No one knows why. All that remains are ruins in southeastern Zimbabwe.

Madagascar's population is also the result of migration. Around A.D. 800, a small group of people sailed in outrigger canoes from islands in Southeast Asia to Madagascar. These people were the Malagasy, and their descendants populated the entire island, later mixing with migrants from the African continent. DNA tests confirm the ancestry of Madagascar is evenly split between Indonesia and East Africa. Despite sharing ancestry, as well as the many political and economic ties to Africa, many Malagasy do not consider themselves to be African.

(t)Tim Gainey/Alamy, (tc)Mary Evans Picture Library/The Image Works, (tc)Africa Media Online/The Image Works, (tcr)Keystone/Hutton Archive/Getty Images, (tr)Per-Anders Pettersson/Getty Images News/Getty Images

Colonial Powers, 1913

- Belgian
- British
- French
- German
- Italian
- Portuguese
- Spanish
- Independent

SPANISH MOROCCO
MADEIRA (Port.)
MOROCCO
IFNI • Agadir
TUNISIA • Tripoli
Mediterranean Sea
CANARY IS. (Sp.)
RÍO DE ORO
ALGERIA
LIBYA
SOUTHWEST ASIA
Cairo •
EGYPT
TROPIC OF CANCER
40°N
20°N
CAPE VERDE (Port.)
FRENCH WEST AFRICA
Red Sea
ANGLO-EGYPTIAN SUDAN
ERITREA
FRENCH SOMALILAND
GAMBIA
PORTUGUESE GUINEA
SIERRA LEONE
TOGO
GOLD COAST
NIGERIA
Lagos
CAMEROON
Fashoda •
Addis Ababa •
ETHIOPIA
BRITISH SOMALILAND
LIBERIA
FERNANDO PÓO (Sp.)
PRÍNCIPE IS. (Port.)
SÃO TOMÉ (Port.)
RÍO MUNI
FRENCH EQUATORIAL AFRICA
UGANDA
BRITISH EAST AFRICA
ITALIAN SOMALILAND
EQUATOR
0°
ATLANTIC OCEAN
CABINDA (Port.)
BELGIAN CONGO
GERMAN EAST AFRICA
Mombasa
PEMBA (U.K.)
ZANZIBAR (U.K.)
SEYCHELLES (U.K.)
ANGOLA
NORTHERN RHODESIA
NYASALAND
MOZAMBIQUE
COMORO IS. (Fr.)
MADAGASCAR
20°S
GERMAN SOUTHWEST AFRICA
SOUTHERN RHODESIA
RÉUNION (Fr.)
TROPIC OF CAPRICORN
WALVIS BAY (U.K.)
BECHUANALAND
Pretoria •
Johannesburg •
SWAZILAND
MAURITIUS (Fr.)
UNION OF SOUTH AFRICA
BASUTOLAND
INDIAN OCEAN
Cape Town •
N W E S
1,000 miles
1,000 kilometers
Lambert Azimuthal Equal-Area projection
40°S

1956
1956
1962
1951
1922
1979
SOUTHWEST ASIA
1975
1960
1960
1960
1993
1965
1960
1974
1960
1960
1960
1956
1958
1960
1977
1961
1960
1960
Never Colonized
1847
1957
1960
1960
1960
1960
1960
1962
1962
1960
1960
1960
1963
1975
1968
1962
1961
1976
ATLANTIC OCEAN
1975
1975
1964
1964
1960
1990
1966
1980
1975
0 1,000 miles
0 1,000 kilometers
1910
1968
1968
1966
Lambert Azimuthal Equal-Area projection

GEOGRAPHY CONNECTION

European control varied from colony to colony. For example, Cape Colony (South Africa) had a strong central government. In Angola, however, the Portuguese had little control over the people who lived in the inland areas.

1. *HUMAN SYSTEMS* Which European country governed Cape Colony?

2. *HUMAN SYSTEMS* Which European country controlled most of inland Southern Africa?

European Influences

Great Zimbabwe had faded away by the time the first Europeans arrived in Southern Africa. In the 1480s, Portuguese explorers, priests, and traders sailed into the Kingdom of Kongo in what is now the northern tip of Angola. At first, relations between the Portuguese and Kongo were peaceful, but it did not last.

By the mid-1500s, an active slave trade was running throughout Southern Africa. Coastal African kings sent raiding parties inland to capture people. The captives were then traded to the Portuguese for firearms and other manufactured goods.

The Portuguese established slave and trading posts on both the east and west coasts of Southern Africa. They either traded or warred with the various kingdoms they encountered, and they shipped much gold, silver, and ivory back to Europe. In the 1600s, other European powers expanded into Africa, causing Portuguese power to wane. By the mid-1700s, Dutch, British, and local African forces had confined the Portuguese to Angola and Mozambique. Only small settlements near the coast remained subject to Portuguese control.

In 1652 the Dutch East India Company established a settlement on the Cape of Good Hope. Dutch settlers took more and more land from the local African inhabitants, expanding well beyond the influence of the Dutch East India Company.

The Dutch were aided in their expansion by disease. As in the Americas, the local population had no immunity to European diseases and many died. After a few generations, the Dutch settlers referred to themselves as Afrikaners, which means "Africans." They were also called Boers, the Dutch word for "farmers."

Most of the Boers used slave labor on their farms. At first, enslaved people were brought in from areas farther north, but later the Boers enslaved local people as well. In order to communicate, the Afrikaners and the Africans developed a pidgin dialect that eventually evolved into Afrikaans, a new language.

At the start of the 1800s, Cape Colony was home to 22,000 whites, 25,000 enslaved blacks, and tens of thousands of free blacks. In 1806 Great Britain seized control of Cape Colony. A year later, Britain outlawed the slave trade in all its colonies. This had a positive effect in that the British and Afrikaners were now prohibited from capturing and enslaving Africans. It did not benefit people already enslaved, who were not freed by the ending of the trade.

European contact with Madagascar was sporadic and varied. The Portuguese, British, and French tried to gain influence on the island. The French established a settlement in 1642, but it did not last. Trade continued, and both the French and the British allied themselves with various local groups and leaders. In 1896 Madagascar became a French colony.

In 1884 Germany's new colonial goals led to the establishment of two colonies on the mainland of Africa. One, German East Africa, was farther north on the continent but did include some of what is today Mozambique. The other, German Southwest Africa, later became the country of Namibia. After losing World War I, Germany was stripped of all its African colonies. German East Africa was divided between Great Britain and Belgium. German Southwest Africa was renamed Southwest Africa, and it was placed under the control of South Africa.

Shaka

One of the most important Africans in the history of Southern Africa was Shaka. He was the son of a Zulu chief and a Langeni princess. Shaka was raised by his mother among the Langeni and the Mthethwa. Both Shaka and his mother were treated cruelly and resented by both groups.

Shaka's father died in 1816, when Shaka was about 30 years old. He took over the Zulu clan his father had led. He reorganized, rearmed, and retrained the Zulu army. He also instituted a draft system.

The new weapons and tactics were devastatingly effective against the other clans. Shaka's army killed hundreds of thousands of other African people. Refugees fled inland and those who remained were incorporated into the Zulu empire.

After his assassination in 1828, the Zulu empire lived on, ruled by various relatives of Shaka. Over time, the Boers and the British began encroaching on Zulu territory. The Boers captured much of southern Zululand, but were forced to give it back. Then, in 1879, the British declared war. The Zulu army of 50,000 soldiers won the first battle in an enormous victory, but it prompted the British to send even more soldiers and supplies. The war ended with the division of the Zulu empire into 13 smaller territories. Even so, Zulu resistance to British rule continued until the first decade of the 1900s.

Cecil Rhodes

One of the most important Europeans in the history of Southern Africa was Cecil Rhodes. He moved from England to Cape Colony (South Africa) in 1870, at the age of 17. He bought up gold and diamond mines throughout the area. By 1891, the company he started, De Beers, produced 90 percent of the world's diamonds.

Shaka created a modern African empire and built an army that fought colonial powers for more than 50 years.

▼ CRITICAL THINKING
1. *Identifying Perspectives* What do you think Shaka thought of the Europeans in Africa?

2. *Assessing* Was Shaka a successful leader? Why or why not?

Tim Gainey/Alamy

THE RHODES COLOSSUS
STRIDING FROM CAPE TOWN TO CAIRO.

In his will, Cecil Rhodes used most of his fortune to establish a scholarship program that would bring students, regardless of race, to Oxford University.

▲ **CRITICAL THINKING**

1. **Identifying Perspectives** Why would Cecil Rhodes want to send students to Oxford rather than to local universities?

2. **Hypothesizing** Why do you think Rhodes specified that scholarships would not be restricted by race?

Rhodes was elected to the parliament of Cape Colony in 1881 and remained in government the rest of his life. He served as prime minister of the colony from 1890 to 1896. A committed imperialist, he used his wealth and power to further his dream of expanding British control to all of Africa.

Rhodes tried to reconcile the Dutch Boers and the British, but under British rule. He expanded British influence north into what is today Botswana, Zambia, and Zimbabwe. The latter two were once combined as the colony of Rhodesia, which was named after Rhodes.

Rhodes's life can be seen as a metaphor for European involvement in Africa. To him, Africa was a resource to be exploited and brought under "civilized" control. Local populations were dealt with ruthlessly—although, in Rhodes's case, generally without violence—and then ignored. The land and riches Africans once controlled were taken over by Europeans.

Independence

After World War I, unrest against colonial rule began to grow across Africa. In 1910 the Union of South Africa was created from four British colonies. Most other countries in Southern Africa were freed from European control in the 1960s and 1970s. Namibia became independent from South Africa in 1990.

Each country had a different road to independence. In Botswana, for example, independence was quick and peaceful. In Angola, however, guerrilla forces fought the Portuguese army for 14 years. Mozambique gained independence in 1976, but fought a civil war until 1992.

In some countries, problems continued after independence. Rhodesia divided into two parts. The northern part became Zambia in 1964 with a black African majority government. In Southern Rhodesia whites took control and declared an independent country in 1965. Black Africans living there did not win the right to vote or run for office until 1978. It came only after a long period of civil war. Once black majority rule was established, the country was renamed Zimbabwe. Robert Mugabe became its first prime minister in 1980. His rule has become more brutal and repressive as political opposition to him has increased.

(t)Mary Evans Picture Library/The Image Works, (b)Keystone/Hulton Archive/Getty Images

TIME LINE ⌄

APARTHEID
and Its Legacy ➡

South Africa's long history of racial discrimination, enshrined in law during the era of apartheid, left deep, long-lasting scars on its people.

CRITICAL THINKING ▶

1. **Describing** How did white settlers secure their power in South Africa?

2. **Analyzing** How has South Africa tried to heal the wounds caused by apartheid?

1750 ➡

1756
Afrikaners begin to import slaves from West Africa, Malaysia, and India.

1948
The National Party comes to power and continues a policy of racial segregation they call apartheid.

THE DIVISIONAL COUNCIL OF THE CAPE
WHITE AREA
BY ORDER SECRETARY
DIE AFDELINGSRAAD VAN DIE KAAP
BLANKE GEBIED
OP LAS SEKRETARIS

1960
Police open fire on protest in Sharpeville; 69 black demonstrators killed

1962
African National Congress (ANC) leader Nelson Mandela is arrested, sentenced to life imprisonment two years later

In South Africa, society had been segregated into blacks, whites, and South African coloured (mixed race) since colonial days. In 1948 the government established a **policy** called apartheid, an Afrikaans word meaning "separateness." Under apartheid, blacks and coloureds were required to live in specific places. They were allowed certain jobs, needed permits to leave their neighborhoods, and were **prohibited** from having social contact with whites. Civil protests and international boycotts during the 1970s and 1980s ended the policy of white minority rule and apartheid. In 1994 South Africa's first black president, Nelson Mandela, was elected. In 1995 he established a Truth and Reconciliation Commission to reveal the abuses that occurred under apartheid and provide restitution to its victims. South Africa had its first black majority government in modern times.

☑ **READING PROGRESS CHECK**

Listing How many different European powers established colonies in Southern Africa?

Population Patterns

GUIDING QUESTION *How have natural resources helped to determine the location of many settlements in Southern Africa?*

Parts of the Southern Africa subregion are heavily populated while others are practically uninhabited. In general, the population increases as one moves from west to east, from the Atlantic to the Indian Ocean. Madagascar follows the same pattern of a west-to-east increase. The explanation for most of the sparsely populated western areas is physical geography. Much of Botswana and Namibia is desert that can support few people.

Throughout history, major cities have often grown up along trade routes. Until the mid-1800s, trade routes mostly involved waterways. Since Southern Africa has almost no navigable rivers, the first trade centers were located along the coast.

A second factor that affected population distribution was natural resources. In 1886 the discovery of gold in the interior of Cape Colony led to a gold rush. One town that was established near the mines, Johannesburg, is now the largest and most important city in South Africa. Many settlements in the subregion

(l)Per-Anders Pettersson/Getty Images News/Getty Images, (r)Alexander Joe/AFP/Getty Images, (b)Africa Media Online/The Image Works

policy an overall plan that establishes goals and determines procedures, decisions, and actions

prohibit to prevent or forbid by authority

1980s
- International campaign to boycott South Africa in protest of apartheid cripples its economy

1976
- More than 600 people killed in clashes between black protesters and government forces during Soweto uprising

1990
- Nelson Mandela released after 27 years in prison

→ **1990**

1994
- First free elections held in post-apartheid South Africa; Mandela elected president as the ANC candidate

1996
- Truth and Reconciliation Commission begins public hearings on crimes committed during apartheid era

→ **2000**

2012
- Capital city of Tshwane (Pretoria) renames streets after leaders of anti-apartheid movement

STEVE BIKO ST
BEATRIX STR
JOHANNES RAMOKHOASE ST
PROES ST

"The township tours around Soweto, for example, seemed to spring up more by accident, arising from demand. Bus loads of tourists bound for other tourist attractions drove through the townships to stop and take a look from the safety of their buses. A few locals decided to offer tours around the townships so they could make some money out of the passing trade. With security tightened for visitors, they can see the huts where people live without water or electricity, and understand what tiny and cramped conditions the people live in."

—Amanda Kendle, "Poverty Tourism: Exploring the Slums of India, Brazil and South Africa," www.vagabondish.com

DBQ *DRAWING INFERENCES*
What name is applied to the urban slums in South Africa?

urbanization the migration of people from rural areas to urban areas

sanitation the disposal of trash and human waste

were established due to the discovery of gold. Pretoria (Tshwane), the administrative capital of South Africa, was first settled by Boers who were attracted not to mineral wealth but to the fertile farmland in the valley and the favorable climate.

Like many other places, Southern Africa has been affected by **urbanization**. People hoping to better their lives migrate to cities in search of jobs. The most populated urban areas are Cape Town, Port Elizabeth, Durban, and Johannesburg in South Africa and Maputo in Mozambique. Some neighborhoods in these cities are wealthy and modern, with expensive homes and fashionable shopping areas. Others are very poor, and many people live without electricity or **sanitation**.

☑ **READING PROGRESS CHECK**

Explaining Why did Johannesburg develop and grow at its location?

Society and Culture Today

GUIDING QUESTION *How do traditions play a role in the lives of the peoples of the subregion?*

The effects of centuries of European influence are still felt in Southern Africa. Colonizing powers divided the continent among themselves with no regard for the traditional territories occupied by the local people. As a result, most of the major ethnic groups—Shona, San, Ambo, Makua, and Ndebele—are distributed in different countries with no defined national territories.

Hundreds of local languages are spoken. However, many countries use the European language of their colonizers to facilitate communication. As a result, French, German, and English are spoken in many places. In South Africa, white Afrikaners form an ethnic group. Their language, Afrikaans, is one of the country's 11 official languages, along with English and 9 local languages such as Zulu.

Christian missionaries worked throughout Africa during colonial times. As a result, Christianity is the most common religion in Southern Africa. Many people still practice traditional religions, however, which involve nature spirits, animal sacrifice and worship, and worshiping the spirits of one's ancestors.

In urban areas, people attend movies, shop in malls, watch television, surf the Internet, and listen to music by African artists. Those living in poverty, both in urban and rural areas, do not have these opportunities. Their leisure activities center on the family and traditional activities such as games, singing, and dancing.

Poverty is an issue throughout the subregion. In Mozambique in 2008, protests against the high prices of necessities such as fuel led to deadly riots. The pattern repeated itself in 2010. When the price of bread rose by 30 percent, there was rioting, Thirteen people were killed and more than 400 were arrested.

Poverty also affects life expectancy, which has been decreasing for decades because of the HIV/AIDS epidemic that was severe in Africa. The situation is improving, however. In 2012, researchers announced that life expectancy in South Africa had increased by five years since 2005. The increase is due to the world's largest drug treatment effort for people with HIV/AIDS.

Education also varies with income. Large cities have colleges and universities, and the children of wealthy families generally get a good education. However, the children of poor families attend school for only a few years, if at all, and often leave school early. Girls, in particular, usually get much less schooling than boys do.

The status of girls and women is an issue in many parts of Southern Africa. In rural areas, women generally do not have the same rights as men. For example, traditional laws often prohibit women from inheriting or even owning property. Most Southern African countries have laws that require gender equality, but traditional laws are often observed instead. To help women, local and international organizations have been working to improve the financial status of African women.

The organizations make small loans to help women set up small businesses, such as growing vegetables or making clothes. These loans are mostly made to women in rural areas, and often lead to women becoming the business leaders in a village or community.

✓ **READING PROGRESS CHECK**

Describing How does tradition affect women in Southern Africa?

Economic Activities

GUIDING QUESTION *What role does mining play in the economy of Southern Africa?*

Most people in Southern Africa are farmers. Over 80 percent of the residents of Angola and Mozambique are agricultural workers. Most of the farms are used for **subsistence farming**. With increased global markets, **commercial farming** has gained in importance. For example, sugar plantations dot the coasts of Angola, South Africa, and Mozambique, while oil palm, cacao, and peanuts are grown in South Africa and Zimbabwe. Fresh flowers and vegetables are grown on commercial farms and exported to European cities.

Control of farmland is a big issue in much of Africa. European colonists claimed much of the best land. A land reform program in Zimbabwe gave land owned by white farmers back to Africans. The result was chaos, violence, and corruption. The land reform led to the failure of Zimbabwe's economy. Zimbabwe's agricultural workforce includes about 50 percent of the country's people.

Mining has always been important to the economy of Southern Africa. One section of the Highveld in South Africa, the Witwatersrand, has been mined since the 1880s. This area has produced about one-third of the gold ever mined in the world. Gold is also found in other parts of South Africa and in Zimbabwe. Most of the world's diamonds come from mines near the Witwatersrand, Botswana, and along the coast near the Namibia–South Africa border. Copper and cobalt are mined in Zimbabwe, Namibia, and Zambia.

Since the 1960s, many countries in Southern Africa have been trying to encourage and support manufacturing. Foreign loans have financed the development of manufacturing. Today, factories produce materials such as paper goods, cement, electric motors, and tractors. Compared to other developing regions, Southern Africa's manufacturing sector is still small, but it is growing.

✓ **READING PROGRESS CHECK**

Explaining From what one area does much of the world's gold come?

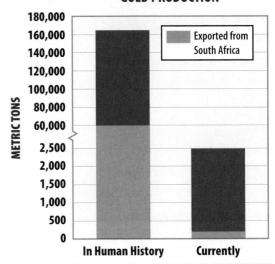

∨ GRAPH SKILLS

GOLD PRODUCTION

Gold mining has long been a central part of South Africa's economy.

▲ **CRITICAL THINKING**

1. ***Using Graphs*** About what percentage of the world's gold exports currently come from South Africa?

2. ***Making Connections*** How has gold influenced South Africa's history?

subsistence farming

small-scale agriculture that provides for the needs of a family or village, but little more

commercial farming

growing large quantities of crops or livestock in order to sell them for a profit

---

LESSON 2 REVIEW

Reviewing Vocabulary

1. ***Applying*** Write a paragraph describing the difference between subsistence farming and commercial farming.

Using Your Notes

2. ***Summarizing*** Use your graphic organizer on Southern Africa's human geography to write a paragraph summarizing how the history of the subregion affected the people and their culture.

Answering the Guiding Questions

3. ***Identifying*** What cultures have influenced life in Southern Africa?

4. ***Making Connections*** How have natural resources helped to determine the location of many settlements in Southern Africa?

5. ***Drawing Inferences*** How do traditions play a role in the lives of the peoples of the subregion?

6. ***Evaluating*** What role does mining play in the economy of Southern Africa?

Writing Activity

7. ***Informative/Explanatory*** In two or three paragraphs, describe how European control in Southern Africa began and ended.

Southern Africa **587**

netw⊙rks

There's More Online!

☑ **IMAGE** LifeStraws

☑ **IMAGE** Poor Urban Sanitation

☑ **IMAGE** Well in Mozambique

☑ **INTERACTIVE SELF-CHECK QUIZ**

☑ **VIDEO** People and Their Environment: Southern Africa

LESSON 3
People and Their Environment: Southern Africa

Reading HELPDESK

Academic Vocabulary

- **cite**
- **administer**

Content Vocabulary

- **shifting cultivation**
- **groundwater**

TAKING NOTES: *Key Ideas and Details*

IDENTIFYING As you read about the environment of Southern Africa, use a graphic organizer like the one below to identify features that are being damaged.

Southern Africa	
Feature	Damage

ESSENTIAL QUESTION • *How do physical systems and human systems shape a place?*

IT MATTERS BECAUSE

Southern Africa is a land of mineral wealth, great natural beauty, and many of the most awe-inspiring animal species on the planet. But if that is to be true a hundred years from now, the governments and people of Southern Africa must increase their focus on conservation and preservation of the environment.

Managing Resources

GUIDING QUESTION *Why are many resources in Southern Africa not managed closely?*

In Southern Africa, as in other parts of the world, poverty is the underlying reason behind much of the failure to manage resources. Most of the people living in Southern Africa are subsistence farmers. In general, these people have little income, poor health care, and little education. The main concern for the people in this region is their families' welfare, not conservation.

Habitat destruction occurs when people collect firewood. While collecting firewood contributes to deforestation, the people have no other choice. Outside of urban areas, people do not have gas or electric stoves, so they use wood for cooking. As the population increases, more pressure is put on forest resources.

Development is not necessarily the major cause of environmental degradation, however. Commercial logging is responsible for most of the deforestation in Southern Africa. The regulations that affect the logging industry have not been changed in 40 years. Furthermore, these regulations are usually ignored. When logging companies are **cited**, or summoned to appear in court, the fines they pay are tiny by today's standards.

Poaching, or illegal hunting, threatens some of Southern Africa's most notable species. Elephant tusks are made of ivory, which is a precious commodity around the world. Many countries have outlawed the importation or sale of ivory, but an illegal market still exists.

(tl)Melanie Stetson Freeman/The Christian Science Monitor/Getty Images, (tc)Courtesy of Vestergaard Frandsen/Newscom, (tr)Per-Anders Pettersson/Getty Images News/Getty Images

Rhinoceroses are another species that may soon be hunted to extinction. Many Asians believe that the horn of the rhinoceros has powerful medicinal properties, even though there is no scientific proof of this claim. In much of Southern Africa, hunting endangered animals, including elephants and rhinoceroses, is illegal. However, because economic conditions are difficult and because it is so profitable, poaching continues to threaten these species.

Access to clean water is another serious issue. In both rural and urban areas, many residents do not have running water in their homes. They either dig their own wells or walk—sometimes for many miles—to collect water from a river or stream. These water sources can be and often are contaminated. During the rainy season, water abounds. However, even unclean water can be difficult to find during winter or in times of drought. Lack of access to clean water has serious health consequences. Many diseases are spread by contaminated water.

In most developed countries, farmers attempt to prevent soil erosion and replenish the nutrients in their soil. Subsistence farmers in less developed countries often cannot and usually do not do this. The result is farmland that becomes unusable for growing crops after just a few years. For example, farmers on the volcanic islands of Comoros face loss of soil fertility and erosion where terraced farming is not practiced.

Shortages of food and water put pressure on communities. What is more likely, though, is that pressure from shortages pushes groups that are already at odds into conflict. For example, both Mozambique and Angola suffered through decades of civil war. The fighting may be over, but shortages of food, water, or other resources could easily drive them back into war.

Resources are not confined by political boundaries. In fact, many of the borders in Southern Africa are defined by rivers—important resources themselves. What starts as a conflict over resources can lead to larger wars between groups, resulting in many casualties and destabilizing the societies and economies of countries where they occur. Resource shortages can thus have effects that extend far beyond the borders of the regions where they occur.

✓ **READING PROGRESS CHECK**

Explaining Why are rhinoceroses endangered?

Human Impact

GUIDING QUESTION *How has environmental degradation affected life in the subregion?*

Human activity is affecting Southern Africa's environment in large and small ways. For example, many subsistence farmers engage in **shifting cultivation**. In this system, farmers cut and then burn brush and trees. After the land is cleared they plant crops in the ash-enriched soil. Once the soil is no longer productive, farmers clear a new plot. They return to the original plot once the fertility is restored. This method of agriculture works if the population is

These women in Mozambique are drawing water from a well.

▲ **CRITICAL THINKING**

1. Analyzing Visuals How might relying on well water affect daily life?

2. Assessing What health risks might relying on well water pose?

cite to summon to appear in a court of law

shifting cultivation
a system in which farmers plant a field for several years until its resources are depleted, then abandon it and clear a new field

Melanie Stetson Freeman/The Christian Science Monitor/Getty Images

Southern Africa has many NGOs that work on environmental issues in the region as a whole. One such group is the Musokotwane Environment Resource Centre for Southern Africa, which is based in Zimbabwe. Its aim is to "inform, motivate, and empower" decision-makers to take positive action on issues such as water resource management, sanitation, and climate change. In 2012 the United Nations Environment Program created the Africa Adaptation Knowledge Network to collect and distribute data for research, information, and strategies for adapting to climate change.

administer to manage; to supervise

The South Africa–based Southern African Development Community **administers** many programs, such as the Regional Environmental Education Programme (SADC REEP), which develops environmental education materials for students and teachers from grade school to the college level. These materials are used in schools in Mauritius, South Africa, Botswana, Lesotho, Swaziland, and Zambia.

The Nature Conservancy, an international organization, has partnered in Southern Africa with a number of local and international organizations. For example, it has teamed with the Zambia Wildlife Authority to reverse the decline in wildlife populations in that country. It is also working with the World Wildlife Fund to protect coral reef ecosystems off Mozambique from climate change. The Conservancy has also partnered with Save the Rhino Trust Namibia to help save the black rhinoceros from extinction. These organizations and many more are working hard to protect Southern Africa's resources and improve the lives of the people who live there. They face two major obstacles.

As discussed above, one of these obstacles is the region's poverty; the other is corruption. In many Southern African countries, national and local governments are corrupt. Officials are more concerned with staying in power and enriching themselves and their friends than with improving conditions for their citizens and their country. Often, this means that help does not reach the people who need it.

As an example, suppose an NGO donates 10,000 propane cookstoves and propane bottles so that people in a rural area can stop cutting firewood. Now imagine that various government officials refuse to distribute the stoves unless they get payoffs from the NGO. Then suppose that 8,000 of the stoves disappear from the warehouse and are sold by friends of a local official. The NGO has spent far more money than it planned, and only 2,000 families got stoves.

Increasingly, NGOs are trying to work directly with the people they want to help, bypassing corrupt officials whenever possible. In addition, new methods and equipment are being developed that will be within the capabilities of even the poorest of people.

☑ **READING PROGRESS CHECK**

Listing What NGOs are working to protect the environment of Southern Africa?

LESSON 3 REVIEW

Reviewing Vocabulary
1. *Identifying* Write a paragraph explaining the effects of shifting cultivation.

Using Your Notes
2. *Summarizing* Use your graphic organizer on managing Southern Africa's resources to write a paragraph summarizing the habitat destruction that has been taking place.

Answering the Guiding Questions
3. *Summarizing* Why are many resources in Southern Africa not managed closely?

4. *Explaining* How has environmental degradation affected life in the subregion?

5. *Identifying* What steps are being taken to combat the environmental challenges?

Writing Activity
6. *Argument* In a paragraph, discuss what you think needs to be done to preserve Southern Africa's resources.

Directions: On a separate sheet of paper, answer the questions below. Make sure you read carefully and answer all parts of the questions.

Lesson Review

Lesson 1

1 *Describing* Describe the location and physical features of the Great Escarpment in Southern Africa.

2 *Hypothesizing* Why do you think many of the most recognizable African mammals live on the savanna? Give examples of the mammals that live there.

3 *Explaining* Southern Africa has vast resources of mineral wealth. Why are these resources often a source of controversy?

Lesson 2

4 *Making Connections* How do Southern Africa's rivers help explain why the first sizable communities developed along the coast?

5 *Hypothesizing* Why do you think many Southern African countries experienced conflict after gaining independence from colonial rule?

6 *Analyzing* Why do people in many countries in Southern Africa continue to use the language of their colonizers?

Lesson 3

7 *Discussing* Discuss how global demand for animal products threatens some of Southern Africa's most notable species.

8 *Identifying Cause and Effect* How might the use of a scarce resource such as river water by one village affect other communities downstream?

9 *Summarizing* What are some factors that have fueled urbanization in Southern Africa?

Critical Thinking

10 *Explaining* Why does subsistence farming in Southern Africa contribute to the mismanagement of resources? Give examples.

11 *Interpreting* Discuss how the life of Cecil Rhodes provides a metaphor for European involvement in Africa.

21st Century Skills

Use the document to answer the following questions.

"Look, I'd rather be free, too, but at least we're not in a zoo anymore."

12 *Identifying Central Issues* What problem does the cartoon address? How do you know?

13 *Hypothesizing* Why do you think the gorillas are in an urban wildlife conservation park rather than their native habitat?

14 *Interpreting* What does the cartoon suggest about efforts to protect African animals?

Robert Mankoff/The New Yorker Collection/www.cartoonbank.com

Need Extra Help?

If You've Missed Question	**1**	**2**	**3**	**4**	**5**	**6**	**7**	**8**	**9**	**10**	**11**	**12**	**13**	**14**
Go to page	576	579	580	585	584	586	588	590	590	588	584	593	593	593

Directions: On a separate sheet of paper, answer the questions below. Make sure you read carefully and answer all parts of the questions.

College and Career Readiness

15 **Problem Solving** You are an environmentalist working for the Nature Conservancy in Madagascar. Use the Internet to research the causes and consequences of severe deforestation. Write a brief report exploring ways to address the problem.

DBQ Analyzing Primary Sources

Use the document to answer the following questions.

Before becoming South Africa's first black president, Nelson Mandela struggled for decades to end apartheid in his country. In 2005 he delivered a speech in London comparing apartheid to poverty today.

PRIMARY SOURCE

"But in this new century, millions of people in the world's poorest countries remain imprisoned, enslaved, and in chains.

They are trapped in the prison of poverty. It is time to set them free.

Like Slavery and Apartheid, poverty is not natural. It is man-made and it can be overcome and eradicated by the actions of human beings.

And overcoming poverty is not a task of charity, it is an act of justice. It is the protection of a fundamental human right, the right to dignity and a decent life."

—Nelson Mandela, February 3, 2005

16 **Identifying Perspectives** Why does Mandela insist that poverty is man-made rather than natural?

17 **Analyzing Arguments** What metaphor does Mandela use to describe the effects of poverty on people? Why does he use this metaphor?

18 **Identifying Perspectives** Why do you think Mandela makes a distinction between charity and justice?

Applying Map Skills

Use the Unit 6 Atlas to answer the following questions.

19 **Places and Regions** Use your mental map of Southern Africa to name the northernmost countries in the subregion.

20 **Environment and Society** Identify areas in Southern Africa where people are most likely to rely on wells to access groundwater. Why might people in these areas have health problems?

21 **Human Systems** Describe how the location of mineral resources has affected the development of cities in Southern Africa. Give examples.

Exploring the Essential Question

22 **Making Connections** Choose one of the countries in Southern Africa. Use what you have learned about its history, politics, art, and economics to create a poster illustrating how human systems have shaped it as a place. Posters should be visual and can include photos, graphs, charts, and maps.

Research and Presentation

23 **Research Skills** Research the legacy of colonial rule in Zimbabwe. What factors led President Robert Mugabe to institute a forced land reform program? How effective was the program? What challenges does the country face today? Write a multi-paragraph summary of your findings and present it to the class.

Writing About Geography

24 **Argument** Using Botswana as an example, write an essay suggesting steps other African countries can take to create a healthy economy. What measures have the people of Botswana taken to strengthen the country's economy? How might other countries benefit from this example? Remember to use standard grammar, spelling, sentence structure, and punctuation in your response.

Need Extra Help?

If You've Missed Question	**15**	**16**	**17**	**18**	**19**	**20**	**21**	**22**	**23**	**24**
Go to page	590	594	594	594	476	476	476	576	584	580

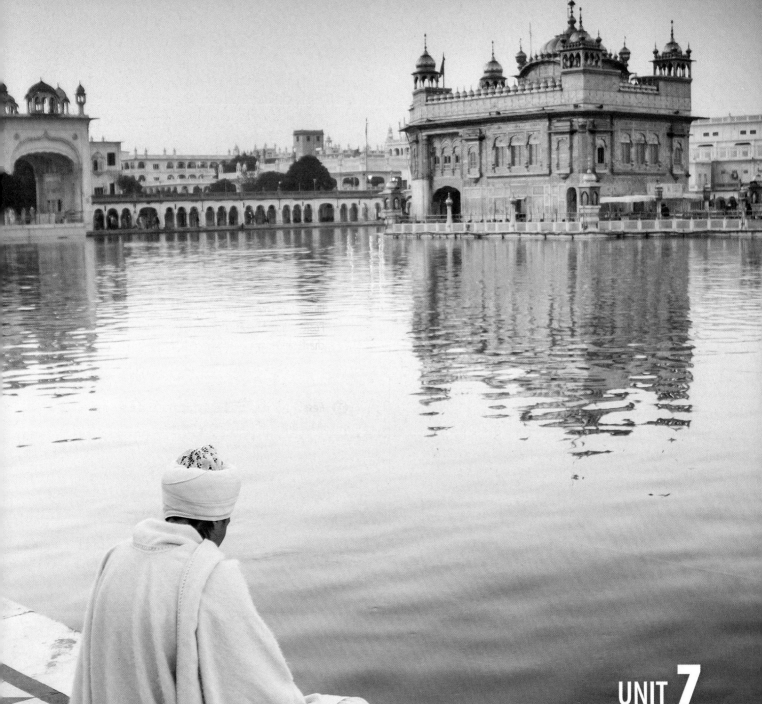

South Asia

Abhinav Sah/Flickr/Getty Images

UNIT **7**

1 Religion These children have been painted as part of a religious festival. Hinduism and Buddhism began in the region, which also has three of the four most populous Muslim countries in the world.

EXPLORE the REGION

Though seemingly isolated from the rest of Asia by mountains, **SOUTH ASIA** has long-practiced religious and cultural traditions that have influenced—and been shaped by—other parts of Asia. With roots in ancient civilizations, the region today is home to vibrant software and film industries, rising economic powers, and countries that play an important role in contemporary international issues.

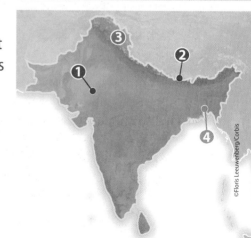

©Floris Leeuwenberg/Corbis

THERE'S MORE ONLINE

(t)SDBWP Sundeep Bhardwaj World Photography 50+ Countries/Flickr/Getty Images,
(bl)Pal Teravagimov Photography/Flickr/Getty Images,
(br)Munir Uz Zaman/AFP/Getty Images

③ Rivers Major rivers that drain the region—such as the Indus, Ganges, and Brahmaputra—flow from their source high in the Himalaya in the north to the floodplains in the south, carrying fertile soil used to grow food for the region's large population.

② Mountains As the Indian subcontinent moves slowly and steadily into Asia, it pushes up the highest mountains in the world, the 1,500-mile- (2,414-km-) long Himalaya, which have individual peaks that reach more than 5 miles (8 km) high.

④ Climate Strong winds traveling across warm seas bring the wet season to South Asia, providing rain vitally needed by the region's farms but also causing flooding in low-lying areas.

South Asia
Physical

KASHMIR
(administered
by Pakistan)

Area ceded by
Pakistan to China,
claimed by India

EAST ASIA

K2
28,250 ft.
(8,611 m)

AKSAI CHIN
area held by China,
claimed by India

**SOUTHWEST
ASIA**

HINDU KUSH · KARAKORAM RANGE

Khyber Pass

ARUNACHAL
PRADESH
largely claimed
by China

JAMMU & KASHMIR
(administered
by India)

Disputed areas

Mt. Dhaulagiri
26,810 ft.
(8,172 m)

Kanchenjunga
28,169 ft.
(8,586 m)

40°N

30°N

Sulaiman Range

Indus R.

Disputed
border

Mt. Everest
29,028 ft.
(8,848 m)

Brahmaputra

H I M A L A Y A

THAR DESERT

GANGETIC PLAIN

Ganges R.

Khasi Hills

Naga
Hills

100°E

ARAVALLI RANGE

Yamuna R.

Indus R.

Gulf of Oman

TROPIC OF CANCER

Gulf of Kachchh

Mizo
Hills

Meghna R.

Sundarbans

Ganges Delta

**SOUTHEAST
ASIA**

VINDHYA RANGE

Narmada R.

Chota Nagpur
Plateau

Kathiawar

Mahanadi R.

Gulf of Khambhat

SATPURA RANGE

20°N

Godavari R.

**Arabian
Sea**

DECCAN
PLATEAU

Bay of Bengal

Krishna R.

EASTERN GHATS

60°E

Preparis
Channels

Andaman
Islands

**Andaman
Sea**

WESTERN GHATS

Kāveri R.

Coromandel Coast

Lakshadweep

Malabar Coast

10°N

Ten Degree
Channel

Palk Strait

Laccadive Sea

**Sri
Lanka**

Nicobar
Islands

Cape
Comorin

N
W   E
S

Maldive
Islands

0°   EQUATOR

500 miles

500 kilometers

Lambert Azimuthal Equal-Area projection

INDIAN OCEAN

Elevations

10,000 ft. (3,000 m)
5,000 ft. (1,500 m)
2,000 ft. (600 m)
1,000 ft. (300 m)
0 ft. (0 m)
Below sea level

—— National boundary
---- Disputed boundary
▲ Mountain peak
■ Pass

Chagos
Archipelago

70°E   80°E   90°E

South Asia
Political

KASHMIR
(administered
by Pakistan)

Area ceded by
Pakistan to China,
claimed by India

AKSAI CHIN
area held by China,
claimed by India

JAMMU & KASHMIR
(administered by India)

Disputed areas

ARUNACHAL
PRADESH
largely claimed
by China

**SOUTHWEST
ASIA**

40°N

30°N

Indus R.

Islamabad
Rawalpindi
Lahore

PAKISTAN

Indus R.

EAST ASIA

NEPAL

Delhi
New Delhi
Jaipur •
Kanpur • Lucknow
Yamuna R.
Ganges R.

Kathmandu

Brahmaputra R.

BHUTAN
Thimphu

100°E

BANGLADESH
Dhaka

Meghna R.

20°N

Gulf of Oman

Karachi

TROPIC OF CANCER   Gulf of Kachchh

Ahmadabad

INDIA

Narmada R.

Kolkata
(Calcutta)
Mahanadi R.

Chittagong

**SOUTHEAST
ASIA**

*Arabian
Sea*

Gulf of Khambhat

Mumbai
(Bombay)

• Pune
Godavari R.
• Hyderabad

Krishna R.

Bay of Bengal

○ National capital
• Major city
- - - Disputed boundary

Bengaluru
(Bangalore)

Kāveri R.

Chennai
(Madras)

*Andaman
Islands*
(India)

*Andaman
Sea*

Lakshadweep
(India)

10°N

Laccadive Sea

Palk Strait

Ten Degree
Channel

N
W   E
S

**SRI
LANKA**
Colombo

• Male

MALDIVES

*Nicobar
Islands*
(India)

UNIT 7
REGIONAL ATLAS

0°   EQUATOR

MAP STUDY

0   500 miles
0   500 kilometers
Lambert Azimuthal Equal-Area projection

INDIAN OCEAN

British Indian Ocean
Territory
(U.K.)

70°E   80°E

1. *Physical Systems* What physical
features would isolate South Asia from
other parts of Asia? What features
would allow contact between this and
other regions?

2. *Human Systems* Which South Asian
countries are involved in disputes over
borders in the region?

South Asia
Climate and Vegetation

Climate
- Tropical rain forest
- Tropical wet/dry
- Semi-arid (steppe)
- Arid (desert)
- Humid subtropical
- Humid continental
- Subarctic
- Tundra and high altitude

SOUTHWEST ASIA

EAST ASIA

40°N

30°N

Islamabad

New Delhi

Kathmandu

Thimphu

100°E

Karachi

TROPIC OF CANCER

Gulf of Oman

Gulf of Kachchh

Ahmadabad

Dhaka

Kolkata (Calcutta)

SOUTHEAST ASIA

20°N

Gulf of Khambhat

Mumbai (Bombay)

Arabian Sea

Bay of Bengal

60°E

Preparis Channels

N
W E
S

Chennai (Madras)

Andaman Sea

10°N

Laccadive Sea

Ten Degree Channel

Palk Strait

0   500 miles

0   500 kilometers

Lambert Azimuthal Equal-Area projection

Colombo

Male

0°   EQUATOR

Vegetation
- Tropical grassland (savanna)
- Desert scrub and desert waste
- Deciduous forest
- Mixed forest (deciduous and coniferous)
- Highland (vegetation varies with elevation)

INDIAN OCEAN

70°E

30°N

TROPIC OF CANCER

20°N

10°N

70°E   80   90°E

600

Resources

Coal	Bauxite
Petroleum	Uranium
Natural gas	Copper
Iron ore	Lead
Zinc	Manganese

SOUTHWEST ASIA

EAST ASIA

40°N

30°N

100°E

TROPIC OF CANCER

Gulf of Oman

Gulf of Kachchh

20°N

Gulf of Khambhat

SOUTHEAST ASIA

Arabian Sea

Bay of Bengal

60°E

Andaman Sea

10°N

Ten Degree Channel

Laccadive Sea

Palk Strait

0°

Land Use

	Commercial farming
	Subsistence farming
	Nomadic herding
	Primarily forest
■	Manufacturing and trade
	Commercial fishing
	Little or no activity

INDIAN OCEAN

70°E

80°E

N
W E
S

0   500 miles
0   500 kilometers
Lambert Azimuthal Equal-Area projection

UNIT 7
REGIONAL ATLAS

MAP STUDY

1. ***Places and Regions*** What generalization can you make to describe the climate of this region?

2. ***Environment and Society*** How do the climate and vegetation maps help explain the presence of nomadic herding in the northwestern part of the region?

40°N

**SOUTHWEST
ASIA**

EAST ASIA

30°N

0 ⊢————————————⊣ 500 miles
0 ⊢————————————⊣ 500 kilometers
Lambert Azimuthal Equal-Area projection

Peshawar
Srinagar
Rawalpindi
Gujranwala
Lahore   Amritsar
Faisalabad   Jalandhar
Multan   Ludhiana

Meerut
Delhi
Faridabad   New
Delhi
Agra   Lucknow
Jaipur
Kanpur   Patna
Allahabad   Varanasi   Guwahati
Hyderabad
Karachi   Dhanbad   Asansol   Dhaka
TROPIC OF CANCER   Gulf of Kachchh   Ahmadabad   Bhopal   Ranchi   Khulna   Chittagong
Gulf of Oman   Indore   Jabalpur   Jamshedpur   Kolkata
Rajkot   Vadodara   (Calcutta)
Surat   Nagpur   Raipur   **SOUTHEAST
ASIA**
Nasik   Aurangabad

*Arabian
Sea*   Gulf of Khambhat   Mumbai   Pune
(Bombay)   Sholapur   *Bay of Bengal*
Hyderabad   Vishakhapatnam
Vijayawada   *Preparis
Channels*

60°E   Bengaluru
(Bangalore)   Chennai
(Madras)   *Andaman
Sea*
Coimbatore
10°N   Madurai   Ten Degree
Kochi   Palk Strait   Channel

Laccadive Sea

Colombo

100°E

EQUATOR   0°

INDIAN OCEAN

70°E   80°E   90°E

UNIT 7
REGIONAL ATLAS

MAP STUDY

1. *Human Systems* Which parts of India have the highest population density? Why do you think this is so?

2. *Human Systems* How do population densities in India compare to those in Pakistan? In Bangladesh?

Cities
(Statistics reflect metropolitan areas.)

■ Over 5,000,000
□ 2,000,000–5,000,000
⊙ 1,000,000–2,000,000

POPULATION	
Per sq. mi.	Per sq. km
1,250 and over	500 and over
250–1,249	100–499
63–249	25–99
25–62	10–24
2.5–24	1–9
Less than 2.5	Less than 1

India

ESSENTIAL QUESTION · *How do physical systems and human systems shape a place?*

Geography Matters...

India is an ethnically diverse country whose population practices a wide range of religions and speaks more than a dozen major languages. Laws have helped to lessen the injustice formerly faced by members of "untouchable" castes, ethnic minorities, and women. But many problems and tensions remain. Despite the challenges of great economic inequality, India's economic power is growing. It has one of the world's largest pools of trained scientists and engineers. India's films, literature, and music have attracted a global following. While some 70 percent of its people live in rural villages, India has three of the world's largest cities.

◀ This woman in Pushkar, India, is wearing traditional dress and jewelry.

©Jens Kalaene/dpa/Corbis

India's population structure

In the third decade of this millennium, India may become the world's most populous country. China is in the lead, but its policies to limit population growth are rigorous, and the population growth rate has declined. India's population growth rate also has fallen. But the population is still large and, unlike China, young. Together, these two countries contain nearly 40 percent of the world's population.

What are the main characteristics of India's population?

India's population growth rate is higher than the world average. In 2012 the number of male children age 4 or under was about 64 million. This age group alone grew by about 8 percent between 1992 and 2012. India's two youngest age groups dominate. Males and females ages 10–24 make up as much as one-third of the country's population.

Over the past decade, India's growth rate has declined. But in 2012 the growth rate, 1.5 percent, led to population increase. This rate still exceeds the world's average annual growth rate of 1.2 percent in 2011. Fertility rates also have dropped. Based on the data, the U.S. Census Bureau predicts that by 2025 India will outrank China as the world's most populous country.

1. Human Systems Why does India have such a large young population?

What are the consequences of this population structure?

The structure and size of India's population produces multiple effects. The population will continue to grow as the large number of young people start families of their own. Additionally, a large population in a poor country leads to poverty. About one-third of the total population lives in poverty. Social services, health care, and education are difficult to provide to a population of this size. With such a young population, India's workforce has expanded, but so too has its need to provide jobs.

At the same time, India faces what some experts think may be a positive effect of its large and growing population. They think a young and growing workforce could fuel India's economy. Much of the young population yearns for educational opportunities and training. However, the country faces a monumental task in providing education and social services as an investment for the future.

2. Places and Regions How might its young, growing population benefit India's future?

How is India responding to its population situation?

To reduce population growth, India's national government has launched an initiative to distribute contraceptives. India's National Family Planning Program has trained service providers to introduce other birth control methods. However, action at the national level may not be enough.

Some Indian states are responding to India's population challenges. They have issued incentives to control the birthrate and offered payments for people who delay having children. In some states, people who have many children are barred from running for political office. However, not all states are implementing programs to slow population growth. This uneven response could thwart efforts to achieve national goals.

3. Human Systems How are governments in India important in controlling population growth?

(l)©2006 Glowimages, Inc.–All rights reserved, (c)Bob Pool/Photographer's Choice RF/Getty Images, (r)©Kapoor Baldev/Sygma/Corbis

1 person is born per second (approx.)

Average household size is 5 PEOPLE

Age Structure

MALE   FEMALE

Median Age: 26.5 years

15-24 years: 18.2%   55-64 years: 6.9%

0-14 years: 28.9%   25-54 years: 40.4%   65 years and over: 5.7%

India total population: 1,220,800,359

Second Highest Population In The World (China Is Higher, U.S. Is Third)

Population Growth 1.3%
89TH IN THE WORLD

Urban Population 30%

Delhi 20.4 Million   Mumbai 20.5 Million   Chennai 5 Million

Largest Cities

Labor Force 498.4 million

53% agriculture   28% services   19% industry

2nd largest in the world

Unemployment Rate 9.90%

Population below the poverty line 29.8%

GEOGRAPHY CONNECTION

Seasonal wind patterns called monsoons influence the subregion's climate.

1. **PHYSICAL SYSTEMS** What generalization can you make to describe the climate of South Asia?

2. **PHYSICAL SYSTEMS** Which monsoon winds are likely to bring wet weather? Which monsoons bring drier weather? Why?

monsoon in Asia, a seasonal wind that brings warm, moist air from the oceans in summer and reverses direction in the winter, bringing cold, dry air from Asia's interior

fluctuate to ebb and flow in waves

Climates, Biomes, and Resources

GUIDING QUESTION *How do monsoons affect India's weather and climate?*

India's climate is influenced by the shape of the subcontinent, its proximity to the Equator, and the Indian Ocean. The difference in temperatures of the air over the ocean and the air over the landmass creates the monsoon winds. These winds cause rainfall and temperature to **fluctuate**, creating wet and dry seasons.

From about mid-June to the end of September, the winds bring hot, wet weather. This occurs during the summer period because heated air rises over the land, which pulls in the moist ocean air from the south and southwest. The monsoon winds cause heavy rains and flooding across the subcontinent during these months. Then, from early October to February, the winds change direction. In these winter months, air from the Asian interior in the north is drawn across the subcontinent toward the ocean. The air is dry and cool compared to the wet season. The weather turns hot and humid with little rainfall from March to mid-June and the winds turn calm. The weather becomes hot with little rainfall but high humidity in the south and dry conditions in the north.

Landforms and location affect the amount of rain that monsoon winds bring to different areas. Different amounts of rain create different biomes. When rain sweeps over the Ganges-Brahmaputra Delta, the Himalaya block rain clouds from moving north. Instead, the rain moves west to the Gangetic Plain. Farmers in this densely populated plain depend on these rains.

When summer monsoons rise over the Western Ghats, the air cools and releases heavy rains that create rain forests in this part of the country. As the

The most important crops on India's farms are rice, wheat, and other grains. Pulses, which include legumes such as chickpeas and lentils, are grown. Large plantations grow commercial agricultural products such as bananas, coffee, tea, and cotton.

Since the 1960s, an effort known as the **green revolution** has sought to increase crop yields in India with high-yielding varieties of crops. These varieties require carefully managed irrigation, fertilizers, and pesticides. The green revolution has stirred controversy, however. Supporters point out that it has increased food production and alleviated hunger in a country that once suffered devastating famines. Opponents argue that excessive use of chemicals pollutes waters, poses health hazards, and leads to pest resistance, requiring farmers to use more and more chemicals to grow crops.

Despite problems of poverty in the country as a whole, India has millions of well-educated, English-speaking people who work in the growing high-technology industry. India is one of the world's largest exporters of high-technology services and software workers. Increasing numbers of schools offer training in technology and engineering. Indian computer professionals are in high demand around the world.

Although India has large oil reserves, it uses more and more energy every year. As more Indians buy cars and trucks, and as electricity serves more homes and businesses, energy consumption increases. According to the U.S. Energy Information Administration, India's oil production held fairly steady at 950,000 barrels of oil per day between 2000 and 2010. However, consumption of oil rose from a little over 2 million barrels a day in 2000 to over 3 million barrels a day in 2010. India must buy the energy from other countries that it does not produce itself.

Many people in India work in **cottage industries**, businesses that employ workers in their homes. These workers work on a variety of tasks. They weave textiles and make shoes, jewelry, woodcarvings, furniture, and bowls. The biggest number of manufacturing jobs comes from India's textile industries, which employ people in turning cotton, jute, wool, silk, and synthetic fibers into fabrics. Nearly every large Indian city has at least one cotton mill.

Heavy industry provides jobs as well. India manufactures iron, steel, cement, and heavy machinery. The Gangetic Plain and parts of eastern India have rich deposits of iron ore, low-grade coal, bauxite, and copper. Many Indians are employed in mining these resources, while others work in India's fishing industry.

✔ **READING PROGRESS CHECK**

Making Connections Why is farming the occupation of so many of India's people?

Connecting Geography to MATH

India's GDP

In 2011 India's gross domestic product (GDP), or the value of the goods and services produced by its people in a year, was $4.421 trillion. India's GDP was the world's fifth highest after the European Union, the United States, China, and Japan. Divide India's GDP by its population of 1.2 billion to figure out the per capita GDP, or the value of goods and services produced per person in a year.

EVALUATING Based on the per capita GDP, do you think the average Indian is wealthy? Explain your answer.

green revolution a program begun in the 1960s to produce higher-yielding, more productive strains of wheat, rice, and other food crops

cottage industry a business that employs workers in their homes

---

LESSON 2 REVIEW

Reviewing Vocabulary

1. ***Drawing Inferences*** Write a paragraph discussing how the green revolution has affected India.

Using Your Notes

2. ***Summarizing*** Use your graphic organizer on India's economy to write a paragraph summarizing the types of work available to most Indians.

Answering the Guiding Questions

3. ***Summarizing*** How did modern India gain its independence?

4. ***Evaluating*** Why are population patterns so important in India?

5. ***Understanding Perspectives*** What is life in Indian villages like?

6. ***Assessing*** What role does agriculture play in the Indian economy?

Writing Activity

7. ***Informative/Explanatory*** Write a paragraph describing the features of a typical Indian village.

WHAT IS THE FUTURE OF KASHMIR?

Pakistanis, Indians, and Kashmiris today all have widely different views about who should govern Kashmir, a region in northern India. The dispute began in 1947 when India and Pakistan became independent from Britain. When Britain withdrew from the region, the Princely states ruled by maharajas were given the choice of joining either India or Pakistan. Hari Singh, the maharaja of Kashmir, wanted Kashmir to be an independent country. Many Pakistanis, however, believed that Kashmir should become part of Pakistan because it included many Muslims. As a result, Pakistanis invaded Kashmir on October 22, 1947. In response, the maharaja signed an agreement to make Kashmir part of India. In return, India agreed to provide military help to force out the Pakistanis. However, India and Kashmir were not successful. Conflict continued until a cease-fire agreement went into effect on January 1, 1949. The United Nations urged Pakistani forces to withdraw until a plebiscite, or vote, could be held to allow Kashmiris to choose between India and Pakistan. However, the Pakistanis did not withdraw and a plebiscite was not held. On July 27, 1949, Pakistan and India signed the Karachi Agreement, which established a line of control (LOC) that divided the territory. Neither Pakistan nor India, however, recognizes the LOC as an international border. Fighting between the two countries intensified when China took parts of northern Kashmir in the 1960s. In 1965 protests in Kashmir against the Indian government stirred unrest in Pakistani cities. Pakistan launched a covert operation, which quickly failed, against Indian forces in Kashmir. The United Nations Security Council became involved and India and Kashmir signed a cease-fire agreement on January 10, 1966. Today, tensions near the LOC continue as both Pakistan and India maintain a large military presence there.

Not only are Pakistan and India divided on who should govern Kashmir, but so are many Kashmiris. Some want to stay part of India. Others are tired of the conflicts and military presence in the area and want to form their own country as the maharaja wanted. Still others support the Pakistani position.

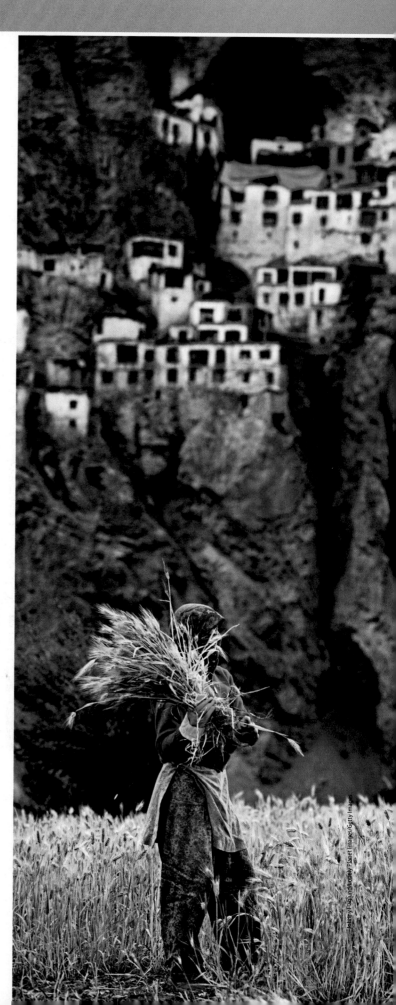

Jeremy Raju/Lonely Planet Images/Getty Images

A New Government is Needed

PRIMARY SOURCE

66 We know that there is armed struggle going on in Kashmir since 1988, and we Kashmiris can sit down to analyse what progress we have made. India and Pakistan had armed conflicts over Kashmir, they were not only indecisive but they created more problems to all the parties to the dispute. In other words by use of force alone we cannot be successful in finding lasting solution. We need to sit down with open mind and see how we can bring peace, stability and prosperity to the region. Peace and stability can only return to Indian Sub-Continent if Kashmir dispute is resolved according to the wishes of the Kashmiri people. 99

—Shabir Choudhry, Director of the Institute of Kashmir Affairs

Support for the Indian Government

PRIMARY SOURCE

66 Government of India has expressed its willingness to accommodate the legitimate political demands of the people of the state of J&K [Jammu and Kashmir]. However, Pakistan sponsored terrorists have terrorised the population and hindered political dialogue by intimidating or silencing voices of moderation that wish to engage in dialogue. The human rights of the people of J&K have been systematically violated by such terror tactics and the kidnappings and killings of innocent people by terrorists.

Jammu & Kashmir is an integral part of India. There can be no compromise on India's unity and integrity. 99

—Ministry of External Affairs, India

What do you think? DBQ

1. *Drawing Conclusions* Why might Kashmiris want both Pakistan and India out of Kashmir?

2. *Identifying Central Issues* On what basis does India defend its claims to Kashmir?

3. *Hypothesizing* Why might it be dangerous for Kashmir to become an independent country?

LESSON 3
People and Their Environment: India

ESSENTIAL QUESTION · *How do physical systems and human systems shape a place?*

Reading **HELP**DESK

Academic Vocabulary

- **environment**
- **prohibit**

Content Vocabulary

- **pesticide**
- **ecotourism**

TAKING NOTES: *Key Ideas and Details*

FORMULATING QUESTIONS As you read about India's people and environment, use a graphic organizer like the one below to list details about India's environment. Then formulate a question about each section.

IT MATTERS BECAUSE

India is among the most biologically diverse countries in the world. Although it has less than 2.5 percent of the Earth's land area, it has nearly 17 percent of its human population. It also has the world's largest population of wild tigers and a number of other rare and endangered species. India's ability to manage its resources while raising the standard of living for its entire population will affect hundreds of millions of people, as well as many species of animals and plants.

Managing Resources

GUIDING QUESTION *Why are many resources in India not managed closely?*

In recent years, India's economy has grown rapidly, and the country has made significant advances in reducing poverty and improving the education and health of its citizens. Despite these advances, India's people still face high levels of poverty and inequality. The government's latest Five-Year Plan (2012–2017) calls for sustainable growth. In the past, India has often favored economic development over resource management and sustainability.

Development policies have put many of India's resources at risk. Forests provide a good example. India's forests are an important natural resource, providing a livelihood for some 250 million people. It is estimated, however, that 40 percent of India's forests are degraded. Deforestation began in the colonial era. The construction of the country's railways, which facilitated moving crops to market, encouraged the development of cash crop agriculture and the clearing of land.

Removal of forests also affects farmland by increasing erosion and flooding. Erosion degrades farmland by washing away the nutrients that make farmland productive. This in turn leads to further degradation because farmers overuse the soil in an effort to increase their harvests. Overuse further depletes nutrients. Erosion caused by wind is also a problem in drier areas of India.

(tl)Andrew Caballero-Reynolds/AFP/Getty Images, (tr)Mike Ledwith/Flickr/Getty Images

No resources are more basic to human life than air and water. Yet these important resources are also threatened in India. Pollution of groundwater is a growing health threat. Much of India's water has high levels of chemicals that make it unsafe to drink because as water supplies are depleted, people draw water from deeper and deeper in the ground, where it is polluted by arsenic, fluoride, and heavy metals. Excessive fertilizer runoff and pollution from human and industrial wastes also contaminate water supplies.

Air pollution is another serious problem. A number of India's cities, including its capital, New Delhi, are among the world's most polluted. Overcrowding in urban areas leads to air pollution as a result of increased use of fossil fuels for transportation and industry.

☑ **READING PROGRESS CHECK**

Summarizing What resources in India are threatened?

Human Impact

GUIDING QUESTION *How has pollution affected the quality of life in India?*

The size and density of India's population creates a significant human impact on the **environment**. Ten percent of the world's population (about half of India's population) lives in the Ganges River valley. Although the Indian economy is growing rapidly, the country also holds a large concentration of people living in poverty, especially in the states of Chhattisgarh, Jharkhand, and Manipur. Large numbers of people cut down trees illegally because they need fuel for cooking. Removal of forests creates many other problems. In addition to soil erosion and flooding, loss of trees can cause temperatures to rise. Loss of forests destroys habitats for wildlife, including birds and other creatures that eat insects, snakes, and rodents.

India's already large cities continue to grow rapidly. Their growing populations have water and energy needs. To supply the residents of northern India with water and power, the Tehri Dam was built in the Himalaya. Located

environment the complex of physical, chemical, and biotic factors (as climate, soil, and living things) that act upon an organism or an ecological community and ultimately determine its survival

⌄ **GRAPH SKILLS**

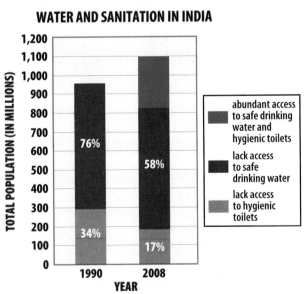

WATER AND SANITATION IN INDIA

TOTAL POPULATION (IN MILLIONS)

1990: 76%, 34%
2008: 58%, 17%

- abundant access to safe drinking water and hygienic toilets
- lack access to safe drinking water
- lack access to hygienic toilets

Source: *New York Times*, March 21, 2012 "Liter by liter, Indians get cleaner water," Amy Yee

India provided more of its people with clean water and better hygiene between 1990 and 2008.

◀ **CRITICAL THINKING**

1. ***Analyzing Visuals*** Which improvement—access to safe drinking water or access to hygienic toilets—affected the greatest number of people?

2. ***Synthesizing*** Why is access to clean water still a problem even though over 80 percent of Indians have it?

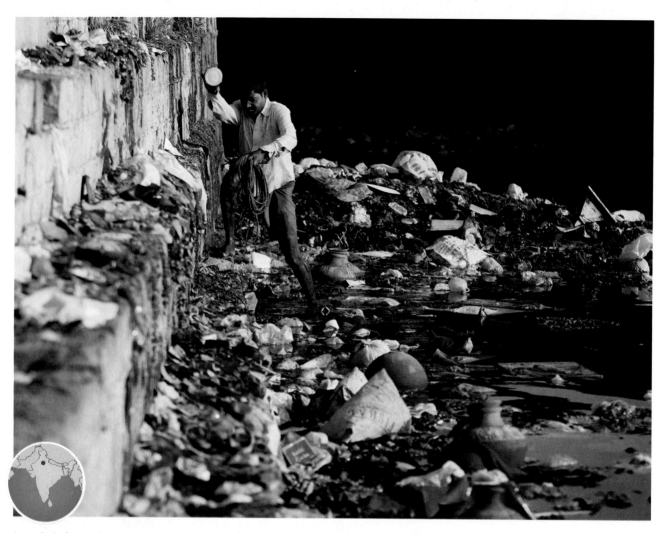

A man looks for recyclable items in the polluted waters of the Yamuna River in New Delhi.

1. Analyzing Visuals What does this photograph indicate about sanitation systems in India's major cities?

2. Identifying Cause and Effect How does India's growing population affect its environment?

on the Bhagirathi River, the Tehri Dam is India's highest dam. The dam was designed to increase the flow of water during the dry season. The poor in India's cities often also lack access to electricity; thus, they rely on biofuels. As cities grow, they increase the demand for timber to be used in construction. This further adds to deforestation.

India is rich in farmland. About half of India's farmers grow their own food, and the other half grow food to sell. The numbers of people who must use the farmland put great pressure on this resource. Much of India's farmland has become less productive because its nutrients have been depleted or because of excessive irrigation with improper drainage. Furthermore, too much plowing can change the structure and composition of the soil. Sometimes it is difficult or impossible to grow food in the soil.

India has made great progress in providing clean water and sanitation. For example, between 1990 and 2008, the number of people who lacked safe drinking water fell by half. But because India's population is so large, tens of millions of people still do not have access to clean water. By 2008, more than half of India's people still did not have toilets and sanitation facilities. Poor sewage and waste disposal threatens water supplies. Overcrowding in urban areas along with lack of sanitation leads to pollution of air and water resources as well as to the spread of disease.

The large population contributes to air pollution as well. Fires used for cooking and heating create smoke. In densely populated cities, construction dust and auto

Andrew Caballero-Reynolds/AFP/Getty Images

emissions create smog that can cause respiratory and heart problems, especially in children and older people. New Delhi has more than 7 million registered vehicles. Each day, another 1,400 cars and trucks get on the city's roads.

One of the worst environmental disasters in history occurred in December 1984 in Bhopal, a city in central India. A dangerous chemical leaked from a factory that made insecticides. It drifted over densely populated areas around the plant and killed between 15,000 and 20,000 people. Hundreds of thousands of others were blinded or suffered respiratory problems or other injuries. By the early twenty-first century, testing found that the soil and water around the factory site was still contaminated. This soil pollution has been blamed for high rates of birth defects and chronic health problems among people living near the factory.

☑ **READING PROGRESS CHECK**

Making Generalizations How did the 1984 Bhopal accident continue to affect people near the site in the early 2000s?

Addressing the Issues

GUIDING QUESTION *What steps has the Indian national government taken to combat pollution?*

Although India's environmental problems are severe, the national government has been involved in environmental protection. In 1976 India's constitution was amended to require the government to protect the environment. India's Supreme Court has made many decisions designed to protect India's air, water, ecosystems, and wildlife. It has considered a proposal to charge an extra tax on diesel cars because of the pollution they cause. In 2011 the Indian Supreme Court **prohibited** the production, distribution, and use of a **pesticide** called endosulfan because of its toxicity and its persistence in the environment.

prohibit to prevent from doing something

pesticide a chemical used to kill insects, rodents, and other pests

Almost 5 percent of India's land area falls within protected areas to ensure that the habitat of Bengal tigers is preserved.

▼ CRITICAL THINKING

1. ***Speculating*** How has the growth of India's human population caused the Bengal tiger to become an endangered species?

2. ***Defending*** Do you think it is important for India's government to protect tigers, elephants, and snow leopards? Defend your position.

Mike Ledwith/Flickr/Getty Images

In 1972 the government of India passed a law to protect wildlife from poaching and smuggling. In 2003 this law was amended to increase punishment and penalties. The law was also expanded to protect plants and ecosystems as well as endangered animals. India has the world's largest population of wild tigers and many other endangered species such as the Asian elephant, the one-horned rhinoceros, and snow leopards. While the African elephant is endangered primarily as a result of poaching of the animals for their ivory tusks, the Asian elephant is at risk due to loss of its natural forest habitat. The clearing of forests for timber and agricultural purposes continues to threaten the viability of this species. To protect its unique biological heritage, India has created a network of 668 protected areas. These include national parks, wildlife sanctuaries, conservation reserves, and community reserves.

ecotourism the practice and business of recreational travel based on a concern for the environment

In 2012 the court restricted **ecotourism** in India's tiger reserves. Although ecotourism is designed to protect wildlife and ecosystems by increasing appreciation, environmentalists argued that tourists were threatening tiger habitat. After a temporary ban on any tourism, the court allowed tourism in the reserves but with far more restrictions.

In 2006 India's government passed the Forest Rights Act. This law gives authority to people living in the forest areas to prohibit forest clearance by large timber and mining companies. This law can trace its origins to a movement in the 1970s among Himalayan women. In what is considered one of India's first environmental movements, the women protected trees that were marked for logging. The women were determined to protect the trees that provided them with firewood and prevented soil erosion.

India continues its efforts to improve the quality of life for its people. The government, however, recognizes that sustainable development and economic growth are related. These principles guided India's Planning Commission as it prepared the near final draft of its twelfth Five-Year Plan. This latest plan states that India must improve living standards "through a growth process which is faster than in the past, more inclusive, and also more environmentally sustainable." The draft went on to say that India cannot "neglect the environmental consequences of economic activity, or allow unsustainable depletion and deterioration of natural resources."

✔ **READING PROGRESS CHECK**

Analyzing What perspective does India's twelfth Five-Year Plan take toward economic growth?

LESSON 3 REVIEW

Reviewing Vocabulary

1. *Hypothesizing* Write a paragraph to form a hypothesis about why India's Supreme Court put a temporary ban on the pesticide endosulfan.

Using Your Notes

2. *Formulating Questions* Choose one of the questions you formulated as you completed your graphic organizer. List three additional questions you might ask to begin researching the answer to the first one.

Answering the Guiding Questions

3. *Identifying Cause and Effect* Why are many resources in India not managed closely?

4. *Finding the Main Idea* How has pollution affected the quality of life in India?

5. *Classifying* What steps has the Indian national government taken to combat pollution?

Writing Activity

6. *Informative/Explanatory* Based on the information covered in this lesson, what environmental challenges are posed by India's rapid population growth?

Directions: On a separate sheet of paper, answer the questions below. Make sure that you read carefully and answer all parts of the questions.

Lesson Review

Lesson 1

1 *Diagramming* Create a diagram to show the following movements that affected India's shape and position: a major tectonic event about 160 million years ago, a major tectonic event about 50 million years ago, and the thrusting up of the Himalaya.

2 *Assessing* What is a monsoon? Discuss the positive and negative effects of monsoons in India.

3 *Identifying Cause and Effect* Identify three types of timber resources in India. Explain the conditions that affect the supply of these resources in India.

Lesson 2

4 *Analyzing Cause and Effect* Discuss the effects of British imperialist policy in India. Explain how the eventual push by Indians for self-rule was in response to this policy.

5 *Summarizing* Provide an overview of the concept of *jati*.

6 *Identifying Central Issues* What is the green revolution? Why has it given rise to controversy?

Lesson 3

7 *Identifying Cause and Effect* What is the approximate percentage of farmers in India that grow their own food or grow food to sell? Create a graphic organizer to show the cause and effect of this pressure on India's farmland.

8 *Summarizing* Summarize efforts taken by the Indian government to address environmental protection since 1976. Explain why these efforts have not achieved all of the desired results.

9 *Evaluating* How did an act passed by India's government in 2006 connect to action among Himalayan women during the 1970s? What does this tell you about the determination of these Himalayan women? Within your response, include the name of the referenced act.

21st Century Skills

Use the document to answer the following questions.

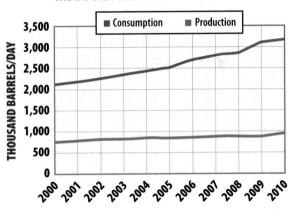

INDIA: OIL PRODUCTION AND CONSUMPTION

Source: U.S. Energy Information Administration

10 *Compare & Contrast* Compare and contrast India's oil production and consumption in 2000 and in 2010.

11 *Using Graphs* Explain whether it would be logical to conclude, based on the data in the graph, that India was reliant on oil imports from 2000 to 2010.

Critical Thinking

12 *Exploring Issues* Suppose you are a geologist who is writing a full-page blog to identify and describe the three major river systems in India and their importance to life in India. Within your blog, include an explanation of at least four types of pollution that could threaten these river systems—and action that might be taken to address the threats.

Need Extra Help?

If You've Missed Question	**1**	**2**	**3**	**4**	**5**	**6**	**7**	**8**	**9**	**10**	**11**	**12**
Go to page	606	608	610	611	611	617	622	623	624	625	625	607

Directions: On a separate sheet of paper, answer the questions below. Make sure that you read carefully and answer all parts of the questions.

Applying Map Skills

Use the Unit 7 Atlas to answer the following questions.

13 *The World in Spatial Terms* Use your mental map of India to describe the spatial relationship among the following: Deccan Plateau, Eastern Ghats, Western Ghats, and the Vindhya and Satpura Ranges.

14 *Places and Regions* Name the location of highest elevation in India.

15 *Human Systems* Identify the plain that is the most densely populated area of India.

Exploring the Essential Question

16 *Making Connections* Use what you have learned about human systems—history, politics, population, society, culture, and economics—to write a one-page essay explaining how the 1984 Bhopal disaster and its aftermath are representative of ways the physical systems and human systems of India have shaped the country.

College and Career Readiness

17 *Reaching Conclusions* Imagine that you work for an organization that is interested in advancing environmental protection efforts in India. Your boss has asked you to research recent legislation and grassroots efforts related to environmental concerns in India. Provide a one-page memo to identify and explain the focus of the legislation and grassroots local efforts that you find the most effective. Your memo should explain why you find these to be the most effective.

Writing About Geography

18 *Informative/Explanatory* Do research and then use standard grammar, spelling, sentence structure, and punctuation to write an article explaining why the Gupta Empire was one of the world's most advanced civilizations of its time—and why it ceased to exist.

DBQ Analyzing Primary Sources

Use the document to answer the following questions.

PRIMARY SOURCE

"*The new constitution gives no scope for retrenchment and therefore no scope for measures of social reform except by fresh taxation, the heavy burden of which on the poor will outweigh all the advantages of any reforms.*

The reformed constitution keeps all the fundamental liberties of person, property, press, and association completely under bureaucratic control. All those laws which give to the irresponsible officers of the Executive Government of India absolute powers to override the popular will, are still unrepealed."

—C. Rajagopalachar, Introduction to *Freedom's Battle: Being a Comprehensive Collection of Writings and Speeches on the Present Situation*, 1922

19 *Understanding Historical Interpretation* What criticisms does the author make of the new constitution?

20 *Interpreting Significance* Explain whether the author's views align with Gandhi's general political and social views regarding the people of India.

Research and Presentation

21 *Gathering Information* With a partner, conduct research to learn more about the slum problem in India. Create a multimedia presentation to explain why and how these slums developed, the living conditions within the slums, and the complex problems that have resulted in them. Propose at least two actions that might ultimately lead to the decline of slums and a better quality of life for those who currently live in slums. Within your presentation, include video and audio clips, maps, photographs, and diagrams or graphs.

Need Extra Help?

If You've Missed Question	**13**	**14**	**15**	**16**	**17**	**18**	**19**	**20**	**21**
Go to page	598	598	598	623	624	611	626	626	615

Pakistan and Bangladesh

networks

There's More Online about Pakistan's and Bangladesh's geography.

CHAPTER **26**

ESSENTIAL QUESTION · *How do physical systems and human systems shape a place?*

Geography Matters...

Pakistan and Bangladesh are lands of contrast. Northern Pakistan is home to K2, the world's second-tallest mountain, which is nearly 5.4 miles (8.7 km) high. Coastal Bangladesh is at sea level. Bangladesh has enough water to grow rice. In Pakistan, farmers must use dry-farming methods to grow crops.

Culturally, Pakistan and Bangladesh have similarities. They are both highly populous countries with strong Muslim cultures. Both share some of their history with neighboring India. Both countries are also struggling to develop their economies to overcome extreme poverty.

◀ Boy in Cox's Bazar, Bangladesh

©Laurence Fordyce/Eye Ubiquitous/Corbis

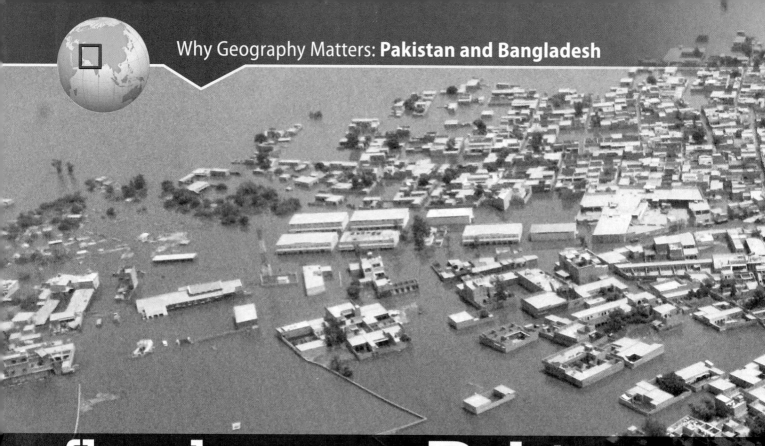

flood-prone Pakistan *and* Bangladesh

Floods are a common and major natural disaster in these two countries. In 2010 more than 18 million Pakistanis experienced terrible floods. In 1984 about 30 million Bangladeshis suffered through floods. Other major floods hit in 1992, 1993, 1995, 1998, 2004, 2005, 2007, and 2012. Why are floods so terrible here? What can Pakistan and Bangladesh do to mitigate them?

THERE'S MORE ONLINE

VIEW a chart about flooding in Pakistan and Bangladesh • *EXPLORE* a map of flood zones in Bangladesh (l)NASA/JSC, (c)Rizwan Tabassum/AFP/Getty Images, (r)©Gideon Mendel For Action Aid/In Pictures/Corbis

Why do floods threaten Pakistan and Bangladesh?

Pakistan and Bangladesh experience at least two floods a year, many of which are devastating. In Pakistan, heavy rains cause the Indus River to overflow its banks. Poor water management adds to the problem. Since population density is high along the river, these floods affect millions of people.

Many factors combine to cause floods in Bangladesh. The country has a very low elevation. Four major rivers—the Ganges, the Brahmaputra, the Jamuna, and the Meghna—flow through Bangladesh's plains, raising the probability of flooding. These floods strike when snow in the Himalaya melts, filling the rivers. The problem is made worse during the wet monsoon season or when cyclones hit.

1. **Physical Systems** In which country do you think flooding occurs more often? Why?

How has flooding harmed these two countries?

Flooding causes a host of problems in Pakistan and Bangladesh. Hundreds of people can die in a flood. In 1988 floods in Bangladesh resulted in the deaths of more than 2,000 people. In 2010 record flooding in Pakistan killed some 1,700 people and destroyed almost 2,000 homes. The economic cost can rise to billions and billions of dollars.

The danger comes from more than the flood waters alone. The water washes away crops, reducing harvests, and stored food becomes wet and spoils more quickly. With reduced food supplies, people suffer from hunger. Ironically, too much water means too little drinkable water. River water is muddy and contaminated by chemicals washed off farmland and by the decomposition of the bodies of dead animals and other waste. This kind of contamination also raises the likelihood of deadly diseases.

2. **Environment and Society** How are these problems affected by high population density in these countries?

What do these countries do to reduce flood damage?

Other countries faced with frequent flooding have taken steps to prevent it. They have built dikes and levees to hold water back, and put limits on building on floodplains or other flood-prone areas. They also have evacuation plans.

In 2005 world leaders met to discuss ways to prevent and recover from disasters like the floods in Pakistan and Bangladesh. They adopted a set of recommendations called the Hyogo Framework for Action. Both Pakistan and Bangladesh have tried to implement some of the recommendations. However, their efforts are hampered by the lack of funding.

3. **Human Systems** What conditions prevent Pakistan and Bangladesh from taking more steps to address flooding?

networks

There's More Online!

☑ **IMAGE** Flood Damage in Bangladesh

☑ **MAP** Mountain Passes of Pakistan

☑ **INTERACTIVE SELF-CHECK QUIZ**

☑ **VIDEO** Physical Geography of Pakistan and Bangladesh

LESSON 1
Physical Geography of Pakistan and Bangladesh

ESSENTIAL QUESTION · *How do physical systems and human systems shape a place?*

Reading HELPDESK

Academic Vocabulary

- **generate**
- **maintain**

Content Vocabulary

- **delta**
- **floodplain**

TAKING NOTES: *Key Ideas and Details*

IDENTIFYING CAUSE AND EFFECT
As you read about the physical geography of Pakistan and Bangladesh, use a graphic organizer like the one below to identify examples of how physical systems shape human systems.

Cause	Effect

IT MATTERS BECAUSE

Pakistan and Bangladesh are two of the ten most populous countries in the world. The landforms and climates of this subregion pose major challenges to the people living there.

Landforms

GUIDING QUESTION *How do landforms in Pakistan and Bangladesh affect human activities?*

Pakistan lies toward the northwestern edge of the Indian subcontinent, and Bangladesh lies toward the northeastern edge. Pakistan borders Afghanistan in Central Asia to the west and China to the north. Bangladesh is almost surrounded by India, but also shares a small border with Myanmar (Burma).

Physically, the two countries have few similarities and many differences. Bangladesh is almost entirely low and flat. Pakistan has large areas of low, flat plains as well as high mountains.

According to the theory of continental drift, the Indian subcontinent is slowly and steadily pushing into the southern edge of Asia. This collision thrust up the Himalaya mountain ranges. The Himalaya cover part of northern Pakistan, as do, farther north, the Karakoram Mountains. Farther west, the Hindu Kush range frames the northwest edge of Pakistan. In Pakistan, these mountain systems range from about 13,000 feet (3,962 m) to more than 19,500 feet (5,944 m) high. A few peaks soar more than 25,000 feet (7,620 m). One peak in the Karakoram, named K2, towers 28,250 feet (8,611 m). It is the second-highest mountain in the world.

These mountains are generally impassable, but the Hindu Kush have several wide, very high passes. The most important is the Khyber Pass in northwest Pakistan along the Afghanistan border. This pass has allowed armies to enter the Indian subcontinent. A highway through the Khunjerab Pass in the Karakoram is being modernized by China as a link to Pakistan.

Three lower mountain ranges run in a north-to-south direction from the western half of Pakistan and toward the Arabian Sea. These are, from north to south, the Salt Range, the Sulaiman Range, and the Kirthar Range.

(tl)Shawkat Khan/AFP/Getty Images, (tcl)NASA/JSC, (tcr)Rizwan Tabassum/AFP/Getty Images, (tr)©Gideon Mendel For Action Aid/In Pictures/Corbis

West of the Kirthar Range is a highland region, the Baluchistan Plateau. The mountainous northern and western regions of Pakistan are prone to earthquakes. A major earthquake in northwestern Pakistan in 2005 killed more than 80,000 people.

To the east is the Indus River valley, which has rich alluvial soil. The Indus and its tributaries form two alluvial plains, the Punjab in the north and the Sind in the south. The Sind includes part of the sandy Thar Desert.

Bangladesh is very flat and low, reaching only about 30 feet (9 m) above sea level. The land is cut deeply by the Ganges and the Brahmaputra. In Bangladesh, these rivers are called the Padma and Jamuna, respectively. After joining, the rivers flow through central Bangladesh to empty into the Bay of Bengal in a **delta** system that is the largest in the world. Islands are scattered along the Bay of Bengal. The land is low, so the rivers often change course, altering the landscape.

Bangladesh has few hilly areas. The Chittagong Hills are the largest and the highest. Mount Keokradong, the highest point in the country, is located here, rising about 4,035 feet (1,230 m) high. Bangladesh also has lowland depressions. The Haor Basin, which covers 9,459 square miles (24,500 sq. km), has hundreds of depressions that form wetlands in the rainy months.

Along the southwest coast of Bangladesh is a massive wetland called the Sundarbans. This region is crossed by many waterways that are part of the Ganges Delta. It is thick with mangrove trees that live in the mixture of salt water and freshwater found in this coastal marshland.

delta an alluvial deposit at a river's mouth that is shaped roughly like the Greek letter *delta* (Δ)

☑ **READING PROGRESS CHECK**

Comparing How are the landforms of Pakistan and Bangladesh similar?

GEOGRAPHY CONNECTION

Pakistan has some of the highest and most rugged mountains in the world.

1. ***PLACES AND REGIONS*** Which pass allows travel into Pakistan through the Karakoram Range?

2. ***ENVIRONMENT AND SOCIETY*** Which pass was most likely used by armies invading South Asia from Central Asia?

Mountain Passes of Pakistan

The Khyber Pass connects Pakistan with Afghanistan. For centuries it has held strategic importance--Persians, Greeks, Mughals, Afghans, and the British all used the pass to gain access to and control of the region. Today it is still a strategic route for military supply operations.

The Gumal Pass is a 4-mile (6-km) gorge with an elevation of 7,500 feet (2,286 m). The name often refers to the entire area covered by the Gomal River. The pass has been an important trade route between Afghanistan and Pakistan and was once used by nomadic Afghan traders called Powindahs.

The Bolan Pass also was used by invaders, travelers, and nomadic groups for many years and links India with Central Asia.

250 miles
250 kilometers
Lambert Azimuthal Equal-Area projection

Elevations

10,000 ft. (3,000 m)
5,000 ft. (1,500 m)
2,000 ft. (600 m)
1,000 ft. (300 m)
0 ft. (0 m)
Below sea level

--- Disputed boundary
--- Line of control
■ Mountain pass
▲ Mountain peak

TAJIKISTAN
TURKMENISTAN
HINDU KUSH
KARAKORAM RANGE
Shandur Pass
Hayal Pass
Noltar Pass
Hispar Pass
Lowari Pass
Babusar Pass
K2 28,250 ft. (8,611 m)
Malakand Pass
Karakar Pass
Kabul
Khyber Pass
Peshawar
AFGHANISTAN
Kohat Pass
Islamabad
Salt Range
HIMALAYA
CHINA
Gumal Pass
Kandahār
Sulaiman Range
Central Brahui Range
30°N
NEPAL
80°E
IRAN
Bolan Pass
PAKISTAN
INDIA
Central Makran Range
Kirthar Range
Indus R.
Karachi
Arabian Sea
TROPIC OF CANCER
60°E
70°E

Powerful cyclones can deliver driving rains to the region and contribute to the causes of flooding. This house in Bangladesh collapsed when the Padma River flooded.

▲ CRITICAL THINKING

1. Comparing and Contrasting Do you think cyclones are more of a problem in Pakistan or Bangladesh? Why?

2. Explaining How is flooding in this region both beneficial and destructive?

floodplain the low-lying land along a river, formed mainly by sediment that has been deposited by floodwaters

generate to produce

maintain to support

Water Systems

GUIDING QUESTION *How are rivers important to the peoples of Pakistan and Bangladesh?*

From their sources high in the Himalaya, three major river systems—the Indus, the Ganges, and the Brahmaputra—supply Pakistan and Bangladesh with water. These three rivers swell with seasonal snowmelt and rains, overflowing their banks and depositing fertile soil on their **floodplains**. People rely on these rivers for farming, transportation, and power. However, the floods can be destructive.

The Indus flows south from the Himalaya through Pakistan, watering wheat and rice crops before emptying into the Arabian Sea. In its upper reaches, the Indus is fed by four tributaries—the Jhelum, Chenab, Ravi, and Sutlej. Together these five rivers give the region its name, Punjab, from a Persian word for "five waters." To the north, the Mangla Dam on the Jhelum and the Tarbela Dam on the Indus are used to **generate** electricity.

The Ganges and Brahmaputra Rivers both retain a flow of water throughout the year. The Ganges is the most important river in South Asia, drawing waters from a basin covering about 400,000 square miles (1 million sq. km). The Brahmaputra is a major waterway for the subregion. Boats can navigate the river inland from the Bay of Bengal some 800 miles (1,290 km). In Bangladesh, these rivers move through lowlands and thus cannot be used for hydroelectric power. Some electricity is generated from rivers that flow through the Chittagong Hills, however.

Depressions in Bangladesh form freshwater wetlands that range from small lakes to marshes. *Haors,* the largest of these features, are bowl-like basins that flood during the wet monsoon. *Beels* are more like saucers that fill with water and then support plant growth. *Baors* are lakes that form after rivers change course. Many *baors* are found in the southwest. In the dry season, they are tapped for irrigation. All these bodies **maintain** fish life that adds to the diet of the people of Bangladesh. They offer potential for developing commercial fisheries, too. Recently, however, commercial shrimp farming has threatened wild fish populations.

Pakistan's lakes are threatened by environmental degradation. The country's largest freshwater lake, Lake Manchar, once supported many fishers. Agricultural runoff in the lake, however, has caused a decline in fish populations.

Karachi, Pakistan's largest city, is located on the Arabian Sea and serves as the country's most important port. Rail lines and highways connect it to the rest of the country, making it possible to move Pakistani exports overseas and to transport imports received at the port around the country. Karachi is also home to a shellfish industry that provides another export for Pakistan.

Both Pakistan and Bangladesh have coastal areas that are vulnerable to flooding when the wet monsoons are strong. Cyclones are a very serious natural hazard in Bangladesh and India. Bangladesh averages 16 cyclones every 10 years. These powerful storms bring high winds and heavy rains. Another coastal danger is tsunamis. These are huge waves caused by underwater earthquakes.

☑ READING PROGRESS CHECK

Explaining Why is Pakistan more able to generate hydroelectric power from the Indus River than Bangladesh can from the Ganges and the Brahmaputra?

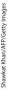

Shawkat Khan/AFP/Getty Images

Climates, Biomes, and Resources

GUIDING QUESTION *What are the most important natural resources in Pakistan and Bangladesh?*

Pakistan and Bangladesh are subject to seasonal monsoons. These winds create three distinct seasons: hot (from late February to June), hot and wet (from June or July until September), and cool (from October to late February). In the hot and wet season, moist monsoon winds from the ocean bring rain. During the cool season, dry, cool monsoon winds from the Asian interior blow across the subcontinent toward the ocean.

Southeastern Bangladesh has a tropical wet climate. Most of Bangladesh receives as much as 60 inches (152 cm) of rain a year. In the wet monsoon months, rain forests absorb great amounts of moisture. In the hot, damp Sundarbans, tropical mangrove forests thrive. The Himalaya block the cold winds blowing from Central Asia. As a result, a band of humid subtropical climate extends across northern Bangladesh. The Hakaluki Hoar wetland in the east is home to rare and endangered fish, bird, and plant species.

Monsoon rains are not nearly as heavy in Pakistan as in Bangladesh. The wettest part of Pakistan, in the northeast, receives about half the rain of Bangladesh. Much of Pakistan only gets a fraction of that. Most rain falls in the hot months and evaporates quickly. Along the lower Indus in southern Pakistan, a desert climate keeps the land arid and windswept. East of the Indus in the Thar Desert, the vegetation is desert scrub. Though irrigation makes it possible to grow wheat in some parts, much of this area remains sandy with little vegetation.

South Asian rivers are important for basic needs, as well as for economic needs. They provide drinking water, alluvial soil, transportation, and hydroelectric power. They also provide fish for local use and export.

Pakistan and Bangladesh have few mineral resources. Of these, Pakistan has large amounts of limestone. It has oil in the north and natural gas in the south. Pakistan has very small amounts of iron ore, copper, coal, and other minerals.

The most abundant resource in Bangladesh, besides water, is natural gas. Most of that is burned by people for heat or used to make fertilizer, rather than exported. Bangladesh has small amounts of oil and coal. Some of these resources are inaccessible because they are buried too deep. Bangladesh does have valuable plant resources in the bamboo that grows in the Chittagong Hills and the high-quality mangrove wood of the Sundarbans.

✓ **READING PROGRESS CHECK**

Explaining In which country would it be easier to irrigate fields? Why?

Analyzing PRIMARY SOURCES

Flooding in Bangladesh

"That is how I felt in Sukkur 10 days ago. Overwhelmed.

People were flowing into the city at a ferocious pace, a ragged river of humanity, with shocked faces and frightened eyes.

They were fleeing on trucks, donkey carts, bicycles and on foot, clutching whatever was precious—electric fans, bedding, pots and pans, chickens and goats. Behind them, a great sinister mass of floodwater was pouring in."

—Jill McGivering, "'Elation' and 'Unease' at Helping Pakistan Flood Child," *BBC News,* September 4, 2010

DBQ *MAKING PREDICTIONS*
What kind of help will these fleeing people need both immediately and in the future?

LESSON 1 REVIEW

Reviewing Vocabulary
1. *Identifying* Write a paragraph explaining how the people of Pakistan and Bangladesh benefit from the floodplain.

Using Your Notes
2. *Describing* Use your graphic organizer on the physical systems of Pakistan and Bangladesh to write a paragraph describing one example of how landforms or waterways shape life in these countries.

Answering the Guiding Questions
3. *Drawing Conclusions* How do landforms in Pakistan and Bangladesh affect human activities?

4. *Explaining* How are rivers important to the peoples of Pakistan and Bangladesh?

5. *Analyzing* What are the most important natural resources in Pakistan and Bangladesh?

Writing Activity
6. *Informative/Explanatory* In a paragraph, discuss how the rivers of Pakistan and Bangladesh are both an advantage and a disadvantage.

networks

There's More Online!

☑ **IMAGE** Microloans in Bangladesh

☑ **IMAGE** Ship-Breaking in Bangladesh

☑ **MAP** Partition of India, 1947

☑ **TIME LINE** Unity and Division

☑ **INTERACTIVE SELF-CHECK QUIZ**

☑ **VIDEO** Human Geography of Pakistan and Bangladesh

LESSON 2
Human Geography of Pakistan and Bangladesh

ESSENTIAL QUESTION · *How do physical systems and human systems shape a place?*

Reading **HELP**DESK

Academic Vocabulary

- **infrastructure**
- **available**

Content Vocabulary

- **total fertility rate**
- **jute**

TAKING NOTES: *Key Ideas and Details*

COMPARING AND CONTRASTING As you read about the human geography of Pakistan and Bangladesh, use a graphic organizer like the one below to identify similarities and differences in the countries' histories.

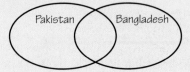

IT MATTERS BECAUSE

Pakistan and Bangladesh share a similar history of Muslim influence, British colonialism, and the pursuit of independence. Both have struggled since independence to build stable governments and overcome poverty.

History and Government

GUIDING QUESTION *How did contemporary Pakistan and Bangladesh form?*

The area we know today as Pakistan has been home to civilizations for thousands of years. One of the earliest settlements here was at Mehrgarh, near the Bolan Pass. Evidence dating back to 7000 B.C. shows that agriculture was being practiced. Around 2500 B.C. a great civilization arose in the Indus River valley. This culture developed a writing system, a strong central government, and a thriving overseas trade. Ruins of the walled city of Mohenjo Daro include evidence of plumbing and other advanced technology.

Beginning in the A.D. 600s, a Buddhist kingdom ruled over what is today Pakistan and Bangladesh for several centuries. Hinduism from India was also practiced. However, the diversity of the two countries suggests that both were crossroads for other groups seeking trade and territorial control. Over the centuries, many peoples crossed or settled in the regions that later became Pakistan and Bangladesh.

Coming of Islam

Muslim invaders and traders brought Islam to southeast Pakistan in the A.D. 700s. About a century later, Muslim traders brought Islam to Bangladesh. Over time, Islam became the majority religion in both areas.

From the 1500s to the 1800s, Pakistan and Bangladesh were part of the Mogul Empire that also ruled much of India. Early in this period, Sikhism, which blends elements of Hinduism and Islam, arose in the Punjab in northwestern India. In the early 1800s, a Sikh named Ranjit Singh established a kingdom in northern Pakistan, but it fell apart after his death.

(t)Bert Brandt/AFP/Getty Images, (tc)Marvin Lichtner/TIME & LIFE Images/Getty Images, (tr)©Michael Blach/Demotix/Corbis

In the middle 1800s, the British included both Pakistan and Bangladesh within their colony of India. Together they formed a single colonial administrative unit called India. The British introduced English, restructured the educational system, built railroads, and developed a civil service. During their rule, tension between Muslims and Hindus in South Asia grew. While Hindus formed a group to campaign for independence, Muslims, who feared domination by the majority Hindus, created a similar organization of their own.

Pakistan Since Independence

By the 1930s, the idea of a Muslim state separate from India had taken hold among South Asia's Muslims. When Hindu and Muslim leaders could not agree on a constitution for a single state, in 1947 the British granted independence to two states based upon the dominant religion. India was formed as a predominantly Hindu state and Pakistan as a predominantly Muslim state. The latter consisted of two sections, known at the time as East Pakistan and West Pakistan. They were separated by about 1,000 miles (1,609 km) of Indian territory.

Independence from Britain was marked by a massive movement of Hindus and Muslims between India and Pakistan. During this time, there were outbreaks of violence between followers of the two religions. Adding to the tension, both countries claimed the area of Kashmir. It was part of India by tradition, but had a majority Muslim population. The two countries fought wars for control of Kashmir in 1948, 1965, and 1999. Tensions over Kashmir and other issues became more worrisome in the 1990s after India and Pakistan both developed nuclear weapons. Relations between the two countries have improved in recent years, but the situation remains uneasy.

Pakistan is a parliamentary republic, but instability and military rule have prevailed since 1971. In the early 2000s, General Pervez Musharraf allied Pakistan with the United States in its war on terror. His rule became increasingly unpopular, however, and he was forced to step down in 2008. Asif Ali Zardari, the widower of former prime minister Benazir Bhutto, was elected president in 2008. He introduced reforms to try to establish civilian government, but poverty and rebel activity in the west plague the country. In 2013 Nawaz Sharif took office as Pakistan's prime minister after a sweeping victory of his Pakistan Muslim League party.

Bangladesh: Independence and After

The people of East Pakistan were culturally different from the people in West Pakistan long before independence from the British. In Bangladesh they are ethnic Bengali and speak Bangla. After independence, West Pakistan wanted to impose a national language, Urdu, on all of Pakistan. Bengali leaders believed that their ethnic majority was treated unfairly by the government, which was dominated by leaders from West Pakistan. They formed a protest movement.

After wins in the 1970–1971 elections, Bengali nationalists pushed for self-rule. Pakistan sent its army to suppress the nationalists, which prompted them to declare independence for Bangladesh—which means "Bengal country." India entered the war on behalf of independent Bangladesh, and Pakistan surrendered.

Bangladesh is also a parliamentary republic. Political and ethnic rivalries have made stable rule difficult. Discontent has continued in recent years.

☑ READING PROGRESS CHECK

Making Connections How was the movement that led to the independence of Bangladesh similar to the movement that led to the formation of Pakistan?

Connecting Geography to **HISTORY**

Natural Disaster

When does a storm create a political movement? Since Pakistan's independence in 1947, many people in East Pakistan were unhappy with domination by those in West Pakistan. In the 1960s, some groups began calling for an independent state. Then a devastating cyclone smashed into East Pakistan in 1970, flooding large areas and killing as many as 500,000 people. The West Pakistan-dominated government was slow to provide desperately needed relief to the suffering Bengalis. The smoldering anger over this slow response contributed to the Bengali declaration of independence the following year and the establishment of Bangladesh.

EVALUATING What effect did the 1970 cyclone have on the history of Bangladesh?

Population Patterns

GUIDING QUESTION *How does high population density affect life in Pakistan and Bangladesh?*

Pakistan, with its population numbering 180 million, is the sixth most populous country in the world. Bangladesh, with a population of 153 million, ranks eighth. In Pakistan, more than one-third of the people are under the age of 15. In Bangladesh, just under one-third are under age 15.

Physical geography shapes settlement in Pakistan, where most of the population lives in the Indus River valley. Although Pakistan is one of South Asia's most urbanized countries, only about 35 percent of the population lives in urban areas. Many people are migrating to cities, however, including Islamabad, Lahore, and Karachi. This growing urban population is straining resources. Migrants to the cities are forced to live in makeshift structures pulled together from scrap material. The **infrastructure** that makes urban life more livable, such as public water and sewage removal, is undeveloped in these poor areas.

In Pakistan the **total fertility rate**, or the average number of children a woman has in her lifetime, is 3.6. This is 50 percent higher than the world average. As a result, its population is growing rapidly, at 2 percent a year.

Bangladesh is the most densely populated country in South Asia, with 3,256 people per square mile (1,257 people per sq. km). The highest population densities occur in cities, such as Dhaka. However, density is generally high throughout the country. Only the Sundarbans in the southwest and the Chittagong Hills in the southeast have lower population densities.

To encourage Bengali women to have fewer children, private and government programs give women small loans to start their own businesses. The programs have achieved some success. While the total fertility rate was 4.4 in 1991, by 2011 it had declined to 2.2, just below the world average. Fertility rates have decreased as women become more educated and have more economic opportunities.

infrastructure system of public services such as power, water and sewage, transportation and communication networks, and schools and health care facilities

total fertility rate the average number of children a woman has in her lifetime

☑ **READING PROGRESS CHECK**

Explaining How did the government of Bangladesh address its high fertility rate?

Bert Brandt/AFP/Getty Images

TIME LINE ⌄

UNITY and Division ➔

The state of Pakistan was born out of a desire of Muslims in India for national unity, but geographic and demographic challenges caused further disunity, leading to the creation of Bangladesh.

▶ **CRITICAL THINKING**

1. ***Explaining*** Why did the Muslim League call for a separate state to be created for Muslims?

2. ***Comparing*** In what ways are the histories of Pakistan and Bangladesh similar?

Muslim League is founded to protect Muslim rights in India under British rule

1906

1900 ➔

1940

Muslim League demands that India be partitioned to create a separate Muslim state called Pakistan

Pakistan is established, but West and East Pakistan are about a thousand miles apart. First governor-general is Mohammed Ali Jinnah.

1947

1948

Pakistan goes to war against India over disputed territory of Kashmir

Society and Culture Today

GUIDING QUESTION *How is life in Pakistan similar to and different from life in Bangladesh?*

Pakistan and Bangladesh became separate countries because of their diverse ethnic heritages. Pakistan is diverse in large part because it experienced invasions and migrations over many centuries. Pakistan has six main ethnic groups: Punjabis, Pashtuns, Sindhis, Sariakis, Muhajirs, and Balochis. Punjabis make up about 45 percent of the country's people. No other group has more than 16 percent. Urdu is the official language of Pakistan, but only 8 percent of the population speaks it. More Pakistanis speak Punjabi than any other language. English, also an official language, is typically spoken by members of the elite, including government officials.

In Bangladesh the majority of the people are Bengali, a term describing both an ethnic and a language group. It is something they share with some of their Hindu neighbors in the Indian state of Bengal. Non-Bengalis, mostly smaller indigenous groups, make up only a small percentage of Bangladesh's population.

Islam is the main religion in both countries. The two countries also have some Hindus and small Christian populations. Most Muslims are Sunnis, but about one-fifth of Pakistan's Muslims are Shias.

Literacy rates in Pakistan and Bangladesh are very low, at less than 60 percent. In Bangladesh, the literacy rate for men is about 10 percentage points higher than for women. In Pakistan, the gap between male and female literacy is much greater. Schooling in Bangladesh is free and required for all children to age ten, but only about half of children actually attend. School is not required in Pakistan, where the educational system is a mix of government-run, Islamic, and private schools.

Lack of health care is a major problem in both countries, mainly because of their large populations and high poverty rates. Spending on health care ranks toward the bottom of priorities, and there are relatively fewer doctors or hospital beds than are needed. Several diseases, including tuberculosis and malaria, have long been problems in both countries. Bangladesh has succeeded in reducing malaria but still faces many public health issues.

(t)Marvin Lichtner/TIME & LIFE Images/Getty Images, (b)Central Intelligence Agency

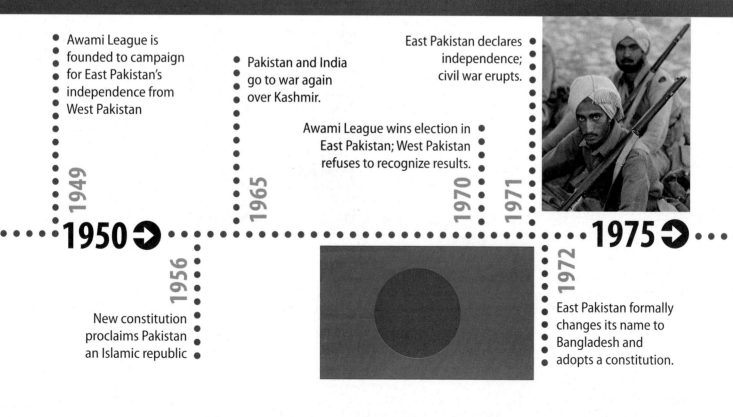

1949 — Awami League is founded to campaign for East Pakistan's independence from West Pakistan

1950 ➔

1956 — New constitution proclaims Pakistan an Islamic republic

1965 — Pakistan and India go to war again over Kashmir.

1970 — Awami League wins election in East Pakistan; West Pakistan refuses to recognize results.

1971 — East Pakistan declares independence; civil war erupts.

1972 — East Pakistan formally changes its name to Bangladesh and adopts a constitution.

1975 ➔

Family and the Status of Women

Family is the social basis in both countries. Extended families live close, often in the same home. Many marriages are arranged, though increasing numbers of educated people choose their own partners. After marriage, the wife typically lives with the husband's family.

In traditional families in Pakistan, women stay in separate parts of the house when nonfamily males visit. This is more common among urban and middle-class Pakistanis than rural or poor people, where women must work to help the family survive. Some Muslim women also wear the burka, a loose garment that covers the face and body, when they are in public.

Both Pakistan and Bangladesh have had women leaders. Benazir Bhutto was twice prime minister of Pakistan. Sheikh Hasina served as prime minister of Bangladesh from 1996 to 2001 and took that office again in 2009.

The Arts

Literature and dance are very important in Bangladesh. In 1913 Bengali Rabindranath Tagore became the first non-European writer to win the Nobel Prize in Literature. The poetry and plays of Kazi Nazrul Islam, "the voice of Bengali nationalism," have inspired poor farmers with themes about the oppression of Muslims. Bangladesh also has developed original and creative traditional dances in classical as well as folk styles.

Music and literature are the richest of all Pakistani art forms. *Qawwali*, a form of devotional singing, is popular. People recite poetry at public *musha'irahs* that are organized like music concerts. The classical music tradition can be traced to the thirteenth-century poet and musician Amir Khosrow, who composed the traditional rhythmic form known as the raga.

✔ **READING PROGRESS CHECK**

Contrasting How are the people of Pakistan more diverse than those of Bangladesh?

These men at a shipyard in Bangladesh, on the Bay of Bengal, work in the ship-breaking industry. They face dangerous conditions, including accidents and exposure to toxic materials.

▼ **CRITICAL THINKING**

1. ***Drawing Conclusions*** Why do people take such dangerous jobs?

2. ***Making Inferences*** Why do you think Bangladesh has a ship-breaking industry, but not a ship-building industry?

Economic Activities

GUIDING QUESTION *What are the dominant economic activities in Pakistan and Bangladesh?*

Pakistan and Bangladesh have traditionally relied on agriculture. Industrial activity is increasing in the twenty-first century. Poverty is widespread. The economic situation has worsened in recent years due mainly to inflation.

About one-fifth of Pakistan's gross domestic product (GDP) comes from agriculture, and about two-fifths of the workforce is in agriculture. Cash crops, including rice, cotton, and sugarcane, bring much-needed income. Pakistan also has a fishing industry, exporting shrimp, lobster, and fish. Despite reforms in the last 50 years, land distribution remains highly unequal. Most farmers continue to use draft animals on small farms. Since the 1980s, service industries have grown. Today, more than half the country's GDP is from the service sector.

Industry constitutes about one-fourth of Pakistan's GDP. The most important are cotton textiles and clothing for export. Other exports include rice, leather, sporting goods, chemicals, and carpets. Most exports are shipped out of the Port of Karachi. Pakistan has become a major trading partner of the United States. Small-scale production, or cottage industries, has played an important role in Pakistan's industrialization. They employ many craftspeople and provide at-home employment for women.

As in India, railways are the principal mode of transportation. The state-owned Pakistan Railways moves both people and cargo throughout Pakistan. Highways are increasing in importance. The Makran Coastal Highway runs along the Arabian Sea and integrates economic activities in the area.

Rural areas in both countries often lack electricity and other modern services. The spread of cell phones has made communications more widely **available** even in rural areas. That is more the case in Pakistan than Bangladesh, however.

available able to be obtained

In Bangladesh most people are sharecroppers. Rice is the major crop. In some areas, farmers can grow three crops a year, alternating rice, wheat, and other crops. **Jute**, a fiber used to make string, rope, and cloth, is also a major cash crop.

jute a fiber used to make string, rope, and cloth

The garment industry has expanded and clothing is now Bangladesh's top export. As in Pakistan, much of Bangladesh's textile manufacturing relies on cottage industries. Dhaka is a center for weaving muslin, a lightweight cotton cloth. Nearly 2 million women work in Bangladesh's garment industry. Other craft goods include jute products, such as upholstery, and leather goods.

Bangladesh is one of the world's largest aquaculture-producing countries. It has many inland fisheries that cultivate fish and shrimp. These proteins are part of a staple diet that also relies heavily on rice and lentils.

☑ **READING PROGRESS CHECK**

Explaining How are cottage industries important in Pakistan and Bangladesh?

LESSON 2 REVIEW

Reviewing Vocabulary

1. *Identifying* Write a paragraph explaining what jute is and why it could be called the "golden crop" of Bangladesh.

Using Your Notes

2. *Comparing and Contrasting* Use your graphic organizer on the human geography of Pakistan and Bangladesh to write a paragraph explaining how their histories are similar and different.

Answering the Guiding Questions

3. *Explaining* How did contemporary Pakistan and Bangladesh form?

4. *Assessing* How does high population density affect life in Pakistan and Bangladesh?

5. *Comparing and Contrasting* How is life in Pakistan similar to and different from life in Bangladesh?

6. *Identifying* What are the dominant economic activities in Pakistan and Bangladesh?

Writing Activity

7. *Argument* In a few paragraphs, discuss challenges Pakistan and Bangladesh face and how they might be tackled.

South ASIA on the BRINK

The nuclear weapons programs of India and Pakistan began in the late 1960s and early 1970s and produced nuclear weapons for both countries. India and Pakistan have over 100 missiles between them. India's Agni-V missile is capable of striking a target more than 3,100 miles (4,989 km) away. Although the two countries' nuclear arsenals are small relative to those of the United States and Russia, they nonetheless pose a threat to the rest of the world.

REGIONAL NUCLEAR STOCKPILES

PAKISTAN

INDIA

1,550 mi (2,500 km)

CHINA 200

INDIA 50

PAKISTAN 60

THREATS

CHINA

KASHMIR DISPUTE

India and Pakistan have both claimed ownership of the northern Kashmir border region. War has erupted over the area twice in the past, and there are worries that if there was ever a third it would go nuclear.

CHINA DISPUTE

China and India too have border disputes. Both countries having nuclear arms raises fears of nuclear war.

TERRORISM

Pakistan's political instability causes some concern. Should its government fall, nuclear materials could fall into the hands of terrorist cells in neighboring countries and within Pakistan itself.

NUCLEAR WINTER

Should both India and Pakistan detonate their arsenals, it could result in a 20% depletion of global ozone levels. The resulting increase in UV radiation could seriously impact human health across the globe.

- ─ ─ **PAKISTANI MAXIMUM RANGE**
- ── **INDIAN MAXIMUM RANGE**

3,107 mi (5,000 km)

Making Connections

1. *Places and Regions* What ongoing regional conflicts increase the risk of nuclear war?

2. *The World in Spatial Terms* According to the map, about how many countries are in range of Pakistan's nuclear weapons?

3. *Environment and Society* What effects could a nuclear conflict have on the environment and the people of the region?

<u>Interact with **Global Connections** Online</u>

netw⬤rks

There's More Online!

☑ **GRAPH** World's Worst Cities for Air Pollution

☑ **IMAGE** Compressed Natural Gas Vehicle

☑ **IMAGE** The Sundarbans

☑ **IMAGE** Water Pollution in Bangladesh

☑ **INTERACTIVE SELF-CHECK QUIZ**

☑ **VIDEO** People and Their Environment: Pakistan and Bangladesh

LESSON 3
People and Their Environment: Pakistan and Bangladesh

ESSENTIAL QUESTION · *How do physical systems and human systems shape a place?*

Reading HELPDESK

Academic Vocabulary

- **principle**
- **accumulate**

Content Vocabulary

- **sustainable development**
- **hydroelectric power**

TAKING NOTES: *Key Ideas and Details*

IDENTIFYING As you read about the environments of Pakistan and Bangladesh, use a graphic organizer like the one below to identify threats to the environment and responses to those threats.

IT MATTERS BECAUSE

Pakistan and Bangladesh are two of the most populous countries in the world. They are also among the world's poorest countries. Their governments need to find solutions to environmental problems to provide their people with health and opportunity. Finding the resources for these solutions is difficult, however.

Managing Resources

GUIDING QUESTION *Why has sustainable development been given a low priority in Pakistan and Bangladesh?*

The economies of Pakistan and Bangladesh are heavily agricultural. The water from their mighty rivers and their fertile soils are very important resources. Agriculture does not generate a great deal of wealth, however. As a result, these countries have huge numbers of people living in poverty and little national income to use to build stronger economies.

In the 1970s and 1980s, people around the world began to notice the environmental damage caused by modern economic activities. Factory smokestacks and vehicle exhaust fouled the air. Chemical and industrial waste poisoned the water. Pumping oil from the ground and mining coal and other resources scarred the landscape. In response, economists and environmentalists began to advocate, or promote, the idea of sustainable development.

When first introduced, **sustainable development** was defined as economic growth that meets the needs of present populations without hampering the ability of people in the future to meet their own needs. Over time, a third goal was added. This was to promote progress in less developed countries and to reduce the gap between those who are wealthy and those who are poor. Sustainable development, then, rests on these **principles**:

- promoting economic development
- protecting the environment
- promoting social fairness

(tl)Tim Laman/National Geographic/Getty Images, (tc)©mhhasan/Demotix/Corbis, (tr)Asif Hassan/AFP/Getty Images

While many people agree on the goals of sustainable development, achieving them is not an easy task. Poorer countries—which need economic development to give their people a better standard of living—have few financial resources to invest in building a more modern economy. For instance, such countries lack the technology or the knowledge needed to create and use more efficient or cleaner production methods.

Less developed countries face another obstacle in achieving sustainable development. For survival, people living in these low-income countries often make short-term choices that can have long-term negative consequences. For example, to gain money by selling valuable wood, they may cut down trees too rapidly to replace them. To expand agricultural output, they may use fertilizers and pesticides that, over time, hurt the soil or leach into the water, damaging drinking supplies.

Both Pakistan and Bangladesh need to develop their economies. One-fourth of Pakistan's people and one-third of Bangladesh's population live in poverty. The rankings of these countries in education and health statistics also reflect their low standards of living. At the same time, both countries suffer from obstacles to development. These include an overreliance on agriculture, a lack of capital to invest, corruption, and the unwise use of resources.

Perhaps the most threatened environmental region in these two countries is the Sundarbans. This wetland area in southwestern Bangladesh and the northeastern coast of India has the world's largest mangrove forest. This forest is threatened due to rising sea levels and the constant floods of salt water from the Bay of Bengal. Mangroves can grow in somewhat salty water, but if the concentration of salt increases too much, they will not thrive. The forest is threatened by action upriver as well. Because of deforestation in the Himalaya, the rivers that flow into Bangladesh now carry more silt and soil than before. Some of this soil reaches the Sundarbans, where it is building up and reducing the easy flow of water on which the trees depend.

sustainable development economic growth that meets the needs of present populations without hampering the ability of people in the future to meet their own needs and that benefits people and societies

principle a fundamental characteristic

The most threatened area in Bangladesh—the Sundarbans— is also the site of the largest mangrove forest in the world. These boats are hauling harvested mangrove for export.

▼ CRITICAL THINKING

1. *Making Connections* What problems threaten this vast forest?

2. *Identifying Cause and Effect* What might be some effects of the death of the mangrove forest?

Tim Laman/National Geographic/Getty Images

Local deforestation is a problem for the mangrove forest as well. About half the trees have been cut down over the years as people seek wood for fuel or clear the land for other uses. The forest faces yet another threat. Wastewater from shrimp farms located just north of the forest is raising concerns about damage it might do to the water quality of the Sundarbans.

Another resource at risk in Pakistan and Bangladesh is water. Most of the population of Pakistan does not have regular access to potable, or drinkable, water. A lack of clean water leads to a host of health problems.

Bangladesh could benefit more than it does from its abundant supply of natural gas by exporting it. Most of it, however, is used locally. Bangladeshis burn natural gas as a fuel, and much of the rest is used in fertilizer factories.

Hydroelectric power, an important source of energy in Pakistan, is also at risk. Pakistan's two hydroelectric power dams in the north are affected by physical geography. The Indus River carries a heavy load of soil as it flows south. While this fertile soil benefits farmers, it **accumulates**, or builds up, behind the dam. This clogs the Tarbela Dam, making it unable to generate power.

✔ **READING PROGRESS CHECK**

Explaining What are the three principles on which sustainable development rests?

Human Impact

GUIDING QUESTION *What kinds of human activity have affected the environment in Pakistan and Bangladesh?*

The threatened Sundarbans is just one example of the way the people of Pakistan and Bangladesh affect the physical environment of these countries. The struggle to survive has created several environmental issues. For example, Pakistan suffers from severe water pollution, which has several different causes. Sewage treatment facilities are scarce, and raw sewage enters the country's rivers and water supply. Industrial pollution is another contributor, as is agricultural runoff of pesticides and fertilizers from farmland. Water pollution is a problem that plagues Bangladesh as well.

With its fairly dry climate, Pakistan has to manage its water. Farmers use irrigation to bring water to areas that receive little rainfall but have fertile soil. However, because of the relatively high temperatures during Pakistan's growing season, some of this water evaporates, leaving minerals and salt in the soil, which makes the land less fertile.

Along with salinity, or saltiness, of the soil, these countries also have problems with soil erosion. This is particularly a problem in Pakistan, where erosion occurs in highlands that are losing their tree cover.

hydroelectric power
electrical energy generated by falling water

accumulate to build up

Agricultural runoff and industrial pollutants in both countries lead to water pollution. People, like this boy in Dhaka, Bangladesh, still rely on fish from the dying rivers for food.

▼ **CRITICAL THINKING**

1. *Explaining* What are other causes of water pollution in these countries?

2. *Identifying Cause and Effect* How do you think this kind of pollution affects humans?

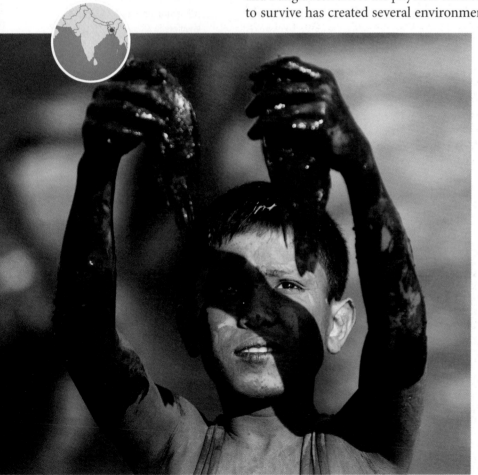

©mhasan/Demotix/Corbis

Bhutan, Maldives, Nepal & Sri Lanka

ESSENTIAL QUESTION · *How do physical systems and human systems shape a place?*

netw●rks

There's More Online about the geography of Bhutan, Maldives, Nepal, and Sri Lanka.

CHAPTER 27

Why Geography Matters
Nepal's Role as a Buffer State

Lesson 1
Physical Geography of Bhutan, Maldives, Nepal & Sri Lanka

Lesson 2
Human Geography of Bhutan, Maldives, Nepal & Sri Lanka

Lesson 3
People and Their Environment: Bhutan, Maldives, Nepal & Sri Lanka

Geography Matters...

The four countries of South Asia profiled in this chapter offer a great variety of natural beauty and human culture and more. They include the world's highest mountain, tropical forests, and beautiful coral reefs. Mountainous Nepal and Bhutan and tropical Sri Lanka and Maldives possess cultures that are centuries old as well as natural wonders that have existed for millions of years.

◄ This Hindu holy man in Nepal wears traditional facial painting.

©Lee Frost/Robert Harding World Imagery/Corbis

Nepal's role *as a* buffer state

Nepal is located high in the rugged Himalaya. Nepal's geographic isolation makes trade and travel difficult. In order to survive, Nepal has forged important political and economic alliances with its neighbors. At times, maintaining friendly relationships with surrounding countries has proved challenging.

Buffer States and Border Disputes in South Asia

International boundary
Province boundary
Disputed boundary

Both China and India have claimed Aksai Chin and Arunachal Pradesh. This boundary dispute has caused tension in the region.

KASHMIR (administered by Pakistan)

Area ceded by Pakistan to China, claimed by India

CHINA

Tibet was divided into Inner Tibet and Outer Tibet, and India and China disputed control of areas to the east and west of Nepal.

JAMMU & KASHMIR (administered by India)

AKSAI CHIN area held by China, claimed by India

TIBET

PAKISTAN

As a result of the Simla Accord, the McMahon Line was drawn in 1914.

30°N

Disputed areas

Disputed border

The small country of Nepal shares borders with India and China. It acts as a buffer state between them.

.Lhasa

New Delhi ✪

NEPAL

INDIA

✪ Kathmandu

BHUTAN

ARUNACHAL PRADESH largely claimed by China

0   250 miles
0   250 kilometers
TROPIC OF CANCER   *Lambert Azimuthal Equal-Area projection*

BANGLADESH

80°E

90°E

MYANMAR (BURMA)

650

What is a buffer state?

Why is Nepal considered a buffer state?

What does Nepal's status as a buffer state mean for its future?

A *buffer* is something that acts as a protective barrier. A *buffer state* is a smaller, independent country sandwiched between larger, more powerful rivals. In the 1800s, Thailand—then called Siam—was a buffer between the British empire in India and the French empire in Southeast Asia. A buffer state serves to prevent conflict between its neighbors. To function as a buffer state, a country must remain neutral and resist pressure from neighboring powers. A buffer state gains trading partners and other advantages from its alliances. It also literally keeps the peace in a region.

A buffer state is vulnerable to instability. It can be caught in the middle if a simmering conflict between its neighbors heats up. Being heavily dependent on other countries, a buffer state also loses the freedom to act as it chooses. A buffer state that is more influenced by one country can lose its independence and become a satellite state.

Small, landlocked Nepal stands between two much larger countries, China and India. In 1775 a Nepalese king described his country's location as "a yam between two boulders." Nepal is considered a buffer state because these two "boulders"—which combined have more than one-third of the world's population—are rivals.

Their hostility largely began with the 1959 Tibetan uprising against China, when the Dalai Lama and other Tibetans took refuge in India. In 1914 British colonial officials had drawn the McMahon Line, which divided Tibet and gave more territory to British-ruled India. India accepted the boundary as legal, but China did not. Since then, the two countries have disagreed about control of areas to the east and west of Nepal and have had an unclear border. In October 1962, Chinese soldiers launched a surprise invasion, crossing the Himalaya and seizing India's territory. Although this Sino-Indian War ended quickly, these quarreling countries have maintained an uneasy peace ever since, putting Nepal's interests at risk.

For years, Nepal benefited from close relations with both China and India. With each country, Nepal shares historical, economic, cultural, and religious bonds. Yet being in the middle means that Nepal must constantly perform a delicate balancing act. Political and economic changes in China and India affect Nepal and its future. Although trading partners, China and India still view one another with suspicion. India's awareness that China is capable of launching a surprise attack led to militarization of the region. Having Nepal as a buffer reduces tensions.

After years of depending heavily on India for trade, Nepal is strengthening its economic ties with China. In 2007, for example, China began constructing a railway extension to link Tibet and Nepal and improve trade. As China and India themselves develop stronger economic ties, there is hope that one day Nepal will become a bridge between them.

1. The World in Spatial Terms How does the location of a country make it a buffer state, and what purpose does a buffer state serve?

2. Places and Regions What physical and political characteristics of a buffer state does Nepal exhibit?

3. Environment and Society In what way does being a buffer state affect Nepal's future?

THERE'S MORE ONLINE

READ a time line of Nepal's history • *SEE* a map of disputed territory claimed by China, India, and Nepal

(l)Bartosz Hadyniak/E+/Getty Images, (c)Ilya S. Savenok/FilmMagic/Getty Images, (r)China Photos/Getty Images News/Getty Images

netw⊙rks

There's More Online!

☑ **IMAGE** Kathmandu, Nepal

☑ **IMAGE** Mount Everest

☑ **INFOGRAPHIC** Ecotourism

☑ **INTERACTIVE**
 SELF-CHECK QUIZ

☑ **VIDEO** Physical Geography of
 Bhutan, Maldives, Nepal & Sri Lanka

LESSON 1
Physical Geography of Bhutan, Maldives, Nepal & Sri Lanka

Reading **HELP**DESK

Academic Vocabulary

- **potential**
- **exploit**

Content Vocabulary

- **aquifer**
- **cowrie shell**

TAKING NOTES: *Key Ideas and Details*

IDENTIFYING As you read about the physical geography of Bhutan, Maldives, Nepal, and Sri Lanka, use a graphic organizer like the one below to give examples of how a landform, water system, or resource affects life in each country.

How Physical Geography Affects Life

Country	Physical Characteristics	Impact on Life

ESSENTIAL QUESTION · *How do physical systems and human systems shape a place?*

IT MATTERS BECAUSE

The northern and southern fringes of South Asia are among the highest and lowest countries in the world. Mountainous Nepal and Bhutan contain the sources of some of South Asia's mighty rivers. Sri Lanka is an island country. Maldives, a collection of low islands, faces the danger of rising sea levels.

Landforms

GUIDING QUESTION *What landforms help shape life in Bhutan, Maldives, Nepal, and Sri Lanka?*

While these four countries are unique, their locations on the fringe of South Asia have meant that their physical and human geographies have been influenced by the South Asian core. Nepal and Bhutan sit on the northern fringe of South Asia, amid the lofty Himalaya. They are bordered by India to the south and China to the north. Far to the south, in the Indian Ocean, sit the island countries of Maldives and Sri Lanka. Sri Lanka is off the southeastern coast of India. Maldives is a collection of islands southwest of India. Its islands stretch over 500 miles (805 km) from north to south and 80 miles (129 km) from east to west. The country's nearly 1,200 islands have a total area of about 115 square miles (298 sq. km).

The Indian subcontinent is a large landmass that is slowly and steadily colliding with the southern edge of Asia. This collision thrust up several mountain ranges, including the Himalaya, which dominate much of Nepal and Bhutan. These ranges spread about 1,500 miles (2,414 km) across the northern edge of South Asia and are hundreds of miles from north to south. The Himalaya are divided into three ranges: the Greater Himalaya to the north, the Lesser Himalaya in the center, and the Outer Himalaya to the south. The Greater Himalaya soar the highest, averaging over 20,000 feet (6,096 m) in elevation. In Nepal, the Greater Himalaya range has eight of the highest mountains in the world. Among them is Mount Everest, the world's highest mountain, which rises to 29,028 feet (8,848 m) above sea level along the border between Nepal and China.

(tl)Richard I'Anson/Lonely Planet Images/Getty Images, (tcl)Bartosz Hadyniak/E+/Getty Images, (tcr)Ilya S. Savenok/FilmMagic/Getty Images, (tr)China Photos/Getty Images News/Getty Images

Nepal has four geographic zones: the Tarai Plain on its southern border, the Churia and Mahabharat Ranges north of it, the Lesser Himalaya north of them, and the Greater Himalaya on its northern rim. The Tarai was originally forested marshland, but much of the plain has been drained for farming. The low Churia Range includes flat valleys that have been cleared of trees for farming. Between the Mahabharat Range and the Lesser Himalaya are more valleys where cities such as Kathmandu are located. Few people live in the Greater Himalaya.

Bhutan has four geographic zones. A narrow lowland called the Duars Plain sits on the southern edge of the country. This area receives heavy rainfall and is thick with vegetation and wild animals. On its northern edge are several passes into the mountains called *dwars*, meaning "doors," which gave their name to the plain. The southern part of the Duars Plain is flat land used for growing rice. A dramatic change in altitude occurs as the Lesser Himalaya and then the Greater Himalaya rise steeply above the plain. Many of Bhutan's people live in the several wide, level valleys that cut through the Lesser Himalaya. The valleys of the Greater Himalaya, found from 12,000 to 18,000 feet (3,658 to 5,486 m) above sea level, are sparsely populated. The Black Mountain range runs north and south from the Greater to the Lesser Himalaya in west-central Bhutan. Few people live in these rugged mountains.

Sri Lanka has two regions. The first is a triangle-shaped mountain range in the south-central part of the island called the Central Highlands. The second is a plain that covers most of the island. The Central Highlands has a few peaks that reach about 7,000 feet (2,134 m), including the island's highest point, Mount Pidurutalagala, with an elevation of 8,281 feet (2,524 m). The plain ranges from sea level at the coast to about 1,000 feet (305 m) high. Many inlets called lagoons lie along the coast, as well as Mannar Island. To the north of the Central Highlands, low ridges cross the plain. To the west, there are higher ridges and deeper valleys. Highland areas are used for growing tea, and the wetter southwest is home to plantations growing rubber plants. Rice is grown in the wet southwest, but irrigation projects have increased rice cultivation in other areas.

Maldives is a scattered collection of about 1,200 islands grouped into 13 atolls with barrier reefs. These atolls are coral reefs formed at the edge of volcanoes that have since sunk beneath the ocean. Colonies of tiny polyps secrete calcium carbonate to form the coral that builds the reefs. Over time, rubble from the reefs and sediment build up low islands. These islands have a very low elevation, mainly reaching no more than 6 feet (1.8 m) above the surrounding water.

Kathmandu, Nepal's capital and most populous city, sits in a valley more than 4,300 feet (1,311 m) above sea level.

◀ **CRITICAL THINKING**

1. *Identifying Cause and Effect* How do the mountains of Nepal and Bhutan influence where people in those countries live?

2. *Hypothesizing* Do you think these two countries have much farmland? Why or why not?

"From Dingboche, Ed Viesturs, Jangbu, and I went ahead of the team to check on the condition of the trail [up Mount Everest]. Reports had reached us that fresh snow had fallen between the settlement of Lobuche and Base Camp, making the final leg of the route impossible for yaks. The yak drivers were refusing to lead their animals through the deep drifts. Hazard pay can tempt the Sherpas [people of Nepal who assist mountain climbers], but they won't risk their yaks for any price. We had nearly one hundred of the animals, but the loads would have to be carried by porters."

—Jamling Tenzing Norgay, *Touching My Father's Soul,* 2002

DBQ *DRAWING INFERENCES*
Why were the Sherpas unwilling to use their yaks?

potential having a capacity that could be developed

aquifer an underground water-bearing layer of porous rock, sand, or gravel

The highest point on the islands is less than 8 feet (2.4 m) above sea level. Barrier reefs off many of these islands protect the sandy beaches from high seas. Many of the atolls are tiny; no island is larger than 5 square miles (13 sq. km). Only about 200 atolls are inhabited; most have fewer than 1,000 people. With little land for growing crops, fishing is a major activity.

✓ **READING PROGRESS CHECK**

Comparing How are the landforms of Bhutan, Maldives, Nepal, and Sri Lanka similar?

Water Systems

GUIDING QUESTION *How are rivers and the sea vital to human life in Bhutan, Maldives, Nepal, and Sri Lanka?*

Three major river systems—the Ganges, the Brahmaputra, and the Indus—fan out across the northern part of the Indian subcontinent. These rivers begin high in the Himalaya and carry fertile soil from mountain slopes onto their floodplains as the rivers swell with seasonal rains. The rivers of Nepal and Bhutan help feed the Ganges and the Brahmaputra.

Several rivers cut through the mountains of Nepal and Bhutan, generally flowing from the north to the south. The chief rivers in Nepal are the Baghmati, the Kosi, the Gandak, and the Karnali. These rivers eventually feed into the Ganges. Some of Nepal's rivers cut deep gorges running north and south through the mountains. Many rivers have fast-flowing streams through rocky channels that make them ideal for white-water rafting. This activity attracts many tourists who are eager for adventure. Nepal's fast rivers have great **potential** for hydroelectric power, but that resource has not been fully developed. The main rivers of Bhutan are the Torsa, the Raidak, the Sankosh, and the Manas. All join the Brahmaputra.

Sri Lanka has a dozen major rivers, most of which flow from the Central Highlands directly to the sea. The exception is the longest river, the Mahaweli, which begins flowing west of the Central Highlands but then follows a northward and eastward path to empty on the northeast coast. The Mahaweli, which swells in the rainy season, is tapped for water to irrigate fields. Rivers that drop from high elevations in the Central Highlands to lower-lying areas give Sri Lanka many spectacular waterfalls. Nearly 20 of them fall more than 348 feet (100 m).

Tiny Maldives has no rivers, but it does have underground aquifers that provide it with freshwater. **Aquifers** are underground water-bearing layers of porous rock, sand, or gravel.

The Indian Ocean surrounds both Maldives and Sri Lanka. The ocean provides rich resources to the people of these islands and affects the climate of all four countries, even far-off Nepal and Bhutan. Ocean waters also represent a threat to low-lying areas of Sri Lanka and to all of Maldives. Powerful cyclones—storms like hurricanes—can bring not only heavy rains but also high seas that can flood Sri Lanka's coastal lowlands and Maldives' low-lying islands.

Another danger comes from earthquakes that occur on the floor of the Indian Ocean. These events can unleash towering waves called tsunamis that can travel and flood far inland. A 2004 tsunami killed more than 31,000 people in Sri Lanka.

The ocean may cause problems for these countries in the future as well. Some climate scientists are forecasting that sea levels may rise as much as several feet during this century. If that does happen, the people of Maldives face a grim future, as the waters could submerge these low islands.

✓ **READING PROGRESS CHECK**

Drawing Inferences Why are aquifers important in Maldives?

Climates, Biomes, and Resources

GUIDING QUESTION *How does topography affect the climate in Nepal and Bhutan?*

As with the rest of South Asia, the monsoons strongly affect the climate of these four countries. However, the locations and landforms of these countries give them different climate zones. Elevation is also an important factor in climate.

Nepal and Bhutan have four climate zones. The Tarai in Nepal and lowlands in Bhutan have a humid subtropical climate. The lower hills and mountains have a warm temperate climate with summer rain and dry winters. The higher mountains have a highland climate with cool temperatures. Areas above 16,000 feet (4,877 m) have an arctic climate with year-round ice. Winters in the Himalaya are long and harsh. Bhutan is directly in the path of wet monsoon winds.

Vegetation varies by elevation and is related to the climate zones. Tropical forests cover the lowlands. Deciduous trees are found at the next elevation level, and conifers appear higher up. Grasses cover the next level, up to about 15,000 feet (4,572 m). Above that is the snow line, where nothing grows. Elephants, rhinoceroses, deer, and tigers roam the lowland areas.

ECOTOURISM IN NEPAL

LOW IMPACT
Ecotourism is responsible travel to areas of natural significance that attempts to conserve the environment and improve the well-being of local people.

INTERNATIONAL
Ecotourism is a growing industry, increasing by 5% each year. It represents 6% of the world's gross domestic product and 11.4% of all consumer spending worldwide.

PRE-TOURIST
Prior to the 1920s, the area was inhabited by the Sherpa people and largely undisturbed by Westerners. Climbing began in the 1920s, and the first successful attempt was in 1953.

TOURIST
Since 1990, climbing has become a popular tourist activity. Unfortunately, these tourists have made a habit of littering the mountain with gear and equipment after use.

ECOTOURIST
Today climbers are encouraged to be more responsible with their waste and unused gear. Since 2008, yearly "eco-climbs" occur to remove trash and raise awareness.

Climbing tourism generates **$500 million** in revenue for Nepal annually. Yet climbers also leave behind **120 tons of trash** each year. **Ecotourists hope to find a balance.**

Ecotourism based on the climate and natural beauty of the Himalaya is a growing industry. Mount Everest, the tallest mountain in the world, is an especially popular destination.

▲ **CRITICAL THINKING**

1. ***Considering Advantages and Disadvantages*** How does Nepal benefit from tourism? What are the negative effects of tourism?

2. ***Hypothesizing*** Why might someone decide to take part in an "eco-climb"?

Maldives has a humid tropical climate with a hot monsoon season and a warm dry season. Coconut palms and breadfruit trees grow on the islands. Tropical fish dart among the coral reefs, and tuna and sea turtles swim in the deeper waters.

Sri Lanka has a similar climate in lowland areas. Temperatures are warm or hot throughout the year. The Central Highlands are much cooler. The country has a tropical wet climate in the west which supports rain forests, but becomes tropical dry in the east with evergreen and deciduous forests and upland grasslands. Drier areas have low bushes and scrub plants. Sri Lanka's forests are teeming with wildlife, including elephants, sloth bears, buffalo, and a variety of birds.

Nepal's rivers have great potential for hydroelectric power, but only a tiny fraction of the country's energy is provided in this way. It does have some power-generating dams, and plans for more are underway. If the country can develop this resource, it could export surplus electricity to India. Mineral resources include deposits of coal, iron ore, copper, and limestone. Nepal's forests contain a variety of trees, but overcutting has resulted in massive soil erosion, prompting the government to implement conservation and reforestation projects.

exploit to make use of a resource

Bhutan has been much more successful in **exploiting** its hydroelectric potential. Electricity is its most valuable export. The Chhukha hydroelectric project began operating in the 1980s. Proceeds from selling power to India paid for the project. Another project completed in 2007 has also been successful, and the country has signed agreements with India to build four more dams. Bhutan has mineral resources, but transportation difficulties have limited its ability to use them.

Bhutan has worked to preserve its forests and has been more successful than Nepal in protecting this resource. While logging is important, it has not resulted in deforestation. Bhutan has taken cautious steps toward promoting tourism. While the government has encouraged tourists, it has also limited their number. The government fears damage to the land or to its culture if too many tourists visit.

Sri Lanka is a major producer of graphite, the material used for the "lead" in pencils. Other mineral resources include precious and semiprecious stones, which are important exports. Among Sri Lanka's most valuable resources, though, are its climate and rich variety of plant and animal life. These resources can attract tourists. Much of the forests have been cut down for farming or for timber, however.

cowrie shell the protective outer covering of a small sea creature, once used as money in Asia and Africa

Most of Maldives's resources derive from its ocean location. Tourism and fishing are the top two economic activities. For many centuries, the people of Maldives took another resource from the sea: **cowrie shells**. These shells were once sold to traders who used them for money in India, China, and Africa.

☑ **READING PROGRESS CHECK**

Explaining How are the climate zones of Nepal and Bhutan related to their landforms?

LESSON 1 REVIEW

Reviewing Vocabulary

1. *Identifying* Write a paragraph defining the term *aquifer* and explaining why aquifers are important to Maldives.

Using Your Notes

2. *Summarizing* Use your graphic organizer on these countries' physical systems to write a paragraph explaining how life in Nepal and Bhutan differs from that in Sri Lanka and Maldives.

Answering the Guiding Questions

3. *Drawing Conclusions* What landforms help shape life in Bhutan, Maldives, Nepal, and Sri Lanka?

4. *Identifying* How are rivers and the sea vital to human life in Bhutan, Maldives, Nepal, and Sri Lanka?

5. *Drawing Conclusions* How does topography affect the climate in Nepal and Bhutan?

Writing Activity

6. *Informative/Explanatory* In a paragraph, explain how the mountains of Nepal or Bhutan or the oceans surrounding Sri Lanka or Maldives both benefit and cause problems for people of those countries.

later than in the past, have fewer children, and live in nuclear families rather than extended families. People are living longer than in the past as well, which creates the need for support from family members and for government pensions and health care.

In traditional cultures, women often are second-class in status. Society in Nepal has long had a caste system similar to that of India, in which social status is set at birth. Males belonging to the highest caste—a tiny minority of the country's people—dominated society. Women, members of the lower castes, Muslims, and others had few rights. The country has tried to change this by making women's equality before the law a fundamental part of the constitution. Women have yet to make much progress. Women are pushing for laws that would give them the rights to own and inherit property.

In Bhutan, women traditionally enjoyed more rights. In fact, women in the family inherited property to ensure their economic security and make sure that they could care for both children and elderly family members. Women were as likely as men to be the head of the household and their work was valued equally with that of men. However, Bhutanese women play a very small role in government and politics—only one-fourth of the seats in the national legislature are held by women, compared to one-third of those in Nepal. In both Sri Lanka and Maldives, less than 10 percent of lawmakers are women.

Women's rights activists say the situation for women is worsening in Maldives. The rise of religious fundamentalism there threatens women's rights. Violence against women is a major problem, and the government is slow to enact a law against it.

The Arts

The artistic spirit is part of daily life in each of these countries. The Buddhist stupas of Nepal and Sri Lanka and the dzongs of Bhutan are works of art and architecture. They also reveal the importance of religion to these cultures. Dance is a popular cultural expression in Bhutan and Nepal and often tells religious or mythical stories. The Newar people of Nepal have a rich architectural and artistic tradition and built many religious shrines. In literature, Michael Ondaatje, who was born in Sri Lanka, won England's prestigious Booker Prize for his novel *The English Patient*.

☑ **READING PROGRESS CHECK**

Making Connections Why is it thought that the Tamils of Sri Lanka originated in southern India?

In Bhutan and Nepal, people fly colorful prayer flags, and prayer wheels twirl on many corners, sending out invocations. Monks chant mantras, or repetitive prayers.

◀ **CRITICAL THINKING**

1. *Drawing Conclusions* Why is religion a strong force in these people's lives? Explain your answer.

2. *Comparing* What do other religions use that is similar to prayer flags? Why do you think Buddhists use prayer flags?

Source: United Nations Development Programme, 2013

The Human Development Index is based on three dimensions: health (life expectancy at birth); education (mean years of schooling and expected years of schooling); and standard of living (gross national income per capita).

▲ **CRITICAL THINKING**

1. *Comparing and Contrasting* Why do you think the four countries of this subregion rank where they do?

2. *Analyzing* Are there any other factors that you think should be included in a measure of a country's development? Why or why not?

recover to return to a previous level of performance after a decline

Economic Activities

GUIDING QUESTION *How important are farming and fishing to the people of Bhutan, Maldives, Nepal, and Sri Lanka?*

The countries of South Asia have relatively few resources and are somewhat remote. As a result, they are not very economically well developed. Nevertheless, all these countries are connected to the world economy as tourist destinations. Some also export goods or services, and all receive development aid from more developed countries. All joined with Bangladesh, India, and Pakistan in 1985 to form the South Asian Association for Regional Cooperation (SAARC).

In the highlands of Nepal, farmers practice terracing, making use of arable land on the steep slopes. They also cultivate the highland valleys and the low-lying Tarai Plain. Almost three-fourths of the workforce is engaged in agriculture and herding. Chief crops are rice, wheat, and corn. Bhutan, like Nepal, also practices terracing to increase its arable area. Farmers in Bhutan also grow wheat and rice, as well as barley, vegetables, and fruit.

Sri Lanka exports tea, rubber, and coconuts, as well as shrimp, lobsters, and fish. Rice is grown for domestic consumption. In 2004 a tsunami in the Indian Ocean caused widespread destruction in Sri Lanka and Maldives, including the devastation of the fishing industry. Fish catches of tuna and bonitos have declined because large-scale fishing operations from other countries are taking more fish and depleting the supply. The people of Maldives export fruits, vegetables, and coconut for food and oil.

Few major industries exist in these countries due to their lack of technical development. Sri Lanka mines and exports graphite. Precious and semiprecious stones are also mined. Food, tea, and rubber processing plants sell products globally. Bhutan has developed hydroelectric projects and sells electricity to India.

Hiking and white-water rafting are popular activities for tourists in Nepal, which also attracts visitors to its religious sites. Bhutan has moved in recent years to expand tourism. To protect the environment and its traditional culture, however, Bhutan restricts the number of tourists who may visit. In Sri Lanka, the tourism industry has **recovered** since relative peace has been restored. Tourism is a mainstay of the economy of Maldives, which attracts visitors with beautiful beaches and colorful coral reefs. The tourism industry was set back by the 2004 tsunami, which damaged some resorts, but it has since rebounded.

☑ **READING PROGRESS CHECK**

Summarizing How important is agriculture, fishing, and industry to the people of these countries?

LESSON 2 REVIEW

Reviewing Vocabulary

1. *Identifying* Write a paragraph explaining how *dzongs*, lamas, mantras, and stupas are related to one another.

Using Your Notes

2. *Considering Advantages and Disadvantages* Use your graphic organizer on one of these countries to write a paragraph explaining the country's greatest advantages and its biggest challenges.

Answering the Guiding Questions

3. *Explaining* What factors influenced the present-day governments of Bhutan, Maldives, Nepal, and Sri Lanka?

4. *Comparing and Contrasting* How do population density maps of Nepal and Bhutan reflect the rugged landscape of the two countries?

5. *Explaining* How do different cultures and religions coexist in Bhutan, Maldives, Nepal, and Sri Lanka?

6. *Comparing and Contrasting* How important are farming and fishing to the people of Bhutan, Maldives, Nepal, and Sri Lanka?

Writing Activity

7. *Argument* In a paragraph, explain what you think can be done to settle conflicts between groups in Sri Lanka or Bhutan and why you think it would work.

East Asia

UNIT **8**

WanRu Chen/Flickr/Getty Images

① Cities East Asia has several cities with many millions of people, where traditional culture is combined with modern skyscrapers and technology.

EXPLORE the REGION

A dynamic region of booming economies and bustling cities, **EAST ASIA** is also home to one of the world's oldest civilizations and to centuries-old traditions. Several countries in the region are major trading partners with the United States, and by virtue of their size and economic output, some countries in East Asia are among the most powerful and influential in the world.

Bob Charlton/Lonely Planet Images/Getty Images

THERE'S MORE ONLINE

(bl)Yann Layma/Stone/Getty Images,
(br)Michael Snell/Alamy

③ Economy   Manufacturing and trade are major economic activities in East Asia, a region that includes four of the most productive economies in the world.

④ Culture   China's culture, thousands of years old, has influenced other countries in the region. But those countries have also maintained their own traditions, like this Korean dance.

② Farming   The people of East Asia have relied on its rivers for thousands of years to bring water and fertile soil to their fields. Rice, long cultivated here, is a major staple of the diet, but China is also a major producer of wheat.

East Asia
Physical

CENTRAL ASIA

Khüiten Peak
14,350 ft.
(4,374 m) ▲

RUSSIA

N
W   E
S

Sea of Okhotsk

ALTAY SHAN

Junggar Basin

TIAN SHAN

Hentiyn Nuruu

Kerulen R.

Da Hinggan Ling

Xiao Hinggan Ling

Songhua R.

Amur R.

PAMIRS

Tarim Basin

TAKLIMAKAN DESERT

Turpan Depression -426 ft. (-130 m) ▼

Altun Shan

G O B I

Manchurian Plain

Liao R.

Changbai Shan

Hokkaidō

Karakoram Range

K2 28,250 ft. (8,611 m) ▲

KUNLUN SHAN

Muztag 25,338 ft. (7,723 m) ▲

Qilian Shan

Huang He (Yellow R.)

Mu Us Desert

Bo Hai

Korea Bay

Shandong Peninsula

Korea Peninsula

Sea of Japan (East Sea)

Honshū

Mt. Fuji 12,388 ft. (3,776 m) ▲

H I M A L A Y A

PLATEAU OF TIBET

Changjiang

Yangtze R.

Salween R.

Mekong R.

Wei R.

Qin Ling

Grand Canal

North China Plain

Yellow Sea

Korea Strait

Shikoku

Kyūshū

140°E

Mt. Everest 29,028 ft. (8,848 m) ▲

Brahmaputra R.

Sichuan Basin

Chang Jiang (Yangtze R.)

Xiang R.

Gan R.

East China Sea

Ryukyu Islands

SOUTH ASIA

80°E

Yunnan Plateau

Xi R.

Red R.

Taiwan Strait

Taiwan

Yü Shan 13,113 ft. (3,997 m) ▲

TROPIC OF CANCER

20°N

Gulf of Tonkin

Salween R.

Mekong R.

Leizhou Peninsula

Hainan

Luzon Strait

PACIFIC OCEAN

Hainan

South China Sea

Philippine Sea

UNIT 8

REFERENCE ATLAS

MAP STUDY

1. *Physical Systems* What physical feature separates Mongolia from China in the southeast?

2. *Human Systems* Which Chinese cities are located along the Chang Jiang?

SOUTHEAST ASIA

EQUATOR 0°

INDIAN OCEAN

0   1,000 miles
0   1,000 kilometers
Lambert Azimuthal Equal-Area projection

100°E

120°E

Elevations

10,000 ft. (3,000 m)	
5,000 ft. (1,500 m)	
2,000 ft. (600 m)	
1,000 ft. (300 m)	
0 ft. (0 m)	
Below sea level	

──── National boundary

⊥⊥⊥⊥ Canal

▲ Mountain peak

▼ Lowest point

East Asia
Political

60°N

RUSSIA

**CENTRAL
ASIA**

*Sea of
Okhotsk*

N
W   E
S

• Ürümqi

Ulaanbaatar ✪

Kerulen R.

MONGOLIA

• Harbin

Songhua R.

Amur R.

• Sapporo

40°N

• Kashi

Liao R.

Shenyang •

**NORTH
KOREA**

*Sea of Japan
(East Sea)*

JAPAN

Beijing ✪

Huang He (Yellow R.)

Tianjin •

Dalian •

*Korea
Bay*

P'yŏngyang ✪

Seoul ✪

Taegu •

**SOUTH
KOREA**

Pusan •

Kōbe •

✪ Tokyo
Nagoya •

• Osaka

Hiroshima •

• Lanzhou

Wei R.

Qingdao •

*Yellow
Sea*

Fukuoka •

• Xi'an

CHINA

Nanjing •

Shanghai •

*East
China
Sea*

140°E

• Lhasa

Chengdu •

Wuhan •

Brahmaputra R.

**SOUTH
ASIA**

80°E

Chongqing •

Chang Jiang (Yangtze R.)

Xiang R.

Gan R.

Fuzhou •

TROPIC OF CANCER

Taiwan Strait

• Taipei ✪

• Kunming

Xi R.

Guangzhou •

Macau •

Hong Kong •

TAIWAN

Kao-hsiung •

20°N

Salween R.

Red R.

*Gulf of
Tonkin*

Luzon Strait

**PACIFIC
OCEAN**

*Bay of
Bengal*

*South
China
Sea*

*Philippine
Sea*

✪ National capital

• Major city

Mekong R.

**SOUTHEAST
ASIA**

EQUATOR   0°

**INDIAN
OCEAN**

0   1,000 miles

0   1,000 kilometers

Lambert Azimuthal Equal-Area projection

100°E

100°E

East Asia
Climate and Vegetation

Climate
- Tropical wet/dry
- Semi-arid (steppe)
- Arid (desert)
- Humid subtropical
- Humid continental
- Subarctic
- Tundra and high altitude

0 — 1,000 miles
0 — 1,000 kilometers
Lambert Azimuthal Equal-Area projection

RUSSIA

CENTRAL ASIA
- Ürümqi
- Kashi
- Ulaanbaatar

Sea of Okhotsk

Sea of Japan (East Sea)

- Beijing
- P'yŏngyang
- Seoul
- Tokyo

Bo Hai
Korea Bay
Korea Strait

Yellow Sea

- Shanghai

East China Sea

SOUTH ASIA
- Lhasa
- Chongqing

Bay of Bengal

- Taipei
- Kao-hsiung

Taiwan Strait

TROPIC OF CANCER

- Hong Kong

Luzon Strait

Gulf of Tonkin

South China Sea

Philippine Sea

PACIFIC OCEAN

SOUTHEAST ASIA

EQUATOR

Vegetation
- Tropical rain forest
- Desert scrub and desert waste
- Temperate grassland
- Deciduous forest
- Coniferous forest
- Mixed forest (deciduous and coniferous)
- Highland (vegetation varies with elevation)

East Asia
Economic Activity

Resources

Coal		Cobalt	
Petroleum		Nickel	
Natural gas		Copper	
Iron ore		Lead	
Tin		Manganese	
Zinc		Gold	
Bauxite		Silver	

Land Use

- Commercial farming
- Subsistence farming
- Nomadic herding
- Primarily forest
- Manufacturing and trade
- Commercial fishing
- Little or no activity

CENTRAL ASIA

RUSSIA

SOUTH ASIA

SOUTHEAST ASIA

Bay of Bengal

INDIAN OCEAN

Sea of Okhotsk

Sea of Japan (East Sea)

Korea Bay

Yellow Sea

East China Sea

South China Sea

Philippine Sea

PACIFIC OCEAN

Gulf of Tonkin

Taiwan Strait

Luzon Strait

TROPIC OF CANCER

EQUATOR

60°N

40°N

140°E

20°N

80°E

100°E

120°E

0°

UNIT 8
REFERENCE ATLAS

MAP STUDY

1. **Physical Systems** Which climate type is found in all of the region's countries?

2. **Human Systems** What is the predominant type of land use in North Korea, South Korea, and Japan?

0   1,000 miles

0   1,000 kilometers

Lambert Azimuthal Equal-Area projection

East Asia
Population Density

RUSSIA

N
W E
S

CENTRAL ASIA

Ürümqi

Qiqihar
Daqing
Jiamusi
Harbin
Changchun
Jilin
Chifeng
Shenyang
Anshan
Hohhot
Huludao
P'yŏngyang
Baotou
Beijing
Tianjin
Dalian
Shijiazhuang
Yantai
Seoul
Taiyuan
Incheon
Daejon
Nagoya
Xining
Jinan
Zibo
Qingdao
Pusan
Kyōto
Gifu
Lanzhou
Handan
Linyi
Gwangju
Osaka
Hiroshima
Zhengzhou
Zaozhuang
Xuzhou
Yellow Sea
Fukuoka
Tianshui
Xi'an
Suzhou
Huaiyin
Nanyang
Huainan
Nanjing
Mianyang
Hefei
Wuxi
Shanghai
Chengdu
Nanchong
Wuhan
Ningbo
East China Sea
Neijiang
Chongqing
Nanchang
Hangzhou
Lupanshui
Changsha
Wenzhou
Guiyang
Hengyang
Fuzhou
Kunming
Quanzhou
Xiamen
Taipei
Nanning
Guangzhou
Shantou
Kao-hsiung
Yulin Macau
Shenzhen
Hong Kong
Zhanjiang

Sea of Okhotsk
Sapporo
Sendai
Tokyo-Yokohama

Sea of Japan (East Sea)

SOUTH ASIA

PACIFIC OCEAN

TROPIC OF CANCER

Luzon Strait

Gulf of Tonkin

Bay of Bengal

South China Sea

Philippine Sea

SOUTHEAST ASIA

EQUATOR

60°N
40°N
20°N
80°E
100°E
120°E
140°E
0°

0   1,000 miles
0   1,000 kilometers
Lambert Azimuthal Equal-Area projection

Cities
(Statistics reflect metropolitan areas.)
- ■ Over 5,000,000
- □ 2,000,000–5,000,000
- ◉ 1,000,000–2,000,000

POPULATION

Per sq. mi.	Per sq. km
1,250 and over	500 and over
250–1,249	100–499
63–249	25–99
25–62	10–24
2.5–24	1–9
Less than 2.5	Less than 1

UNIT 8
REFERENCE ATLAS

MAP STUDY

1. *Human Systems* Is it fair to say that East Asia is a region of huge cities? Why or why not?

2. *Places and Regions* How does the population density in western China compare to the population density in eastern China?

676

China and Mongolia

networks

There's More Online about the geography of China and Mongolia.

CHAPTER 28

ESSENTIAL QUESTION · *How do physical systems and human systems shape a place?*

Geography Matters...

China is one of the largest and most influential countries in the world. With a civilization thousands of years old, this country of geographical contrasts has more than a billion people.

China's mammoth economy exports goods globally. However, environmental problems resulting from the country's rapid growth have given rise to domestic and international challenges.

The center of a vast empire, Mongolia was later ruled by China. A land of windswept plains and mountains, today independent Mongolia faces challenges of urbanization and development.

◄ This woman stands in the Forbidden City located in the center of Beijing, China.

Jason Hosking/Taxi/Getty Images

Why Geography Matters: **China and Mongolia**

China's growing energy demands

Since 1978 China has been shifting from an agriculture-based economy to an industrial economy. Today, Chinese workers manufacture clothes, shoes, cars, bicycles, ships, planes, washing machines, televisions, computers, cement, steel, iron, toys, and many other products for export. China's surging industrial growth has resulted in an increasing demand for energy, which is needed to power thriving cities and bustling factories from Guangzhou to Shenyang.

Why is China consuming more energy resources?

Since the 1970s, China's consumption of oil, natural gas, coal, and other sources of energy has climbed steadily. One reason for this increase in demand is China's population. With more than 1.3 billion people, China has the largest population in the world. An even more important factor is that China has developed its economy through industrialization, which causes energy demand to increase. Following this path of development, China will continue to require more energy to generate electricity for homes, schools, and businesses and to fuel cars, trucks, and buses. As of 2013, China had the second-largest economy in the world, behind the United States.

1. Human Systems What has caused increased consumption of energy in China?

How can China meet the growing demand for energy?

Currently, China uses more of the world's energy than any other country. In 2011, for example, China was the largest consumer of coal and the second-largest consumer of oil in the world. To meet its demand for energy that will keep its economy growing, China must supply enough affordable oil, coal, and other types of energy to its people and industries. To supplement its own reserves of oil, coal, and natural gas, China must also import these fuels. In addition, the Chinese are investing in products and technologies—such as solar power, LED lighting, fuel-efficient cars, and advanced coal technology—that either use or provide energy more efficiently. By improving energy efficiency, China can shrink its energy consumption and make sure it has adequate supplies of energy for the future.

2. Places and Regions How can China ensure continued economic growth?

What challenges does China face in supplying its future energy needs?

China uses coal to generate nearly 80 percent of its electricity. In 2010 its demand for coal caused a 75-mile traffic jam involving 10,000 trucks carrying coal from Inner Mongolia. The government's twelfth Five-Year Plan lays out guidelines for reducing consumption, increasing efficiency, and developing alternative sources such as solar, nuclear, water, and wind. China—one of the world's largest producers of electricity from wind power—can harness the wind energy created by its long coastline and large landmass. With many rivers and mountains, China has become the world's leading producer of hydroelectric power, or energy from flowing water.

3. Environment and Society How could developing new energy sources help China meet its energy demands?

(l)View Stock/Getty Images, (c)GFC Collection/age fotostock, (r)Lou Linwei/Alamy

YESTERDAY ca. 1980s

As China industrialized, it relied heavily on fossil fuels and non-renewable resources such as oil, natural gas, and coal. Emissions from vehicles remained low, as many relied on animal transportation in rural areas or bicycles in urban areas.

*
- 85% COAL
- OIL
- HYDROELECTRIC
- 14%
- 1%

TODAY

While China is making strides in the implementation of green technology, it continues to have issues with pollution due to a heavy reliance on fossil fuels. Greater access to cars and trucks has also increased emissions.

- 70% COAL
- OIL
- 19%
- HYDROELECTRIC
- 6%
- NATURAL GAS
- 4%
- NUCLEAR
- 1%

TOMORROW

In the future, China has planned to increase its usage of solar, nuclear, and wind energy sources. However, as its population continues to grow, China's ability to reduce overall emissions remains in question.

*
- 65% COAL
- OIL
- 22%
- HYDROELECTRIC
- 8%
- NUCLEAR
- 3%
- RENEWABLES
- 2%

*Data for Yesterday and Tomorrow are approximations.

networks

There's More Online!

☑ **IMAGE** Giant Panda Eating Bamboo

☑ **IMAGE** Terraced Rice Fields

☑ **MAP** China's River: Huang He

☑ **MAP** East Asian Monsoons

☑ **MAP** Physical Geography of China and Mongolia

☑ **INTERACTIVE SELF-CHECK QUIZ**

☑ **VIDEO** Physical Geography of China and Mongolia

Reading **HELP**DESK

Academic Vocabulary

- **symbol**

Content Vocabulary

- **range**
- **loess**
- **monsoon**
- **typhoon**

TAKING NOTES: *Key Ideas and Details*

IDENTIFYING China's landforms and bodies of water are regularly affected by weather systems. As you read, use a graphic organizer like the one below to note the main landforms, water systems, and climate regions.

China's Physical Geography		
Landforms	Water Systems	Climate Regions

LESSON 1

Physical Geography of China and Mongolia

ESSENTIAL QUESTION · *How do physical systems and human systems shape a place?*

IT MATTERS BECAUSE

China is a large land of many contrasts: high mountains, grassy steppes, broad plains, tropical rain forests, and cold, dry deserts. Great rivers nourish the soil to support agriculture that feeds millions, but monsoons can cause dangerous floods. Mongolia has large areas of upland plateau that are deserts, as well as mountainous regions and basins. Most of its land consists of extensive grasslands used for livestock herding.

Landforms

GUIDING QUESTION *What physical features dominate China and Mongolia, and how have they affected human geography?*

East Asia spans the snowy peaks of the Himalaya in the west all the way to the Japanese islands in the east. China makes up the majority of the landmass in this subregion. Rivers, hills, and plains occupy the east. The southwest contains high plateaus, and the northwest stretches wide into dry deserts. Mongolia is mostly an upland plateau with mountains in the north and south interrupted by basins.

The landforms of China can be divided into two parts: the mountains and plateaus of the west and the plains and hills of the east. In far western China, a mountain **range** called the Pamirs forms a node from which several other ranges radiate. This remote interior region includes the Kunlun Shan and Tian Shan ranges. Between them stretches the Tarim Basin, characterized by deserts and salt marshes. A desert, the Taklimakan, covers most of the Tarim Basin. Another massive desert, the Gobi, is found in north-central China and southern Mongolia. It is the source of frequent dust storms that plague these areas. Rain seldom falls in the Gobi—less than 3 inches (7.5 cm) a year.

The Altay Shan range, farther north, lies along part of the border between China and Mongolia. China is separated from South Asia by the Himalaya, the world's highest mountains and the location of the world's tallest peak, Mount Everest, at 29,028 feet (8,848 m) on the border of China and Nepal. The country's lowest point, the Turpan Depression, is found in western China near Mongolia and lies 426 feet (130 m) below sea level.

(tl)©Nature Picture Library/Alamy, (tr)Rodrigo A. Torres/Glowimages

The highest plateau in East Asia is the Plateau of Tibet (or the Plateau of Xizang), which forms a large part of China's southwest. It has an average elevation of 16,000 feet (4,875 m). North and east of the Plateau of Tibet lie other rugged highlands, although with lower elevations. The mountains, plateaus, and great deserts in China's west and south have formed natural barriers that helped protect its inhabitants from land invasion for thousands of years, but also isolated China from peaceful contact with foreigners. For centuries China's isolation served to protect the country, allowing it to develop its unique culture.

The eastern part of China is characterized by vast lowland plains and some hills. It contains most of China's population because the land is conducive to farming and settlement. The Manchurian Plain (or Northeast Plain) and the North China Plain are two of the most important of the numerous plains areas.

In addition to the southern upland plateau and deserts, Mongolia has areas of mountains in the north and the south with basins in between. Some contain remnants of extinct volcanoes, also found on the far eastern plateau.

range a group of mountains

✓ **READING PROGRESS CHECK**

Identifying Which mountain ranges originate in the Pamirs?

Water Systems

GUIDING QUESTION *How do China's major rivers influence human systems?*

China's major rivers originate on the Plateau of Tibet and flow eastward, down from the plateau and across the lowland plains to the Pacific Ocean. In northern

GEOGRAPHY CONNECTION

China's immense size covers many types of landforms and physical features.

1. *THE WORLD IN SPATIAL TERMS*
   Approximately how wide is the Plateau of Tibet?

2. *PLACES AND REGIONS* Which desert spans both Mongolia and China? What is the elevation of this desert?

Physical Geography of China and Mongolia

KAZAKHSTAN, KYRGYZSTAN, TIAN SHAN, TAKLIMAKAN DESERT, KUNLAN SHAN, Qaidam Basin, QILIAN SHAN, PLATEAU OF TIBET, HIMALAYA, NEPAL, BHUTAN, Mt. Everest 29,028 ft. (8,848 m), TROPIC OF CANCER, BANGLADESH, INDIA, MYANMAR (BURMA), Bay of Bengal, LAOS, THAILAND, VIETNAM, Hainan, South China Sea, Gonga Shan 24,790 ft. (7,556 m), Sichuan Basin, Chang Jiang (Yangtze R.), CHINA, Huang He (Yellow R.), Beijing, Bo Hai, Grand Canal, North China Plain, Yellow Sea, NORTH KOREA, SOUTH KOREA, East China Sea, JAPAN, Sea of Japan (East Sea), TAIWAN, PACIFIC OCEAN, ALTAY SHAN, Khüiten Peak 14,350 ft. (4,374 m), Selenge R., Kerulen R., MONGOLIA, GOBI, Manchurian Plain, Turpan Depression -426 ft. (-130 m)

Elevations
- 10,000 ft. (3,000 m)
- 5,000 ft. (1,500 m)
- 2,000 ft. (600 m)
- 1,000 ft. (300 m)
- 0 ft. (0 m)
- Below sea level

— National boundary
⊥⊥⊥⊥ Canal
▲ Mountain peak
▼ Lowest point

0 — 500 miles
0 — 500 kilometers
Lambert Azimuthal Equal-Area projection

China and Mongolia **681**

Bamboo is an important plant to people and animals.

▲ **CRITICAL THINKING**

1. *Identifying Central Issues* Why is bamboo so important to China?

2. *Analyzing* What characteristics of bamboo make it so useful?

loess fine, yellowish-brown topsoil made up of particles of silt and clay, carried by the wind

symbol a sign or image that stands for an idea

China, the major river system is the Huang He, or Yellow River. The river gets its name from the sediment load of yellowish-brown topsoil it carries. This fine topsoil is called **loess**, and it covers large areas in northern China. The loess was created by wind erosion that deposited the fine soil at the end of the last ice age. When the soil is eroded by the Huang He, the yellow loess is carried into the water. The combination of nutrient-rich loess and water flowing into the North China Plain makes this area ideal for farming wheat and soybeans. Over the years, though, flooding of the Huang He has killed hundreds of thousands of people, earning it the nickname "China's sorrow."

The longest of central China's rivers is the Chang Jiang, or Yangtze River, which flows for 3,434 miles (5,525 km) through dramatic gorges as it comes off the Plateau of Tibet and then across the wide plains of eastern China. It empties into the ocean near Shanghai. More than half of China's rice and other grains are watered by the Chang Jiang, which is also a major transportation route. The recent construction of the Three Gorges Dam helps control flooding along the lower portions of the river and provides water for irrigation and hydroelectric power.

The Xi River and its tributaries form the largest major river system in southern China. This river system is also known as the Pearl River. The ports of Guangzhou, Hong Kong, and Macau benefit from fertile soil deposited there to form a great delta. This area is undergoing some of the most rapid development in China.

China's Grand Canal is the world's longest human-made waterway. Its construction was begun in the 400s B.C. and has been expanded and rebuilt. Today it carries goods and people from Beijing to Hangzhou, some 1,000 miles (1,609 km).

Most of Mongolia's rivers are located in the north and are difficult to harness for transportation or hydroelectric power. The Orhon is the longest, at almost 700 miles (1,130 km). Lakes in Mongolia are often saltwater or temporary.

☑ **READING PROGRESS CHECK**

Summarizing What process makes the North China Plain so fertile?

Climates, Biomes, and Resources

GUIDING QUESTION *What defines the climates, natural resources, and biomes of China?*

China's climates are extremely diverse, ranging from tropical in the south to subarctic in the north. Climates are shaped by the interaction of latitude and physical features such as mountains, highlands, and coastal regions. Each type of climate supports different vegetation. Temperate, humid forests dominate the south and east, while the north and west are characterized by dry highlands and grasslands.

A humid subtropical climate, with hot summers and heavy rains, characterizes the midlatitudes of much of eastern China and the island of Taiwan. Deciduous trees and evergreens with broad leaves both flourish here. The tree-like grass bamboo grows in many warmer areas. This fast-growing plant—some species can grow as much as one foot per day—is also hardy and versatile. It is used for everything from medicine, food, and decoration to building material. It also provides the only source of food for two rare mammals: the giant panda and the smaller red panda. Other native plants with important economic value are the mulberry tree, whose leaves are food for silkworms, and the tea bush. Tea, silk, and bamboo have long been important in China and have come to stand as **symbols** of Chinese culture, as has the panda.

©Nature Picture Library/Alamy

Deserts have formed in China's north and west where surrounding mountains create a rain shadow, blocking the moist coastal winds. These deserts stretch across Mongolia into inland northern China. Deserts—such as the Gobi and the Taklimakan—and the steppe lands experience temperature extremes. Daytime temperatures can fall as much as 55°F (30.5°C) overnight. Temperatures in the Gobi range from 100°F to –30°F (38°C to –34°C), although they average 73°F (23°C) in summer and 0°F (–18°C) in winter. The large steppe climate regions east of the Gobi display a natural vegetation of grasses, with few trees.

The climate in mountainous areas changes according to elevation. In most cases, the higher the elevation, the cooler the temperature is. For example, the average annual high temperature on the Plateau of Tibet is a cool 58°F (14°C), because the elevation is very high—averaging about 16,000 feet (4,875 m). Lower mountain slopes bloom with meadows, grass, flowers, and trees. No trees grow above the timberline; only mosses and colorful lichens survive here.

Lush rain forest covers much of Hainan, an island off China's southern coast. This tropical wet climate has high temperatures year-round. In summer, the rainy **monsoons**, or seasonal winds, blow in. Palms and tropical hardwoods grow alongside tropical fruit trees and broad-leaved evergreens.

Seasonal weather patterns in China are created by the interactions that occur between the air mass above the continent and the air mass above the Pacific Ocean. The meeting of these air masses results in monsoons that blow in the same direction for about half the year before switching directions. Heat and humidity from the Pacific Ocean arrive in China during the summer monsoon, which blows from southeast to northwest. As a result, more than 80 percent of the subregion's

monsoon in Asia, a seasonal wind that brings warm, moist air from the oceans in summer and cold, dry air from land in winter

GEOGRAPHY CONNECTION

In China, monsoons blowing westward bring warm, moist air in the summer. Monsoons blowing eastward bring cold, dry air in the winter.

1. *PHYSICAL SYSTEMS* Winds originating on the Plateau of Tibet have what characteristics?

2. *ENVIRONMENT AND SOCIETY* How do summer monsoon rains affect farms near the coast?

East Asian Monsoons

Winter monsoons
Summer monsoons
Extent of summer monsoon precipitation

There are many areas of rich farmland in China. In the photo, rice fields have been planted on terraced slopes in Guangxi.

▲ **CRITICAL THINKING**

1. Analyzing What is the benefit of terracing?

2. Synthesizing What geographic conditions make China such a successful crop producer?

typhoon a violent tropical storm that forms in the western Pacific Ocean, usually in late summer

annual rainfall occurs in heavy downpours from April through October. Farmers depend on these summer monsoons to water their crops. If monsoons come late or bring less rainfall, crops fail. When too much rain comes, flooding can result. From November to March, the winter monsoon brings cold, dry arctic air across China as it blows from the land onto the ocean.

Large, violent storms with high winds, known as **typhoons**, result when warm, humid air over the Pacific Ocean moves onto land. Typhoons are also called tropical cyclones and are similar to hurricanes. These intense spiral storms with their storm surges can raise coastal waters up to 20 feet (6 m) above normal, creating serious flooding along the coasts. As with hurricanes in the Atlantic and Caribbean, typhoon season peaks between late August and October.

Far from the ocean, Mongolia has a continental climate with long, cold winters and short, cool-to-hot summers. Mostly arid to semi-arid, annual rainfall in desert areas is less than 4 inches (10 cm), with 14 inches (35 cm) in the northern forested mountains.

China has a wealth of mineral resources, including iron ore, tin, tungsten, and gold. The South China Sea and the Taklimakan contain large petroleum deposits. Northeastern China has abundant coal deposits. Other natural resources include natural gas, mercury, aluminum, lead, zinc, and uranium. Mongolia has a number of minerals, including coal, gold, and copper, as well as oil and uranium deposits.

China is the world's leading producer of rice. The "rice bowl" in southern China yields two harvests of rice per year. Agricultural income as a percentage of gross domestic product (GDP) equaled 10 percent in 2011. In fact, China is the world leader in the gross value of its farm output for rice, wheat, potatoes, corn, peanuts, tea, millet, barley, apples, cotton, oilseed, pork, and fish. East Asia, including China, boasts the world's biggest deep-sea fishing industries.

☑ **READING PROGRESS CHECK**

Making Connections How can monsoons affect China's agricultural output?

LESSON 1 REVIEW

Reviewing Vocabulary

1. *Using Context Clues* Write a sentence explaining how loess affects agriculture in China.

Using Your Notes

2. *Comparing and Contrasting* How do the water systems in China affect its climates compared to the effects of weather systems?

Answering the Guiding Questions

3. *Identifying Central Issues* What physical features dominate China and Mongolia, and how have they affected human geography?

4. *Analyzing* How do China's major rivers influence human systems?

5. *Synthesizing* What defines the climates, natural resources, and biomes of China?

Writing Activity

6. *Informative/Explanatory* In a paragraph, explain the possible effects of typhoons and monsoons on the Chinese people and their economy.

Rodrigo A. Torres/Glowimages

(tl)©Hulton-Deutsch Collection/Corbis, (tcl)Catherine Henriette/AFP/Getty Images, (tcl)Martin Bureau/AFP/Getty Images, (tcr)Justin Guariglia/National Geographic/Getty Images, (trl)Lane Oatey/Getty Images

net**w**orks

There's More Online!

☑ **GRAPHS** Grain Production in China, 1950–1970

☑ **GRAPHS** U.S. Exports to China and Imports from China

☑ **IMAGE** Chinese Schoolgirl

☑ **IMAGE** Mongolian Flag

☑ **TIME LINE** Modern China

☑ **INTERACTIVE SELF-CHECK QUIZ**

☑ **VIDEO** Human Geography of China and Mongolia

Reading **HELP**DESK

Academic Vocabulary

- **impact**
- **institute**
- **dominate**

Content Vocabulary

- **dynasty**
- **aborigine**
- **ideogram**
- **atheist**
- **commune**
- **merchant marine**
- **dissident**
- **economic sanctions**
- **Special Economic Zone (SEZ)**

TAKING NOTES: *Key Ideas and Details*

DESCRIBING As you read about the human geography of China, use a graphic organizer like the one below to identify how history, population, culture, and economics have created the China of today.

LESSON 2
Human Geography of China and Mongolia

ESSENTIAL QUESTION · *How do physical systems and human systems shape a place?*

IT MATTERS BECAUSE

China's long history and geography have shaped its culture. The ideas on which its government is based have had long-lasting effects. The country's economic life reveals how it fulfills the daily needs of its citizens and interacts with the larger world. Mongolia, once the world's largest land empire, today faces modern challenges.

History and Government

GUIDING QUESTION *What influence did ancient Chinese history have on government, culture, and daily life in modern times?*

Many powerful transformations have occurred during China's history. Under leaders ranging from nomadic warriors to long-ruling **dynasties**, the country has endured profound political and cultural changes.

China's culture spans more than 5,000 years. Archaeological evidence indicates it began in the Wei River valley. Historical records were begun when invaders established a dynasty, or ruling family, around 1766 B.C. This Shang dynasty arose on the North China Plain. They faced attacks by nomads from Central Asia, rebellions by local nobles, and natural disasters. In China, dynasties were believed to rule under the "mandate of heaven," or the approval of the gods and goddesses. When the people suffered, it was assumed that the dynasty had lost its mandate. The Shang dynasty ruled for over 700 years, coming to an end in 1046 B.C.

The Zhou (JOH) dynasty then took control of the region, and continued for the next 800 years. Under its rule, trade grew, Chinese culture spread, and the making of iron tools began. Crossbows, ox-drawn plows, and horseback riding were introduced, as were widespread irrigation and other efforts to control water. This helped to increase crop yields.

China's most famous teacher and philosopher, Confucius (or Kongfuzi), lived during the Zhou dynasty. He founded a system of thought called Confucianism. It is based on discipline and proper moral conduct, and continues to have an **impact** in China and other Asian civilizations to the present day.

dynasty a ruling house or continuing family of rulers

impact an effect on something

Around the same time, a thinker named Laozi (or Lao-Tzu) helped found another major Chinese philosophy called Daoism. This philosophy emphasizes harmony with nature and simple living. Daoism is still followed in modern China.

Qin Shihuangdi, the first emperor of the Qin dynasty (221–206 B.C.), was the first to unify the country. To ward off attacks from peoples from Central Asia, he joined together and added sections to existing defensive walls, thus establishing the Great Wall of China. Traders and missionaries under the Han (206 B.C.–A.D. 220) and Tang (A.D. 618–907) dynasties spread Chinese culture throughout East Asia. The explorer Zheng He (JUHNG HUH) sailed as far as the coast of East Africa under the Ming dynasty in the early 1400s. From the mid-1600s to the early 1900s, the Qing dynasty ruled China.

By the 1600s, Europeans had set up trade routes to China and other parts of East Asia to gain access to the region's silk and tea trade. China rejected these efforts. In the 1800s, frustration prompted some European countries to force China to open more ports. Later, European governments and Japan divided large areas of China into spheres of influences, or areas over which they had exclusive trading rights.

China Today

The Chinese dynasties ended in 1911 after a revolution led by Sun Yat-sen. A tumultuous period followed in which warlords competed for control. When Sun died in 1925, Chiang Kai-shek came to power and formed the Nationalist government of the Republic of China. Chiang's rival, however, a Communist named Mao Zedong, won the support of China's farmers. Civil war erupted, and the Communists eventually took power in 1949. They set up the People's Republic of China on the mainland. The Republic of China was on Taiwan, where Nationalists had fled after the war.

institute to organize and establish

In the 1950s, the Communist government **instituted** the "Great Leap Forward." Small-scale farms were replaced by large government farms as part of economic and social change. However, the economy faltered and the new farms failed to produce enough food, causing millions to die of starvation.

Despite these failures, Mao still dreamed of a classless society. He believed that only an atmosphere of constant revolutionary fervor could enable the Chinese to

©Hulton-Deutsch Collection/Corbis

TIME LINE ⌄

Modern CHINA ➜

Since the mid-twentieth century, China has been transformed from an impoverished agrarian society to the economic powerhouse of Asia. However, these profound changes have come at a great cost.

▶ **CRITICAL THINKING**

1. **Describing** How did China try to speed up its change from an agrarian society to an industrial society?

2. **Analyzing** What challenges has China faced during its transformation?

Mao Zedong proclaims the establishment of the People's Republic of China.

1949

Cultural Revolution begins with the aim of strengthening communism in China by rooting out capitalist and traditionalist elements.

1966

1958 Great Leap Forward begins a massive government effort to industrialize China.

China issues Open Door Policy to encourage foreign investment and development of a market economy.

1986

achieve the final stage of communism. In 1966 Mao launched the Great Proletarian Cultural Revolution in which urban and educated Chinese were sent to work on rural farms. The goal was to cleanse Chinese society of elements thought to be guilty of embracing capitalist ideas.

In the late 1970s, Deng Xiaoping, with other new government leaders, encouraged modernization and limited capitalism. They made some businesses and farms available for private ownership. Officials also began to allow foreign businesses and technology into the country. Free-market influences and the modernization they represented have gradually opened China's economy and society further.

Taiwan and Mongolia

Tensions have remained high between the Republic of China on Taiwan and the People's Republic of China on the mainland. Both have desired to reunite since the 1950s, but they cannot agree on how this might occur. Taiwan has embraced democracy and industry, but China has not reached the same levels of change. In recent years, Taiwan has invested billions of dollars in factories located on the mainland. China receives key computer and electronics parts from Taiwan. Such business interests have created an intermingling of the Chinese and Taiwanese economies.

Early inhabitants of Mongolia were the Xiongnu, a contemporary rival empire of the Qin and Han dynasties in China. Later, under Genghis Khan, the Mongol Empire with its capital in Karakorum took over much of China, Russia, and Central Asia. After the empire fell, the Qing claimed it as the Chinese province of Outer Mongolia. When China's Qing dynasty was overthrown in 1911, the modern country of Mongolia was created. The Mongolians had been loyal to the Qing dynasty, but they refused to be part of the new China. From 1924 to 1991, the Soviet Union pressured Mongolia to become and remain a Communist state. When Soviet communism collapsed in the 1990s, however, Mongolia adopted a democratic constitution.

☑ **READING PROGRESS CHECK**

Describing How did the Great Leap Forward affect China?

Mongolia became an independent country in 1921. The current flag was adopted in 1992, when Mongolia became a democracy.

▲ **CRITICAL THINKING**

1. *Drawing Conclusions* Why do you think Mongolians did not want to be part of the People's Republic of China?

2. *Making Inferences* Why might the Soviet Union have wanted Mongolia to be a Communist state?

(t)©Digital Archive Japan/Alamy, (c)Catherine Henriette/AFP/Getty Images, (cr)Martin Bureau/AFP/Getty Images, (b)Yulia Zhukova/age fotostock

1989 Troops open fire on pro-democracy demonstrators in Beijing's Tiananmen Square.

1990 ➤

1997 After more than 150 years of British rule, the United Kingdom returns Hong Kong to Chinese control.

2001 China joins the World Trade Organization.

2006 Three Gorges Dam is completed, providing hydroelectric power and flood control on the Chang Jiang.

2008 China hosts Summer Olympics in Beijing.

2011 China overtakes Japan to become world's second-largest economy.

2010 ➤

Population Patterns

GUIDING QUESTION *How has the migration of rural dwellers to urban areas affected China's population?*

The Han dynasty gave its name to what is today the largest ethnic group in China. Ethnic Han Chinese comprise approximately 92 percent of China's more than 1.3 billion people. About 55 ethnic groups make up the remaining 8 percent.

Non-Chinese ethnic groups governed by China have their own distinct cultures and histories. Tibet, once a Buddhist kingdom, was taken over by China in 1959. Tensions between the Chinese and the Tibetans continue to simmer.

Taiwan and China share a long history. Chinese people who immigrated to Taiwan hundreds of years ago make up the majority of the inhabitants there. Only about 2 percent of Taiwan's people are **aborigines**, or original inhabitants.

aborigine an area's original inhabitant

Like China, Mongolia has a largely homogeneous population. Most inhabitants of Mongolia are ethnic Mongolians. About 80 percent are Khalka Mongols who speak Mongolian. Other Mongolians belong to different linguistic groups.

Although China covers a large territory, 90 percent of the people live on just one-sixth of the land. Much of this is in the fertile valleys and plains of the Huang He, Chang Jiang, and Xi, China's three great rivers. River valleys and coastal plains are the location of many large urban centers, such as Shanghai, Beijing, Tianjin, and Guangzhou. In contrast, Mongolia claims a population density of only 5 people per square mile (2 people per sq. km) over its vast steppe regions. Most of Mongolia's 2.7 million people live in the north-central part of the country.

While many Chinese still live and work on farms, the migration of millions of people to quickly growing urban areas continues to increase. This urbanization has resulted in overcrowding in cities. Because so many people have left the countryside to live in cities, farms are experiencing labor shortages. To remedy this, China has built dozens of new agricultural towns in remote areas. Such towns provide more social services to attract rural people, with a higher quality of life.

The Chinese government has reduced population growth. A policy of one child per married couple was begun in 1979. This policy has reduced the number of people born since then. While the current population of China is 1.36 billion, it is no longer growing as rapidly. China now adds about 5.3 million people to its population per year. If the one-child policy is maintained, population growth is projected to decline within the next 10 years, and the population is likely to begin to decrease. This is already resulting in an aging population in which older people are a higher proportion of the population.

☑ **READING PROGRESS CHECK**

Summarizing What ethnicity and occupation does a typical Chinese person have, and where do most Chinese people live?

Society and Culture Today

GUIDING QUESTION *What cultural beliefs, practices, and distinctive achievements and attitudes characterize China?*

China's government has encouraged efforts to increase literacy among the general population. In the past, only wealthy people learned to read and write. Efforts to improve the literacy rate suffered a setback during the upheavals of the Cultural Revolution. At that time, people suspected of being enemies of Mao's form of communism were persecuted. Schools and factories were closed, and educated people were seen as a threat. After Mao's death, these policies were reversed. The government began emphasizing education again, and literacy rates are increasing.

Mandarin, a dialect of Han Chinese, **dominates** spoken language in China. Written Chinese uses **ideograms**, symbolic pictures that stand for ideas, unlike Western languages, which use letters to represent sounds. Compared to letters, ideograms have both an advantage and a disadvantage. The advantage is that people who do not speak the same dialect can still communicate through writing. The disadvantage is that Chinese writing is difficult to learn, since there are thousands of ideograms. However, the written language has been a major factor in China's cultural unity.

Life expectancy has risen to over 70 years due to better health care. As a general rule, Communist governments pay for medical treatment for their people. Economic reforms, however, have resulted in the Chinese government covering fewer services than before.

Some Chinese are identified as **atheists**, or those who do not believe in a deity, but others practice Buddhism, Daoism, and Islam. The government controls religious freedom. Since 1962, ruling Chinese officials have restricted the practice of Buddhism in Tibet. In 2012 the government began stationing officials in Tibetan monasteries to oversee what is taught there. Mongolia's people are largely Tibetan Buddhist (Lamaist), but some are Muslim or atheist or shamanist.

The Arts

China has a rich history of artistic and literary accomplishments. These include ancient poetry, which describes nature's beauty, and traditional opera, which features costumes and martial arts. Under the Tang dynasty, potters created fine porcelain, known as "china" in the West. During the Cultural Revolution, the government nearly destroyed traditional arts, but it did not crush all artistic endeavors. China is now enjoying a resurgence in the arts.

dominate to exert influence over

ideogram a pictorial character or symbol that represents a specific meaning or idea

atheist a person who believes there is no god

ANALYZING PRIMARY SOURCES

The Spread of Rap and Hip-Hop in China

Chinese youth interpret rap music in their own way by changing its language and instrumentation. They have made it unique to their culture.

❝ . . . Chinese students admitted to not fully understanding the lyrics of the American songs they loved. But the love for hip-hop was not solely for the lyrics. It was the music, the emotions conveyed in the lyrics, that made hip-hop such a unifying force that overcame any barrier—linguistic or geographic.

Still, the music did not come so easily. 'You can't rap in Chinese, it's a tonal language.' 'You can't rap in China, they'll censor you.' . . . all sentiments that kept Chinese hip-hop from taking off with a running start. But hip-hop could not be stopped. . . .

To this day, hip-hop still comes head-to-head with censorship and doubts on the ability to rap in Chinese dialects and creativity. Yet, these obstacles provide listeners with the best form of hip-hop music Chinese youth can offer. ❞

— Carla Amurao, "From West to Far East: Rappin' and Rockin' the House," *Tavis Smiley: China Week, PBS California,* July 2011

Hip-hop dancing started in the early 1970s in the United States in the Bronx in New York City. It has become increasingly popular among youth in China's major cities, as the country has opened its borders to other cultural influences.

DBQ ▲ CRITICAL THINKING

1. **Identifying Central Issues** Describe the difficulties that people may face creating rap music in China.

2. **Speculating** How do you think the Internet and social media have influenced the growth of hip-hop in China?

PHOTO: Lane Oatey/Getty Images; TEXT: Tavis Smiley on PBS.org

Today in China, females are receiving more education and gaining places in the workforce.

▲ **CRITICAL THINKING**

1. Analyzing How has the position of women and girls improved in China?

2. Drawing Conclusions What long-term effects might the one-child policy have on China's families?

commune a collective farming community

merchant marine a country's shipping fleet that engages in trade

dissident a citizen who speaks out against government policies

Family and the Status of Women

The most important social unit in China's history has been the family. It monitors the behavior of its members and their welfare and provides work. Communism changed family bonds, weakening them in favor of loyalty to the state. The government provided jobs. Husbands and wives were sometimes separated in different regions to work. Family size was generally limited to one child.

Economic growth has brought changes to Chinese families. While most rural families remain traditional, families in urban areas are changing. Extended families are decreasing, while nuclear families are increasing. People are marrying and having their only child later in life. Divorce rates are climbing.

Traditionally, Chinese women have had a low position in society. Men have had all the power in the family, society, and government. Modern governments have worked to improve women's status, with some success. Women participate in the Chinese workforce at a relatively high level. However, few have achieved powerful positions in business or government.

The traditional preference for boys, combined with the one-child policy, has resulted in the death of many female infants through purposeful neglect. Girls are put up for adoption at high rates. Population figures from 2012 show that there are 93 women for every 100 men in China. In contrast, there are 102 women for every 100 men in the United States.

☑ **READING PROGRESS CHECK**

Making Connections What features of the Chinese written language have helped unify the culture?

Economic Activities

GUIDING QUESTION *How do domestic activities, government policy, trade, and international relations interact to form modern China's economy?*

A little more than one-third of China's workforce is involved in agriculture, making it the largest segment of the economy. No country produces more rice, wheat, and tea than China. China's farmers also produce soybeans, cotton, and silk.

Since 1949, the agricultural sector has seen many changes enforced by the government. The Great Leap Forward organized farmers into **communes**, or large farming communities. The government decided what and how much to grow, and commune members shared work and products. The results were disastrous. Crop production plunged, and starvation resulted during this time.

Chinese leaders instituted new policies in the 1980s. Such policies encouraged the development of small family farms. Farmers could sell and keep the profits when they produced extra crops or livestock.

At the beginning of the twenty-first century, Chinese farms began operating under different conditions. China's growing urban population now consumes crops through markets supplied by the country's farms as well as imports. Farmers' earnings have improved substantially as a result. These reforms, however, created a surplus of farmworkers who had previously been employed by larger collective farms. Many of these workers have moved to cities.

Market reforms have increased the growth of China's economy at a strong rate. In 2010 China became the world's largest exporter. However, challenges exist. Areas on the coast are wealthier than the interior agricultural regions. This economic gap brings new migrants from the countryside to the cities.

Justin Guariglia/National Geographic/Getty Images

China's waterways provide important routes for the transportation of goods to seaports like Shanghai near the outlet of the Chang Jiang. Large, oceangoing ships can navigate to the port of Wuhan in central China. The **merchant marine** fleet is used for commercial transport. Major ports are Hong Kong, Tianjin, and Guangzhou.

Both China and the United States seek increased trade, but obstacles exist. The United States wants increased trade to take advantage of China's growing economy. China's undervalued currency, established to keep export prices down, has resulted in a flood of inexpensive Chinese goods to other countries.

Although China's economy is strong and its global trade aspirations ambitious, positive political reforms have lagged. Human rights issues are of particular concern. **Dissidents**, or citizens who speak out against government policies, are harshly treated. In 1989, when Chinese students held a protest for democratic reforms in Beijing's Tiananmen Square, government troops responded brutally. China's major trade partners—the United States, Japan, and the European Union—have encouraged China to respect human rights. They imposed **economic sanctions**, or trade restrictions, on China in response to the events at Tiananmen Square. When China released several imprisoned dissidents, the United States relented on some sanctions.

The United States hopes that trade may stimulate democratic reforms in China. The U.S. Congress granted full trading privileges to China in 2000. The next year, China became a member of the World Trade Organization (WTO), an international body overseeing trade agreements and disputes.

Hong Kong's robust economy is due in large part to its proximity to Shenzhen, a **Special Economic Zone (SEZ)** that welcomes foreign investment. Such areas experience only minimal interference by the government. China has established SEZs, and uses the income to expand manufacturing. Taiwan has one of the strongest export-based economies in the world. Textiles, plastics, and electronic goods are specialties. Taiwan's depends heavily on exports. When demand decreased in 2009, it weakened the economy, which began to improve in 2010.

Mongolia's extensive grasslands are used for grazing huge numbers of livestock—sheep, goats, cattle, camels, and horses. Only 1 percent of its land can be farmed. Land is often government-owned, although efforts to privatize it have boosted economic growth. In the last few decades mining has become a major activity, with coal, copper, and fluorite as exports. Some textile and food production takes place, and tourism is on the rise. In addition, foreign investors have shown interest in the country's mineral deposits, which include gold, copper, and oil found in the Gobi.

☑ **READING PROGRESS CHECK**

Inferring How has life changed for Chinese farmers since the 1980s?

U.S. EXPORTS TO CHINA

U.S. IMPORTS FROM CHINA

Source: http://www.census.gov/foreign-trade/balance/c5700.html

China is one of the United States's leading trading partners.

▲ **CRITICAL THINKING**

1. *Drawing Inferences* Describe the U.S.-China trade relationship.

2. *Predicting* How will these graphs look in 10 years? Explain.

economic sanctions trade restrictions

Special Economic Zone (SEZ) a relatively small district in China that is fully open to global commerce

---

LESSON 2 REVIEW

Reviewing Vocabulary

1. *Identifying* What is an ideogram and how is it used?

2. *Summarizing* Who are dissidents? What role did they play in the Tiananmen protests of 1989?

Using Your Notes

3. *Interpreting* Use your graphic organizer to write a paragraph about how China's culture has affected the way it carries out economic activities.

Answering the Guiding Questions

4. *Drawing Conclusions* What influence did ancient Chinese history have on government, culture, and daily life in modern times?

5. *Evaluating* How has the migration of rural dwellers to urban areas affected China's population?

6. *Identifying Perspectives* What cultural beliefs, practices, and distinctive achievements and attitudes characterize China?

7. *Assessing* How do domestic activities, government policy, trade, and international relations interact to form modern China's economy?

Writing Activity

8. *Argument* In a paragraph, take the Chinese government's position and defend their policies, economic reforms, and attitude toward human rights.

ARE LIMITS ON GROWTH BEST FOR CHINA?

The high population growth rate has long been a concern in China. When the People's Republic of China came to power in 1949, its leaders wanted to slow population growth in order to reduce poverty and to stabilize the country. People were exposed to new family planning methods, and birth control was encouraged. However, once the population neared 1 billion in the 1970s, leaders wanted people to have no more than two children. By 1980, a one-child policy was put in place and continues today.

Most people must follow the one-child policy and are often rewarded for doing so. However, there are exceptions. Some minority groups, people whose first child has disabilities, and people in rural areas may have more than one child. Also, people may pay a fine for each additional child. However, those who knowingly go against the policy are required to pay large fines. Some have reported that if the fine is not paid, families can lose their homes or jobs.

Demographers and others think that China should end the one-child policy. These critics warn that if it continues, China will have a shortage of workers and fewer people available to take care of an aging population.

China's National Population and Family Planning Commission manages the one-child policy. Their goal is to make sure that the country can sustain development. To meet this goal, they monitor the effects of population changes on the economy, society, natural resources, and the environment. The Family Planning Commission is currently reexamining the one-child policy. They will only make changes if the revisions benefit the continued growth of the country.

Benefits of Population Limits

PRIMARY SOURCE

❝ Henan has much to teach the world in family planning, but it is a hard lesson to learn. Officials from Africa and India come to study what we are doing in China, but I'm not sure that they can apply it the same way. That's because they don't have a Communist party so it is difficult for them to take such strong steps. ❞

—Liu Shaojie, Vice Director of the Population Commission in Henan province, China, quoted in *The Guardian*, October 25, 2011

End Population Limits

PRIMARY SOURCE

❝ Everyone has blindly accepted the fact that population control has helped China's economy, but it has never been proven. . . . In fact, by 2030, no matter what policy China adopts the population will start to shrink. And I have never seen a country with a shrinking population and sound economic development. Western countries may have shrinking populations, but they have immigrants to make up their labour force. China does not have large immigration, so aiming for a zero or negative birth rate is very risky. ❞

—Liang Zhongtang, a former committee member of the National Family Planning Commission, quoted in *The Telegraph*, September 25, 2010

❝ We have been discussing the one-child policy since 2000. . . . It is just a matter of finding the right solution. Making the jump to two children is only a matter of time now. . . . If China sticks to the one-child policy, we are looking at a situation as bad as the one in southern Europe. Old people will make up a third of the population by 2050. ❞

—Li Jiamin, a specialist in population studies at Nankai University, quoted in *The Telegraph*, October 31, 2012

What do you think? DBQ

1. **Drawing Conclusions** Why might Chinese people hesitate to speak out against China's one-child policy?

2. **Identifying Central Issues** According to critics of the one-child policy, what are the risks of keeping the policy as it is?

3. **Hypothesizing** Why might Chinese people in urban areas and Chinese people in rural areas have different opinions on the one-child policy?

continues to occur. Thousands of dams built rapidly during the 1950s and 1960s are now damaged and at risk of failure.

To deal with these problems, China has invested more in flood-control projects. The Three Gorges Dam created a huge reservoir almost 400 miles (644 km) long. The project has been controversial due to the relocation of over a million people. Countless farms, villages, and historic temples were submerged. Natural ecosystems have been damaged with some animal and plant species now endangered or extinct. Changing the flow of the river may interfere with migratory paths of fish species. Pollutants in the soil and chemicals in abandoned factories may leach into the soil as the water rises in the reservoir. Supporters of the project point to the energy created by the dam and its commercial benefits. The dam will help to ease China's future power shortages that could potentially be crippling. The reservoir behind the dam holds an enormous amount of water for farming.

☑ **READING PROGRESS CHECK**

Understanding Relationships How does China's physical geography influence flooding?

THREE GORGES DAM
Harnessing the Yangtze

THE NEW RESERVOIR - HIGHS AND LOWS

SF to LA length of reservoir

trade ships can now reach Chongqing

356 major landslides have occurred

1.5 million people relocated

Circle size relative to city's population

EAST CHINA SEA

CHINA

YANGTZE RIVER

CHONGQING
p 29 million

WUHAN
p 10 million

NANJING
p 8 million

SHANGHAI
p 23 million

SHOCKING NUMBERS

Avg. Coal Plant	Hoover Dam	Three Gorges
667 Mwtz	**2080** Mwtz	**1.21** Gwtz
Powers **60** Homes	Powers **100** Homes	Powers **200** Homes

POWERING CHINA

18.2 MILLION HOMES ARE POWERED BY THE THREE GORGES DAM.

SUPPLIES 11% OF CHINA'S TOTAL POWER NEEDS.

EQUALS 10 TO 15 FEWER NUCLEAR REACTORS.

The Three Gorges Dam was a major engineering project. The dam was designed to serve several purposes: to prevent flooding along the Yangtze; to generate electricity; to increase the water supply for agriculture, industry, and homes; and to improve navigation along the river.

▲ **CRITICAL THINKING**

1. Analyzing What are the benefits of the Three Gorges Dam for China?

2. Formulating Questions Write three questions that you would ask the engineers of the Three Gorges Dam.

Addressing the Issues

GUIDING QUESTION *How are the Chinese people and government addressing the effects of industrial and economic growth on the environment?*

Industrial and economic growth has created very real benefits for the Chinese people. This growth has also created a host of environmental problems. The impact of China's economic boom on the environment and human health has been enormous. Urban areas are plagued by serious air pollution created by old technology in transportation and industry. One major reason for such severe pollution is China's heavy use of the country's coal supply. In addition to its abundance, coal is relatively cheap, making it attractive. But coal produces high amounts of several types of pollution. Blowing coal dust in the northern industrial areas worsens air pollution and causes many people to suffer from lung diseases. These same problems affect the inhabitants of Mongolia's few cities, where coal is burned for heating and cooking in homes and by factories.

Burning coal creates not only air pollution, but also acid rain. It has become a major problem in China and has also affected other areas of Asia. Mercury from the burning of coal has been found in the Pacific Ocean and as far away as the western United States. Ashes that remain in the coal furnaces after burning occurs, called bottom ash, must be removed and stored in large ponds that can leak or spill.

Like other rapidly urbanizing countries, China has trouble disposing of wastes. About one-third of China's population lacks access to clean water. Many tons of sewage are discharged annually into the Chang Jiang—30 billion tons in 2006. Seventy percent of the country's lakes and rivers are polluted. Two-thirds of China's cities lack clean water, forcing millions of people to boil their drinking water to make it safe to consume.

Pollution has made cancer the leading cause of death in China. Industrial waste is mostly to blame. One metal company in Shenyang spewed clouds of sulfur dioxide and other hazardous chemicals into the atmosphere for many years. City residents affected by health problems from the plant's toxic emissions prompted the government, for the first time, to shut down a state-run factory. In early 2013, a state of emergency was declared in Beijing due to terrible air quality. Mongolia's capital, Ulaanbaatar, is also among cities with the worst air pollution in the world.

Efforts to regulate polluters have been slow for many reasons. In China, one reason is the industries' reluctance to support stricter laws. More importantly, the overwhelming focus on economic growth encourages governments at both the national and local levels to ignore pollution regulations. Record-high pollution levels in Beijing in 2012 caused temporary emergency measures. The Beijing government shut down 103 factories and took 30 percent of government vehicles off the road. Despite such measures, air quality remains extremely hazardous.

China's large population and thriving industry depend on huge quantities of lumber. Each year China cuts down thousands of acres of forests, but also plants many trees to replace this resource. However, the replacement forests

Increasingly, air pollution is a major health concern in today's large Chinese cities.

▲ CRITICAL THINKING

1. **Identifying Cause and Effect** Why is coal creating a health problem in China?

2. **Analyzing** How effectively has China dealt with air pollution?

narvikk/E+/Getty Images

Each year, China meets its demand for lumber by cutting down thousands of acres of forest.

▲ **CRITICAL THINKING**

1. *Identifying Cause and Effect* What are the effects of the huge demand for lumber in China?

2. *Exploring Issues* Why does reforestation not completely solve the problems of deforestation?

are much less diverse than the original forests. In addition, reforestation efforts have not kept up with demand for timber. China has become a major importer of lumber, pulp, and paper. Clear-cutting timber has led to soil erosion, which in turn, leads to deforestation. Trees and other types of living ground cover help to slow runoff from rain. When they are destroyed, large-scale soil erosion and flooding occur.

A succession of unusually heavy rains in the late 1990s caused flooding of the Chang Jiang and the Huang He. The consequences of such flooding included the deaths of thousands, widespread destruction of property, and damage to the landscape.

Disasters such as these prompted China to plant trees on millions of acres of deforested riverbanks. A major dam was constructed along the Huang He to control flooding. Nature and wetland reserves and wildlife protection zones have also been created.

Western China is experiencing desertification. Much of the vegetation has been depleted by over-grazing and soil erosion, leaving the land bare and dry. Desertification has worsened sandstorms and dust storms that travel from the Gobi through southern Mongolia, northern and western China, and to the Pacific Ocean.

Traditional Chinese medicines often utilize products derived from rare or exotic animals. Deer antlers, rhinoceros horns, and bear gallbladders contain desirable substances that are often worth incredible sums of money. As wealth grows in the country, the demand for these products has increased. Because China has little wildlife remaining, many animal products are imported. The consequences of this increased demand have affected exotic and endangered animal populations worldwide.

Some areas in China are protected, however. These include forests at high altitudes and bamboo groves in the western Sichuan province, where pandas need the bamboo to survive. A surviving population of Siberian tigers was found in the far northeast in 1997. Wildlands there and in northwestern Yunnan have been protected as well.

✓ **READING PROGRESS CHECK**

Identifying How has industrialization affected the health of Chinese and Mongolian people living in cities?

LESSON 3 REVIEW

Reviewing Vocabulary

1. *Analyzing Text Structure* How does nuclear power differ from hydroelectric or fossil fuel power sources?

Using Your Notes

2. *Organizing* Using your graphic organizer, list three environmental issues in China and efforts to address them.

Answering the Guiding Questions

3. *Analyzing* How do China and Mongolia use fossil fuels as energy sources?

4. *Hypothesizing* What special challenges come with globalization, and how will they affect China's future?

5. *Exploring Issues* How are the Chinese people and government addressing the effects of industrial and economic growth on the environment?

Writing Activity

6. *Narrative* In two or three paragraphs, discuss the possible effects of environmental problems on a fictional Chinese person. Be sure to include relevant details from the lesson.

©Reuters/Corbis

Directions: On a separate sheet of paper, answer the questions below. Make sure you read carefully and answer all parts of the questions.

Lesson Review

Lesson 1

1 *Describing* What are some ways in which the Chang Jiang influences human systems in China?

2 *Explaining* Why do monsoons form in China? How do they change during the summer and winter months?

3 *Evaluating* China has a wealth of natural resources. How important is the production of rice to the Chinese economy?

Lesson 2

4 *Summarizing* Describe some of the challenges facing farms and cities in China as a result of urbanization.

5 *Considering Advantages and Disadvantages* What are some of the advantages and disadvantages of using ideograms for written language?

6 *Hypothesizing* Why do you think the creation of farming communes during the Great Leap Forward led to starvation?

Lesson 3

7 *Making Connections* Why has China's growing economic prosperity created more demand for electric power?

8 *Making Generalizations* How has the balance of power in East Asia shifted since the end of the Cold War?

9 *Explaining* How has burning coal affected the health of urban populations in China and Mongolia? Give examples.

Critical Thinking

10 *Analyzing* Why has globalization led to the rise of regional identities in China?

11 *Hypothesizing* China currently has a population with more boys than girls. What problems might this imbalance cause in coming years?

21st Century Skills

Use the cartoon below to answer the following questions.

PRIMARY SOURCE

MADE IN CHINA

12 *Identifying Perspectives* What do you think the "Made in China" sign means to the cartoonist?

13 *Explaining Continuity and Change* What does the cartoon suggest about how China's booming economy will affect its tradition of isolation?

College and Career Readiness

14 *Decision Making* Imagine that you have been asked to advise the Chinese government on the construction of a new dam. Research the controversy surrounding the Three Gorges Dam. Write a two-page report summarizing the controversy and offering your recommendations. Should China go forward with the new dam project? Explain.

Need Extra Help?

If You've Missed Question	**1**	**2**	**3**	**4**	**5**	**6**	**7**	**8**	**9**	**10**	**11**	**12**	**13**	**14**
Go to page	682	683	684	688	689	690	694	695	697	695	690	690	695	696

Michael Shaw/The New Yorker Collection/www.cartoonbank.com

Directions: On a separate sheet of paper, answer the questions below. Make sure you read carefully and answer all parts of the questions.

DBQ Analyzing Primary Sources

Use the document to answer the following questions.

China has taken significant measures to deal with threats to its environment. But these measures haven't kept pace with the rapid growth of its economy.

PRIMARY SOURCE

"*Given what scientists now predict about the timing of climate change, the greening of China will probably come too late to prevent more dramatic warming, and with it the melting of Himalayan glaciers, the rise of the seas, and the other horrors Chinese climatologists have long feared.*

It's a dark picture. Altering it in any real way will require change beyond China—most important, some kind of international agreement that transforms the economics of carbon. At the moment China is taking green strides that make sense for its economy. 'Why would they want to waste energy?' Deborah Seligsohn of the World Resources Institute asked, adding that 'if the U.S. changed the game in a fundamental way—if it really committed to dramatic reductions—then China would look beyond its domestic interests and perhaps go much further.' Perhaps it would embrace more expensive and speedier change. In the meantime China's growth will blast onward, a roaring fire that throws off green sparks but burns with ominous heat."

—Bill McKibben, "Can China Go Green?" *National Geographic,* June 2011

15 *Interpreting* Why does the author say that the greening of China is a "dark picture"?

16 *Problem Solving* What does Deborah Seligsohn think the United States should do to help China go green?

17 *Analyzing* How does the description of China's economy as a "roaring fire that throws off green sparks" help you understand the problem?

Applying Map Skills

Use the Unit 8 Atlas to answer the following questions:

18 *Environment and Society* What features of Mongolia's geography help explain why some areas have lower density populations than others?

19 *Human Systems* Locate the area that was submerged by the construction of the Three Gorges Dam. To what neighboring towns and cities did people most likely relocate?

20 *Physical Systems* Use your mental map of China to describe the location of the Gobi and Taklimakan deserts.

Exploring the Essential Question

21 *Making Connections* Write a paragraph describing how China's three great rivers have shaped population patterns. How have rivers influenced the location of urban centers?

Research and Presentation

22 *Understanding Relationships* Use the Internet and other resources to research the debate over human rights in China. What are the major issues? How effective have economic sanctions been at encouraging positive political reforms? Use your findings to create a poster that includes photos and quotes and present it to the class.

Writing About Geography

23 *Informative/Explanatory* Use standard grammar, spelling, sentence structure, and punctuation to write three paragraphs describing the problems of erosion and deforestation in China. How has economic growth fueled these problems?

Need Extra Help?

If You've Missed Question	**15**	**16**	**17**	**18**	**19**	**20**	**21**	**22**	**23**
Go to page	696	696	694	672	672	672	688	691	697

McKibben, William/National Geographic Stock

Japan

ESSENTIAL QUESTION · *How do physical systems and human systems shape a place?*

Why Geography Matters
Japan's Aging Population

Lesson 1
Physical Geography of Japan

Lesson 2
Human Geography of Japan

Lesson 3
People and Their Environment: Japan

Geography Matters...

The rugged island country of Japan, punctuated by volcanic mountains formed over millions of years, is home to a population that is ethnically homogenous compared to most industrialized countries. With its ancient culture and complex traditions, Japan also has a leading world economy that is highly advanced technologically. One of Asia's most successful democracies, Japan has become a strong ally of the United States and a force of stability in the region.

◄ Nobu Matsuhisa is the chef and owner of one of Tokyo's most famous restaurants.

Jeremy Sutton-Hibbert/Alamy

Japan's aging population

After a post–World War II baby boom, Japan experienced declining birthrates and slower overall population growth. The current population is declining and aging. Aging populations mean that a smaller percentage is working to support those in retirement. This could pose a threat to Japan's economic future.

How is the population of Japan changing?

In Japan, the fertility rate is declining as women marry later, delay having children, or never marry at all. (Most Japanese babies are born to married women.) At the same time, life spans are lengthening. The Japanese have the longest life spans in the world. In 1950, on average, Japanese women lived to age 63 and Japanese men to 60. Today Japanese women are expected to live to age 86 and Japanese men to 79. In 2012 more than 50,000 Japanese were centenarians, or people who are more than 100 years old. That number is projected to be at least six times as great by 2030. The combination of lower fertility rates and lengthening life spans has caused the population of Japan to decline and age.

1. Human Systems What effect is Japan's current fertility rate having on its population?

What are the long-term effects of an aging population?

As Japan's population decreases and ages, the workforce shrinks. Statistics indicate that the number of Japanese workers peaked in 1999 and has shrunk 2.5 percent since then. Therefore, there is a possibility of a shortage of workers in the future, as well as an increase in the dependency ratio. The dependency ratio refers to the ratio of economically dependent persons to economically active persons. As the dependency ratio increases, more retired people are drawing on government services, such as pensions and health care, and fewer people are paying taxes to fund those services. Interestingly, the lengthening life span of the Japanese does not necessarily mean increased sickness and debility. Japan's men appear to be healthier longer than men in other countries, and its women are generally vigorous for much of old age.

2. Human Systems Write a paragraph describing how the increased dependency ratio may change the way the Japanese people live in the future.

How are people in Japan and across the world finding solutions?

Many Japanese men—almost one-third of the workforce—continue to work beyond the usual retirement age. If more Japanese women also worked longer, the dependency ratio and the burden on the younger workforce would be reduced. Internationally, some governments are taking steps to make it easier for people over 65 to continue working. Others are considering raising the retirement age from 65 to 67—or possibly higher—and even linking the retirement age to life expectancy. Other options include increasing taxes to pay for government pensions and social security systems or reducing payouts in these systems.

3. Human Systems Why is raising the retirement age in Japan and other countries an important response to aging populations?

(l)UFO RF/a.collectionRF/Getty Images, (c)Haruyoshi Yamaguchi/Bloomberg/Getty Images, (r)Ryan McVay/Getty Images

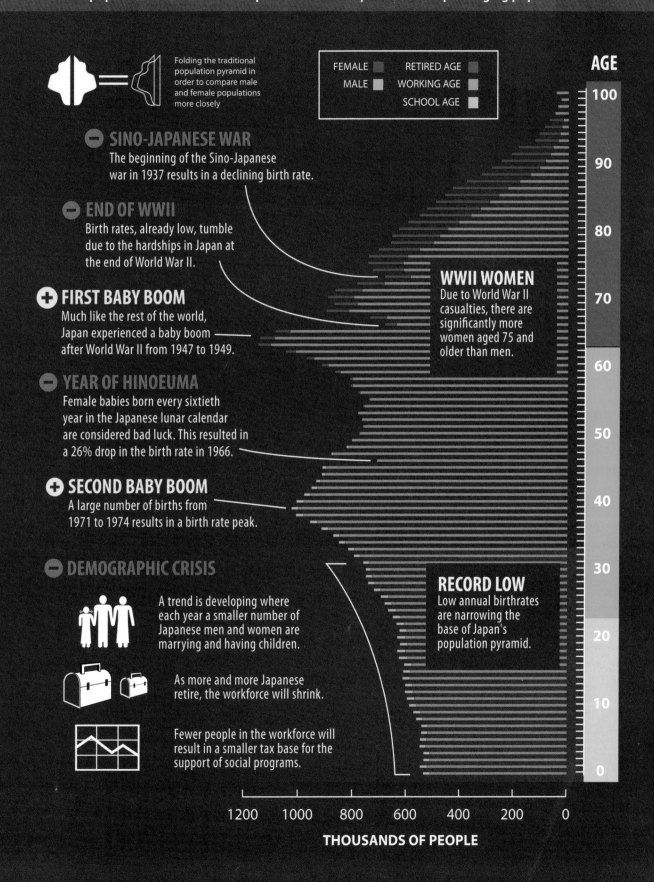

Folding the traditional population pyramid in order to compare male and female populations more closely

FEMALE | **RETIRED AGE**
MALE | **WORKING AGE**
SCHOOL AGE

AGE

⊖ SINO-JAPANESE WAR
The beginning of the Sino-Japanese war in 1937 results in a declining birth rate.

⊖ END OF WWII
Birth rates, already low, tumble due to the hardships in Japan at the end of World War II.

⊕ FIRST BABY BOOM
Much like the rest of the world, Japan experienced a baby boom after World War II from 1947 to 1949.

⊖ YEAR OF HINOEUMA
Female babies born every sixtieth year in the Japanese lunar calendar are considered bad luck. This resulted in a 26% drop in the birth rate in 1966.

⊕ SECOND BABY BOOM
A large number of births from 1971 to 1974 results in a birth rate peak.

⊖ DEMOGRAPHIC CRISIS

A trend is developing where each year a smaller number of Japanese men and women are marrying and having children.

As more and more Japanese retire, the workforce will shrink.

Fewer people in the workforce will result in a smaller tax base for the support of social programs.

WWII WOMEN
Due to World War II casualties, there are significantly more women aged 75 and older than men.

RECORD LOW
Low annual birthrates are narrowing the base of Japan's population pyramid.

1200 1000 800 600 400 200 0

THOUSANDS OF PEOPLE

netw⊙rks
There's More Online!

- ☑ **IMAGE** Fuji Fishing Harbor
- ☑ **IMAGE** Japanese Macaques
- ☑ **MAP** Physical Map of Japan
- ☑ **INTERACTIVE SELF-CHECK QUIZ**
- ☑ **VIDEO** Physical Geography of Japan

LESSON 1
Physical Geography of Japan

ESSENTIAL QUESTION · *How do physical systems and human systems shape a place?*

Reading **HELP**DESK

Academic Vocabulary

- **consist**
- **affect**
- **predominant**

Content Vocabulary

- **archipelago**
- **tsunami**

TAKING NOTES: *Key Ideas and Details*

SUMMARIZING As you read about the physical geography of Japan, use a graphic organizer like the one below to take notes on the climate.

IT MATTERS BECAUSE

The islands of Japan form an arc of 1,500 miles (2,400 km) across the North Pacific Ocean, rocky evidence of the Ring of Fire and its powerful volcanic action. Japan's waterways have been essential to transportation and agriculture. Although much of Japan is mountainous, its people have benefited from proximity to the ocean as a rich natural resource.

Landforms

GUIDING QUESTION *How has volcanic activity changed the land?*

Although there are more than 6,800 islands in its **archipelago**, most of Japan's land area **consists** of four main islands: Hokkaidō, Honshū, Shikoku, and Kyūshū. The islands of Japan formed during the last 15 to 20 million years. They were thrust upward and layered by the force of volcanic action from the Pacific Ring of Fire, the volcanic zone that runs along the edges of the Pacific. The resulting landscape of mostly rugged mountains **affects** where people live. The majority of the population dwells in the Japanese coastal lowlands.

Japan's location in the Ring of Fire causes some 50 Japanese volcanoes to remain active and frequent earthquakes. Japan's tallest peak, Mount Fuji, is a volcano not currently active. In a typical year more than 1,000 small earthquakes occur, with major quakes occurring less frequently but with more destruction and often causing deaths. When a powerful undersea earthquake occurs nearby, a major **tsunami**, or huge wave, can form, bringing terrible consequences. Tsunamis can travel more than 250 miles per hour (400 km per hour) and be more than 30 feet (10 m) high. In 2011 an earthquake off the coast of Honshū rocked the island and launched a series of tsunamis that flooded the coastline in massive waves. More than 19,000 people died, property damage was enormous, and a serious accident occurred at a nuclear power station. The cleanup and reconstruction effort was predicted to take ten years and cost upwards of $150 billion.

☑ **READING PROGRESS CHECK**

Summarizing How does the Ring of Fire affect Japan?

(tlc)©Creatas/PunchStock, (tc)©Jose Fuste Raga/Corbis

Waterways

GUIDING QUESTION *What makes the waterways in Japan unique?*

A distinctive feature of Japan is its short, swift rivers that flow from the mountains, often plunging over cliffs as stunning waterfalls. Typically, Japanese rivers rise in forested mountains, flow through steep valleys, and then cross alluvial plains to empty into the sea. The lower regions of these rivers are often used for farming rice. In the coastal lowlands, where rivers slow and pass through cities, some waterways are used for transportation, including carrying passenger water buses. These uses often involve damming the rivers or altering their original courses. Increasing demand for water—for agriculture, industry, and drinking—remains a challenge. The lack of natural or easily constructed reservoirs makes it difficult to contain rapid runoff from rainfall.

Japan's longest rivers, the Shinano and the Tone, flow from mountain heights on Honshū and can be very destructive when they flood. The Shinano is 225 miles (360 km) long. The Tone is 200 miles (322 km) long and supplies Tokyo with drinking water. On Hokkaidō, the Ishikari and the Teshio flow into the Sea of Japan (East Sea). The Yoshino River on Shikoku flows through a deep gorge and is known for its rapids.

Japan's largest lake, Lake Biwa, is in the central part of Honshū. It was created from a depression along a fault. Many of Japan's coastal lakes are simply drowned river valleys that reached the sea and were dammed over time by accumulating silt and sandbars.

✔ **READING PROGRESS CHECK**

Identifying Name three ways in which Japan's waterways have changed over time and identify whether the processes were natural or caused by humans.

archipelago a group or chain of islands

consist to be composed of

affect to have an effect on

tsunami a huge wave resulting from undersea earthquake or volcanic activity that gets higher and higher as it approaches the coast

GEOGRAPHY CONNECTION

More than four-fifths of Japan's surface area consists of mountains.

1. *PHYSICAL SYSTEMS* Why does Japan experience frequent earthquakes and occasional tsunamis?

2. *PLACES AND REGIONS* Name the four major bodies of water that border the islands of Japan.

Physical Map of Japan

Sea of Okhotsk

Hokkaidō

NORTH KOREA

Sea of Japan (East Sea)

EURASIAN PLATE

Honshū

JAPAN

PACIFIC PLATE

SOUTH KOREA

Korea Strait

PACIFIC OCEAN

Tokyo

Shikoku

Kyūshū

PHILIPPINE PLATE

Elevations

10,000 ft. (3,000 m)
5,000 ft. (1,500 m)
2,000 ft. (600 m)
1,000 ft. (300 m)
0 ft. (0 m)
Below sea level

✪ Capital city
▲ Active volcano
■ Site of major earthquake
〰 Tectonic plate boundary

250 miles
250 kilometers
Lambert Azimuthal Equal-Area projection

Japan **705**

The Japanese tradition of viewing cherry tree *(sakura)* blossoms in springtime dates back a thousand years. The beautiful tree and its flowers symbolize the natural cycle of human life and became revered as a symbol of good luck. Today cherry blossom festivals are held throughout Japan, timed to match the local blooms. People picnic under the blossoming trees and hold tea ceremonies, welcoming spring together with food, drink, and song.

SPECULATING What effect might *hanami* celebrations have on Japanese society as a whole?

predominant present as the strongest or main characteristic

The Japanese macaque lives farther north than any other wild primate.

▶ **CRITICAL THINKING**

1. *Analyzing Visuals* How is the Japanese macaque well adapted to its natural environment?

2. *Contrasting* How does the appearance of the Japanese macaque differ from most other monkeys?

Climates, Biomes, and Resources

GUIDING QUESTION *How does the climate in Japan change from north to south?*

Japan generally has a mild climate and abundant rainfall. Nevertheless, there are startling contrasts in climate and biomes. The northern latitudes of Hokkaidō are much colder and experience deep winter snows and howling winds, while far to the south, Okinawa has a subtropical climate.

Climate Regions and Biomes

The **predominant** influences on Japan's climate are latitude, the proximity of the great Asian landmass to the west, the mountainous terrain, and ocean currents. The combination of these factors makes Japan monsoonal, or influenced by seasonal winds that bring either precipitation or dry air. The cold winter monsoon drops rain or snow on the western slopes of Japan's mountains, leaving the eastern region drier. The warmer summer monsoon reverses this action, bringing warm rains to the eastern slopes and dry air to the west. The warm waters of the Japan Current, which is similar to the flow of the Atlantic Gulf Stream, provide moisture for the summer monsoon. The cold Kuril Current is responsible for dense fogs off Hokkaidō. Tropical storms called typhoons, similar to hurricanes, generally occur in late summer, forming over warm tropical waters. They bring torrential rains, high winds, and frequently great destruction to Japanese islands.

Northern Japan has a humid continental climate with warm summers and cold, snowy winters. Southern Japan has a humid subtropical climate with hot summers and heavy rain. Most of Japan is forested, or was once. Semitropical rain forests in the south consist of broad-leaved evergreen trees. Transitioning north, broad-leaved evergreens mix with deciduous trees, which eventually give way to conifers in the far north. Typical Japanese trees include beeches, maples, oaks, and spruces. A variety of cherry trees, with blossoms that are a symbol of Japan, are planted throughout the country. Much of the original vegetation of Japan is gone, however, supplanted by agriculture and foreign flora.

Despite great concentrations of people in Japan, large land mammals and birds still thrive in isolated mountain regions. Bears, wild boars, deer, and antelope are common in remote mountains. The Japanese macaque, a kind of wild monkey, also dwells as far north as Honshū. Ocean animals along the coasts

©Creatas/PunchStock

include sea turtles, sea snakes, abundant water birds, fish, whales, dolphins, and porpoises. The rich sea life is made possible, in part, by the meeting of the warm and cold currents.

Natural Resources

Despite extensive mountains and poor soil, Japanese agriculture makes the most of reliable rainfall and irrigation from rivers to grow most of the rice the country needs. The main crop is rice, cultivated in flooded paddies and principally on small farms. Other crops include a variety of grains, vegetables, fruits, potatoes, and tea. Dairy cattle are raised on Hokkaidō. Beef cattle are raised in feed lots, where they are carefully nurtured to provide high-fat meat.

Because Japan is an island country, the resources of the sea have always been an important source of food for the Japanese. Modern sources come from both deep sea fishing and human-made seafood farms. Japan is a world leader in harvesting and importing fish. The country consumes some 7.5 billion tons of fish annually, which amounts to nearly 15 percent of the world's catch, and reaps about $14 billion a year from the fishing industry. Ocean pollution and overfishing, however, have jeopardized the fisheries on which Japan depends. Such overfishing has seriously depleted marine life.

The mining industry is small and declining, with the exception of gold. Limestone is still widely quarried. Coal was historically the most important mineral mined in Japan. However, various factors—such as foreign competition, high production costs, and increasing reliance on oil—have lessened the importance of the industry. The last coal mine in Japan closed in 2002. Most of Japan's limited oil and natural gas resources come from Niigata Prefecture several hundred miles north of Tokyo. Since the beginning of this century, copper and iron mines have virtually ceased production, and these minerals are imported.

This busy fishing harbor lies at the base of Mount Fuji, Japan's highest mountain at 12,388 feet (3,776 m). There are about 200,000 vessels in Japan's entire fishing fleet.

▲ **CRITICAL THINKING**

1. *Analyzing Visuals* What details in the photo reflect how Japan's physical geography shapes its human geography?

2. *Hypothesizing* How may Japan's economy be affected if fisheries continue to decline?

✔ **READING PROGRESS CHECK**

Determining Importance Which of Japan's natural resources are most closely related to Japan's being an island country? What is threatening these resources?

LESSON 1 REVIEW

Reviewing Vocabulary

1. *Making Connections* How is a tsunami different from an ordinary ocean wave?

Using Your Notes

2. *Listing* Use your graphic organizer from this lesson to list the characteristics of Japan's climate.

Answering the Guiding Questions

3. *Examining* How has volcanic activity changed the land?

4. *Describing* What makes the waterways in Japan unique?

5. *Summarizing* How does the climate in Japan change from north to south?

Writing Activity

6. *Informative/Explanatory* Write a paragraph describing how latitude affects the climate and biomes of Japan.

©Jose Fuste Raga/Corbis

THE TOHOKU **Earthquake** AND **Tsunami**

The Tohoku earthquake of 2011 was the most powerful recorded earthquake ever to hit Japan. The earthquake and resulting tsunami caused incredible damage in Japan, including a meltdown at the Fukushima Daiichi Nuclear Power plant. Japan and the world will feel the effects of these disasters for years to come.

EARTHQUAKE SITE

9.0 Magnitude

NUCLEAR PLANT FAILURE

 Fukushima Daiichi 200 Rads into area

EARTHQUAKE TYPE

Massive swell

Upward force — Sea floor

Fault Line

Continental Plate

FAULTING

EXTENT OF THE TSUNAMI

112 mph

33 ft (11.65 m)

RELIEF EFFORTS

In total, 116 countries and 28 international organizations offered assistance to Japan after the earthquake and tsunami.

As of early 2012, over 5 billion dollars for relief and recovery had been donated from people in Japan and across the globe.

700 million

Private donations from the United States exceeded 700 million dollars, the third highest U.S. donations total for any overseas disaster.

WORLD IMPACT

4–10 inches off axis

At a magnitude of 9.0, the earthquake was so strong it shifted Earth on its axis by estimates of 4 to 10 inches (10–25 cm).

High tsunami waters from the Tohoku earthquake hit several other places, such as Hawaii, Peru, Chile, Oregon and California.

As of early 2013, twelve pieces of Japanese debris from the tsunami, including a 20-foot (6.1-m) boat, had washed onto U.S. and Canadian shores.

Making Connections

1. **The World in Spatial Terms** How did the location of the Tohoku earthquake contribute to the destructive power of the tsunami?

2. **Physical Systems** Outside of Japan, the tsunami impacted areas in the southern and eastern Pacific Ocean the most. Explain why areas in the western Pacific were largely spared.

3. **Environment and Society** Numerous images and video of the Tohoku earthquake and subsequent tsunami were posted on the Internet almost immediately following the disaster. How do you think these visuals contributed to international relief efforts?

*Interact with **Global Connections** Online*

networks
There's More Online!

☑ **GRAPH** Japanese Balance of Trade with the United States

☑ **IMAGE** Commodore Perry in Japan

☑ **IMAGE** Japan and the West

☑ **IMAGE** Port of Tokyo

☑ **TIME LINE** Shifting Power

☑ **INTERACTIVE SELF-CHECK QUIZ**

☑ **VIDEO** Human Geography of Japan

Reading HELP DESK

Academic Vocabulary

- **approximate**
- **generation**
- **unique**
- **overseas**

Content Vocabulary

- **clan**
- **samurai**
- **acculturation**
- **trade surplus**

TAKING NOTES: *Key Ideas and Details*

IDENTIFYING As you read about the human geography of Japan, use a graphic organizer like the one below to identify major turning points in Japan's history.

Turning Points in Japan's History

```
  400
Japan unites        1868   1945

  1100  1542  1854  1931
```

LESSON 2
Human Geography of Japan

ESSENTIAL QUESTION · *How do physical systems and human systems shape a place?*

IT MATTERS BECAUSE

Japan's early human geography reflects influences from ancient China and Korea. The Japanese civilization that developed from these two countries then isolated itself well into the 1800s. Since then, the forces of modernization and globalization have influenced Japanese society. Yet the traditional and the modern exist comfortably together.

History and Government

GUIDING QUESTION *How do ancient and modern traditions influence life in Japan?*

Japan's history combines tradition with transformation. Japan has maintained its cultural traditions while leaping toward modernization over the last century.

Japan Through the Ages

Because they are so close, China and Korea had a significant impact on Japan. Once ruled by many **clans,** in the A.D. 400s Japan united under the Yamato dynasty. Yamato rulers adopted China's philosophy, writing system, art, sciences, and governmental structure. Korean scholars also influenced the Japanese.

Japan was ruled by dynasties for centuries. Japanese society evolved, developing a government and establishing an emperor as absolute monarch. In the 1100s, warring armies brought about the shogunate, a feudal society under the control of a shogun, or military ruler. Professional warriors known as **samurai** supported the shogun's rule, although the emperor was the official ruler of Japan.

The first documented contact with the West occurred in 1542. A Portuguese ship sailing to China was blown off course and landed in Japan. Roman Catholic missionaries and European traders followed, raising concerns that military conquest might follow. To prevent this, shoguns restricted foreigners. They expelled all but a few European merchants who traded at Nagasaki.

During the 1800s, the United States worked to open Japan to trade. In 1854, following a show of strength, Commodore Matthew C. Perry of

(tl)©Bettmann/Corbis, (tcl)©The Print Collector/Corbis, (tc)©Christian Kober/AWL Images/Getty Images, (tcr)©Popperfoto/Getty Images, (tr)©HO/Reuters/Corbis the U.S. Navy negotiated opening Japanese ports to U.S. ships. A trade agreement followed that benefited the United States. This action eventually sparked a samurai rebellion that returned full authority to the emperor.

A Changing Government

The return of authority to the emperor was a turning point in Japanese history. Called the Meiji Restoration, this period—which lasted from 1868 to 1912— involved the rapid modernization of Japanese society, including its government, economy, military, education, and legal systems. Japan became a modern country set on building an empire. In a war with China, Japan conquered the island of Formosa (Taiwan). Japan also fought Russia for control of Korea and gained rights to Manchuria and the Russian island of Sakhalin. By the time Japan took control of Korea in 1910, it had become the most powerful empire in Asia.

During World War I, Japan sided with the Allies, profiting from exports to Allied countries and expanding its influence in Asia. Following the war, Japan continued to prosper and extend its reach. Though Japan steered toward a democracy, military leaders gained influence over the government. In 1931 they invaded northeast China; in 1937 they invaded northern China. By 1939, Japan had joined with Germany and Italy as the Axis Powers. Japan's 1941 attack on Pearl Harbor drew the United States into World War II, in which some 3 million Japanese lost their lives. The war ended when Japan surrendered after the United States dropped atomic bombs on Hiroshima and Nagasaki in 1945.

The American occupation of Japan that followed stripped the empire of its territories and military. It also set the country back on the road to democracy. Out of the ruins, a vibrant economy emerged. Several decades later, Japan

clan a family group

samurai a professional warrior of preindustrial Japan

Commodore Perry's steamships, anchored in Edo Bay (now Tokyo Bay) and bristling with huge guns, convinced the Japanese government to sign a trade treaty with the United States and open two ports and a trade agreement.

▼ **CRITICAL THINKING**

1. **Identifying Central Issues** What caused the Japanese to end years of isolation abruptly?

2. **Defending** Provide a main reason to justify Commodore Perry's pressure on Japan to open trade with the United States. Then provide a main reason to justify Japanese resistance to opening their country to foreign influences.

©Bettmann/Corbis

became a global economic power. Today, Japan has the third-largest economy in the world, surpassed only by China and the United States. As a constitutional monarchy, Japan has an emperor who is the head of state.

☑ **READING PROGRESS CHECK**

Sequencing Information How has the role of the Japanese emperor changed over time?

Population Patterns

GUIDING QUESTION *How does population density influence life in parts of Japan?*

The majority of Japanese—99 percent—belong to the same ethnic group. They are descendants of migrating people from Asia who pushed out most of the indigenous people of Japan, the Ainu (EYE•noo). Some Ainu still live on Hokkaidō today.

Since most of Japan is mountainous, there is limited land area suitable for habitation. Most people are concentrated in the lowlands along the seacoasts, or in valleys and plains. The urban concentrations are great, and Tokyo-Yokohama is one of the largest megacities in the world, with more than 37 million people. The population density in this urban area is 11,300 people per square mile (4,300 people per sq. km). The average population density in Japan is about 907 people per square mile (350 per sq. km). The lowest density is on Hokkaidō.

approximate close to, but not exact

Approximately four-fifths of Japanese live in urban areas, and a majority of these people live near the Tokaido megalopolis. This corridor is a 750-mile- (1,200-km-) long urbanization zone and is concentrated along rail lines. Extreme urbanization affects many aspects of Japanese life. This includes not only the lifestyles of the Japanese, but architecture, transportation, housing, and daily activities. In Tokyo, crowded streets with immense skyscrapers and numerous commercial areas are located within easy walking distance of subway stations. Bicycles are widely used by Japanese of all ages to ride to

©The Print Collector/Corbis

TIME LINE ∨

Shifting
POWER →

Although Japan's monarchy has endured longer than that of any other country, the real power in Japan has shifted from emperors to shoguns to the people. The emperor plays an important symbolic role in Japan, but wields little power.

▶ **CRITICAL THINKING**

1. Describing In Japan's history, how has power passed from one leader to another?

2. Analyzing What role has the military played in Japan's government in the past?

Japan's capital moves to Nara, the emperor emerges as leader, and Buddhism becomes the state religion.

710

Minamoto Yoritomo becomes first shogun, or military leader, and establishes capital at Kamakura

1185

700 →

1300 →

794

Capital moves to Heian-kyo; powerful Fujiwara family dominates political and cultural life

1392

During the Muromachi period, provincial warlords hold real power in Japan.

712

school or to commute to work or subway stations. These stations link to a system of railroads. This system includes a network of long-distance, high-speed electric bullet trains. In 2012 a Japanese railroad company unveiled plans for a magnetic levitation train. It would float above the tracks, levitated by the force of powerful magnets. This train is expected to travel at 311 miles per hour (500 km per hour). Due to its emphasis on convenient public transportation, Tokyo has developed very differently from cities that are heavily reliant on privately owned vehicles. It has no slums and is considered environmentally friendly.

✔ READING PROGRESS CHECK

Summarizing What is the largest urban area in the world?

Society and Culture Today

GUIDING QUESTION *How do traditions influence family life and art?*

The written form of Japanese began around A.D. 400, when Chinese writing was first introduced and adapted to Japanese. The Japanese borrowed words from Chinese and have more recently borrowed thousands of English words for concepts that do not exist in Japanese.

Japan's indigenous religion, Shinto, is often practiced in a mix with Buddhism and even Christianity. Shinto emphasizes reverence for nature and is polytheistic. While it is unusual for Japanese children to receive formal religious instruction, many households have both a Buddhist altar and a Shinto altar.

Modern Japanese society has a high regard for education. By law, children must attend school until age 15. Many begin at a very young age to focus on getting into the best schools and eventually the best universities. Adults often continue to seek instruction long after formal schooling is completed, studying

(l)Christian Kober/AWL Images/Getty Images, (r)©HO/Reuters/Corbis, (b)Popperfoto/Getty Images

1600

Tokugawa period begins, initiating a prolonged period of peace.

1868

In response to pressure from Western nations, Japan begins rapidly modernizing during Meiji Restoration.

1900 →

1912

Yoshihito becomes emperor; Japan becomes more democratic as more men gain the vote.

1926

Hirohito becomes emperor; militarism increases, eventually leading to World War II.

1989

Akihito becomes emperor; period of sluggish economic performance and political turmoil

2006

Prince Hisahito is born and becomes third in line for the Japanese throne.

"Composed of the Japanese words *oru* (to fold) and *kami* (paper), origami has a rich and complex history that spans culture, class and geography.

Paper was first invented in China around 105 A.D., and was brought to Japan by monks in the sixth century. Handmade paper was a luxury item only available to a few, and paper folding in ancient Japan was strictly for ceremonial purposes, often religious in nature.

By the Edo period (1603–1868), paper folding came to be regarded as a new form of art that was enabled by the advent of paper both mass-produced and more affordable."

—"Between the Folds," *PBS Independent Lens,* November 30, 2009

DBQ **ANALYZING PRIMARY SOURCES** Why did the advent of mass-produced paper change origami from a ceremonial practice to a new form of art?

generation a group of individuals born and living at the same time

unique unlike anything else

acculturation cultural modification of an individual, group, or people by adapting to or borrowing traits from another culture

anything from foreign languages to technology. The Japanese are also avid travelers and tourists.

In addition to a high standard of living, the Japanese enjoy corresponding general good health and the longest life expectancy in the world. National health insurance covers all citizens. However, high life expectancy combined with a low birthrate means the Japanese population is aging disproportionately, which strains health and social services. A smaller workforce is a future issue for Japan.

Family and the Status of Women

Family in Japan remains a source of stability but is undergoing changes. Families are smaller than in the past. Many children are still taught the importance of being part of a group, as opposed to being an individual. They are also taught to revere their ancestors. But as new **generations** are more likely to be raised in urban settings, rather than in villages, they tend to have more modern views and changing values. In a technologically advanced and consumer-driven society, less social conformity and increasing individual choice is apparent in ordinary Japanese people. Young Japanese are among the world's major users of technology.

Some of the changes in Japanese families are due to the changing role of women in society. There is growing dissatisfaction with traditional roles in which women are expected to bear major responsibility for the household tasks. Women make up about two-fifths of Japanese workers. As they enter the workforce in larger numbers, they often delay marriage and childbearing. This has contributed to the trend of smaller families and Japan's declining population. Female role models, such as Japan's first female foreign minister, have helped to change perceptions. Many women work only part-time jobs, however, and those in careers often find it difficult to win promotion to high-level positions.

The Arts

The Japanese have developed their own art forms. Traditional Japanese art rejects what is showy and emphasizes delicacy, exquisite forms, and simplicity. Poets, writers, and artisans throughout the centuries have portrayed the beauty of the seasons, the changing environment, and the rugged mountains through poetry, theater, painting, pottery, embroidery, and silk screen.

Among the enduring and **unique** art forms of Japan is origami. Origami uses folded paper to depict the shapes of objects of nature, such as animals and birds. The Japanese tea ceremony is also a time-honored tradition. Still practiced as a hobby today, the tea ceremony combines literary and artistic traditions into a ritual of great artistry. Another tradition that remains popular is the classical Kabuki theater. It blends music, dancing, costumes, elaborate makeup, and well-timed special effects to bring characters to life.

A taste for literature among educated Japanese has a long history. Lady Murasaki Shikibu, a Japanese noblewoman, wrote one of the world's first novels, *The Tale of Genji*, around A.D. 1010. It tells the story of an imperial prince in the emperor's court. Another well-known form of Japanese poetry, the haiku, is famous for its concise and exact use of words and nature imagery that appeals to the senses.

Japan's rich artistic history does not hinder the Japanese from enjoying and excelling at art and cultural forms that developed in other places. Such **acculturation** is apparent in many ways. Western classical music is popular, and major Japanese cities may have more than one symphony orchestra. Western-style painting, sculpture, architecture, and cinema are also very popular.

✔ **READING PROGRESS CHECK**

Identifying List five of Japan's unique art forms and describe two of them.

Economic Activities

GUIDING QUESTION *How have industry and trade transformed the country?*

Japan's industrial economy is based on a mixed market system. Such a system emphasizes private ownership of the means of production and distribution. The government plays a key role, however. It carefully coordinates with industry and regulates all sectors of the economy, including imports.

Industry and Manufacturing

Japan has seen extraordinary economic growth since World War II. High demand for trade goods around the world has made it a global economic power since the 1980s. Lacking ample natural resources, such as fossil fuels and minerals, Japan has focused on trade to sustain its industries. It has highly skilled and educated workers and newly advanced technology. This has helped Japan to become a leading producer of consumer goods such as cars, electronic devices, and computers.

Telecommunications industries, particularly mobile communications leaders NTT and KDDI, rank among the largest Japanese corporations. Other huge conglomerates include Canon, which makes office and consumer products, and Honda, which builds automobiles and motorcycles. Nippon Steel is Japan's leading steel producer. Mitsubishi is a major multinational company with diverse business interests. These and many other important companies earn billions in annual profits.

Apart from their economic role, Japanese corporations play an important societal role. Although this practice is beginning to change, most employees expect to stay with their company until retirement. This culture of lifetime employment creates very strong bonds between company and workers. Politeness, sensitivity, and good manners are basic elements in the workplace. Executives and their families are ranked socially by the importance of the company for which they work.

Trade

To sustain its economy, Japan imports raw materials such as fossil fuels and minerals from other countries. Japanese industries turn these resources into finished products that they sell in huge quantities **overseas**. For example, China, the United States, South Korea, Taiwan, Singapore, Thailand, and Germany all buy Japanese cars and high-tech devices.

overseas relating to a foreign country, especially one across the ocean

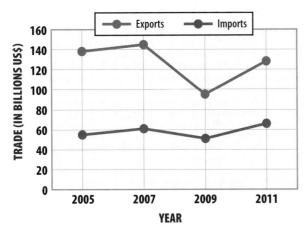

Japanese Balance of Trade with the United States

Source: http://tse.export.gov/TSE/MapDisplay.aspx

⌄ GRAPH SKILLS

Japan relies heavily on exports to drive economic growth.

◄ CRITICAL THINKING

1. *Analyzing Visuals* Describe the difference between Japan's exports to and imports from the United States.

2. *Constructing a Thesis* Give a logical reason for why Japan's imports and exports both dipped in 2009. Explain your reasoning.

This freighter is docked at the Port of Tokyo. Japan's economy depends on imports of raw materials and exports of finished products.

▲ **CRITICAL THINKING**

1. Finding the Main Idea Why is Japan's economy so reliant on trade?

2. Evaluating Why are Japanese finished goods in high demand globally?

trade surplus earning more money from export sales than spending for imports

Although the global demand for Japanese goods is high, the Japanese government protects local industries from foreign competition. It places tariffs, or taxes, on the manufactured goods imported from foreign countries (but not on raw materials). This restricts foreign sales in Japan and helps tip the balance of trade in Japan's favor. Balance of trade is the difference between a country's imports and exports. Japan has a **trade surplus** in its balance of trade because the value of its exports is greater than the value of its imports. This makes Japan wealthy. However, it lowers profits for trading partners, such as the United States, which import from Japan more than they export. Efforts to persuade Japan to drop tariff protections on finished products have yet to be completely successful.

In the 1990s, an economic downturn occurred in Japan. This was brought about, in part, by poor banking practices. As banks failed, production slowed, exports dropped, and unemployment rose. The Japanese government took steps to reform unsafe economic practices. By 2003, the situation had stabilized and was improving. The recession that began in 2008, however, reduced global demand for Japanese goods and slowed the economy. Things began to improve again in 2013.

☑ **READING PROGRESS CHECK**

Inferring What might a Japanese official say to defend the practice of placing tariffs on finished goods imported from overseas countries?

LESSON 2 REVIEW

Reviewing Vocabulary

1. Identifying Write several sentences indicating the role of the clan and the samurai in Japan's early history.

Using Your Notes

2. Sequencing Use your graphic organizer to write a paragraph describing the major turning points in Japan's history.

Answering the Guiding Questions

3. Drawing Conclusions How do ancient and modern traditions influence life in Japan?

4. Evaluating How does population density influence life in parts of Japan?

5. Making Connections How do traditions influence family life and art?

6. Synthesizing How have industry and trade transformed the country?

Writing Activity

7. Informative/Explanatory In a paragraph, discuss the advantages and disadvantages of living in Tokyo-Yokohama, the largest urban area in the world.

Toshifumi Kitamura/AFP/Getty Images

(tl)Sue Flood/The Image Bank/Getty images, (tr)Gyro Photography/amanaimagesRF/Getty Images

networks

There's More Online!

☑ **GRAPH** Total Whale Catch

☑ **IMAGE** Aquaculture

☑ **IMAGE** Bluefin Tuna

☑ **MAP** Japan's Nuclear Power Plants

☑ **INTERACTIVE SELF-CHECK QUIZ**

☑ **VIDEO** People and Their Environment: Japan

Reading **HELP**DESK

Academic Vocabulary

- **sustain**
- **widespread**
- **issue**
- **strategy**

Content Vocabulary

- **aquaculture**
- **supertrawler**
- **chlorofluorocarbon (CFC)**

TAKING NOTES: *Key Ideas and Details*

ORGANIZING Use a graphic organizer like the one below to take notes on Japan's environment as you read the lesson.

Japan: Managing Resources

Human Activity	Impact on Environment	Addressing the Issues
Build nuclear reactor		

LESSON 3
People and Their Environment: Japan

ESSENTIAL QUESTION · *How do physical systems and human systems shape a place?*

IT MATTERS BECAUSE

Because of its economic strength, Japan has an impact on the world disproportionate to its size. The choices the Japanese make about energy, ocean harvesting, and consumption resonate far beyond their island home.

Managing Resources

GUIDING QUESTION *What resources are at risk in Japan?*

Because of its limited supply of fossil fuels, Japan has long depended on foreign oil imports. To reduce this dependence, the Japanese people turned to nuclear technology to provide electrical energy. Today Japan has more than 50 nuclear reactors with the capacity to supply the country with some 30 percent of its energy. A series of nuclear accidents at Japanese reactors, however, exposed people to radiation. The first accident occurred in 1978. A more recent one happened in 2011.

Following the earthquake and resulting tsunami in 2011, a damaged nuclear reactor had a meltdown. The perils of operating nuclear reactors in a region prone to earthquakes and tsunamis were made clear. In response, Japan temporarily shut down most of its reactors. This prompted calls to reverse Japan's reactor-building program and focus instead on renewable energy sources. Japanese businesses and industries resisted such a change, however. They pointed out that rising fossil fuel prices could cripple the Japanese economy and add substantially to pollution. Japan continues searching for answers and alternatives that will allow it to **sustain** its energy needs safely. Investment in solar power, wind, biomass, and small hydroelectric and geothermal projects has expanded. Increasingly, Japanese companies are building wind turbines and developing solar cell technology. There is some discussion of building floating wind turbines to be anchored in the sea because Japan has limited land area for wind projects.

☑ **READING PROGRESS CHECK**

Identifying What are the pros and cons of using nuclear power in Japan?

Japan's Nuclear Power Plants

Kashiwazaki-Kariwa is the world's largest nuclear power plant. Of Japan's 50 nuclear power plants, only 2 are still online in the wake of the 2011 accident. New guidelines and new safety measures are being planned to increase plant safety and ease public concerns.

The Fukushima Daiichi nuclear power plant sustained heavy damage when a 9.0 magnitude earthquake occurred in the Pacific Ocean on March 11, 2011. The tsunami generated by the quake destroyed the power supply and cooling system for the nuclear reactors. A considerable amount of radioactive material was released into the environment.

- ■ Nuclear power plant
- ✪ Capital city
- • City

0 — 250 miles
0 — 250 kilometers
Lambert Azimuthal Equal-Area projection

GEOGRAPHY CONNECTION

Nuclear energy has been a priority in Japan since 1973. Following the earthquake and tsunami that caused a nuclear accident in 2011, however, this policy came under review.

1. **THE WORLD IN SPATIAL TERMS** Why do you think Japan's nuclear power plants are located along the coasts?

2. **ENVIRONMENT AND SOCIETY** What are located within a few hundred miles of the largest clusters of nuclear power plants?

sustain to continue without interruption

widespread found over a large area

Human Impact

GUIDING QUESTION *What human activities have affected the physical environment of Japan?*

Japan's rapid industrialization brought with it severe and **widespread** environmental problems. Economic growth took priority over health and safety. The result was increasing air and water pollution and declining populations of aquatic animals.

Following World War II and into the 1960s, Japan's major industries—iron and steel, cement, paper and pulp—expanded without regard to pollution. Power plants vented noxious pollutants into the air. Industry and agriculture tainted rivers, lakes, and surrounding ocean with chemicals and raw sewage. Air and water quality suffered, creating health risks. In the 1970s, smog alerts were frequent. Acid rain became a serious problem. It damaged lakes, crops, and trees; threatened animal species; and corroded buildings and bridges. By the mid-1990s, Japan was the world's fourth-largest emitter of carbon dioxide.

Starting in the 1970s, Japan began adopting stricter rules to protect air and water quality. Today, it has some of the strictest regulations in the world. Yet despite improvements in the levels of soot and smoke emissions from factory smoke stacks, motor vehicle emissions remain a serious problem. Diesel engines, in particular, foul the air of Japan's major cities. Regulations to further tighten control of emissions from diesel engines and buses have been passed. Japanese industries have been at work on fuel-cell engines, which produce no dangerous emissions.

The problem of acid rain is particularly troubling. Even as Japan has tightened regulations on emissions, the acidity of the rain has remained largely unchanged. Researchers believe that massive sulfur dioxide emissions in China contribute to acid rain in Japan. China and Japan signed a preliminary treaty to address acid rain in 1994. Evidence suggests, however, that acid rain that originates in China will reach disastrous levels in Japan by 2020.

For centuries, the Japanese have depended on the abundance of the sea for food. Fish are an integral part of the Japanese diet, and the Japanese consume large amounts of fish and other seafood. Japan's most internationally famous food specialty, sushi, usually includes thin slices of raw fish called sashimi. Overfishing—resulting from catching too many fish of particular species—has seriously depleted fish stocks around the Japanese islands and around the world. Water pollution has also contributed to declining fish stocks.

Some 40,000 people use Japan's Inland Sea—a narrow body of salt water surrounded by Honshū, Shikoku, and Kyūshū—for fishing and **aquaculture**, which is the farming of seafood. Wild fish stocks such as red sea bream, anchovies, and mackerel have declined dramatically in this area in recent decades. The number of jellyfish—which negatively affect other species—is increasing in these waters because the plankton on which they feed thrive in polluted water.

Globally, the huge Japanese demand for seafood has contributed to declining fish stocks in fisheries in the Atlantic and the Pacific. Many countries, including Japan, rely on powerful deep-sea **supertrawlers**. These huge ships are equipped to serve as modern factories at sea, harvesting distant fisheries. Harvesting huge catches using supertrawlers leads to overfishing. It also causes the killing and waste of tons of unwanted sea life. To protect fisheries from foreign commercial fishing, many countries claim an economic zone extending to 200 nautical miles off their shores. Public opposition to supertrawlers is gaining strength globally, particularly as numerous fish stocks are rapidly declining.

Despite a 1986 international ban on commercial whaling, Japan remains the largest consumer of whale meat in the world. Centuries of overhunting by many countries caused serious declines in whale populations globally, severely endangering many species. By the 1960s, only Japan, Russia, and a few other countries continued to hunt whales on a large scale. Following the 1986 ban, Japan, Iceland, and Norway used a loophole for scientific research to continue to capture whales. Hundreds are still killed and eaten each year. Japan and the other whale-hunting countries are widely criticized for their whaling practices.

☑ **READING PROGRESS CHECK**

Understanding Relationships Explain why fish stocks in Japan's Inland Sea have declined in recent years.

Mature bluefin tuna reach 6.5 feet (2 m) and live up to 15 years in the wild. These endangered and valuable fish are prized for food in Japan even as their stocks have plummeted worldwide. A premium fish sold for $1.76 million in 2013.

▲ **CRITICAL THINKING**

1. *Making Connections* Why did fish become a staple in the Japanese diet?

2. *Analyzing Ethical Issues* Is it ethical to consume endangered species? Explain your answer.

aquaculture the cultivation of seafood

supertrawler an ocean-going factory ship with facilities for processing and freezing fish

---

Total Whale Catch

NUMBER OF WHALES — YEAR

Legend: ■ Total ■ Japan ▨ Iceland ■ Norway

Y-axis: 40,000 / 35,000 / 30,000 / 25,000 / 20,000 / 15,000 / 10,000 / 5,000 / 0

X-axis: 1946 1956 1966 1976 1986 1996 2009

Source: http://awionline.org/sites/default/files/uploads/flash/whaling_stats/main.swf.

⌄ GRAPH SKILLS

Whales have been hunted and eaten by the Japanese, Icelanders, and Norwegians for centuries.

◄ **CRITICAL THINKING**

1. *Analyzing Visuals* What was the largest total number of whales captured by Japan, Iceland, and Norway after 1946?

2. *Constructing a Thesis* Give a logical reason for why all three countries' whale hunting dipped sharply in 1986.

Sue Flood/The Image Bank/Getty images

This fish farm is located in Ago Bay, along the south coast of Honshū. Aquaculture provides the Japanese with about 22 percent of their seafood.

▲ **CRITICAL THINKING**

1. Assessing Why would it be advantageous to create a fish farm in a bay?

2. Making Predictions Do you think the number of fish farms will increase in the future? Why or why not?

issue a topic for debate and discussion

chlorofluorocarbon (CFC) a chemical substance, found mainly in liquid coolants, that damages the Earth's protective ozone layer

strategy a plan of action to achieve an aim

Addressing the Issues

GUIDING QUESTION *How are environmental issues being addressed in Japan?*

Japan ignored the environmental problems caused by rapid economic growth for many years. In the 1970s, however, the Japanese government began to encourage industries to address environmental **issues** more responsibly.

Today Japan is a leader in addressing environmental issues and supporting industries that profit from sustainable growth. For example, the Japanese have taken steps to reduce emissions of **chlorofluorocarbons (CFCs)**. CFCs were widely used in industry in the liquid coolants for refrigerators and air conditioners during the mid-1900s. Released into the atmosphere, they deplete Earth's stratospheric ozone layer that protects the planet from harmful ultraviolet solar radiation. Japan is working toward a complete phase-out of CFCs.

Japan is also working to mitigate climate change. In 1997 Japan hosted an international meeting under the United Nations Framework Convention on Climate Change. The treaty that resulted is known as the Kyoto Protocol. It defined ways to reduce the carbon emissions that contribute to global warming. It was ratified by 37 industrial countries, including Japan (but not the United States). The treaty set binding targets to reduce greenhouse gas emissions, including carbon dioxide. Nevertheless, by 2007 carbon dioxide emissions in Japan had increased by more than 10 percent from 1990 levels.

One **strategy** Japan has used to offset declining fish stocks and other sea products has been to increase aquaculture. Fish farmers raise fish by floating fish cages in the sea or anchoring them to the sea floor. Floats and ropes are used to cultivate shellfish such as oysters, scallops, pearl oysters, and abalone, as well as kelp. Although aquaculture production has declined somewhat since 1990, nearly 1.2 million tons of fish and seaweed were farmed in Japan in 2010.

Another strategy for maintaining fish populations has been to prevent overfishing by establishing quotas that limit fish harvests. Like the ban on whaling, however, quotas have limited effectiveness.

✓ **READING PROGRESS CHECK**

Inferring Has Japan been generally successful in addressing environmental problems? Explain your answer.

LESSON 3 REVIEW

Reviewing Vocabulary

1. Making Connections Write a paragraph describing the relationship of supertrawlers to Japanese aquaculture.

Using Your Notes

2. Transferring Information Use your graphic organizer to identify a human activity that has been destructive to Japan's environment and write a paragraph about ways in which the issue is being addressed.

Answering the Guiding Questions

3. Differentiating What resources are at risk in Japan?

4. Assessing What human activities have affected the physical environment of Japan?

5. Drawing Conclusions How are environmental issues being addressed in Japan?

Writing Activity

6. Argument Write an introductory paragraph for a personal essay that will argue whether you think Japan is doing enough to address environmental issues and why.

Gyro Photography/amanaimagesRF/Getty Images

Directions: On a separate sheet of paper, answer the questions below. Make sure you read carefully and answer all parts of the questions.

Lesson Review

Lesson 1

1 *Identifying Cause and Effect* How does the Pacific Ring of Fire affect human settlement in Japan?

2 *Explaining* What combination of factors makes Japan monsoonal?

3 *Assessing* Generally, where does Japan rank in oil production and oil exporting in the world? Why?

Lesson 2

4 *Describing* How did the United States "open up" Japan in 1854?

5 *Explaining* Explain whether it would be accurate to state that the emperor is the government official in Japan who makes or approves the major governmental decisions for the country.

6 *Identifying Central Issues* Where does Japan rank in life expectancy among all countries of the world? Explain the reasons that have greatly contributed to this ranking and discuss problems that are arising in relation to it.

Lesson 3

7 *Explaining* What events in 2011 caused Japan to reconsider its reliance on nuclear energy? Why?

8 *Interpreting Significance* Explain the significance of economic zones to fisheries.

9 *Evaluating* Explain whether it would be accurate to state that the level of carbon dioxide emissions in Japan in 2007 poses an irony to its role relative to the Kyoto Protocol.

21st Century Skills

Use the cartoon below to answer the questions that follow.

PRIMARY SOURCE

"Before you do something you may regret, I think you should know that I contain six parts of mercury per million."

10 *Using Primary Sources* How does the cartoon provide a whimsical look at a serious issue plaguing Japan and other countries around the world? Why might this cartoon be more effective than long lists of statistics in attracting the attention of some readers?

11 *Drawing Conclusions* What effect do you think the artist hoped to gain through this cartoon? Explain whether you think this effect is evident in Japan.

Critical Thinking

12 *Exploring Issues* You are preparing a Web site about the current economy of Japan—and the connection of the current economy to the country's history, physical geography (including natural resources), human geography (including population patterns), and environmental concerns. Lay out the main page of the Web site, including at least 10 links to information for the content detailed above. For each link, provide a short paragraph to summarize the main content that will be accessed through the link.

Lee Lorenz/The New Yorker Collection/www.cartoonbank.com

Need Extra Help?

If You've Missed Question	**1**	**2**	**3**	**4**	**5**	**6**	**7**	**8**	**9**	**10**	**11**	**12**
Go to page	704	706	707	710	712	714	717	719	720	721	721	701

Directions: On a separate sheet of paper, answer the questions below. Make sure you read carefully and answer all parts of the questions.

DBQ Analyzing Primary Sources

Use the excerpts to answer the questions that follow.

PRIMARY SOURCE

"*Based on a detailed investigation of all the facts, and supported by the testimony of the surviving Japanese leaders involved, it is the Survey's opinion that certainly prior to 31 December 1945, and in all probability prior to 1 November 1945, Japan would have surrendered even if the atomic bombs had not been dropped, even if Russia had not entered the war, and even if no invasion had been planned or contemplated.*"

—United States Bombing Survey, July 1, 1946, President's Secretary's File, Truman Papers

"*. . . There are voices which assert that the bomb should never have been used at all. I cannot associate myself with such ideas. Six years of total war have convinced most people that had the Germans or Japanese discovered this new weapon, they would have used it upon us to our complete destruction with the utmost alacrity. . . . Future generations will judge these dire decisions, and I believe that if they find themselves dwelling in a happier world from which war has been banished, and where freedom reigns, they will not condemn those who struggled for their benefit amid the horrors and miseries of this gruesome and ferocious epoch.*"

—Winston Churchill, "Why Should We Fear for Our Future?" House of Commons, August 16, 1945

13 **Understanding Historical Interpretation** Explain whether these quotes provide statements that are essentially in agreement with one another.

14 **Interpreting Significance** Explain whether you think Churchill's anticipation of potential future events related to weapons of mass destruction were valid.

Exploring the Essential Question

15 **Sequencing** Conduct research on Japan's history and the periods in which the island country was influenced by or isolated from other countries. Use your research to create a time line of at least five entries explaining major milestones.

Applying Map Skills

Refer to the Unit 8 Atlas to answer the following questions.

16 **Places and Regions** Identify the major body of water nearest Ōsaka.

17 **The World in Spatial Terms** Use your mental map of Japan to describe the spatial relationship among the islands of Hokkaidō, Honshū, Kyūshū, and Shikoku.

18 **Human Systems** Which two cities in Japan are the most densely populated?

College and Career Readiness

19 **Examining Information** You are located in Washington, D.C., and wish to host a traditional Japanese cherry blossom festival. Conduct research to determine the connection between Japan and the cherry blossom trees in your location, as well as the optimal time of the year to have a festival. Write a report explaining why you have chosen this time of the year and the activities that will be included in the experience.

Research and Presentation

20 **Gathering Information** With a partner, conduct research to learn more about the Shinkansen and the development of the bullet train and the magnetic levitation train in Japan. Create a multimedia presentation to detail development of the bullet train and the magnetic levitation train—and the importance of rapid mass transit to Japan. Include audio, video, and maps in your presentation.

Writing About Geography

21 **Argument** Use standard grammar, spelling, sentence structure, and punctuation to write an editorial to support the pro or con side of this statement: "Supertrawlers should be banned." Focus your argument on the effects of such a ban. Support your statements with information from the chapter and address opposing arguments in your editorial.

Need Extra Help?

If You've Missed Question	**13**	**14**	**15**	**16**	**17**	**18**	**19**	**20**	**21**
Go to page	722	722	710	672	672	676	706	713	719

North Korea and South Korea

ESSENTIAL QUESTION · *How do physical systems and human systems shape a place?*

netw⬤rks

There's More Online about the geography of North Korea and South Korea.

CHAPTER 30

Geography Matters...

The Korean Peninsula is composed of North Korea and South Korea. The people of the peninsula share a language, a history, and a culture. War and politics, however, divided a single country into two. Political philosophies and decisions about resources have resulted in two very different economies and standards of living. Prosperity and scarcity threaten the environment in different ways. Urbanization, the increased use of technology and consumer goods, and the loss of rural culture threaten South Korea. Famine, erosion, and lack of freedom threaten North Korea.

◀ This young Korean woman wears traditional dress.

©Topic Photo Agency/Corbis

complementarity: *two* Koreas

North Korea and South Korea share a peninsula, a history, and a culture, but they are enemies. The two countries are in a situation known as a complementarity in which two places are economically interdependent. North Korea has more natural resources and raw materials, while South Korea is more developed industrially and economically. Economic cooperation could occur if the countries were not bitterly divided.

©Topic Photo Agency/Corbis

How is life in both countries similar on the peninsula?

North Korea occupies the northern half of the Korean Peninsula and South Korea occupies the southern half. North Korea's rugged, mountainous terrain makes farming difficult, but yields coal and other valuable mineral resources such as gold, iron, copper, and zinc. Although North Korea has invested in mining, as well as other industries, it is one of the world's poorest countries. In contrast, South Korea's economy is one of the strongest in Asia. Like North Korea, South Korea is mountainous. However, fertile plains in the west and south are suitable for growing rice, barley, soybeans, corn, and other crops. Since the 1940s, South Korea's economy has been shifting from agricultural to industrial. Today, South Korea is the largest shipbuilder in the world. Its modern factories also manufacture cars, electronics, chemicals, and steel.

1. Places and Regions How are North Korea and South Korea similar and different?

How has a shared history affected both countries?

North Korea and South Korea have common bonds and the ability to satisfy each other's economic needs. However, these two countries have political differences that make cooperation difficult. Communist North Korea is a dictatorship. It has ties to China and Russia. South Korea is a republic with ties to the United States, Western Europe, and Japan. Since the Korean War ended in 1953, relations between the two countries have remained tense. North Korea has threatened South Korea with military strikes and South Korea has vowed to strike back. In 2010 North Korea fired on South Korean forces stationed on Yeonpyeong Island. Despite hostile relations, the two countries have been able to find some common ground. For instance, North Korea has been plagued by serious food shortages brought on by drought and flooding. To help solve this problem, North Korean leaders have accepted emergency food aid from South Korea. With few of its own resources, South Korea supports its growing industries by buying raw materials from North Korea.

2. Human Systems What prevents North Korea and South Korea from working together?

In a global economy, what do the two countries have in common?

To help one another, North Korea and South Korea can continue to work toward sharing resources. In the late 1990s, South Korea pursued the "Sunshine Policy" of economic cooperation, sharing aid and resources with North Korea. Later, the leaders of North Korea and South Korea attended inter-Korean summits to address political and economic issues. As a result, the two countries were able to launch a joint economic venture in 2006: the Kaesong Industrial Complex in North Korea. For this project, South Korea supplied money and technology. North Korea supplied land and labor. More than 100 South Korean-owned companies employ about 47,000 North Koreans. Other inter-Korean economic projects have been proposed, such as a natural gas pipeline from Russia to South Korea via North Korea. There is also a proposal to extend the Trans-Siberian Railway through North Korea to South Korea.

3. The World in Spatial Terms What are the possible benefits of economic cooperation between North Korea and South Korea?

(l)©Christophe Boisvieux/Corbis, (c)©Getty Images News/Getty Images, (r)Chung Sung-Jun/Getty Images News/Getty Images

THERE'S MORE ONLINE

EXPLORE a map of the Korean War • *COMPARE* the populations of North Korea and South Korea

networks

There's More Online!

- ☑ **CHART** Average Year-Round Temperatures in Select Korean Cities
- ☑ **IMAGE** Rural Town in South Korea
- ☑ **MAP** Waterways of the Korean Peninsula
- ☑ **INTERACTIVE SELF-CHECK QUIZ**
- ☑ **VIDEO** Physical Geography of North Korea and South Korea

LESSON 1
Physical Geography of North Korea and South Korea

ESSENTIAL QUESTION · *How do physical systems and human systems shape a place?*

Reading HELPDESK

Academic Vocabulary

- **vary**
- **cooperative**

Content Vocabulary

- **islet**
- **aquaculture**

TAKING NOTES: *Key Ideas and Details*

COMPARING As you read about the physical geography of the Korean Peninsula, use a graphic organizer like the one below to compare the landforms, climates, and natural resources of North Korea and South Korea.

Physical Geography

Landforms   Climates   Natural Resources

IT MATTERS BECAUSE

The Korean Peninsula lies between China and Japan, at the center of one of the world's most economically and strategically important regions—East Asia. The peninsula's rugged terrain was a barrier to invaders for many centuries, and the population of the peninsula is quite homogeneous. However, the peninsula's long coastline allowed contact with other civilizations. The two Koreas have great economic and geopolitical importance today.

Landforms

GUIDING QUESTION *How does the land on the Korean Peninsula change from north to south?*

The Korean Peninsula is about half the size of California. It is bordered by Russia and China to the north and surrounded by water on the other three sides. To the east is the Sea of Japan, which the Koreans call the East Sea. To the south, the Korea Strait lies between South Korea and Japan. To the southwest lies the East China Sea, which joins the Yellow Sea on the western coast. The Koreans call this the West Sea. After World War II, the peninsula was split into the Democratic People's Republic of Korea (North Korea) and the Republic of Korea (South Korea). The two countries have been divided by the Korean Demilitarized Zone (DMZ) since 1953. The DMZ is a strip of unoccupied land about 6 miles (9.7 km) wide.

The peninsula is very mountainous. The T'aebaek Mountains begin in North Korea and stretch along the Sea of Japan following the east coast into South Korea. Both North Korea and South Korea are mountainous, but North Korea is more mountainous with highlands and plateaus across its northern and eastern areas. North Korea has some of the highest peaks. Mount Paektu, the highest peak on the peninsula, is on the border between North Korea and China. Most of the rivers on the peninsula originate in the T'aebaek range and flow westward. This creates large river valleys on the western side of the peninsula. Most people live in these lowlands or along the coastline.

(tl)Topic Photo Agency/age fotostock, (tc)©Topic Photo Agency/Corbis, (tr)©Christophe Boisvieux/Corbis

The coastline has many small inlets and bays, especially in the south. There are more than 3,000 islands off the coast of South Korea. The two largest, Cheju and Ulleungdo, were formed by volcanic lava. Cheju, in the East China Sea south of the Korean Peninsula, is beloved by honeymooners, and is known as the "Hawaii of Korea." Ulleungdo, located to the north in the Sea of Japan, has fine fishing and forests. South Korea also claims a group of rocky **islets** that Koreans call the Tok Islands. Japan also claims them.

islet a very small island

☑ **READING PROGRESS CHECK**

Specifying Which country of the Korean Peninsula is more mountainous?

Waterways

GUIDING QUESTION *Which major waterways drain the land in the peninsula?*

Most of the peninsula's major rivers are in South Korea. The Han, the Kŭm, and the Naktong all begin in the T'aebaek Mountains. The northernmost of these rivers, the Han, flows through South Korea's capital, Seoul. Both the Han and the Kŭm flow west to reach the Yellow Sea. South Korea's longest river, the Naktong, flows south for 325 miles (523 km) to the Korea Strait.

North Korea's longest river is the Yalu, which forms most of the border between North Korea and China. The source of the Yalu is on the southern slope of Mount Paektu. The Yalu flows some 500 miles (800 km) to Korea Bay, an inlet of the Yellow Sea, on the western coast of North Korea. Several other rivers in North Korea—the Ch'ŏngch'ŏn, Taedong, Chaeryŏng, and Yesŏng—also drain into Korea Bay. These large river valleys are important farming areas in this mountainous land.

Waterways of the Korean Peninsula

GEOGRAPHY CONNECTION

The major rivers of the Korean Peninsula flow to the west.

1. **ENVIRONMENT AND SOCIETY** Why are the major cities of the Korean Peninsula located along rivers?

2. **THE WORLD IN SPATIAL TERMS** On which river is P'yŏngyang located?

They have also been important to the settlement of the peninsula. Both South Korea's capital, Seoul, and P'yŏngyang, the capital of North Korea, began as river ports.

Rivers are no longer the main transportation routes in the two Koreas. However, they are still used for irrigation and for generating hydroelectricity. Water levels **vary** throughout the year. Rivers are shallow except during the summer rainy season. This means that the water supply and ability to generate power vary seasonally, as well.

vary to exhibit or undergo change

☑ **READING PROGRESS CHECK**

Identifying Which country on the Korean Peninsula has the longest river?

Climates, Biomes, and Resources

GUIDING QUESTION *How does the climate change from north to south?*

The Korean Peninsula lies close to the landmass of Asia. Because of this, much of it has a continental climate of cold winters and warm or hot summers. The north is closer to the mainland, so it has the greater temperature extremes. The southern part of the peninsula has a humid subtropical climate, with temperatures moderated by the surrounding seas. Monsoon winds bring heavy rains to the entire peninsula in the summer.

Climate Regions and Biomes

North Korea's winter lasts from December to March, and is generally cold and snowy. The average temperatures in January range from about –10°F (–23°C) in the northern interior to 20°F (–7°C) in the south. Mean July temperatures are above the 60s°F (about 20°C) in most of the country. South Korea's winter temperatures also vary from north to south. Seoul, in the northwest, has average January temperatures in the low 20s°F (–5°C), while in Pusan, in the southeast, winter temperatures are in the mid-30s°F (about 2°C). In summer, it is generally hot with heavy rain. During August, the warmest month, the temperature is typically in the high 70s°F (about 25°C) throughout South Korea.

The Korean Peninsula typically receives between 35 and 60 inches (89 and 152 cm) of precipitation each year. About 60 percent falls as rain between June and September, the summer monsoon season. The northern plateau is the driest area, receiving about 24 inches (61 cm) of precipitation per year. Typhoons may bring heavy rains in the late summer, especially along the southern coast. The South Korean island of Cheju is the wettest area, receiving more than 70 inches (178 cm) per year.

There are many coniferous trees, such as the Siberian fir, spruce, and Korean pine, especially in the northern highlands. Much of the lowlands have been deforested in both countries, but small pine groves mixed with oaks, lindens, maples, and birches can be found in some areas. The humid subtropical regions of the south support over 4,500 known species of plants, including needle-leaved deciduous trees and broad-leaved evergreens, such as camellias and camphor trees.

Due to deforestation, large predators such as tigers, leopards, lynx, and bears are nearly gone from the peninsula. Deer are the most common large mammals. The peninsula

This small rural town in South Korea is surrounded by fields that have been terraced for growing crops.

▼ **CRITICAL THINKING**

1. ***Comparing*** What resources do both countries have?

2. ***Drawing Conclusions*** How does terracing like this increase crop yields?

Topic Photo Agency/age fotostock

is home to many species of birds. These include wild pigeons, herons, and cranes. The Demilitarized Zone (DMZ) serves as an important nature preserve, where waterfowl, migratory birds, and mammals can flourish undisturbed by humans.

Natural Resources

Neither country on the peninsula has a great deal of farmland because of the mountainous landscape. North Korea has increased agricultural production through the use of fertilizers and irrigation. Since 1958, North Korea's farms have been administered as **cooperatives**, or communal farms. The country has more than 3,000 cooperatives. About 300 families work each farm on about 1,200 acres (500 ha) of land. These farms deliver produce to the government, which distributes food through state-owned stores. South Korea also has little arable land, but recent improvements in processes and commercialization of the farming industry have led to increased production.

In both countries, rice is the most important crop. North Korea also grows corn, wheat, barley, potatoes, and fruit. The warmer climate areas, particularly in the southern regions of South Korea, are suitable for growing a variety of citrus fruits. These areas also produce vegetables, pears, and persimmons. Ginseng is an important export for South Korea. In North Korea, livestock is raised on land that cannot be used to grow crops. Livestock production, especially of poultry, has steadily increased over time. The raising of livestock is also important in South Korea. South Korea also has a large dairy industry. After rice, the top agricultural products of South Korea are pork, beef, and milk.

Both countries rely heavily on fish and seafood. In North Korea, deep-sea fishing supplies pollack, sardines, mackerel, and other fish. About one-fourth of the country's fish comes from **aquaculture**. South Korea is one of the world's major deep-sea fishing countries. Significant fish production there also comes from coastal fisheries and inland aquaculture.

Most of the peninsula's known mineral resources are in North Korea. Iron ore and coal are especially important. North Korea has the world's largest magnesite deposits, too. Magnesite is used to make many items, including fertilizers and rubber. The main source of electric power in North Korea is hydroelectric. While South Korea has one of the world's largest supplies of graphite and tungsten, it has very few other minerals. It must import all its petroleum and most metals. More than half of South Korea's electricity is generated through thermal electric plants. However, nuclear power generation is becoming more important in that country.

✔ **READING PROGRESS CHECK**

Identifying Which country on the Korean Peninsula has the warmest winters?

Connecting Geography to SCIENCE

The DMZ

The Demilitarized Zone, or DMZ, created as a buffer zone between warring North Korea and South Korea, has become an important nature preserve. Bordered by armies and landmines, the DMZ has been untouched by humans since 1953. Former farmland has been overgrown or slowly replaced by trees, wetlands, and estuaries. These habitats provide a home to rare species, including Asiatic black bears, Siberian musk deer, and the red-crowned crane, one of the rarest birds in the world. Naturalists are working on ways to protect the nature preserve if the two Koreas are ever reunited.

DRAWING CONCLUSIONS Why might reunification of Korea pose a threat to habitats and species in the DMZ?

cooperative an organization, often a farm, whose members work together and share expenses and profits

aquaculture the cultivation of seafood

LESSON 1 REVIEW

Reviewing Vocabulary

1. *Describing* Write several sentences to describe how cooperative farming works in North Korea.

Using Your Notes

2. *Comparing and Contrasting* Use your graphic organizer to write a paragraph comparing and contrasting the physical geography of North Korea and South Korea.

Answering the Guiding Questions

3. *Drawing Conclusions* How does the land on the Korean Peninsula change from north to south?

4. *Identifying* Which major waterways drain the land in the peninsula?

5. *Analyzing* How does the climate change from north to south?

Writing Activity

6. *Argument* In three paragraphs, formulate an argument about whether you think life would be easier in North Korea or South Korea, based only on physical geography and resources.

networks

There's More Online!

LESSON 2
Human Geography of North Korea and South Korea

ESSENTIAL QUESTION · *How do physical systems and human systems shape a place?*

Reading HELPDESK

Academic Vocabulary

- **principle**
- **isolation**
- **regime**
- **authority**
- **ethnicity**

Content Vocabulary

- **coup**
- **cultural divergence**

TAKING NOTES: *Key Ideas and Details*

LISTING As you read about the human geography of North Korea and South Korea, use a graphic organizer like the one below to list details about the history, population, culture, and economic activities of the two countries.

	North Korea	South Korea
History		
Population		
Culture		
Economic Activities		

IT MATTERS BECAUSE

Korea's ancient culture has influenced and been influenced by the cultures of China and Japan. The tense relationship between North Korea and South Korea and their different political systems have gained the two countries different major allies—China and the United States, respectively.

History and Government

GUIDING QUESTION *How has the decision to divide the Korean Peninsula into North Korea and South Korea affected the development of each country?*

Before China expanded into Korea in 108 B.C. during the Han dynasty, the Korean people mostly belonged to village communities that were not united. Eventually three united kingdoms emerged that shared similar cultures, but were also rivals. These kingdoms traded with China and were greatly influenced by Chinese culture.

Buddhism was introduced to the peninsula from China in the fourth century A.D. It was widely adopted by ruling kingdoms. From Korea, Buddhism was introduced to Japan in the sixth century A.D. Confucianism came to the peninsula from China about a thousand years ago. At first, Korean rulers promoted Buddhism as the religion for personal enlightenment. Confucianism was adopted as a standard for political **principles**. Later, Buddhism was suppressed, and Confucianism provided the basis for a complex governmental bureaucracy.

The Silla dynasty drove out the Chinese and united Korea in A.D. 668. Other Korean dynasties came to power, which were largely supported by the Chinese and influenced by Chinese culture. The government of the Koryo dynasty, for example, used China's government as a model.

After resisting Mongol invaders, Korea became known as the Hermit Kingdom because of its **isolation**. In 1871 Korea declared an official policy of isolation, but Japanese warships arrived in 1876. This "gunboat diplomacy" by Japan forced Korea to open its ports for trade. After the Russo-Japanese War ended in 1905, Korea became a Japanese protectorate.

(tl)Jon Choi Jin/AP Images, (tcl)DEA/G. Dagli Orti/De Agostini/Getty Images, (tcl)Interim Archives/Archive Photos/Getty Images, (tcr)Mark Ralston/AFP/Getty Images, (tr)SeongJoon Cho/Bloomberg/Getty Images

Japan annexed Korea in 1910, making it a colony. Japanese occupation deprived the Korean people of many rights and freedoms. The Japanese attempted to assimilate Koreans into Japanese life and culture. They used Korea's resources to grow their own economy. Anti-Japanese feelings increased the desire to gain independence.

Korea Divided

After Japan's defeat in World War II, the Korean Peninsula was jointly occupied by the Soviet Union and the United States. North of the 38th parallel, or line of latitude, the Soviets set up a Communist government. The new country that emerged was the Democratic People's Republic of Korea. The south was administered by the United States. The Republic of Korea was established there, with Seoul as its capital.

In June 1950, North Korean forces invaded South Korea. The conflict that followed became known as the Korean War. United Nations forces, mainly troops from the United States, came to the aid of the Republic of Korea (South Korea), while the Soviet Union and China supported the Democratic People's Republic of Korea (North Korea) with troops and equipment. Fighting concluded in 1953 with an armistice, or cease-fire, and the establishment of the Korean Demilitarized Zone (DMZ) separating the two countries.

United States military support of South Korea has continued since 1950. Soviet support for North Korea ceased with the fall of the Soviet Union and the end of the Cold War in 1991. Today, North Korea remains isolated from the global community. Its main supporter is China.

Kim Il Sung became the first premier under the North Korean Communist **regime** in 1948. His son succeeded him, and his grandson became premier in 2011. It is a family dynasty that rules the north. South Korea has become a democratic government. The first president of South Korea, Syngman Rhee, was elected in 1948. Strict governmental control, a military **coup**, and meeting economic over social needs continued until 1993, when civilian **authority** was restored. In 2012 South Korea elected a woman as its president for the first time.

Korean Governments Today

The Korean Workers' Party controls North Korea's elections and provides lists of approved candidates—usually only one for each position. This legislative body meets for only one or two weeks a year. Most decisions are made by the 15-member Presidium of the Supreme People's Assembly (SPA).

principle a rule or code of conduct

isolation state of being set or kept apart from others

regime a form of government

coup an overthrow of the government

authority the power to influence or command thought, opinion, or behavior

North Korean soldiers celebrate their country's successful nuclear bomb test on February 13, 2013. Experts believe that North Korea's nuclear operations are being carefully planned to avoid detection.

▼ CRITICAL THINKING

1. *Interpreting* What may underground nuclear testing reveal about North Korea's intentions?

2. *Making Inferences* Why might North Korea want to hide its nuclear development?

Jon Chol Jin/AP Images

"The invasion of Southern Korea by the Reds [Communists] was not altogether a surprise but it was shocking to us all just the same....I hope the UN forces will drive the Reds beyond the Korean territory and that they will unify the Korean Government under a competent person."

—Philip Jaisohn, letter dated September 14, 1950, to U.S. Lieutenant General John R. Hodge

DBQ *IDENTIFYING CENTRAL ISSUES* What did the author hope would be accomplished by the Korean War?

Since 1987, South Korea's government has had a structure like that of the United States. Power is balanced among legislative, executive, and judicial branches. South Korea has a multiparty system, although two parties usually dominate.

In the early twenty-first century, relations between North Korea and South Korea seemed to improve. South Korean industrial firms opened plants in Kaesong, North Korea. The plants employed people from both countries. Leaders of both countries met in 2000 and in 2007. In 2010 hundreds of South Koreans crossed the border to meet North Korean relatives for the first time since the division.

There were also setbacks in relations, however. In December 2007 South Korea elected a conservative president. His policies offended North Korea, and North Korea closed the Kaesong facility in 2009 and 2013. In 2010 a North Korean submarine sank a South Korean warship, killing 46 sailors. In 2012 North Korea conducted two rocket-test fires in violation of international agreements. A successful underground nuclear test was carried out by North Korea in 2013. The United Nations and many countries around the world condemned the test. It also strained relations between North Korea and China, its closest ally.

✔ **READING PROGRESS CHECK**

Explaining What 2010 incident increased distrust of North Korea by South Koreans?

Population Patterns

GUIDING QUESTION *How do settlement patterns and density differ in the two countries?*

Almost all of those who live on the Korean Peninsula can trace their heritage to China and Central Asia. The people all share a common Korean **ethnicity**. The peninsula is one of the most ethnically homogeneous regions in the world. The number of foreigners—mostly from Japan, the United States, and China—in South Korea is growing, especially in major urban areas.

DEA Picture Library/age fotostock

FOCUS
of Rival Interests ➔

Although often troubled by its larger neighbors China and Japan, Korea maintained political and cultural independence for more than a thousand years.

▶ **CRITICAL THINKING**

1. *Explaining* How did geography play a role in the relationship between Korea, China, and Japan?

2. *Analyzing* How did Korea lose its independence in the early twentieth century?

• First Korean kingdom, Choson, emerges in northern Korean Peninsula

300 B.C. ➔

300 B.C.

• Kingdom of Silla defeats rival kingdoms, uniting entire Korean Peninsula

668

108 B.C.

China conquers Choson and turns it into a Chinese colony.

935

Kingdom of Koryo overthrows Silla; takes control of Korean Peninsula

The countries' populations have had different growth rates. North Korea's population more than doubled between 1953 and 1993. The estimated 2011 population of North Korea was less than half that of South Korea's nearly 50 million. Food shortages, adequate health services, and living standards have been problems. Most of the rural population lives in the coastal lowlands and river valleys. Inland areas are very sparsely settled due to mountains and lack of arable land. Government expansion of industry has relocated people from rural to urban areas. P'yŏngyang is the largest city and capital of North Korea.

South Korea's population also doubled in the second half of the twentieth century. In the twenty-first century, although the birthrate has fallen, population density is high in urban areas. About 80 percent of South Korea's population is urban. Roughly half of that population lives in the country's seven largest cities. The capital city of Seoul is by far the largest city. Satellite cities encircle Seoul to form an extremely large urban region. These satellite cities include Anyang, Songnam, Suwŏn, Inch'ŏn, and Puch'on.

☑ **READING PROGRESS CHECK**

Summarizing Describe the changes to South Korea's population since the 1950s.

Society and Culture Today

GUIDING QUESTION *What are the major influences on the culture and art in each country?*

People of both countries speak Korean. They use the alphabet known as hangul in South Korea and as *Choson muntcha* in North Korea. The heritage of Buddhism, Confucianism, shamanism, and Christianity are important in South Korea. North Korea is atheist, and religion is not permitted. Despite their shared ethnicity, the political divisions between the two countries have created **cultural divergence**, or contrasts in the development of culture.

ethnicity relating to races or large groups of people classed according to common traits and customs

cultural divergence a separation of people or societies, with regard to beliefs, values, and customs, due to a division under different political systems

(t)DEA/G. Dagli Orti/De Agostini/Getty Images, (b)Interim Archives/Archive Photos/Getty Images

---

1391
Choson dynasty begins; it rules Korea until 1910.

1500s
Korea suffers devastating invasions by Japan.

1500 ➔

1636
After Mongol invasion, Korea seeks isolation from other countries.

1860s
Korea resists attempts by Great Britain, France, and United States to open Korea to trade.

1910
Japan annexes Korea.

1945 ➔

1945
Korea becomes free from Japan, but divided into two separate countries at the 38th parallel.

While North Korea and South Korea share similar cultural traits, there are also major differences between them. Government-provided medical care is free in North Korea. There is at least one clinic in every village. However, there are shortages of physicians and equipment. Most people in South Korea have medical insurance. There, the basic health care needs of the citizens are generally met. The life expectancy at birth for North Korean males was 61.2 years in 2009; for females it was 66.5 years. In South Korea, life expectancy at birth for males in 2008 was 76.5 years; for females it was 83.3 years.

Both countries value education. In North Korea, students receive a free and compulsory education for 11 years, starting with a year of preschool. Some older students are required to work while they are in school. The emphasis of all education is on science and technology. Institutions of higher education offer an additional two to six years of education. The most important institution of higher education is Kim Il-Sung University in P'yŏngyang.

Nine years of primary and middle school are compulsory for South Koreans. Nearly all who graduate from middle school go to high school or technical school. About 80 percent of high school graduates continue their education at a college or university. High school students undergo strenuous preparation for the highly competitive entrance exams to top universities. Most of the prestigious universities are in Seoul. In the twenty-first century, more and more South Koreans have pursued higher education abroad, with many studying in the United States.

Family and the Status of Women

After World War II, the occupying Soviets destroyed the lineage records of North Koreans. Since awareness of ancestry is an important component of Confucian practice, the fabric of Korean tradition was destroyed in that country. People are encouraged to spend their leisure time in activities that support the communist state. An extensive internal security apparatus monitors people's activities. The Korean Central News Agency censors newspapers. Government controls radio and television broadcasts. Internet is restricted to a few people, although illegal cell phone and Internet connections are more common in recent years.

Confucian traditions still affect family life in South Korea, and ancestral worship is central. People regard the most recent generation of ancestors as very much part of their family celebrations and rituals. South Koreans celebrate the first 100 days of a baby's life, marriages, and 61st birthdays with special family gatherings to honor their ancestors.

The roles of women in South Korea have changed as the extended nuclear family began to decline during rapid urbanization in the late twentieth century. Increasingly, women have become active in South Korea's economy. This has led to a decrease in family size and a lower birthrate. Women have gained equal rights in property ownership. In 2012 the first female president was elected in South Korea. Women in North Korea do not enjoy the same kinds of equality as those in South Korea. Many have entered the workforce out of economic necessity. Increased participation in the economy has given women more of a voice in decision making inside and outside the family.

The Arts

Cultural distinctions are noticeable in the arts. North Korea's Communist government heavily influences the country's art. In South Korea, Western culture has had a strong influence. Writers, artists, and dancers in North Korea are assigned to work for state-run theaters, orchestras, and other institutions. Their role is to promote traditional arts, commemorate the revolution, and express the superiority of Korean culture. The government maintains museums and archaeological sites to support its communist goals.

Mark Ralston/AFP/Getty Images

These North Korean women are performing a traditional fan dance.

◀ **CRITICAL THINKING**

1. *Analyzing* Why do you suppose the North Korean government supports traditional dance?

2. *Comparing and Contrasting* How are the goals of art similar and different in the two Koreas?

In South Korea, dancers, singers, and musicians preserve traditions. Dances with masked dancers and folk songs are still performed. Symphony orchestras, modern theater, rock concerts, and art galleries are also enjoyed by South Koreans.

✔ **READING PROGRESS CHECK**

Evaluating What are advantages of North Korea's health care system?

Economic Activities

GUIDING QUESTION *What industries have grown in importance in North Korea and South Korea?*

North Korea has a command economy in which the state plans and controls all economic production. Today, the economy's main industries produce metal products, machinery, military equipment, and chemicals. In its initial years, the focus was on development of heavy industry. Later it worked to improve technology and infrastructure. It was not until the 1970s and 1980s that North Korea's government began to pay attention to the production of agricultural products and consumer goods.

In the 1990s, a series of natural disasters occurred. These included floods and drought, which contributed to the problems of widespread starvation and malnutrition originally caused by government mismanagement. Subsequently, the North Korean government's highest priority has been to solve what it calls its "food problem."

Since the 1960s, South Korea has transformed itself from one of the world's poorest states to a highly industrialized society. Government and business

These workers at a South Korean aircraft factory are building a helicopter. The same factory hopes to start constructing aircraft for the U.S. Air Force.

▲ **CRITICAL THINKING**

1. ***Making Connections*** Why are exports so important to South Korea's economy?

2. ***Drawing Conclusions*** Why did North Korea's goal of economic self-reliance fail?

leaders developed a strategy focused on exports. They targeted specific industries for development, beginning with textiles and light manufacturing, and then moving to heavy industries such as iron, steel, chemicals, ship building, and automobiles. Manufacturing occurs in the Seoul region. Heavy industry is located in the south with access to seaports. High-tech industries, such as aerospace, electronics, and information technology, have thrived. Industries with global consumers, including smart phones, computers, and tablets, are a specialty.

Since the 1950s, the government of North Korea has stressed self-reliance as a goal in its economic planning. Nonetheless, it had trade relations with the Soviet Union and its Eastern European satellites, as well as with China. The severe economic hardships of the 1990s forced North Korea to accept foreign aid, seek foreign investment, and expand its range of trade partners. Now North Korea exports live animals and agricultural products, textiles and apparel, machinery, and mineral fuels and lubricants. It imports beverages, food and other agricultural products, mineral fuels, machinery, and textiles. Its major trading partners include China, South Korea, Russia, Japan, and Thailand.

Exports have been important to South Korea's economic growth. Many industries produce goods such as machinery, electronics, textiles, automobiles, and footwear for export around the world. South Korea's trade partners include the United States, Japan, the European Union, and the countries of Southeast Asia. South Korea imports raw materials, such as textile fibers, metal ores, and mineral fuels. The raw materials are converted into high-value exports known for their high quality.

✓ **READING PROGRESS CHECK**

Identifying Which country is a major trading partner with both countries?

LESSON 2 REVIEW

Reviewing Vocabulary

1. ***Understanding Relationships*** Write a paragraph discussing the reasons for the cultural divergence of North Korea and South Korea.

Using Your Notes

2. ***Listing*** Use your graphic organizer on the human geography of North Korea and South Korea to describe how physical and human systems have shaped these two countries.

Answering the Guiding Questions

3. ***Making Connections*** How has the decision to divide the Korean Peninsula into North Korea and South Korea affected the development of each country?

4. ***Contrasting*** How do settlement patterns and density differ in the two countries?

5. ***Comparing and Contrasting*** What are the major influences on the culture and art in each country?

6. ***Identifying*** What industries have grown in importance in North Korea and South Korea?

Writing Activity

7. ***Informative/Explanatory*** In a paragraph, explain how the Kaesong facility might have strengthened the relationship between the two Koreas had it continued to operate.

SeongJoon Cho/Bloomberg/Getty Images

networks

There's More Online!

- ☑ **CHART** South Korean Exports
- ☑ **IMAGE** Drift-Net Fishing
- ☑ **INFOGRAPHIC** Facing Empty Oceans
- ☑ **INTERACTIVE SELF-CHECK QUIZ**
- ☑ **VIDEO** People and Their Environment: North Korea and South Korea

LESSON 3
People and Their Environment: North Korea and South Korea

ESSENTIAL QUESTION · *How do physical systems and human systems shape a place?*

Reading HELPDESK

Academic Vocabulary

- **impact**
- **affect**

Content Vocabulary

- **deforestation**
- **drift-net fishing**
- **fertilizer**

TAKING NOTES: *Key Ideas and Details*

IDENTIFYING CAUSE AND EFFECT
As you read about the human geography of the Korean Peninsula, use a graphic organizer like the one below to identify causes of environmental problems.

Managing Resources	
Feature	Damage
Air	
Water	
Forests	
Fish	

IT MATTERS BECAUSE

The environments of North Korea and South Korea are affected by the air quality of other countries. The actions of North Koreans and South Koreans, in turn, affect their own air and water, as well the air and water of other countries.

Managing Resources

GUIDING QUESTION *What resources are at risk on the Korean Peninsula?*

Air pollution is a serious problem, especially in South Korea's cities. Because 80 percent of the population lives in cities, this problem has an **impact** on millions of people. Dust and pollutants that are carried by the winds from China and Mongolia are a big source of the problem. Pollutants from automobiles and industries also contribute to the poor air quality. A 2012 study found that South Korea had more pollutants in its air than either Japan or the United States.

South Korea's large cities also have problems with sewage and sanitation. Sanitation systems built after the Korean War are failing because of the impact of the country's rapid urbanization. Sanitation companies are unable to keep up with and properly dispose of the increasing amount of waste materials. Instead, they pump untreated sewage directly into the Pacific Ocean. Sewer systems also overflow and pollute local rivers and streams. Pollution of the seas and other bodies of water has **affected** fish supplies.

Population density has also had a negative effect on garbage disposal and sanitation systems in Seoul and other South Korean cities. These cities have many rules for garbage disposal and recycling, but they are not enforced or followed. On some streets, uncollected garbage is left in stacks.

North Korea also continues to experience its own problems with water and air pollution. A 2004 report found that almost 30,000 cubic meters, or about 7,925 gallons, of industrial wastewater was being poured every day into the Taedong River. This river flows through the middle of

(tl)Borut Furlan/age fotostock, (tc)Getty Images News/Getty Images, (tr)Chung Sung-Jun/Getty Images News/Getty Images

impact a significant or major effect

affect to produce an effect upon

North Korea's most densely populated city and its capital, P'yŏngyang. In North Korea, too, coal is used to heat most homes and provides power for most industries. Coal burning is a major cause of air pollution and acid rain.

✔ **READING PROGRESS CHECK**

Identifying What are two effects of pollution on natural resources in the Koreas?

Human Impact

GUIDING QUESTION *What human activities have affected the physical environment of North Korea and South Korea?*

deforestation the loss or destruction of forests, mainly from logging or farming

Deforestation has affected the entire Korean Peninsula. Forests once covered about two-thirds of South Korea and most of the western lowlands of North Korea. These forests are now nearly gone due to deforestation. Trees are cut for fuel for heating and cooking. Additionally, much land is also cleared

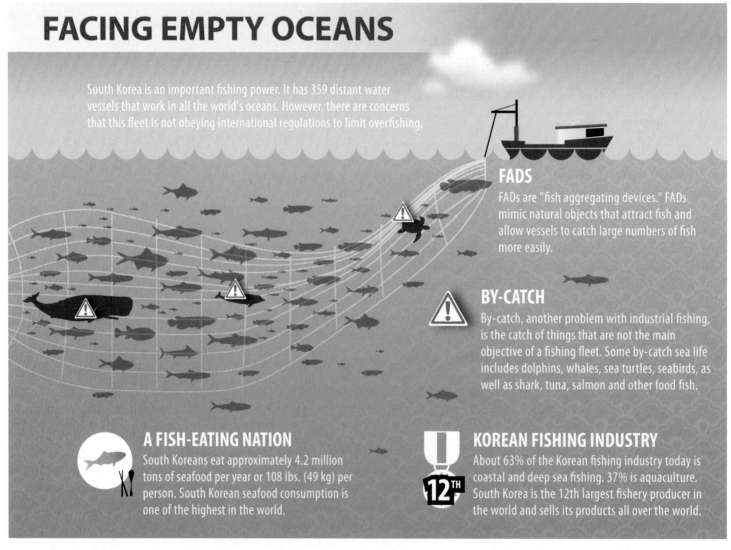

FACING EMPTY OCEANS

South Korea is an important fishing power. It has 359 distant water vessels that work in all the world's oceans. However, there are concerns that this fleet is not obeying international regulations to limit overfishing.

FADS
FADs are "fish aggregating devices." FADs mimic natural objects that attract fish and allow vessels to catch large numbers of fish more easily.

BY-CATCH
By-catch, another problem with industrial fishing, is the catch of things that are not the main objective of a fishing fleet. Some by-catch sea life includes dolphins, whales, sea turtles, seabirds, as well as shark, tuna, salmon and other food fish.

A FISH-EATING NATION
South Koreans eat approximately 4.2 million tons of seafood per year or 108 lbs. (49 kg) per person. South Korean seafood consumption is one of the highest in the world.

KOREAN FISHING INDUSTRY
About 63% of the Korean fishing industry today is coastal and deep sea fishing. 37% is aquaculture. South Korea is the 12th largest fishery producer in the world and sells its products all over the world.

Techniques commonly used by the South Korean fishing industry threaten the sustainability of fish populations in the region.

▲ **CRITICAL THINKING**

1. *Making Predictions* What could happen to South Korea's fishing industry if fishing practices are not changed?

2. *Constructing a Thesis* Do you think fish aggregating devices and drift-net fishing should be banned? Why or why not?

for cultivation and settlement as the population density increases in forested areas. The resulting loss of habitat has caused a decrease in the populations of many animals that once lived on the peninsula.

Deforestation has several terrible effects. For example, deforestation has led to soil erosion in North Korea. Trees and bushes hold soil in place during the heavy seasonal rains. Without them, soil washes down the steep hillsides of the mountainous terrain. This depletes nutrients that keep the soil healthy and make the land fertile for raising crops. Soil loss makes growing food more of a challenge for a country with ongoing food shortages.

Deforestation also leads to flooding. Aside from the direct loss of lives and property that floods can cause, flooding can also pollute drinking water. This occurs when a buildup of mud and debris is washed into the water supply. In 2012 severe floods polluted drinking water supplies in parts of North Korea. As a result, an estimated 50,000 families were left without potable, or drinkable, water.

The isolation and secrecy of the North Korean government make it difficult to monitor the environmental problems facing the country. A 2004 report by the United Nations Environment Programme makes it clear that the country faces multiple environmental problems. Data from 2012 indicate that the average temperature in North Korea had increased by almost 2°C as a result of deforestation. During this period, the world average temperature increase was only about 0.7°C.

While Japan has the world's largest fleet of drift-net fishing vessels, South Korea and Taiwan also have sizable drift-net operations. Environmentalists, the United States Coast Guard, and the Canadian Department of Fisheries and Wildlife are concerned about **drift-net fishing**, or fishing with huge nets. Drift-net fishing is condemned by many countries because it is so destructive. Sometimes called "walls of death," the huge nets are suspended vertically in the ocean and allowed to drift with the currents. Some of these nets are as much as 37 miles (60 km) long.

They are intended to trap squid and fish. However, sea mammals, such as dolphins, porpoises and whales, also die in these nets. So do turtles and birds. Animals like these die in such numbers that the nets are considered a threat to the marine ecosystem. It is estimated that about 80 percent of the animals caught in the nets are not even their intended targets.

In 1992 the United Nations created a resolution to end drift-net fishing. At the same time, the Convention for the Conservation of Anadromous Stocks (fish that migrate from salt water to freshwater) in the North Pacific Ocean was formed. The governments of the United States, Canada, Russia, and Japan signed this convention. South Korea agreed to it in 2003. However, there are few resources to monitor compliance with this agreement, and it is not clear that such fishing has declined.

✔ **READING PROGRESS CHECK**

Hypothesizing What might be the long-term effect of drift-net fishing on the South Korean fishing industry if it is not controlled?

Although unintended, drift-net fishing kills sea life. Several countries continue to practice drift-net fishing even though it violates a United Nations resolution.

▲ **CRITICAL THINKING**

1. ***Drawing Conclusions*** Do you think South Korea will really stop drift-net fishing while other countries still practice it? Why or why not?

2. ***Identifying Cause and Effect*** Why is drift-net fishing so threatening to the marine ecosystem?

drift-net fishing the use of fishing nets of great length and depth

Borut Furlan/age fotostock

Addressing the Issues

GUIDING QUESTION *How are environmental issues being addressed in North Korea and South Korea?*

In 2012 North Korea invited scientists from eight countries to a conference in P'yŏngyang to review the condition of the country's environment. Scientists found barren landscapes on the brink of collapse, a result of decades of environmental degradation. They developed plans for restoring the environment and increasing food security in North Korea.

Deforestation has reached alarming levels in North Korea. According to one study, North Korea was the third most deforested country in the world, after Nigeria and Indonesia. However, North Korea has also started a series of reforestation programs. Such programs are planned for Kandong province, located outside P'yŏngyang. Other programs intend to reforest the western lower hillsides and eastern highlands.

South Korea, which has succeeded in restoring forests to nearly two-thirds of the country, is also attempting to promote reforestation in North Korea. South Korea's former president, Lee Myung-bak, pledged to send seedlings to North Korea. Many private South Korean citizens have donated trees to North Korea as well. Increasing forest coverage is a good step, but these young forests are very different from the old-growth forests that were cut down. They are much less diverse and offer minimal habitat for animals.

fertilizer a substance, such as manure or a chemical mixture, used to make soil more fertile

The North Korean government has also encouraged the use of **fertilizer** to restore soil nutrients to help the trees grow back healthy. The government has provided fertilizer to farmers. However, there is evidence that, rather than use the fertilizer, farmers are selling it to cover their costs.

South Korea is also working on increasing its energy efficiency. In the twenty-first century, the government announced a plan to change every light-bulb in every public building to energy-efficient LED lights.

An ambitious stimulus package launched in 2009 allocated money to research low-carbon technologies, build high-speed railways, and expand bus lanes. The effort also supports building one million "green" homes and improving the efficiency of one million existing homes. While South Korea has yet to solve all of its environmental problems, the country is making substantial progress.

☑ **READING PROGRESS CHECK**

Predicting How might an effective reforestation program improve life for North Koreans?

LESSON 3 REVIEW

Reviewing Vocabulary
1. ***Explaining*** Write a paragraph explaining the impact of deforestation on North Korea's agriculture.

Using Your Notes
2. ***Identifying Cause and Effect*** Use your graphic organizer on the effects of the Korean Peninsula's environmental problems to write a paragraph about the cause and effect of one type of pollution on the peninsula.

Answering the Guiding Questions
3. ***Evaluating*** What resources are at risk on the Korean Peninsula?

4. ***Analyzing*** What human activities have affected the physical environment of North Korea and South Korea?

5. ***Summarizing*** How are environmental issues being addressed in North Korea and South Korea?

Writing Activity
6. ***Informative/Explanatory*** Use reliable Internet sources to locate an article about environmental issues in North Korea or South Korea over the past year. Write a paragraph summarizing the main points of the article.

Directions: On a separate sheet of paper, answer the questions below. Make sure you read carefully and answer all parts of the questions.

Lesson Review

Lesson 1

1 *Comparing* Why do you think the capitals of North Korea and South Korea began as river ports?

2 *Explaining* How did the creation of the Demilitarized Zone affect wildlife in the area?

3 *Drawing Conclusions* Why does mountainous terrain make livestock production and fishing particularly important on the Korean Peninsula?

Lesson 2

4 *Summarizing* How has North Korea's population changed since the middle of the twentieth century?

5 *Contrasting* Describe how life expectancy in South Korea differs from life expectancy in North Korea.

6 *Describing* What are some methods the North Korean government uses to control free speech and limit individual freedoms?

Lesson 3

7 *Explaining* How does deforestation contribute to food shortages in North Korea?

8 *Identifying Cause and Effect* What are some problems caused by population density in South Korea's cities?

9 *Analyzing* Why do you think South Korea has taken steps to address the problem of deforestation in North Korea?

Critical Thinking

10 *Making Predictions* How might reunification on the Korean Peninsula affect the economy of South Korea?

11 *Hypothesizing* South Korea has taken steps to improve energy efficiency. How might these steps help improve air quality in urban areas?

21st Century Skills

Use the graph to answer the following questions.

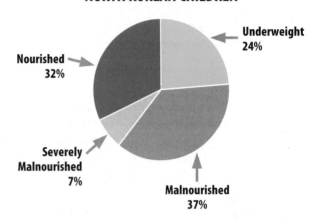

NORTH KOREAN CHILDREN

Underweight 24%

Nourished 32%

Severely Malnourished 7%

Malnourished 37%

Source: http://han-schneider.org

12 *Creating and Using Graphs, Charts, Diagrams, and Tables* According to the circle graph, what percentage of children in North Korea faces problems with nourishment?

13 *Creating and Using Graphs, Charts, Diagrams, and Tables* Why do you think the graph includes the percentage of children who are underweight?

14 *Identifying Perspectives and Differing Interpretations* Based on the graph, how severe do you think the problem of malnutrition is among North Korean children?

College and Career Readiness

15 *Problem Solving* Imagine you have been hired by the United Nations World Food Program to find solutions to the problem of food shortages in North Korea. Use the Internet to research the causes and consequences of the problem. How have natural disasters affected the availability of food in that country? What role has the government played? Write a paragraph outlining steps the North Korean government can take to solve the problem.

Need Extra Help?

If You've Missed Question	1	2	3	4	5	6	7	8	9	10	11	12	13	14	15
Go to page	728	729	729	733	734	734	739	737	740	735	740	734	734	734	734

Directions: On a separate sheet of paper, answer the questions below. Make sure you read carefully and answer all parts of the questions.

DBQ Analyzing Primary Sources

Use the document to answer the following questions.

People who have fled North Korea report that the country's economy relies on forced labor. Even children are forced to work, as a former student describes below.

PRIMARY SOURCE

"When I was between 11 and 15 years old I had to work on the government farm almost every day. . . . We finished class at 1 P.M. and had to rush back home to eat lunch because the school didn't provide food for the students. The school would announce that we'd have to meet back at the school field and bring our own farm tools. . . . They forced everyone, even the small children, to work. I felt bad because this didn't benefit our family and I had many responsibilities to do for my family but the government forced me to work for them. I was always very exhausted as a child."

—Human Rights Watch, "North Korea: Economic System Built on Forced Labor," June 13, 2012

16 *Drawing Conclusions* What does the use of children as forced labor imply about the North Korean economy?

17 *Interpreting* The students were told to bring their own tools from home. How does this detail help you understand the problem?

18 *Analyzing* What features of the farm in the description show that it is a cooperative?

Applying Map Skills

Use the Unit 8 Atlas to answer the following questions.

19 *Human Systems* Compare population density in North and South Korea. Which country is more urbanized?

20 *Places and Regions* Use your mental map of East Asia to identify the major rivers in North Korea and South Korea. Which river forms the border between North Korea and China?

21 *Environment and Society* How has the geography of the Korean Peninsula affected agricultural activity?

Exploring the Essential Question

22 *Analyzing* How has the division of Korea into two countries affected human systems on the Korean Peninsula? Write a paragraph describing the changes, including population density, use of resources, and culture.

Research and Presentation

23 *Research Skills* Using the Internet, research the development of the North Korean nuclear program. How have other countries responded to these nuclear tests? Create a time line showing your findings and share it with the class.

Writing About Geography

24 *Narrative* Use the Internet to research and write a two-page essay describing the Demilitarized Zone. How does the DMZ create a buffer between North and South Korea? What features of the DMZ make it an important nature preserve?

Need Extra Help?

If You've Missed Question	16	17	18	19	20	21	22	23	24
Go to page	733	733	729	676	672	672	730	731	729

Southeast Asia and the Pacific World

UNIT **9**

| **Chapter 31** | **Chapter 32** | **Chapter 33** |
| Southeast Asia | Australia and New Zealand | Oceania |

Alex Linghorn/Photolibrary/Getty Images

travelstock44/age fotostock

1 Diversity Southeast Asia and the Pacific World is a diverse region. Some ethnic groups have maintained their traditions while other cultures have blended together.

EXPLORE the REGION

Stretching from the eastern border of India east nearly to South America, and from China in the north to the Southern Ocean, **SOUTHEAST ASIA** and the **PACIFIC WORLD** is a vast region dominated by the sea. With populous countries like Indonesia, countries with large land areas like Australia, and small island countries with few people, this region shows great variety of land-forms, people, and cultures.

THERE'S MORE ONLINE

3 Environment The region has dense rain forests and unique plants and animals. Various countries are taking steps to protect these natural wonders from the threats they face.

4 Water Almost all of the countries in the region have coastlines, and some countries rely on rivers. Fishing and shipping are important activities in the region.

(t)Purestock/PunchStock,
(bl)Nora Carol Photography/Flickr/Getty Images,
(br)Jonathan Saruk/Getty Images News/Getty Images

2 Landforms The region has many different landforms, from soaring, glacier-covered mountains to low islands that rise just a few feet above the sea.

PACIFIC OCEAN

N
W · E
S

Hkakabo Razi
19,295 ft.
(5,881 m)

EAST ASIA

TROPIC OF CANCER

20°N

Irrawaddy R.
Salween R.
Red R.
Chao Phraya R.
Mekong R.

Gulf of Tonkin

South China Sea

Luzon Strait

Luzon

Philippine Sea

Mariana Islands

Guam

M I C R O N E S I A

Wake Island

Marshall Islands

Bay of Bengal

Mt. Pinatubo
5,770 ft.
(1,759 m)

Andaman Sea

Gulf of Thailand

Malay Peninsula

Sulu Sea

Mindanao

Palau

Caroline Islands

For eastern Oceania, see inset below

Strait of Malacca

Celebes Sea

EQUATOR

0°

Sumatra

Borneo

Sulawesi

Moluccas

M E L A N E S I A

Bismarck Archipelago

Nauru

Gilbert Islands

Greater Sunda Islands

Banda Sea

Puncak Jaya
16,535 ft.
(5,040 m)

New Guinea

Mt. Wilhelm
14,762 ft.
(4,500 m)

Tuvalu

Sunda Strait

Java Sea

Bali

Flores Sea

Timor Sea

Arafura Sea

Solomon Islands

Wallis and Futuna

Krakatau

Java

Lesser Sunda Islands

Timor

Cape York Peninsula

Coral Sea

Vanuatu

Fiji

Gulf of Carpentaria

Great Barrier Reef

New Caledonia

TROPIC OF CAPRICORN

Tonga

Elevations

10,000 ft. (3,000 m)
5,000 ft. (1,500 m)
2,000 ft. (600 m)
1,000 ft. (300 m)
0 ft. (0 m)
Below sea level

— National boundary
▲ Mountain peak
▼ Lowest point

Great Sandy Desert

WESTERN PLATEAU

Macdonnell Ranges

Central Lowlands

GREAT DIVIDING RANGE

Gibson Desert

Great Victoria Desert

Lake Eyre (dry) ▼
-52 ft.
(-16 m)

GREAT ARTESIAN BASIN

PACIFIC OCEAN

Darling R.

Nullarbor Plain

Murray R.

0   1,000 miles
0   1,000 kilometers

Mercator projection

40°S

Great Australian Bight

Mt. Kosciuszko
7,310 ft.
(2,228 m)

Tasman Sea

North Island

Cook Strait

Bass Strait

South Island

Southern Alps

Chatham Islands

Tasmania

Mt. Cook
12,349 ft.
(3,764 m)

120°E   140°E   160°E   180°

MAP STUDY

1. *Physical Systems* What large islands make up the country of Indonesia?

2. *Environment and Society* Where are most major cities in Australia located? What about the physical geography of the country explains these settlement patterns?

EQUATOR

Line Islands

0   1,000 miles
0   1,000 kilometers

Mercator projection

N
W · E
S

0°

Gilbert Islands

Phoenix Islands

Tuvalu

Tokelau

Samoa Islands

P O L Y N E S I A

PACIFIC OCEAN

Marquesas Islands

Wallis and Futuna

Fiji

Niue

Society Islands

Tuamotu Archipelago

Tonga

Cook Islands

Tahiti

20°S

TROPIC OF CAPRICORN

Pitcairn Island

180°   160°W   140°W

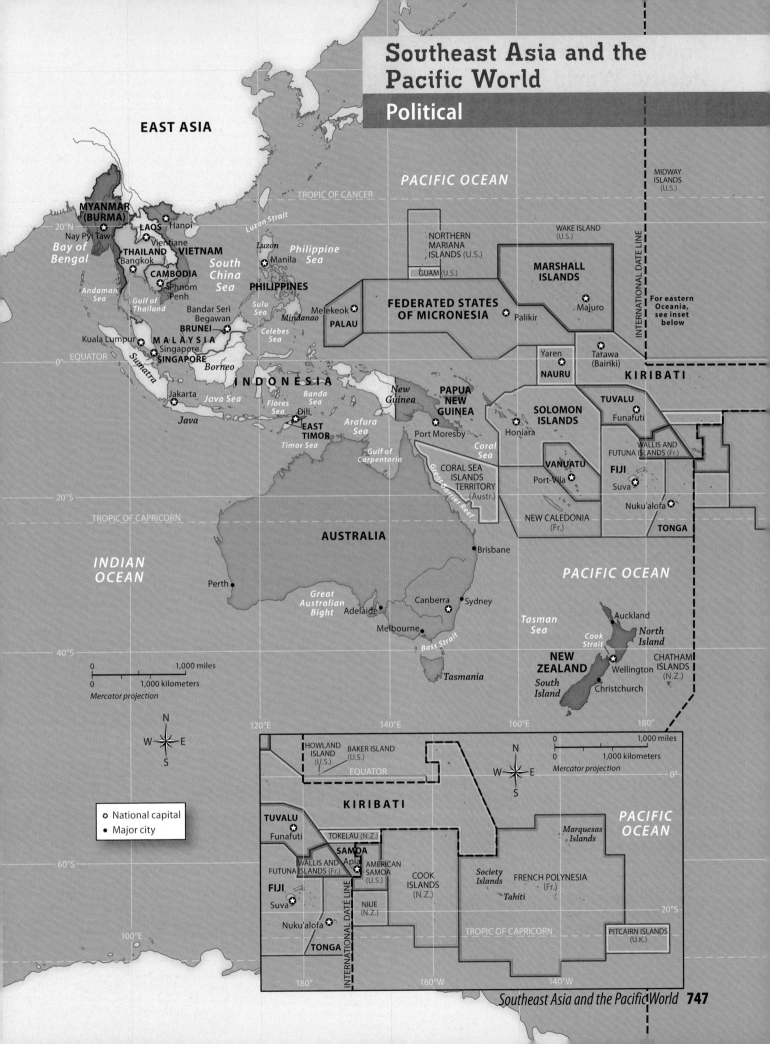

Southeast Asia and the Pacific World

Political

EAST ASIA

PACIFIC OCEAN

MIDWAY ISLANDS (U.S.)

TROPIC OF CANCER

WAKE ISLAND (U.S.)

MYANMAR (BURMA)
Nay Pyi Taw
Hanoi
Luzon Strait
Bay of Bengal
LAOS
Vientiane
THAILAND
Bangkok
VIETNAM
Luzon
Philippine Sea
Manila
NORTHERN MARIANA ISLANDS (U.S.)
GUAM (U.S.)
MARSHALL ISLANDS
Majuro

INTERNATIONAL DATE LINE

For eastern Oceania, see inset below

Andaman Sea
CAMBODIA
Sphnom Penh
South China Sea
PHILIPPINES
Mindanao
FEDERATED STATES OF MICRONESIA
Palikir
Gulf of Thailand
Bandar Seri Begawan
Sulu Sea
Melekeok
PALAU

BRUNEI
Celebes Sea
Kuala Lumpur
MALAYSIA
Singapore
SINGAPORE
Borneo
Yaren
NAURU
Tarawa (Bairiki)
KIRIBATI

EQUATOR
Sumatra
INDONESIA
Jakarta
Java Sea
Banda Sea
New Guinea
PAPUA NEW GUINEA
SOLOMON ISLANDS
Honiara
TUVALU
Funafuti
WALLIS AND FUTUNA ISLANDS (Fr.)

Java
Flores Sea
Dili
EAST TIMOR
Timor Sea
Arafura Sea
Port Moresby
Coral Sea
VANUATU
Port-Vila
FIJI
Suva

Gulf of Carpentaria
CORAL SEA ISLANDS TERRITORY (Austr.)
Great Barrier Reef
Nuku'alofa

20°S
TROPIC OF CAPRICORN
NEW CALEDONIA (Fr.)
TONGA

INDIAN OCEAN
AUSTRALIA
Brisbane
PACIFIC OCEAN

Perth
Great Australian Bight
Adelaide
Canberra
Sydney
Tasman Sea
Auckland
North Island

Melbourne
Bass Strait
Cook Strait
NEW ZEALAND
Wellington
CHATHAM ISLANDS (N.Z.)

40°S
Tasmania
South Island
Christchurch

0   1,000 miles
0   1,000 kilometers
Mercator projection

120°E   140°E   160°E   180°

N
W   E
S

HOWLAND ISLAND (U.S.)
BAKER ISLAND (U.S.)
EQUATOR

0   1,000 miles
0   1,000 kilometers
Mercator projection

N
W   E
S

KIRIBATI

PACIFIC OCEAN

TUVALU
Funafuti
TOKELAU (N.Z.)
Marquesas Islands

SAMOA
Apia
AMERICAN SAMOA (U.S.)
WALLIS AND FUTUNA ISLANDS (Fr.)

FIJI
Suva
NIUE (N.Z.)
COOK ISLANDS (N.Z.)
Society Islands
FRENCH POLYNESIA (Fr.)
Tahiti

Nuku'alofa

20°S

Legend	
⊛ National capital	
• Major city	

60°S
TROPIC OF CAPRICORN
PITCAIRN ISLANDS (U.K.)

TONGA

100°E   180°   160°W   140°W

INTERNATIONAL DATE LINE

Southeast Asia and the Pacific World

Climate and Vegetation

EAST ASIA

PACIFIC OCEAN

TROPIC OF CANCER

20°N
Nay Pyi Taw
Hanoi
Gulf of Tonkin
Luzon Strait
South China Sea
Bay of Bengal
Philippine Sea
Manila
Andaman Sea
Phnom Penh
Gulf of Thailand
Sulu Sea
Celebes Sea
Strait of Malacca
0° EQUATOR
Singapore
Majuro
Java Sea
Flores Sea
Banda Sea
Dili
Timor Sea
Arafura Sea
Port Moresby
Honiara
Funafuti
Gulf of Carpentaria
Great Barrier Reef
Coral Sea
Suva
20°S
TROPIC OF CAPRICORN
Brisbane
PACIFIC OCEAN
Perth
Great Australian Bight
Canberra
Sydney
Tasman Sea
Cook Strait
INDIAN OCEAN
40°S
Bass Strait
Wellington

Climate
- Tropical rain forest
- Tropical wet/dry
- Semi-arid (steppe)
- Arid (desert)
- Humid subtropical
- Marine west coast
- Mediterranean

0 — 1,000 miles
0 — 1,000 kilometers
Mercator projection

120°E   140°E   160°E   180°

MAP STUDY

1. **Physical Systems** Which parts of the region would you expect to have ample rainfall? Which would you expect to be dry?

2. **The World in Spatial Terms** What generalization can you make about the location of centers of manufacturing in the region?

Vegetation
- Tropical rain forest
- Tropical grassland (savanna)
- Desert scrub and desert waste
- Temperate grassland
- Deciduous forest
- Mixed forest (deciduous and coniferous)

TROPIC OF CANCER
20°N
EQUATOR
20°S
TROPIC OF CAPRICORN
40°S
100°E   120°E   140°E   160°E   180°

Southeast Asia and the Pacific World

Economic Activity

EAST ASIA

PACIFIC OCEAN

TROPIC OF CANCER

20°N

Bay of Bengal

Andaman Sea

Gulf of Thailand

South China Sea

Luzon Strait

Philippine Sea

Sulu Sea

Celebes Sea

EQUATOR

0°

Java Sea

Flores Sea

Banda Sea

Arafura Sea

Timor Sea

Gulf of Carpentaria

Coral Sea

For eastern Oceania, see inset below

TROPIC OF CAPRICORN

PACIFIC OCEAN

Tasman Sea

Cook Strait

Great Australian Bight

Great Barrier Reef

N W E S

Resources

Coal	Nickel
Petroleum	Copper
Natural gas	Lead
Iron ore	Manganese
Tin	Gold
Zinc	Silver
Bauxite	Platinum
Cobalt	Gems
Uranium	

Land Use

Commercial farming
Subsistence farming
Livestock raising
Primarily forest
■ Manufacturing and trade
Commercial fishing
Little or no activity

0   1,000 miles
0   1,000 kilometers
Mercator projection

40°S

INDIAN OCEAN

120°E   140°E   160°E   180°

EQUATOR

N W E S

0   1,000 miles
0   1,000 kilometers
Mercator projection

PACIFIC OCEAN

20°S

TROPIC OF CAPRICORN

180°   160°W   140°W

Southeast Asia and the Pacific World
Population Density

PACIFIC OCEAN

EAST ASIA

TROPIC OF CANCER

Mandalay

Hanoi
Haiphong

20°N
Yangon
(Rangoon)
Bay of
Bengal

Luzon Strait

South
China
Sea

Philippine
Sea

Bangkok

Manila

Phnom Penh

Andaman
Sea

Ho Chi Minh
City

Gulf of
Thailand

Sulu
Sea

Davao

For eastern
Oceania,
see inset
below

Medan

Kuala Lumpur

Celebes
Sea

EQUATOR
Singapore

0°

Palembang

Java Sea
Jakarta
Semarang
Bandung
Surabaya

Ujung
Pandang

Banda
Sea

Flores
Sea

INDIAN
OCEAN

Timor
Sea

Arafura
Sea

Gulf of
Carpentaria

Coral Sea

Great Barrier Reef

PACIFIC OCEAN

20°S

TROPIC OF CAPRICORN

Brisbane

Perth

Great
Australian
Bight

Adelaide

Newcastle
Sydney

Auckland

Melbourne

Tasman
Sea

Cook
Strait

40°S

Bass Strait

Cities
(Statistics reflect metropolitan areas.)
- ■ Over 5,000,000
- ☐ 2,000,000–5,000,000
- ⊙ 1,000,000–2,000,000

POPULATION

Per sq. mi.		Per sq. km
1,250 and over		500 and over
250–1,249		100–499
63–249		25–99
25–62		10–24
2.5–24		1–9
Less than 2.5		Less than 1

0   1,000 miles

0   1,000 kilometers

Mercator projection

120°E   140°E   160°E   180°

EQUATOR

0   1,000 miles

0   1,000 kilometers

Mercator projection

0°

PACIFIC
OCEAN

20°S

TROPIC OF CAPRICORN

180°   160°W   140°W

MAP STUDY

1. *Human Systems* Which countries in Southeast Asia are the most densely populated?

2. *Human Systems* Why do you think the heaviest concentration of population in Australia is along the eastern coast?

Southeast Asia

ESSENTIAL QUESTION • *How do physical systems and human systems shape a place?*

Geography Matters...

Southeast Asia is one of the most diverse regions on Earth. All the world's religions, political systems, and economic systems can be found here—and so can hundreds of ethnic groups speaking hundreds of different languages. The physical geography is just as diverse, with towering volcanoes, broad river deltas, and a wide variety of tropical biomes and climates. The region includes a mainland area and two large island archipelagoes.

This young woman in Bali, Indonesia, wears an intricate gold headdress.

©Dallas and John Heaton/Free Agents Limited/Corbis

emerging markets *in* Southeast Asia

*A country that is moving toward a more complex economy as the result of rapid industrialization, economic growth, and integration into the global economic system is an emerging market. In **Southeast Asia**, emerging markets are booming. Indonesia, Malaysia, and Vietnam—three of Southeast Asia's roaring "tigers"—are among the fastest-growing economies in the region.*

Why are businesses investing in Indonesia, Malaysia, and Vietnam?

During the late twentieth century, Southeast Asia underwent an economic transformation. From the 1970s to the 1990s, traditionally agricultural countries began to industrialize. Today, Indonesia, Malaysia, and Vietnam manufacture shoes, clothing, electronics, and many other goods, most of which are exported around the world. Industrialization has led to economic growth in Southeast Asia. Between 1967 and 1997, Indonesia's gross domestic product (GDP) grew 6 percent per year. Since adopting a constitution in 1992 that encouraged the development of a market economy, Vietnam has experienced impressive economic growth. Vietnam's economy is growing about 7 percent each year. Both Vietnam and Malaysia have some of the lowest unemployment rates in the world. Thriving industries have boosted international trade and attracted European, Chinese, and American businesses to the region.

1. **Human Systems** What economic changes have taken place in these countries?

How did these three countries become classified as emerging markets?

Indonesia, Malaysia, and Vietnam share characteristics that helped them develop strong economies. First, each country has abundant natural resources. For example, Indonesia has oil, tin and other minerals, timber, coal, and natural gas. Second, these countries have plenty of highly skilled workers and low labor costs. Although several Southeast Asian countries raised the minimum wage in 2013, workers here still earn much less per hour compared to workers in Europe, North America, and many other parts of the world. Third, the location of these countries spurs trade. Southeast Asia is strategically located between the South China Sea and the Indian Ocean. The Strait of Malacca—a choke point that connects the two bodies of water—is one of the most important shipping routes in the world. Finally, the governments of Indonesia, Malaysia, and Vietnam adopted policies that foster trade and business. These policies include building new roads, airports, and ports; improving education; and streamlining processes to make it easier for people to start businesses.

2. **Places and Regions** What four factors led to Indonesia, Malaysia, and Vietnam becoming classified as emerging markets?

What are some risks of rapid economic growth?

Indonesia, Malaysia, and Vietnam have impressively expanded their economies, yet the widening gap between the rich and the poor in these countries often creates tensions. At times, economic inequality has led to labor unrest, corruption, violent clashes, and political and social instability. In 2011, for instance, copper miners at the Grasberg Mine in Indonesia went on strike for higher wages and large protests demanding political reforms erupted in Malaysia. Rapid growth can also involve uneven economic development. Indonesia, for example, has relied heavily on its oil and natural gas resources for economic growth, but its manufacturing sector has fallen behind. In addition, these emerging market countries have developed greater vulnerability to the ripple effects of the global recession. Even the most vibrant emerging markets can be negatively affected by a struggling worldwide economy.

3. **Human Systems** What are the economic challenges these countries face?

(l)Hoang Dinh Nam/AFP/Getty Images, (c)Iryna Rasko/age fotostock, (r)Achmad Ibrahim/AP Images

THERE'S MORE ONLINE

READ a chart of Southeast Asia's major exports • **VIEW** a graph showing GDP of Southeast Asian countries

networks

There's More Online!

☑ **IMAGE** Mekong River Delta

☑ **IMAGE** Population Density in Southeast Asia

☑ **MAP** Volcanoes in Southeast Asia

☑ **INTERACTIVE SELF-CHECK QUIZ**

☑ **VIDEO** Physical Geography of Southeast Asia

LESSON 1
Physical Geography of Southeast Asia

Reading HELPDESK

Academic Vocabulary

- **access**
- **differentiate**

Content Vocabulary

- **tsunami**
- **cyclone**
- **typhoon**
- **biodiversity**

TAKING NOTES: *Key Ideas and Details*

IDENTIFYING As you read about the physical geography of Southeast Asia, use a diagram like the one below to note similarities and differences between the features of the mainland and those of the islands and archipelagoes.

Mainland Both Islands and Archipelagoes

ESSENTIAL QUESTION · *How do physical systems and human systems shape a place?*

IT MATTERS BECAUSE

Few regions have such diverse physical features as Southeast Asia. The Pacific Ocean dominates this subregion where the combined landmass of island countries is greater than the combined landmass of countries on the mainland. Southeast Asia's physical geography also shows the effects of the Pacific Ring of Fire. Most Southeast Asian countries are mountainous, and many have volcanoes and are prone to earthquakes.

Landforms

GUIDING QUESTION *What are the major physical characteristics of Southeast Asia?*

The 11 countries of Southeast Asia can be divided into two areas: the countries of the mainland in the west, and island archipelagoes in the east. Cambodia, Vietnam, Thailand, Myanmar (Burma), Laos, and the city-state of Singapore make up most of the subregion's mainland. Malaysia is partly on the Malay Peninsula and partly on the island of Borneo. In the east, Indonesia, East Timor, Brunei, and the Philippines form island countries that stretch across the western half of the Pacific Ocean.

It is important to realize how large Southeast Asia's island countries are. Their landmass is actually larger than that of the subregion's mainland countries. Indonesia consists of more than 13,000 islands that span a distance of 1,000 miles (1,609 km) from north to south and 3,000 miles (4,828 km) from west to east—an area wider than the continental United States. The Philippines is a close second to Indonesia, with more than 7,000 islands and an area close in size to the state of Arizona. Countries on the mainland are also large. Myanmar is larger than the country of France, while Laos is almost as large as the entire United Kingdom.

Southeast Asia also includes some of the world's tiniest countries. On the Malay Peninsula, the tiny city-state of Singapore, at 433 square miles (697 sq. km), is approximately a third of the size of Rhode Island. Brunei, which borders Malaysia on the island of Borneo, is a little smaller than Delaware.

(tl)Universal Images Group/Getty Images, (tr)Bartosz Hadyniak/the Agency Collection/Getty Images

Southeast Asia's position along the Pacific Ocean places it within the Ring of Fire, a belt of volcanoes and tectonic plate boundaries that surrounds the Pacific. About 75 percent of the world's volcanoes are located along the Ring of Fire. Most of the world's strongest earthquakes also occur here because of the moving and colliding of Earth's tectonic plates. As tectonic plates move, they produce earthquakes and volcanoes. In much of Southeast Asia, earthquakes and volcanic eruptions are common. Volcanic eruptions are more typical in the subregion's island countries. One exception, however, is the large island of Borneo. Borneo is often called a mini-continent because it is so stable. This stability makes Borneo's physical geography an exception to the pattern of the rest of the subregion. Although it has mountains, Borneo has no active volcanoes and rarely experiences earthquakes.

The rest of Indonesia, in contrast, has the largest concentration of active volcanoes in the world. In 2010 Mount Merapi erupted, sending hot gases high into the air and dropping hot ash and debris on nearby villages. As Merapi's explosions grew more violent, Indonesia's government evacuated nearby villages and cut off **access** to them. It was two weeks before villagers were able to return.

Eruptions in Indonesia have changed world history. In 1815 Mount Tambora sent so much ash into the sky that the year was dubbed "the year without a summer" because ash blocked so much solar energy. Crops failed and millions went hungry in places as far away as the United States, Europe, and northern Africa. In 1883 Krakatau exploded, killing some 36,000 people. Indonesia's most destructive volcanic eruption is thought to have occurred about 73,000 years ago, when Mount Toba exploded. That eruption changed weather patterns worldwide for the next 20 years.

Some of Southeast Asia's earthquakes occur under water. When this happens, they can produce **tsunamis**, dangerous huge waves that can flood inland areas with little warning. This is what happened in 2004, when an underwater

access a way to approach or enter a place

tsunami a dangerous large ocean wave caused by an underwater earthquake, a volcanic eruption, or a sudden displacement of the ocean floor

GEOGRAPHY CONNECTION

Southeast Asia is located along the Ring of Fire, an area of frequent volcanic eruptions and earthquakes.

1. ***THE WORLD IN SPATIAL TERMS***
What countries that are not part of Southeast Asia are also affected by the Ring of Fire around the Pacific Ocean?

2. ***ENVIRONMENT AND SOCIETY***
In what ways are people affected by volcanic eruptions?

Volcanoes in Southeast Asia

▲ Active volcano
〰 Tectonic plate boundary

Large earthquakes under the sea (those greater than 7.0 on the Richter Scale) may cause the ocean floor to move vertically, displacing water and changing the waves at the surface. At sea, waves may only be a few inches taller than usual. When these waves reach a coast, however, they may be many feet taller than usual, producing a tsunami. In 1883 earthquakes preceding Krakatau's eruption produced a 98-foot- (30-m-) high tsunami.

IDENTIFYING CAUSE AND EFFECT
What is the connection between undersea earthquakes and tsunamis?

differentiate to distinguish or recognize as different

earthquake in Indonesia triggered a devastating tsunami in the Indian Ocean. The tsunami generated a series of waves, killing more than 300,000 people.

The same tectonic forces that produce volcanoes also produced the mountains for which this subregion is famous. Not all Southeast Asian countries are as vulnerable to volcanic eruptions as Indonesia. All, however, are known for their mountains and hills. These mountains have folded together into a fanlike shape as the Indian subcontinental plate has collided with the Eurasian plate over the past 50 million years. Mountain ranges on the mainland run from north to south and form a natural backbone for most mainland countries in this region. Mountains also form natural borders. For example, the border between Thailand and Myanmar is marked by a mountain range. So is Thailand's border with China. Another major mountain range divides Laos and Vietnam. The island archipelagoes and the Malay Peninsula are also mountainous. Many elevations in the region exceed 10,000 feet (3,000 m).

✔ **READING PROGRESS CHECK**

Explaining How does the Ring of Fire affect the landforms of Southeast Asia?

Waterways

GUIDING QUESTION *What are the major waterways of Southeast Asia?*

In much of Southeast Asia, everyday life revolves around water. Rivers provide water for crops and serve as navigable arteries for trade. Much of the subregion's population lives along the rivers. This is especially the case on the mainland, where the population has become concentrated in certain areas. The Pacific Ocean is also ideal for transportation, which has made it the basis of international trade networks and much of the subregion's tourism. Southeast Asia's waterways have made it one of the most accessible regions in the world. This accessibility has shaped the region's economy and its politics. Not surprisingly, waterways have historically been important to Southeast Asian cultures as well.

Southeast Asia's most dominant waterway, the Pacific Ocean, holds the archipelagoes of the eastern half of the region. But Southeast Asia's mainland countries also depend on the ocean for trade, fishing, and tourism. Rivers on the islands are mostly short, steep, and fast-flowing. The vast majority of Southeast Asia's population is concentrated on the islands of Indonesia and the Philippines.

The mainland has five major rivers: the Mekong (sometimes called the Danube of Southeast Asia), the Red River, the Irrawaddy, the Salween, and the Chao Phraya. Three of these rivers—the Irrawaddy, the Salween, and the Mekong—originate in the mountains of Tibet in China. Southeast Asia's river systems **differentiate** the farming economies of the lowland valley and delta areas from the hill and mountain communities that separate the river valleys.

The Mekong River is the longest river in Southeast Asia and the seventh longest in Asia. It flows south for some 2,600 miles (4,200 km) across the mainland to the South China Sea. As it progresses, the Mekong passes through or along the borders of five Southeast Asian countries: Myanmar, Laos, Thailand, Cambodia, and Vietnam. It forms Laos's border with Myanmar and part of its border with Thailand. Access to the Mekong and its tributaries is the basis of several international disputes. The countries most affected by the Mekong—Laos, Thailand, Cambodia, and Vietnam—have formed a committee to study water projects involving the river.

By contrast, the mainland's other four major rivers are shorter and each is mostly (though not entirely) contained within a single country. The Irrawaddy and the Salween form part of the drainage system for Myanmar's western mountains. The Chao Phraya is the shortest of the four rivers and drains

Thailand's western mountains. Thailand's capital, Bangkok, is located on the Chao Phraya's delta. The Red River, though it has a smaller drainage area than the other rivers, provides Vietnam with a second river delta (in addition to that of the Mekong). Both of these deltas provide rich soil for farming.

☑ **READING PROGRESS CHECK**

Explaining How do the major waterways of Southeast Asia affect population distribution?

Climates, Biomes, and Resources

GUIDING QUESTION *How does climate affect human activities in Southeast Asia?*

Southeast Asia's climate is tropical and subtropical. Its seasonal changes are based more on rainfall than on temperature. Most of Southeast Asia, like many tropical oceanic regions, divides the year into just two seasons: the dry season and the rainy monsoon season. These seasons are created by the changing wind patterns over the Indian and Pacific Oceans. For most of the mainland, the rainy season is from May to September when the monsoon winds blow from the Indian Ocean. Between November and March, the winds blow in a reverse direction and make the season drier. Some parts of Malaysia, Indonesia, and the Philippines have a rainy season that lasts all year, however. They are surrounded by the Pacific and Indian Oceans so that the winds bring rains no matter which way they are blowing. During the rainy season, rainfall is quite heavy. Most of Southeast Asia gets more than 60 inches (150 cm) of rain each year. Some areas get two or three times that amount. For example, Ranong, on the coast of Thailand, receives some 160 inches (400 cm) of rain each year.

During the rainy season, some weather systems become severe tropical storms. These storms are called **cyclones** in the Indian Ocean and **typhoons** in the western Pacific Ocean. They are the same kind of storm as Atlantic hurricanes. In late 2013, Typhoon Haiyan struck in what meteorologists refer to as the "typhoon belt" of the Philippines. Haiyan was among the strongest storms ever recorded. Besides devastating winds, it produced torrential rains that caused flash floods which destroyed villages. Thousands died and many more were left homeless.

Climate Regions and Biomes

Overall, the average annual temperature in most Southeast Asian countries is around 80°F (27°C). But temperatures in Southeast Asia, like temperatures around the world, tend to vary by elevation. Mountainous regions are differentiated from river lowland deltas not only by their different landforms, but also by their cooler temperatures. Air temperatures typically drop by about 0.9°F (0.5°C) for every 328 feet (100 m) climbed.

Some tourists to Southeast Asia head for the cooler highlands. Other tourists go to the coastal areas to experience the beaches and coral reefs. They find that the heat of these areas is somewhat mitigated by the sea breezes.

This satellite image helps show population density in Southeast Asia.

▲ **CRITICAL THINKING**

1. **Comparing** In what areas of Southeast Asia is the population most dense? What geographic features are common in areas of high population density?

2. **Identifying** Do you see a connection between Southeast Asia's waterways and the borders of countries? What countries seem to have been affected by the locations of rivers?

cyclone a spiral-shaped tropical storm with winds of at least 74 miles (119 km) per hour and with heavy rain; occurs in the western South Pacific or Indian Oceans

typhoon a cyclone that occurs in the western North Pacific Ocean (west of the International Date Line)

Universal Images Group/Getty Images

shatter belt a region where political alliances are constantly splintering and fracturing based on ethnicity

buffer zone a neutral area serving to separate powerful countries or those that are hostile to each other

united hundreds of ethnic groups into one country. The British united Malaysia's nine sultanates with the indigenous groups of northern Borneo. The Dutch colonized indigenous people on hundreds of islands who spoke different languages. The colony became Indonesia. Thailand remained an independent country called Siam because it served as a buffer zone between the British and the French colonies.

Independence and Movements for Change

The era of European colonialism did not last. The first independence movement began in the Philippines. Following the Spanish-American War, the Filipino revolutionaries targeted the U.S. military. In World War II the Philippines were invaded by Japan. World War II ended in 1945, and the Philippines became independent in 1946. The last U.S. military base in the Philippines was closed in 1992 following the eruption of the Mount Pinatubo volcano.

The end of World War II started the independence movement in Southeast Asia. Britain's colonies in the region gained independence: Myanmar in 1948; Malaysia and Singapore in the 1960s. Indonesia declared independence in 1946. French Indochina had a strong colonial administration that resisted independence movements. After 1946 the colonies of France became flash points for struggles between groups supporting either democratic or communist leaders.

Independence from colonial powers did not always lead to peace. Cambodia became independent in 1953. In the 1970s it fell to the Khmer Rouge, a brutal regime that murdered at least 1.5 million Cambodians. In Vietnam, France fought to keep control until it was finally defeated at Dien Bien Phu in 1954. Vietnam was divided between communist forces in the north and a U.S.-supported government in the south. U.S. support for the democratic Republic of Vietnam ended in 1975. In 1976 North and South Vietnam were unified under a communist government.

Cambodia and Vietnam were not the only countries in the region to have internal conflicts. Thailand had a military coup in 2006, followed by mass protests in the streets from 2008 to 2010. When Thailand held elections in 2011, the new government began reforming the Thai constitution to be more democratic.

AFP/Getty Images

TIME LINE ⌄

COLONIZATION
and Independence ➜

After decades or even centuries of colonization, Southeast Asian countries won their independence, but liberation brought problems of its own.

▶ CRITICAL THINKING

1. *Describing* How did the United States gain territory in Southeast Asia?

2. *Comparing* How were the experiences of colonization and independence similar in Laos and Vietnam?

Dutch colonists begin to settle Indonesia in order to control spice trade

1602

1521 ➜

1521 First Spaniards arrive in what they later name the Philippines in honor of the heir to the Spanish throne.

Ho Chi Minh founds Indo-Chinese Communist Party in Vietnam.

1930

1850s–1890s French colonial rule established in Indochina; United States wins Philippines from Spain

The country also struggles with an armed separatist revolt in its southern provinces, where the population is ethnically Malay rather than Thai. In Burma, a military coup overthrew the government in 1988 and renamed the country Myanmar. A military junta has ruled ever since, although democratic elections were held in 1990. After the election, the junta put the leader of the winning opposition party, Aung San Suu Kyi, under house arrest. In 2007 the junta killed 13 people and arrested thousands more who participated in protests led by pro-democracy activists and Buddhist monks. Elections were held in 2010, and Myanmar is now slowly reforming its government.

Several Southeast Asian countries such as Indonesia and the Philippines have guerrilla independence movements. The Philippines has struggled to achieve a peace accord with an ethnic group called the Moro in the south and is threatened by Maoist guerrillas throughout the entire country. Indonesia made peace with its separatists in 2005, though it still faces an independence movement in Papua. East Timor is an example of a local independence movement that succeeded. East Timor was invaded by Indonesia after declaring independence from Portugal in 1975. Over the next 24 years, hundreds of thousands of East Timorese died fighting for independence. In 1999 a majority of East Timorese voted for independence in an election supervised by the United Nations (UN). The Indonesian military violently retaliated, however. Later that year, Australian-led UN peacekeeping troops stopped the bloodshed. In 2002 East Timor became independent. East Timor continued to struggle with a rebel group for several more years, but most rebels surrendered in 2008 after unsuccessfully trying to attack East Timor's president and prime minister.

Today, Southeast Asia's political map is diverse. The subregion includes communist governments, emerging democracies, and recent military dictatorships. There is also a sultanate (Brunei) and a city-state (Singapore).

☑ READING PROGRESS CHECK

Explaining How did European colonialism affect the countries of Southeast Asia?

Keystone/Hulton Archive/Getty Images

- Japan invades and occupies most of Southeast Asia during World War II.

1941–1945

- France reimposes protectorate over Indochina; communist guerrillas rise up against the French in Vietnam

1946

- After independence, civil war occurs in Laos; Vietnam divided at the 17th parallel, followed by war

1954

1940 ➔

1954 ➔

1935
- United States creates plan for Philippine independence, eventually granted in 1946

1948
Burma becomes independent; U Nu becomes first prime minister

1953
Cambodia becomes independent, led by King Norodom Sihanouk.

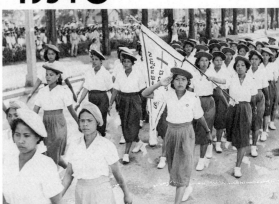

Population Patterns

GUIDING QUESTION *How has Southeast Asia's physical geography influenced its population patterns?*

Southeast Asia's mainland countries are not densely populated. The soil on the volcanic islands tends to be more productive than soil in the region's mainland. River valleys and flood plains of major rivers, such as the Mekong, are the exception with rich alluvial soil. Islands and river plains are where the population density is greatest. Southeast Asia's largest population center is located on Java, an island in Indonesia. Almost 150 million people live on this one island in this country that is entirely made up of islands. More than half of all Indonesians live on Java, and more than half of the people on Java work as farmers.

Despite Indonesia's agrarian economy, it is the fourth most populous country in the world. After Indonesia, the next most populous country in Southeast Asia is the Philippines. Southeast Asia is becoming more urban as it industrializes. Each country has at least one large city: Kuala Lumpur in Malaysia; Bangkok in Thailand; Manila in the Philippines; Yangon(Rangoon) in Myanmar; Hanoi and Ho Chi Minh City in Vietnam; and the city-state of Singapore. All of these cities have become a thriving part of the global economy.

☑ **READING PROGRESS CHECK**

Explaining What parts of Southeast Asia have the highest population density?

Society and Culture Today

GUIDING QUESTION *Why is Southeast Asia one of the world's most culturally diverse areas?*

Southeast Asia encompasses hundreds of different ethnic groups. Most ethnic groups have their own language. The greatest ethnic diversity is found in the island countries. For example, Indonesia's ethnic groups include Javanese, Madurese, Sundanese, and Balinese. In addition, while Malays predominate in Malaysia, they have also emigrated to nearby areas. As a result, there are enclaves of Malay minorities throughout Indonesia, and Indonesia's official language is a modified form of Malay.

The borders of Southeast Asia's mainland countries match the locations of each country's majority ethnic group. In Myanmar, which is also called Burma, Burmans form the majority. The Thai dominate Thailand. Cambodia's population is mostly Khmer, and Vietnam's population is primarily Vietnamese. Each of these countries has other, smaller ethnic groups that live in the mountains. In addition to indigenous ethnic groups, Southeast Asian countries are home to minority ethnic groups that originally emigrated from other countries. For example, Malaysia and Singapore have large Indian communities. However, the largest minority ethnic group in the region is Chinese. About 32 million Chinese live in Southeast Asia, primarily in urban areas. Nearly one-fourth of Malaysia's population is Chinese. The population of Singapore is about 77 percent Chinese.

Angkor Wat is a Cambodian temple complex built in the twelfth century. It was built as a Hindu temple, but was later turned into a Buddhist temple.

▼ **CRITICAL THINKING**

1. *Drawing Conclusions* Why do you think that the Angkor Wat temple continued to be important to Cambodians even after local religious traditions had changed?

2. *Hypothesizing* Why do you think an image of Angkor Wat appears on the Cambodian national flag?

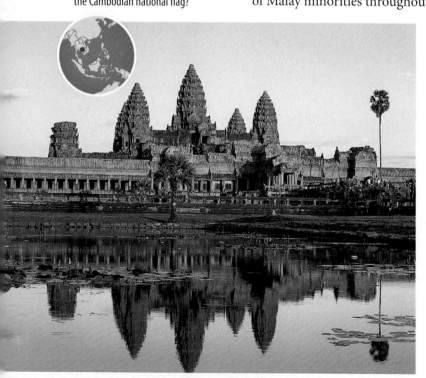

Digital Vision/Punchstock

Southeast Asia's religions reflect the waves of immigration that have reached this subregion. The most common religion in Southeast Asia is Islam. Muslim traders who traveled to the region in the twelfth and thirteenth centuries introduced the religion to Southeast Asia. Islam is the dominant religion in Southeast Asia's largest country, Indonesia. About 90 percent of Indonesian people are Muslim. So are most Malaysians. Malaysia's South Asian minority, though, tend to be Hindu, while Chinese Malaysians are mostly Buddhist. Buddhism predominates in the following mainland countries as well: Myanmar, Thailand, and Cambodia. Vietnam historically served as a religious crossroads for Buddhism, Catholicism, Daoism, Confucianism, and ancestor worship. In the last census, however, more than 80 percent of the people of Vietnam, a communist country, claimed no religion. In the Philippines, where Spanish missionaries accompanied colonists, most people are Catholic.

Family and the Status of Women

Women in Southeast Asia have traditionally been responsible for raising families but are also a major part of the workforce. This produces a double burden for women who must both work outside the home and still keep up with the daily household responsibilities. In rural areas, women often work as farmers. In cities, women frequently work in factories, especially in industries such as textiles, food processing, and electronics.

Increasingly, Southeast Asian women are rising to positions of leadership as activists for change. In Myanmar, activist Aung San Suu Kyi leads a democratic opposition party that often wins national elections. But Aung San Suu Kyi has not become president of Myanmar. Instead, she has spent much of her life under house arrest because of her outspoken opposition to the military government. She was awarded the Nobel Peace Prize in 1991 for her work to bring democracy to Myanmar. After the government liberalized policies, she was elected to the lower house of the Myanmar parliament in 2012.

In 1991 a young Indonesian woman and law student, Dita Sari, began leading factory workers to strike for higher wages and better working conditions. She was jailed and beaten, but was later released as Indonesia became more democratic. Today Sari is still a union leader, but instead of leading strikes, she lobbies Indonesia's parliament for labor law reforms.

The Arts

The arts in Southeast Asia have been heavily influenced by religion, which can be seen through the architectural style of Buddhist and Hindu temples and monuments. Chinese and Indian styles are prevalent in Southeast Asian ceramics and bronze, as well as architecture. Elaborate Chinese-style pagodas and Indian-style *wats*, or temples, dot the landscape. Traditional crafts such as weaving and other textile techniques such as batik are still practiced.

✔ **READING PROGRESS CHECK**

Identifying What are the dominant religions of Southeast Asia?

This woman is selling produce in a market in Bukittinggi on the Indonesian island of Sumatra. Originally the site of a Dutch fort and now an important commercial center, the town reflects a multiethnic heritage.

▲ **CRITICAL THINKING**

1. *Drawing Conclusions* How do you think that people of many different ethnic groups, speaking many different languages, are able to coexist peacefully and conduct transactions with each other in urban markets?

2. *Considering Advantages and Disadvantages* What are some of the advantages and disadvantages of Indonesia's ethnic diversity?

Tibor Bognar/age fotostock

Workers are drilling in preparation for the construction of a railroad that will run from China into several Southeast Asian mainland countries.

▲ CRITICAL THINKING

1. Identifying Cause and Effect
Why do you think Southeast Asian countries are cooperating with China to build a railroad? How do you think the new railroad will affect trade networks in the region?

2. Making Connections What characteristics of Southeast Asia's geography make the construction of railroads a challenge?

emerging market an economy that may not have been very strong in the recent past, but that is in transition to becoming a stronger market

free port a place where goods can be unloaded, stored, or reshipped free of import duties

Economic Activities

GUIDING QUESTION *How have Southeast Asia's location and natural resources contributed to its economic development?*

Southeast Asia is as diverse economically as it is politically. It holds one of the world's most prosperous cities, Singapore, and a wealthy sultanate, Brunei. The subregion also includes some countries that are among the world's poorest, such as Myanmar. During the 1980s, the industrializing countries of Southeast Asia enjoyed an economic boom. This boom was based on advantages of location, natural resources, inexpensive labor, and increased foreign investment. Some countries that did not have strong economies in the past—such as Indonesia, Malaysia, and Vietnam—are now considered to be **emerging markets**, ripe for foreign investment.

Today agriculture remains Southeast Asia's leading economic activity. Most people in the subregion make their living as farmers. More than half of the subregion's arable land is used to grow rice. Thailand and Vietnam are among the world's top exporters of rice. Indonesia and Myanmar are also two of the region's major rice producers. Farmers in Southeast Asia also grow cassava, yams, corn, bananas, sugarcane, coffee, coconuts, and spices. Thailand, Indonesia, and Malaysia—the world's "rubber belt"—have many rubber plantations. Palm oil, a product of oil palm trees, is an important cash crop in Malaysia and Indonesia. Forestry and logging have also become important in Malaysia, the Philippines, Indonesia, and Thailand. Forests include teak and ebony trees, which are very valuable. The high value results in illegal logging and export to other countries. Along the coasts and rivers, many people make a living by fishing.

Resources, Power, and Industry

Southeast Asia is rich in mineral resources. Thailand, Malaysia, and Indonesia are three of the world's leading producers of tin. Iron ore is mined in Malaysia and the Philippines. Brunei, Malaysia, and Indonesia also produce oil. Indonesia once exported much of its oil. Now that it is industrializing, however, it has begun to import more oil than it exports. Malaysia uses some of its resources for industries such as manufacturing electronics, cement, chemicals, and processed foods. Indonesia's industries focus on textile and garment manufacturing.

Economic Integration

Southeast Asia has long been the crossroads of major ocean trade routes. Today, most shipping between Europe and East Asia passes through the Strait of Malacca, near Singapore. The city-state of Singapore was originally founded as a British trading colony. Not only does Singapore have a strategic location at the crossing of trade routes, but it has a large, deep natural harbor. Historically, this harbor was used by British warships to displace the economic dominance of the Dutch in the region. Today, its location enables Singapore to prosper as a **free port**, a place where goods can be unloaded, stored, and reshipped free of import duties. Its port is the largest container port in the world. These large metal containers may be stacked on ships for crossing oceans, and then carried by rail or truck after they reach ports. Singapore has also attracted foreign investors, especially in technology, consumer electronics, and pharmaceuticals. Singapore's efforts to build its economy have paid off. Its population has the highest standard of living in the subregion, and its per-capita income levels are comparable to those of the United States and Switzerland.

Hoang Dinh Nam/AFP/Getty Images

Human Impact

GUIDING QUESTION *Why do urban growth and industrialization create environmental problems in Southeast Asia?*

Industrialization and economic development in Southeast Asia have resulted in the pollution and the destruction of natural environments. Southeast Asians—like people everywhere—affect their environment. Increased manufacturing and industrialization create jobs and raise standards of living. They also produce industrial waste, however. Growing populations and crowded conditions in cities such as Bangkok, Manila, and Jakarta raise concerns about adequate housing, water supplies, sanitation, and traffic control. Bangkok, for example, has become overheated as a result of increased industrial development. In recent years, Bangkok's heat, humidity, and pollution levels have increased at levels higher than the global average. Its levels are also higher than the surrounding rural areas, making it what is known as an urban heat island.

Urbanization and industrialization also put a strain on shared local resources. Such resources include rivers and water supplies. When different communities share a river, the communities upstream have an advantage because the water flows there first. Currently, Southeast Asian communities face a dispute over how to best manage the shared Mekong River. The Mekong originates in China, and the Chinese are building a series of hydroelectric dams across it. Downstream farmers worry, however, about what will happen if the water level in the Mekong drops. Cambodians are concerned about the Tonle Sap lake, which gets its water from the Mekong. Vietnamese farmers fear that Mekong Delta rice paddies could be invaded by salt water from the sea if the Mekong's water levels drop.

The dumping of toxic waste from newly developed industries has become a widespread problem. In some cases, pollution extends into rural areas. This includes the subregion's national parks. In one of Thailand's national parks, for example, most of the freshwater wells have been contaminated by poor waste disposal. Water pollution is also severe in Thailand's coastal areas. Activities related to agriculture, as well as offshore oil and natural gas exploration, have resulted in the loss of over half of Thailand's mangrove forests.

One industry that produces substantial water pollution is mining. At Indonesia's largest gold mine, waste is dumped into the Ajkwa River that flows through part of Papua province, contaminating the surrounding water systems. Even the fishing industry causes some water pollution. Southeast Asian fishers use poisons and explosives to capture certain kinds of fish to supply Asian restaurants and the world's aquarium industry. Many of these fish come from the coral reefs. Like rain forests, coral reefs are a rich source of biodiversity. Also like rain forests, the reefs are being threatened by economic development. The live reef-fish trade generates huge profits, but current fishing methods are destroying the reefs. Once the reefs are gone, local communities will lose a primary source of food.

Deforestation has become a source of air pollution. Land is often cleared by fire, especially in Indonesia, where rain forests are being cleared to make room for palm oil plantations. Forest fires cause air pollution in rural areas, and this pollution often makes its way to cities

This river in Vietnam has been polluted by drainage from mines. Water pollution has become a serious problem in parts of Southeast Asia.

▼ **CRITICAL THINKING**

1. *Explaining* How does water pollution affect human populations?

2. *Problem Solving* What steps might governments in Southeast Asia take to reduce water pollution?

Terry Whittaker/FLPA/age fotostock

These children in Thailand are planting tree seedlings. Several Southeast Asian countries have introduced reforestation programs to conserve forests.

▲ **CRITICAL THINKING**

1. Drawing Conclusions Do you think that reforestation programs will be enough to save Southeast Asia's forests? Why or why not?

2. Drawing Inferences What might be the purpose of teaching children to plant trees?

sustainable development
technological and economic growth that does not deplete the human and natural resources of a given area

reforestation the planting and cultivating of new trees in an effort to restore a forest that has been reduced by fire or by cutting

as well. Since the late 1990s, Indonesian forest fires have created pollution and respiratory problems for people as far away as mainland Malaysia and Singapore. The Malaysian government has been forced on occasion to declare a state of emergency. Singapore's government often advises residents who have heart or respiratory conditions to avoid going outside.

☑ **READING PROGRESS CHECK**

Identifying What environmental challenges are the results of human economic activities in Southeast Asia?

Addressing the Issues

GUIDING QUESTION *How are people and governments addressing environmental issues in Southeast Asia?*

Many Southeast Asian countries are shifting their economic policies to focus on **sustainable development**. This strategy means encouraging technological and economic growth that relies on renewable resources. For example, to prevent further loss of rain forests, Thailand, Indonesia, the Philippines, and Malaysia have limited certain timber exports. They have also introduced **reforestation** programs. In Laos, the government has encouraged some highland farmers to resettle on more fertile ground and marked some of its forests as protected areas.

Southeast Asian governments are also starting to develop ideas to reduce the impact of urban growth on the surrounding environment. Scientists are trying to come up with solutions to urban heat islands. One proposal is to create "green zones," areas within a city that are granted special environmental protection. Another idea is to ban the construction of tall buildings near the sea so that winds would be able to blow farther into the city and disperse more air pollution.

Southeast Asia's environmental issues have attracted widespread interest. The Association of Southeast Asian Nations (ASEAN) has tried to address some of these issues. In 2002 ASEAN countries formed an agreement on Transboundary Haze Pollution. Indonesia refused to ratify it. However, countries around the world are still pressuring Indonesia to conserve its rain forests. In 2012 Indonesia announced a two-year ban on forest clearing.

☑ **READING PROGRESS CHECK**

Identifying How are Southeast Asian governments working to limit deforestation?

Eco Images/Universal Images Group/Getty Images

LESSON 3 REVIEW

Reviewing Vocabulary

1. Identifying Write a paragraph explaining what sustainable development is and why it is important.

Using Your Notes

2. Summarizing Use your graphic organizer on the causes and effects of environmental challenges in Southeast Asia to write a paragraph explaining how economic development is connected to environmental challenges.

Answering the Guiding Questions

3. Identifying Cause and Effect How has the management of natural resources impacted the environment in Southeast Asia?

4. Drawing Conclusions Why do urban growth and industrialization create environmental problems in Southeast Asia?

5. Identifying How are people and governments addressing environmental issues in Southeast Asia?

Writing Activity

6. Informative/Explanatory In a paragraph, explain how Southeast Asia's urban and rural areas differ from each other in terms of their effects on the environment.

Directions: On a separate sheet of paper, answer the questions below. Make sure that you read carefully and answer all parts of the questions.

Lesson Review

Lesson 1

1 *Specifying* Discuss three examples of volcanic eruptions in Indonesia and how they have changed world history.

2 *Assessing* In most of Southeast Asia, the soil is poor. What causes the soil in some areas to be richer and more productive? With mostly poor soil, why does plant growth remain strong throughout most of Southeast Asia?

3 *Drawing Conclusions* Explain whether it would be accurate to state that Southeast Asia's rain forests are an example of biodiversity.

Lesson 2

4 *Identifying Cause and Effect* Discuss an example of a successful independence movement in which a Southeast Asian country gained independence from a colonial power.

5 *Summarizing* Write a summary about the religions practiced in Southeast Asia. Specify the most common religion and briefly explain how it came to be practiced in the region.

6 *Drawing Conclusions* How do Singapore and Myanmar represent the opposite extremes of economic development in Southeast Asia?

Lesson 3

7 *Analyzing* What is the greatest threat to rain forests in Indonesia and Malaysia? Why does the possibility of eliminating this threat pose a challenge to these countries?

8 *Considering Advantages and Disadvantages* Explain the advantages and disadvantages of manufacturing and industrialization in Southeast Asia.

9 *Describing* Describe two activities that Southeast Asian governments have undertaken to reduce the impact of urban growth on the surrounding environment.

21st Century Skills

Use the graph below to answer the questions that follow.

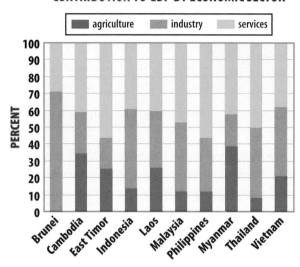

CONTRIBUTION TO GDP BY ECONOMIC SECTOR

Source: The CIA World Factbook.

10 *Using Charts* Within which countries are the contributions to GDP by industry the greatest? What is the percentage for each? Why is industry more successful there?

11 *Creating Diagrams* Explain whether Brunei's agricultural contribution to GDP seems logical when considering the size of the country.

Critical Thinking

12 *Making Connections* Write a one-page essay explaining the role of agriculture in Southeast Asia's economy. Identify specific crops and other economic activities. Explain where agriculture ranks overall economically in relation to the other activities.

College and Career Readiness

13 *Clear Communication* Suppose you want to publish biographies of Aung San Suu Kyi and Dita Sari. Write a one-page proposal about each of these women and her connection to Southeast Asia. In your proposal, explain each woman's significance in the region's recent history.

Need Extra Help?

If You've Missed Question	❶	❷	❸	❹	❺	❻	❼	❽	❾	❿	⓫	⓬	⓭
Go to page	755	758	758	760	763	764	767	769	770	771	771	764	763

Directions: On a separate sheet of paper, answer the questions below. Make sure that you read carefully and answer all parts of the questions.

DBQ **Analyzing Primary Sources**

Read the excerpts and answer the following questions.

There was great division of opinion regarding the way the United States entered and participated in the war in Vietnam.

PRIMARY SOURCE

"*I know we oughtn't to be there, but I can't get out. . . . I just can't be the architect of surrender.***"**

—President Lyndon B. Johnson, commenting on Vietnam to Senator Eugene McCarthy, February 1966

"*I saw courage both in the Vietnam War and in the struggle to stop it. I learned that patriotism includes protest, not just military service. But you don't have to go half way around the world or march on Washington to learn about bravery or love of country. Again and again, in the causes that define our nation, we have seen the uncommon courage that is common to the American people.***"**

—John Kerry, announcement speech, Patriots Point, South Carolina, September 2, 2003 (as prepared for delivery)

14 *Understanding Historical Interpretation* Explain how the quote by President Johnson indicates his attitude toward withdrawal of U.S. troops from Vietnam. How does this quote help you to assess why the United States remained involved in combat in Vietnam for such a long period of time?

15 *Interpreting Significance* Explain whether Kerry's statement indicates that he is aligned with Johnson's attitudes and action regarding Vietnam. How does Kerry's statement indicate a widespread attitude regarding U.S. involvement in Vietnam?

Writing About Geography

16 *Informative/Explanatory* Use standard grammar, spelling, sentence structure, and punctuation to write a one-page essay discussing family and the current status of women in Southeast Asia.

Applying Map Skills

Refer to the Unit 9 Atlas to answer the following questions.

17 *Places and Regions* Identify the five major rivers of Southeast Asia's mainland countries. Of these rivers, which is the shortest? Which is the longest?

18 *The World in Spatial Terms* Use your mental map of Southeast Asia to describe the spatial relationship between Thailand and Myanmar. Explain whether travel on foot would be easy or difficult between these two countries.

19 *Human Systems* Write a general statement to compare and contrast the populations of the Philippines and Indonesia with the populations of other countries in Southeast Asia.

Exploring the Essential Question

20 *Organizing* Select one of the countries from the chapter. Then suppose you are developing a website to provide information explaining how physical systems and human systems have shaped that country. Create the home page, including relevant links. For each link, write a two-to-three paragraph summary of the information to be included.

Research and Presentation

21 *Gathering Information* With a partner, conduct research to learn more about the Association of Southeast Asian Nations (ASEAN). Create a multimedia presentation to explain the following: members, mission, and growth of the organization, as well as its past and present activities. Explain whether you think the organization is effective currently—and whether you think the organization will be effective in the future. Within your presentation, include video, maps, photographs, and diagrams or graphs.

Need Extra Help?

If You've Missed Question	**14**	**15**	**16**	**17**	**18**	**19**	**20**	**21**
Go to page	772	772	763	746	746	746	765	765

Australia and New Zealand

ESSENTIAL QUESTION · *How do physical systems and human systems shape a place?*

netw⊙rks

There's More Online about the geography of Australia and New Zealand.

CHAPTER 32

Geography Matters...

Australia and New Zealand have rich cultural heritages influenced by indigenous peoples, English settlers, and more recent immigrants.

The subregion faces many challenges. Environmental threats include climate change, deforestation, and the introduction of invasive plant and animal species which have no natural predators. The countries of Australia and New Zealand are making strides in addressing both cultural and environmental issues, but more work is needed. With the help of individuals, companies, organizations, and governments, change is moving forward.

◀ This young man has the Australian flag painted on his face.

©Marianna Massey/Corbis

Australia *and* New Zealand: indigenous peoples

Australia and New Zealand are home to indigenous peoples, or the people who lived in a place before it was conquered by colonial societies. Australia's Aborigines (A•buh•RIH•juh•neez) and New Zealand's Maori (MAWR•ee) have their own languages, customs, and traditions. This made it more difficult for them to live among European settlers. In the twentieth century, Aborigines and Maori protested unfair treatment. They pressed for their political, economic, social, cultural, and civil rights.

Andrew Holt/Photographer's Choice/Getty Images

774

(l)Holger Leue/Lonely Planet Images/Getty Images, (c)Charles O. Cecil/Alamy, (r)©Matthew McKee/Eye Ubiquitous/Corbis

How do the indigenous peoples in Australia and New Zealand differ?

Both Australian Aborigines and the Maori migrated from Southeast Asia. The Aboriginal people arrived in Australia about 40,000 to 60,000 years ago. Over time, these nomadic hunters and gatherers dispersed across the entire continent. They divided into different groups and developed distinct cultures with as many as 700 languages.

Unlike the Aborigines, the Maori arrived in New Zealand sometime in the A.D. 1200s. The Maori remained in close contact with one another in a relatively small geographic area and were isolated from other societies. Therefore, Maori culture is less diffuse than Australia's Aboriginal culture. For instance, Maori today speak dialects of the same ancestral language and share similar beliefs. Although Aboriginal and Maori societies developed differently, British colonization greatly affected both peoples.

1. The Uses of Geography How did geography affect the development of Aboriginal and Maori cultures?

Why have both populations been denied basic rights?

European explorers visited Australia and New Zealand in the 1600s. The British government established a colony for convicts in Australia in 1788. Free settlers soon followed. In the 1820s, Europeans settled in New Zealand. Colonization, however, caused friction. Indigenous peoples and European colonists competed for control of valuable resources such as land and water. Social, political, and cultural differences between colonists and indigenous peoples also led to conflicts. Aborigines and Maori frequently endured discrimination. The Australian government believed the Aborigines would not survive colonization and placed many in reserve stations. It provided them with food and clothing but restricted their movements. The government also forcibly removed about 100,000 Aboriginal children from their families in order to absorb them into white society. These children are referred to as the "Stolen Generations." In New Zealand in 1840, Maori chiefs and British representatives signed the Treaty of Waitangi, but many of the Maori rights (such as land ownership) granted by the treaty were ignored, resulting in many years of war and conflict.

2. Human Systems What caused Aborigines and Maori to be denied their rights?

What steps are being taken to restore the rights of these indigenous peoples?

In the twentieth century, Australia's Aborigines and New Zealand's Maori began to pressure their governments to restore their rights. Activists fought to preserve their land, language, and way of life through political protests. Eventually, they regained land, fishing, and other basic rights. Children belonging to the "Stolen Generations" received a national apology in 2008. The Australian government later adopted a formal policy aimed at preserving Aboriginal languages. In the 1990s, New Zealand's government began to honor the Treaty of Waitangi. The government transferred land and control of important fisheries and paid millions of dollars in claim settlements to the Maori. New Zealand made Maori an official language and established Maori-language schools and a Maori-language television station. These indigenous peoples, however, still suffer disproportionately from poor health and lower life expectancies due to poverty, unemployment, and other factors.

3. Environment and Society What role has the environment played in helping Aborigines and Maori regain their rights?

THERE'S MORE ONLINE

WATCH a video about the lives of Aborigines • *SEE* an image of protests by Aborigines and Maori

netw⊙rks

There's More Online!

- ☑ **IMAGE** Australian Outback
- ☑ **IMAGE** Hopetoun Falls, Australia
- ☑ **IMAGE** Kings Canyon, Australia
- ☑ **INFOGRAPHIC** Life in the Great Barrier Reef
- ☑ **INTERACTIVE SELF-CHECK QUIZ**
- ☑ **VIDEO** Physical Geography of Australia and New Zealand

LESSON 1
Physical Geography of Australia and New Zealand

ESSENTIAL QUESTION · *How do physical systems and human systems shape a place?*

Reading **HELP**DESK

Academic Vocabulary

- **resource**
- **dominate**
- **benefit**

Content Vocabulary

- **bush**
- **sunken mountains**
- **atoll**
- **caldera**
- **lagoon**
- **artesian well**
- **coral**

TAKING NOTES: *Key Ideas and Details*

IDENTIFYING As you read about the geography of Australia and New Zealand, use a graphic organizer like the one below to note the characteristics of the land, water, and biomes of the subregion.

Australia and New Zealand		
Land	Water	Biomes

IT MATTERS BECAUSE

Australia and New Zealand are part of the diverse South Pacific region, which includes sandy beaches, snow-capped mountains, volcanoes, stunning reefs, grasslands, and deserts. Among these varied physical features are unique species of wildlife that attract tourists and scientists alike.

Landforms

GUIDING QUESTION *How do the landforms of Australia and New Zealand influence the economies of these countries?*

Like an island, Australia is surrounded by water. However, geographers classify Australia as a continent because of its massive size. Australia lies between the Pacific Ocean and Indian Ocean in the Southern Hemisphere, and it has a largely flat terrain.

A chain of hills and mountains known as the Great Dividing Range interrupts the level landscape. The peaks of the range stretch along the eastern coast from the Cape York Peninsula to the island of Tasmania. The Western Plateau, a low area of flat land in central and western Australia, covers almost two-thirds of the continent. Few people live in the Western Plateau. It is made up of three deserts: the Great Sandy, the Gibson, and the Great Victoria Deserts. These areas are collectively called the Outback. Many Australians also call these sparsely inhabited areas the **bush**.

South of Great Victoria Desert is the Nullarbor Plain. The name is from the Latin *nullus arbor,* meaning "no tree." On the southern edge of this dry, treeless landscape, giant cliffs tower above the Great Australian Bight, a part of the Indian Ocean. The Western Plateau and the Great Dividing Range are separated by the Central Lowlands. This area of arid grassland and desert stretches across the east-central part of Australia. The large expanses of dry regions make most of Australia unsuitable for agriculture. As a result, it relies upon the extraction of mineral **resources** for much of its industry and exports.

(l)Beanstock Images/Radius Images/Getty Images, (c)©Digital Vision/Getty Images

New Zealand is about 1,000 miles (1,600 km) southeast of Australia. It is primarily two islands: North Island and South Island. Both islands are dotted with sandy beaches, emerald hillsides, and snow-tipped mountains. North Island's northern region includes golden beaches, ancient forests, and rich soils. A central plateau of volcanic stone features hot springs and several active volcanoes. Mount Ruapehu (roo•uh•PAY•hoo), North Island's highest point, is an active volcano.

The snowy peaks of the Southern Alps run along South Island's western edge. New Zealand's highest peak, Mount Cook, rises to 12,349 feet (3,764 m). Glaciers are responsible for carving out many of New Zealand's lakes and rivers. New Zealand also contains **sunken mountains** like those in the Marlborough Sounds and Fiordland that have sunk into the sea. Rugged cliffs, deep fjords, and coastal caves dot the western coast. The Canterbury Plains, New Zealand's flattest and most fertile land, lie on the eastern coast where livestock and agriculture flourish.

New Zealand has **atolls**, or ring-shaped islands formed by the buildup of coral reefs on the rim of submerged volcanoes. In the New Zealand territory of Tokelau, the Atafu atoll has 19 islets that rise to 15 feet (5 m) above sea level and enclose a lagoon. There is also a **caldera**, or large volcanic crater. Reporoa, a caldera on New Zealand's North Island in the Taupo volcanic zone, formed about 230,000 years ago. The last activity recorded was in 2005. **Lagoons** are shallow lakes that are at times connected to a river, another lake, or the sea. New Zealand has mainly coastal lagoons. These inland bodies of water are important to the survival of certain species of flora and fauna.

✔ **READING PROGRESS CHECK**

Identifying What three deserts make up Australia's Outback?

Water Systems

GUIDING QUESTION *What is the effect of water on life in Australia and New Zealand?*

Although surrounded by water, Australia is the driest inhabited continent on Earth. Freshwater is unevenly distributed, unreliable, and seasonal. Seventy percent of Australia is described as arid or semi-arid with limited precipitation.

bush a wild or sparsely inhabited region

resource a source or supply from which benefit can be gained

sunken mountains a high mountain range that has become submerged

atoll a ring-shaped island formed by the buildup of a coral reef on the rim of a submerged volcano

caldera a large volcanic crater

lagoon a shallow lake that is intermittently connected to a river, another lake, or the sea

Kings Canyon (left) and Hopetoun Falls (right) are examples of the diversity of Australia's landscape.

▼ **CRITICAL THINKING**

1. *Analyzing Visuals* Describe the climate of Kings Canyon.

2. *Analyzing Visuals* Describe the climate of Hopetoun Falls.

(l)Beanstock Images/Radius Images/Getty Images, (r)©Digital Vision/Getty Images

artesian well a well that brings pressurized water to the surface without pumping

In Australia, rivers and lakes in the Central Lowlands fill with water after heavy rainfall. Because rains are infrequent, most rivers and lakes remain dry much of the year. In the southeast, however, the Murray River and the Darling River supply water that supports irrigated farming. A large amount of pressurized underground water, known as the Great Artesian Basin, lies beneath the lowlands. The water that gushes from these **artesian wells**, or wells from which pressurized water flows to the surface, is too salty for humans or crops. Ranchers use it to water livestock, however.

Australia's main water problem is that people and water are not located in the same place. Most of the rainfall is in the tropical north of Australia. However, the population is growing along the east coast and across most of southern Australia. Today these areas receive less rainfall than they did 50 years ago.

In contrast, New Zealand has an abundance of freshwater. It is ranked fourth out of 30 countries that belong to the Organization for Economic Co-operation and Development (OECD) for its per-capita renewable freshwater resource. Both North Island and South Island are bisected by hills and mountains covered with snow. These mountains support fast-flowing, snow-fed rivers and streams. Fifty-one percent of New Zealand's total length of rivers and streams are catchments, or places where water is collected. Theses catchments have predominantly natural land cover such as native bush, alpine rock, and clumps of grass. The remainder of New Zealand's rivers have been modified by agriculture, plantation forestry, or urban development.

Australia and New Zealand have striking landscapes with mountains, rock formations, coral reefs, abundant sea life, and beaches. Although they are isolated geographically, tourism is a growing part of the subregion's economies. In Australia, many people live close enough to a beach to visit it regularly. Some people go to the beach for sun and surfing. Others go to parasail, fish, snorkel, scuba dive, and comb the beaches for shells. People come from around the world to visit the Great Barrier Reef, the world's largest coral reef. New Zealanders and visitors to the country also enjoy recreational use of coastal beaches, rivers, and lakes. As in Australia, many people go to the beach for sun and surfing. But in Auckland, New Zealand's most populous city, it seems that everyone sails. Perfect sailing beaches can be found only 15 minutes from the central business district.

✔ **READING PROGRESS CHECK**

Explaining How is water from the Great Artesian Basin used for agriculture?

Climates, Biomes, and Resources

GUIDING QUESTION *How do the climates of Australia and New Zealand affect the biomes and resources of each?*

The climates of Australia and New Zealand differ by region. In Australia, which is about two-thirds the size of the United States, the climate varies widely from temperate zones to desert regions. New Zealand's climate varies from warm subtropical to cool temperate climates.

Australia's climate includes tropical wet/dry and humid subtropical climates in the north and northeast; deserts in the interior; and midlatitude temperate areas along the eastern, southern, and southwestern coasts. Differences in rainfall cause these significant changes in climate and vegetation. The Outback covers almost two-thirds of Australia. The Indian and Pacific Oceans surround Australia, and the winds coming off the water create erratic shifts from one extreme to the other.

Australia also has a tropical savanna across its low latitude areas in the north. During the dry season, the savanna is prone to wildfires. In the rainy season, the rain falls in heavy bursts accompanied by thunderstorms and monsoons. The wet season lasts from December to March.

Marsupials—such as the kangaroo, koala, and agile wallaby—**dominate** the savanna. They owe their survival to Australia's early isolation after the breakup of the early supercontinent Gondwana (which later collided with other landmasses to form Pangaea). Before more advanced mammals could move in to Australia and replace the marsupials, the landmass separated from Antarctica and India.

Australia also has more than 900 wetlands that provide a natural source of water and protection against floods. They also support a variety of animals such as the giant barred frog and the ringtail possum, as well as migratory birds. The Australian government and several nonprofit organizations work to protect the wetlands.

Along Australia's northeastern coast lies the Great Barrier Reef. Its habitat hosts multitudes of creatures. It has been shrinking by a rate of about 20 percent every 100 years. The United Nations has designated it a World Heritage Site.

dominate to exercise the main influence on; to control

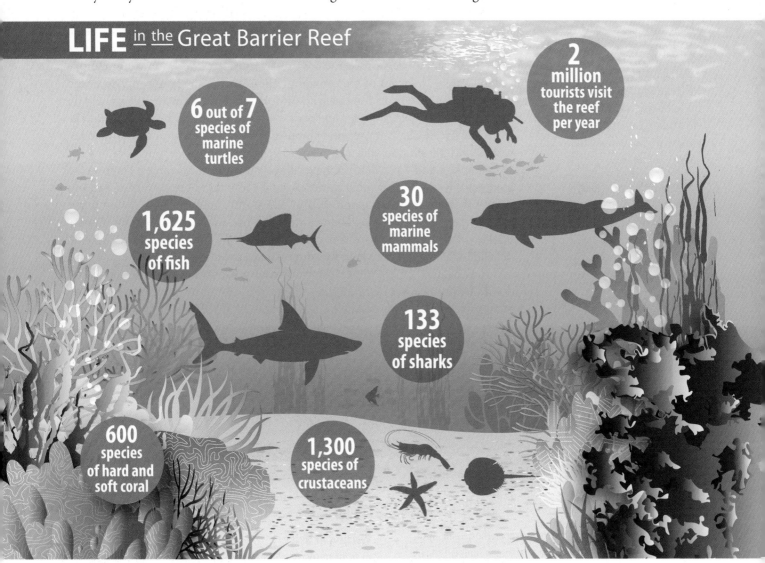

LIFE in the Great Barrier Reef

6 out of **7** species of marine turtles

2 million tourists visit the reef per year

1,625 species of fish

30 species of marine mammals

133 species of sharks

600 species of hard and soft coral

1,300 species of crustaceans

The Great Barrier Reef is a chain of coral reefs located off Australia's northeastern coast.

▲ **CRITICAL THINKING**

1. Assessing What factors threaten the health of the Great Barrier Reef?

2. Constructing Arguments What steps could Australia take to protect the Great Barrier Reef?

The Outback

The Outback, or *the bush*, is an important symbol of the national identity of Australians. It features prominently in Australian literature, painting, popular music, film, and food. What is so special about the bush? It was unique to Australia and very different from the European landscapes known to the early settlers. The Aborigines' skill at surviving in the bush became legendary. The idea of the Outback as integral to Australian national identity was reinforced in 1958 with the publication of *The Australian Legend*, which looked at what defined the typical Australian.

ANALYZING Why might living in regions such as the Outback shape the national identity of Australians?

coral the limestone skeleton of a tiny sea animal

benefit to promote well-being; to be useful

Its name suggests a single reef, but the Great Barrier Reef is actually a string of about 2,900 small reefs. Formed from **coral**, the limestone skeletons of a tiny sea animal, it stretches for 1,250 miles (2,000 km).

Much of New Zealand has a marine west coast climate. Ocean winds warm the land in winter and cool it in summer. Temperatures range from 65°F to 85°F (18°C to 29°C) in summer and 35°F to 55°F (2°C to 13°C) in winter.

Differences in geography also cause variations in climate. Mountainous areas exposed to western winds generally receive more rainfall than other areas. While the country as a whole averages 25 to 60 inches (64 to 152 cm) of rain annually, the Southern Alps on South Island have an average annual rainfall of 315 inches (800 cm).

New Zealand's isolation gives rise to unique plant life. Almost 90 percent of the country's indigenous plants are only found there. Manuka, a small shrub, carpets land where prehistoric volcanic eruptions destroyed ancient forests. To address erosion in deforested areas, several tree species have been imported.

The natural resources of both Australia and New Zealand enrich the economies of both countries. Water resources provide fishing and a means of transportation and trade. Only 6 percent of Australia's land is arable, but farmers grow wheat, barley, fruit, and sugarcane. In the more arid regions, ranchers raise cattle, sheep, and chickens. Aboriginal land rights have been an ongoing debate in the country.

New Zealand's fertile volcanic soil greatly **benefits** its economy. Agriculture accounts for 12 percent of the economy. About half of the land supports crops and livestock. New Zealand's sheep and wool products dominate exports, and its forests yield valuable timber. The country's rivers and dams produce abundant hydroelectric power. New Zealand also uses geothermal energy, created by water heated underground in volcanic fields, to generate power.

Australia and New Zealand have abundant mineral resources. Deposits of gold and silver as well as other minerals can be found in both countries. Mining is an important industry. One-fourth of the world's bauxite, used in aluminum production, is in Australia. The country also has some of the world's highest quality opals as well as deposits of coal, iron ore, lead, zinc, and gold. Nickel and petroleum are found mainly in western offshore sites.

New Zealand's main minerals include coal, gold, silver, iron, limestone, clay, dolomite, pumice, salt, serpentinite, zeolite, and bentonite. However, one of its most prized minerals is *pounamu*. *Pounamu* is a type of jade or greenstone that is treasured for its beauty and its spiritual significance by the Maori.

☑ **READING PROGRESS CHECK**

Describing What are some natural resources in Australia and New Zealand?

LESSON 1 REVIEW

Reviewing Vocabulary
1. ***Describing*** Write a paragraph describing the water resources of Australia and New Zealand.

Using Your Notes
2. ***Summarizing*** Use your graphic organizer on the geography of Australia and New Zealand to write a paragraph summarizing the characteristics of the land, water, and biomes of the subregion.

Answering the Guiding Questions
3. ***Explaining*** How do the landforms of Australia and New Zealand influence the economies of these countries?

4. ***Analyzing*** What is the effect of water on life in Australia and New Zealand?

5. ***Identifying*** How do the climates of Australia and New Zealand affect the biomes and resources of each?

Writing Activity
6. ***Informative/Explanatory*** In a paragraph, discuss the land of Australia and New Zealand and how it affects daily life in the countries.

networks

There's More Online!

☑ **IMAGE** The Lost Generations

☑ **MAP** Australia and New Zealand

☑ **MAP** Patterns of European Settlement

☑ **TIME LINE** Migration and Settlement

☑ **INTERACTIVE SELF-CHECK QUIZ**

☑ **VIDEO** Human Geography of Australia and New Zealand

Reading **HELP**DESK

Academic Vocabulary

- **accompany**
- **impact**

Content Vocabulary

- **Aborigine**
- **clan**
- **boomerang**
- **Maori**
- **dominion**
- **national identity**
- **Strine**
- **pidgin**

TAKING NOTES: *Key Ideas and Details*

IDENTIFYING As you read about the human geography of Australia and New Zealand, use a graphic organizer like the one below to identify the factors that influenced the subregion's culture.

Indigenous Peoples	Migration	European Colonization	Power Struggles

(tl)©Jason Edwards/National Geographic Society/Corbis, (tcl)Hulton Archive/Getty Images, (tcl)North Wind Picture Archives/The Image Works, (tcr)©Cannon Collection/Australian Picture Library/Corbis, (tr)©Corbis

LESSON 2
Human Geography of Australia and New Zealand

ESSENTIAL QUESTION · *How do physical systems and human systems shape a place?*

IT MATTERS BECAUSE

Long isolated from most of the rest of the world, Australia and New Zealand gave rise to unique indigenous cultures and wildlife. The Aborigines were the first human inhabitants of the subregion. Later, British colonists brought their culture and traditions, introducing European influences.

History and Government

GUIDING QUESTION *How did migration and geography influence the history and governments of Australia and New Zealand?*

The indigenous peoples and the settlers who arrived as part of the colonization of Australia and New Zealand have made them culturally diverse countries. These distinct groups of inhabitants have contributed to the rich cultural fabric of Australia and New Zealand and have shaped the modern countries of today.

The early inhabitants of Australia, the **Aborigines**, have the world's oldest surviving culture. The first of these nomadic hunters and gatherers is believed to have arrived in Australia about 40,000 to 60,000 years ago from Southeast Asia. Scientists believe that they migrated to the area over land bridges during the Ice Age, when ocean levels that were much lower than today had exposed land on the continental shelves. These early Aborigines led a nomadic life and used well-traveled routes to reach water and seasonal food sources. Family groups called **clans** traveled together within their territories, carrying what they needed to survive, such as baskets, bowls, spears, and sticks for digging. To hunt, Aboriginal men used a heavy throwing stick, called a **boomerang**, which soars and curves in flight. Women and children gathered plants and seeds.

Migration was **accompanied** by increased trade among the islands of the South Pacific. Research suggests that sometime in the A.D. 1200s the **Maori** left eastern Polynesia and settled the islands of New Zealand. Maori farmers lived in villages and grew traditional root crops such as taro and yams, which they had brought from their Polynesian homeland.

Aborigine the indigenous people of Australia

clan a group of close-knit, interrelated families

boomerang an Australian throwing stick that soars and curves in flight and returns near the thrower

accompany to go with

Maori the indigenous people of New Zealand

impact an effect

dominion a largely self-governing country within the British Empire

European Exploration and Settlement

Europeans began exploring the South Pacific in the 1500s. Captain James Cook, a British sailor, arrived in Tahiti in 1769. He then sailed southwest and, in October 1769, reached New Zealand. Over the next six months, Cook charted all of New Zealand's coasts. He decided to take a different route home. He crossed the Tasman Sea westward instead of going east around Cape Horn. In 1770 Cook came upon the southeast coast of Australia. He claimed the lands of New Zealand and Australia for Great Britain.

In 1788 Great Britain began to use Australia as a penal colony for convicts from overcrowded British prisons. By the early 1850s, free British settlers were establishing settlements along the east coast. Sheep were introduced and settlers profited from exporting wool to Britain. Another source of wealth was gold, discovered in the 1850s. Escaped convicts known as bushrangers, however, made life for the free settlers difficult. Bushrangers supported themselves by stealing from free settlers. To combat the problem, strict laws were put in effect.

During the same time, the British and other Europeans established settlements in New Zealand, which offered fishing and rich soil for farming. By the end of the 1800s, raising livestock, primarily sheep, had become a major part of the economy.

The arrival of Europeans in Australia had a disastrous **impact** on indigenous peoples. Many were forcibly removed from their land and denied basic rights. They resisted the Europeans, and conflicts were common. Violence and European diseases steadily reduced the Aboriginal population. In the mid-1800s, authorities placed many Aborigines on reserves, or separate areas.

In New Zealand, British settlement brought hardship to the Maori as well. In 1840 Britain and Maori groups of North Island signed the Treaty of Waitangi. The treaty purported to protect the rights of the Maori, but was used as a basis for British annexation of North Island. The Maori mounted an armed resistance against British rule that lasted for 15 years, but ended up losing most of their land.

Independence

In 1901 the Australian colonies decided to form a federation known as the Commonwealth of Australia. The new country was a **dominion**, a largely self-governing country within the British Empire. Its government blended a U.S.-style federal system with a British-style parliamentary democracy.

New Zealand became self-governing as a colony in 1853. The colony's Parliament permitted voting to men who owned property. This requirement effectively disqualified most Maori from voting since they did not own property. In 1867 New Zealand passed the Maori Representation Act, which ensured that the Maori were represented in Parliament. In 1893 New Zealand became the first country in the world to legally recognize women's right to vote, including Maori women. In 1907 New Zealand became a self-governing dominion using a British parliamentary system. Beginning in 1975 the voting laws changed, permitting Maori and people of European descent to vote in the same elections.

Both Australia and New Zealand are constitutional monarchies with parliamentary systems. They have written constitutions. The head of state, a largely ceremonial role, is the British monarch. The people elect members of a parliament. The leader of the political party with the majority of votes for parliament is the prime minister, or head of the government. A movement in Australia known as republicanism hopes to establish a republican form of government in which the British monarch would not be head of state. Instead, officials of the state would be citizens who are directly or indirectly elected or appointed.

☑ **READING PROGRESS CHECK**

Sequencing When did the first inhabitants of Australia and New Zealand arrive?

Population Patterns

GUIDING QUESTION *How have migration and an aging population affected Australia's and New Zealand's population patterns?*

Both New Zealand and Australia are multicultural countries. Immigrants from around the world move to both countries to make a new life. Over 7 million people have migrated to Australia since 1945. Of today's population, 44 percent were either born or have a parent who was born in another country. As a reult, many languages are spoken in the country, although English is the dominant language. In New Zealand, most early immigrants were British with some Dutch immigrants arriving in the 1950s. By the mid-1970s policy changes began to make New Zealand a more welcoming place for other immigrants. From the 1980s to today, Pacific Islanders, Asians, Africans, and immigrants from the Middle East have immigrated and now call New Zealand home.

Numerous groups make up the indigenous Aborigines of Australia. One such group is the Arrente, who have lived in central Australia for about 20,000 years. Another group is the Palawa, who have lived on the island of Tasmania for about 32,000 years. After years of harsh treatment and isolation in the Outback and other isolated areas, the Aborigines are now demanding more opportunities. In 1967 the Australian government finally recognized the Aborigines as citizens. Today, about two percent of the population is Aborigine. Growing numbers of Aborigines, particularly youth, are moving to cities.

Approximately 7 percent of New Zealand's population is Maori. The majority of the country's population is descended from British settlers. Asians and Pacific Islanders, attracted by the growing economy, have increased the diversity of New Zealand's society.

GEOGRAPHY CONNECTION

Europeans settled parts of Australia at different times.

1. *HUMAN SYSTEMS* Which areas of Australia were the last to be settled by Europeans?

2. *THE WORLD IN SPATIAL TERMS* Where were most areas of Aboriginal resistance located?

Patterns of European Settlement

Areas of European Settlement
- by 1830
- by 1850
- by 1880
- by 1930
- No European settlement by 1930
- Area of Aboriginal resistance

Mercator projection

Australia's physical geography affects the distribution of its people. Few people live in the dry central plateaus and deserts of the interior. Most people in Australia live along the southeastern, eastern, and southwestern coasts. These areas have mild climates, fertile soils, and access to the sea. The largest Australian cities are Sydney and Melbourne. Both have more than 4 million residents and both are major commercial ports.

About 85 percent of the inhabitants of New Zealand live in urban areas, mostly located along the coast. New Zealand's ports of Auckland, Christchurch, and Wellington are the country's largest cities. Both Wellington, the capital, and Auckland are located on North Island, where about 75 percent of all New Zealanders live.

The aging population of Australia and New Zealand is of growing concern for the governments of both countries. In Australia, the government expects record increases in the number of people ages 65 and over between now and the year 2021. By 2051, models indicate that between 24 and 26 percent of the population is expected to be over the age of 65. Even though New Zealand's population is aging, its birthrates have increased. As a result, the total number of young people is increasing.

An aging population means that there will be more and more people in need of long-term health care and other assistance. Without an increase in the birthrate, the ratio of persons in the labor force will decline in relation to the number of retired persons. Thus, the government and people of Australia may have to make economic choices in the future. Spending on retirement support, education, or infrastructure are among those choices. Young immigrants from countries such as Great Britain, Taiwan, China, and South Korea are also affecting the population dynamics of Australia.

✓ **READING PROGRESS CHECK**

Analyzing What are the concerns about the aging population in Australia and New Zealand?

TIME LINE ⌄

MIGRATION
and Settlement ➔

The history of Australia and New Zealand has been deeply affected by the relations between the indigenous peoples—the Aborigines in Australia and the Maori in New Zealand—and the Europeans who settled there.

▶ **CRITICAL THINKING**

1. **Describing** What are the origins of the major groups of people in Australia and New Zealand?

2. **Explaining** What are some reasons why British people immigrated to Australia?

- First Aborigines arrive in Australia from Southeast Asia.

c. 40,000 B.C.

40,000 B.C. ➔

- Aborigines spread throughout Australia and Tasmania.

c. 20,000 B.C.

A.D. 1200s

Maori arrive in New Zealand from different parts of eastern Polynesia.

1642
- Dutch explorer Abel Tasman is the first European known to reach New Zealand.

©Jason Edwards/National Geographic Society/Corbis

Society and Culture Today

GUIDING QUESTION *How have immigration and migration shaped society and culture in Australia and New Zealand?*

The people of Australia and New Zealand feel a strong **national identity** toward their countries. Australia and New Zealand blend both European and indigenous elements in their cultures. In recent years, Asian influences have also increased in the region. While daily life in much of Australia and New Zealand may resemble that in Western countries, there is diversity in lifestyles across the subregion.

English is the major language spoken in both Australia and New Zealand. Australian English, called **Strine**, has a unique vocabulary made up of Aboriginal words, terms used by early settlers, and slang created by modern Australians.

Because of the rather large population of Maori in New Zealand, Maori is also spoken in certain places. Two percent of Australians—the Aborigines—speak Aboriginal languages. Aborigines also speak **pidgin** English.

Most Australians and New Zealanders, especially those in cities, have access to quality medical care and other social services. The rugged terrain and isolation of some parts of Australia can make access to health care difficult. However, modern technology allows doctors to reach and treat patients through the use of two-way radios, mobile clinics, computer displays, and by air ambulance.

Aborigines and Maori have historically received lower levels of health care, education, and government benefits. Many Aborigines are poor and suffer from malnutrition and unemployment. In recent years, the governments and private organizations of both countries have been working to make up for past injustices. The courts have recognized the claims of the Aborigine and Maori peoples to government assistance, land, and natural resources.

Both Australia and New Zealand provide free compulsory education. Literacy rates are over 99 percent, and many students attend universities. Students in Australia's remote Outback receive and submit work using the Internet, mail, and two-way radios.

national identity the sense of a being part of the whole of a country including its culture, traditions, language, and politics

Strine Australian English that has a unique vocabulary made up of Aboriginal words, terms used by early settlers, and slang created by modern Australians

pidgin a grammatically simplified form of a language

(l)Hulton Archive/Getty Images, (r)North Wind Picture Archives/The Image Works, (b)©Cannon Collection/Australian Picture Library/Corbis

---

1769
British captain James Cook explores New Zealand coastline.

1788
Great Britain begins to send convicts to penal colony in New South Wales, Australia.

➔ 1800

1815
First British missionaries arrive in New Zealand with goal of converting Maori to Christianity.

Gold rush in Victoria draws thousands to Australia, increasing the population by 300 percent.

1850s

1855
Increasing Chinese immigration leads to anti-Chinese legislation in Australian colony of Victoria.

1888
Severe restrictions on Chinese immigration enacted throughout Australia

➔ 1900

Homes Are Sought For These Children

A GROUP OF TINY HALF-CASTE AND QUADROON CHILDREN at the Darwin half-caste home. The Minister for the Interior (Mr Perkins) recently appealed to charitable organisations in Melbourne and Sydney to find homes for the children and rescue them from becoming outcasts.

I like the little girl in Centre of group, but if taken by anyone else, any of the others would do, as long as they are strong

Children of the Stolen Generations were forcibly removed from their families, who had no right of appeal.

▲ CRITICAL THINKING

1. Analyzing Visuals What effect do you think removal had on the children of the Stolen Generations?

2. Speculating Why do you think the Australian government removed Aboriginal children from their families?

Today Aborigines are also known as the First Australians, though they lost their land and were marginalized by European settlement. The Australian government tried to suppress the Aborigines' culture to make them more English by removing Aboriginal children from their homes and placing them with foster parents or in boarding schools. The children were not allowed contact with their families. These children are known as the Stolen Generations. About 100,000 Australian Aboriginal and Torres Strait Islander children were removed from their families between 1910 and 1971.

There have been recent improvements in relations between Aborigines and Australians of European descent. In 2008 Australian prime minister Kevin Rudd apologized to the Aborigines for policies and laws of previous governments that were responsible for the Stolen Generations. In the same year, a documentary presented the history of the Aborigines from 1778 through 1993. It showed the experiences of Aborigines during the British settlement of Australia.

In New Zealand, relations between the Maori and the Europeans were on a more equal basis. However, the Maori were still disadvantaged. In the 1970s and 1980s, the Maori become more politically active. They sought the return of land and compensation from the British government for the loss of access to natural resources. These were items promised to the Maori with the Treaty of Waitangi in 1840. By 1998, the Maori had achieved substantial compensation from the government of New Zealand. Eventually, the government also apologized for the injustices inflicted on the Maori.

For many years, both the Aborigines and the Maori suffered racism from British settlers. With increased integration into the national identity of each country and a better understanding of the cultures that make up these national identities, some think that racism will become a thing of the past.

Australians and New Zealanders have strong family ties. Most people in Australia live in a nuclear family of parents and children, but some live as extended families. In New Zealand, families are also made up of mainly parents and children. The Maori tradition has strong ties with extended families. The word for extended family in the Maori language is *whānau*.

Australian women are active in both working at home and in jobs outside the home. Women work both full and part time while they have young children. Many women have balanced full-time work with caring for a family. Between 1996 and 2005, working mothers with a child under five increased from 46 percent to 52 percent. This suggests more women are in full- and part-time work.

In New Zealand, the role of women has shifted to greater opportunities. After high school, some women find jobs in offices, shops, the health services, and as teachers. Others go on to study at a university. Some women may choose to stay home as wives and mothers.

Australians and New Zealanders traditionally used the arts—art, music, dance, and storytelling—to pass on knowledge from generation to generation. Aborigines recorded their past in rock paintings and used songs to share information about watering holes and landmarks. Maori artisans developed skills in canoe making, basketry, and woodcarving.

©Corbis

Movies have been filmed in New Zealand, including *The Lord of the Rings* trilogy. New Zealand has an innovative and world-renowned film industry. It is also the birthplace of several famous directors and actors including Peter Jackson, Anna Paquin, and Russell Crowe. Australian actors include Cate Blanchett, Hugh Jackman, and Nicole Kidman.

☑ **READING PROGRESS CHECK**

Analyzing How might racism become a thing of the past in Australia and New Zealand?

Economic Activities

GUIDING QUESTION *What are the characteristics of the economies of Australia and New Zealand?*

Australia and New Zealand have a close economic relationship. They have signed trade agreements such as the Australia–New Zealand Closer Economic Relations Trade Agreement (ANZCERTA) of 1983. The objective of these agreements is to improve their economic relationship by eliminating trade barriers.

Australia has a diverse economy. This includes agriculture, mining, steel, industrial and transportation equipment, food processing, and chemicals. The country also has natural resources such as gold, iron ore, copper, coal, and uranium that invite foreign investments and are a significant portion of the country's exports. Australia also exports energy and food. Agricultural products include wheat, barley, sugarcane, fruit, cattle, sheep, and poultry.

New Zealand's main industries are food processing, wood and paper products, textiles, machinery, transportation equipment, banking and insurance, tourism, and mining. Agriculture is a significant part of New Zealand's economy. Agricultural products include wool, dairy products, lamb and mutton, beef, fish, wheat, barley, potatoes, pulses, fruit, and vegetables. New Zealand also has many natural resources. These include natural gas, iron ore, sand, coal, timber, hydropower, gold, and limestone. New Zealand is a significant exporter of dairy products, meat, wool, wood products, fish, and machinery.

Both countries have large service sectors. About 75 percent of people in Australia and 71 percent of the people in New Zealand work in services ranging from government agencies to banking and tourism. Service industries are major contributors to the economies of both countries and increase their gross domestic product (GDP). Australia is ranked nineteenth in the world by GDP and New Zealand is ranked sixty-fifth.

☑ **READING PROGRESS CHECK**

Evaluating What economic activities can be found in Australia and New Zealand?

LESSON 2 REVIEW

Reviewing Vocabulary

1. *Identifying* Name four elements of the cultures of the peoples of Australia and New Zealand and describe each in a sentence.

Using Your Notes

2. *Analyzing* Use your graphic organizer about the human geography of Australia and New Zealand to write a paragraph about how physical systems affect both countries.

Answering the Guiding Questions

3. *Drawing Conclusions* How did migration and geography influence the history and governments of Australia and New Zealand?

4. *Evaluating* How have migration and an aging population affected Australia's and New Zealand's population patterns?

5. *Making Connections* How have immigration and migration shaped society and culture in Australia and New Zealand?

6. *Explaining* What are the characteristics of the economies of Australia and New Zealand?

Writing Activity

7. *Informative/Explanatory* In a paragraph, discuss how the people of Australia and New Zealand, both past and present, have shaped the subregion.

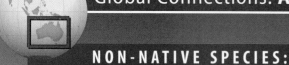

NON-NATIVE SPECIES:

RABBITS
— IN —
AUSTRALIA

In 1859 a man named Thomas Austin freed 24 rabbits in Australia. By 1920 there were 10 billion rabbits spread across the continent. By 1990, due to successful eradication attempts, the population was down to 600 million.

RABBIT POPULATION

- ⬤ ABUNDANT
- ⬤ COMMON
- ⬤ RARE
- — RABBIT FENCE

BARWIN PARK, VICTORIA
Original site of Thomas Austin's rabbit release.

"The introduction of a few rabbits could do little harm and might provide a touch of home, in addition to a spot of hunting."

—Thomas Austin

WHY DID THEY BECOME A PROBLEM?

- 🐰 Rabbits reproduce at an extremely rapid rate.

- 🐰 They can live in a variety of environments.

- 🐰 They have few natural predators in Australia.

- 🐰 Diseases and parasites that control the populations in Europe do not exist in Australia.

DAMAGING EFFECTS ⚠

Rabbits are particularly damaging to the natural environment of Australia. They graze on plants, eating them down to the ground. Rabbits also eat young plants and seedlings.

Rabbit burrows damage plants' roots and loosen the soil, leading to erosion. Rabbits compete with other burrowing animals for food.

ATTEMPTED SOLUTIONS

DISEASE
A disease (Myxomatosis) common in South American rabbits was introduced in 1950 and killed most of the rabbits. However, the disease is carried by mosquitoes so this solution didn't work in arid areas where mosquitoes don't thrive.

RABBIT-PROOF FENCES
About 2,023 miles (3,236 km) of rabbit-proof fences were built in Australia. They failed because parts were poorly built and rabbits climbed over or dug under the fences.

INTEGRATED RABBIT CONTROL
The best solution seems to be using a variety of tools at once. Combined techniques include poisoning, fencing, and burrow destruction and fumigation.

Making Connections

1. *Places and Regions* Why do you think rabbit species found in Australia are most commonly found in the south?

2. *Physical Systems* Few natural rabbit predators exist in Australia. Why would the introduction of a non-native predator species only add further problems?

3. *Human Systems* The threat of an accidental introduction of non-native species is still real, especially in seaports and airports. Write a one-page proposal recommending ways Australia can protect itself against such accidental intrusions.
*Interact with **Global Connections** Online*

networks
There's More Online!

- ☑ **GRAPHS** Levels of Coral Bleaching in the Great Barrier Reef
- ☑ **IMAGE** European Rabbit
- ☑ **IMAGE** Kaingaroa Forest
- ☑ **IMAGE** Whale Rescue
- ☑ **MAP** How Invasive Species Get to Australia
- ☑ **INTERACTIVE SELF-CHECK QUIZ**
- ☑ **VIDEO** People and Their Environment: Australia and New Zealand

LESSON 3
People and Their Environment: Australia and New Zealand

ESSENTIAL QUESTION · *How do physical systems and human systems shape a place?*

Reading HELPDESK

Academic Vocabulary

- **adapt**
- **compound**

Content Vocabulary

- **station**
- **copra**
- **grazier**

TAKING NOTES: *Key Ideas and Details*

IDENTIFYING As you read about the people and the environment of Australia and New Zealand, use a graphic organizer like the one below to identify problems faced by the people and the different ways in which groups and individuals are trying to address these problems.

IT MATTERS BECAUSE

Australia and New Zealand have strikingly beautiful landscapes. Both countries are also rich in natural resources. However, overuse and misuse have created some threats to the natural habitats in this subregion, which both countries are trying to correct. Government, national and international organizations, and individuals are working toward new ways to protect and preserve their environment.

Managing Resources

GUIDING QUESTION *What environmental impact did European practices of land management have in Australia and New Zealand?*

The remote geographic locations and challenging environments of Australia and New Zealand influence how people earn their living. Agriculture is by far the most significant economic activity. As you have read, for example, Australia and New Zealand export large quantities of farm products. Australia is the world's leading producer of wool. However, the lack of rainfall in Australia makes agriculture challenging.

Although less than 5 percent of Australians work in agriculture, much of the country's vast land area is devoted to raising livestock. Because of the generally dry climate, animals need large areas to find enough vegetation to eat. As a result, some Australian ranches, called **stations**, are huge. They can be as large as 6,000 square miles (15,540 sq. km), roughly the size of Connecticut. The impact of long-term livestock grazing in Australia can be the degradation of many of the natural ecosystems in the country.

Due to Australia's dry climate, less than 10 percent of its land is arable, or suitable for growing crops. Irrigation, fertilizers, and modern technology help Australian farmers make more productive use of the land.

In New Zealand, more than half of the land is used for agriculture. In addition to grazing, animals are fed **copra**, or dried coconut meat, which is mainly imported from countries such as Fiji and Papua New Guinea. Ranchers, known as **graziers** in New Zealand, raise sheep, cattle, and

(tl)Mitch Reardon/Lonely Planet Images/Getty Images, (tc)Tony Ashby/AFP/Getty Images, (tr)George Clerk/E+/Getty Images red deer. The country has nearly 20 times more livestock than people. Unlike Australia, New Zealand has some of the most fertile soil in the region. This allows farmers to grow wheat, barley, potatoes, and fruits.

Although new plants and animals have helped both countries develop strong agricultural industries, the impact to the region has been detrimental in some cases. In Australia, native plants and animals were well **adapted**, or fitted, to life on an isolated continent. Since European settlement, these native plants and animals have had to compete for habitat, food, and shelter with exotic plants and animals that were introduced into the environment from other places. For example, 24 European rabbits were introduced to Australia in 1859 for hunting purposes. But rabbits had few natural predators in Australia, so by 1926, there were about 10 billion rabbits in the country. These animals destroyed native plants, exposing the soil to wind and water. The resulting soil erosion caused so much damage that some farms were abandoned. Feral foxes are also a problem. They have caused the decline of several species of native animals, and they prey on newborn lambs.

The introduction of exotic plants in Australia has also created problems. While some of these exotic plants were introduced intentionally, most were introduced accidentally. Many of them have become weeds. For example, the prickly pear was introduced in the 1900s to establish a cochineal dye industry. The prickly pear soon became an invasive species, turning 15,000 square miles (40,000 sq. km) of farmland into a sea of weeds. The weed was eventually eradicated.

Introduced species have also harmed native species in New Zealand. Feral cats and dogs have made New Zealand's native kiwi, a flightless bird, an endangered species. Weasels and ferrets introduced to control rabbits have killed native birds.

☑ **READING PROGRESS CHECK**

Explaining How do people in Australia and New Zealand use the land?

Human Impact

GUIDING QUESTION *How has land use in Australia and New Zealand affected their respective environments?*

Another issue of concern is the protection of forest, soil, and freshwater resources. In Australia, many forests have been cleared for farms and grazing lands. In New Zealand, forests are also converted to pastureland for livestock. Between 2000 and 2010, some 74,000 acres (30,000 ha) of forest were converted to pasture. This leaves little protection against soil erosion. Deforestation also causes soil salinity as well as increased risk of fire and floods. However, pasture continues to dominate New Zealand land use.

In Australia, soil erosion is **compounded** by overgrazing in arid areas and by the country's worst drought in over a century. Soil conservation in the region is closely linked to reducing deforestation. Australia and New Zealand are aware of the problems and are developing plans to use forest resources while reducing damage to the environment.

Land use has significantly modified the physical and chemical nature of rivers in both countries. This has caused unforeseen consequences. For example, drought, salt, irrigation, and agricultural runoff threaten Australia's freshwater sources.

The introduction of the European rabbit to Australia and New Zealand caused serious damage to the environment.

▲ **CRITICAL THINKING**

1. *Making Connections* How could a small mammal cause damage to the environment?

2. *Problem Solving* What steps could the governments of Australia and New Zealand take to deal with the rabbit problem?

station an Australian ranch

copra dried coconut meat fed to animal herds

grazier a New Zealand rancher

adapt to make fit to changing circumstances

compound to add to

Mitch Reardon/Lonely Planet Images/Getty Images

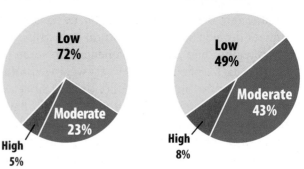

Note: Inshore = less than 6.2 miles (10 km) from the coast
Source: Australian Institute of Marine Science.

Warmer water temperatures can result in coral bleaching. In 1998 and 2002 massive bleaching events occurred in the Great Barrier Reef.

▲ **CRITICAL THINKING**

1. **Differentiating** In which year did inshore reefs suffer higher levels of coral bleaching?

2. **Analyzing** Which type of reef has suffered higher levels of coral bleaching?

The fertile Murray-Darling River Basin is one of the world's largest drainage basins. However, the use of water for irrigation and the increasing needs of growing city populations have dramatically reduced the rivers' flows. Large areas within the basin are also at risk from increasing soil salinity. One of the major causes of increasing salinity has been the replacement of native vegetation—which has deep root structures—with pastures and non-native shallow-rooted crops. Deep root structures prevent rainwater from reaching deep underground and causing salt to rise to the surface soil and water. In increased soil salinity, plants are unable to grow. Salinity causes building foundations, fences, and roads to crumble. Water salinity causes water to be too salty for consumption by humans and wildlife.

Agricultural runoff, chemical fertilizers, and organic waste also threaten the oceans that surround the subregion. Toxic waste, in particular, endangers Australia's Great Barrier Reef and other coral reefs. Tourists, boaters, divers, oil shale mining, and rising water temperatures increasingly stress these coral environments.

Some scientists argue that increases in Earth's temperatures could be devastating. If polar ice caps melt, ocean levels will rise. This would cause extensive flooding. Rising ocean temperatures also affect certain types of plankton and algae that grow in warm waters, causing overgrowth and the choking out of other life forms. The breakdown of the relationship between coral and the algae that provide it with nutrients causes coral bleaching. Scientists in the region are studying global warming and are hoping to discover causes, predict consequences, and provide solutions.

Pollution also affects marine life. This is especially true for the tiny organisms that make up coral reefs. Algae—on which these organisms thrive—and plankton are key parts of the ocean's food web. A food web refers to the interlinking chains of predators and their food sources in an ecosystem. As these tiny living things are destroyed, larger plants and animals that rely on them for food also die off.

☑ **READING PROGRESS CHECK**

Explaining How does human activity impact the land and waters of Australia and New Zealand?

mountains can create areas that are nearly inaccessible. On many high islands, however, mountain areas have greater population density. The interior of Papua New Guinea, for example, is densely populated. This pattern is much different from Fiji, where the majority of the population centers are on the coast.

The settlement patterns of Oceania were limited by the physical geography of the region. Isolated valleys, such as those on the larger islands of Melanesia, created an environment of cultural differences between people living on the coast and those living inland. Even today the islands of Melanesia are characterized by a diversity of languages and customs.

Volcanoes shaped another type of islands, **low islands**. Low islands, such as many of the Marshall Islands, are ring-shaped islands known as atolls. They are formed by the buildup of **coral reefs** on the rim of submerged volcanoes. Atolls encircle lagoons—or shallow pools of clear water—and usually rise only a few feet above sea level. Low islands have poor soil and few natural resources. The landscape of low islands increased interaction between people since lack of physical barriers allowed for more uniform languages and cultures to develop and spread.

✔ **READING PROGRESS CHECK**

Describing What is the relationship between atolls and lagoons?

high island an island with mountain ranges and volcanic soils

low island an island formed by the buildup of a coral reef on the rim of a submerged volcano; sometimes known as an atoll

coral reef a reef made up of fragments of corals, coral sands, algal and other organic deposits, and the solid limestone resulting from their consolidation

from VOLCANO to ATOLL

Phase 1
VOLCANIC ISLAND

As magma erupts through the Earth's crust and is cooled by ocean waters, a volcanic island is slowly formed.

Phase 2
SINKING ISLAND

Once the volcano becomes dormant, the island begins to sink. As it sinks, a coral island forms around it. A small, shallow lagoon forms between the coral island and the mountain.

Phase 3
BARRIER REEF

As the island sinks further, the coral fringes form a barrier reef. The lagoon between the reef and the shore gets larger and deeper.

Phase 4
ATOLL

Once the island disappears below the waves, it becomes a seamount. The barrier reef becomes an atoll, circling an open lagoon.

SEAMOUNT

LAGOON

RISING MAGMA

CORAL ISLAND

MANTLE PLUME

OCEAN CRUST

HOT SPOT

A volcanic island forms when magma erupts from deep within the Earth. Over millions of years, a volcanic island will become an atoll.

▲ **CRITICAL THINKING**

1. Identifying What tectonic forces are required for volcanic islands to form?

2. Identifying Cause and Effect Describe how atolls are formed.

The landscapes of low islands (left) and high islands (right) are strikingly different.

▲ **CRITICAL THINKING**

1. *Speculating* Why are low islands more vulnerable to rises in ocean temperature?

2. *Identifying* What landforms are typically found on high islands?

converse reversed in order, relation, or action; the other way around

Water Systems

GUIDING QUESTION *How do the bodies of water surrounding the islands of Oceania affect ways of life and settlement?*

Oceania extends across millions of square miles in the Pacific. Occupying about a third of the Earth's surface, the Pacific is the largest body of water on the planet. It also has the greatest biodiversity of all the oceans. The Equator marks the division between the North Pacific and the South Pacific.

The Pacific Ocean displays diverse landforms, both above and below its surface. Some can be seen as they rise above the ocean, while others are submerged. The underwater mountain ranges vary from isolated ranges to complex systems that run for thousands of miles. Many of the ranges have exposed volcanic summits. The summits are visible as islands or chains of islands that dot the ocean. One of the longest mountain ranges forms the Hawaiian Islands of Polynesia. Its highest peak, Mauna Kea, rises nearly 13,796 feet (4,205 m) above sea level.

The world's greatest ocean depth has been measured in the Mariana Trench. Located near Guam, it reaches a maximum depth of 36,198 feet (11,033 m). Like the Mariana Trench, most trenches in the western Pacific are next to island chains. Other deep trenches in Oceania reach depths in excess of 32,000 feet (9,754 m).

Asian migrants settled Oceania in family groups along island coasts. They survived on fish, turtles, and shrimp, as well as breadfruit and coconuts. Over time, they cultivated root crops including taro and yams. They also raised livestock such as chickens and pigs. Well-built canoes made lengthy voyages possible. Thus trade gradually developed between islands. To make trading easier, people on some islands used long strings of shell pieces as money. On a few islands, shell money is still exchanged today for canned goods or vegetables at markets.

Important trade routes cross through the ocean surrounding the subregion. The majority of the exports moving from west to east and from north to south are manufactured goods on their way to markets. **Conversely**, most of the exports that are shipped from east to west and from south to north are raw materials. The small islands of Oceania depend heavily on trade—especially for basic necessities, such as foodstuffs and fuels. There are exceptions, however. Papua New Guinea generally exports more than it imports.

The bodies of freshwater in the subregion vary depending on the island type. High islands have sources of freshwater that support agriculture. The low islands typically have no freshwater sources other than places that catch rainwater.

☑ **READING PROGRESS CHECK**

Describing What landforms are found in the Pacific Ocean?

(l)Reinhard Dirscherl/Waterframe/Getty Images, (r)Peter Hendrie/Lonely Planet Images/Getty Images

Climates, Biomes, and Resources

GUIDING QUESTION *How do the various climates on the islands of Oceania affect their biomes?*

Most of Oceania has a tropical wet climate and is warm year-round. The dry season features cloudless skies, but the wet season brings constant rain and high humidity. High islands are high enough to force warm, moist air to rise. It then cools and condenses. Low islands do not have this effect, so there is less rainfall.

The great expanses of open water in the Pacific Ocean influence wind and pressure patterns. These are reflected in the climatic conditions. A generally windless area called the doldrums occupies a narrow band near the Equator where the direct rays of the sun cause air to rise vertically instead of blowing horizontally. As the air rises, air from the north and south is drawn in by the low surface pressure, creating the trade winds. The flow also causes typhoons to form.

The amount of rainfall on the islands creates both arid and wet climate regions, depending on the island type and location. Only shrubs and grasses grow on dry, low islands. These islands have only a small proportion of arable land. Palms and other trees appear on islands with more rainfall.

Hot, steamy rain forests thrive where heavy rains drench island interiors. The summit of Mount Waialeale on Hawaii's island of Kauai is one of the wettest places on Earth. Rainfall averages about 450 inches (1,143 cm) of rain each year. The rich soils of high islands support a diverse group of plants and animals.

The isolation of many of the islands in Oceania makes for the presence of endemic species. Species that are endemic to a particular location are found only in that place. The intentional and accidental introduction of new species to the islands of Oceania has resulted in negative changes to the ecosystem.

Island type and location also influence resources. Low islands have poor soils and few resources beyond the sea. These mainly consist of coconut oil, dried coconut (copra), bananas, and seabed minerals. The volcanic materials of the islands of American Samoa include an abundance of pumice and pumicite. Fiji, also volcanic, has timber, gold, and a potential for offshore oil. Fiji also generates hydropower.

High islands have rich soils that support agriculture. Products grown for export include sugar, coffee, and cocoa. The high islands also have rain forests with diverse flora and fauna. The islands with forests, such as Samoa, harvest hardwoods. Islands without forests have to import lumber. In addition to forests, the Solomon Islands have a variety of other natural resources. These resources include fish and other marine animals, gold, bauxite, phosphates, lead, zinc, and nickel. Tourism helps to **offset** the lack of natural resources on many of the islands, especially the mineral-poor low islands.

offset to compensate for or serve to counterbalance

✔ **READING PROGRESS CHECK**

Naming What are some of the natural resources of Oceania?

LESSON 1 REVIEW

Reviewing Vocabulary

1. Describing Describe the significance of coral reefs to atolls.

Using Your Notes

2. Contrasting Using your graphic organizer from the lesson, contrast the soils of the low islands with those of the high islands.

Answering the Guiding Questions

3. Explaining How do the islands of Oceania affect settlement?

4. Identifying Cause and Effect How do the bodies of water surrounding the islands of Oceania affect ways of life and settlement?

5. Synthesizing How do the various climates on the islands of Oceania affect their biomes?

Writing Activity

6. Narrative Write a paragraph describing the type of activities you might participate in during a vacation to one of the islands in Oceania. Include a discussion of how the landforms and bodies of water relate to the type of activities available.

netw⬤rks

There's More Online!

- ☑ **IMAGE** Hotel in Fiji
- ☑ **IMAGE** Oceanic Art
- ☑ **IMAGE** Subsistence Farming in Oceania
- ☑ **MAP** Oceania in World War II
- ☑ **TIME LINE** Colonization and Independence
- ☑ **INTERACTIVE SELF-CHECK QUIZ**
- ☑ **VIDEO** Human Geography of Oceania

Reading **HELP**DESK

Academic Vocabulary

- **nevertheless**
- **bulk**

Content Vocabulary

- **kinship group**
- **trust territory**
- **pidgin**

TAKING NOTES: *Key Ideas and Details*

IDENTIFYING As you read about the human geography of Oceania, use a graphic organizer like the one below to record how the different island groups were settled.

Human Geography of Oceania
Settlement

Melanesia → [   ]
Micronesia → [   ]
Polynesia → [   ]

LESSON 2
Human Geography of Oceania

ESSENTIAL QUESTION · *How do physical systems and human systems shape a place?*

IT MATTERS BECAUSE

Migrations of people among hundreds of islands over many generations shaped societies in Oceania. European colonization had a profound impact as well. Today, the islands in Oceania are reshaping themselves as independent countries.

History and Government

GUIDING QUESTION *How has life on the islands of Oceania changed over time?*

Hundreds of indigenous people lived on the many islands in the Pacific Ocean when European explorers began arriving in the 1500s. The islands had been their homes for thousands of years. Many of these cultures practiced religious beliefs that connected them to the land and sea. In many of the societies, the chief was the person of highest status. Complex systems of ranked lineages and **kinship groups** had developed. Many societies, such as that of the Marshall Islands, were matrilineal, meaning that family history was traced through the mother's family. Powerful chiefs were recognized for their ability to communicate with the gods and secure food and other valuables for the people.

Europeans brought far-reaching changes to the peoples of Oceania. When Europeans settled in Oceania in the 1800s, they developed plantations for growing sugarcane, pineapples, and other tropical products that were sold around the world. Europeans also brought new diseases that resulted in epidemics among the indigenous populations. In order to maintain a workforce, plantation owners recruited workers from East Asia and South Asia. In just a few years many of the islands became culturally diverse and dependent on the global demands for sugar, the major crop. Meanwhile, Europeans further altered traditional beliefs and customs by establishing schools, sponsoring missions, and converting people to Christianity.

During the late 1800s and early 1900s, Britain, France, Germany, Spain, and the United States struggled for control of various Pacific islands. These countries wanted to acquire or expand their influences in the region and gain new sources of raw materials. Colonial powers did not have a strong physical presence in many of the islands. In some places, islanders ignored the laws of European countries that they felt did not apply to them.

(tl)©Stapleton Collection/Corbis, (tcl)LAPI/Roger-Viollet/The Image Works, (tc)National Archives and Records Administration (NWDNS-80-G-396229), (tcr)Mike Robinson/Alamy, (tr)©Michele Westmorland/Corbis

The two world wars dramatically changed Oceania's political geography. After World War I, most of Germany's territories came under Japanese colonial rule. During World War II, those same islands were the sites of fierce battles. The United States and its allies were fighting to drive the Japanese from these islands. After Japan's defeat in the war, its Pacific territories were turned over to the United States as **trust territories**. These are designated by the United Nations to be temporarily governed by another country.

Beginning in the 1960s, islands in Oceania, also called the Pacific Islands, began gaining independence. In 1962 Samoa, formerly Western Samoa, became the first Pacific Island to win freedom, after rule by Germany and New Zealand. Since the 1970s most of these islands—including Palau, the Marshall Islands, and the Federated States of Micronesia—have become independent countries. Most Pacific Islands achieved independence by the end of the 1900s.

Many islands in Oceania have a dual government system that includes the government imposed upon them by colonial powers and their traditional, indigenous systems of government. In some countries, the traditional systems have been integrated into the constitutional structure. However, in other countries, such as Papua New Guinea, indigenous governance systems have been largely sidelined in the formal government structures.

The Solomon Islands represents one example of cultural rebirth occurring in the subregion. Traditional beliefs are starting to resurface and guide decision making. These countries are looking within their borders for input on economic and social issues. Social structure is a high priority. People in the Solomon Islands are once again valuing equality for all people. They are also building a strong relationship to the land by practicing traditional farming and marketing locally grown foods rather than relying on imported, processed foods.

☑ **READING PROGRESS CHECK**

Explaining Describe how the arrival of Europeans affected the population of Oceania.

Population Patterns

GUIDING QUESTION *How and why did people spread across Oceania?*

The islands of Oceania were probably first settled by peoples from Asia around 40,000 years ago. Waves of migrants from Asia continued to arrive over many centuries. People already living there moved from island to island and settled into three major island groups: Melanesia, Micronesia, and Polynesia.

Melanesia includes Papua New Guinea, Fiji, the Solomon Islands, as well as New Caledonia. Melanesian cultures differ greatly, even among groups living on the same island. Some Melanesians are more closely related to Polynesians than to other Melanesians. The earliest people of Melanesia were the Papuans. They were hunters and gatherers that adapted to the rain forests on New Guinea. Some estimates date their arrival on the island to 40,000 years ago. Eventually the early Melanesians domesticated root crops and sugarcane. They may have also kept domesticated pigs as long as 9,000 years ago. In areas of the highlands of New Guinea, water and irrigation systems were in place 5,000 years ago.

Arriving by sea from Southeast Asia, the Austronesian peoples settled in Melanesia about 4,000 years ago. Pottery, tools, and shell ornaments date their arrival in the islands. The languages they spoke were similar to languages used in the Solomon Islands, Vanuatu, and New Caledonia. Language was used to test theories of migration among people in the region. Distances were great across ocean waters. Migration between the islands meant the people were accomplished mariners. The migration theory explained both when and how the islands were first inhabited. It also explained the cultural differences and similarities in this region.

Connecting Geography to **SCIENCE**

Carbon Dating

Archaeologists, geologists, anthropologists, and botanists all work together to determine when a given island was first inhabited. Carbon-14 dating is a tool used to determine the age of items such as bone, cloth, wood, and plant fibers. This method works on archaeological artifacts that have a biological origin and that are less than 50,000 years old. Scientists look at the ratio of carbon-12 to carbon-14 as a means for dating the objects. The ratio of the two carbons is the same in every living thing at the moment of death. The carbon-14 begins to decay at death and is not replaced. By looking at the ratio between the two carbons and comparing it to the ratio in a living organism, it is possible to determine the date of the object with precision.

IDENTIFYING Describe some of the objects that geographers and archaeologists can use from an island to determine when people first arrived.

kinship group a group of people related by blood or marriage

trust territory a dependent area that the United Nations placed under the temporary control of another country

Micronesia includes the Federated States of Micronesia, Nauru, Kiribati, Guam, and the Mariana Islands. The first people to settle in Micronesia are thought to have arrived by ocean canoes some 2,000 to 3,500 years ago from Southeast Asia and Melanesia. The islands in this region are more densely populated. A larger population suggested that more migrants arrived and that they had skills necessary to develop farming and trading economies. Language has again been used as the key to dating the arrival of the first people to these islands.

Polynesia includes Samoa, Tonga, Tuvalu, and French Polynesia. The ancestors of present-day Polynesians may have come to the islands 6,000 to 8,000 years ago. The western islands—Wallis and Futuna, Samoa, and Tonga—were settled first. These first Polynesians grew traditional crops and domesticated animals. Women gathered plants and were weavers. Many Polynesians share a similar language and culture. Today the largest population of Polynesians lives in the Samoan Islands.

Asian communities also exist in the subregion. Chinese, Japanese, and South Asian traders and laborers settled parts of Oceania during the 1800s. Today their descendants live in such places as French Polynesia and Fiji.

The population of Oceania's islands varies considerably, and many of the islands are not large enough to support people. Population is not evenly distributed among the countries of the subregion. Papua New Guinea has the largest population, with about 7 million people. Nauru, the world's smallest republic, has a population of about 10,000 people. The total land area of Oceania's 25,000 islands is 551,059 square miles (1,427,246 sq. km) spread across a vast Pacific Ocean expanse of over 20 million square miles (52 million sq. km). Urban population varies greatly. For example, the capital city of Papua New Guinea, Port Moresby, has a population of nearly 300,000. In contrast, Apia, the capital city of Samoa, has a population of approximately 36,000.

☑ **READING PROGRESS CHECK**

Identifying Who first settled Melanesia and from where did they come?

©Stapleton Collection/Corbis

TIME LINE ⌄

COLONIZATION
and Independence ➜

Beginning in the 1800s, European countries and the United States colonized the islands of Oceania, but those islands became independent states in the twentieth century.

▶ **CRITICAL THINKING**

1. *Explaining* How did the islands of Oceania change after contact with Europeans was made?

2. *Describing* When did the pace of independence quicken in Oceania?

1600s–1700s
European navigators first contact the peoples of Oceania.

1870s
Population of the Ellice Islands (now Tuvalu) declines dramatically.

1600 ➜

1830s
European missionaries arrive in Fiji, Samoa, and Tonga to convert the population to Christianity.

1880s
Germany takes control over part of New Guinea, Marshall Islands, and Nauru.

806

Society and Culture Today

GUIDING QUESTION *What is life like on the islands of Oceania today?*

Today, societies in Oceania have been shaped by a variety of cultures. The subregion's countries display a blend of European, Asian, and indigenous traditions. The traditional lifestyle of indigenous people is difficult to find in the twenty-first-century world.

Before the era of modern transportation and advanced communications, vast distances of open ocean separated the peoples of the subregion from the rest of the world. As a result, isolated groups developed their own languages and belief systems without outside interference. Of the world's some 6,000 languages, 1,200 are spoken in Oceania. Some are spoken by only a few hundred people.

European colonization brought European languages to the subregion. Today French is widely spoken in Oceania. In many areas, varieties of **pidgin** English are spoken as well. This is a blend of English and indigenous words that do not follow rules of grammar, but do allow better communication among different groups.

pidgin a blend of English and indigenous words to form a new language

The peoples of the subregion practice several different religions. These include Christianity, Hinduism, Islam, and Buddhism. Sometimes the religions are combined with traditional religious beliefs. Christianity is the most widely practiced religion in Oceania and was introduced by Europeans. Hinduism was introduced to Fiji when thousands of Indians immigrated to work on plantations. Over 30 percent of the population of Fiji is Hindu.

The quality of education varies throughout Oceania. In the Solomon Islands, missionary schools provided primary education until the mid-1970s. Today secondary schools and universities are common in the Solomon Islands, Fiji, and Papua New Guinea. Differences in education throughout the subregion can be seen in the varying rates of literacy. For example, over 93 percent of Fiji's population is literate, but only 57 percent of the inhabitants of Papua New Guinea are literate.

(cl)LAPI/Roger-Viollet/The Image Works, (cr)Mike Robinson/Alamy, (b)National Archives and Records Administration (NWDNS-80-G-396229)

1941–1945
Japanese forces occupy Kiribati, Marshall Islands, Micronesia, Nauru, Palau, Papua New Guinea, and Tuvalu.

1962
Western Samoa becomes first colony in Oceania to become independent.

1970s
Fiji, Tonga, Papua New Guinea, Tuvalu, Federated States of Micronesia, Kiribati, and Marshall Islands become independent.

1900 ➜

1899s
Germany buys Palau from Spain. Germany and the United States divide Samoa between them.

1946
United States begins nuclear weapons testing in Marshall Islands; islanders are forced to evacuate.

2000 ➜

1994
Republic of Palau becomes independent state.

Many Pacific Islanders suffer from their countries' poor economies and low standards of living. Health care varies from country to country. On remote islands, electricity, schools, and hospitals are difficult services. As independent countries, they qualify for international assistance and are making improvements.

Family and the Status of Women

The traditional family structure has consisted of many generations living in the same village and often in the same house. This structure includes large families with children being cared for by many family members. Contemporary life today has changed throughout Oceania due to modernization and loss of traditional values. Many young people migrate to Australia, the United States, and France.

The status of women is of great concern in the subregion. The status of women varies from strong matrilineal leadership to victims of violence long accepted as a part of a culture. In Oceania, there are cultural practices that make it difficult to protect women against violence. For example, in Melanesia, there is the practice of early arranged marriages and forced marriages. The practice of burning and scarring brides also occurs among some ethnic groups in the region even though it is against the law.

The Arts

The ocean, fish, and many types of plants and animals are the subjects for the arts of Oceania. Carving wood is an art mastered by many people. Skilled crafts, such as making canoes and building houses, are a tradition. Art and architecture are closely related to each other. Buildings are carved with elaborate designs. Detailed designs are also important in the symbolic art of body tattooing in Oceania. Creating decorative patterns on the skin by scarification, the intentional scarring of the skin by incision or burning, and tattooing are practiced by some culture groups.

☑ **READING PROGRESS CHECK**

Explaining How did Hinduism become one of the religions practiced in Oceania?

Skilled crafts are a tradition in Oceania with artwork on buildings, canoes, and on bodies in the form of tattoos.

▼ **CRITICAL THINKING**

1. *Identifying* What are some of the common themes found in traditional Oceanic art?

2. *Explaining* What type of artwork is part of ritual practices in Oceania?

Economic Activities

GUIDING QUESTION *What economic opportunities are available to the people of Oceania?*

The remote geographic locations and challenging environments of Oceania influence how people there earn a living. Agriculture continues to be an important economic activity in the subregion, but new industries are contributing to national economies. Tourism has gained in importance as people looking for scuba diving in pristine waters select the islands of the Pacific for their environments. Niche markets, such as call centers and web development, have increased in Oceania.

Although many islands in Oceania lack abundant natural resources, some have excellent resources for tourism. The tourism industry is still a relatively new industry in many parts of Oceania. French Polynesia, Guam, the Northern Mariana Islands, and Fiji have all experienced recent growth in this industry, and it has become vital to their economies.

Much of Oceania lacks arable land, which in turn limits agriculture. Farmers on smaller islands use traditional

©Michele Westmorland/Corbis

farming and fishing. On some islands, the village or island government owns the land and rents or leases it to farmers.

Some islands have rich volcanic soil and ample rainfall. On many of the islands, Europeans had once claimed the most productive land. Land settlements in recent years have returned the best land to indigenous people.

Most of the islands in Oceania have only small amounts of mineral resources and lack large manufacturing centers. Papua New Guinea, Fiji, and New Caledonia have mineral deposits. Papua New Guinea's rich deposits of gold and copper have only recently been exploited. Nauru's deposits of phosphates are depleted due to excessive mining. Manufacturing in Oceania is mostly limited to small-scale enterprises like clothing production. Many of the larger islands with forests have sawmills that produce lumber for export or domestic use. Some of the larger islands have facilities that process agricultural products, such as coconut oil and cane sugar.

Trade between Oceania and other parts of the world has increased because of improvements in transportation and communications, as well as the creation of new trade agreements. Some island countries are too small, too poor, or too rugged to have well-developed road or rail systems. **Nevertheless**, some governments are improving these systems. Cargo ships and planes move imports and exports to and from the subregion. Principal trading partners of Oceania include the United States, Japan, and Australia. Commercial airlines and cruise ships bring travelers. Air transport and interisland shipping are the vital lines for transportation. Many of the island groups in Oceania have international airports. Extensive road networks are limited to the larger islands.

Cellular, digital, satellite communications, and the Internet are becoming common in the subregion. In recent decades, improved transportation and communications links have increased trade between Oceania and other parts of the world. The subregion's agricultural and mining products earn the **bulk** of its export income. Countries also export handicrafts—such as baskets, floor mats, ceremonial masks, and pottery—made by island craftspeople.

Holger Leue/LOOK/Getty Images

Farming in Oceania frequently relies on animal power and human labor.

▲ **CRITICAL THINKING**

1. *Speculating* What are some of the crops that might be grown by traditional farming practices in the Solomon Islands and other parts of Oceania?

2. *Contrasting* How were the plantations established by the Europeans different from traditional farming?

nevertheless despite; in spite of

bulk the greater part of something

✔ **READING PROGRESS CHECK**

Making Connections What factors contribute to the lack of adequate networks of roads on many of the islands in Oceania?

LESSON 2 REVIEW

Reviewing Vocabulary

1. *Summarizing* Define *trust territory* and explain how it relates to Oceania.

Using Your Notes

2. *Describing* Use your notes from the graphic organizer to describe the order of settlement for Melanesia, Micronesia, and Polynesia.

Answering the Guiding Questions

3. *Identifying Cause and Effect* How has life on the islands of Oceania changed over time?

4. *Explaining* How and why did people spread across Oceania?

5. *Describing* What is life like on the islands of Oceania today?

6. *Examining* What economic opportunities are available to the people of Oceania?

Writing Activity

7. *Informative/Explanatory* Write a paragraph describing the negative impact Europeans had on the population and cultures of Oceania.

WHO OWNS THE
HIGH SEAS?

In 1994 the Law of the Sea Treaty came into force. This treaty defines which countries own parts of the oceans. According to the treaty, each country's territory extends about 12 nautical miles from its coasts. Beyond this boundary a country may establish an exclusive economic zone (EEZ) that extends 200 nautical miles from shore. When two countries' ocean territories overlap, a border is set. The areas beyond these descriptions and the EEZs, however, belong to no one. They are the high seas. The high seas can be accessed by anyone with the means to do so. These areas are rich in fish and other natural resources, such as oil and minerals. People travel through the high seas, and countries mine resources and fish there.

People have different ideas about how the high seas and the natural resources within them should be managed and protected. More than 150 countries have signed the Law of the Sea Treaty, which gives the International Seabed Authority (ISA) jurisdiction over the high seas. However, some countries refuse to sign the treaty. In these countries, some fear that the ISA will impose new tax laws on travel in the oceans and reduce free movement on the waters. Another concern is that fewer economic opportunities will exist if the ISA regulates the high seas. They also worry that future projects, such as mining the sea floor, will be restricted. Other critics of the treaty think the high seas should be privately owned or even feel that no one should own and regulate these areas. Still others oppose the treaty because they want more regulation of the high seas. They do not think that the ISA and the United Nations have the authority to protect the oceans. They feel the ISA has not done enough to protect the waters from the existing threats of overfishing, pollution, pirating, and drug trafficking.

Dan Rafla/Aurora/Getty Images

Problems of the Law of the Sea Treaty

PRIMARY SOURCE

❝ One of the primary missions of the United States Navy for over two centuries has been to maintain freedom of the seas for all. As a Navy veteran, I am offended to think that the Senate and the Chief of Naval Operations would even consider ceding any part of that mission to the United Nations. . . .

There is no guarantee that the treaty will remain what it is at the time of ratification. Under its terms, its content can later be changed by an amendment process that does not require the approval of the United States government. This undermines U.S. sovereignty and, to put it bluntly, is unconstitutional. ❞

—Colin Hanna, president of Let Freedom Ring, quoted in "Kill the Law of the Sea Treaty," *U.S. News*, May 10, 2012

Support More Regulation

PRIMARY SOURCE

❝ We have the Law of the Sea [Treaty] but it does not have enough teeth to actually do anything. We also have the regional fisheries management organisation but this is not a global treaty and they are regional treaties. We see the high seas as the last frontier to be tackled. It is nobody's territory but everybody's territory to exploit. . . .

So we need to bridge this gap somehow and there is no framework yet or a treaty for this purpose. Eventually all the fish caught has to land somewhere and it becomes somebody's business and therefore this dilemma needs to be solved. ❞

—Gustavo Fonseca, head of natural resources quoted in "Oceans Special: GEF Rolls Out Investment Project to Address Issues for Areas in the High Seas," June 16, 2012

What do you think? DBQ

1. *Drawing Conclusions* Why might people want to make a plan for regulation and management of the high seas?

2. *Identifying Central Issues* According to critics of the treaty, what are the risks of regulation of the high seas?

3. *Hypothesizing* What might be some benefits and drawbacks of private ownership of the high seas?

(l)MC3 Dylan McCord/US Navy/DoD, (r)Courtesy of Doug Helton, NOAA/NOS,ORR/ERD

networks

There's More Online!

☑ **CHART** United States
Nuclear Tests July 1945
through September 1992

☑ **IMAGE** Easter Island

☑ **MAP** Global Sea Level Trends

☑ **INTERACTIVE SELF-CHECK QUIZ**

☑ **VIDEO** People and Their
Environment: Oceania

Reading HELPDESK

Academic Vocabulary

- **nuclear**
- **grant**

Content Vocabulary

- **climate change**
- **ocean warming**

TAKING NOTES: *Key Ideas and Details*

IDENTIFYING Use a graphic organizer like the one below to take notes as you read about the issues that relate to people and their environment in Oceania.

Human Impacts
on the Environment in Oceania

Deforestation	
Rising Sea Levels	
Rising Ocean Temperatures	

LESSON 3

People and Their Environment: Oceania

ESSENTIAL QUESTION · *How do physical systems and human systems shape a place?*

IT MATTERS BECAUSE

The islands of Oceania account for the emission of less than 1 percent of greenhouse gases that contribute to climate change. They are, nonetheless, among the most vulnerable to the adverse effects of greenhouse gases.

Managing Resources

GUIDING QUESTION *How are the people of Oceania using their islands' natural resources?*

The protection of forest, soil, and freshwater resources is a major concern throughout Oceania. The management of resources includes reforestation plans and implementation of "tabu," or no-take zones to control overfishing. Countries with valuable timber resources, such as Papua New Guinea, are developing plans to use forest resources without damaging the environment.

Methods used to improve agriculture production have a long history in Oceania and continue to be implemented to increase output. The labor-intensive practice of terracing is used to grow crops on the sides of hills. Terraces built into a slope increase the amount of available land for cultivating crops. Terracing helps reduce irrigation runoff and erosion. Irrigation practices such as simple flooding are used to slow down the flow of water to help trap sediment and control erosion. Crop diversification is another innovative agricultural practice used in Oceania. Planting a diversity of crops helps preserve the soil and increase productivity.

Overseas companies also have a stake in the management of Oceania's natural resources. Foreign companies are actively mining and logging on the islands and deep seabeds of the subregion. The investment in tourism by foreign companies is also a factor in managing the natural resources of the subregion. As new resorts are built, the environment is affected. Fiji has established a joint effort with hotel and resort owners to create a tourist sector that builds energy-efficient hotels, cuts carbon emissions, and reduces the impact on the environment.

(tl)©Mago World Image/age fotostock, (tcl)©Bettmann/Corbis, (tcr)Kyodo/AP Images, (tr)Kyodo/Newscom

Oceania's environmental challenges include managing its freshwater resources. Many coral atolls and volcanic islands hold only limited supplies of freshwater. The lack of clean drinking water keeps the standard of living low. It also poses barriers to economic growth in some countries. Improvement will come with better management of runoff and the construction of additional sanitation facilities.

The low islands of Oceania are in particular danger. **Climate change** can cause sea levels to rise and produce extreme weather. Managing the resources of these islands will help reduce their vulnerability. Programs include the protection of mangroves and seagrass habitats on these islands. In some cases, infrastructures have been relocated farther inland.

Many of the countries in Oceania have developing economies. This means that implementing programs can be difficult. Programs continue to be initiated, however, through grassroots organizations, government efforts, and international organizations. Environmental education is also an important factor in managing natural resources. Understanding issues such as climate change will help empower people to make changes and welcome new methods of balancing their needs with those of the ecosystems.

climate change any significant change in the measures of climate lasting for an extended period of time

☑ **READING PROGRESS CHECK**

Summarizing What issues are the most important in regard to managing the natural resources of Oceania?

Human Impact

GUIDING QUESTION *How have humans—both in Oceania and elsewhere—affected the environments of the islands of Oceania?*

The natural resources of Oceania, as in other regions, have not always been well managed. Today the subregion faces many environmental problems. Conservation efforts, however, are gaining recognition.

GEOGRAPHY CONNECTION

In recent decades, global sea levels have been rising.

1. *THE WORLD IN SPATIAL TERMS* In what areas of the globe have sea levels risen the most?

2. *THE WORLD IN SPATIAL TERMS* List three areas in which sea levels have fallen.

Global Sea Level Trends

Regional trends from 1992 to 2013

-10 -8 -6 -4 -2 0 2 4 6 8 10
Falling   Sea Level Trends (mm/yr)   Rising

nuclear a weapon whose destructive power derives from an uncontrolled nuclear reaction

The testing of **nuclear** weapons has had major effects on parts of the subregion's environment. In the late 1940s and 1950s, the United States and other countries with nuclear capability carried out aboveground testing of nuclear weapons in the South Pacific. The dangers of such testing were gravely underestimated at the time. In 1954 the United States exploded a nuclear device on Bikini Atoll in the Marshall Islands. The people of Bikini Atoll had been moved to safety. However, those living downwind of the explosion on Rongelap Atoll were exposed to massive doses of radiation. This exposure resulted in deaths, illnesses, and genetic abnormalities.

Like other world regions, Oceania is threatened by climate change. One resulting impact is an apparent increase in the frequency and severity of the El Niño weather pattern. Climate and weather in Oceania are highly sensitive to seasonal El Niño weather patterns that can cause both droughts and powerful storms. Another impact of climate change that many scientists consider imminent is that continued increases in Earth's temperatures could cause a devastating rise in sea level. If ice caps and glaciers on land experience increased melting, the resulting rise in the level of ocean waters would cause flooding on many of Oceania's islands.

ocean warming rise in the temperature of the ocean water

Rising sea levels not only cause loss of land but also contaminate the limited freshwater sources of the low islands with salt water. This reduces the amount of freshwater available for agricultural purposes as well as human consumption. In addition, **ocean warming** affects certain types of plankton and algae that grow in tropical waters, causing overgrowth and the choking out of other life forms. Coral reefs are also destroyed by ocean warming. Scientists in the region are studying global warming. They hope to discover its causes, predict its consequences, and provide solutions.

Pollution also poses a threat to the waters of Oceania and is damaging to the coral reefs. Agricultural runoff and inadequate sanitation cause pollution that further threatens freshwater supplies. Agricultural runoff, chemical fertilizers, and organic waste also threaten the subregion's oceans.

Famous for its giant carved statues, Easter Island no longer has the natural vegetation that once covered the land.

Deforestation and the loss of native species are not new to the islands of Oceania. Easter Island, also called Rapa Nui, had already lost most of its native

▼ **CRITICAL THINKING**

1. **Contrasting** Describe how the vegetation of Easter Island has changed since the arrival of Europeans.

2. **Describing** What are commonly held theories for causes of deforestation on Easter Island?

plant species by the late 1700s. When European explorers first arrived on Easter Island in the late 1700s, deforestation had already taken place. They found a grassy island devoid of plants greater than 10 feet (3 m) in height. The island at one time had been covered with a subtropical forest dominated by large palms, an ecosystem entirely different from what the Europeans observed.

The exact cause of Easter Island's deforestation is not certain. However, a few theories prevail. Some scientists believe that uncontrolled populations of rodents were the cause. Others blame overexploitation of the available resources or the island's natural fragility for the deforestation. The loss of forest caused soil erosion that limited the inhabitants' ability to grow crops.

Present-day causes of deforestation include mining and logging. Although mining is beneficial to Papua New Guinea's economy it has had negative impacts on the rain forests there. Mining operations cause destruction of forest habitats and pollute waterways and oceans. The process of removing the nontarget material from the ore and the discarded waste rock can cause damage to the environment. The process of separating gold in alluvial mining releases mercury into streams and soils, which creates another source of contamination. Pollutants can cause localized fish kills and loss of vegetation in areas. Pollution in rivers and streams is harmful to agriculture as well as to animals and humans.

☑ **READING PROGRESS CHECK**

Describing What are the pros and cons of mining operations in Papua New Guinea?

Addressing the Issues

GUIDING QUESTION *How are the people of Oceania reacting to the changes in their environments?*

The U.S. nuclear testing in Oceania was stopped. The effects of radiation exposure and environmental damage, however, have continued. Today the atolls affected by the testing remain off-limits to human settlement. Recent studies, however, offer hope for eventual environmental recovery. In the 1990s, the United States government provided $90 million to help decontaminate Bikini Atoll. It also set up a $45 million trust fund to provide **grants** for blast survivors from Rongelap Atoll and their offspring. The nuclear legacy also has had political effects. Antinuclear activism is a major factor in regional politics. French plans to conduct nuclear tests on an atoll in French Polynesia caused antinuclear demonstrations. The international outcry led to an early halt to the tests.

Countries within the subregion are also addressing other environmental concerns. The Nauru Agreement Concerning Cooperation in the Management of Fisheries of Common Interest (also known as the Parties to the Nauru Agreement, or PNA) was created to manage tuna populations sustainably. The agreement requires catch reporting and maintenance of logbooks. It also mandates the installation of electronic position and data transfer devices on the fishing vessels. The members include eight Pacific Island countries that together control 25 percent of the world's supply of tuna. The PNA members are Federated States of Micronesia, Kiribati, Marshall Islands, Nauru, Palau, Papua New Guinea, Solomon Islands, and Tuvalu.

Coral reefs require clean water and sunlight to survive. The U.S. Environmental Protection Agency (EPA) has established programs to protect, restore, and maintain water quality around the coral reefs. Hawaii's Department of Land and Natural Resources (DLNR) reported that nearly 25 percent of all living coral was lost between 1994 and 2006 on Maui. Grants offered by the EPA aim at controlling pollution, managing watersheds, research, monitoring, and education.

Analyzing
PRIMARY SOURCES

Rising Sea Levels Cause Plans for Relocation

Kiribati is making plans to relocate its entire population before the islands disappear under rising ocean water.

"With the Australian Government the Pacific countries have the Australia–Pacific Technical Colleges (APTC) scheme which provides international standard qualification training thus allowing access for our people to the international labour markets. We also have a pilot program, the Kiribati–Australia Nursing Initiative (KANI) under which a number of our student nurses are trained to international standards in Australian institutions. We acknowledge with gratitude these innovative programs by our two development partners and indeed invite consideration of these models by other partner countries."

—Kiribati president Anote Tong, statement to the Second Session of Global Platform for Disaster Risk-Reduction Conference, June 2009

DBQ *DRAWING CONCLUSIONS*
What initiatives does the statement discuss regarding plans for relocation of people from Kiribati?

grant a sum of money given to a person or organization for a particular purpose

Between 1945 and 1992, the United States carried out more than 1,000 nuclear tests. The vast majority were conducted underground, but more than 100 were done in the Pacific region.

▶ CRITICAL THINKING

1. Hypothesizing Why do you think the United States chose Pacific islands and atolls as locations for nuclear tests?

2. Interpreting Significance What do you suppose were the long-term effects of these nuclear tests on the region?

United States Nuclear Tests July 1945 through September 1992	
Location	**Number of tests**
Bikini	23
Christmas Island	24
Enewetak	43
Johnston Island	12
Pacific	4
Total Pacific	106

Source: United States Nuclear Tests July 1945 through September 1992, U.S. Department of Defense, December 2000

Another joint effort is in the area of sugar production. For decades sugar has been a leading export for Fiji's economy. Sugar production has, however, also damaged the ecosystems of the islands of Fiji. The Sustainable Sugar program was created by the World Wildlife Fund (WWF). The program is working with local people, companies, and the government to reduce the negative effects of sugarcane production on the coral reefs. WWF also promotes an approach to conservation and development in the region that recognizes the rights of local people to manage and benefit from use of natural resources. The approach is called Community Based Natural Resource Management (CBNRM). CBNRM supports community-based initiatives and empowerment of communities. The program seeks legislation and partnerships with public and private sectors in the use of natural resources. The goal is to create sustainable development and allow communities to benefit economically. Emphasis is also placed on maintaining traditional customs and values that have protected the natural resources for centuries.

Climate change is a major concern for all the islands of Oceania. Many of the low islands are already planning strategies to cope with rising sea levels. The plans include moving human populations to the remaining areas after loss of land due to rising sea levels. Another plan is for emigration to nearby countries. Climate change will have severe effects on the economies of the subregion as populations relocate. It will also impact the subregion's culture as people move to countries such as Australia, New Zealand, and the United States. In those countries, emigrants would not own land and might lose their social structures and traditional ways of life.

☑ **READING PROGRESS CHECK**

Describing What are some measures that have been taken to address environmental concerns?

LESSON 3 REVIEW

Reviewing Vocabulary

1. Making Connections Define climate change and ocean warming. Explain how they relate to each other.

Using Your Notes

2. Summarizing Use your graphic organizer to write a description of the major causes of deforestation in Oceania.

Answering the Guiding Questions

3. Exploring Issues How are the people of Oceania using their islands' natural resources?

4. Evaluating How have humans—both in Oceania and elsewhere—affected the environments of the islands of Oceania?

5. Discussing How are the people of Oceania reacting to the changes in their environments?

Writing Activity

6. Argument Suppose you are a citizen of Oceania. Write a short letter to your local government regarding environmental issues on your island and actions you would like the government to take.

Directions: On a separate sheet of paper, answer the questions below. Make sure that you read carefully and answer all parts of the questions.

Lesson Review

Lesson 1

1 *Classifying* Explain the criteria for classifying the three groups of Oceania's islands. Within your explanation, include the names of these three groups.

2 *Making Connections* What is the relationship between atolls and the development of uniform languages and culture?

3 *Identifying Cause and Effect* How do climate conditions in the doldrums contribute to typhoons?

Lesson 2

4 *Explaining* What conditions gave rise to the number of languages spoken in Oceania? Would it be accurate to state that nearly half the languages spoken in the world are in Oceania?

5 *Problem Solving* Why is health care not uniform across all the islands of Oceania? Briefly discuss the importance of health care, and suggest an idea that might help to make it accessible across the islands.

6 *Describing* What is the most important economic activity in Oceania? Describe conditions that support as well as pose threats to this activity.

Lesson 3

7 *Considering Advantages and Disadvantages* How have countries outside of Oceania helped boost the economies of some countries of Oceania but also caused damage? Discuss the efforts of Fiji to balance economic and environmental interests.

8 *Drawing Conclusions* Discuss the impact of nuclear testing on the Oceania subregion. Can the destruction caused by nuclear testing be reversed? Explain.

9 *Identifying Central Issues* Describe the environmental issues tied to gold mining and production in Papua New Guinea. Explain why these issues have not been resolved.

21st Century Skills

Use the cartoon below to answer the questions that follow.

PRIMARY SOURCE

"Gentlemen, it's time we gave some serious thought to the effects of global warming."

10 *Using Primary Sources* Explain how this cartoon illustrates a key issue in understanding why the management of resources is important to the low islands of Oceania.

11 *Analyzing* Based on the above issue, discuss the potential impacts to freshwater resources and agriculture.

Critical Thinking

12 *Making Connections* What is the relationship of "tabu" to overfishing? How does implementation of "tabu" show the rise of the countries of Oceania to overcome challenges from the past under colonialism? Write a one-page essay to explain your responses.

Exploring the Essential Question

13 *Organizing* Create a Venn diagram to compare and contrast three countries in Oceania. Include the following information: physical geography; climate and resources; population patterns; society and culture today; economic activities; managing resources; and addressing issues related to human impact. After you complete your Venn diagram, write a summary to explain it.

Need Extra Help?

If You've Missed Question	1	2	3	4	5	6	7	8	9	10	11	12	13
Go to page	800	801	803	807	808	808	812	814	815	817	817	812	797

Mick Stevens/The New Yorker Collection/www.cartoonbank.com

Directions: On a separate sheet of paper, answer the questions below. Make sure that you read carefully and answer all parts of the questions.

Applying Map Skills

Refer to the Unit 9 Atlas to answer the following questions.

14 *Places and Regions* Identify the largest island in Melanesia.

15 *The World in Spatial Terms* Use your mental map of Oceania to describe the spatial relationship among the Marshall Islands, Palau, Vanuatu, Fiji, and Tonga.

16 *Human Systems* Of all the islands of Oceania, which has the largest population? What is the approximate population? What are the likely causes that led to this island having the greatest population?

College and Career Readiness

17 *Decision Making* Suppose you are a college student and your professor is researching subsistence farming. She has asked you to research subsistence farming in the Solomon Islands and explain how it is being practiced currently and whether it seems to have met the goals and needs of the people living there. Additionally, your professor has asked you to express and support an opinion about whether subsistence farming has been a worthwhile activity in the Solomon Islands—or whether another type of farming should replace it. Conduct research, and then prepare a summary of your findings to submit to your professor.

Writing About Geography

18 *Narrative* Suppose you are traveling at low altitude through airspace above the high islands of Oceania. Describe the scenery below using standard grammar, spelling, sentence structure, and punctuation.

DBQ Analyzing Primary Sources

Use the document to answer the following questions.

Dr. Elizabeth Lindsey, an explorer, author, filmmaker, and speaker, is the first female to become a Fellow of the National Geographic Society. She received the Visionary Award from the United Nations in 2010 for contributions in intercultural engagement and understanding.

PRIMARY SOURCE

"*'As a child, I was cared for by three old Hawaiian women while my parents worked. These elders were revered in our community for their mastery in ancient traditions. They told me that I would travel far to keep the voices of the ancestors alive and that it would take the wisdom of these elders to return the world to balance.'*"

—Dr. Elizabeth Lindsey, www.elizabethlindsey.com

19 *Interpreting Significance* How does the quote help you understand the attitude of Hawaiians toward elders in their communities? Does this attitude accurately reflect the general attitude of most Americans toward senior citizens today?

20 *Drawing Inferences* What types of actions do you think Dr. Lindsey and others are taking—or should take—to return the world to balance?

Research and Presentation

21 *Gathering Information* With a partner, conduct research to learn additional information about the Mariana Trench. Create a multimedia presentation to explain the following: location of the trench, tectonic activity that formed the trench, relevance of the trench to Oceania, likelihood that the trench will give rise to future earthquakes, recent newsworthy events relevant to the trench. Within your presentation, include maps, photographs, and graphs. Additionally, develop sketched or computer-generated diagrams.

Need Extra Help?

If You've Missed Question	**14**	**15**	**16**	**17**	**18**	**19**	**20**	**21**
Go to page	746	746	750	808	800	818	818	802

Antarctica:

the land of ICE

Earth's fifth-largest continent, the frigid land of Antarctica is found at the bottom of the world. Since its discovery in 1820 it has fascinated explorers and scientists.

(t) Danita Delimont/Gallo Images/Getty Images; Dean Lewins/Newscom

 THERE'S MORE ONLINE

LAND, CLIMATE & ENVIRONMENT

Antarctica is located at the southernmost point on Earth. It is surrounded by the Antarctic Ocean, also known as the Southern Ocean. At 5.5 million square miles (14.2 million sq. km) Antarctica is the Earth's fifth-largest continent. Almost completely covered by ice, it has no indigenous inhabitants and very limited plant and animal life. The only humans to live here are researchers and scientists, and even they live here only temporarily.

SOUTHERN OCEAN

WIND AND ICE

3.1 mi
(4776 m)
Thickest known piece of ice: (at Terre Adélie)

An ice shelf is a thick, floating sheet of ice that forms in coastal areas where a glacier meets a body of water.

-126° F
(-88° C)
Lowest recorded temperature

203 mph
(327 kmph)
Highest recorded wind speed

Weddell Sea

Ronne Ice Shelf

ANTARCTIC PENINSULA

WEST ANTARCTICA

ELLSWORTH LAND

Amundsen Sea

MARIE BYRD LAND

NASA Goddard Space Flight Center Image by Reto Stöckli (land surface, shallow water, clouds). Enhancements by Robert Simmon (ocean color, compositing, 3D globes, animation). Data and technical support: MODIS Land Group; MODIS Science Data Support Team, Dave Pape; (bkgrd) rolfo/Flickr Open/Getty Images

Fimbul
Ice Shelf

**QUEEN MAUD
LAND**

Comparing Area:
Antarctica and
the United States*

5,500,000 mi²
(14,245,000 km²)

3,619,969 mi²
(9,375,720 km²)

*Contiguous United States

**ENDERBY
LAND**

**60-
80%**
OF THE EARTH'S
**FRESH
WATER**

Amery
Ice Shelf

**AMERICAN
HIGHLAND**

West
Ice Shelf

• **SOUTH
POLE**

**EAST
ANTARCTICA**

Shackleton
Ice Shelf

TRANSANTARCTIC MOUNTAINS

**WILKES
LAND**

Ross
Ice Shelf

Though there is no
sand, Antarctica's arid
climate means it is
considered a desert. The
continent receives very little
precipitation and what little it
receives always falls as snow.

**VICTORIA
LAND**

Ross
Sea

ANTARCTIC DISCOVERIES

The frigid temperatures and isolation of Antarctica make sustaining plant and animal life difficult. The majority of animal life is found in the Southern Ocean. All warm-blooded animals in the Antarctic (seals, whales, penguins, etc.) rely on a thick layer of blubber for insulation. Most plant life is concentrated in the milder climate of West Antarctica, particularly the Antarctic Peninsula. There are no trees. Most of the vegetation consists of algae, lichens and mosses.

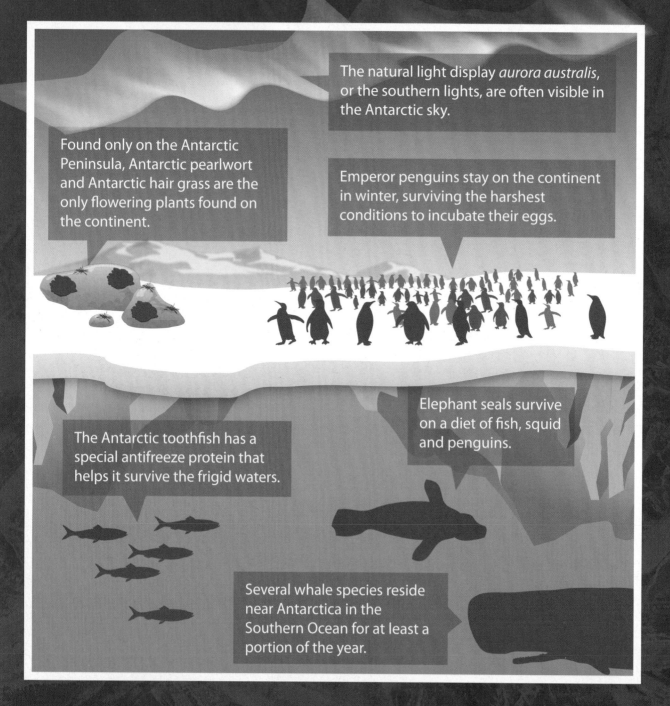

The natural light display *aurora australis*, or the southern lights, are often visible in the Antarctic sky.

Found only on the Antarctic Peninsula, Antarctic pearlwort and Antarctic hair grass are the only flowering plants found on the continent.

Emperor penguins stay on the continent in winter, surviving the harshest conditions to incubate their eggs.

Elephant seals survive on a diet of fish, squid and penguins.

The Antarctic toothfish has a special antifreeze protein that helps it survive the frigid waters.

Several whale species reside near Antarctica in the Southern Ocean for at least a portion of the year.

(bkgrd) rolfo/Flickr Open/Getty Images

SOUTHERN
OCEAN

1772-74

Weddell
Sea

Ronne
Ice Shelf

Graham
Land

Marie
Byrd
Land

Amundsen
Sea

Ross
Ice Shelf

Ross
Sea

Queen Maud
Land

South
Pole

Amery
Ice Shelf

Wilkes
Land

Victoria
Land

1911

1901

1838

1840s

TIMELINE OF ANTARCTIC DISCOVERY

James Cook
1772-74
First to cross the
Antarctic Circle.

Charles Wilkes
1838
Explored the eastern
coast of Antarctica.

James Ross
1840s
Charted much of
Antarctica's coastline.

Robert F. Scott
1901
Begins first inland
exploration.

Roald Amundsen
1911
First to reach the
South Pole.

In the 1770s British captain James Cook became the first person to cross the Antarctic Circle. Cook never saw land, but discoveries were soon made in the early 1800s by scientists, explorers, and whalers. As interest in the new continent grew, several nations and individuals sponsored scientific explorations.

SCIENTIFIC RESEARCH

As interests grew in Antarctica, countries such as Argentina, Australia, Great Britain, Chile, France, New Zealand and Norway began to lay claim to sections of the continent. In 1959, 12 countries negotiated the Antarctic Treaty to preserve Antarctica for peaceful scientific research and to put all territorial claims on hold. Since then, other countries have established research programs in Antarctica and 26 new countries have signed on. Today, there are around 69 research stations operated by 30 countries that serve as bases for scientists to study physical geography, climate and wildlife. The population of these stations ranges from 4,000 people in the summer to 1,000 people in the winter.

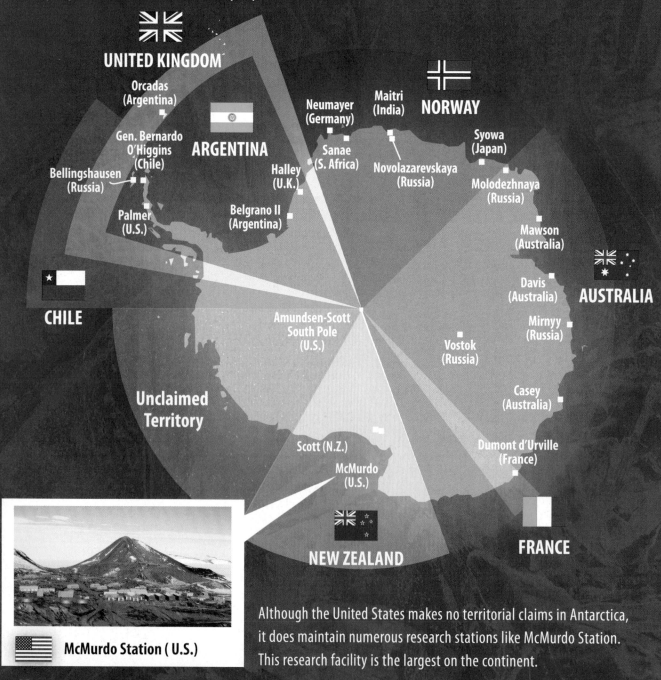

UNITED KINGDOM

Orcadas (Argentina)

Neumayer (Germany)

Maitri (India)

NORWAY

Gen. Bernardo O'Higgins (Chile)

ARGENTINA

Syowa (Japan)

Sanae (S. Africa)

Bellingshausen (Russia)

Halley (U.K.)

Novolazarevskaya (Russia)

Molodezhnaya (Russia)

Palmer (U.S.)

Belgrano II (Argentina)

Mawson (Australia)

Davis (Australia)

AUSTRALIA

CHILE

Amundsen-Scott South Pole (U.S.)

Mirnyy (Russia)

Vostok (Russia)

Unclaimed Territory

Casey (Australia)

Scott (N.Z.)

Dumont d'Urville (France)

McMurdo (U.S.)

FRANCE

NEW ZEALAND

McMurdo Station (U.S.)

Although the United States makes no territorial claims in Antarctica, it does maintain numerous research stations like McMurdo Station. This research facility is the largest on the continent.

NHPA/SuperStock; (bkgrd) rolfo/Flickr Open/Getty Images

New Frontiers

Because Antarctica's environment remains largely untouched by humans, it is a valuable location for scientific research. Compared to other places on Earth, for example, Antarctica's atmosphere, soil, and water contain little to no trace of pollution. As a result, scientists are able to use samples from Antarctica as a base for analyzing other parts of the world.

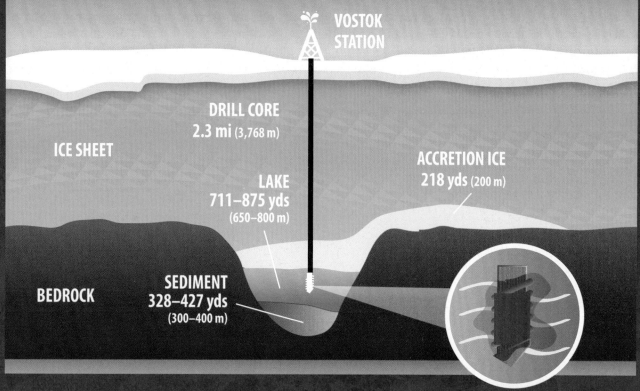

VOSTOK STATION

DRILL CORE
2.3 mi (3,768 m)

ICE SHEET

ACCRETION ICE
218 yds (200 m)

LAKE
711–875 yds
(650–800 m)

BEDROCK

SEDIMENT
328–427 yds
(300–400 m)

Lake Vostok

Lake Vostok, a subglacial lake, was discovered in the early 1970s. In 2012 scientists began probing the lake for study using deep drilling techniques. From this study scientists have found bacteria not known to exist anywhere else on Earth.

Impacts

Some scientists have raised concerns that antifreeze used in Lake Vostok drilling could harm the life forms discovered. Scientists must take great care to preserve Antarctica's pristine environment.

Thinking Geographically

1. **Analyzing Information** What characteristics of Antarctica's land and climate make sustaining life difficult?

2. **Identifying** What are some common characteristics shared by animal life and plant life found in and around Antarctica?

3. **Drawing Conclusions** What potential problems could result from nations' making territorial claims in Antarctica?

4. **Exploring Issues** Even though natural resource deposits have been found in Antarctica, many countries have agreed not to mine them. What do you think are the pros and cons of mining in the Antarctic?

5. **Expository** Research the type of work being done by scientists at one of the Antarctic stations. How could the work of these scientists impact your life? Write an essay detailing your findings.

TERMS

animism—belief that spirits inhabit natural objects and forces of nature

atheism—disbelief in the existence of any god

monotheism—belief in one God

polytheism—belief in more than one god

secularism—belief that life's questions can be answered apart from religious belief

sect—a subdivision within a religion that has its own distinctive beliefs and/or practices

tenet—a belief, doctrine, or principle believed to be true and held in common by members of a group

A *religion* is a set of beliefs in an ultimate reality and a set of practices used to express those beliefs. Religion is a key component of culture.

Each religion is defined and set apart from other religions by its own special beliefs, celebrations, and worship styles. Most religions also have their own sacred texts, sacred symbols, and sacred sites. All of these aspects of religion help to unite followers of that faith regardless of where in the world they live.

The religions examined in this World Religions Handbook all have these sacred elements, celebrations, and worship styles. Examining these characteristics provides insight into each of these religions.

World Religions Today

Legend:
- Roman Catholic
- Protestant
- Christian (Mixed)
- Eastern Churches
- Sunni Muslim
- Shia Muslim
- Hinduism
- Buddhism
- Confucianist or Taoist
- Traditional religions
- ✡ Judaism
- ☬ Sikhism
- Uninhabited

We study religion because it is an important component of culture, shaping how people interact with one another, dress, and eat. It stands at the core of a religious culture.

The diffusion of religion throughout the world has been caused by a variety of factors including migration, missionary work, trade, and war. Buddhism, Christianity, and Islam are the three major religions that spread their beliefs through missionary activities. Religions such as Hinduism, Sikhism, and Judaism are associated with a particular culture group. Followers are usually born into these religions. Sometimes close contact and differences in beliefs have resulted in conflict between religious groups.

Percentage of World Population

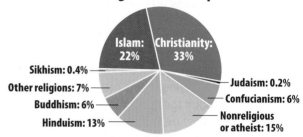

Note: Total exceeds 100% because numbers were rounded.
Sources: www.cia.gov, The World Factbook 2006; www.adherents.com.

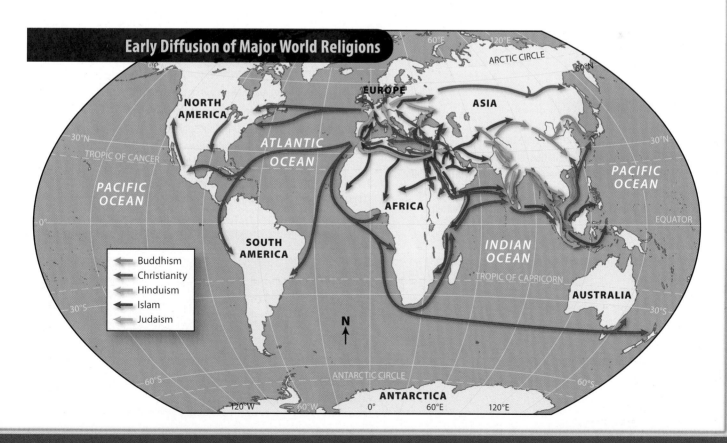

Early Diffusion of Major World Religions

Buddhism

Siddhartha Gautama, known as the Buddha ("the Awakened") after his enlightenment at the age of 35, was born some 2,500 years ago in what is now Nepal. The Buddha's followers adhere to his teachings (dharma, meaning "divine law"), which aim to end suffering in the world. Buddhists call this goal Nirvana; and they believe that it can be achieved only by understanding the Four Noble Truths and by following the 4th Truth, which says that freedom from suffering is possible by practicing the Eightfold Path. Through the Buddha's teachings, his followers come to know the impermanence of all things and reach the end of ignorance and unhappiness.

Over time, as Buddhism spread throughout Asia, several branches emerged. The largest of these are Theravada Buddhism, the monk-centered Buddhism which is dominant in Sri Lanka, Burma, Thailand, Laos, and Cambodia; and Mahayana, a complex, more liberal variety of Buddhism that has traditionally been dominant in Tibet, Central Asia, Korea, China, and Japan.

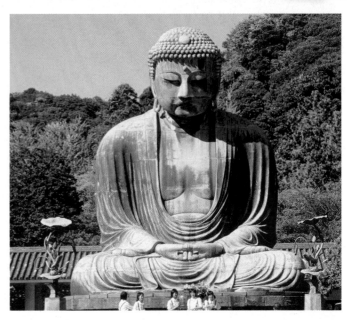

Statue of the Buddha, Siddhartha Gautama

Sacred Text ▾

For centuries the Buddha's teachings were transmitted orally. For Theravada Buddhists, the authoritative collection of Buddhist texts is the Tripitaka ("three baskets"). These texts were first written on palm leaves in a language called Pali. This excerpt from the *Dhammapada,* a famous text within the Tripitaka, urges responding to hatred with love:

> " *Never in this world is hate*
> *Appeased by hatred.*
> *It is only appeased by love—*
> *This is an eternal law.* "
> —*Dhammapada 1.5*

Sacred Symbol ▾

The *dharmachakra* ("wheel of the law") is a major Buddhist symbol. Among other things, it signifies the overcoming of obstacles. The eight spokes represent the Eightfold Path—right view, right intention, right speech, right action, right livelihood, right effort, right mindfulness, right concentration—that is central for all Buddhists.

PHOTOS: (t) Author's Image/Punchstock; (b) ©ArmandsPharyos/Alamy; TEXT: From SOURCES OF INDIAN TRADITION SECOND EDITION, VOL I, From the Beginning to 1800 By Ainslie T. Embree. Copyright © 1988 Columbia University Press. Reprinted with permission of the publisher.

Sacred Site ▲

Buddhists believe that Siddhartha Gautama achieved enlightenment beneath the Bodhi Tree in Bodh Gayā, India. Today, Buddhists from around the world flock to Bodh Gayā in search of their own spiritual awakening.

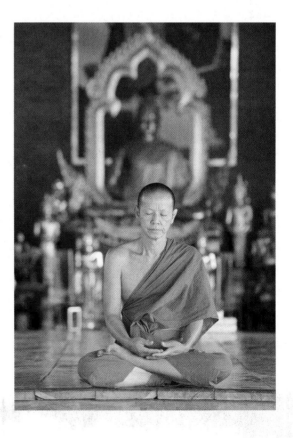

Worship and Celebration ▶

The ultimate goal of Buddhists is to achieve Nirvana, the enlightened state in which individuals are free from ignorance, greed, and suffering. Theravada Buddhists believe that monks are most likely to reach Nirvana because of their lifestyle of renunciation, moral virtue, study, and meditation.

(t) Robert Nickelsberg/Getty News Images/Getty Images; (b) Ingolf Pompe/LOOK/Getty Images

Christianity

Christianity claims more members than any of the other world religions. It dates its beginning to the death of Jesus in A.D. 33 in what is now Israel. It is based on the belief in one God and on the life and teachings of Jesus. Christians believe that Jesus, who was born a Jew, is the son of God and is fully divine and human. Christians regard Jesus as the Messiah (Christ), or savior, who died for humanity's sins. Christians feel that people are saved and achieve eternal life by faith in Jesus.

The major forms of Christianity are Roman Catholicism, Eastern Orthodoxy, and Protestantism. All three are united in their belief in Jesus as savior, but have developed their own individual theologies.

Sacred Text ▾

The Christian Bible is the spiritual text for all Christians and is considered to be inspired by God. This excerpt, from Matthew 5:3-12, is from Jesus' Sermon on the Mount.

Stained glass window depicting Jesus

" *Blessed are the poor in spirit, for theirs is the kingdom of heaven.*
Blessed are those who mourn, for they shall be comforted.
Blessed are the meek, for they shall inherit the earth.
Blessed are those who hunger and thirst for righteousness, for they shall be satisfied.
Blessed are the merciful, for they shall obtain mercy.
Blessed are the pure in heart, for they shall see God.
Blessed are the peacemakers, for they shall be called sons of God.
Blessed are those who are persecuted for righteousness' sake, for theirs is the kingdom of heaven.
Blessed are you when men revile you and persecute you and utter all kinds of evil against you falsely on my account.
Rejoice and be glad, for your reward is great in heaven, for so men persecuted the prophets who were before you. "

Sacred Symbol ▾

Christians believe that Jesus died for their sins. His death redeemed those who follow his teachings. The statue *Christ the Redeemer*, located in Rio de Janeiro, Brazil, symbolizes this fundamental belief.

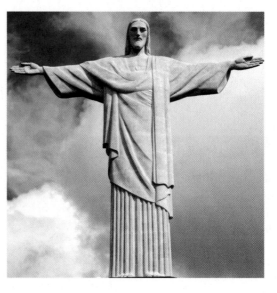

PHOTOS: (t) ©Tom Grill/Corbis; (b) Antonello/Flickr/Getty Images; TEXT: Scripture quotations are from The Holy Bible, English Standard Version® (ESV®), Copyright © 2001 by Crossway, a publishing ministry of Good News Publishers. Used by permission. All rights reserved.

Sacred Site ▸

The Gospels affirm that Bethlehem was the birthplace of Jesus. Consequently, it holds great importance to Christians. The Church of the Nativity is located in the heart of Bethlehem. It houses the spot where Christians believe Jesus was born.

Worship and Celebration ▾

Christians celebrate many events commemorating the life and death of Jesus. Among the most widely known and observed are Christmas, Good Friday, and Easter. Christmas is often commemorated by attending church services to celebrate the birth of Jesus. As part of the celebration, followers often light candles.

(t) ©Atlantide Phototravel/Corbis; (b) Remi Benali/The Image Bank/Getty Images

Confucianism

Confucianism began more than 2,500 years ago in China. Although considered a religion, it is actually a philosophy. It is based upon the teachings of Confucius, which are grounded in ethical behavior and good government.

The teachings of Confucius focused on three areas: social philosophy, political philosophy, and education. Confucius taught that relationships are based on rank. Persons of higher rank are responsible for caring for those of lower rank. Those of lower rank should respect and obey those of higher rank. Eventually his teachings spread from China to other East Asian societies.

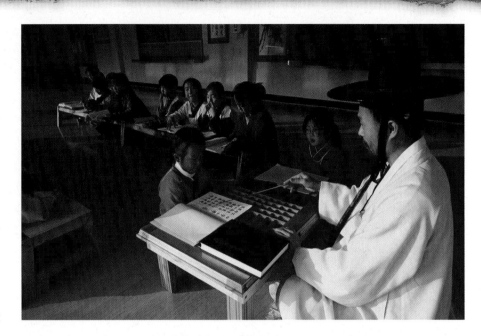

Students study Confucianism, Chunghak-dong, South Korea

The Analects

Sacred Text ▾

Confucius was famous for his sayings and proverbs. These teachings were gathered into a book called the *Analects* (see image above) after Confucius's death. Below is an example of Confucius's teachings:

Confucius said:

❝ *To learn and to practice what is learned time and again is pleasure, is it not? To have friends come from afar is happiness, is it not? To be unperturbed when not appreciated by others is gentlemanly, is it not?* ❞

Sacred Symbol ▾

Yin-yang symbolizes the harmony offered by Confucianism. The light half represents *yang,* the creative, firm, strong elements in all things. The dark half represents *yin,* the receptive, yielding, weak elements. The two act together to balance one another.

(t) ©Michel Setboun/Corbis; (c) ©Bettmann/Corbis; (b) Blank Archives/Archive Photos/Getty Images

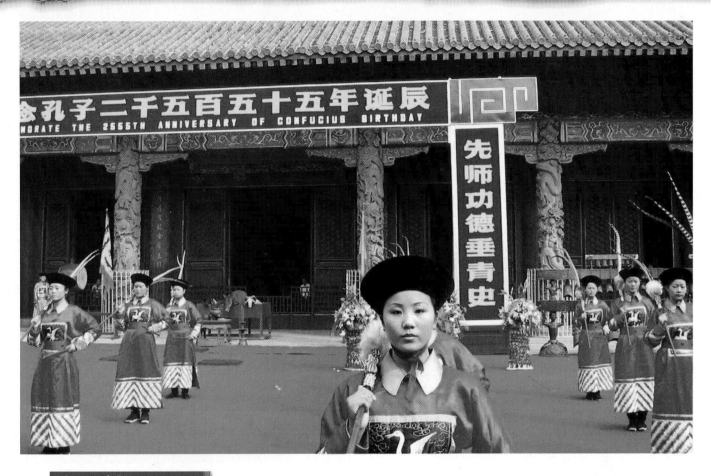

先师功德垂青史

念孔子二千五百五十五年诞辰

COMMEMORATE THE 2555TH ANNIVERSARY OF CONFUCIUS BIRTHDAY

Sacred Site ▲

The temple at Qufu is a group of buildings dedicated to Confucius. It is located on Confucius's ancestral land. It is one of the largest ancient architectural complexes in China. Every year followers gather at Qufu to celebrate the birthday of Confucius.

Worship and Celebration ▶

Confucianism does not have a god or clergy, but there are temples dedicated to Confucius, the spiritual leader. Those who follow his teachings see Confucianism as a way of life and a guide to ethical behavior and good government.

(t) STR/AFP/Getty Images; (b) Photodisc/Getty Images

Hinduism

Hinduism is the oldest of the world's major living religions. It developed among the cultures in India as they spread out over the plains and forests of the subcontinent. It has no single founder or founding date. Hinduism is complex; it has numerous sects and many different practices among its followers. Most Hindus believe in one god, whose qualities are represented by various divinities. Among the more famous Hindu divinities are Brahma, Vishnu, and Shiva, who represent respectively the creative, sustaining, and destructive forces in the universe. Major Hindu beliefs are reincarnation, karma, and dharma.

Hindus believe the universe contains several heavens and hells. According to the concept of rebirth or reincarnation, which is central to their beliefs, souls are continually reborn. In what form one is reborn is determined by the good and evil actions performed in his or her past lives. Those acts are karma. A soul continues in the cycle of rebirth until release is achieved.

Sacred Text ▼

The Vedas consist of hymns, prayers, and speculations composed in ancient Sanskrit. They are the oldest religious texts in an Indo-European language. The Rig Veda, Sama Veda, Yajur Veda, and Atharva Veda are the four great Vedic collections. Together, they make up one of the most significant and authoritative Hindu religious texts.

Statue of Vishnu

> *Now, whether they perform a cremation for such a person or not,*
> *people like him pass into the flame,*
> *from the flame into the day,*
> *from the day into the fortnight of the waxing moon,*
> *from the fortnight of the waxing moon into the six months when the sun moves north,*
> *from these months into the year, from the year into the sun,*
> *from the sun into the moon, and from the moon into the lightning.*
> *Then a person who is not human—he leads them to Brahman.*
> *This is the path to the gods, the path to Brahman.*
> *Those who proceed along this path do not return to this human condition.* »
>
> —The Chandogya Upanishad 4:15.5

Sacred Symbol ▼

One important symbol of Hinduism is actually a symbol for a sound. "Om" is a sound that Hindus often chant during prayer, mantras, and rituals.

PHOTOS: (t) Dorling Kindersley/Getty Images; (b) Image Source/Getty Images; TEXT: THE EARLY UPANISADS: ANNOTATED TEXT AND TRANSLATION by Olivelle (1998) 11 lines from "Chandogya Upanisad" p.227 © 1998 by Patrick Olivelle. By permission of Oxford University Press, USA.

Sacred Site

Hindus believe that when a person dies his or her soul is reborn. This is known as reincarnation. Many Hindus bathe in the Ganges and other sacred rivers to purify their soul and to be released from rebirth.

Worship and Celebration

Holi is a significant North Indian Hindu festival celebrating the triumph of good over evil. As part of the celebration, men, women, and children splash colored powders and water on each other. In addition to its religious significance, Holi also celebrates the beginning of spring.

(t) Mark Downey/Getty Images; (b) Poras Chaudhary/The Image Bank/Getty Images

Islam

Followers of Islam, known as Muslims, believe in one God, whom they call Allah. The word *Allah* is Arabic for "god." The founder of Islam, Muhammad, began his teachings in Makkah (Mecca) in A.D. 610. Eventually the religion spread throughout much of Asia, including parts of India to the borders of China, and a substantial portion of Africa. According to Muslims, the Quran, their holy book, contains the direct word of God, revealed to their prophet Muhammad sometime between A.D. 610 and A.D. 632. Muslims believe that God created nature and without his intervention, there would be nothingness. God serves four functions: creation, sustenance, guidance, and judgment.

Central to Islamic beliefs are the Five Pillars. These are affirmation of the belief in Allah and Muhammad as his prophet; group prayer; tithing, or the giving of money to charity; fasting during Ramadan; and a pilgrimage to Makkah once in a lifetime if physically and financially able. Within Islam, there are two main branches, the Sunni and the Shia. The differences between the two center on a criterion for legitimate Muslim rule. The Shia believed that the rulers should descend from Muhammad. The Sunni believed that the rulers need only be followers of Muhammad. Most Muslims are Sunni.

The Dome of the Rock, Jerusalem

The Quran

Sacred Text ▾

The sacred text of Islam is the Quran. Preferably, it is written and read only in Arabic, but translations have been made into many languages. The excerpt below is a verse repeated by all Muslims during their five daily prayers.

❝ *In the Name of Allah, the Compassionate,*
    the Merciful,
Praise be to Allah, the Lord of the World,
The Compassionate, the Merciful,
Master of the Day of Judgment,
Only You do we worship, and only You
Do we implore for help.
Lead us to the right path,
The path of those you have favored
Not those who have incurred
Your wrath or
Have gone astray. ❞

—The Quran

Sacred Symbol ▾

Islam is often symbolized by the crescent moon. It is an important part of Muslim rituals, which are based on the lunar calendar.

PHOTOS: (t) Design Pics/Natural Selection Ralph Curtin; (c) ©GOODSHOOT/Alamy; (b) Martin Moos/Lonely Planet Images/Getty Images; TEXT: Excerpt from The Koran Selected Suras. Copyright © 1958 by The George Macy Companies, Inc. All rights reserved under Pan American and International Copyright Conventions. Used by permission of Dover Publications, Inc.

Sacred Site ▸

Makkah is a sacred site for all Muslims. One of the Five Pillars of Islam states that all who are physically and financially able must make a hajj, or pilgrimage, to the holy city once in their life. Practicing Muslims are also required to pray facing Makkah five times a day.

Worship and Celebration ▾

Ramadan is a month-long celebration commemorating the time during which Muhammad received the Quran from Allah. It is customary for Muslims to fast from dawn until sunset all month long. Muslims believe that fasting helps followers focus on spiritual rather than bodily matters and creates empathy for one's fellow men and women. Ramadan ends with a feast known as Eid-al-Fitr, or Feast of the Fast.

(t) Nabeel Turner/Stone/Getty Images; (b) Fuse/Getty Images

Judaism

Judaism is a monotheistic religion. In fact, Judaism was the first major religion to believe in one God. Jews trace their national and religious origins back to God's call to Abraham and the revelation of the Torah to Moses at Mount Sinai. Jews believe that they have a covenant with God, who expects them to pursue justice and live ethical lives and will one day usher in a new era of universal peace.

Over time Judaism has separated into branches, including Orthodox, Reform, Conservative, and Reconstructionist. Orthodox Jews are the most traditional of all the branches.

El Ghriba Synagogue, Jerba, Tunisia

The Torah scroll

Sacred Text ▾

The Torah is the five books of Moses, which tell the story of the origins of the Jews and explain Jewish laws. The remainder of the Hebrew Bible contains the writings of the prophets, Psalms, and ethical and historical works.

> ❝ *I am the Lord your God, who brought you out of the land of Egypt, out of the house of slavery; you shall have no other gods before me.* ❞
>
> —Exodus 20:2

Sacred Symbol ▾

The menorah is used in the celebration of Hanukkah, commemorating the rededication of the Temple of Jerusalem following the Maccabees' victory over the Syrian Greeks.

PHOTOS: (t) Juergen Ritterbach/Fonline/Getty Images; (c) JUPITERIMAGES/Comstock Images/Alamy; TEXT: Exodus 20:2-3 From The Torah. Copyright © 2004 by Kuperard, an imprint of Bravo Ltd. English translation copyright © 1917, 1945, 1955 by The Jewish Publication Society of America. Published by Henry Holt and Company, LLC.

Sacred Site ◀

The Western Wall is what remains of the structure surrounding the Second Jerusalem Temple, built after the Jews' return from the Babylonian captivity. It is considered a sacred spot in Jewish religious tradition. Prayers are offered at the wall morning, afternoon, and evening.

Worship and Celebration

The day-long Yom Kippur service ends with the blowing of the ram's horn (shofar). Yom Kippur is the holiest day in the Jewish calendar. During Yom Kippur, Jews do not eat or drink for 25 hours. The purpose is to reflect on the past year, repent for one's sins, and gain forgiveness from God. It falls in September or October, ten days after Rosh Hashanah, the Jewish New Year.

(t) Paul Chesley/Stone/Getty Images; (b) Geoff Manasse/Getty Images

Sikhism

Sikhism emerged in the mid-1500s in the Punjab, in northwest India, rising from the religious experience and teachings of Guru Nanak. The religion exhibits influences from Islam and Hinduism, but it is distinct from both. Sikh traditions teach that Nanak encountered God directly and was commissioned by Him to be His servant.

Sikhs ("students, disciples") believe in one almighty god who is formless and without qualities (*nirguna*) but can be known through meditation and heard directly. Sikhism forbids discrimination on the basis of class, color, religion, caste, or gender. While over 80 percent of the world's 23 million Sikhs live in the Punjab, Sikhism has spread widely as many Sikhs have migrated to new homes around the world.

Sikh men often wear long beards and cover their heads with turbans.

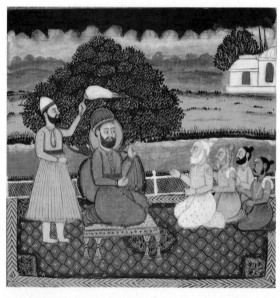

Guru Nanak

Sacred Text ▾

The great authoritative sacred text for Sikhs is the Adi Granth ("Principal Book," also known as the Guru Granth Sahib). Compiled from the mid-1500s through the 1600s, it includes contributions from Sikh gurus and from some persons also claimed as saints by Hindus and Muslims, such as Namdev, Ravidas, and Kabir.

> ❝ *Enshrine the Lord's Name within your heart. The Word of the Guru's Bani prevails throughout the world, through this Bani, the Lord's name is obtained.* ❞
> —Guru Amar Das, page 1066

Sacred Symbol ▾

The sacred symbol of the Sikhs is the *khanda*. It is composed of four traditional Sikh weapons: the *khanda* or double-edged sword (in the center), from which the symbol takes its name; the *cakkar* (disk), and two curved daggars (*kirpan*) representing temporal and spiritual power, respectively Piri and Miri.

(t) British Library/Robana/Hulton Fine Art Collection/Getty Images; (tr) ©PhotosIndia/Corbis; (b) Chirag Pithadaya/Alamy

Sacred Site ▸

Amritsar is the spiritual capital of Sikhism. The Golden Temple (*Harimandir Sahib*) in Amritsar is the most sacred of Sikh shrines.

Worship and Celebration ▾

Vaisakhi is a significant Punjabi and Sikh festival in April celebrating the new year and the beginning of the harvest season. Celebrations often take place along riverbanks with participants dancing and wearing brightly colored clothes.

(t) Adrian Pcpe/Photographer's Choice RF/Getty Images; (b) Don MacKinnon/Stringer/Getty Images News/Getty Images

Indigenous Religions

Native Americans The beliefs of most Native Americans center on the spirit world; however, the rituals and practices of individual groups vary. Most Native Americans believe in a Great Spirit who, along with other spirits, influences all aspects of life. These spirits make their presence known primarily through acts of nature.

The rituals, prayers, and ceremonies of Native Americans are often centered on health and good harvest and hunting. Rituals used to mark the passage through stages of life, including birth, adulthood, and death, are passed down as tribal traditions. Religious ceremonies often focus on important points in the agricultural and hunting seasons. Prayers, which are offered in song and dance, also concentrate on agriculture and hunting themes as well as health and well-being.

Rituals are passed down from generation to generation. These Native Americans are performing a ritual dance.

There are many different Native American groups throughout the United States and Canada. This Pawnee is wearing traditional dress during a celebration in Oklahoma.

Totem poles, like this one in Alaska, were popular among the Native American peoples of the Northwest Coast. They were often decorated with mythical beings, family crests, or other figures. They were placed outside homes.

(t) Ira Block/National Geographic Stock; (bl) ©David Ball/Corbis; (br) ML Harris/The Image Bank/Getty Images

Assessment

Reviewing Vocabulary

Match the following terms with their definitions.

1. sect
2. monotheism
3. polytheism
4. animism
5. atheism

a. belief that spirits inhabit natural objects and forces of nature
b. belief in one God
c. a subdivision within a religion that has its own distinctive belief and/or practices
d. belief in more than one god
e. disbelief in the existence of any god

Reviewing the Main Ideas

World Religions

6. Which religion has the most followers worldwide?

7. On a separate sheet of paper, make a table of the major world religions. Use the chart below to get you started.

Name	Founder	Geographic distribution	Sacred sites
Buddhism			
Christianity			
Confucianism			
Hinduism			
Islam			
Judaism			
Sikhism			
Indigenous			

Buddhism

8. According to Buddhism, how can the end of suffering in the world be achieved?

9. What is Nirvana? According to Buddhists, who is most likely to achieve Nirvana and why?

Christianity

10. In what religion was Jesus raised?

11. Why do Christians regard Jesus as their savior?

Confucianism

12. What is Confucianism based on?

13. What does yin-yang symbolize?

Hinduism

14. Where did Hinduism develop?

15. What role do Hindus believe karma plays in reincarnation?

Islam

16. What are the two branches of Islam? What is the main difference between the two groups?

17. What role does Makkah play in the Islamic faith?

Judaism

18. What is the Torah?

19. What is the purpose of Yom Kippur?

Sikhism

20. Where do most Sikhs live? Why?

21. What religions have contributed to the Adi Granth?

Indigenous Religions

22. Why would local religions feature sacred stories about the creation of people, animals, and plant life?

23. Which of the indigenous religions has the largest membership?

Problem-Solving Activity

24. **Research Project** Use library and Internet sources to research the role of food and food customs in one of the world's major religions. Create a presentation to report your findings to the class.

Gazetteer

A Gazetteer (GA·zuh·TIHR) is a geographic index or dictionary. It shows latitude and longitude for cities and certain other places. This Gazetteer lists most of the world's largest independent countries, their capitals, and several important geographic features. The page numbers tell where each entry can be found on a map in this book. As an aid to pronunciation, many entries are spelled phonetically.

A

Abidjan (AH·bee·JAHN) Capital and port city of Côte d'Ivoire, Africa. 5°N 4°W (p. 477)

Abu Dhabi (AH·boo DAH·bee) Capital of the United Arab Emirates, on the Persian Gulf. 24°N 54°E (p. 361)

Abuja (ah·BOO·jah) Capital of Nigeria. 8°N 9°E (p. 477)

Accra (AH·kruh) Capital and port city of Ghana. 6°N 0° longitude (p. 477)

Aconcagua (AH·kohn·KAH·gwah) Highest peak of the Andes and of the Western Hemisphere, in western Argentina near the Chilean border. 32°S 76°W (p. 168)

Addis Ababa (AHD·dihs AH·bah·BAH) Capital of Ethiopia. 9°N 39°E (p. 477)

Adriatic (AY·dree·A·tihk) **Sea** Arm of the Mediterranean Sea between the Balkan Peninsula and Italy. (p. 244)

Aegean (ee·JEE·uhn) **Sea** Arm of the Mediterranean Sea between Greece and Turkey. (p. 242)

Afghanistan Country in Central Asia, west of Pakistan. (p. 361)

Ahaggar Mountains Highest plateau region in the central Sahara. (p. 360)

Albania Country on the east coast of the Adriatic Sea, south of Serbia and Montenegro. (p. 243)

Algeria Country in North Africa. (p. 361)

Algiers (al·JIHRZ) Capital of Algeria. 37°N 3°E (p. 361)

Alps Mountain system extending through central Europe. (p. 242)

Altay Shan Mountain system between western Mongolia and China and between Kazakhstan and southern Russia. (p. 672)

Amazon River River flowing through Peru and Brazil in South America and into the Atlantic Ocean. (p. 168)

Amman Capital of Jordan. 32°N 36°E (p. 361)

Amsterdam Capital of the Netherlands. 52°N 5°E (p. 243)

Amu Dar'ya River in Turkmenistan in central and western Asia. (p. 360)

Amur River River in northeast Asia. (p. 672)

Andes Mountain system along western South America. (p. 168)

Andorra (an·DAWR·uh) Country in southern Europe, between France and Spain. (p. 243)

Angola (ang·GOH·luh) Country in Africa, south of the Democratic Republic of the Congo. (p. 477)

Ankara (AHN·kuh·ruh) Capital of Turkey. 40°N 33°E (p. 361)

Antananarivo (AHN·tah·NAH·nah·REE·voh) Capital of Madagascar. 19°S 48°E (p. 477)

Antigua Island in the West Indies, part of independent Antigua and Barbuda. 18°N 61°W (p. 169)

Apennines (A·puh·NYNZ) Mountain range in central Italy. (p. 242)

Appalachian Mountains Mountain system in eastern North America. (p. 112)

Arabian Sea Part of the Indian Ocean between India and the Arabian Peninsula. (p. 361)

Aral Sea Inland sea between Kazakhstan and Uzbekistan. (p. 360)

Argentina Country in South America, east of Chile. (p. 169)

Arkansas River River in south-central United States, emptying into the Mississippi River. (p. 112)

Armenia (ahr·MEE·nee·uh) Southeastern European country between the Black and Caspian Seas. (p. 361)

Ashkhabad (ASH·kuh·BAD) Capital of Turkmenistan. 40°N 58°E (p. 361)

Asmara (az·MAHR·uh) Capital of Eritrea. 16°N 39°E (p. 477)

Astana Capital of Kazakhstan. 52°N 72°E (p. 361)

Asunción (ah·SOON·SYOHN) Capital of Paraguay. 25°S 58°W (p. 169)

Athens Capital of Greece. 38°N 24°E (p. 243)

Atlas Mountains Mountain range on the northern edge of the Sahara. (p. 360)

Australia Country and continent southeast of Asia. (p. 747)

Austria Country in central Europe, east of Switzerland. (p. 243)

Azerbaijan (A·zuhr·by·JAHN) European-Asian country on the Caspian Sea. (p. 361)

B

Baghdad Capital of Iraq. 33°N 44°E (p. 361)

Bahamas Independent state comprising a chain of islands, cays, and reefs southeast of Florida and north of Cuba. 24°N 76°W (p. 169)

Bahrain (bah·RAYN) Independent state in the western Persian Gulf. (p. 361)

Baku Capital of Azerbaijan. 40°N 50°E (p. 361)

Balkan Mountains Mountain range extending across central Bulgaria to the Black Sea. (p. 242)

Balkan Peninsula Peninsula in southeastern Europe bordered on the west by the Adriatic Sea. (p. 242)

Baltic Sea Arm of the Atlantic Ocean in northern Europe that connects with the North Sea. (p. 242)

Bamako (BAH·mah·KOH) Capital of Mali. 13°N 8°W (p. 477)

Bangkok Capital of Thailand. 14°N 100°E (p. 5)

Bangladesh (BAHNG·gluh·DESH) Country in South Asia, bordered by India and Myanmar. (p. 599)

Bangui (bahng·GEE) Capital of the Central African Republic. 4°N 19°E (p. 477)

Banjul Capital of Gambia. 13°N 17°W (p. 477)

Barbados Island country between the Atlantic Ocean and the Caribbean Sea. 14°N 59°W (p. 169)

Barbuda Island in the West Indies, part of independent Antigua and Barbuda. 18°N 62°W (p. 169)

Barents Sea Part of the Arctic Ocean, north of Norway and Russia. (p. 242)

Bay of Bengal Part of the Indian Ocean between eastern India and Southeast Asia. (p. 598)

Beijing Capital of China. 40°N 116°E (p. 5)

Beirut (bay·ROOT) Capital of Lebanon. 34°N 36°E (p. 361)

Belarus (BEE·luh·ROOS) Eastern European country west of Russia. (p. 243)

Belgium (BEHL·juhm) Country in northwestern Europe, south of the Netherlands. (p. 243)

Belgrade Capital of Serbia. 45°N 21°E (p. 243)

Belize (buh·LEEZ) Country in Central America. (p. 169)

Belmopan (BEHL·moh·PAHN) Capital of Belize. 17°N 89°W (p. 169)

Benin (buh·NEEN) Country in western Africa. (p. 477)

Ben Nevis Peak in the highlands region of the Grampian Mountains in Scotland. 54°N 5°W (p. 242)

Bering Sea Part of the north Pacific Ocean, extending between the United States and Russia. (p. 112)

Berlin Capital of Germany. 53°N 13°E (p. 243)

Bern Capital of Switzerland. 47°N 7°E (p. 243)

Bhutan (boo·TAHN) Country in the eastern Himalaya, northeast of India. 27°N 91°E (p. 599)

Bishkek Capital and largest city of Kyrgyzstan. 43°N 75°E (p. 361)

Bissau (bih·SOW) Capital of Guinea-Bissau. 12°N 16°W (p. 477)

Black Sea Sea between Europe and Asia. (p. 242)

Bloemfontein (BLOOM·FAHN·TAYN) Judicial capital of the Republic of South Africa. 29°S 26°E (p. 477)

Bogotá (BOH·goh·TAH) Capital of Colombia. 5°N 74°W (p. 169)

Bolivia Republic in west-central South America. (p. 169)

Bosnia-Herzegovina (BAHZ·nee·uh HERT·suh·goh·VEE·nuh) Southeastern European country between Serbia and Croatia. (p. 243)

Bosporus Strait between European and Asian Turkey, connecting the Sea of Marmara with the Black Sea. (p. 413)

Botswana (baht·SWAH·nuh) Country in Africa, north of the Republic of South Africa. (p. 477)

Brahmaputra River River that begins in Tibet, flows through northeast India and Bangladesh, and empties into the Bay of Bengal. (p. 598)

Brasília (bruh·ZIHL·yuh) Capital of Brazil. 16°S 48°W (p. 169)

Bratislava (BRAH·tuh·SLAH·vuh) Capital and largest city of Slovakia. 48°N 17°E (p. 243)

Brazil Largest country in South America, in east-central South America. (p. 169)

Brazzaville (BRA·zuh·VIHL) Capital of Congo. 4°S 15°E (p. 477)

Brunei (bru·NY) Country on the northern coast of the island of Borneo. (p. 747)

Brussels Capital of Belgium. 51°N 4°E (p. 243)

Bucharest (BOO·kuh·REHST) Capital of Romania. 44°N 26°E (p. 243)

Budapest Capital of Hungary. 48°N 19°E (p. 243)

Buenos Aires (BWAY·nuhs AR·eez) Capital of Argentina. 34°S 58°W (p. 169)

Bujumbura (BOO·juhm·BUR·uh) Capital of Burundi. 3°S 29°E (p. 477)

Bulgaria (BUHL·GAR·ee·uh) Country in southeastern Europe, south of Romania. (p. 243)

Burkina Faso (bur·KEE·nuh FAH·soh) Country in western Africa, south of Mali. (p. 477)

Burundi (bu·ROON·dee) Country in central Africa at the northern end of Lake Tanganyika. (p. 477)

C

Cairo (KY·roh) Capital of Egypt. 31°N 32°E (p. 5)

Cambodia (kam·BOH·dee·uh) Country in Southeast Asia, south of Thailand. (p. 747)

Cameroon (KA·muh·ROON) Country in West Africa, on the northeast shore of the Gulf of Guinea. (p. 477)

Canada Country in northern North America. (p. 113)

Canberra Capital of Australia. 35°S 149°E (p. 747)

Cape Town Legislative capital of the Republic of South Africa. 34°S 18°E (p. 5)

Cape Verde Republic consisting of a group of volcanic islands in the Atlantic Ocean. 15°N 26°W (p. 477)

Caracas (kah·RAH·kahs) Capital of Venezuela. 11°N 67°W (p. 169)

Caribbean (KAR·uh·BEE·uhn) **Sea** Part of the Atlantic Ocean, bounded by the West Indies, South America, and Central America. (p. 170)

Carpathian Mountains Mountain range in eastern Europe in Slovakia and Romania. (p. 242)

Caspian (KAS·pee·uhn) **Sea** Salt lake between Europe and Asia. (p. 242)

Caucasus (KAW·kuh·suhs) **Mountains** Mountain range in southwestern Russia. (p. 360)

Central African Republic Country in central Africa, south of Chad. (p. 477)

Central Siberian Plateau Tableland area in Siberia. (p. 242)

Chad Country in north-central Africa. (p. 477)

Chang Jiang (CHAHNG JYAHNG) River in north-central and eastern China, also known as the Yangtze River. (p. 672)

Chao Phraya (chow PRY·uh) River in Thailand, flowing south into the Gulf of Thailand. (p. 746)

Chile (CHIH·lee) Western South American country, along the Pacific Ocean. (p. 169)

China (People's Republic of China) Country in eastern and central Asia. (p. 673)

Chişinău (KEE·shee·NOW) Capital and largest city of Moldova. 47°N 29°E (p. 243)

Colombia Republic in northern South America. (p. 169)

Colombo Capital of Sri Lanka. 7°N 80°E (p. 599)

Colorado Plateau Highlands region in the western United States. (p. 112)

Colorado River River in the western United States that flows through the Grand Canyon. (p. 112)

Columbia Plateau Flat plains area primarily in western Washington State in the United States. (p. 112)

Comoros (KAH·muh·ROHZ) **Islands** Island country in the Indian Ocean between the island of Madagascar and Africa. 13°S 43°E (p. 477)

Conakry (KAH·nuh·kree) Capital of Guinea. 10°N 14°W (p. 477)

Congo, Democratic Republic of the African country on the Equator, north of Zambia and Angola. (p. 477)

Congo, Republic of the Country in Equatorial Africa. (p. 477)

Congo River River that runs through the Democratic Republic of the Congo. (p. 476)

Copenhagen (KOH·puhn·HAY·guhn) Capital of Denmark. 56°N 12°E (p. 243)

Costa Rica (KAWS·tah REE·kuh) Central American country, south of Nicaragua. (p. 169)

Côte d'Ivoire (KOHT dee·VWAHR) West African country, south of Mali. (p. 477)

Croatia (kroh·AY·shuh) Southeastern European country on the Adriatic Sea. (p. 243)

Cuba Island country southeast of Florida. 21°N 80°W (p. 169)

Cyprus Island country in the eastern Mediterranean Sea, south of Turkey. 35°N 31°E (p. 243)

Czech (CHEHK) **Republic** Central European country south of Germany and Poland. (p. 243)

D

Dakar Capital of Senegal. 15°N 17°W (p. 6)

Damascus Capital of Syria. 34°N 36°E (p. 361)

Danube River River in Europe that begins in Germany and flows into the Black Sea. (p. 242)

Dardanelles Strait between European and Asian Turkey, connecting the Sea of Marmara with the Aegean Sea. (p. 413)

Dar es Salaam (DAHR EHS suh·LAHM) Commercial capital of Tanzania. 7°S 39°E (p. 477)

Darling River River in southeast Australia. (p. 746)

Dead Sea A landlocked body of salt water located between Israel and Jordan. (p. 360)

Deccan Plateau The peninsula of India south of the Narmada River. (p. 598)

Denmark Country in northwestern Europe, between the Baltic and North Seas. (p. 243)

Dhaka Capital of Bangladesh. 24°N 90°E (p. 599)

Djibouti (juh·BOO·tee) Country in East Africa, on the Gulf of Aden. (p. 477)

Dnieper (NEE·puhr) **River** River that begins in Russia, flows through Belarus and Ukraine, and then drains into the Black Sea. (p. 242)

Dniester (NEE·stuhr) **River** River in south-central Europe that begins in Ukraine and flows southeast to the Black Sea. (p. 242)

Dodoma (doh·DOH·MAH) Political capital of Tanzania. 7°S 36°E (p. 477)

Doha (DOH·hah) Capital of Qatar. 25°N 51°E (p. 361)

Dominica Island republic in the West Indies, lying in the center of the Lesser Antilles. 15°N 61°W (p. 169)

Dominican Republic Republic occupying the eastern two-thirds of Hispaniola Island in the West Indies. 19°N 70°W (p. 169)

Don River River in southwestern Russia. (p. 242)

Drakensberg (DRAH·kuhnz·BUHRG) Range Mountain range in South Africa. (p. 476)

Dublin Capital of Ireland. 53°N 6°W (p. 243)

Dushanbe (doo·SHAM·buh) Capital and largest city of Tajikistan. 39°N 69°E (p. 361)

E

East Timor (Timor-Leste) Island country in the Indonesian archipelago, northwest of Australia. (p. 747)

Eastern Ghats Mountain range in India. (p. 598)

Ecuador (EH·kwuh·DAWR) Country in South America, south of Colombia. (p. 169)

Egypt Country in northern Africa on the Mediterranean Sea. (p. 361)

Elbe River River in central Europe. (p. 243)

Elburz Mountains Mountain range in northern Iran parallel to the shore of the Caspian Sea. (p. 360)

El Salvador (ehl SAL·vuh·DAWR) Country in Central America, southwest of Honduras. (p. 169)

Equatorial Guinea (EE·kwuh·TOHR·ee·uhl GIH·nee) Country in western Africa, south of Cameroon. (p. 477)

Eritrea (EHR·uh·TREE·uh) Country in northeast Africa, north of Ethiopia. (p. 477)

Estonia (eh·STOH·nee·uh) Northern European country on the Baltic Sea. (p. 243)

Ethiopia (EE·thee·OH·pee·uh) Country in eastern Africa, north of Somalia and Kenya. (p. 477)

Euphrates (yu·FRAY·teez) **River** River in southwestern Asia that flows through Syria and Iraq and joins the Tigris River. (p. 360)

F

Fiji (FEE·jee) Country comprising an island group in the southwest Pacific Ocean. 19°S 175°E (p. 747)

Gazetteer

Finland Country in northern Europe, east of Sweden. (p. 243)

France Country in western Europe. (p. 243)

Freetown Capital and port city of Sierra Leone, in western Africa. 9°N 13°W (p. 477)

French Guiana Overseas department of France on the northeast coast of South America. (p. 169)

G

Gabon (ga·BOHN) Country in western Africa, on the Atlantic Ocean. (p. 477)

Gaborone (GAH·boh·ROH·nay) Capital of Botswana, in southern Africa. 24°S 26°E (p. 477)

Gambia Country in western Africa. (p. 477)

Ganges (GAN·JEEZ) **River** River in northern India and Bangladesh that flows into the Bay of Bengal. (p. 598)

Gangetic (gan·JEH·tic) **Plain** A fertile plains region in northern India traversed by the Ganges River. (p. 598)

Georgetown Capital of Guyana. 8°N 58°W (p. 169)

Georgia Asian/European country bordering the Black Sea, south of Russia. (p. 361)

Germany (Federal Republic of Germany) Country in north-central Europe. (p. 243)

Ghana (GAH·nuh) Country in western Africa, on the Gulf of Guinea. (p. 477)

Gobi Desert in Central Asia. (p. 672)

Godavari River River in central India. (p. 598)

Gran Chaco Region in south-central South America located in Paraguay, Bolivia, and Argentina. (p. 168)

Great Britain Kingdom in western Europe comprising England, Scotland, and Wales. (p. 242)

Great Dividing Range Chain of hills and mountains, on Australia's eastern coast. (p. 746)

Great Plains Rolling treeless area of central North America. (p. 112)

Great Salt Lake Large saltwater lake in Utah in the United States that has no outlet. (p. 112)

Great Slave Lake A lake in the south-central mainland of the Northwest Territories in Canada. (p. 112)

Greece Country in southern Europe, on the Balkan Peninsula. (p. 243)

Greenland Island in the northwestern Atlantic Ocean. 74°N 40°W (p. 112)

Grenada Island in the self-governing West Indies. 17°N 61°W (p. 169)

Guam Island in the western Pacific. It is an unincorporated United States territory. 13°N 144°E (p. 747)

Guatemala (GWAH·tuh·MAH·luh) Country in Central America, south of Mexico. (p. 169)

Guatemala Capital of Guatemala and the largest city in Central America. 15°N 91°W (p. 169)

Guinea (GIH·nee) West African country on the Atlantic coast. 11°N 12°W (p. 477)

Guinea-Bissau (GIH·nee bih·SOW) West African country on the Atlantic coast. 12°N 20°W (p. 477)

Gulf of Aden Arm of the Indian Ocean between the Arabian Peninsula and Africa. (p. 360)

Gulf of Mexico Gulf on the southern coast of North America. (p. 112)

Gulf of Thailand Inlet of the South China Sea. (p. 746)

Guyana Republic in northern South America. (p. 169)

H

Hainan (HY·NAHN) Island province of China in the South China Sea. 19°N 109°E (p. 672)

Haiti (HAY·tee) Republic occupying the western third of Hispaniola Island in the West Indies. 19°N 72.25°W (p. 169)

Hanoi Capital of Vietnam. 21°N 106°E (p. 747)

Harare (huh·RAH·ray) Capital of Zimbabwe. 18°S 23°E (p. 477)

Havana Capital of Cuba. 23°N 82°W (p. 169)

Helsinki Capital of Finland. 60°N 24°E (p. 243)

Himalaya (HIH·muh·LAY·uh) Mountain system in South Asia, bordering the Indian subcontinent on the north. (p. 598)

Hindu Kush Mountain range in Central Asia. (p. 360)

Honduras (hahn·DUR·uhs) Central American republic. (p. 169)

Hong Kong Administrative district and port in southern China. 22°N 115°E (p. 7)

Huang He (HWAHNG HUH) Major river in central China, also known as the Yellow River. (p. 672)

Hudson Bay Inland sea in east-central Canada. (p. 112)

Hungary Central European country, south of Slovakia. (p. 243)

I

Iberian (eye·BIHR·ee·uhn) **Peninsula** Peninsula in southwestern Europe. (p. 242)

Iceland Island country between the north Atlantic and Arctic Oceans. 65°N 20°W (p. 243)

India South Asian country south of China. (p. 599)

Indochina Southeast peninsula of Asia. (p. 3)

Indonesia (IHN·duh·NEE·zhuh) Group of islands that forms the Southeast Asian country of the Republic of Indonesia. 5°S 119°E (p. 747)

Indus River River in Asia that rises in Tibet and flows through Pakistan to the Arabian Sea. (p. 598)

Iran (ih·RAHN) Southwest Asian country, formerly called Persia. (p. 361)

Iraq (ih·RAHK) Southwest Asian country, south of Turkey. (p. 361)

Ireland Island west of England, occupied by the Republic of Ireland and by Northern Ireland. 54°N 8°W (p. 243)

Irrawaddy River River in central Myanmar formed by the confluence of the Mali and Nmai Rivers. (p. 746)

Irtysh River River in northeast Kazakhstan and the western part of Russia, in Asia. (p. 242)

Gazetteer

Islamabad (ihs·LAH·muh·BAHD) Capital of Pakistan. 34°N 73°E (p. 599)

Israel (IHZ·ree·uhl) Country in Southwest Asia, south of Lebanon. (p. 361)

Isthmus of Panama Narrow strip of land that forms the link in Central America between North America and South America. (p. 168)

Italy Southern European country, south of Switzerland and east of France. (p. 243)

J

Jakarta Capital of Indonesia. 6°S 107°E (p. 5)

Jamaica (juh·MAY·kuh) Island country in the West Indies. 18°N 78°W (p. 169)

Japan Country in East Asia, consisting of four main islands of Hokkaidō, Honshū, Shikoku, and Kyūshū, plus thousands of small islands. 37°N 134°E (p. 673)

Jerusalem (juh·ROO·suh·luhm) Capital of Israel and a holy city for Jews, Christians, and Muslims. 32°N 35°E (p. 361)

Jordan Country in Southwest Asia. (p. 361)

Juba Capital of South Sudan. 5°N 31°E (p. 477)

Jutland Peninsula extending north from Germany. (p. 242)

K

K2 (Godwin Austen) Himalayan mountain in Jammu and Kashmir. 35°N 76°E (p. 598)

Kabul Capital of Afghanistan. 35°N 69°E (p. 361)

Kalahari Desert Plateau and part desert located in the southern part of Africa. (p. 476)

Kamchatka Peninsula Peninsula in northeast Russia, in Asia. (p. 242)

Kampala (kahm·PAH·lah) Capital of Uganda. 0° latitude 32°E (p. 477)

Kara Sea Arm of the Arctic Ocean north of Russia. (p. 242)

Kathmandu (KAT·MAN·DOO) Capital of Nepal. 28°N 85°E (p. 599)

Kazakhstan (KA·zak·STAN) Large Asian country south of Russia, bordering the Caspian Sea. (p. 361)

Kenya (KEH·nyuh) Country in eastern Africa, south of Ethiopia. (p. 477)

Khartoum Capital of Sudan. 16°N 33°E (p. 477)

Khyber Pass Mountain pass between Afghanistan and Pakistan. 34°N 71°E (p. 360)

Kiev See **Kyiv (Kiev)**

Kigali (kee·GAH·lee) Capital of Rwanda, in central Africa. 2°S 30°E (p. 477)

Kilimanjaro Highest mountain in Africa, located in Tanzania. 3°S 37°E (p. 476)

Kingston Capital of Jamaica. 18°N 77°W (p. 169)

Kinshasa (kihn·SHAH·suh) Capital of the Democratic Republic of the Congo. 4°S 15°E (p. 5)

Kiribati (KIHR·uh·BAS) One of the two Federated States of Micronesia. 5°S 170°W (p. 747)

Korean Peninsula Peninsula on which both North and South Korea are located. (p. 672)

Kosovo (KAW·saw·VOH) Country in southeastern Europe, between Serbia and Montenegro. (p. 243)

Krishna River River of the Deccan Plateau in south India. (p. 598)

Kuala Lumpur (KWAH·luh LUM·PUR) Capital of Malaysia. 3°N 102°E (p. 747)

Kunlun Shan Mountain ranges in western China on the north edge of the Plateau of Tibet. (p. 672)

Kuwait (ku·WAYT) Country between Saudi Arabia and Iraq, on the Persian Gulf. (p. 361)

Kyiv (Kiev) (KEE·EHF) Capital of Ukraine. 50°N 31°E (p. 243)

Kyrgyzstan (KIHR·gih·STAN) Small Central Asian country on China's western border. (p. 361)

L

Labrador Sea Part of the Atlantic Ocean south of Baffin Bay off the coast of Newfoundland. (p. 112)

Lagos Port city of Nigeria. 6°N 3°E (p. 477)

Lake Baikal Lake in southern Siberia, Russia. It is the largest freshwater lake in Eurasia. (p. 242)

Lake Chad Reservoir located in Chad. (p. 476)

Lake Erie One of the Great Lakes of the United States and Canada. (p. 112)

Lake Huron One of the Great Lakes of the United States and Canada. (p. 112)

Lake Malawi Lake in southeast Africa. (p. 476)

Lake Michigan One of the Great Lakes of the United States and Canada. (p. 112)

Lake Ontario The easternmost and smallest of the Great Lakes of the United States and Canada. (p. 112)

Lake Superior One of the Great Lakes of the United States and Canada. (p. 112)

Lake Tanganyika Lake in east-central Africa. (p. 476)

Lake Titicaca Lake on the border between Peru and Bolivia. Highest navigable lake in the world. (p. 168)

Lake Victoria Freshwater lake in Tanzania and Uganda. (p. 476)

Lake Volta Reservoir located in Ghana. (p. 476)

Lake Winnipeg Lake in south-central Manitoba, Canada. (p. 112)

Laos (LOWS) Southeast Asian country, south of China and west of Vietnam. (p. 747)

La Paz (lah PAHZ) Administrative capital of Bolivia, and the highest capital in the world. 17°S 68°W (p. 169)

Latvia (LAT·vee·uh) Northeastern European country on the Baltic Sea, west of Russia. (p. 243)

Lebanon (LEH·buh·nuhn) Country on the Mediterranean Sea, south of Syria. (p. 361)

Lena River River in east-central Russia. (p. 242)

Lesotho (luh·SOH·toh) Country in southern Africa. (p. 477)

Liberia (ly·BIHR·ee·uh) West African country, south of Guinea. 7°N 10°W (p. 477)

Libreville (LEE·bruh·VIHL) Capital and port city of Gabon. 1°N 9°E (p. 477)

Libya (LIH·bee·uh) North African country on the Mediterranean Sea, west of Egypt. (p. 361)

Liechtenstein (LIHK·tuhn·STYN) Small country in central Europe. (p. 243)

Lilongwe (lih·LAWNG·gway) Capital of Malawi. 14°S 34°E (p. 477)

Lima (LEE·muh) Capital of Peru. 12°S 77°W (p. 169)

Lisbon Capital of Portugal. 39°N 9°W (p. 243)

Lithuania (LIH·thuh·WAY·nee·uh) European country on the Baltic Sea, west of Belarus. (p. 243)

Ljubljana (lee·OO·blee·AH·nuh) Capital of Slovenia. 46°N 14°E (p. 243)

Llanos Vast plains in northern South America. (p. 168)

Loire River River in Europe that rises in southeastern France and empties into the Bay of Biscay. (p. 242)

Lomé (loh·MAY) Capital and port city of Togo in Africa. 6°N 1°E (p. 477)

London Capital of the United Kingdom, on the Thames River. 52°N 0° longitude (p. 243)

Luanda Capital of Angola. 9°S 13°E (p. 477)

Lusaka Capital of Zambia. 15°S 28°E (p. 477)

Luxembourg (LUHK·suhm·BUHRG) European country between France, Germany, and Belgium. (p. 243)

M

Macau (muh·KOW) Administrative district and port in southern China. (p. 673)

Macedonia (MA·suh·DOH·nee·uh) Republic in southeastern Europe, north of Greece. Macedonia also refers to a geographic region in the Balkan Peninsula. (p. 243)

Mackenzie River River in the western portion of the Northwest Territories in Canada. (p. 112)

Madagascar (MA·duh·GAS·kuhr) Island in the Indian Ocean, southeast of Africa. (p. 477)

Madrid Capital of Spain. 40°N 4°W (p. 243)

Malabo (mah·LAH·boh) Capital of Equatorial Guinea. 4°N 9°E (p. 477)

Malawi (muh·LAH·wee) Southeastern African country, south of Tanzania and east of Zambia. (p. 477)

Malaysia (muh·LAY·zhuh) Federation of states in Southeast Asia on the Malay Peninsula and the island of Borneo. (p. 747)

Maldives (MAWL·DEEVZ) Island country in the Indian Ocean near South Asia. 5°N 42°E (p. 599)

Mali Country in western Africa, south of Algeria. (p. 361)

Malta An independent state consisting of three islands in the Mediterranean Sea. 36°N 15°E (p. 243)

Managua (mah·NAH·gwah) Capital of Nicaragua. 12°N 86°W (p. 169)

Manila (muh·NIH·luh) Capital and port city of the Republic of the Philippines. 15°N 121°E (p. 747)

Marshall Islands Independent group of atolls and reefs in the western Pacific Ocean. 11°N 108°E (p. 747)

Maseru (MA·suh·ROO) Capital of Lesotho, in southern Africa. 29°S 27°E (p. 477)

Masqat Capital of Oman. 23°N 59°E (p. 361)

Mato Grosso Plateau Highlands area in southwest Brazil. (p. 168)

Mauritania (MAWR·uh·TAY·nee·uh) West African country, north of Senegal. (p. 361)

Mauritius (maw·RIH·shuhs) Island country in the Indian Ocean east of Madagascar. 21°S 58°E (p. 477)

Mbabane (EHM·bah·BAH·nay) Capital of Swaziland, in southeastern Africa. 26°S 31°E (p. 477)

Mediterranean Sea Inland sea enclosed by Europe, Asia, and Africa. (p. 244)

Mekong River River in Southeast Asia that flows south through Laos, Cambodia, and Vietnam. (p. 746)

Melekeok New capital of Palau. 7°N 134°E (p. 747)

Meseta The plains of central Spain. (p. 242)

Mexico Country in North America, south of the United States. (p. 169)

Mexico City Capital and most populous city of Mexico. 19°N 99°W (p. 169)

Minsk Capital of Belarus. 54°N 28°E (p. 243)

Mississippi River River in the central United States that rises in Minnesota and flows southeast into the Gulf of Mexico. (p. 112)

Missouri River River in the central United States that joins the Mississippi River. (p. 112)

Mogadishu (MAH·guh·DIH·shoo) Capital and major seaport of Somalia, in eastern Africa. 2°N 45°E (p. 477)

Moldova (mahl·DOH·vuh) European country between Ukraine and Romania. (p. 243)

Monaco (MAH·nuh·KOH) Independent principality in southern Europe, on the Mediterranean. (p. 243)

Monaco Capital of Monaco. 44°N 8°E (p. 243)

Mongolia (mahn·GOHL·yuh) Country in Asia between Russia and China. (p. 673)

Monrovia (muhn·ROH·vee·uh) Capital and major seaport of Liberia, in western Africa. 6°N 11°W (p. 477)

Mont Blanc The highest mountain of the Alps, in southeastern France. 46°N 7°E (p. 242)

Montenegro (mahn·tuh·NEH·groh) European country between the Adriatic Sea and Serbia. (p. 243)

Montevideo (MAHN·tuh·vuh·DAY·oh) Capital of Uruguay. 35°S 56°W (p. 169)

Morocco (muh·RAH·koh) Country in northwestern Africa on the Mediterranean Sea and the Atlantic Ocean. (p. 361)

Moscow Capital of Russia. 56°N 38°E (p. 243)

Mount Ararat Mountain in eastern Turkey. 39°N 44°3 (p. 360)

Mount Elbrus Highest point in the Caucasus Mountains. 43°N 42°E (p. 242)

Mount Everest (EHV·ruhst) Highest mountain in the world, in the Greater Himalaya mountain range between Nepal and Tibet. 28°N 87°E (p. 598)

Gazetteer

Mount Fuji Peak in south-central Honshū, Japan. It is the highest peak in Japan. 35°N 138°E (p. 672)

Mount Logan Peak in northwest Canada. 60°N 140°W (p. 112)

Mount McKinley Highest peak in North America, located in Denali National Park in Alaska. 63°N 151°W (p. 112)

Mount Pinatubo Active volcanic mountain in the Philippines. 15°N 170°E (p. 746)

Mount Whitney Peak in the Sierra Nevada range in central California. 36°N 118°W (p. 112)

Mozambique (MOH•zuhm•BEEK) Country in southeastern Africa, south of Tanzania. (p. 477)

Murray River River in Australia. (p. 746)

Myanmar (MYAHN•MAHR) Country in Southeast Asia, south of China, also called Burma. (p. 747)

N

Nairobi Capital of Kenya. 1°S 37°E (p. 477)

Namib Desert Arid region along the coast of Namibia in southwestern Africa. (p. 476)

Namibia (nuh•MIH•bee•uh) Country in southwestern Africa, on the Atlantic Ocean. (p. 477)

Narmada River River in central India that flows into the Gulf of Khambat in the Arabian Sea. (p. 598)

Nassau (NA•saw) Capital of the Bahamas. 25°N 77°W (p. 169)

Nauru (nah•OO•roo) One of the two Federated States of Micronesia. 32°S 166°E (p. 747)

Nay Pyi Taw New capital of Myanmar. 20°N 96°E (p. 747)

N'Djamena (uhn•jah•MAY•nah) Capital of Chad. 12°N 15°E (p. 477)

Nepal (nuh•PAWL) Mountain country between India and China. (p. 599)

Netherlands Western European country on the North Sea. (p. 243)

New Delhi Capital of India. 29°N 77°E (p. 599)

New Zealand Major island country in the south Pacific, southeast of Australia. 42°S 175°E (p. 747)

Niamey (nee•AH•may) Capital and commercial center of Niger, in western Africa. 14°N 2°E (p. 477)

Nicaragua (NIH•kuh•RAH•gwuh) Republic in Central America. (p. 169)

Nicosia (NIH•kuh•SEE•uh) Capital of Cyprus. 35°N 33°E (p. 243)

Niger (NY•juhr) Landlocked country in western Africa, north of Nigeria. (p. 361)

Nigeria (ny•JIHR•ee•uh) Country in western Africa, south of Niger. (p. 477)

Niger River River in western Africa. (p. 476)

Nile River Longest river in the world, flowing north and east through eastern Africa. (p. 476)

Northern European Plain Plain that sweeps across western and central Europe into Russia and includes most of European Russia. (p. 242)

North Korea Asian country in the northernmost part of the Korean Peninsula. (p. 673)

North Sea Arm of the Atlantic Ocean extending between the European continent on the south and east and Great Britain on the west. (p. 242)

Norway Country on the Scandinavian Peninsula. (p. 243)

Nouakchott (nu•AHK•SHAHT) Capital of Mauritania. 18°N 16°W (p. 361)

Nullarbor Plain Dry, treeless area that lies south of the Great Victorian Desert in Australia. (p. 746)

O

Ob' River River in western Russia. (p. 242)

Ohio River Major river in the midwestern United States, emptying into the Mississippi River. (p. 112)

Oman (oh•MAHN) Country on the Arabian Sea and the Gulf of Oman. (p. 361)

Orinoco River River in Venezuela. (p. 168)

Oslo Capital of Norway. 60°N 11°E (p. 243)

Ottawa Capital of Canada. 45°N 76°W (p. 114)

Ouagadougou (WAH•gah•DOO•goo) Capital of Burkina Faso, in western Africa. 12°N 2°W (p. 477)

P

Pakistan South Asian country on the Arabian Sea, northwest of India. (p. 599)

Palau (puh•LOW) Island country in the western Pacific Ocean. 7°N 135°E (p. 747)

Pamirs Mountainous region of Central Asia. (p. 360)

Pampas Plains area of South America. (p. 168)

Panama (PA•nuh•MAH) Republic in south Central America, on the Isthmus of Panama. (p. 169)

Panama Capital of Panama. 9°N 79°W (p. 169)

Papua New Guinea (PA•pyuh•wuh noo GIH•nee) Independent island country in the south Pacific Ocean. 7°S 142°E (p. 747)

Paraguay (PAR•uh•GWY) Country in South America, north of Argentina. (p. 169)

Paraguay River River in south-central South America. (p. 168)

Paramaribo (PAR•uh•MAR•uh•BOH) Capital and port city of Suriname. 6°N 55°W (p. 169)

Paraná River River in southeast central South America. (p. 168)

Paris Capital and river port of France. 49°N 2°E (p. 243)

Patagonia Plateau region of South America primarily in Argentina. (p. 168)

Peace River River in western Alberta, Canada. (p. 112)

Persian Gulf Arm of the Arabian Sea between Iran and Saudi Arabia. (p. 360)

Peru Country in South America, south of Ecuador and Colombia. (p. 169)

Philippines (FIH•luh•PEENZ) Country in the Pacific Ocean, southeast of China. (p. 747)

Phnom Penh (NAHM PEHN) Capital of Cambodia. 12°N 106°E (p. 747)

Podgorica (PAWD·GAWR·eet·sah) Capital of Montenegro. 42°N 19°E (p. 243)

Poland Country on the Baltic Sea in eastern Europe. 52°N 18°E (p. 243)

Po River River in northern Italy that flows to the Adriatic Sea. (p. 243)

Port-au-Prince (POHRT·oh·PRIHNTS) Capital of Haiti. 19°N 72°W (p. 169)

Port Moresby (MOHRZ·bee) Capital of Papua New Guinea. 10°S 147°E (p. 747)

Porto-Novo (POHR·toh·NOH·voh) Capital and port city of Benin, in western Africa. 7°N 3°E (p. 477)

Portugal (POHR·chih·guhl) Country on the Iberian Peninsula, south and west of Spain. (p. 243)

Prague (PRAHG) Capital of the Czech Republic. 50°N 15°E (p. 243)

Pretoria See **Tshwane (Pretoria)**

Puerto Rico Island in the West Indies. It is a self-governing commonwealth in union with the United States. 18°N 66°W (p. 169)

P'yŏngyang (PYAWNG·YAHNG) Capital of North Korea. 39°N 126°E (p. 673)

Pyrenees Mountain range extending along the border of France and Spain. (p. 242)

Q

Qatar (KAH·tuhr) Country on the southwestern shore of the Persian Gulf. (p. 361)

Qin Ling Mountain range in northern China. (p. 672)

Quito (KEE·toh) Capital of Ecuador. 0° latitude 79°W (p. 169)

R

Rabat Capital of Morocco. 34°N 7°W (p. 361)

Red River River in the south-central United States, emptying into the Mississippi River. (p. 112)

Red River River in Vietnam that empties into the South China Sea. (p. 746)

Red Sea Inland sea between the Arabian Peninsula and northeast Africa. (p. 360)

Reykjavík (RAY·kyuh·VIHK) Capital of Iceland. 64°N 22°W (p. 243)

Rhine River River in western Europe that flows to the North Sea. (p. 242)

Rhône River River in Switzerland and France. (p. 242)

Riga Capital of Latvia. 57°N 24°E (p. 243)

Río de la Plata Estuary of the Paraná and Uruguay Rivers between Uruguay and Argentina. (p. 168)

Rio Grande/Río Bravo del Norte River forming part of the boundary between the United States and Mexico. (p. 112)

Riyadh (ree·YAHD) Capital of Saudi Arabia. 25°N 47°E (p. 361)

Rocky Mountains An extensive mountain system in western North America. (p. 112)

Romania (ru·MAY·nee·uh) Country in eastern Europe, south of Ukraine. (p. 243)

Rome Capital of Italy. 42°N 13°E (p. 243)

Rub' al-Khali Desert region in the southern Arabian Peninsula, also called the Empty Quarter. (p. 360)

Russia Largest country in the world, covering parts of Europe and Asia. (p. 243)

Rwanda (roo·AHN·dah) Country in Africa, south of Uganda. (p. 477)

S

Sahara Vast region of deserts and oases in North Africa. (p. 360)

St. Lawrence River River in southern Quebec and southeast Ontario, Canada. (p. 112)

St. Lucia Independent island state in the Caribbean Sea. 13°N 60°W (p. 169)

St. Vincent Principal island of St. Vincent and the Grenadines, south of St. Lucia. 13°N 61°W (p. 169)

Samoa Group of independent islands in the southwest Pacific Ocean. 13°S 172°W (p. 747)

Sanaa (sa·NAH) Capital of Yemen. 15°N 44°E (p. 361)

San José Capital of Costa Rica. 10°N 84°W (p. 169)

San Marino (SAN muh·REE·noh) Small European country, located on the Italian Peninsula. (p. 243)

San Salvador (san SAL·vuh·DAWR) Capital of El Salvador. 14°N 89°W (p. 169)

Santiago Capital of Chile. 33°S 71°W (p. 169)

Santo Domingo (SAN·tuh duh·MIHNG·goh) Capital of the Dominican Republic. 19°N 70°W (p. 169)

São Francisco River River in eastern Brazil flowing into the Atlantic Ocean. (p. 168)

São Tomé and Príncipe (SOWN tuh·MAY and PRIHN·sih·pee) Small island country in the Gulf of Guinea off the coast of central Africa. 1°N 7°E (p. 477)

Sarajevo (SAR·uh·YAY·voh) Capital of Bosnia and Herzegovina. 43°N 18°E (p. 243)

Saskatchewan River River in south-central Canada that flows into Lake Winnipeg. (p. 112)

Saudi Arabia (SOW·dee uh·RAY·bee·uh) Country on the Arabian Peninsula. (p. 361)

Scandinavia A peninsula in northern Europe. (p. 242)

Sea of Japan (East Sea) Branch of the Pacific Ocean between Japan and the Korean Peninsula. (p. 672)

Sea of Marmara Sea in northwest Turkey. (p. 413)

Sea of Okhotsk An inlet of the Pacific Ocean on the eastern coast of Russia. (p. 672)

Seine (SAYN) **River** French river that flows through Paris and into the English Channel. (p. 243)

Senegal (SEH·nih·GAWL) Country on the coast of western Africa, on the Atlantic Ocean. (p. 477)

Seoul (SOHL) Capital of South Korea. 38°N 127°E (p. 5)

Serbia (SUHR·bee·uh) European country between Macedonia and Hungary. (p. 243)

Seychelles (say•SHEHLZ) Small island country in the Indian Ocean near East Africa. 6°S 56°E (p. 477)

Siberia An area in the region of north-central Asia, primarily in Russia. (p. 242)

Sierra Leone (see•EHR•uh lee•OHN) Country in western Africa, south of Guinea. (p. 477)

Sierra Madre del Sur Mountain range along the coast of southern Mexico. (p. 168)

Sierra Madre Occidental Mountain range running parallel to the Pacific Ocean coast in Mexico. (p. 168)

Sierra Madre Oriental Mountain range running parallel to the Gulf of Mexico coast in Mexico. (p. 168)

Sierra Nevada Mountain range in eastern California in the United States. (p. 112)

Sinai Peninsula Peninsula in northeast Egypt between the Gulf of Suez and the Gulf of Aqaba. (p. 360)

Singapore Multi-island country in Southeast Asia near the tip of the Malay Peninsula. 2°N 104°E (p. 747)

Skopje (SKAW•pyeh) Capital of the Republic of Macedonia. 42°N 21°E (p. 243)

Slovakia (sloh•VAH•kee•uh) Central European country south of Poland. (p. 243)

Slovenia (sloh•VEE•nee•uh) Small central European country on the Adriatic Sea, south of Austria. (p. 243)

Sofia Capital of Bulgaria. 43°N 23°E (p. 243)

Solomon Islands Independent island group in the west Pacific Ocean. 8°S 159°E (p. 747)

Somalia (soh•MAH•lee•uh) Country in East Africa, on the Gulf of Aden and the Indian Ocean. (p. 477)

South Africa Country at the southern tip of Africa. (p. 477)

South China Sea Part of the Pacific Ocean extending from Japan to the tip of the Malay Peninsula. (p. 672)

South Korea Country in Asia on the Korean Peninsula between the Yellow Sea and the Sea of Japan. (p. 673)

South Sudan Country in eastern Africa, west of Ethiopia. (p. 477)

Spain Country on the Iberian Peninsula. (p. 243)

Sri Lanka (sree LAHNG•kuh) Island country in the Indian Ocean south of India. 9°N 83°E (p. 599)

Stockholm Capital of Sweden. 59°N 18°E (p. 243)

Strait of Gibraltar Passage connecting Mediterranean Sea to the Atlantic Ocean. (p. 360)

Strait of Hormuz Strait between the northern tip of Oman, the southeastern Arabian Peninsula, and the southern coast of Iran. (p. 360)

Strait of Malacca Ocean trade route running between Indonesia and Malaysia, near Singapore. (p. 746)

Sucre (SOO•kray) Constitutional capital of Bolivia. 19°S 65°W (p. 169)

Sudan Northeast African country on the Red Sea. (p. 477)

Suriname Republic in South America. (p. 169)

Suva Capital of Fiji. 18°S 177°E (p. 747)

Swaziland (SWAH•zee•LAND) South African country west of Mozambique. (p. 477)

Sweden Northern European country on the eastern side of the Scandinavian Peninsula. (p. 243)

Switzerland (SWIHT•suhr•luhnd) European country in the Alps, south of Germany. (p. 243)

Syr Dar'ya River in west-central Asia in Kyrgyzstan, Uzbekistan, and Kazakhstan. (p. 360)

Syria (SIHR•ee•uh) Country in Asia on the eastern side of the Mediterranean Sea. (p. 361)

T

Taipei (TY•PAY) Capital of Taiwan. 25°N 122°E (p. 673)

Taiwan (TY•WAHN) Island country off the southeast coast of China, claimed by China. 24°N 122°E (p. 673)

Tajikistan (tah•JIH•kih•STAN) Central Asian country north of Afghanistan. (p. 361)

Taklimakan Desert in western China. (p. 3)

Tallinn (TA•luhn) Capital and largest city of Estonia. 59°N 25°E (p. 243)

Tanzania (TAN•zuh•NEE•uh) East African country on the coast of the Indian Ocean. (p. 477)

Tashkent Capital of Uzbekistan. 41°N 69°E (p. 5)

Tasman Sea Part of the south Pacific Ocean between Australia and New Zealand. (p. 746)

Taurus Mountains Mountain range in southern Turkey. (p. 360)

Tbilisi (tuh•BEE•luh•see) Capital of the Republic of Georgia. 42°N 45°E (p. 361)

Tegucigalpa (tuh•GOO•suh•GAL•puh) Capital of Honduras. 14°N 87°W (p. 169)

Tehran (TAY•RAN) Capital of Iran. 36°N 52°E (p. 5)

Thailand (TY•LAND) Southeast Asian country south of Myanmar. (p. 747)

Thames (TEHMZ) **River** River in southern England that flows into the North Sea. (p. 243)

Thar Desert Region of sandy desert in northwest India and southeast Pakistan. (p. 598)

Thimphu (thihm•POO) Capital of Bhutan. 28°N 90°E (p. 599)

Tian Shan Mountain range in western China. (p. 672)

Tierra del Fuego Archipelago off southern South America. 54°N 68°W (p. 168)

Tiranë (tih•RAH•nuh) Capital of Albania. 42°N 20°E (p. 243)

Togo West African country between Benin and Ghana, on the Gulf of Guinea. (p. 477)

Tokyo Capital of Japan. 36°N 140°E (p. 5)

Tonga South Pacific island country. 20°S 175°W (p. 747)

Trinidad and Tobago (TRIH•nih•DAD tuh•BAY•goh) Independent republic comprising the islands of Trinidad and Tobago, located in the Atlantic Ocean off the northeast coast of Venezuela. 11°N 61°W (p. 169)

Tripoli Capital of Libya. 33°N 13°E (p. 361)

Tshwane (Pretoria) Administrative capital of the Republic of South Africa. 26°S 28°E (p. 477)

Tunis Capital of Tunisia. 37°N 10°E (p. 361)

Tunisia (too•NEE•zhuh) North African country on the Mediterranean Sea between Libya and Algeria. (p. 361)

Turkey Country in southeastern Europe and western Asia. (p. 361)

Turkmenistan (tuhrk•MEH•nuh•STAN) Central Asian country on the Caspian Sea. (p. 361)

Tuvalu Independent island group in the western Pacific Ocean. 8°S 178°E (p. 747)

U

Uganda (oo•GAHN•duh) East African country south of Sudan. (p. 477)

Ukraine (yoo•KRAYN) Large eastern European country west of Russia, on the Black Sea. (p. 243)

Ulaanbaatar (OO•LAHN•BAH•TAWR) Capital of Mongolia. 48°N 107°E (p. 673)

United Arab Emirates Country of seven states on the eastern side of the Arabian Peninsula. (p. 361)

United Kingdom Country in western Europe made up of England, Scotland, Wales, and Northern Ireland. (p. 243)

United States Country in North America located between Canada and Mexico. (p. 113)

Ural Mountains Mountain range in Russia that marks the traditional boundary between European Russia and Asian Russia. (p. 242)

Ural River River in eastern Europe and western Asia, originating in the Ural Mountains. (p. 242)

Uruguay (UR•uh•GWY) South American country, south of Brazil on the Atlantic Ocean. (p. 169)

Uzbekistan (uz•BEH•kih•STAN) Central Asian country south of Kazakhstan. (p. 361)

V

Vanuatu (VAN•WAH•TOO) Country made up of islands in the Pacific Ocean, east of Australia. 17°S 170°W (p. 747)

Vatican (VA•tih•kuhn) **City** Headquarters of the Roman Catholic Church, located in the city of Rome, Italy. 42°N 13°E (p. 243)

Venezuela Republic in northern South America. (p. 169)

Verkhoyanski Mountains Mountain range in northeastern Russia, just east of the Lena River. (p. 242)

Vienna Capital of Austria. 48°N 16°E (p. 243)

Vientiane (vyehn•TYAHN) Capital of Laos. 18°N 103°E (p. 747)

Vietnam Southeast Asian country, east of Laos and Cambodia. (p. 747)

Vindhya Range Mountain range in central India. (p. 598)

Vistula River River in southwestern Poland that flows north into the Baltic Sea. (p. 243)

Volga River River in western Russia that flows south into the Caspian Sea. (p. 242)

W

Warsaw Capital of Poland. 52°N 21°E (p. 243)

Washington, D.C. Capital of the United States, near the Atlantic coast. 39°N 77°W (p. 114)

Wellington Capital of New Zealand. 41°S 175°E (p. 747)

Western Ghats Mountain range in southern India. (p. 598)

Western Sahara Territory in northwest Africa. (p. 361)

West Siberian Plain Area of flat land that stretches from the Arctic Ocean to the grasslands of Central Asia. (p. 242)

Windhoek (VIHNT•HUK) Capital of Namibia, in southwestern Africa. 22°S 17°E (p. 477)

X

Xi (SHEE) **River** River in southeast China, known in its upper course as the Hongshui. (p. 672)

Y

Yablonovyy Range Mountain range in southern Russia. (p. 242)

Yamoussoukro (YAH•muh•SOO•kroh) Second capital of Côte d'Ivoire, in western Africa. 7°N 6°W (p. 477)

Yangon (Rangoon) Former capital of Myanmar. 17°N 96°E (p. 750) See also **Nay Pyi Taw**

Yaoundé (yown•DAY) Capital of Cameroon, in western Africa. 4°N 12°E (p. 477)

Yellow Sea Large inlet of the Pacific Ocean between northeast China and the Korean Peninsula. (p. 672)

Yemen (YEH•muhn) Country on the Arabian Peninsula, south of Saudi Arabia. (p. 361)

Yenisey River River in western Russia that flows north into the Kara Sea. (p. 242)

Yerevan (YEHR•uh•VAHN) Former capital and largest city of Armenia. 40°N 44°E (p. 361)

Yucatán Peninsula Peninsula including parts of southeastern Mexico, Belize, and Guatemala in Central America. (p. 168)

Yukon River River in the Yukon Territory, Canada. (p. 112)

Z

Zagreb Capital and largest city of Croatia. 46°N 16°E (p. 243)

Zagros Mountains Mountain system in southern and southwestern Iran. (p. 360)

Zambezi River River in south-central Africa. (p. 476)

Zambia (ZAM•bee•uh) Country in south-central Africa, east of Angola. (p. 477)

Zimbabwe (zihm•BAH•bwee) Country in south-central Africa, southeast of Zambia. (p. 477)

Gazetteer

GLOSSARY/GLOSARIO

- Content vocabulary terms in this glossary are words that relate to geography content. They are highlighted yellow in your text.
- Words below that have an asterisk (*) are academic vocabulary terms. They help you understand your school subjects and are **boldface** in your text.

*abandon • alluvial plain

ENGLISH	A	ESPAÑOL

*abandon to give up completely (p. 137)

aborigine/Aborigine an area's original inhabitants (p. 688); the indigenous people of Australia (p. 781)

absolute location the exact position of a place on the Earth's surface (p. 17)

*abundant present in large amounts; plentiful (p. 558)

*access a way to approach or enter (pp. 295, 755)

*accompany to be together with something (p. 781)

accretion slow process in which an oceanic plate slides under a continental plate, creating debris that can cause continents to grow (p. 47)

acculturation cultural modification of an individual, group, or people by adapting to or borrowing traits from another culture (p. 714)

*accumulate to build up (p. 644)

*accurate free from error (p. 303)

*achieve to carry out successfully; to accomplish (p. 259)

acid deposition wet or dry airborne acids that fall to the ground (p. 286)

acid rain precipitation carrying large amounts of dissolved acids, which kills wildlife and damages buildings, forests, and crops (p. 135)

*acquire to gain possession (p. 341)

*adapt to make fit to changing circumstances (p. 791)

*adequate satisfactory or acceptable (p. 406)

*administer to manage; to supervise (p. 592)

*advocate to publicly recommend or support (p. 151)

*affect to have an effect on; to produce an effect upon (pp. 704, 737)

agribusiness an industry engaged in agriculture on a large scale, sometimes including the manufacture and distribution of farm supplies (p. 281)

alluvial plain floodplain, such as the Gangetic Plain in India, on which flooding rivers have deposited silt (p. 606)

*abandonar dejar por completo (pág. 137)

aborígenes habitantes originarios de una región (pág. 688); indígenas de Australia (pág. 781)

localización absoluta ubicación exacta de un lugar en la superficie terrestre (pág. 17)

*abundante presente en grandes cantidades; copioso (pág. 558)

*acceso ruta para acercarse o entrar a un lugar (págs. 295, 755)

*acompañar estar junto a alguien o algo (pág. 781)

acreción proceso lento en el cual una placa oceánica se desliza por debajo de una placa continental, de modo que se forman detritos que originan el crecimiento de los continentes (pág. 47)

aculturación cambio cultural de un individuo, grupo o pueblo que se presenta cuando este se adapta a las características de otra cultura o las adquiere (pág. 714)

*acumular amontonar (pág. 644)

*preciso sin errores (pág. 303)

*lograr realizar exitosamente; cumplir (pág. 259)

deposición ácida partículas ácidas húmedas o secas que caen al suelo (pág. 286)

lluvia ácida precipitación con grandes cantidades de ácidos disueltos, que destruye la vida silvestre y daña edificaciones, bosques y cultivos (pág. 135)

*adquirir obtener la posesión de algo (pág. 341)

*adaptarse ajustarse a las circunstancias cambiantes (pág. 791)

*adecuado satisfactorio o aceptable (pág. 406)

*administrar dirigir; supervisar (pág. 592)

*defender recomendar o apoyar públicamente (pág. 151)

*afectar tener un efecto en algo; influir en algo (págs. 704, 737)

agroindustria industria relacionada con la agricultura a gran escala, que en ocasiones incluye la manufactura y distribución de insumos agrícolas (pág. 281)

llanura aluvial planicie en la cual las inundaciones fluviales depositan limo, como la llanura del Ganges en la India (pág. 606)

Glossary/Glosario

ENGLISH   ESPAÑOL

alluvial soil rich soil made up of sand and mud deposited by running water (p. 369)

suelo aluvial suelo fértil compuesto de arena y lodo depositados por una corriente de agua (pág. 369)

*alter to change partly (p. 120)

*alterar cambiar parcialmente (pág. 120)

*alternative different from the usual or regular (p. 236)

*alternativa diferente de lo usual o común (pág. 236)

altiplano Spanish for "high plain," a region in Peru and Bolivia encircled by the Andes (p. 218)

altiplano llanura alta; región de Perú y Bolivia rodeada por los Andes (pág. 218)

*altitude height above sea level (p. 576)

*altitud altura sobre el nivel del mar (pág. 576)

animist pertaining to the religious beliefs of animism in which nature, such as animals and mountains, has spirits (p. 491)

animista relacionado con las creencias religiosas del animismo, según las cuales los elementos de la naturaleza, como los animales y las montañas, tienen espíritu (pág. 491)

*annual occurring once a year (p. 610)

*anual que ocurre una vez al año (pág. 610)

*approximate close to but not exact (p. 712)

*aproximado cercano pero no exacto (pág. 712)

*approximately about or nearly (p. 514)

*aproximadamente casi o alrededor de (pág. 514)

aquaculture the cultivation of seafood (pp. 148, 719, 729)

acuicultura cultivo de especies acuáticas (págs. 148, 719, 729)

aqueduct a channel or pipeline for carrying a large quantity of flowing water (p. 138)

acueducto canal o tubería que transporta una gran cantidad de agua corriente (pág. 138)

aquifer underground water-bearing layers of porous rock, sand, or gravel (pp. 54, 380, 654)

acuífero estrato de roca permeable, arena o grava que almacena agua subterránea (págs. 54, 380, 654)

archipelago a group or chain of islands (pp. 197, 704)

archipiélago grupo o cadena de islas (págs. 197, 704)

*area a geographical region; the amount of space that the surface of a place covers (p. 484)

*área región geográfica; espacio que ocupa la superficie de un lugar (pág. 484)

arid excessively dry (p. 434)

árido excesivamente seco (pág. 434)

artesian well a well that brings pressurized water to the surface without pumping (p. 778)

pozo artesiano pozo por el cual el agua sometida a presión asciende a la superficie, sin necesidad de bombearla (pág. 778)

*assume to gain or acquire (p. 418)

*asumir contraer o adquirir (pág. 418)

atheist a person who does not believe in God (p. 689)

ateo persona que no cree en Dios (pág. 689)

atmosphere a thin layer of gases that surrounds the Earth (p. 42)

atmósfera capa delgada de gases que rodea la Tierra (pág. 42)

atoll a ring-shaped island formed by the buildup of a coral reef on the rim of a submerged volcano (p. 777)

atolón isla en forma de anillo compuesta por la acumulación de arrecifes de coral alrededor de un volcán sumergido (pág. 777)

*attribute to explain by indicating a cause (p. 204)

*atribuir explicar indicando una causa (pág. 204)

*authority power to influence or command thought, opinion, or behavior (pp. 88, 731)

*autoridad poder de influenciar o dominar el pensamiento, la opinión o el comportamiento (págs. 88, 731)

autocracy system of government in which one person rules with unlimited power and authority (p. 88)

autocracia sistema de gobierno en el cual una persona dirige con poder y autoridad ilimitados (pág. 88)

*available able to be obtained (p. 639)

*disponible que se puede utilizar (pág. 639)

avalanche a large mass of ice, snow, and rock that slides down a mountainside (p. 273)

avalancha gran masa de hielo, nieve y roca que se desliza por una ladera (pág. 273)

Glossary/Glosario

ENGLISH	ESPAÑOL
average daily temperature the average of the daily high temperature and the overnight low; often used for comparison across climate regions (p. 69)	**temperatura media diaria** promedio de la temperatura máxima diurna y la temperatura mínima nocturna; por lo general se usa para hacer comparaciones entre las regiones climáticas (pág. 69)
axis an imaginary line that runs through the center of the Earth between the North and South Poles (p. 60)	**eje** línea imaginaria que atraviesa el centro de la Tierra del Polo Norte al Polo Sur (pág. 60)

B

Balkanization division of a region into smaller hostile regions (p. 320)	**balcanización** división de una región en zonas más pequeñas, que tienen relaciones hostiles (pág. 320)
basin an area drained by a river and its tributaries (p. 554)	**cuenca** área drenada por un río y sus tributarios (pág. 554)
bedouin member of the nomadic desert peoples of North Africa and Southwest Asia (p. 375)	**beduino** miembro de los pueblos nómadas de los desiertos de África del Norte y el Sudoeste Asiático (pág. 375)
***benefit** to gain (p. 486); to promote well-being; to be useful (p. 780)	***beneficiar** ganar (pág. 486); promover el bienestar; ser útil (pág. 780)
biodiversity biological diversity in an environment as indicated by numbers of different species of plants and animals (p. 199); the diverse life forms in a habitat or ecosystem (p. 758)	**biodiversidad** diversidad biológica de un medioambiente expresada mediante la cantidad de especies vegetales y animales diferentes (pág. 199); las diversas formas de vida de un hábitat o un ecosistema (pág. 758)
biome major type of ecological community defined primarily by distinctive plant and animal groups (p. 69)	**bioma** área extensa ocupada por una comunidad ecológica definida principalmente a partir de las especies vegetales y animales que predominan allí (pág. 69)
biofuel fuel created from living matter, such as trees (p. 568)	**biocombustible** combustible que se obtiene de materia orgánica, como los árboles (pág. 568)
biosphere the part of the Earth where life exists (p. 42)	**biosfera** parte de la Tierra donde hay vida (pág. 42)
birthrate number of births per year for every 1,000 people (p. 82)	**tasa de natalidad** número de nacimientos al año por cada 1,000 habitantes (pág. 82)
black market illegal trade of scarce or illegal goods, usually sold at high prices (p. 346)	**mercado negro** comercio ilegal de productos ilícitos o escasos, que por lo general se venden a precios altos (pág. 346)
boomerang an Australian throwing stick that soars and curves in flight and returns near the thrower (p. 781)	**bumerán** utensilio de madera originario de Australia que al ser lanzado hace un movimiento curvo y vuelve a quien lo lanzó (pág. 781)
brain drain the loss of highly educated and skilled workers to other countries (p. 225)	**fuga de talentos** migración de trabajadores altamente calificados e instruidos hacia otros países (pág. 225)
break-of-bulk act of unloading, transferring, or distributing part or all of a shipment (p. 258)	**descarga** acción de descargar, transferir o distribuir parte de un cargamento o su totalidad (pág. 258)
buffer zone a neutral area serving to separate powerful nations or nations that are hostile to each other (p. 759)	**zona de amortiguación** área neutral que separa naciones poderosas o que tienen relaciones hostiles (pág. 759)
***bulk** the greater part of something (p. 809)	***grueso** la mayor parte de algo (pág. 809)
bush a sparsely inhabited region (p. 776)	**monte** región poco poblada (pág. 776)

C

caldera a large volcanic crater (p. 777)	**caldera** gran cráter volcánico (pág. 777)
canopy top layer of a rain forest, where the tops of tall trees form a continuous layer of leaves (p. 558)	**dosel** capa superior de una selva tropical, donde las copas de los árboles más altos forman una cubierta de hojas (pág. 558)

Glossary/Glosario

ENGLISH	ESPAÑOL
cap-and-trade a method for managing pollution in which a limit is placed on emissions and businesses or countries can buy and sell emissions allowances (p. 287)	**comercio de derechos de emisión** método de control de la polución según el cual se establece un límite a las emisiones, y las empresas o países pueden comprar y vender permisos de emisión (pág. 287)
capitalism a system in which factors of production are privately owned (p. 95)	**capitalismo** sistema en el que los medios de producción son de propiedad privada (p. 95)
carrying capacity the population that an area will support without undergoing deterioration (p. 522); the maximum population of any given species that an environment can sustain (p. 546)	**capacidad de carga** población que puede soportar un medioambiente sin deteriorarse (pág. 522); población máxima de una especie dada que puede soportar un medioambiente (pág. 546)
cash crop farm product grown to be sold or traded rather than used by the farm family (pp. 181, 560)	**cultivo comercial** producto agrícola que se cultiva para venta o intercambio, no para el consumo de las familias que lo cultivan (págs. 181, 560)
cataract a large waterfall (p. 507)	**catarata** cascada de gran tamaño (pág. 507)
central place theory geographical theory that seeks to explain the number, size, and location of human settlements in an urban system (p. 104)	**teoría de los lugares centrales** teoría geográfica que busca explicar el número, tamaño y ubicación de los asentamientos humanos en un sistema urbano (pág. 104)
cereal any grain like barley, oats, or wheat, grown for food (p. 457)	**cereal** grano que se cultiva para el consumo, como la cebada, la avena o el trigo (pág. 457)
*__challenge__ to arouse or stimulate especially by presenting with difficulties (p. 340)	*__desafiar__ retar o estimular especialmente planteando dificultades (pág. 340)
chernozem (cher•nuh•ZYAWM) rich, black topsoil found in the Northern European Plain, especially in Russia and Ukraine (p. 336)	*__chernozem__* tierra negra y rica en humus que se encuentra en la llanura noreuropea, particularmente en Rusia y Ucrania (pág. 336)
chinook a seasonal warm wind that blows down the Rockies in late winter and early spring (p. 147)	*__chinook__* viento estacional cálido y seco que sopla desde las montañas Rocosas a finales de invierno y comienzos de primavera (pág. 147)
chlorofluorocarbon (CFC) a chemical substance, found mainly in liquid coolants, that damages the Earth's protective ozone layer (p. 720)	**clorofluorocarbono (CFC)** sustancia química empleada principalmente en los refrigerantes líquidos, que deteriora la capa de ozono que protege la Tierra (pág. 720)
choke point a strategic, narrow waterway between two larger bodies of water (p. 443)	**punto de estrangulamiento** vía fluvial reducida entre dos cuerpos de agua más grandes que tiene fines estratégicos (pág. 443)
*__circumstance__ an event or fact that accompanies or determines another (p. 497)	*__circunstancia__ suceso o hecho que se relaciona con otro o lo determina (pág. 497)
*__cite__ to summon to appear in a court of law (p. 588)	*__citar__ convocar para comparecer ante un tribunal (pág. 588)
city-state an independently governed community consisting of a city and the surrounding lands, notably present in ancient Greece (p. 299)	**ciudad-estado** comunidad política independiente que incluye una ciudad y el territorio circundante, propia de la antigua Grecia (pág. 299)
clan a large group of people descended from the same ancestor; a family group (pp. 494, 710); a group of close-knit, interrelated families (p. 781)	**clan** grupo numeroso de personas que descienden de un mismo antepasado; grupo familiar (págs. 494, 710); grupo de familias estrechamente relacionadas (pág. 781)
clear-cutting the removal of all trees in a stand of timber (pp. 134, 664)	**tala generalizada** corte de todos los árboles de un área (págs. 134, 664)
climate weather patterns typical for an area over a long period of time (p. 60)	**clima** patrones de tiempo atmosférico característicos de un área en un periodo largo (pág. 60)

Glossary/Glosario

ENGLISH	ESPAÑOL
climate change any significant change in the measures of climate lasting for an extended period of time (p. 813)	**cambio climático** cualquier cambio significativo del clima en un periodo prolongado (pág. 813)
*coincide to happen in the same place and at the same time (p. 447)	***coincidir** ocurrir en el mismo lugar y al mismo tiempo (pág. 447)
Cold War the power struggle between the Soviet Union and the United States after World War II (p. 277)	**Guerra Fría** conflicto de poder entre Estados Unidos y la Unión Soviética después de la Segunda Guerra Mundial (pág. 277)
command economy a system of resource management in which decisions about production and distribution of goods and services are made by a central authority (p. 95)	**economía planificada** sistema de la administración de recursos en el cual una autoridad central toma las decisiones relacionadas con la producción y distribución de bienes y servicios (pág. 95)
commercial farming growing large quantities of crops or livestock in order to sell them for a profit (p. 587)	**agricultura comercial** producción de grandes cantidades de cultivos o ganado para obtener ganancias (pág. 587)
commune a collective farming community whose members share work and products (p. 690)	**comuna** comunidad agrícola colectiva cuyos miembros comparten el trabajo y los bienes que producen (pág. 690)
communism society based on equitable distribution of wealth, land, and public ownership of production (p. 276)	**comunismo** sociedad basada en la distribución equitativa de la riqueza, la tierra y la propiedad pública de los medios de producción (pág. 276)
*community people with common interests living in a particular area (p. 82)	***comunidad** personas con intereses comunes que viven en un área específica (pág. 82)
complementarity relationship between two places in which one produces something the other needs, resulting in an exchange (p. 302)	**complementariedad** relación entre dos lugares en la cual uno produce algo que el otro necesita, de modo que se produce un intercambio (pág. 302)
*compound to make something more extreme or intense by adding something to it (p. 791)	***combinar** hacer que un elemento sea más extremo o intenso agregando algo a este (pág. 791)
*comprehensive covering completely or broadly; inclusive (p. 279)	***exhaustivo** que abarca completamente o ampliamente; inclusivo (pág. 279)
*comprise to make up; to constitute (p. 317)	***comprender** conformar; constituir (pág. 317)
*concentrated less dilute (p. 469)	***concentrado** menos diluido (pág. 469)
condensation the process of excess water vapor changing into liquid water when warm air cools (p. 51)	**condensación** proceso mediante el cual el exceso de vapor de agua se convierte en agua líquida cuando el aire caliente se enfría (pág. 51)
*conflict a competition or struggle (p. 126)	***conflicto** competencia o lucha (pág. 126)
conflict diamonds diamonds that are mined in war-torn areas and are used to finance one or more parties involved (p. 534)	**diamantes de guerra** diamantes que se extraen de zonas devastadas por la guerra y con los cuales se financia una o más partes involucradas (pág. 534)
conic projection a map created by projecting an image of Earth onto a cone placed over part of an Earth model (p. 15)	**proyección cónica** mapa que se crea proyectando una imagen de la Tierra sobre un cono situado en una parte de un modelo terrestre (pág. 15)
coniferous referring to vegetation having cones and needle-shaped leaves, including many evergreens, that keep their foliage throughout the winter (p. 71)	**conífera** vegetación con conos y hojas en forma de aguja, como muchos árboles de hoja perenne, que mantiene su follaje durante el invierno (pág. 71)
connectivity the directness of routes linking pairs of places (p. 102)	**conectividad** efectividad de las rutas que conectan dos lugares (pág. 102)
conquistador Spanish for "conqueror"; Spanish soldier who participated in conquest of indigenous peoples of Latin America (p. 181)	**conquistador** soldado español que participó en la conquista de los pueblos indígenas de América Latina (pág. 181)
*consequence the result of an action (p. 694)	***consecuencia** resultado de una acción (pág. 694)

ENGLISH

*consist to be composed of or made up of (pp. 270, 704)

*constant unchanging (p. 51)

*constitute to compose or form (p. 507)

continental relating to or characteristic of a continent (p. 256)

continental drift the theory that the continents were once joined and then slowly drifted apart (p. 45)

continental shelf part of a continent that extends out underneath the ocean (p. 43)

continentality effect of extreme variation in temperature and very little precipitation within the interior portions of a landmass (p. 340)

*contribute to give or supply (p. 136)

*controversy the presence of opposing views (p. 148)

*converse reversed in order, relation, or action (p. 802)

*convert to change from one system, use, or method to another (p. 535)

*cooperation a common effort (p. 211)

*cooperative an organization, often a farm, whose members work together and share expenses and profits (p. 729)

copra dried coconut meat fed to animal herds (p. 790)

coral the limestone skeleton of a tiny sea animal (p. 780)

coral reef a reef made up of fragments of corals, coral sands, algal, and other organic deposits, and the solid limestone resulting from their consolidation (p. 801)

cordillera parallel chains or ranges of mountains (p. 218)

core innermost layer of the Earth made up of a super-hot but solid inner core and a super-hot liquid outer core (p. 44)

Coriolis effect the resulting diagonal movement, either north or south, of prevailing winds caused by the Earth's rotation (p. 66)

*corporate formed into an association and endowed by law with the rights and liabilities of an individual (p. 188)

*corresponding showing a direct connection between two things (p. 340)

cottage industry a business that employs workers in their homes (pp. 205, 617)

coup an overthrow of the government (p. 731)

cowrie shell the protective outer covering of a small sea creature, once used as money in Asia and Africa (p. 656)

ESPAÑOL

*consistir estar compuesto o formado por (págs. 270, 704)

*constante que no cambia (pág. 51)

*constituir componer o formar (pág. 507)

continental relacionado con un continente o característico de este tipo de formación (pág. 256)

deriva continental teoría según la cual los continentes alguna vez estuvieron unidos y luego se separaron lentamente (pág. 45)

plataforma continental parte de un continente que se extiende por debajo del océano (pág. 43)

continentalidad efecto de la variación extrema de la temperatura y la precipitación escasa en las áreas interiores de un continente (pág. 340)

*contribuir dar o suministrar (pág. 136)

*controversia presencia de puntos de vista opuestos (pág. 148)

*contrario opuesto en orden, relación o acción (pág. 802)

*convertir cambiar de un sistema, uso o método a otro (pág. 535)

*cooperación esfuerzo colectivo (pág. 211)

*cooperativa organización, por lo general una granja, cuyos miembros trabajan juntos y comparten gastos y ganancias (pág. 729)

copra pulpa seca del coco que sirve de alimento para los rebaños (pág. 790)

coral esqueleto calcáreo de un animal marino diminuto (pág. 780)

arrecife de coral arrecife compuesto por fragmentos de corales, arena de coral, algas y otros depósitos orgánicos, y la piedra caliza que se forma cuando estos elementos se consolidan (pág. 801)

cordillera cadenas de montañas paralelas (pág. 218)

núcleo centro de la Tierra compuesto por un núcleo interno sólido y un núcleo externo líquido, ambos extremadamente calientes (pág. 44)

efecto Coriolis movimiento diagonal de los vientos predominantes, en dirección norte o sur, causado por la rotación de la Tierra (pág. 66)

*corporativo que se ha convertido en una asociación a la cual la ley ha otorgado los derechos y responsabilidades civiles de un individuo (pág. 188)

*correspondiente que muestra una relación directa entre dos elementos (pág. 340)

industria artesanal empresa en la cual las personas trabajan desde su hogar (págs. 205, 617)

golpe de Estado derrocamiento de un gobierno (pág. 731)

cauri caparazón de una criatura marina pequeña que se usaba como moneda en Asia y África (pág. 656)

ENGLISH

ESPAÑOL

*create to bring into being or cause to exist (p. 45)

*crear dar vida o hacer que exista (pág. 45)

*crucial vitally important (p. 65)

*crucial de vital importancia (pág. 65)

crust outer layer of the Earth, a hard rocky shell forming Earth's surface (p. 44)

corteza capa exterior de la Tierra, rocosa y sólida, que forma la superficie terrestre (pág. 44)

cultural boundary a geographical boundary between two different cultures (p. 90)

frontera cultural límite geográfico entre dos culturas diferentes (pág. 90)

cultural diffusion the spread of culture traits, material and non-material, from one culture to another (p. 80)

difusión cultural expansión de características culturales, materiales e inmateriales, de una cultura a otra (pág. 80)

cultural divergence a separation of people or societies, with regard to beliefs, values, and customs, due to a division under different political systems (p. 733)

divergencia cultural separación de personas o sociedades con respecto a sus creencias, valores y costumbres debido a la división causada por sistemas políticos diferentes (pág. 733)

culture way of life of a group of people who share similar culture traits, including beliefs, customs, technology, and material items (p. 78)

cultura modo de vida de un grupo de personas que comparten características culturales similares, tales como creencias, costumbres, tecnología y elementos materiales (pág. 78)

*culture the customary beliefs, social forms, and material traits of a racial, religious, or social group (p. 180)

*cultura creencias usuales, actitudes sociales y rasgos materiales de un grupo racial, religioso o social (pág. 180)

culture hearth a center where cultures developed and from which ideas and traditions spread outward (pp. 80, 416)

centro cultural centro donde se desarrollaron las culturas y desde los cuales se difundieron las ideas y tradiciones (págs. 80, 416)

culture region division of the Earth in which people share a similar way of life, including language, religion, economic systems, and values (p. 79)

región cultural división en la cual las personas comparten un modo de vida similar, tal como el idioma, la religión, los sistemas económicos y los valores (pág. 79)

cuneiform wedge-shaped symbols that were pressed into clay tablets (p. 416)

cuneiforme símbolos en forma de cuña inscritos en tablas de arcilla (pág. 416)

current cold or warm stream of seawater that flows in the oceans, generally in a circular pattern (p. 65)

corriente movimiento del agua marina fría o cálida en un patrón circular (pág. 65)

cyclone a storm with heavy rains and high winds that blow in a circular pattern around an area of low atmospheric pressure (p. 609)

ciclón tormenta con lluvias y vientos fuertes que circula alrededor de un área con baja presión atmosférica (pág. 609)

cyclone, typhoon a spiral-shaped tropical storm in which the winds have reached speeds of at least 74 miles per hour (119 km per hour). Storms in the Northwestern Pacific, west of the International Date Line, are called typhoons; storms in the Southwest Pacific or Indian Oceans are referred to as cyclones. (p. 757)

tifón (ciclón) tormenta tropical en forma de espiral cuyos vientos han alcanzado una velocidad mínima de 74 millas por hora (119 km/h). Las tormentas del Pacífico Noroeste, al oeste de la línea internacional de cambio de fecha, se llaman tifones; las del Pacífico Sur y el océano Índico se conocen como ciclones. (pág. 757)

cylindrical projection a map created by projecting Earth's image onto a cylinder (p. 15)

proyección cilíndrica mapa que se crea proyectando la imagen de la Tierra en un cilindro (pág. 15)

czar ruler of Russia until the 1917 revolution; originally from Latin word *Caesar*, title of Roman emperors (p. 341)

zar gobernador de Rusia hasta la revolución de 1917; se deriva de la palabra latina *Caesar*, título de los emperadores romanos (pág. 341)

D

death rate number of deaths per year for every 1,000 people (p. 82)

tasa de mortalidad número de fallecimientos en un año por cada 1,000 habitantes (pág. 82)

deciduous falling off or shed seasonally or periodically; trees such as oak and maple, which lose their leaves in autumn (p. 71)

caducifolio relacionado con la vegetación que muda según la estación o periódicamente; árboles que pierden sus hojas en otoño, como los robles y los arces (pág. 71)

*decline to become less in amount (p. 344)

*disminuir reducir en número o cantidad (pág. 344)

Glossary/Glosario

ENGLISH	ESPAÑOL
deforestation the loss or destruction of forests, mainly for logging or farming (pp. 188, 738, 766)	**deforestación** pérdida o destrucción de bosques principalmente por la explotación maderera y agrícola (págs. 188, 738, 766)
delta a triangular section of land formed by sand and silt carried downriver (p. 487); a triangular-shaped area of silt and sediment deposit found at the mouth of a river (p. 578); an alluvial deposit at a river's mouth that is shaped roughly like the Greek letter delta (Δ) (p. 631)	**delta** terreno triangular que se forma donde una corriente fluvial deposita arena y limo (pág. 487); terreno triangular en la desembocadura de un río formado por un depósito de limo y sedimento (pág. 578); depósito aluvial en la desembocadura de un río cuya forma se parece a la letra griega delta (Δ) (pág. 631)
democracy system of government in which leaders rule with consent of the citizens (p. 89)	**democracia** sistema de gobierno según el cual los líderes dirigen con el consentimiento de los ciudadanos (pág. 89)
demographic transition the model that uses birthrates and death rates to show how populations in countries or regions change over time (p. 82)	**transición demográfica** modelo que se vale de las tasas de natalidad y mortalidad para mostrar la manera como cambia la población de los países o regiones con el paso del tiempo (pág. 82)
*__demonstrate__ to clearly show the value of (p. 547)	*__demostrar__ mostrar claramente el valor de algo (pág. 547)
*__derive__ to obtain something from (pp. 66, 368)	*__derivar__ obtener algo de otro elemento (págs. 66, 368)
desalination the removal of salt from seawater to make it usable for drinking and farming (pp. 53, 447)	**desalinización** proceso que elimina la sal del agua marina para que sea potable y apta para la agricultura (págs. 53, 447)
desertification process in which arable land becomes desert (p. 404); the destruction of land in arid and semiarid climates (p. 496)	**desertización** proceso en el cual la tierra cultivable se convierte en un desierto (pág. 404); destrucción del suelo en áreas de climas áridos y semiáridos (pág. 496)
devolution the granting of self-rule to local and regional authorities (p. 277)	**descentralización** proceso de devolver la autonomía a las autoridades regionales y locales (pág. 277)
dialect local form of a language used in a particular place or by a certain group (p. 203)	**dialecto** variedad local de una lengua que comparte un grupo específico o se habla en un lugar particular (pág. 203)
*__differentiate__ to recognize features that make similar regions, communities, or people different from each other (p. 756)	*__diferenciar__ reconocer las características que hacen que regiones, comunidades o personas similares se diferencien unas de otras (pág. 756)
dike large bank of earth and stone that holds back water (p. 272)	**dique** gran muro de tierra y piedra que retiene el agua (pág. 272)
*__diminish__ to make less or cause to appear less (p. 159)	*__reducir__ hacer menos o hacer parecer menos (pág. 159)
dissident a citizen who speaks out against government policies (p. 691)	**disidente** ciudadano que manifiesta su oposición con respecto a las políticas gubernamentales (pág. 691)
*__distinct__ recognizably different (p. 70)	*__distinto__ evidentemente diferente (pág. 70)
*__diverse__ differing from one another (p. 180)	*__diverso__ que se diferencia de otro (pág. 180)
divide a high point or ridge that determines the direction rivers flow (p. 121)	**divisoria** terreno alto o cresta que determina la dirección de las corrientes fluviales (pág. 121)
*__document__ to write to provide information (p. 443)	*__documentar__ escribir para suministrar información (pág. 443)
doldrums a frequently windless area near the Equator (p. 66)	**calmas ecuatoriales** área con vientos muy suaves cercana al ecuador (pág. 66)
domesticate to adapt plants and animals from the wild for human use (pp. 372, 489)	**domesticar** adaptar especies vegetales y animales silvestres para el beneficio humano (págs. 372, 489)
*__dominate__ to have a commanding position in (p. 611); to exert influence over (pp. 689, 779)	*__dominar__ tener una posición de mando (pág. 611); ejercer influencia (págs. 689, 779)
dominion a partially self-governing country with close ties to another country (p. 150); a largely self-governing country within the British Empire (p. 782)	**dominio** país con autonomía parcial que tiene vínculos estrechos con otra nación (pág. 150); país autónomo que formaba parte del Imperio británico (pág. 782)

Glossary/Glosario

863

ENGLISH

ESPAÑOL

doubling time the number of years it takes for a population to double in size (p. 83)

tiempo de duplicación número de años que tarda una población en doblar su tamaño (pág. 83)

drift-net fishing the use of fishing nets of great length and depth (p. 739)

pesca con redes de deriva uso de redes de pesca de gran tamaño y profundidad (pág. 739)

dry farming a farming method used in dry regions in which crops are grown that rely only on the natural precipitation (p. 127)

agricultura de secano método agrícola que se emplea en regiones secas, en el cual se siembran cultivos que dependen solo de las precipitaciones naturales (pág. 127)

dune a mound or ridge of sand formed by the wind (p. 435)

duna colina o montículo de arena que se forma por el viento (pág. 435)

*dynamic energetic, characterized by constant change and progress (p. 765)

*dinámico enérgico, que se caracteriza por el cambio y el progreso constantes (pág. 765)

dynasty a ruling house or continuing family of rulers, especially in China (p. 685)

dinastía casa real o serie de gobernantes de una misma familia, particularmente en China (pág. 685)

dzong a fortified monastery that also served as an administrative and commercial center (p. 660)

jong fortaleza-monasterio que también servía como centro administrativo y comercial (pág. 660)

E

e-commerce buying and selling on the Internet (p. 541)

comercio electrónico compra y venta de productos por Internet (pág. 541)

economic sanction a restriction placed on trade (p. 691)

sanción económica restricción al comercio (pág. 691)

*economy an ordered system for the production, distribution, and consumption of goods and services (p. 317)

*economía sistema ordenado para la producción, la distribución y el consumo de bienes y servicios (pág. 317)

ecotourism the practice and business of recreational travel based on concern for the environment (pp. 205, 264, 624)

ecoturismo práctica y empresa relacionadas con los viajes recreativos que se basan en el cuidado del medioambiente (págs. 205, 264, 624)

elevation the height of a land surface above the level of the sea (p. 18)

altura (altitud) distancia de un punto en la superficie terrestre con respecto al nivel del mar (pág. 18)

El Niño a periodic reversal of the pattern of ocean currents and water temperatures in the mid-Pacific region (p. 67)

El Niño cambio periódico en los patrones de las corrientes marinas y la temperatura del agua en la región del Pacífico Medio (pág. 67)

embargo a ban on trade (p. 421)

embargo prohibición al comercio (pág. 421)

*emerge to rise from an obscure position or condition; to become visible (p. 251)

*emerger aparecer desde una posición o condición de oscuridad; hacerse visible (pág. 251)

emerging market an economy that may not have been very strong in the recent past, but that is in transition to becoming a stronger market (p. 764)

mercado emergente economía que no fue sólida en el pasado reciente, pero que se encuentra en el proceso de convertirse en un mercado más fuerte (pág. 764)

emigrate to leave one's own country to settle permanently in another (p. 153)

emigrar dejar el país natal para establecerse permanentemente en otro (pág. 153)

*emphasis importance or special consideration (p. 327)

*énfasis importancia o consideración especial (pág. 327)

enclave a distinct territorial or cultural area that is within a foreign territory (p. 463)

enclave territorio o grupo cultural que se encuentra dentro de otro territorio (pág. 463)

*energy usable power (p. 198)

*energía fuerza aprovechable (pág. 198)

*enhance to improve or increase (p. 490)

*realzar mejorar o aumentar (pág. 490)

*enormous gigantic; exceedingly large (p. 52)

*enorme gigantesco; extremadamente grande (pág. 52)

*ensure to make sure or certain (p. 288)

*asegurar garantizar o hacer que algo sea seguro (pág. 288)

Glossary/Glosario

ENGLISH

entrepôt commercial center where goods are received (p. 258)

*environment natural surroundings (p. 534); the complex of physical, chemical, and biotic factors (such as climate, soil, and living things) that act upon an organism or an ecological community and ultimately determine its survival (p. 621)

equinox one of two days (about March 21 and September 23) on which the sun is directly above the Equator, making day and night equal in length (p. 61)

erosion the movement of weathered rock and material by wind, glaciers, and moving water (p. 49); the wearing away of soil (p. 546)

escarpment a steep cliff or slope between a higher and lower land surface (pp. 219, 507, 576)

*estimate to judge approximately the size or value of something (p. 437)

*ethnic of or relating to large groups of people classed according to common traits and customs (p. 321)

ethnic cleansing the expelling from a country or genocide of an ethnic group (p. 319)

ethnic group group of people who share common ancestry, language, religion, customs, or place of origin (p. 79)

*ethnicity relating to races or large groups of people classed according to common traits and customs (p. 732)

eutrophication process by which a body of water becomes too rich in dissolved nutrients, leading to plant growth that depletes oxygen (p. 137)

evaporation the process of converting liquid into vapor, or gas (p. 51)

*eventually in the end, especially after a long delay (p. 63)

*exceed to be greater than (p. 412)

exclave a territory that belongs to a particular political unit but is separated from it and surrounded by another political unit (p. 462)

*exhibit to demonstrate or show publicly (p. 198)

*exploit to make use of a resource (p. 656)

*export a commodity sent from one country to another for purposes of trade (p. 517)

*expose to put someone at risk for harm or injury (p. 469)

extended family household made up of several generations of family members (p. 183)

*external arising outside of (p. 49)

*extract to remove (pp. 158, 665)

ESPAÑOL

almacén aduanero centro de comercio donde se reciben mercancías (pág. 258)

*medioambiente entorno natural (pág. 534); unión de factores físicos, químicos y bióticos (como el clima, el suelo y los seres vivos) que actúan sobre un organismo o una comunidad ecológica y básicamente determinan su supervivencia (pág. 621)

equinoccio uno de dos días (hacia el 21 de marzo y el 23 de septiembre) en los cuales el sol se encuentra directamente sobre el ecuador, de modo que la duración del día y la noche es igual (pág. 61)

erosión proceso por el cual el viento, los glaciares y el agua en movimiento transportan roca y material desgastados (pág. 49); desgaste del suelo (pág. 546)

escarpe acantilado o pendiente entre una superficie más alta y una más baja (págs. 219, 507, 576)

*calcular determinar el tamaño o el valor aproximado de algo (pág. 437)

*étnico perteneciente o relativo a grupos de personas extensos con características y costumbres en común (pág. 321)

limpieza étnica expulsión de un país o genocidio de un grupo étnico (pág. 319)

grupo étnico grupo de personas que tienen la misma ascendencia, lengua, religión, lugar de origen y costumbres (pág. 79)

*etnicidad relativo a etnias o extensos grupos de personas con características y costumbres en común (pág. 732)

eutrofización proceso por el cual en un cuerpo de agua se presenta un aumento excesivo de nutrientes disueltos, lo cual da lugar a un crecimiento de plantas que agota el oxígeno(pág. 137)

evaporación proceso por el cual un líquido se convierte en vapor o gas (pág. 51)

*a la larga al final, especialmente luego de una demora (pág. 63)

*exceder ser más grande que algo (pág. 412)

exclave territorio que pertenece a una unidad política específica pero que está separado de esta y se encuentra rodeado por otra unidad política (pág. 462)

*manifestar exponer o mostrar públicamente (pág. 198)

*explotar hacer uso de un recurso (pág. 656)

*exportación conjunto de mercancías que se envía de un país a otro con fines comerciales (pág. 517)

*exponer poner a alguien en riesgo, de manera que puede resultar lastimado o perjudicado (pág. 469)

familia extensa grupo familiar compuesto por varias generaciones de parientes (pág. 183)

*externo que se manifiesta hacia el exterior (pág. 49)

*extraer remover (págs. 158, 665)

Glossary/Glosario

865

F

ENGLISH

*factor something that actively contributes to the production of a result (p. 306)

fall line a boundary in the eastern United States where the higher land of the Piedmont drops to the lower Atlantic Coastal Plain (p. 121)

fault a crack or break in Earth's crust (pp. 47, 506)

faulting process of cracking that occurs when the folded land cannot be bent any further (p. 47)

*feature to have as a characteristic or as a prominent attribute (p. 177)

federal system form of government in which powers are divided between the national government and state or provincial government (p. 87)

feeder stream a tributary to a larger river (p. 424)

fertilizer a chemical or natural substance added to soil or land to increase its fertility; a substance (as manure or a chemical mixture) used to make soil more fertile (pp. 402, 740)

First Nation one of the indigenous peoples of Canada who are neither Inuit nor Métis (p. 149)

fishery an area in which fish or sea animals are caught (p. 148); a place for catching fish, the fishing industry (p. 546)

fjord (fee•YAWRD) a long, steep-sided glacial valley now filled by seawater (p. 250)

floodplain the low-lying land along a river, formed mainly by sediment that has been deposited by floodwaters (p. 632)

*fluctuate to ebb and flow in waves (p. 608)

*fluctuation a shift from a previous condition (p. 33)

*focus to concentrate attention or effort (p. 279)

foehn (FUHRN) a dry wind that blows from the leeward sides of mountains, sometimes melting snow and causing avalanches; term used mainly in Europe (p. 273)

fold a bend in layers of rock, sometimes caused by plate movement (p. 47)

foreclosure legal proceeding in which a borrower's rights to a property are relinquished due to his or her inability to make payments on the loan (p. 131)

formal region a region defined by a common characteristic, such as production of a product (p. 30)

fossil fuel a resource formed in the Earth by plant and animal remains (p. 124)

ESPAÑOL

*factor algo que contribuye activamente a la producción de un resultado (pág. 306)

línea de descenso frontera en el este de Estados Unidos donde la meseta de Piedmont se une con la llanura litoral atlántica (pág. 121)

falla fractura o ruptura de la corteza terrestre (págs. 47, 506)

formación de fallas proceso de fractura que ocurre cuando no se pueden formar más pliegues en los estratos rocosos (pág. 47)

*distinguirse tener una característica o un atributo prominente (pág. 177)

sistema federal forma de gobierno en la cual el poder se divide entre el gobierno nacional y el gobierno estatal o provincial (pág. 87)

afluente tributario de un río grande (pág. 424)

fertilizante sustancia química o natural que se agrega al suelo para que sea más fértil; sustancia (como el estiércol o una mezcla química) que se usa para que el suelo sea más fértil (págs. 402, 740)

Primera Nación uno de los pueblos indígenas de Canadá excluyendo a los inuits y los métis (pág. 149)

pesquería área donde se atrapan peces y animales marinos (pág. 148); lugar donde se pesca, la industria pesquera (pág. 546)

fiordo valle glacial largo y empinado que está cubierto de agua marina (pág. 250)

planicie de inundación tierra baja a lo largo de un río formada por los sedimentos que depositan las inundaciones fluviales (pág. 632)

*fluctuar subir y bajar en ondas (pág. 608)

*fluctuación cambio de una condición anterior (pág. 33)

*enfocar concentrar la atención o el esfuerzo (pág. 279)

foehn viento seco que sopla a sotavento de las montañas, y que a veces derrite la nieve y causa avalanchas (pág. 273)

pliegue ondulación de las capas de roca que a veces se produce por el movimiento de las placas tectónicas (pág. 47)

ejecución hipotecaria procedimiento legal mediante el cual se retiran los derechos de una propiedad a un prestatario que no puede pagar las cuotas del préstamo adquirido (pág. 131)

región formal región definida por una característica común, como la manufactura de un producto (pág. 30)

combustible fósil recurso de la Tierra formado por restos de plantas y animales (pág. 124)

ENGLISH	ESPAÑOL
free enterprise a system in which private individuals or groups have the right to own property or businesses and make a profit with limited government interference (p. 94)	**libre empresa** sistema según el cual individuos o grupos privados tienen derecho a aposeer propiedades o empresas y obtener ganancias de los mismos con una interferencia gubernamental mínima (pág. 94)
free port a place where goods can be unloaded, stored, or reshipped free of import duties (p. 764)	**zona franca** lugar en el cual se pueden descargar, almacenar y reenviar mercancías sin pagar impuestos de importación (pág. 764)
free trade zone an area of a country in which trade restrictions do not apply (p. 186)	**zona de libre comercio** zona de un país en la cual no se aplican las restricciones comerciales (pág. 186)
***function** a special purpose (p. 102)	***función** propósito especial (pág. 102)
functional region a central place and the surrounding territory linked to it (p. 30)	**región funcional** lugar central y el territorio circundante que se vincula a este (pág. 30)
***furthermore** besides; in addition (p. 446)	***además** aparte de, adicionalmente (pág. 446)

G

ENGLISH	ESPAÑOL
***generate** to produce (p. 632); to bring into existence (p. 666)	***generar** producir (pág. 632); traer a la existencia (pág. 666)
***generation** a group of individuals born and living at the same time (pp. 714, 766)	***generación** grupo de individuos que nacen y viven en un mismo periodo de tiempo (págs. 714, 766)
geographic information systems (GIS) computer programs that process and organize details about places on Earth and integrate those details with satellite images and other pieces of information (p. 23)	**sistemas de información geográfica (SIG)** programas informáticos que procesan y organizan detalles sobre lugares de la Tierra y los integran con imágenes satelitales y otra información (pág. 23)
geometric boundary a boundary that follows a geometric pattern (p. 90); a fixed limit or extent that typically follows straight lines (p. 373)	**frontera geométrica** límite o frontera que tiene un patrón geométrico (pág. 90); frontera fija o extensión territorial que generalmente sigue una línea recta (pág. 373)
geopolitics government policy as it is influenced by physical, human, and economic geography (p. 444)	**geopolítica** políticas gubernamentales que se ven influidas por la geografía física, humana y económica (pág. 444)
geothermal energy a form of energy conversion that captures heat energy from within Earth (p. 251)	**energía geotérmica** forma de conversión de energía que toma energía calorífica del interior de la Tierra (pág. 251)
geyser a spring that throws forth intermittent jets of heated water and steam (p. 251)	**géiser** manantial que expulsa columnas intermitentes de agua caliente y vapor (pág. 251)
glaciation a process by which glaciers form and spread (p. 250)	**glaciación** proceso de formación y expansión de los glaciares (pág. 250)
glacier a large body of ice that moves across the surface of the Earth (p. 49)	**glaciar** enorme masa de hielo que se desliza por la superficie terrestre (pág. 49)
glasnost Russian term for a new openness in areas of politics, social issues, and media; part of Gorbachev's reform plans (p. 343)	**glasnost** término ruso para la apertura en temas políticos, asuntos sociales y medios de comunicación; parte de los planes de reforma de Gorbachov (pág. 343)
global positioning system (GPS) a navigational system that can determine absolute location by using satellites and receivers on Earth (p. 21)	**sistema de posicionamiento global (GPS)** sistema de navegación que puede determinar una localización absoluta mediante satélites y receptores instalados en la Tierra (pág. 21)
***goal** the aim of an activity (p. 544)	***meta** objetivo de una actividad (pág. 544)
***grant** a sum of money given to a person or organization for a particular purpose (p. 815)	***beca** suma de dinero que se le da a una persona u organización para un propósito determinado (pág. 815)
grazier a New Zealand rancher (p. 790)	**grazier** ganadero de Nueva Zelanda (pág. 790)

Glossary/Glosario

867

Glossary/Glosario

ENGLISH	ESPAÑOL
great circle route an imaginary line that follows the curve of the Earth and represents the shortest distance between two points (p. 14)	**ruta del círculo máximo** línea imaginaria que sigue la curvatura de la Tierra y presenta la distancia más corta entre dos puntos (pág. 14)
green revolution a program begun in the 1960s to produce higher-yielding, more productive strains of wheat, rice, and other food crops (p. 617)	**revolución verde** programa que inició en la década de 1960 con el fin de producir variedades mejoradas y más rentables de trigo, arroz y otros cultivos (pág. 617)
greenhouse effect the capacity of certain gases in the atmosphere to trap heat, thereby warming the Earth (p. 63)	**efecto invernadero** capacidad de algunos gases presentes en la atmósfera de retener el calor y así calentar la Tierra (pág. 63)
griot traditional oral historians, storytellers, singers, and musicians of West Africa (p. 540)	**griot** historiadores orales, narradores, cantantes y músicos tradicionales de África Occidental (pág. 540)
gross domestic product (GDP) the value of goods and services produced within a country in a year (p. 185)	**producto interno bruto (PIB)** valor de los bienes y servicios producidos en un país durante un año (pág. 185)
groundwater water located underground within the Earth that supplies wells and springs (pp. 54, 590)	**agua subterránea** agua ubicada debajo de la superficie terrestre que forma pozos y manantiales (págs. 54, 590)
guest worker a foreign laborer living and working temporarily in another country (p. 278)	**trabajador invitado** persona que vive y trabaja temporalmente en un país distinto a su país de origen (págs. 278)

H

ENGLISH	ESPAÑOL
habitat area with conditions suitable for certain plants or animals to live (p. 522)	**hábitat** área con condiciones favorables para la supervivencia de ciertas plantas y animales (pág. 522)
hajj in Islam, the yearly pilgrimage to Makkah that Muslims must make at least once in a lifetime (p. 441)	**hach** en el islam, la peregrinación a La Meca que deben realizar los musulmanes por lo menos una vez en su vida (pág. 441)
*hamper to make difficult; to impede (p. 566)	***obstaculizar** dificultar; impedir (pág. 566)
harmattan a hot, dry wind that blows from the northeast or east in the western Sahara (p. 488)	**harmattan** viento caliente y seco que proviene del nordeste o el este y que sopla en el Sahara Occidental (pág. 488)
headwaters the source of a stream or river (p. 121)	**cabecera** fuente de un arroyo o un río (pág. 121)
hieroglyphics an ancient Egyptian writing system in which pictures and symbols represent words or sounds (p. 372)	**jeroglífico** antiguo sistema de escritura egipcio en el cual las imágenes y símbolos representan palabras o sonidos (pág. 372)
high island an island with mountain ranges and volcanic soils (p. 800)	**isla alta** isla con cordilleras y suelos volcánicos (pág. 800)
Holocaust the mass murder of 6 million Jews by Germany's Nazi regime during World War II (p. 277)	**Holocausto** exterminio masivo de 6 millones de judíos ejecutado por el régimen nazi de Alemania durante la Segunda Guerra Mundial (pág. 277)
hot spring a spring whose water issues at a temperature higher than that of its surroundings (p. 251)	**fuente termal** manantial cuya agua brota a una temperatura mayor que la de su entorno (pág. 251)
*hub the center of an activity or region (p. 561)	***núcleo** centro de una actividad o región (pág. 561)
hurricane a large, powerful windstorm that forms over warm ocean waters (p. 122)	**huracán** tormenta con vientos torrenciales que se forma sobre aguas oceánicas cálidas (pág. 122)
hydroelectric power electrical energy generated by falling water (p. 644)	**energía hidroeléctrica** energía eléctrica producida por agua en movimiento (pág. 644)
hydrosphere the watery areas of the Earth, including oceans, lakes, rivers, and other bodies of water (p. 42)	**hidrosfera** capa de agua de la Tierra que incluye océanos, lagos, ríos y otros cuerpos de agua (pág. 42)

ENGLISH	ESPAÑOL

Ibadhism a conservative form of Islam distinct from Sunni and Shia sects (p. 441)

ibadismo forma conservadora del islamismo que se distingue de las sectas suníes y chiíes (pág. 441)

ideogram a pictorial character or symbol that represents a specific meaning or idea (p. 689)

ideograma carácter o símbolo pictórico que representa un significado o idea específicos (pág. 689)

*ignorance lack of knowledge, education, or awareness (p. 188)

*ignorancia falta de conocimiento, educación o conciencia (pág. 188)

*immigrate to change residence from a country to begin living permanently in another country (p. 127)

*inmigrar dejar de residir en un país para establecerse de forma permanente en otro (pág. 127)

*impact an effect on something; a significant or major effect (pp. 685, 737, 782)

*impacto efecto sobre algo; efecto importante o considerable (págs. 685, 737, 782)

imperialism the actions by which one country is able to extend power to control another country (p. 611)

imperialismo acciones mediante las cuales un país extiende su poder para controlar a otro (pág. 611)

*implement to carry out or accomplish by concrete measures (p. 229)

*implementar llevar a cabo o realizar algo con medidas concretas (pág. 229)

*incentive something that motivates one to act (p. 96)

*incentivo lo que motiva a alguien a actuar (pág. 96)

*incorporate to blend or thoroughly combine (p. 514)

*incorporar mezclar o combinar por completo (pág. 514)

indigenous native to a place (p. 513)

indígena nativo de un lugar (pág. 513)

industrial capitalism an economic system in which business leaders use profits to expand their companies (p. 276)

capitalismo industrial sistema económico en el cual los líderes de negocios usan las ganancias para expandir sus compañías (pág. 276)

Industrial Revolution beginning in the 1700s, rapid major change in the economy with the introduction of power-driven machinery (p. 276)

Revolución Industrial cambio rápido e importante en la economía que se dio con la introducción de la mecanización en el siglo XVIII (pág. 276)

*inevitable incapable of being avoided or evaded (p. 178)

*inevitable que no puede evitarse o evadirse (pág. 178)

infrastructure the set of systems that affect how well a place or organization operates, such as telephone or transportation systems, within a country (p. 539)

infraestructura conjunto de sistemas dentro de un país que influyen en el buen funcionamiento de un lugar u organización, como los sistemas de telefonía o transporte (pág. 539)

*infrastructure public services or systems such as power, water and sewage, transportation and communication networks, and schools and health care facilities (p. 636)

*infraestructura servicios públicos o sistemas como la electricidad, el acueducto y el alcantarillado, las redes de comunicación y transporte, y las instalaciones de escuelas e instituciones sanitarias (pág. 636)

*institute to organize and establish (p. 686)

*instituir organizar y establecer (pág. 686)

*integrate to blend into a functioning whole (p. 256)

*integrar incorporar algo en un conjunto de partes (pág. 256)

*interdependence a condition in which people or groups rely on each other, rather than only relying on themselves (p. 765)

*interdependencia situación en la que las personas o los grupos dependen el uno del otro, no solo de sí mismos (pág. 765)

*intermittent occurring at irregular intervals; occasional (p. 561)

*intermitente que ocurre a intervalos irregulares; ocasional (pág. 561)

*internal existing or lying within (p. 21)

*interno que está en la parte de adentro (pág. 21)

internally displaced person a refugee within his or her own country (p. 567)

persona desplazada interna refugiado dentro de su propio país (pág. 567)

*intervene to take action to change what is happening (p. 521); to come between so as to prevent or alter a course of events (p. 767)

*intervenir actuar para cambiar lo que sucede (pág. 521); interponerse con el fin de evitar o alterar el curso de los hechos (pág. 767)

Inuit a member of the Arctic native peoples of North America; once known as Eskimo (p. 149)

inuit miembro de los pueblos árticos de América del Norte, conocido en el pasado como esquimal (pág. 149)

Glossary/Glosario

869

ENGLISH	ESPAÑOL
invasive species non-indigenous or non-native species that threatens ecosystems, habitats, or other species (p. 261)	**especie introducida** especie no nativa que amenaza ecosistemas, hábitats y otras especies (pág. 261)
islet a very small island (p. 727)	**islote** isla muy pequeña (pág. 727)
*****isolate** to place or keep by itself (p. 285)	*****aislar** dejar algo solo y separado (pág. 285)
*****isolation** set or kept apart from others (p. 730)	*****aislamiento** efecto de separar de otros (pág. 730)
*****issue** a vital or unsettled matter (p. 300)	*****conflicto** asunto vital o sin resolver (pág. 300)
*****issue** a topic for debate and discussion (p. 720)	*****asunto** tema de debate y discusión (pág. 720)
isthmus a narrow strip of land connecting two larger land areas (p. 197)	**istmo** franja delgada de tierra que conecta dos territorios más grandes (pág. 197)

J

jati a group that defines one's occupation and social position (p. 611)	*jati* En la India, grupo que define la ocupación y la posición social de una persona (pág. 611)
jazz musical form that developed in the United States in the early 1900s, blending African rhythms and European harmonies (p. 129)	*jazz* género musical que se desarrolló en Estados Unidos a principios del siglo XX, que mezcla ritmos africanos y armonías europeas (pág. 129)
jute a fiber used to make string, rope, and cloth (p. 639)	*yute* fibra que se usa para hacer cuerdas, sogas y telas (pág. 639)

K

karma in Hindu belief, the sum of good and bad actions in one's present and past lives (p. 614)	**karma** en la creencia hindú, la suma de las acciones buenas y malas de la vida presente y las vidas pasadas de una persona (pág. 614)
karst terrain dominated by limestone bedrock and characterized by rocky ground, caves, sinkholes, underground rivers, and the absence of surface streams and lakes (p. 314)	**carst** terreno dominado por un lecho de roca caliza que se caracteriza por suelo rocoso, cavernas, sumideros, ríos subterráneos y la ausencia de arroyos y lagos superficiales (pág. 314)
kibbutz a communal farm or settlement in Israel (p. 394)	**kibutz** granja comunal o asentamiento en Israel (pág. 394)
kinship group a group of people related by blood or marriage (p. 804)	**grupo familiar** grupo de personas relacionadas por lazos sanguíneos o matrimonio (pág. 804)
Kyoto Protocol an amendment to the international treaty on climate change designed to reduce the amount of greenhouse gases emitted by specific countries (p. 287)	**Protocolo de Kioto** enmienda al tratado internacional sobre el cambio climático diseñada para reducir la cantidad de gases de efecto invernadero que emiten países específicos (pág. 287)

L

lagoon a shallow lake that is intermittently connected to a river, another lake, or the sea (p. 777)	**laguna** lago superficial conectado intermitentemente a un río, otro lago o el mar (pág. 777)
lama Buddhist monk (p. 657)	**lama** monje budista (pág. 657)
land bridge a strip of land that connects two larger landmasses, enabling migration of plants and animals to new areas (p. 176)	**puente intercontinental** franja de tierra que conecta dos continentes más grandes, lo que permite la migración de plantas y animales a otras zonas (pág. 176)
land subsidence the sinking or settling of land to a lower level in response to various natural and human-caused factors (p. 189)	**subsidencia del terreno** hundimiento o desplazamiento de la tierra a un nivel inferior como respuesta a varios factores naturales o humanos (pág. 189)
language family group of related languages that have all developed from one earlier language (p. 78)	**familia lingüística** grupo de lenguas relacionadas que se desarrollaron a partir de una misma lengua (pág. 78)

Glossary/Glosario

ENGLISH	ESPAÑOL
latifundia in Latin America, large agricultural estates owned by families or corporations (p. 205)	**latifundios** en América Latina, grandes propiedades agrícolas cuyos dueños son familias o corporaciones (pág. 205)
leeward being in or facing the direction toward which the wind is blowing (p. 68)	**sotavento** estar de frente a la dirección en la que sopla el viento (pág. 68)
less developed country a country that, according to the United Nations, exhibits the lowest indicators of socioeconomic development (p. 97)	**país menos desarrollado** país que, según las Naciones Unidas, muestra los indicadores de desarrollo socioeconómico más bajos (pág. 97)
lingua franca a common language used among people with different native languages (p. 514)	**lengua franca** lenguaje común que se usa entre personas que tienen diferentes lenguas nativas (pág. 514)
*****link** a connecting structure (p. 425)	*****vínculo** estructura conectora (pág. 425)
lithosphere uppermost layer of the Earth that includes the crust, continents, and ocean basins (p. 42)	**litosfera** capa más exterior de la Tierra que incluye la corteza, los continentes y las cuencas oceánicas (pág. 42)
llanos (LAH•nohs) fertile grasslands found in inland areas of Colombia and Venezuela (p. 219)	**llanos** praderas fértiles que se encuentran en zonas del interior de Colombia y Venezuela (pág. 219)
loess (LEHS) fine, yellowish, brownish topsoil made up of particles of silt and clay, carried and deposited by the wind (pp. 270, 682)	**loess** capa superior del suelo fina, amarillenta y marrón, compuesta de partículas de limo y arcilla, llevada y depositada por el viento (págs. 270, 682)
low island an island known as an atoll (p. 801)	**isla baja** isla conocida como atolón (pág. 801)
Loyalist an American colonist who remained loyal to the British government (p. 152)	**leal** colono norteamericano que permanecía fiel al gobierno británico (pág. 152)

M

ENGLISH	ESPAÑOL
magma molten rock that is located below Earth's surface (p. 45)	**magma** roca fundida ubicada debajo de la superficie de la Tierra (pág. 45)
*****maintain** to support (p. 632)	*****mantener** sostener (pág. 632)
*****major** greater in importance or interest (p. 80)	*****principal** de mayor importancia o interés (pág. 80)
mantle thick middle layer of the Earth's interior structure consisting of hot rock that is dense but flexible (p. 44)	**manto** capa gruesa intermedia del interior de la estructura terrestre formada por roca caliente densa pero flexible (pág. 44)
mantra sacred words or phrases that are repeated in prayers or chants (p. 660)	**mantra** palabras o frases sagradas que se repiten en oraciones o cantos (pág. 660)
Manufacturing Belt a concentrated region of manufacturing industries in the northeastern and midwestern United States (p. 127)	**Manufacturing Belt (cinturón manufacturero)** región de industrias manufactureras concentradas en el noroeste y medio oeste de Estados Unidos (pág. 127)
Maori the indigenous people of New Zealand (p. 781)	**maorí** pueblo indígena de Nueva Zelanda (pág. 781)
map projection a mathematical formula used to represent the curved surface of the Earth on the flat surface of a map (p. 15)	**proyección cartográfica** fórmula matemática que se usa para representar la superficie curva de la Tierra en la superficie plana de un mapa (pág. 15)
maquiladora in Mexico, a manufacturing plant owned by a foreign company (p. 186)	**maquiladora** en México, planta manufacturera de propiedad de una compañía extranjera (pág. 186)
market economy an economic system based on free enterprise, in which businesses are privately owned and production and prices are determined by supply and demand (p. 94)	**economía de mercado** sistema económico que se basa en la libertad de empresa, en el cual los negocios son de propiedad privada y la producción y los precios los determinan la oferta y la demanda (pág. 94)

Glossary/Glosario

ENGLISH	ESPAÑOL
marsh a wetland typically covered with grasses (p. 424)	**marisma** humedal usualmente cubierto de hierba (pág. 424)
massif a body of mountain ranges formed by fault-line activity (p. 294)	**macizo** cuerpo de cordilleras formado por la actividad de las líneas de falla (pág. 294)
matriarchal family ruled by a woman such as a mother, grandmother, or aunt (p. 203)	**matriarcal** familia controlada por una mujer, como la madre, la abuela o la tía (pág. 203)
Mediterranean climate a climate marked by warm, dry summers and cool but not cold winters (p. 579)	**clima mediterráneo** clima que se caracteriza por veranos cálidos y secos e inviernos frescos, mas no fríos (pág. 579)
megacity a great city that is made up of several large and small cities (p. 182)	**megaciudad** ciudad enorme formada por varias ciudades grandes y pequeñas (pág. 182)
megalopolis a large population concentration made up of several large and many smaller cities, such as the area between Boston and Washington, D.C. (p. 129)	**megalópolis** gran concentración de población formada por varias ciudades grandes y pequeñas, como el área entre Boston y Washington, D. C. (pág. 129)
meltwater water formed by melting snow and ice (p. 327)	**agua de fusión** agua que se forma cuando la nieve y el hielo se derriten (pág. 327)
mercantilism the theory or practice of merchant or trading pursuits (p. 611)	**mercantilismo** teoría o práctica de las actividades comerciales y mercantiles (pág. 611)
merchant marine a country's fleet of ships that engages in trade (p. 691)	**marina mercante** flota de barcos de un país que realiza actividades comerciales (pág. 691)
mestizo refers to people of mixed indigenous and European descent (p. 181)	**mestizo** persona descendiente de la mezcla de indígenas y europeos (pág. 181)
metropolitan area region that includes a central city and its surrounding suburbs (p. 103)	**área metropolitana** región que incluye una ciudad central y los suburbios que la rodean (pág. 103)
midnight sun continuous daylight, a time when the sun is visible at midnight during the summer in either the Arctic or Antarctic (p. 62)	**sol de medianoche** luz solar continua, periodo en el cual el sol es visible a la medianoche durante el verano en el Polo Ártico o el Polo Antártico (pág. 62)
*migrate to move from one place to another (p. 253)	***migrar** desplazarse de un lugar a otro (pág. 253)
migration the movement of people from place to place (p. 82)	**migración** movimiento de personas de un lugar a otro (pág. 82)
minifundia in Latin America, small farms that produce food chiefly for family use (p. 205)	**minifundios** en América Latina, granjas pequeñas que producen alimentos principalmente para consumo familiar (pág. 205)
mistral a strong northerly wind from the Alps that can bring cold air to southern France (p. 273)	**mistral** viento fuerte del norte que sopla desde los Alpes y lleva aire frío al sur de Francia (pág. 273)
mixed economy a system of resource management in which the government supports and regulates enterprise through decisions that affect the marketplace (p. 95)	**economía mixta** sistema de administración de recursos en el cual el gobierno mantiene y regula las empresas mediante decisiones que influyen en el mercado (pág. 95)
mixed forest forest with both coniferous and deciduous trees (p. 71)	**bosque mixto** bosque con árboles coníferos y caducifolios (pág. 71)
monarchy a form of autocracy with a hereditary king or queen exercising supreme power (p. 88)	**monarquía** forma de autocracia en donde un rey o reina herederos ejercen el poder supremo (pág. 88)
*monitor to watch closely, evaluate (p. 428)	***monitorizar** observar de cerca, evaluar (pág. 428)
monoculture the cultivation or growth of a single crop over a wide area for a consecutive number of years (p. 233)	**monocultivo** cultivo de un solo producto en un área extensa durante varios años consecutivos (pág. 233)
monotheism belief in one God (p. 395)	**monoteísmo** creencia en un solo dios (pág. 395)
monsoon in Asia, a seasonal wind that brings warm, moist air from the oceans in summer and cooler, dry air from inland in winter (pp. 436, 607, 683)	**monzón** en Asia, viento estacional que transporta aire cálido y húmedo desde los océanos en verano, y aire fresco y seco desde el interior en invierno (págs. 436, 607, 683)

Glossary/Glosario

ENGLISH	ESPAÑOL
montane referring to a mountainous area (p. 557)	**montañoso** relativo a un área con montañas (pág. 557)
moraine piles of rocky debris left by melting glaciers (p. 50)	**morrena** pilas de desechos rocosos que deja un glaciar cuando se derrite (pág. 50)
more developed country a country that has a highly developed economy and advanced technological infrastructure relative to other less developed nations (p. 97)	**país más desarrollado** país con un alto grado de desarrollo económico y una infraestructura tecnológica avanzada en relación con otros países menos desarrollados (pág. 97)
moshav a cooperative settlement of small individual farms in Israel (p. 394)	**moshav** asentamiento de carácter cooperativo formado por pequeñas granjas individuales en Israel (pág. 394)
mosque in Islam, a house of worship (p. 396)	**mezquita** en el islam, casa de culto (pág. 396)
mujahideen Islamic guerrilla fighters (p. 461)	**muyahidín** combatiente de la guerrilla islámica (pág. 461)

N

ENGLISH	ESPAÑOL
national identity the sense of a being part of the whole of a country, including its culture, traditions, language, and politics (p. 785)	**identidad nacional** sentido de pertenencia a un país, incluyendo su cultura, tradiciones, lenguaje y política (pág. 785)
nationalism belief in the right of each people to be an independent nation (p. 373)	**nacionalismo** creencia en el derecho de un pueblo a ser una nación independiente (pág. 373)
natural boundary a fixed limit or extent defined along physical geographic features such as mountains and rivers (p. 90); a boundary created by a physical feature, such as a mountain, river, or strait (p. 416)	**frontera natural** límite fijo o extensión definida a lo largo de formaciones geográficas como montañas y ríos (pág. 90); frontera creada por una formación física como una montaña, un río o un estrecho (pág. 416)
natural increase the growth rate of a population; the difference between birthrate and death rate (p. 82)	**crecimiento natural** tasa de crecimiento de una población; diferencia entre la tasa de natalidad y la tasa de mortalidad (pág. 82)
natural vegetation plant life that grows in a certain area if people have not changed the natural environment (p. 69)	**vegetación natural** vida vegetal que crece en ciertas áreas donde los seres humanos no han alterado el medioambiente natural (pág. 69)
*__navigable__ able to be traveled by boat (p. 556)	*__navegable__ que se puede atravesar en barco (pág. 556)
*__neutral__ not favoring either side in a quarrel, contest, or war (p. 612)	*__neutral__ que no favorece a ningún lado en una disputa, contienda o guerra (pág. 612)
*__nevertheless__ despite a situation or comment (p. 809)	*__sin embargo__ a pesar de una situación o comentario (pág. 809)
newly industrialized country a country that has begun transitioning from primarily agricultural to primarily manufacturing and industrial activity (p. 95)	**país recientemente industrializado** país que ha comenzado a hacer la transición de principalmente agrícola a principalmente manufacturero e industrial (pág. 95)
nomad a member of a wandering pastoral people (p. 374)	**nómada** miembro de un pueblo pastoril errante (pág. 374)
*__nonetheless__ in spite of; nevertheless (p. 435)	*__no obstante__ a pesar de; sin embargo (pág. 435)
*__nuclear__ of, relating to, or using the atomic nucleus, atomic energy, the atom bomb, or atomic power (p. 694)	*__nuclear__ relativo al uso de los núcleos atómicos, la energía atómica o la bomba atómica (pág. 694)
*__nuclear__ a weapon whose destructive power derives from an uncontrolled nuclear reaction (p. 814)	*__nuclear__ arma cuyo poder de destrucción se deriva de una reacción nuclear sin control (pág. 814)
nuclear family a family unit made up of a husband, wife, and children (p. 494)	**familia nuclear** unidad familiar compuesta por el esposo, la esposa y los hijos (pág. 494)
nuclear wastes by-products of producing nuclear power and weapons (p. 352)	**residuos radiactivos** subproductos de la producción de energía y armas nucleares (pág. 352)

Glossary/Glosario

ENGLISH

O

oasis small area in a desert where water and vegetation are found (p. 70)

*obtain to gain or acquire usually by planning or effort (p. 33)

*obvious easily discovered, seen, or understood (p. 205)

*occur to come into existence, happen (p. 264)

ocean warming rise in the temperature of the ocean water (p. 814)

*offset to counterbalance or to compensate for something else (p. 803)

old-growth forest complex forest that has developed over a long period of time and is relatively untouched by human activity (p. 158)

oligarchy system of government in which a small group holds power (p. 88)

oral tradition the practice of passing down stories from generation to generation by word of mouth (p. 495)

organic farming the use of natural substances rather than chemical fertilizers and pesticides to enrich the soil and grow crops (p. 665)

*output something produced; a mineral, agricultural, or industrial production (p. 401)

overfarming situation in which land is repeatedly farmed so that soil nutrients are depleted (p. 517)

overfishing harvesting fish to the point that species are depleted and the value of the fishery reduced (p. 148)

overgrazing grazing so heavily that the vegetation is damaged and the ground erodes (p. 404)

*overlap to partly cover (p. 69)

*overseas relating to a foreign country, especially one across the ocean (p. 715)

oxisol a thick, weathered soil of the humid tropics that is largely depleted of fertility and nutrients (p. 233)

P

pampas grassy, treeless plains of southern South America (p. 219)

panchayat village council (p. 614)

*parallel two items that are side by side and having the same distance between them (p. 530)

*participate to take part in (p. 418)

*partner one associated with another especially in an action (p. 377)

ESPAÑOL

O

oasis área pequeña en un desierto donde hay agua y vegetación (pág. 70)

*obtener ganar o adquirir, usualmente mediante la planeación o el esfuerzo (pág. 33)

*obvio fácil de descubrir, ver o entender (pág. 205)

*ocurrir acontecer, suceder (pág. 264)

calentamiento oceánico aumento de la temperatura del agua del océano (pág. 814)

*compensación para contrarrestar o para compensar algo más (pág. 803)

bosque primario bosque complejo que se ha desarrollado durante un periodo largo y relativamente no ha sido influenciado por la actividad humana (pág. 158)

oligarquía sistema de gobierno en el cual un grupo pequeño ejerce el poder (pág. 88)

tradición oral transmisión de historias de generación en generación mediante la palabra hablada (pág. 495)

agricultura orgánica uso de sustancias naturales en lugar de pesticidas y fertilizantes químicos para cultivar y fertilizar el suelo (pág. 665)

*producto algo producido; producción industrial, agrícola o minera (pág. 401)

sobrexplotación agraria situación en la que la tierra se cultiva repetidamente, de modo que se agotan los nutrientes del suelo (pág. 517)

sobrepesca pescar hasta el punto de agotar las especies y reducir el valor de la pesca (pág. 148)

sobrepastoreo pastoreo intensivo que daña la vegetación y erosiona el suelo (pág. 404)

*superponer cubrir parcialmente (pág. 69)

*ultramar relativo a un país extranjero, en especial al otro lado del océano (pág. 715)

oxisol suelo grueso y meteorizado de los trópicos húmedos que no es fértil ya que los nutrientes se han agotado (pág. 233)

P

pampa llanura austral de América del Sur, cubierta de hierba pero desprovista de árboles (pág. 219)

panchayat concejo comunal (pág. 614)

*paralelos dos objetos que están lado a lado a la misma distancia entre sí (pág. 530)

*participar tomar parte en algo (pág. 418)

*socio persona asociada con otra para tomar parte en una actividad (pág. 377)

874

ENGLISH

pastoralism the raising of animals for food (p. 415)

patois a dialect used in everyday speech that blends elements of several languages (p. 203)

patriarchal a family that is headed by a male family member (p. 494)

peninsula a portion of land nearly surrounded by water (p. 434)

perceptual region a region defined by popular feelings and images rather than by objective data (p. 30)

perestroika (PEHR•uh•STROY•kuh) in Russian, "restructuring"; part of Gorbachev's plan for reforming Soviet economy and government (p. 343)

permafrost permanently frozen layer of soil beneath the surface of the ground (pp. 72, 339)

*persistent continuing, existing, or acting for a long time (p. 319)

pesticide chemical used to kill crop-damaging insects, rodents, and other pests (pp. 353, 402, 623)

*phenomenon a fact or event of scientific interest that can be scientifically explained or described (p. 221)

phosphate natural mineral containing chemical compounds often used in fertilizers (p. 371)

pidgin a grammatically simplified form of a language (p. 785); a blend of English and an indigenous language (p. 807)

planar projection a map created by projecting an image of the Earth onto a geometric plane (p. 15)

plantation a large commercial farm (p. 561)

plate tectonics the term scientists use to describe the activities of continental drift and magma flow, which create many of Earth's physical features (p. 45)

poaching illegal hunting (p. 523)

polder low-lying area from which seawater has been drained to create new land (p. 272)

*policy an overall plan that establishes goals and determines procedures, decisions, and actions (p. 585)

pollution hot spot a location where pollution and other human activities have led to the degradation, or even death, of an ecosystem (p. 307)

population density the average number of people living on a square mile or square kilometer of land (p. 85)

population distribution the variations in population that occur across a country, a continent, or the world (p. 85)

ESPAÑOL

pastoreo cría de animales para el consumo (pág. 415)

patois dialecto usado en la expresión cotidiana, en el que se mezclan elementos de varios idiomas (pág. 203)

patriarcal familia dirigida por uno de sus integrantes de sexo masculino (pág. 494)

península porción de tierra que está prácticamente rodeada por agua (pág. 434)

región perceptiva región que se define mediante sentimientos e imágenes populares en vez de información objetiva (pág. 30)

perestroika en ruso, "reestructuración"; parte del plan de Gorbachov para reestructurar el gobierno y la economía soviéticos (pág. 343)

permafrost capa de suelo permanentemente congelada debajo de la superficie (págs. 72, 339)

* persistente durable, que existe o actúa por un tiempo largo (pág. 319)

pesticida sustancia química que se usa para matar insectos, roedores y otras plagas que dañan las cosechas (págs. 353, 402, 623)

*fenómeno hecho o suceso de interés científico que se puede describir o explicar científicamente (pág. 221)

fosfato mineral natural que contiene compuestos químicos y se usa con frecuencia en los fertilizantes (pág. 371)

pidgin tipo de lengua gramaticalmente simplificada (pág. 785); mezcla de inglés y una lengua indígena (pág. 807)

proyección planar mapa que se crea proyectando una imagen de la Tierra en un plano geométrico (pág. 15)

plantación granja comercial grande (pág. 561)

tectónica de placas término con el cual los científicos describen las actividades de la deriva continental y el flujo de magma, que crearon muchas de las formaciones físicas de la Tierra (pág. 45)

caza furtiva caza ilegal (pág. 523)

pólder área baja de la cual se ha drenado el agua del mar para crear tierra nueva (pág. 272)

*política plan general que establece metas y determina procedimientos, decisiones y acciones (pág. 585)

punto de emisión de contaminantes lugar donde la contaminación y otras actividades humanas han llevado a la degradación, o incluso la muerte, de un ecosistema (pág. 307)

densidad de población número promedio de personas que viven en una milla cuadrada o un kilómetro cuadrado de territorio (pág. 85)

distribución de la población cambios en la población que ocurren en un país, un continente o el mundo (pág. 85)

Glossary/Glosario

ENGLISH

population pressure the sum of factors within a population that reduce the ability of an environment to support the population, therefore resulting in migration or population decline (p. 202)

population pyramid a diagram that shows the distribution of a population (p. 84)

*portion share of a whole (p. 660)

*pose to come to attention as; to set forth or offer for consideration (p. 369)

postindustrial economy that emphasizes services and technology rather than industry and manufacturing (p. 131)

*potential possible or likely (p. 490); having a capacity that could be developed (p. 654)

prairie an inland grassland area (p. 71)

precipitation moisture that falls to the Earth as rain, sleet, hail, or snow (p. 51)

*predictable occurring in a way that is expected (p. 61)

*predominant the main or strongest (pp. 221, 706)

*preliminary something done in preparation for something more important (p. 406)

prevailing wind wind in a region that blows in a fairly constant directional pattern (p. 66)

*primary of first rank, importance, or value (p. 30)

primate city a city that dominates a country's economy, culture, and government and in which population is concentrated; usually the capital (p. 182)

*principal most important, consequential, or influential (p. 374)

*principle fundamental characteristic (p. 642); a rule or code of conduct (p. 730)

*priority something given or meriting attention before competing alternatives (p. 566)

privatization a change to private ownership of state-owned companies and industries (p. 346)

*prohibit to forbid somebody from doing something through law or rule (pp. 523, 585)

*prohibit to prevent from doing something (p. 623)

*promote to help something grow or develop (p. 306)

prophet person believed to be a messenger from God (p. 396)

ESPAÑOL

presión demográfica conjunto de factores dentro de una población que reducen la capacidad de un medioambiente para sustentarla, lo que resulta en la migración o la disminución de la población (pág. 202)

pirámide demográfica diagrama que muestra la distribución de una población (pág. 84)

*porción parte de un todo (pág. 660)

*plantear presentar; exponer o poner a consideración (pág. 369)

posindustrial economía que se enfoca más en los servicios y la tecnología que en la industria y la manufactura (pág. 131)

*potencial posible o probable (pág. 490); que tiene una capacidad que se puede desarrollar (pág. 654)

pradera área de pastizal en el interior (pág. 71)

precipitación agua que cae a la Tierra en forma de lluvia, aguanieve, granizo o nieve (pág. 51)

*predecible que ocurre de la forma esperada (pág. 61)

*predominante lo prevaleciente o más fuerte (págs. 221, 706)

*preliminar algo que se hace en preparación para algo más importante (pág. 406)

viento predominante viento que sopla en un patrón direccional bastante constante en una región (pág. 66)

*primordial de primer rango, importancia o valor (pág. 30)

ciudad principal ciudad que domina la economía, la cultura y el gobierno de un país y en la cual se concentra la mayor parte de su población; por lo general es la capital (pág. 182)

*principal más importante, relevante o influyente (pág. 374)

*principio característica fundamental (pág. 642); norma o código de conducta (pág. 730)

*prioridad algo a lo que se presta más atención que a otras alternativas con las que compite (pág. 566)

privatización cambio de propiedad al sector privado de las industrias y empresas públicas (pág. 346)

*prohibir impedir que alguien haga algo mediante leyes o normas (págs. 523, 585)

*prohibir impedir la ejecución de algo (pág. 623)

*promover ayudar al crecimiento o desarrollo de algo (pág. 306)

profeta persona considerada mensajera de Dios (pág. 396)

Q

qanat an underground canal first built by the ancient Persians (p. 417)

Quebecois Quebec's French-speaking inhabitants (p. 150)

qanat canal subterráneo inventado por los antiguos persas (pág. 417)

quebequenses habitantes de habla francesa de Quebec (pág. 150)

Glossary/Glosario

ENGLISH	ESPAÑOL
quipu (KEE•poo) knotted cords of various lengths and colors used by the Inca to keep financial records (p. 223)	**quipu** cuerdas anudadas de varias longitudes y colores que los incas usaban para llevar sus registros financieros (pág. 223)

R

*radical fundamental or extreme; drastic (p. 352)	***radical** básico o extremo; drástico (pág. 352)
radioactive material material contaminated by residue from the generation of nuclear energy or the testing of nuclear weapons (pp. 352, 469)	**material radiactivo** material contaminado por residuos de la generación de energía nuclear o de pruebas de armas nucleares (págs. 352, 469)
rain shadow result of a process by which dry areas develop on the leeward sides of mountain ranges (pp. 68, 579)	**sombra pluviométrica** resultado de un proceso mediante el cual se desarrollan áreas secas en los lados de sotavento de las cordilleras (págs. 68, 579)
*range a series of things in a row; a series of mountains (pp. 390, 680)	***cadena** serie de eslabones entrelazados; (montañas) cordillera (págs. 390, 680)
*recover to return to a previous level of performance after a decline (p. 662)	***recuperar** retornar a un nivel previo de desempeño después de un declive (pág. 662)
reforestation the planting and cultivating of new trees in an effort to restore a forest where the trees have been cut down or destroyed (pp. 212, 326, 770)	**reforestación** plantación y cultivo de nuevos árboles para restaurar un bosque al que se le cortaron o fueron destruidos (págs. 212, 326, 770)
*regime a form of government (p. 731)	***régimen** forma de gobierno (pág. 731)
*regulate to govern or direct according to rule (p. 95)	***regular** gobernar o dirigir según las normas (pág. 95)
reincarnation rebirth in new bodies or forms of life (p. 614)	**reencarnación** acción de volver a nacer en un cuerpo o forma de vida nuevos (pág. 614)
relative location location in relation to other places (p. 17)	**localización relativa** localización con relación a otros lugares (pág. 17)
*relevant having to do with the matter at hand (p. 381)	***pertinente** que tiene que ver con el asunto que se trata (pág. 381)
relief the variation in elevation across an area of Earth's land (p. 19)	**relieve** variación en la elevación de una superficie terrestre (pág. 19)
*rely to be dependent (p. 401)	***confiar** depender de algo (pág. 401)
remote sensing any technique used to measure, observe, or monitor a subject or process without physically touching the object under observation (p. 25)	**detección remota** cualquier técnica que se usa para medir, observar o monitorizar un sujeto o proceso sin tocarlo físicamente (pág. 25)
Renaissance rebirth; the period in European civilization characterized by a surge of interest in classical learning and values (p. 300)	**Renacimiento** acción de volver a nacer; periodo de la civilización europea caracterizado por el resurgimiento del interés en el aprendizaje y los valores clásicos (pág. 300)
reservoir a natural or artificial lake used as a source of water (p. 532)	**embalse** lago natural o artificial que se usa como fuente de agua (pág. 532)
*resource a usable stock or supply; a source from which benefit can be gained (pp. 298, 776)	***recursos** existencias o provisiones que se pueden usar; fuente de la cual se puede obtener beneficio (págs. 298, 776)
*restrain to hold down or back; to limit (p. 460)	***contener** reprimir; limitar (pág. 460)
*reverse the opposite (p. 456)	***inverso** opuesto (pág. 456)
revolution in astronomy, the Earth's yearly trip around the sun, taking 365¼ days (p. 61)	**revolución** en astronomía, el giro anual de la Tierra alrededor del Sol, el cual toma 365¼ días (pág. 61)
rift valley a valley formed by the separation of tectonic plates (p. 391)	**fosa tectónica** valle formado por la separación de las placas tectónicas (pág. 391)

ENGLISH	ESPAÑOL

river plain a plain formed by the deposit of sediment over a long period of time by one or more rivers (p. 533)

llanura fluvial llanura que se forma por el depósito de sedimentos de uno o más ríos durante un largo periodo (pág. 533)

Russification in nineteenth- and twentieth-century Russia and the Soviet Union, a government program that required everyone in the empire to speak Russian and to become a Christian; assignment of some Russian-speaking people to non-Russian ethnic regions (p. 342)

rusificación durante los siglos XIX y XX en Rusia y la Unión Soviética, programa gubernamental que exigía que todos los habitantes del imperio hablaran ruso y se convirtieran al cristianismo; asignación de algunos hablantes de ruso a regiones étnicas no rusas (pág. 342)

S

Sahel the band of land extending from Senegal to Somalia; also called the African Transition Zone (p. 484)

Sáhel franja de tierra que se extiende desde Senegal hasta Somalia; también se conoce como la zona de transición africana (pág. 484)

samurai a professional warrior of preindustrial Japan (p. 710)

samurái guerrero profesional del Japón preindustrial (pág. 710)

sanitation the disposal of waste products (pp. 493, 586)

recolección de basuras eliminación de desechos (págs. 493, 586)

satellite a country controlled by another country, notably Eastern European countries controlled by the Soviet Union by the end of World War II (p. 342)

país satélite país controlado por otro, en particular los países de Europa del Este controlados por la Unión Soviética al final de la Segunda Guerra Mundial (pág. 342)

*scope the extent of an activity or influence (p. 537)

*alcance extensión de una actividad o influencia (pág. 537)

sedimentation the action or process of forming or depositing sediment (p. 211)

sedimentación acción o proceso por el cual se depositan sedimentos (pág. 211)

seismic relating to or caused by an earthquake (p. 176)

sísmico relacionado con o causado por un terremoto (pág. 176)

separatism the breaking away of one part of a country to create a separate, independent country (p. 151)

separatismo doctrina que propugna la separación de una parte de un país para crear un país independiente (pág. 151)

*series a number of things of the same type following one after the other in space or time (p. 144)

*serie conjunto de cosas del mismo tipo que suceden unas a otras en el tiempo o en el espacio (pág. 144)

shamal a northwesterly wind in the Persian Gulf (p. 436)

shamal viento del noroeste que sopla sobre el golfo Pérsico (pág. 436)

shari'ah Islamic law derived from the Qur'an and the teachings of Muhammad (p. 441)

sharia ley islámica derivada del Corán y las enseñanzas de Mahoma (pág. 441)

shatter belt a region where political alliances are constantly splintering and fracturing based on ethnicity (p. 319); a region caught between stronger political groups, which is under constant stress and is sometimes fragmented as the result of conflicts between strong rivals (p. 759)

zona de conflicto región donde las alianzas políticas se rompen frecuentemente por diferencias étnicas (pág. 319); región rodeada por grupos políticos más fuertes, bajo tensión constante y que algunas veces está fragmentada como resultado de conflictos entre rivales fuertes (pág. 759)

sheikhdom territory ruled by an Arab tribal leader (p. 438)

reino de unjeque territorio gobernado por un líder de una tribu árabe (pág. 438)

Shia a branch of Islam that regards Muhammad's son-in-law Ali and the imams as his rightful successors (p. 441)

chiismo rama del islam que considera al yerno de Mahoma, Alí, y a los imanes, como los sucesores legítimos de Mahoma (pág. 441)

*shift to change the place, position, or direction of (p. 120)

*mover cambiar de lugar, posición o dirección (pág. 120)

shifting cultivation a system in which farmers plant a field for several years until its resources are depleted, then abandon it and clear a new field (p. 589); a form of agriculture in which an area of ground is cleared of vegetation, cultivated for a few years, and then abandoned in order to move to a new area (p. 768)

agricultura itinerante sistema en el cual los agricultores cultivan un campo durante varios años hasta que agotan sus recursos, luego lo abandonan y abren claros en otro campo (pág. 589); forma de agricultura en la que se abren claros en la vegetación de un área, se cultiva durante unos años y luego se abandona para hacer lo mismo en un área nueva (pág. 768)

Glossary/Glosario

ENGLISH	ESPAÑOL
*significant important (p. 271)	***significativo** importante (pág. 271)
*similar comparable (p. 78)	***similar** comparable (pág. 78)
simoom a hot, dry, suffocating wind that blows from time to time in the Arabian Peninsula (p. 436)	**simún** viento caliente, seco y sofocante que sopla de vez en cuando en la península arábiga (pág. 436)
site the specific location of a place, including its physical setting (p. 29)	**emplazamiento** ubicación específica de un lugar que incluye su configuración física (pág. 29)
situation the geographic position of a place in relation to other places or features of a larger region (p. 29)	**situación** posición geográfica de un lugar en relación con otros lugares o formaciones de una región más grande (pág. 29)
smog haze caused by the interaction of ultraviolet solar radiation with chemical fumes from automobile exhausts and other pollution sources (p. 136)	**esmog** niebla ocasionada por la interacción de la radiación solar ultravioleta con gases químicos que salen de los tubos de escape de los automóviles y otras fuentes de contaminación (pág. 136)
solstice one of two days (about June 21 and December 22) on which the sun's rays strike directly on the Tropic of Cancer or Tropic of Capricorn, marking the beginning of summer or winter (p. 62)	**solsticio** uno de los dos días (cerca del 21 de junio y el 22 de diciembre) en los que los rayos del sol caen directamente sobre el Trópico de Cáncer o el Trópico de Capricornio, lo que marca el comienzo del verano o el invierno (pág. 62)
*source a point of origin (p. 580)	***fuente** punto de origen (pág. 580)
spatial perspective a way of looking at the human and physical patterns on Earth and their relationships to one another (p. 26)	**perspectiva espacial** forma de mirar los patrones físicos y humanos de la Tierra y las relaciones entre ellos (pág. 26)
Special Economic Zones (SEZ) relatively small districts that are fully open to global commerce (p. 691)	**zonas económicas especiales (ZEE)** distritos relativamente pequeños completamente abiertos al comercio global (pág. 691)
*sphere a globe-shaped body (p. 40)	***esfera** cuerpo en forma de globo (pág. 40)
spreading process by which magma wells up between oceanic plates and pushes the plates apart (p. 47)	**expansión** proceso mediante el cual el magma brota entre las placas oceánicas y las separa empujándolas (pág. 47)
*stable not changing or fluctuating (p. 354)	***estable** que no cambia ni fluctúa (pág. 354)
stateless nation an ethnic group without a formal state (p. 397)	**nación sin estado** grupo étnico sin un estado formal (pág. 397)
station an Australian ranch (p. 790)	**estación** rancho australiano (pág. 790)
steppe wide, grassy plains of Eurasia; also, similar semiarid grassy areas elsewhere (p. 457)	**estepa** pradera amplia de Euroasia; también, praderas semiáridas similares en otros lugares (pág. 457)
*strategy a plan or method; a plan of action to achieve an aim (pp. 443, 720)	***estrategia** plan o método; plan de acción para lograr un objetivo (págs. 443, 720)
*stress pressure or strain (p. 496)	***estrés** presión o tensión (pág. 496)
Strine Australian English that has a unique vocabulary made up of Aboriginal words, terms used by early settlers, and slang created by modern Australians (p. 785)	*strine* inglés australiano compuesto por un vocabulario único que se toma de palabras aborígenes, términos de los primeros colonos y jerga de los australianos modernos (pág. 785)
*structure something constructed or arranged in a definite pattern of organization (p. 104)	***estructura** algo que se construye o se organiza con un patrón definido (pág. 104)
stupa a dome-shaped structure that serves as a Buddhist shrine (p. 660)	**estupa** estructura en forma de domo que se usa como templo budista (pág. 660)
subcontinent a large landmass that is part of a continent but still distinct from it, such as India (p. 606)	**subcontinente** gran masa de tierra que forma parte de un continente pero se distingue de él, como la India (pág. 606)

ENGLISH	ESPAÑOL
subduction process by which oceanic plates dive beneath continental plates, often causing mountains to form on land (p. 46)	**subducción** proceso que ocurre cuando una placa oceánica se mueve debajo de una placa continental, lo que ocasiona que se formen montañas en la superficie (pág. 46)
***subsequent** occurring or being carried out at a time after something else (p. 381)	***subsecuente** que ocurre o se realiza después de otra cosa (pág. 381)
subsistence farming farming that provides the basic needs of a family with little surplus (pp. 495, 587)	**agricultura de subsistencia** tipo de agricultura que proporciona lo suficiente para cubrir las necesidades básicas de una familia, con poco excedente (págs. 495, 587)
Sunbelt a mild climate region in the southern and southwestern portions of United States (p. 128)	***Sunbelt* [Cinturón del Sol]** región de clima templado en el sur y el suroeste de Estados Unidos (pág. 128)
sunken mountains high mountain range that has become submerged (p. 777)	**montañas sumergidas** cordillera de montañas altas que se han sumergido (pág. 777)
Sunni a branch of Islam that regards the first four successors of Muhammad as his rightful successors (p. 441)	**sunismo** rama del islam que considera legítimos a los cuatro primeros sucesores de Mahoma (pág. 441)
supertrawler an oceangoing factory ship with facilities for processing and freezing fish (p. 719)	**súper arrastrero** buque factoría de pesca de arrastre con instalaciones para procesar y congelar el pescado (pág. 719)
***survive** to manage to stay alive (p. 509)	***sobrevivir** seguir viviendo (pág. 509)
***sustain** to give support to (p. 413); to continue without interruption (p. 717)	***sostener** dar apoyo (pág. 413); proseguir sin interrupción (pág. 717)
***sustainable** a method of harvesting or using a resource so that the resource is not depleted or permanently damaged (p. 212)	***sustentable** método de agricultura o uso de un recurso de manera que el recurso no se agote ni se dañe permanentemente (pág. 212)
sustainable development technological and economic growth that does not deplete the human and natural resources of a given area (p. 188)	**desarrollo sustentable** crecimiento económico y tecnológico que no agota los recursos humanos y naturales de un área dada (pág. 188)
sustainable development economic growth that meets the needs of present populations without hampering the ability of people in the future to meet their own needs and that benefits poor people and societies (p. 642); technological and economic growth that does not deplete the human and natural resources of a given area (p. 770)	**desarrollo sostenible** desarrollo económico que satisface las necesidades de la población actual sin obstaculizar la capacidad de las generaciones futuras de satisfacer sus propias necesidades y que beneficia a las sociedades y gentes pobres(pág. 642); crecimiento económico y tecnológico que no agota los recursos naturales y humanos de un área dada (pág. 770)
***symbol** a sign or image that stands for an idea (p. 682)	***símbolo** signo o imagen que representa una idea (pág. 682)
syncretism a blending of beliefs and practices from different religions into one faith (p. 182)	**sincretismo** mezcla de creencias y prácticas de diferentes religiones en una sola fe (pág. 182)

T

Taliban from Arabic for "seeker" or "student"; name of a fundamentalist Sunni Muslim group, active in Afghanistan, which controlled the Afghan government from 1996 to 2001 (p. 461)	**talibán** del árabe que significa "estudiante" o "investigador"; nombre de un grupo fundamentalista musulmán suní activo en Afganistán, que controló el gobierno afgano de 1996 a 2001 (pág. 461)
tar sands sand or sandstone naturally impregnated with petroleum (p. 148)	**arenas de alquitrán** arena o arenisca impregnada naturalmente de petróleo (pág. 148)
thematic map a map that emphasizes a single idea or a particular kind of information about an area (p. 20)	**mapa temático** mapa que se enfoca en una sola idea o un tipo particular de información sobre un área (pág. 20)
theocracy system of government in which the officials are regarded as divinely inspired (p. 88)	**teocracia** sistema de gobierno en donde se considera que los funcionarios están inspirados por una divinidad (pág. 88)

Glossary/Glosario

ENGLISH

*theory a plausible general principle offered to explain (p. 41)

timberline elevation above which it is too cold for trees to grow (p. 147)

total fertility rate the average number of children a woman has in her lifetime (p. 636)

*trace to follow or study in detail or step by step (p. 149)

trade surplus earning more money from export sales than spending for imports (p. 716)

traditional economy a system in which tradition and custom control all economic activity; exists in only a few parts of the world today (p. 94)

*transform to change in form or appearance (p. 393)

transition zone an area in which the properties of the land undergo a radical change (p. 484)

*transmit to send from one person, thing, or place to another (p. 24)

*trend a general movement (p. 84)

tributary a smaller river or stream that feeds into a larger river (p. 121)

*trigger to initiate, actuate, or set off as if by pulling a trigger (p. 342)

trust territory a dependent area that the United Nations placed under the temporary control of a foreign country (p. 805)

tsunami a Japanese term used for a huge sea wave caused by an undersea earthquake (p. 609); a huge wave that gets higher and higher as it approaches the coast (p. 704); a dangerous large ocean wave caused by an undersea earthquake, a volcanic eruption, or a sudden displacement of the ocean floor (p. 755)

tungsten an extremely rare heavy-metal element essential in high-tech industry (p. 298)

typhoon a violent tropical storm that forms in the Pacific Ocean, usually in late summer (pp. 684, 757)

ESPAÑOL

*teoría principio general verosímil que se ofrece para explicar algo (pág. 41)

límite del bosque altitud por encima de la cual hace demasiado frío para que los árboles crezcan (pág. 147)

tasa de fertilidad total número promedio de hijos que una mujer tiene a lo largo de la vida (pág. 636)

*rastrear seguir o estudiar en detalle o paso por paso (pág. 149)

superávit comercial situación en la que el valor de las ganancias obtenidas por concepto de las exportaciones es mayores que el de las importaciones (pág. 716)

economía tradicional sistema en el cual la tradición y las costumbres controlan las actividades económicas; en la actualidad, existe solo en algunas partes del mundo (pág. 94)

*transformar cambiar de forma o apariencia (pág. 393)

zona de transición área en la cual las propiedades de la tierra sufren un cambio radical (pág. 484)

*transmitir enviar desde una persona, cosa o lugar a otra (pág. 24)

*tendencia movimiento general (pág. 84)

tributario río o arroyo pequeño que desemboca en uno más grande (pág. 121)

*desencadenar iniciar, impulsar o provocar (pág. 342)

territorio en fideicomiso área dependiente que las Naciones Unidas pone bajo el control temporal de un país extranjero (pág. 805)

tsunami término japonés que describe una ola gigantesca ocasionada por un terremoto submarino (pág. 609); ola gigantesca que se vuelve más grande a medida que se acerca a la costa (pág. 704); ola gigante peligrosa ocasionada por un terremoto submarino, una erupción volcánica o un desplazamiento repentino del lecho marino (pág. 755)

tungsteno elemento metálico pesado extremadamente raro y esencial en la industria de alta tecnología (pág. 298)

tifón tormenta tropical violenta que se forma en el océano Pacífico, usualmente al final del verano (págs. 684, 757)

U

*undergo to go through, experience (p. 261)

Underground Railroad a network of safe houses in the United States that helped thousands of enslaved people escape to freedom (p. 126)

understory a lower layer of the rain forest (p. 558)

uneven development condition in which some places do not benefit as much as others from social and economic advancement (p. 227)

*someterse a experimentar, padecer (pág. 261)

Tren Clandestino red de casas seguras en Estados Unidos que ayudó a escapar a miles de personas esclavizadas (pág. 126)

sotobosque capa inferior de la selva tropical (pág. 558)

desarrollo desigual condición por la que algunos lugares no se benefician de los avances sociales y económicos tanto como otros (pág. 227)

Glossary/Glosario

Glossary/Glosario

ENGLISH	ESPAÑOL
*unique being the only one; without a like or an equal; distinctive; unlike anything else (pp. 87, 454, 714)	***único** singular; sin igual; exclusivo; como ningún otro (págs. 87, 454, 714)
unitary system form of government in which powers are divided between the national government and state or provincial governments (p. 87)	**sistema unitario** forma de gobierno en la que los poderes se dividen entre el gobierno nacional y los gobiernos estatales o provinciales (pág. 87)
urban sprawl spreading of urban developments on undeveloped land near a city (p. 102)	**expansión urbana** diseminación de los desarrollos urbanos hacia las áreas sin desarrollo cercanas a una ciudad (pág. 102)
urbanization the relocation of people from rural areas to urban areas (p. 586)	**urbanización** traslado de las personas de las áreas rurales a las áreas urbanas (pág. 586)
*utilize to make use of (p. 694)	***utilizar** usar (pág. 694)

V

ENGLISH	ESPAÑOL
*vary to exhibit or undergo change (p. 728)	***variar** mostrar o experimentar cambios (pág. 728)
vertical climate zone a climate zone that occurs as elevation increases, with its own natural vegetation and crops (p. 178)	**zona climática vertical** zona climática que se presenta a medida que aumenta la altitud, con su propia vegetación y cultivos naturales (pág. 178)
*virtually almost; nearly (p. 460)	***prácticamente** casi; por poco (pág. 460)
*voluntary of one's own choice or consent (p. 236)	***voluntario** por elección o consentimiento propio (pág. 236)

W

ENGLISH	ESPAÑOL
wadi in the desert, a streambed that is dry except during a heavy rain (p. 369)	**uadi** en el desierto, un cauce seco excepto durante una temporada lluviosa (pág. 369)
water cycle regular movement of Earth's water from ocean to air to ground and back to the ocean (p. 51)	**ciclo del agua** movimiento regular del agua de la Tierra, desde el océano hasta el aire, luego a la tierra y de nuevo al océano (pág. 51)
weather condition of the atmosphere in one place during a short period of time (p. 60)	**tiempo atmosférico** condición de la atmósfera en un lugar durante un corto periodo (pág. 60)
weathering chemical or physical processes that break down rocks into smaller pieces (p. 49)	**meteorización** proceso químico o físico que rompe las rocas en pedazos más pequeños (pág. 49)
welfare state a state that assumes primary responsibility for the social welfare of its citizens (p. 259)	**estado de bienestar** estado que asume la responsabilidad primaria del bienestar social de sus ciudadanos (pág. 259)
*widespread covering a wide area; prevalent (pp. 228, 718)	***extendido** que cubre un área extensa; generalizado (págs. 228, 718)
windward being in or facing the direction from which the wind is blowing (p. 68)	**barlovento** estar en la dirección desde donde viene el viento (pág. 68)
world cities cities generally considered to play an important role in the global economic system (p. 105)	**ciudades globales** ciudades que desempeñan un papel importante en el sistema económico mundial (pág. 105)

Z

ENGLISH	ESPAÑOL
ziggurat a large temple built by the Sumerians (p. 420)	**zigurat** templo grande construido por los sumerios (pág. 420)

Index

Index

Index

Index

Index

Index

Index

N

Index

Index

Index

Index

Index

Index

Index

Index

Index

Index

Index